RF and Microwave Power Amplifier Design

About the Author

Andrei Grebennikov received his Dipl. Eng. degree in radio electronics from the Moscow Institute of Physics and Technology, Moscow, Russia, in 1980, and his Ph.D. in radio engineering from the Moscow Technical University of Communications and Informatics, Moscow, Russia, in 1991. He obtained long-term academic and industrial experience working with the Moscow Technical University of Communications and Informatics, also in Moscow, the Institute of Microelectronics (Singapore), M/A-COM (Cork, Ireland), Infineon Technologies (Munich, Germany, and Linz, Austria), and Bell Labs, Alcatel-Lucent (Dublin, Ireland) as an Engineer, Researcher, Lecturer, and Educator. He has lectured as a Guest Professor with the University of Linz, Linz, Austria, and presented short courses and tutorials as an Invited Speaker at the IEEE International Microwave Symposia, European and Asia-Pacific Microwave Conferences, the Institute of Microelectronics (Singapore), the Motorola Design Centre (Penang, Malaysia), the Tomsk State University of Control Systems and Radioelectronics (Tomsk, Russia), and the Aachen Technical University (Aachen, Germany). He is an author and co-author of more than 100 papers, holds 30 European and U.S. patents and patent applications, and has authored six books dedicated to RF and microwave circuit design.

RF and Microwave Power Amplifier Design

Andrei Grebennikov

Second Edition

McGraw Hill Education

New York Chicago San Francisco
Athens London Madrid
Mexico City Milan New Delhi
Singapore Sydney Toronto

Cataloging-in-Publication Data is on file with the Library of Congress.

RF and Microwave Power Amplifier Design, Second Edition

1 2 3 4 5 6 7 8 9 0 QVS/QVS 1 2 0 9 8 7 6 5 4

ISBN 978-0-07-182862-8
MHID 0-07-182862-1

This book is printed on acid-free paper.

Sponsoring Editor
Michael McCabe

Editorial Supervisor
Donna M. Martone

Production Supervisor
Pamela A. Pelton

Acquisitions Coordinator
Amy Stonebraker

Project Manager
Harleen Chopra,
Cenveo® Publisher Services

Copy Editor
Cenveo Publisher Services

Proofreader
Cenveo Publisher Services

Indexer
Jeff Evans

Art Director, Cover
Jeff Weeks

Composition
Cenveo Publisher Services

Contents

Preface xi
Acknowledgment xv

1 Two-Port Network Parameters and Passive Elements **1**
1.1 Traditional Network Parameters 1
1.2 Scattering Parameters 7
1.3 Interconnections of Two-Port Networks 11
1.4 Practical Two-Port Networks 16
1.4.1 Single-Element Networks 16
1.4.2 π- and T-Type Networks 17
1.5 Three-Port Network with Common Terminal 20
1.6 Lumped Elements 23
1.6.1 Inductors 23
1.6.2 Capacitors 27
1.7 Transmission Line 29
References 34

2 Nonlinear Circuit Design Methods **37**
2.1 Frequency-Domain Analysis 37
2.1.1 Trigonometric Identities 38
2.1.2 Piecewise-Linear Approximation 40
2.1.3 Bessel Functions 45
2.2 Time-Domain Analysis 47
2.3 Newton-Raphson Algorithm 50
2.4 Quasilinear Method 54
2.5 Harmonic Balance Method 58
2.6 X-Parameters 62
References 69

3 Nonlinear Active Device Modeling **71**
3.1 Power MOSFETs 72
3.1.1 Small-Signal Equivalent Circuit 72
3.1.2 Determination of Equivalent Circuit Elements 75
3.1.3 Nonlinear I-V Models 79
3.1.4 Nonlinear C-V Models 84
3.1.5 Charge Conservation 90

3.1.6 Gate-Source Resistance 91
3.1.7 Temperature Dependence 92
3.2 MESFETs and HEMTs 94
3.2.1 Small-Signal Equivalent Circuit 94
3.2.2 Determination of Equivalent Circuit Elements 99
3.2.3 Curtice Quadratic Nonlinear Model 102
3.2.4 Materka-Kacprzak Nonlinear Model 104
3.2.5 Chalmers (Angelov) Nonlinear Model 105
3.2.6 IAF (Berroth) Nonlinear Model 108
3.2.7 Model Selection 109
3.3 BJTs and HBTs 110
3.3.1 Small-Signal Equivalent Circuit 110
3.3.2 Determination of Equivalent Circuit Elements 112
3.3.3 Equivalence of Intrinsic π- and T-Type Topologies 116
3.3.4 Nonlinear Bipolar Device Modeling 117
References 123

4 Impedance Matching 127
4.1 Main Principles 127
4.2 Smith Chart 131
4.3 Matching with Lumped Elements 136
4.3.1 Analytic Design Technique 136
4.3.2 Bipolar UHF Power Amplifier 149
4.3.3 MOSFET VHF High-Power Amplifier 153
4.4 Matching with Transmission Lines 156
4.4.1 Analytic Design Technique 156
4.4.2 Equivalence between Circuits with Lumped and Distributed Parameters 165
4.4.3 Narrow-Band Microwave Power Amplifier 169
4.4.4 Broadband UHF High-Power Amplifier 170
4.5 Types of Transmission Lines 174
4.5.1 Coaxial Line 174
4.5.2 Stripline 176

4.5.3 Microstrip Line 178
4.5.4 Slotline 181
4.5.5 Coplanar Waveguide 183
References 185

5 Power Transformers, Combiners, and Couplers ... 187
5.1 Basic Properties 187
5.1.1 Three-Port Networks 188
5.1.2 Four-Port Networks 189
5.2 Transmission-Line Transformers and Combiners 191
5.3 Baluns 204
5.4 Wilkinson Power Dividers/Combiners 210
5.5 Branch-Line Hybrid Couplers 221
5.6 Coupled-Line Directional Couplers 229
References 236

6 Power Amplifier Design Fundamentals 241
6.1 Main Characteristics 242
6.2 Power Gain and Stability 249
6.3 Stabilization Circuit Technique 252
6.3.1 Frequency Domains of BJT Potential Instability 252
6.3.2 Frequency Domains of MOSFET Potential Instability 258
6.3.3 Some Examples of Stabilization Circuits 262
6.4 Basic Classes of Operation: A, AB, B, and C 267
6.5 Load Line and Output Impedance 274
6.6 Classes of Operation Based on Finite Number of Harmonics 278
6.7 Mixed-Mode Class B and Nonlinear Effect of Collector Capacitance 281
6.8 Load-Pull Characterization 286
6.9 Linearity 290
6.10 Push-Pull and Balanced Power Amplifiers 298
6.10.1 Basic Push-Pull Configurations 298
6.10.2 Balanced Power Amplifiers 303
6.11 Bias Circuits 306
6.12 Practical Aspect of RF and Microwave Power Amplifiers 313
References 319

7 **High-Efficiency Power Amplifiers** **323**
- 7.1 Overdriven Class B 323
- 7.2 Class-F Circuit Design 327
  - 7.2.1 Idealized Class-F Mode 329
  - 7.2.2 Class F with Maximally Flat Waveforms 333
  - 7.2.3 Class F with Quarterwave Transmission Line 338
  - 7.2.4 Effect of Saturation Resistance 342
  - 7.2.5 Load Networks with Lumped and Distributed Parameters 344
  - 7.2.6 Design Examples of Class-F Power Amplifiers 348
- 7.3 Inverse Class F 354
  - 7.3.1 Idealized Inverse Class-F Mode 356
  - 7.3.2 Inverse Class F with Quarterwave Transmission Line 359
  - 7.3.3 Load Networks with Lumped and Distributed Parameters 361
  - 7.3.4 Design Examples of Inverse Class-F Power Amplifiers 363
- 7.4 Class E with Shunt Capacitance 373
  - 7.4.1 Optimum Load-Network Parameters 374
  - 7.4.2 Effect of Saturation Resistance, Finite Switching Time, and Nonlinear Shunt Capacitance 381
  - 7.4.3 Optimum, Nominal, and Off-Nominal Class-E Operation 385
  - 7.4.4 Load Network with Transmission Lines 387
  - 7.4.5 Practical Class-E Power Amplifiers ... 390
- 7.5 Class E with Finite DC-Feed Inductance 394
  - 7.5.1 General Analysis and Optimum Circuit Parameters 395
  - 7.5.2 Parallel-Circuit Class E 402
  - 7.5.3 Load Networks with Transmission Lines 405
- References 408

8 **Broadband Power Amplifiers** **413**
- 8.1 Bode-Fano Criterion 414
- 8.2 Matching Networks with Lumped Elements 416

8.3 Matching Networks with Mixed Lumped and Distributed Elements 427
8.4 Matching Networks with Transmission Lines 431
8.5 Power Amplifiers with Lossy Compensation Networks 443
8.5.1 Lossy Match Design Techniques 444
8.5.2 Practical Examples 450
8.6 Broadband Class-E Power Amplifiers 456
8.6.1 Reactance Compensation Technique 456
8.6.2 Broadband Class E with Shunt Capacitance 469
8.6.3 Broadband Parallel-Circuit Class E 477
8.6.4 Monolithic Microwave Broadband Class-E Power Amplifiers 484
8.6.5 Broadband CMOS Class-E Power Amplifiers 487
8.7 Practical Broadband RF and Microwave Power Amplifiers 490
References 497

9 Linearization and Efficiency Enhancement Techniques 501
9.1 Feedforward Amplifier Architecture 501
9.2 Predistortion Linearization 509
9.3 Outphasing Power Amplifiers 517
9.4 Envelope Tracking 532
9.5 Switched-Path and Variable-Load Power Amplifiers 540
9.6 Monolithic HBT and CMOS Power Amplifiers for Handset Applications 551
References 565

10 High-Efficiency Doherty Power Amplifiers 573
10.1 Historical Aspect and Conventional Doherty Architectures 573
10.1.1 Basic Structures 575
10.1.2 Operation Principle 580
10.1.3 Offset Lines 583
10.1.4 Linearity 585
10.1.5 Series-Connected Load 587

10.2 Efficiency Improvement 587
10.3 Asymmetric Doherty Amplifiers 591
10.4 Multistage Doherty Amplifiers 594
10.5 Inverted Doherty Amplifiers 601
10.6 Integration 604
10.7 Digitally Driven Doherty Amplifier 611
10.8 Multiband and Broadband Capability 613
10.8.1 Dual-Band Parallel Doherty Architecture 614
10.8.2 Tri-Band Inverted Doherty Configuration 621
References 630

Index 635

Preface

The main objective of this book is to present all relevant information required for RF and microwave power amplifier design, including well-known historical and recent novel schematic configuration, theoretical approaches, circuit simulation results, and practical implementation techniques. This comprehensive book can be very useful for lecturing to promote the systematic way of thinking with analytical calculations, circuit simulation, and practical verification, thus making a bridge between theory and practice of RF and microwave engineering. As it often happens, a new result is the well-forgotten old one. Therefore, the demonstration of not only new results based on new technologies or circuit schematics is given, but some sufficiently old ideas or approaches are also introduced that could be very useful in modern design practice or could contribute to appearance of new general architectural ideas and specific circuit design techniques. As a result, this book is intended for and can be recommended to *university-level professors* as a comprehensive reference material to help in lecturing for graduates and postgraduates students, to *researchers and scientists* to combine the theoretical analysis with practical design and to provide a sufficient basis for innovative ideas and circuit design techniques, and to *practicing designers and engineers* as the book contains numerous well-known and novel practical circuits, architectures, and theoretical approaches with detailed description of their operational principles and applications.

In Chap. 1, the two-port networks are introduced to describe the behavior of linear and nonlinear circuits. To characterize the nonlinear properties of the bipolar or field-effect transistors, their equivalent circuit elements are expressed through the impedance *Z*-parameters, admittance *Y*-parameters, or hybrid *H*-parameters. On the other hand, the transmission *ABCD*-parameters are very important in the design of the distributed circuits such as a transmission line or cascaded elements, whereas the scattering *S*-parameters are widely used to simplify a measurement procedure.

The main purpose of Chap. 2 is to present widely used nonlinear circuit design techniques to analyze nonlinear power amplifier circuits.

In general, there are several approaches to analyze and design these nonlinear circuits, depending on their main specifications, for example, an analysis in the time domain when it is necessary to determine the transient circuit behavior, or in the frequency domain to provide an improvement of the power and spectral performances when such parasitic effects as instability and spurious emissions must be eliminated or minimized. By using the time-domain technique, it is quite easy to describe the circuit by differential equations, whereas the frequency-domain analysis is more explicit when a relatively complex circuit can be reduced to one or more sets of immittances at each harmonic component. The dynamic *X*-parameters are introduced as a novel way to build behavioral models for power amplifiers that include long-term memory effects.

In Chap. 3, all necessary steps to provide an accurate device modeling procedure starting with the determination of the device small-signal equivalent circuit parameters are described and discussed. A variety of nonlinear models for MOSFET, MESFET, HEMT, and bipolar devices including HBTs, which are very prospective for modern microwave monolithic integrated circuits of power amplifiers, are presented. In order to highlight the advantages or drawbacks of one nonlinear device model over the other, a comparison of the measured and modeled voltage-current and voltage-capacitance characteristics and a frequency range of model application are provided.

A concept of the impedance matching and impedance-matching technique, which is very important when designing power amplifiers, is described and presented in Chap. 4. First, the main principles and impedance-matching tools such as the Smith chart are described, providing a starting point of the matching design procedure. As an engineering solution in general depends on the different circuit requirements, the designer should choose the optimum solution among a variety of the matching networks, including either lumped elements and transmission lines or both of them. To simplify and visualize the matching design procedure, an analytical approach, which allows the parameters of the matching circuits using simple equations to be calculated, and the Smith chart traces are discussed and illustrated with several examples of the narrowband and broadband RF and microwave power amplifiers using bipolar or MOSFET devices. Finally, design formulas and curves are given for several types of transmission lines, such as stripline, microstrip line, slotline, and coplanar waveguide.

Chapter 5 describes the basic properties of three- and four-port networks, as well as a variety of different power combiners, impedance transformers, and directional couplers for RF and microwave power amplifier applications. Therefore, for power combining in view of insufficient power performance of the active devices, it is best to use the coaxial cable combiners with ferrite core to combine the output powers of RF power amplifiers intended for wideband applications.

As the device output impedance for high output power levels is usually too small, to match this impedance with a standard 50-Ω load, it is necessary to use the coaxial cable transformers with specified impedance transformation. For narrowband applications, the *N*-way Wilkinson combiners are widely used due to the simplicity of their practical realization. Because the size of the power combiners should be very small at microwave frequencies, the commonly used hybrid microstrip combiners including different types of the microwaves hybrids and directional couplers are described and analyzed.

Chapter 6 represents the fundamentals of the power amplifier design, which is generally a complicated procedure when it is necessary to provide simultaneously accurate active device modeling, effective impedance matching depending on the technical requirements and operation conditions, stability in operation, and ease in practical implementation. Therefore, initially the key definitions of different power gains and stability are introduced. For a stable operation of the power amplifier, it is necessary to evaluate the frequency domains where the active device may be potentially unstable. To avoid parasitic oscillations, the stabilization circuit technique for different frequency domains from low frequencies to high frequencies close to the device transition frequency is analyzed and discussed. One of the key parameters of the power amplifier is its linearity, which is very important for many TV and cellular applications. Therefore, the relationships between the output power, 1-dB gain compression point, third-order intercept point, and third- and higher-order intermodulation distortions are given and illustrated for different active devices. The basic classes of the power amplifier operation, namely Classes A, AB, B, and C, are introduced, analyzed, and illustrated. The device biasing conditions and examples of the bias circuits for MOSFET and bipolar devices to improve linearity or to increase efficiency are shown and discussed. Also, the concept of push-pull amplifiers and their circuit design using balanced transistors is given. In a final section, the several practical examples of power amplifiers using MOSFET, MESFET, and bipolar devices in the different frequency ranges and for different output powers are shown and discussed.

Modern commercial and military communication systems require high-efficiency long-term operating conditions. Chapter 7 describes in detail the possible load-network solutions to provide a high-efficiency power amplifier operation based on using overdriven Class-B, Class-F, inverse Class-F, and Class-E operation modes depending on the technical requirements. In Class-F power amplifiers analyzed in the frequency domain, the fundamental and harmonic load impedances are optimized by short-circuit termination and open-circuit peaking to control the voltage and current waveforms at the drain of the device to obtain maximum efficiency. In Class-E power amplifiers analyzed in the time domain, an efficiency improvement is achieved by realizing the on/off switching operation with special current and voltage waveforms so that high voltage and high current do not exist at the same time.

In many telecommunication, radar or testing systems, the transmitters operate in a very wide frequency range. Chapter 8 describes the power amplifier design based on a broadband concept that provides some advantages when there is no need to tune the resonant-circuit parameters. However, there are many factors that restrict the frequency bandwidth depending on the active device parameters. As a result, it is sufficiently easy to provide multioctave amplification from very low frequencies up to ultrahigh frequencies using the power MOSFET devices when lossy gain compensation is provided. At higher frequencies when the device input impedance is significantly smaller and influence of its internal feedback and parasitic parameters is substantially higher, it is necessary to use multisection matching networks with lumped and distributed elements. As an alternative, the parallel-circuit Class-E load-network configuration can be easily implemented in the broadband high-efficiency power amplifier design. A variety of broadband power amplifiers using in different frequency ranges are presented and described.

In modern telecommunication systems, it is very important to realize both high-efficiency and linear operation of the power amplifiers. Chapter 9 describes a variety of techniques and approaches that can improve the power amplifier performance. To increase efficiency over power backoff range, the outphasing and envelope-tracking power amplifier architectures, as well as the switched-path and variable-load power amplifier configurations are discussed and analyzed. To improve linearity of the operation, the feedforward linearizing technique and different types of predistortion linearization circuit schematics are demonstrated and explained. Finally, the design and implementation of the monolithic integrated circuits of the high-efficiency GaAs HBT and CMOS power amplifiers for handset application are shown and described.

Chapter 10 describes the historical aspect of the Doherty approach to the power amplifier design and modern trends in Doherty amplifier design techniques using multistage and asymmetric multiway architectures. To increase efficiency over the power-backoff range, the switchmode Class-E, conventional Class-F, or inverse Class-F operation mode by controlling the second and third harmonics can be used in the load network. The Doherty amplifier with a series connected load and inverted Doherty architectures are also described and discussed. Finally, examples of the lumped Doherty amplifier implemented in monolithic microwave integrated circuits, digitally driven Doherty technique, and broadband capability of the two-stage Doherty amplifier are given.

Acknowledgment

The author wishes to thank his wife, Galina Grebennikova, for performing important numerical calculation and computer artwork design, and for her constant encouragement, inspiration, support, and assistance.

RF and Microwave Power Amplifier Design

CHAPTER 1

Two-Port Network Parameters and Passive Elements

The two-port equivalent circuits are widely used in RF and microwave circuit design to describe the electrical behavior of both active devices and passive networks [1–4]. The two-port network impedance *Z*-parameters, admittance *Y*-parameters, or hybrid *H*-parameters are very important to characterize the nonlinear properties of the active devices, bipolar or field-effect transistors. The transmission *ABCD*-parameters of a two-port network are very convenient for designing the distributed circuits like transmission lines or cascaded elements. The scattering *S*-parameters are useful to characterize linear circuits and required to simplify the measurement procedure. Transmission lines are widely used in matching circuits in power amplifiers, directional couplers, power combiners, and dividers. Monolithic implementation of lumped inductors and capacitors is usually required at microwave frequencies and for portable devices.

1.1 Traditional Network Parameters

The basic diagram of a two-port nonautonomous transmission system can be represented by the equivalent circuit shown in Fig. 1.1, where V_S is the independent voltage source, Z_S is the source impedance, *LN* is the linear time-invariant two-port network without independent source, and Z_L is the load impedance. The two independent phasor currents I_1 and I_2 (flowing across input and output terminals) and phasor voltages V_1 and V_2 characterize such a two-port network. For autonomous oscillator systems, in order to provide an appropriate analysis in the frequency domain of the two-port network in the negative one-port representation, it is sufficient to set the source impedance to infinity. For a power amplifier design, the elements of the matching or resonant circuits, which are assumed to be linear or

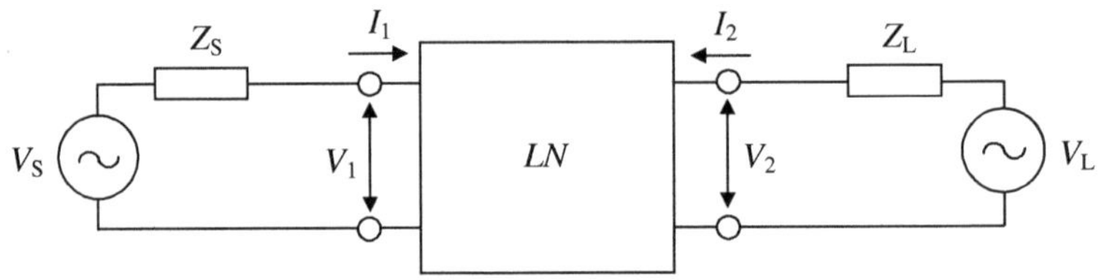

FIGURE 1.1 Basic diagram of two-port nonautonomous transmission system.

appropriately linearized, can be found among the *LN*-network elements, or additional two-port linear networks can be used to describe their frequency domain behavior.

For a two-port network, the following equations can be considered to be imposed boundary conditions:

$$V_1 + Z_S I_1 = V_S \tag{1.1}$$

$$V_2 + Z_L I_2 = V_L \tag{1.2}$$

Suppose that it is possible to obtain a unique solution for the linear time-invariant circuit shown in Fig. 1.1. Then, the two linearly independent equations, which describe the general two-port network in terms of circuit variables V_1, V_2, I_1, and I_2, can be expressed in a matrix form as

$$[M][V]+[N][I]=0 \tag{1.3}$$

or

$$\left.\begin{aligned} m_{11}V_1 + m_{12}V_2 + n_{11}I_1 + n_{12}I_2 = 0 \\ m_{21}V_1 + m_{22}V_2 + n_{21}I_1 + n_{22}I_2 = 0 \end{aligned}\right\} \tag{1.4}$$

The complex 2×2 matrices $[M]$ and $[N]$ in Eq. (1.3) are independent of the source and load impedances Z_S and Z_L and voltages V_S and V_L, respectively, and they depend only on the circuit elements inside the *LN* network.

If matrix $[M]$ in Eq. (1.3) is nonsingular with $|M| \neq 0$, this matrix equation can be rewritten in terms of $[I]$ as

$$[V] = -[M]^{-1}[N][I] = [Z][I] \tag{1.5}$$

where $[Z]$ is the open-circuit impedance two-port network matrix. In a scalar form, matrix Eq. (1.5) is given by

$$V_1 = Z_{11}\, I_1 + Z_{12}\, I_2 \tag{1.6}$$

$$V_2 = Z_{21}\, I_1 + Z_{22}\, I_2 \tag{1.7}$$

where Z_{11} and Z_{22} are the open-circuit driving-point impedances, and Z_{12} and Z_{21} are the open-circuit transfer impedances of the two-port network. The voltage components V_1 and V_2 due to the input current I_1 can be found by setting $I_2 = 0$ in Eqs. (1.6) and (1.7), thus resulting in an open-output terminal. Similarly, the same voltage components V_1 and V_2 are determined by setting $I_1 = 0$ when the input terminal becomes open-circuited.

The resulting driving-point impedances can be written as

$$Z_{11} = \left.\frac{V_1}{I_1}\right|_{I_2=0} \qquad Z_{22} = \left.\frac{V_2}{I_2}\right|_{I_1=0} \tag{1.8}$$

whereas the two transfer impedances are

$$Z_{21} = \left.\frac{V_2}{I_1}\right|_{I_2=0} \qquad Z_{12} = \left.\frac{V_1}{I_2}\right|_{I_1=0} \tag{1.9}$$

Dual analysis can be used to derive the short-circuit admittance matrix when the current components I_1 and I_2 are considered as outputs caused by V_1 and V_2. If matrix $[N]$ in Eq. (1.3) is nonsingular with $|N| \neq 0$, this matrix equation can be rewritten in terms of $[V]$ as

$$[I] = -[N]^{-1}[M][V] = [Y][V] \tag{1.10}$$

where $[Y]$ is the short-circuit admittance two-port network matrix. In a scalar form, matrix Eq. (1.10) is written as

$$I_1 = Y_{11}\, V_1 + Y_{12}\, V_2 \tag{1.11}$$

$$I_2 = Y_{21}\, V_1 + Y_{22}\, V_2 \tag{1.12}$$

where Y_{11} and Y_{22} are the short-circuit driving-point admittances, and Y_{12} and Y_{21} are the short-circuit transfer admittances of the two-port network. In this case, the current components I_1 and I_2 due to the input voltage source V_1 are determined by setting $V_2 = 0$ in Eqs. (1.11) and (1.12), thus creating a short output terminal. Similarly, the same current components I_1 and I_2 are determined by setting $V_1 = 0$ when input terminal becomes short-circuited.

As a result, the two driving-point admittances are

$$Y_{11} = \left.\frac{I_1}{V_1}\right|_{V_2=0} \qquad Y_{22} = \left.\frac{I_2}{V_2}\right|_{V_1=0} \tag{1.13}$$

whereas the two transfer admittances are

$$Y_{21} = \left.\frac{I_2}{V_1}\right|_{V_2=0} \qquad Y_{12} = \left.\frac{I_1}{V_2}\right|_{V_1=0} \tag{1.14}$$

In some cases, an equivalent two-port network representation can be redefined in order to express the voltage source V_1 and output current I_2 in terms of the input current I_1 and output voltage V_2. If the submatrix

$$\begin{bmatrix} m_{11} & n_{12} \\ m_{21} & n_{22} \end{bmatrix}$$

given in Eq. (1.4) is nonsingular, then

$$\begin{bmatrix} V_1 \\ I_2 \end{bmatrix} = -\begin{bmatrix} m_{11} & n_{12} \\ m_{21} & n_{22} \end{bmatrix}^{-1} \begin{bmatrix} n_{11} & m_{12} \\ n_{21} & m_{22} \end{bmatrix} \begin{bmatrix} I_1 \\ V_2 \end{bmatrix} = [H] \begin{bmatrix} I_1 \\ V_2 \end{bmatrix} \tag{1.15}$$

where $[H]$ is the hybrid two-port network matrix. In a scalar form, it is best to represent matrix Eq. (1.15) as

$$V_1 = h_{11} I_1 + h_{12} V_2 \tag{1.16}$$

$$I_2 = h_{21} I_1 + h_{22} V_2 \tag{1.17}$$

where h_{11}, h_{12}, h_{21}, and h_{22} are the hybrid H-parameters. The voltage source V_1 and current component I_2 are determined by setting $V_2 = 0$ for the short output terminal in Eqs. (1.16) and (1.17) as

$$h_{11} = \left.\frac{V_1}{I_1}\right|_{V_2=0} \qquad h_{21} = \left.\frac{I_2}{I_1}\right|_{V_2=0} \tag{1.18}$$

where h_{11} is the driving-point input impedance and h_{21} is the forward current transfer function. Similarly, the input voltage source V_1 and output current I_2 are determined by setting $I_1 = 0$ when input terminal becomes open-circuited as

$$h_{12} = \left.\frac{V_1}{V_2}\right|_{I_1=0} \qquad h_{22} = \left.\frac{I_2}{V_2}\right|_{I_1=0} \tag{1.19}$$

where h_{12} is the reverse voltage transfer function and h_{22} is the driving-point output admittance.

The transmission parameters, often used for passive device analysis, are determined for the independent input voltage source V_1 and input current I_1 in terms of the output voltage V_2 and output current I_2. In this case, if the submatrix

$$\begin{bmatrix} m_{11} & n_{11} \\ m_{21} & n_{21} \end{bmatrix}$$

given in Eq. (1.4) is nonsingular, one can obtain

$$\begin{bmatrix} V_1 \\ I_1 \end{bmatrix} = -\begin{bmatrix} m_{11} & n_{11} \\ m_{21} & n_{21} \end{bmatrix}^{-1} \begin{bmatrix} m_{12} & n_{12} \\ m_{22} & n_{22} \end{bmatrix} \begin{bmatrix} V_2 \\ -I_2 \end{bmatrix} = [ABCD] \begin{bmatrix} V_2 \\ -I_2 \end{bmatrix} \tag{1.20}$$

where $[ABCD]$ is the forward transmission two-port network matrix. In a scalar form, Eq. (1.20) can be written as

$$V_1 = AV_2 - BI_2 \tag{1.21}$$

$$I_1 = CV_2 - DI_2 \tag{1.22}$$

where A, B, C, and D are the transmission parameters. The voltage source V_1 and current component I_1 are determined by setting $I_2 = 0$ for the open output terminal in Eqs. (1.21) and (1.22) as

$$A = \left. \frac{V_1}{V_2} \right|_{I_2=0} \qquad C = \left. \frac{I_1}{V_2} \right|_{I_2=0} \tag{1.23}$$

where A is the reverse voltage transfer function and C is the reverse transfer admittance. Similarly, the input independent variables V_1 and I_1 are determined by setting $V_2 = 0$ when the output terminal is short-circuited as

$$B = \left. \frac{V_1}{I_2} \right|_{V_2=0} \qquad D = \left. \frac{I_1}{I_2} \right|_{V_2=0} \tag{1.24}$$

where B is the reverse transfer impedance and D is the reverse current transfer function. The reason that a minus sign is associated with I_2 in Eqs. (1.20) to (1.22) is that historically the input signal for transmission networks is considered as flowing to the input port whereas the output current is flowing to the load. The direction of the current $-I_2$ entering the load is shown in Fig. 1.2.

The parameters describing the same two-port network through different two-port matrices (impedance, admittance, hybrid, or transmission) can be cross-converted, and the elements of each matrix can be expressed by the elements of other matrices. For example, Eqs. (1.11) and (1.12) for the Y-parameters can be easily

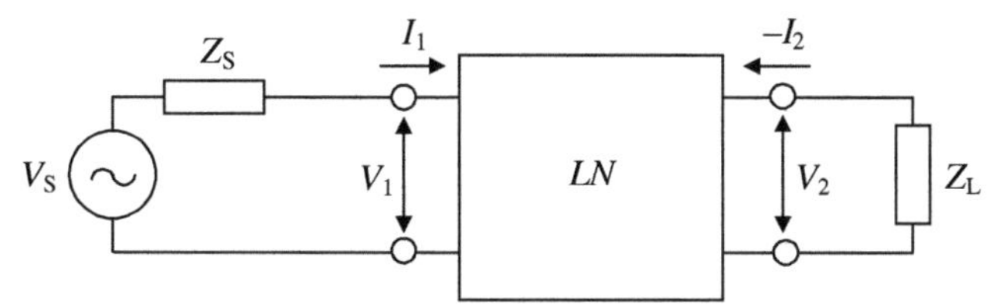

FIGURE 1.2 Basic diagram of loaded two-port transmission system.

solved for the independent input voltage source V_1 and input current I_1 as

$$V_1 = -\frac{Y_{22}}{Y_{21}} V_2 + \frac{1}{Y_{21}} I_2 \tag{1.25}$$

$$I_1 = -\frac{Y_{11} Y_{22} - Y_{12} Y_{21}}{Y_{21}} V_2 + \frac{Y_{11}}{Y_{21}} I_2 \tag{1.26}$$

By comparing the equivalent Eqs. (1.21) and (1.22) and Eqs. (1.25) and (1.26), the direct relationships between the elements of the transmission *ABCD*-matrix and admittance *Y*-matrix are written as

$$A = -\frac{Y_{22}}{Y_{21}} \qquad B = -\frac{1}{Y_{21}} \tag{1.27}$$

$$C = -\frac{\Delta Y}{Y_{21}} \qquad D = -\frac{Y_{11}}{Y_{21}} \tag{1.28}$$

where $\Delta Y = Y_{11}Y_{22} - Y_{12}Y_{21}$.

A summary of the relationships between the impedance *Z*-parameters, admittance *Y*-parameters, hybrid *H*-parameters, and transmission *ABCD*-parameters is shown in Table 1.1, where $\Delta Z = Z_{11}Z_{22} - Z_{12}Z_{21}$ and $\Delta H = h_{11}h_{22} - h_{12}h_{21}$.

	[*Z*]	[*Y*]	[*H*]	[*ABCD*]
[*Z*]	$\begin{matrix} Z_{11} & Z_{12} \\ Z_{21} & Z_{22} \end{matrix}$	$\begin{matrix} \frac{Y_{22}}{\Delta Y} & -\frac{Y_{12}}{\Delta Y} \\ -\frac{Y_{21}}{\Delta Y} & \frac{Y_{11}}{\Delta Y} \end{matrix}$	$\begin{matrix} \frac{\Delta H}{h_{22}} & \frac{h_{12}}{h_{22}} \\ -\frac{h_{21}}{h_{22}} & \frac{1}{h_{22}} \end{matrix}$	$\begin{matrix} \frac{A}{C} & \frac{AD-BC}{C} \\ \frac{1}{C} & \frac{D}{C} \end{matrix}$
[*Y*]	$\begin{matrix} \frac{Z_{22}}{\Delta Z} & -\frac{Z_{12}}{\Delta Z} \\ -\frac{Z_{21}}{\Delta Z} & \frac{Z_{11}}{\Delta Z} \end{matrix}$	$\begin{matrix} Y_{11} & Y_{12} \\ Y_{21} & Y_{22} \end{matrix}$	$\begin{matrix} \frac{1}{h_{11}} & -\frac{h_{12}}{h_{11}} \\ \frac{h_{21}}{h_{11}} & \frac{\Delta H}{h_{11}} \end{matrix}$	$\begin{matrix} \frac{D}{B} & -\frac{AD-BC}{B} \\ -\frac{1}{B} & \frac{A}{B} \end{matrix}$
[*H*]	$\begin{matrix} \frac{\Delta Z}{Z_{22}} & \frac{Z_{12}}{Z_{22}} \\ -\frac{Z_{21}}{Z_{22}} & \frac{1}{Z_{22}} \end{matrix}$	$\begin{matrix} \frac{1}{Y_{11}} & -\frac{Y_{12}}{Y_{11}} \\ \frac{Y_{21}}{Y_{11}} & \frac{\Delta Y}{Y_{11}} \end{matrix}$	$\begin{matrix} h_{11} & h_{12} \\ h_{21} & h_{22} \end{matrix}$	$\begin{matrix} \frac{B}{D} & \frac{AD-BC}{D} \\ -\frac{1}{D} & \frac{C}{D} \end{matrix}$
[*ABCD*]	$\begin{matrix} \frac{Z_{11}}{Z_{21}} & \frac{\Delta Z}{Z_{21}} \\ \frac{1}{Z_{21}} & \frac{Z_{22}}{Z_{21}} \end{matrix}$	$\begin{matrix} -\frac{Y_{22}}{Y_{21}} & -\frac{1}{Y_{21}} \\ -\frac{\Delta Y}{Y_{21}} & -\frac{Y_{11}}{Y_{21}} \end{matrix}$	$\begin{matrix} -\frac{\Delta H}{h_{21}} & -\frac{h_{11}}{h_{21}} \\ -\frac{h_{22}}{h_{21}} & -\frac{1}{h_{21}} \end{matrix}$	$\begin{matrix} A & B \\ C & D \end{matrix}$

TABLE 1.1 Relationships between *Z*-, *Y*-, *H*- and *ABCD*-Parameters

1.2 Scattering Parameters

The concept of incident and reflected voltage and current parameters can be illustrated by the one-port network shown in Fig. 1.3, where the network impedance Z is connected to the signal source V_S with the internal impedance Z_S. In a common case, the terminal current I and voltage V consist of incident and reflected components (assume their rms values). When the load impedance Z is equal to the conjugate of source impedance expressed as $Z = Z_S^*$, the terminal current becomes the incident current, which is calculated from

$$I_i = \frac{V_S}{Z_S^* + Z_S} = \frac{V_S}{2 \operatorname{Re} Z_S} \tag{1.29}$$

The terminal voltage defined as the incident voltage can be determined from

$$V_i = \frac{Z_S^* V_S}{Z_S^* + Z_S} = \frac{Z_S^* V_S}{2 \operatorname{Re} Z_S} \tag{1.30}$$

Consequently, the incident power, which is equal to the maximum available power from the source, can be obtained by

$$P_i = \operatorname{Re}(V_i I_i^*) = \frac{|V_S|^2}{4 \operatorname{Re} Z_S} \tag{1.31}$$

The incident power can be presented in a normalized form using Eq. (1.30) as

$$P_i = \frac{|V_i|^2 \operatorname{Re} Z_S}{|Z_S^*|^2} \tag{1.32}$$

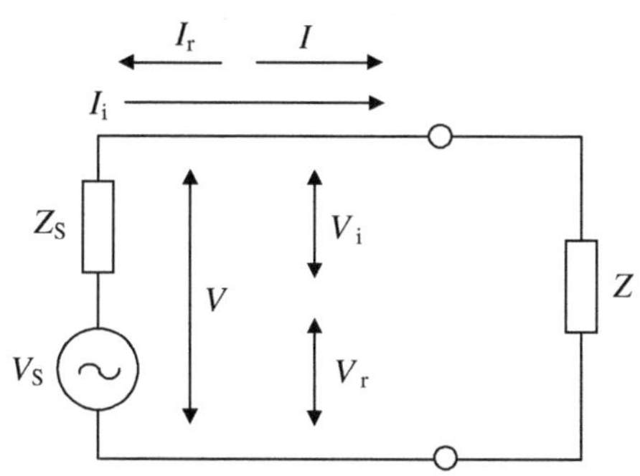

FIGURE 1.3 Incident and reflected voltages and currents.

This allows the normalized incident voltage wave a to be defined as the square root of the incident power P_i by

$$a = \sqrt{P_i} = \frac{V_i\sqrt{\text{Re}\,Z_S}}{Z_S^*} \tag{1.33}$$

Similarly, the normalized reflected voltage wave b defined as the square root of the reflected power P_r can be written as

$$b = \sqrt{P_r} = \frac{V_r\sqrt{\text{Re}\,Z_S}}{Z_S} \tag{1.34}$$

The incident power P_i can be expressed by the incident current I_i and the reflected power P_r can be respectively expressed by the reflected current I_r as

$$P_i = |I_i|^2\,\text{Re}Z_S \tag{1.35}$$

$$P_r = |I_r|^2\,\text{Re}Z_S \tag{1.36}$$

As a result, the normalized incident voltage wave a and reflected voltage wave b can be given by

$$a = \sqrt{P_i} = I_i\sqrt{\text{Re}\,Z_S} \tag{1.37}$$

$$b = \sqrt{P_r} = I_r\sqrt{\text{Re}\,Z_S} \tag{1.38}$$

The parameters a and b also can be called the *normalized incident* and *reflected current waves* or simply the *normalized incident* and *reflected waves*, respectively, because the normalized current waves and the normalized voltage waves represent the same parameters.

The voltage V and current I related to the normalized incident and reflected waves a and b can be written as

$$V = V_i + V_r = \frac{Z_S^*}{\sqrt{\text{Re}\,Z_S}}a + \frac{Z_S}{\sqrt{\text{Re}\,Z_S}}b \tag{1.39}$$

$$I = I_i - I_r = \frac{1}{\sqrt{\text{Re}\,Z_S}}a - \frac{1}{\sqrt{\text{Re}\,Z_S}}b \tag{1.40}$$

where

$$a = \frac{V + Z_S I}{2\sqrt{\text{Re}\,Z_S}} \qquad b = \frac{V - Z_S^* I}{2\sqrt{\text{Re}\,Z_S}} \tag{1.41}$$

The source impedance Z_S is often purely real and, therefore, is used as the normalized impedance. In microwave design technique, the characteristic impedance of the passive two-port networks, including the transmission lines and connectors, is considered as real and equal to 50 Ω. This is very important for measuring *S*-parameters when all transmission lines, source, and load should have the same real impedance. When $Z_S = Z_S^* = Z_0$, where Z_0 is the characteristic impedance, the ratio of the normalized reflected wave and the normalized incident wave for a one-port network is called the *reflection coefficient* Γ defined as

$$\Gamma = \frac{b}{a} = \frac{V - Z_S^* I}{V + Z_S I} = \frac{V - Z_S I}{V + Z_S I} = \frac{Z - Z_S}{Z + Z_S} \tag{1.42}$$

where $Z = V/I$.

For a two-port network shown in Fig. 1.4, the normalized reflected waves b_1 and b_2 can also be represented by the normalized incident waves a_1 and a_2, respectively, as

$$b_1 = S_{11} a_1 + S_{12} a_2 \tag{1.43}$$

$$b_2 = S_{21} a_1 + S_{22} a_2 \tag{1.44}$$

or, in a matrix form,

$$\begin{bmatrix} b_1 \\ b_2 \end{bmatrix} = \begin{bmatrix} S_{11} & S_{12} \\ S_{21} & S_{22} \end{bmatrix} \begin{bmatrix} a_1 \\ a_2 \end{bmatrix} \tag{1.45}$$

where the incident waves a_1 and a_2 and the reflected waves b_1 and b_2 for complex source and load impedances Z_S and Z_L are given by

$$a_1 = \frac{V_1 + Z_S I_1}{2\sqrt{\mathrm{Re}\, Z_S}} \qquad a_2 = \frac{V_2 + Z_L I_2}{2\sqrt{\mathrm{Re}\, Z_L}} \tag{1.46}$$

$$b_1 = \frac{V_1 - Z_S^* I_1}{2\sqrt{\mathrm{Re}\, Z_S}} \qquad b_2 = \frac{V_2 - Z_L^* I_2}{2\sqrt{\mathrm{Re}\, Z_L}} \tag{1.47}$$

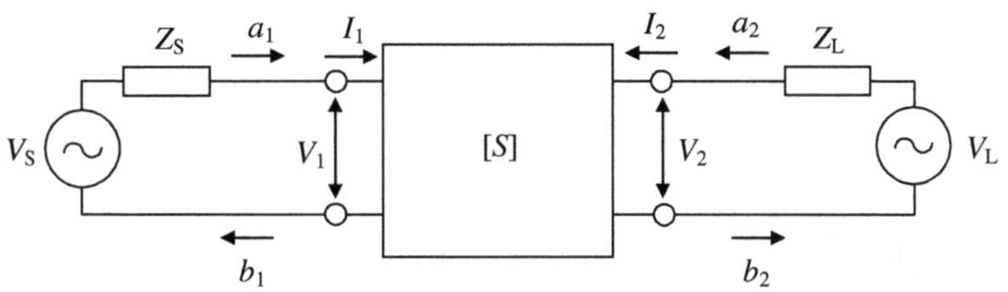

FIGURE 1.4 Basic diagram of S-parameter two-port network.

where S_{11}, S_{12}, S_{21}, and S_{22} are the S-parameters of the two-port network.

From Eq. (1.45), it follows that if $a_2 = 0$, then

$$S_{11} = \left.\frac{b_1}{a_1}\right|_{a_2=0} \qquad S_{21} = \left.\frac{b_2}{a_1}\right|_{a_2=0} \tag{1.48}$$

where S_{11} is the reflection coefficient and S_{21} is the transmission coefficient for ideal matching conditions at the output terminal when there is no incident power reflected from the load.

Similarly,

$$S_{12} = \left.\frac{b_1}{a_2}\right|_{a_1=0} \qquad S_{22} = \left.\frac{b_2}{a_2}\right|_{a_1=0} \tag{1.49}$$

where S_{12} is the transmission coefficient and S_{22} is the reflection coefficient for ideal matching conditions at the input terminal.

To convert S-parameters to the admittance Y-parameters, it is convenient to represent Eqs. (1.46) and (1.47) as

$$I_1 = (a_1 - b_1)\frac{1}{\sqrt{Z_0}} \qquad I_2 = (a_2 - b_2)\frac{1}{\sqrt{Z_0}} \tag{1.50}$$

$$V_1 = (a_1 + b_1)\sqrt{Z_0} \qquad V_1 = (a_2 + b_2)\sqrt{Z_0} \tag{1.51}$$

where it is assumed that the source and load impedances are real and equal to $Z_S = Z_L = Z_0$.

Substituting Eqs. (1.50) and (1.51) to Eqs. (1.11) and (1.12) results in

$$\frac{a_1 - b_1}{\sqrt{Z_0}} = Y_{11}(a_1 + b_1)\sqrt{Z_0} + Y_{12}(a_2 + b_2)\sqrt{Z_0} \tag{1.52}$$

$$\frac{a_2 - b_2}{\sqrt{Z_0}} = Y_{21}(a_1 + b_1)\sqrt{Z_0} + Y_{22}(a_2 + b_2)\sqrt{Z_0} \tag{1.53}$$

which can then be respectively converted to

$$-b_1(1 + Y_{11}Z_0) - b_2Y_{12}Z_0 = -a_1(1 - Y_{11}Z_0) + a_2Y_{12}Z_0 \tag{1.54}$$

$$-b_1Y_{21}Z_0 - b_2(1 + Y_{22}Z_0) = a_1Y_{21}Z_0 - a_2(1 - Y_{22}Z_0) \tag{1.55}$$

Equations (1.54) and (1.55) can be easily solved for the reflected waves b_1 and b_2 as

$$\begin{aligned} &b_1[(1 + Y_{11}Z_0)(1 + Y_{22}Z_0) - Y_{12}Y_{21}Z_0^2] \\ &\quad = a_1[(1 - Y_{11}Z_0)(1 + Y_{22}Z_0) + Y_{12}Y_{21}Z_0^2] - 2a_2Y_{12}Z_0 \end{aligned} \tag{1.56}$$

$$b_2\,[(1+Y_{11}Z_0)(1+Y_{22}Z_0)-Y_{12}Y_{21}Z_0^2]$$

$$=-2\,a_1\,Y_{21}Z_0+a_2\,[(1+Y_{11}Z_0)(1-Y_{22}Z_0)+Y_{12}Y_{21}Z_0^2] \tag{1.57}$$

Comparing equivalent Eqs. (1.43) and (1.44) and Eqs. (1.56) and (1.57) gives the following relationships between the scattering S-parameters and admittance Y-parameters:

$$S_{11}=\frac{(1-Y_{11}Z_0)(1+Y_{22}Z_0)+Y_{12}Y_{21}Z_0^2}{(1+Y_{11}Z_0)(1+Y_{22}Z_0)-Y_{12}Y_{21}Z_0^2} \tag{1.58}$$

$$S_{12}=\frac{-2\,Y_{12}Z_0}{(1+Y_{11}Z_0)(1+Y_{22}Z_0)-Y_{12}Y_{21}Z_0^2} \tag{1.59}$$

$$S_{21}=\frac{-2\,Y_{21}Z_0}{(1+Y_{11}Z_0)(1+Y_{22}Z_0)-Y_{12}Y_{21}Z_0^2} \tag{1.60}$$

$$S_{22}=\frac{(1+Y_{11}Z_0)(1-Y_{22}Z_0)+Y_{12}Y_{21}Z_0^2}{(1+Y_{11}Z_0)(1+Y_{22}Z_0)-Y_{12}Y_{21}Z_0^2}. \tag{1.61}$$

The relationships between S-parameters with Z-, H-, and $ABCD$-parameters can be obtained in a similar fashion. Table 1.2 shows the conversions between S-parameters and Z-, Y-, H-, and $ABCD$-parameters for the simplified case when the source impedance Z_S and the load impedance Z_L are equal to the characteristic impedance Z_0 [5].

1.3 Interconnections of Two-Port Networks

When analyzing the behavior of a particular electric circuit in terms of the two-port network parameters, it is often necessary to define the parameters of a combination of the two or more internal two-port networks. For example, the negative feedback amplifier circuit consists of an active two-port network representing the amplifier stage, which is connected in parallel with a passive feedback two-port network. In general, the two-port networks can be interconnected using parallel, series, series-parallel, or cascade connections.

To characterize the resulting two-port networks, it is necessary to take into account which currents and voltages are common for individual two-port networks. The most convenient set of parameters is one, for which the common currents and voltages represent a simple linear combination of the independent variables. For the interconnection shown in Fig. 1.5(*a*), the two-port networks Z_a and Z_b are connected in series for both the input and output terminals. Therefore, the currents flowing through these terminals are equal when

$$I_1=I_{1a}=I_{1b} \qquad I_2=I_{2a}=I_{2b} \tag{1.62}$$

S-Parameters through *Z*-, *Y*-, *H*-, *ABCD*-Parameters	***Z*-, *Y*-, *H*-, *ABCD*-Parameters through S-Parameters**
$S_{11}=\dfrac{(Z_{11}-Z_0)(Z_{22}+Z_0)-Z_{12}Z_{21}}{(Z_{11}+Z_0)(Z_{22}+Z_0)-Z_{12}Z_{21}}$	$Z_{11}=Z_0\dfrac{(1+S_{11})(1-S_{22})+S_{12}S_{21}}{(1-S_{11})(1-S_{22})-S_{12}S_{21}}$
$S_{12}=\dfrac{2Z_{12}Z_0}{(Z_{11}+Z_0)(Z_{22}+Z_0)-Z_{12}Z_{21}}$	$Z_{12}=Z_0\dfrac{2S_{12}}{(1-S_{11})(1-S_{22})-S_{12}S_{21}}$
$S_{21}=\dfrac{2Z_{21}Z_0}{(Z_{11}+Z_0)(Z_{22}+Z_0)-Z_{12}Z_{21}}$	$Z_{21}=Z_0\dfrac{2S_{21}}{(1-S_{11})(1-S_{22})-S_{12}S_{21}}$
$S_{22}=\dfrac{(Z_{11}+Z_0)(Z_{22}-Z_0)-Z_{12}Z_{21}}{(Z_{11}+Z_0)(Z_{22}+Z_0)-Z_{12}Z_{21}}$	$Z_{22}=Z_0\dfrac{(1-S_{11})(1+S_{22})+S_{12}S_{21}}{(1-S_{11})(1-S_{22})-S_{12}S_{21}}$
$S_{11}=\dfrac{(1-Y_{11}Z_0)(1+Y_{22}Z_0)+Y_{12}Y_{21}Z_0^2}{(1+Y_{11}Z_0)(1+Y_{22}Z_0)-Y_{12}Y_{21}Z_0^2}$	$Y_{11}=\dfrac{1}{Z_0}\dfrac{(1-S_{11})(1+S_{22})+S_{12}S_{21}}{(1+S_{11})(1+S_{22})-S_{12}S_{21}}$
$S_{12}=\dfrac{-2Y_{12}Z_0}{(1+Y_{11}Z_0)(1+Y_{22}Z_0)-Y_{12}Y_{21}Z_0^2}$	$Y_{12}=\dfrac{1}{Z_0}\dfrac{-2S_{12}}{(1+S_{11})(1+S_{22})-S_{12}S_{21}}$
$S_{21}=\dfrac{-2Y_{21}Z_0}{(1+Y_{11}Z_0)(1+Y_{22}Z_0)-Y_{12}Y_{21}Z_0^2}$	$Y_{21}=\dfrac{1}{Z_0}\dfrac{-2S_{21}}{(1+S_{11})(1+S_{22})-S_{12}S_{21}}$
$S_{22}=\dfrac{(1+Y_{11}Z_0)(1-Y_{22}Z_0)+Y_{12}Y_{21}Z_0^2}{(1+Y_{11}Z_0)(1+Y_{22}Z_0)-Y_{12}Y_{21}Z_0^2}$	$Y_{22}=\dfrac{1}{Z_0}\dfrac{(1+S_{11})(1-S_{22})+S_{12}S_{21}}{(1+S_{11})(1+S_{22})-S_{12}S_{21}}$
$S_{11}=\dfrac{(h_{11}-Z_0)(1+h_{22}Z_0)-h_{12}h_{21}Z_0}{(h_{11}+Z_0)(1+h_{22}Z_0)-h_{12}h_{21}Z_0}$	$h_{11}=Z_0\dfrac{(1+S_{11})(1+S_{22})-S_{12}S_{21}}{(1-S_{11})(1+S_{22})+S_{12}S_{21}}$
$S_{12}=\dfrac{2h_{12}Z_0}{(h_{11}+Z_0)(1+h_{22}Z_0)-h_{12}h_{21}Z_0}$	$h_{12}=\dfrac{2S_{12}}{(1-S_{11})(1+S_{22})+S_{12}S_{21}}$
$S_{21}=\dfrac{-2h_{21}Z_0}{(h_{11}+Z_0)(1+h_{22}Z_0)-h_{12}h_{21}Z_0}$	$h_{21}=\dfrac{-2S_{21}}{(1-S_{11})(1+S_{22})+S_{12}S_{21}}$
$S_{22}=\dfrac{(h_{11}+Z_0)(1-h_{22}Z_0)+h_{12}h_{21}Z_0}{(h_{11}+Z_0)(1+h_{22}Z_0)-h_{12}h_{21}Z_0}$	$h_{22}=\dfrac{1}{Z_0}\dfrac{(1-S_{11})(1-S_{22})-S_{12}S_{21}}{(1-S_{11})(1+S_{22})+S_{12}S_{21}}$
$S_{11}=\dfrac{AZ_0+B-CZ_0^2-DZ_0}{AZ_0+B+CZ_0^2+DZ_0}$	$A=\dfrac{(1+S_{11})(1-S_{22})+S_{12}S_{21}}{2S_{21}}$
$S_{12}=\dfrac{2(AD-BC)Z_0}{AZ_0+B+CZ_0^2+DZ_0}$	$B=Z_0\dfrac{(1+S_{11})(1+S_{22})-S_{12}S_{21}}{2S_{21}}$
$S_{21}=\dfrac{2Z_0}{AZ_0+B+CZ_0^2+DZ_0}$	$C=\dfrac{1}{Z_0}\dfrac{(1-S_{11})(1-S_{22})-S_{12}S_{21}}{2S_{21}}$
$S_{22}=\dfrac{-AZ_0+B-CZ_0^2+DZ_0}{AZ_0+B+CZ_0^2+DZ_0}$	$D=\dfrac{(1-S_{11})(1+S_{22})+S_{12}S_{21}}{2S_{21}}$

Table 1.2 Conversions between S-Parameters and *Z*-, *Y*-, *H*-, and *ABCD*-Parameters

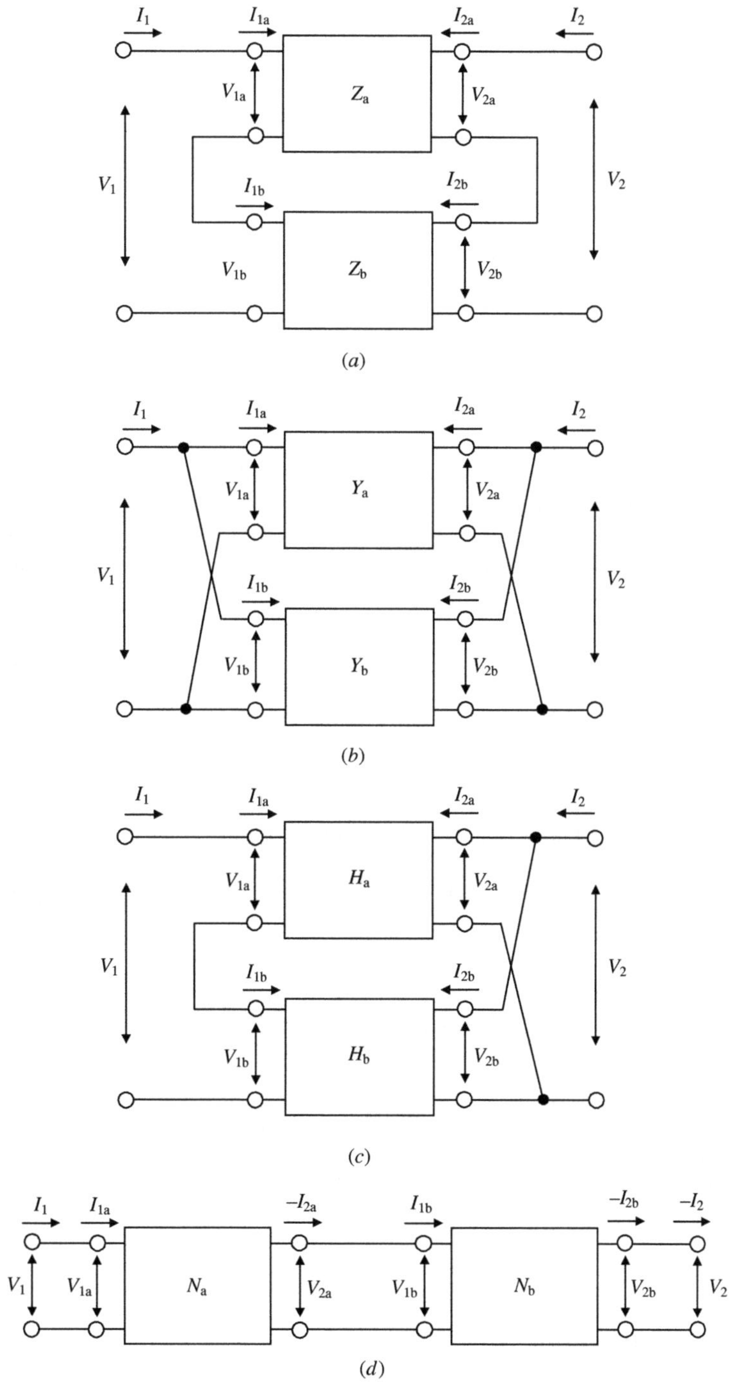

FIGURE 1.5 Different interconnections of two-port networks.

or, in a matrix form,

$$[I]=[I_a]=[I_b] \tag{1.63}$$

The terminal voltages of the resulting two-port network represent the corresponding sums of the terminal voltages of the individual two-port networks when

$$V_1 = V_{1a} + V_{1b} \qquad V_2 = V_{2a} + V_{2b} \tag{1.64}$$

or, in a matrix form,

$$[V]=[V_a]+[V_b] \tag{1.65}$$

The currents are common components both for the resulting and individual two-port networks. Consequently, to describe the properties of such a circuit, it is most convenient to use the impedance matrices. For each two-port network Z_a and Z_b, we can write using Eq. (1.62), respectively,

$$[V_a]=[Z_a][I_a]=[Z_a][I] \tag{1.66}$$

$$[V_b]=[Z_b][I_b]=[Z_b][I] \tag{1.67}$$

Adding both sides of Eqs. (1.66) and (1.67) yields

$$[V]=[Z][I] \tag{1.68}$$

where

$$[Z]=[Z_a]+[Z_b]=\begin{bmatrix} Z_{11a}+Z_{11b} & Z_{12a}+Z_{12b} \\ Z_{21a}+Z_{21b} & Z_{22a}+Z_{22b} \end{bmatrix} \tag{1.69}$$

The circuit shown in Fig. 1.5(*b*) is composed of the two-port networks Y_a and Y_b connected in parallel, where the common components for both resulting and individual two-port networks are the input and output voltages, respectively,

$$V_1 = V_{1a} = V_{1b} \qquad V_2 = V_{2a} = V_{2b} \tag{1.70}$$

or, in a matrix form,

$$[V]=[V_a]=[V_b] \tag{1.71}$$

Consequently, to describe the circuit properties in this case, it is convenient to use the admittance matrices that give the resulting matrix equation in the form

$$[I]=[Y][V] \tag{1.72}$$

where

$$[Y]=[Y_a]+[Y_b]=\begin{bmatrix} Y_{11a}+Y_{11b} & Y_{12a}+Y_{12b} \\ Y_{21a}+Y_{21b} & Y_{22a}+Y_{22b} \end{bmatrix} \tag{1.73}$$

The series connection of the input terminals and parallel connection of the output terminals are characterized by the circuit in Fig. 1.5(*c*), which shows a series-parallel connection of two-port networks. The common components for this circuit are the input currents and the output voltages. As a result, it is most convenient to analyze the circuit properties using hybrid matrices. The resulting two-port hybrid matrix is equal to the sum of the two individual hybrid matrices written as

$$[H]=[H_a]+[H_b]=\begin{bmatrix} h_{11a}+h_{11b} & h_{12a}+h_{12b} \\ h_{21a}+h_{21b} & h_{22a}+h_{22b} \end{bmatrix} \tag{1.74}$$

Figure 1.5(*d*) shows the cascade connection of the two individual two-port networks. For such an approach using the one-by-one interconnection of the two-port networks, the output voltage and the output current of the first network are equal to the input voltage and the input current of the second one, respectively, when

$$V_1=V_{1a} \qquad I_1=I_{1a} \tag{1.75}$$

$$V_{2a}=V_1 \qquad -I_{2a}=I_{1b} \tag{1.76}$$

$$V_{2b}=V_2 \qquad -I_{2b}=-I_2 \tag{1.77}$$

In this case, it is convenient to use a system of *ABCD*-parameters given by Eqs. (1.21) and (1.22). As a result, for the first individual two-port network shown in Fig. 1.5(*d*),

$$\begin{bmatrix} V_{1a} \\ I_{1a} \end{bmatrix}=\begin{bmatrix} A_a & B_a \\ C_a & D_a \end{bmatrix}\begin{bmatrix} V_{2a} \\ -I_{2a} \end{bmatrix} \tag{1.78}$$

or, using Eqs. (1.75) and (1.76),

$$\begin{bmatrix} V_1 \\ I_1 \end{bmatrix}=\begin{bmatrix} A_a & B_a \\ C_a & D_a \end{bmatrix}\begin{bmatrix} V_{1b} \\ I_{1b} \end{bmatrix} \tag{1.79}$$

Similarly, for the second individual two-port network,

$$\begin{bmatrix} V_{1b} \\ I_{1b} \end{bmatrix} = \begin{bmatrix} A_b & B_b \\ C_b & D_b \end{bmatrix} \begin{bmatrix} V_{2b} \\ -I_{2b} \end{bmatrix} = \begin{bmatrix} A_b & B_b \\ C_b & D_b \end{bmatrix} \begin{bmatrix} V_2 \\ -I_2 \end{bmatrix} \quad (1.80)$$

Then, substituting matrix Eq. (1.80) to matrix Eq. (1.79) yields

$$\begin{bmatrix} V_1 \\ I_1 \end{bmatrix} = \begin{bmatrix} A_a & B_a \\ C_a & D_a \end{bmatrix} \begin{bmatrix} A_b & B_b \\ C_b & D_b \end{bmatrix} \begin{bmatrix} V_2 \\ -I_2 \end{bmatrix} = \begin{bmatrix} A & B \\ C & D \end{bmatrix} \begin{bmatrix} V_2 \\ -I_2 \end{bmatrix} \quad (1.81)$$

Consequently, the transmission matrix of the resulting two-port network obtained by the cascade connection of the two or more individual two-port networks is determined by multiplying the transmission matrices of the individual networks. This important property is widely used in the analysis and design of transmission networks and systems.

1.4 Practical Two-Port Networks

1.4.1 Single-Element Networks

The simplest networks, which include only one element, can be constructed by a series-connected admittance Y, as shown in Fig. 1.6(*a*), or by a parallel-connected impedance Z, as shown in Fig. 1.6(*b*).

The two-port network, consisting of the single series admittance Y, can be described in a system of the admittance Y-parameters as

$$I_1 = YV_1 - YV_2 \quad (1.82)$$

$$I_2 = -YV_1 + YV_2 \quad (1.83)$$

or, in a matrix form,

$$[Y] = \begin{bmatrix} Y & -Y \\ -Y & Y \end{bmatrix} \quad (1.84)$$

which means that $Y_{11} = Y_{22} = Y$ and $Y_{12} = Y_{21} = -Y$. The resulting matrix is a singular matrix with $|Y| = 0$. Consequently, it is impossible to

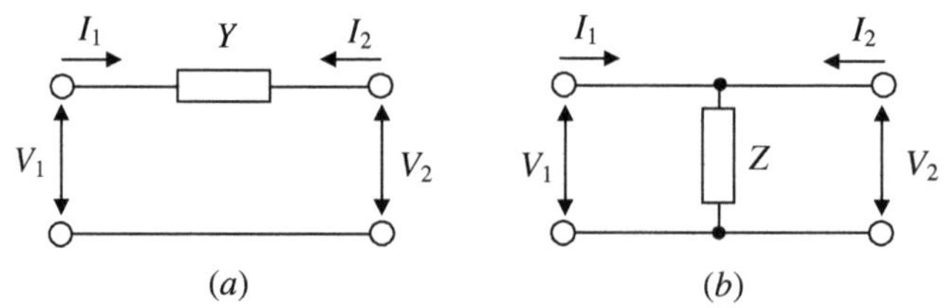

FIGURE 1.6 Single-element networks.

determine such a two-port network with the series admittance Y-parameters through a system of the impedance Z-parameters. However, by using the hybrid H-parameters and the transmission $ABCD$-parameters, it can be respectively described by

$$[H]=\begin{bmatrix} 1/Y & 1 \\ -1 & 0 \end{bmatrix} \qquad [ABCD]=\begin{bmatrix} 1 & 1/Y \\ 0 & 1 \end{bmatrix} \tag{1.85}$$

Similarly, for a two-port network with a single parallel impedance Z,

$$[Z]=\begin{bmatrix} Z & Z \\ Z & Z \end{bmatrix} \tag{1.86}$$

which means that $Z_{11} = Z_{12} = Z_{21} = Z_{22} = Z$. The resulting matrix is a singular matrix with $|Z| = 0$. In this case, it is impossible to determine such a two-port network with the parallel impedance Z-parameters through a system of the admittance Y-parameters. By using the H- and $ABCD$-parameters, the two-port network can be described by

$$[H]=\begin{bmatrix} 0 & 1 \\ -1 & 1/Z \end{bmatrix} \qquad [ABCD]=\begin{bmatrix} 1 & 0 \\ 1/Z & 1 \end{bmatrix}. \tag{1.87}$$

1.4.2 π- and T-Type Networks

The basic configurations of a two-port network, which usually describe the electrical properties of the active devices, can be represented in the form of a π-circuit shown in Fig. 1.7(*a*) or in the form of a T-circuit shown in Fig. 1.7(*b*). Here, the π-circuit includes the current source $g_m V_1$ and the T-circuit includes the voltage source $r_m I_1$.

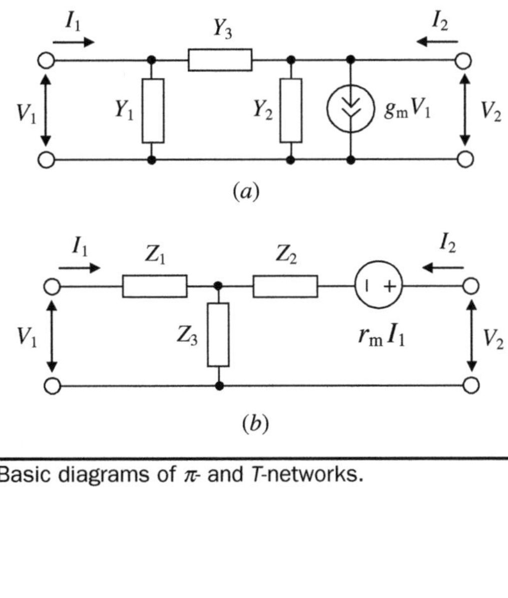

FIGURE 1.7 Basic diagrams of π- and T-networks.

By writing the two loop equations using Kirchhoff's current law or applying Eqs. (1.13) and (1.14) for the π-circuit, we obtain

$$I_1 - (Y_1 + Y_3)V_1 + Y_3\ V_2 = 0 \tag{1.88}$$

$$I_2 + (g_m - Y_3)V_1 + (Y_2 + Y_3)\ V_2 = 0 \tag{1.89}$$

Equations (1.88) and (1.89) can be rewritten as matrix Eq. (1.3) with

$$[M] = \begin{bmatrix} 1 & 0 \\ 0 & 1 \end{bmatrix} \quad \text{and} \quad [N] = \begin{bmatrix} -(Y_1 + Y_3) & Y_3 \\ -g_m + Y_3 & -(Y_2 + Y_3) \end{bmatrix}$$

Because matrix $[M]$ is nonsingular, such a two-port network can be described by a system of the admittance Y-parameters as

$$[Y] = -[M]^{-1}[N] = \begin{bmatrix} Y_1 + Y_3 & -Y_3 \\ g_m - Y_3 & Y_2 + Y_3 \end{bmatrix} \tag{1.90}$$

Similarly, for a two-port network in the form of a T-circuit using Kirchhoff's voltage law or applying Eqs. (1.8) and (1.9), we obtain

$$[Z] = -[M]^{-1}[N] = \begin{bmatrix} Z_1 + Z_3 & Z_3 \\ r_m + Z_3 & Z_2 + Z_3 \end{bmatrix} \tag{1.91}$$

If $g_m = 0$ for a π-circuit and $r_m = 0$ for a T-circuit, their corresponding matrices in a system of $ABCD$-parameters can be written as for π-circuit

$$[ABCD] = \begin{bmatrix} 1 + \dfrac{Y_2}{Y_3} & \dfrac{1}{Y_3} \\ Y_1 + Y_2 + \dfrac{Y_1 Y_2}{Y_3} & 1 + \dfrac{Y_1}{Y_3} \end{bmatrix} \tag{1.92}$$

for T-circuit

$$[ABCD] = \begin{bmatrix} 1 + \dfrac{Z_2}{Z_3} & Z_1 + Z_2 + \dfrac{Z_1 Z_2}{Z_3} \\ \dfrac{1}{Z_3} & 1 + \dfrac{Z_1}{Z_3} \end{bmatrix} \tag{1.93}$$

When the impedances of a T-circuit and the admittances of a π-circuit are properly interrelated, these two circuits become equivalent regarding the effect on any other two-port network. For example, for a π-circuit shown in Fig. 1.8(*a*), we can write

$$I_1 = Y_1\ V_{13} + Y_3\ V_{12} = Y_1\ V_{13} + Y_3\ (V_{13} - V_{23})$$

$$= (Y_1 + Y_3)V_{13} - Y_3 V_{23} \tag{1.94}$$

$$I_2 = Y_2 V_{23} - Y_3 V_{12} = Y_2 V_{23} - Y_3 (V_{13} - V_{23})$$

$$= -Y_3 V_{13} + (Y_2 + Y_3) V_{23} \tag{1.95}$$

Solving Eqs. (1.94) and (1.95) for voltages V_{13} and V_{23} yields

$$V_{13} = \frac{Y_2 + Y_3}{Y_1Y_2 + Y_1Y_2 + Y_1Y_2}\ I_1 + \frac{Y_3}{Y_1Y_2 + Y_1Y_2 + Y_1Y_2} I_2 \tag{1.96}$$

$$V_{23} = \frac{Y_3}{Y_1Y_2\ + Y_1Y_2 + Y_1Y_2}\ I_1 + \frac{Y_1 + Y_3}{Y_1Y_2 + Y_1Y_2 + Y_1Y_2} I_2 \tag{1.97}$$

Similarly, for a T-circuit shown in Fig. 1.8(*b*),

$$V_{13} = Z_1 I_1 + Z_3 I_3 = Z_1 I_1 + Z_3 (I_1 + I_2)$$

$$= (Z_1 + Z_3)\ I_1 + Z_3 I_2 \tag{1.98}$$

$$V_{23} = Z_2\ I_2 + Z_3\ I_3 = Z_2\ I_2 + Z_3 (I_1 + I_2)$$

$$= Z_3 I_1 + (Z_2 + Z_3)\ I_2 \tag{1.99}$$

and the equations for currents I_1 and I_2 can be obtained by

$$I_1 = \frac{Z_2 + Z_3}{Z_1Z_2 + Z_1Z_2 + Z_1Z_2} V_{13} - \frac{Z_3}{Z_1Z_2 + Z_1Z_2 + Z_1Z_2} V_{23} \tag{1.100}$$

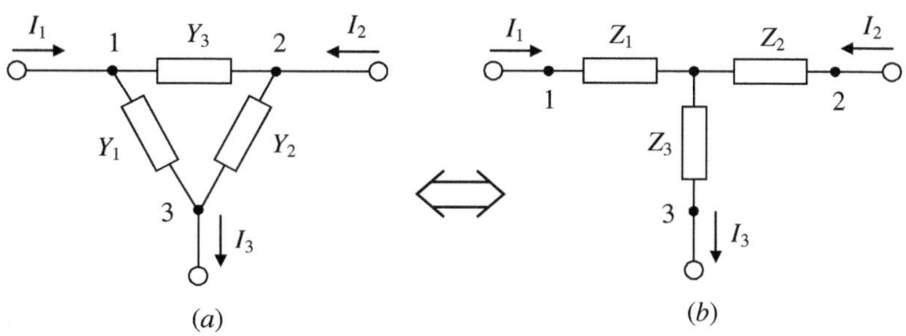

Figure 1.8 Equivalence of π- and T-circuits.

T- to π- Transformation	π- to *T*- Transformation
$Y_1 = \frac{Z_2}{Z_1Z_2 + Z_2Z_3 + Z_1Z_3}$	$Z_1 = \frac{Y_2}{Y_1Y_2 + Y_2Y_3 + Y_1Y_3}$
$Y_2 = \frac{Z_1}{Z_1Z_2 + Z_2Z_3 + Z_1Z_3}$	$Z_2 = \frac{Y_1}{Y_1Y_2 + Y_2Y_3 + Y_1Y_3}$
$Y_3 = \frac{Z_3}{Z_1Z_2 + Z_2Z_3 + Z_1Z_3}$	$Z_3 = \frac{Y_3}{Y_1Y_2 + Y_2Y_3 + Y_1Y_3}$

TABLE 1.3 Relationships between π- and *T*-Circuit Parameters

$$I_2 = -\frac{Z_3}{Z_1Z_2 + Z_1Z_2 + Z_1Z_2} V_{13} + \frac{Z_1 + Z_3}{Z_1Z_2 + Z_1Z_2 + Z_1Z_2} V_{23} \tag{1.101}$$

To establish a *T*- to π- transformation, it is necessary to equate the coefficients for V_{13} and V_{23} in Eqs. (1.100) and (1.101) to the corresponding coefficients in Eqs. (1.94) and (1.95). Similarly, to establish a π- to *T*- transformation, it is necessary to equate the coefficients for I_1 and I_2 in Eqs. (1.98) and (1.99) to the corresponding coefficients in Eqs. (1.96) and (1.97). The resulting relationships between the admittances for a π-circuit and the impedances for a *T*-circuit are given in Table 1.3.

1.5 Three-Port Network with Common Terminal

The concept of a two-port network with two independent sources can generally be extended to any multiport networks. Figure 1.9 shows the three-port network, where all three independent sources are connected to a common point. The three-port network matrix Eq. (1.3) can be described in a scalar form as

$$\left.\begin{aligned} m_{11}V_1 + m_{12}V_2 + m_{13}V_3 + n_{11}I_1 + n_{12}I_2 + n_{13}I_3 = 0 \\ m_{21}V_1 + m_{22}V_2 + m_{23}V_3 + n_{21}I_1 + n_{22}I_2 + n_{23}I_3 = 0 \\ m_{31}V_1 + m_{32}V_2 + m_{33}V_3 + n_{31}I_1 + n_{32}I_2 + n_{33}I_3 = 0 \end{aligned}\right\} \tag{1.102}$$

If matrix [*N*] in Eq. (1.102) is nonsingular when $|N| \neq 0$, this system of three equations can be rewritten in admittance matrix representation in terms of the voltage matrix [*V*] similarly to a two-port network by

$$\begin{bmatrix} I_1 \\ I_2 \\ I_3 \end{bmatrix} = \begin{bmatrix} Y_{11} & Y_{12} & Y_{13} \\ Y_{21} & Y_{22} & Y_{23} \\ Y_{31} & Y_{32} & Y_{33} \end{bmatrix} \begin{bmatrix} V_1 \\ V_2 \\ V_3 \end{bmatrix} \tag{1.103}$$

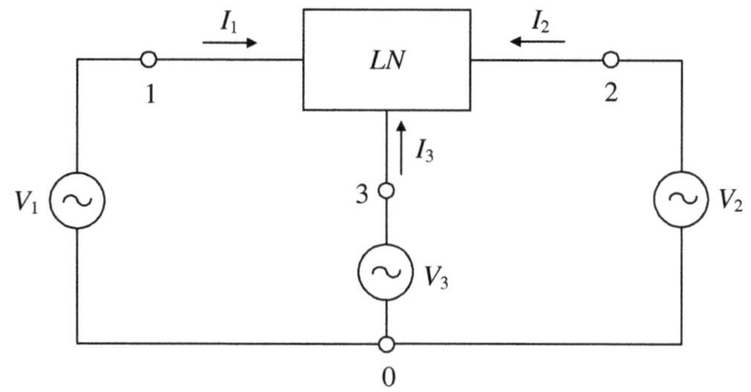

FIGURE 1.9 Basic diagram of three-port network with common terminal.

The matrix [Y] in Eq. (1.103) is the indefinite admittance matrix of the three-port network and represents a singular matrix with two important properties:

The sum of all terminal currents entering the circuit is equal to zero, that is, $I_1 + I_2 + I_3 = 0$.
All terminal currents entering the circuit depend on the voltages between circuit terminals, which makes the sum of all terminal voltages equal to zero, that is, $V_{13} + V_{32} + V_{21} = 0$.

According to the first property, adding the left- and right-hand parts of matrix Eq. (1.103) results in

$$(Y_{11} + Y_{21} + Y_{31})V_1 + (Y_{12} + Y_{22} + Y_{32})V_2 + (Y_{13} + Y_{23} + Y_{33})\,V_3 = 0 \tag{1.104}$$

Because all terminal voltages (V_1, V_2, and V_3) can be set independently from each other, Eq. (1.104) can be satisfied only if any column sum is identically zero,

$$\left.\begin{aligned} Y_{11} + Y_{21} + Y_{31} &= 0 \\ Y_{12} + Y_{22} + Y_{32} &= 0 \\ Y_{13} + Y_{23} + Y_{33} &= 0 \end{aligned}\right\} \tag{1.105}$$

The neither terminal currents will neither decrease nor increase with the simultaneous change of all terminal voltages by the same magnitude. Consequently, if all terminal voltages are equal to a nonzero value when $V_1 = V_2 = V_3 = V_0$, a lack of the terminal currents occurs when $I_1 = I_2 = I_3 = 0$. For example, from the first row of the matrix Eq. (1.103), it follows that $I_1 = Y_{11}V_1 + Y_{12}V_2 + Y_{13}V_3$, and we can write

$$0 = (Y_{11} + Y_{12} + Y_{13})\,V_0 \tag{1.106}$$

which results due to the nonzero value V_0 in

$$Y_{11} + Y_{12} + Y_{13} = 0 \tag{1.107}$$

Applying the same approach to other two rows results in

$$\left.\begin{array}{l} Y_{11} + Y_{12} + Y_{13} = 0 \\ Y_{21} + Y_{22} + Y_{23} = 0 \\ Y_{31} + Y_{32} + Y_{33} = 0 \end{array}\right\} \tag{1.108}$$

Consequently, by using Eqs. (1.105) through (1.108), the indefinite admittance Y-matrix of a three-port network can be rewritten by

$$[Y] = \begin{bmatrix} Y_{11} & Y_{12} & -(Y_{11} + Y_{12}) \\ Y_{21} & Y_{22} & -(Y_{21} + Y_{22}) \\ -(Y_{11} + Y_{21}) & -(Y_{12} + Y_{22}) & Y_{11} + Y_{12} + Y_{21} + Y_{22} \end{bmatrix} \tag{1.109}$$

By selecting successively terminals 1, 2, and 3 as the datum terminal, the corresponding three two-port admittance matrices of the initial three-port network can be obtained [6]. In this case, the admittance matrices of the bipolar transistor will correspond to a common-emitter configuration shown in Fig. 1.10(*a*), a common-base configuration shown in Fig. 1.10(*b*), and a common-collector configuration shown in Fig. 1.10(*c*), respectively. If the common-emitter device is treated as a two-port network characterized by four Y-parameters (Y_{11}, Y_{12}, Y_{21}, and Y_{22}), the two-port matrix of the common-collector configuration with grounded collector terminal is simply obtained by deleting the second row and the second column in matrix Eq. (1.109). For the common-base configuration with grounded base terminal, the first row and the first column are deleted because the emitter terminal is considered the input terminal, and the remaining rows and columns are interchanged.

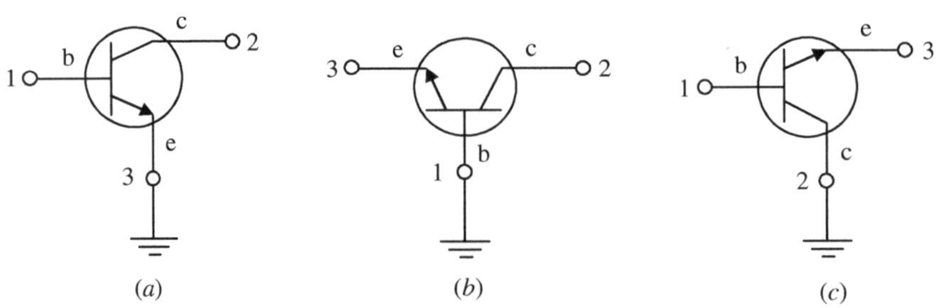

Figure 1.10 Bipolar transistors with different common terminals.

	Y-Parameters	Z-Parameters
Common emitter (source)	$\begin{matrix} Y_{11} & Y_{12} \\ Y_{21} & Y_{22} \end{matrix}$	$\begin{matrix} Z_{11} & Z_{12} \\ Z_{21} & Z_{22} \end{matrix}$
Common base (gate)	$\begin{matrix} Y_{11}+Y_{12}+Y_{21}+Y_{22} & -(Y_{12}+Y_{22}) \\ -(Y_{21}+Y_{22}) & Y_{22} \end{matrix}$	$\begin{matrix} Z_{11}+Z_{12}+Z_{21}+Z_{22} & -(Z_{12}+Z_{22}) \\ -(Z_{21}+Z_{22}) & Z_{22} \end{matrix}$
Common collector (drain)	$\begin{matrix} Y_{11} & -(Y_{11}+Y_{12}) \\ -(Y_{11}+Y_{21}) & Y_{11}+Y_{12}+Y_{21}+Y_{22} \end{matrix}$	$\begin{matrix} Z_{11} & -(Z_{11}+Z_{12}) \\ -(Z_{11}+Z_{21}) & Z_{11}+Z_{12}+Z_{21}+Z_{22} \end{matrix}$

TABLE 1.4 *Y*- and *Z*-parameters of active device with different common terminal

A similar approach can be applied to the indefinite three-port impedance network. This allows the *Z*-parameters of the impedance matrices of the common-base and the common-collector configurations through known impedance *Z*-parameters of the common-emitter configuration of the transistor to be determined. Parameters of the three-port network, which can describe the electrical behavior of the three-port bipolar or field-effect transistor configured with different common terminals, are given in Table 1.4.

1.6 Lumped Elements

Generally, passive RF and microwave lumped or integrated circuits are designed based on the lumped elements, distributed elements, or combination of both types of elements. Distributed elements represent any sections of the transmission lines of different lengths, types, and characteristic impedances. The basic lumped elements are inductors and capacitors that are small in size in comparison with the transmission-line wavelength λ, and usually their linear dimensions are less than $\lambda/10$ or even $\lambda/16$. In applications where lumped elements are used, their basic advantages are small physical size and low production cost. However, their main drawbacks are lower quality factor and power-handling capability compared with distributed elements.

1.6.1 Inductors

Inductors are lumped elements that store energy in a magnetic field. Lumped inductors can be realized using several different configurations such as a short-section of a strip conductor or wire, a single loop, or a spiral. The printed high-impedance microstrip-section inductor is usually used for low inductance values, typically less than 2 nH and often meandered to reduce the component size. The printed microstrip single-loop inductors are not very popular due to their limited inductance per unit area. The approximate expression for the microstrip short-section inductance in free space is given by

$$L(\text{nH}) = 0.2\times10^{-3}l\left[\ln\left(\frac{l}{W+t}\right)+1.193+\frac{W+t}{3l}\right]K_g \tag{1.110}$$

where the conductor length l, conductor width W, and conductor thickness t are in microns, and the term K_g accounts for the presence of a ground plane defined as

$$K_g = 0.57 - 0.145\ln\frac{W}{h} \qquad \text{for}\frac{W}{h} > 0.05 \tag{1.111}$$

where h is the spacing from ground plane [7, 8].

Spiral inductors can have a circular configuration, a rectangular (square) configuration shown in Fig. 1.11(*a*), or an octagonal configuration shown in Fig. 1.11(*b*), if the technology allows 45° routing. The circular geometry is superior in electrical performance, whereas the rectangular shapes are easy to lay out and fabricate. Printed inductors are based on using thin-film or thick-film Si or GaAs fabrication processes, and the inner conductor is pulled out to connect with other circuitry through a bondwire, an air bridge, or by using multilevel crossover metal. The general expression for spiral inductor, which is also valid for its planar integration within accuracy of around 3%, is based on a Wheeler formula and can be obtained as

$$L\ (\text{nH}) = \frac{K_1 n^2 d_{avg}}{1+K_2\rho} \tag{1.112}$$

where n is the number of turns, $d_{avg} = (d_{out} + d_{in})/2$ is the average diameter, $\rho = (d_{out} + d_{in})/(d_{out} - d_{in})$ is the fill ratio, d_{out} is the outer diameter in μm, d_{in} is the inner diameter in μm, and the coefficients K_1 and K_2 are layout dependent as follows: square – $K_1 = 2.34$, $K_2 = 2.75$, hexagonal – $K_1 = 2.33$, $K_2 = 3.82$, and octagonal – $K_1 = 2.25$, $K_2 = 3.55$ [9, 10].

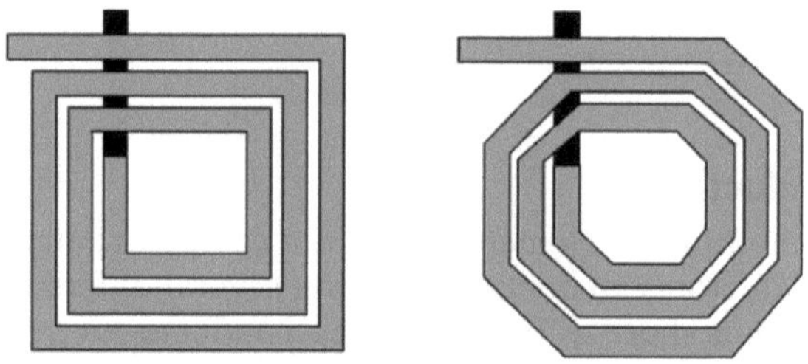

(*a*) (*b*)

FIGURE 1.11 Spiral inductor layouts.

In contrast to the capacitors, high-quality inductors cannot be readily available in a standard complementary metal-oxide semiconductor (CMOS) technology. Therefore, it is necessary to use special techniques to improve the inductor electrical performance. By using a standard CMOS technology with only two metal layers and a heavily doped substrate, the spiral inductor will have a large series resistance compared with three-four metal layer technologies, and the substrate losses become a very important factor due to a relatively low resistivity of silicon. A major source of substrate losses is the capacitive coupling when current is flowing not only through the metal strip but also through the silicon substrate. Another important source of substrate losses is the inductive coupling when the magnetic field penetrates deeply into the silicon substrate due to the planar inductor structure, inducing current loops and related losses. However, the latter effects are particularly important for large-area inductors and can be overcome by using silicon micromachining techniques [11].

The simplified equivalent circuit for the CMOS spiral microstrip inductor is shown in Fig. 1.12, where L_s models the self and mutual inductances, R_s is the series coil resistance, C_{ox} is the parasitic oxide capacitance from the metal layer to the substrate, R_{si} is the resistance of the conductive silicon substrate, C_{si} is the silicon substrate parasitic capacitance, and C_c is the parasitic coupling capacitance [12]. The parasitic silicon substrate capacitance C_{si} is sufficiently small and in most cases it can be neglected. Such a model shows an accurate agreement between simulated and measured data within 10% across a variety of inductor geometries and substrate dopings up to 20 GHz [13]. At frequencies well below the inductor self-resonant frequency ω_{SRF}, the coupling capacitance C_c between metal segments due to

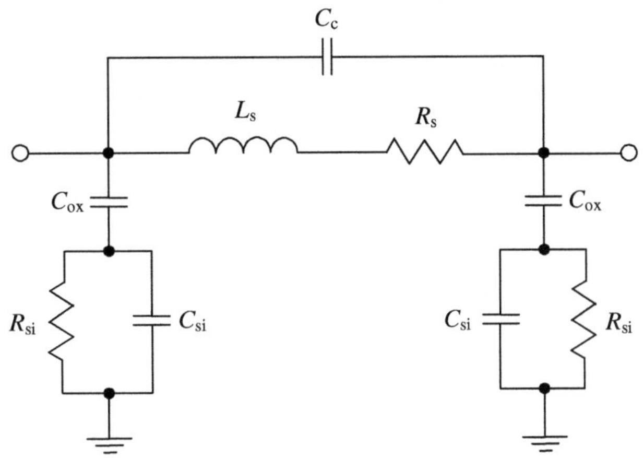

Figure 1.12 Equivalent circuit of a square spiral inductor.

fringing fields in both the dielectric and air regions can also be neglected as the relative dielectric constant of the oxide is small enough [14]. In this case, if one side of the inductor is grounded, the self-resonant frequency of the spiral inductor can approximately be calculated from

$$\omega_{\mathrm{SRF}} = \frac{1}{\sqrt{L_s C_{ox}}} \sqrt{\frac{L_s - R_s^2 C_{ox}}{L_s - R_{si}^2 C_{ox}}} \tag{1.113}$$

At frequencies higher than the self-resonant frequency ω_{SRF}, the inductor exhibits a capacitive behavior. The self-resonant frequency ω_{SRF} is limited mainly by the parasitic oxide capacitance C_{ox}, which is inversely proportional to the oxide thickness between the metal layer and substrate. The frequency, at which the inductor quality factor Q is maximal, can be obtained as

$$\omega_{\mathrm{Q}} = \frac{1}{\sqrt{L_s C_{ox}}} \sqrt{\frac{R_s}{2R_{si}}} \left(\sqrt{1 + \frac{4R_{si}}{3R_s}} - 1 \right)^{0.5} \tag{1.114}$$

The inductor metal conductor series resistance R_s can be easily calculated at low frequencies as the product of the sheet resistance and the number of squares of the metal trace. However, at high frequencies, the skin effect and other magnetic field effects will cause a nonuniform current distribution in the inductor profile. In this case, a simple increase in a diameter of the inductor metal turn does not necessarily reduce correspondingly the inductor series resistance. For example, for the same inductance value, the difference in resistance between the two inductors, when one of which has a two times wider metal strip, is only a factor of 1.35 [15]. Moreover, at very high frequencies, the largest contribution to the series resistance does not come from the longer outer turns, but from the inner turns. This phenomenon is a result of the generation of circular eddy currents in the inner conductors, whose direction is such that they oppose the original change in magnetic field. On the inner side of the inner turn, coil current and eddy current flow in the same direction so the current density is larger than average. On the outer side, both currents cancel and the current density is smaller than average. As a result, the current in the inner turn is pushed to the inside of the conductor.

In hybrid or monolithic applications, bondwires are used to interconnect different components such as lumped elements, planar transmission lines, solid-state devices, and integrated circuits. These bondwires, which are usually made of gold or aluminium, have 0.5- to 1.0-mil diameters and their lengths are electrically short compared to the operating wavelength. To characterize the electrical behaviour of the bondwires, simple formulas in terms of their inductances and series resistances can be used. As a first-order approximation, the parasitic capacitance associated with bondwires can be neglected. In this case, the bondwire inductance can be estimated for $l >> d$, where l is the bondwire length in μm and d is the bondwire diameter in μm, as

$$L(\text{nH}) = 0.2 \times 10^{-3} l \left(\ln \frac{4l}{d} + 0.5 \frac{d}{l} - 1 + C \right) \tag{1.115}$$

where $C = 0.25 \times \tanh(4\delta/d)$ is the frequency-dependent correction factor, which is a function of bondwire diameter and its material's skin depth δ [9, 16].

1.6.2 Capacitors

Capacitors are lumped elements that store energy due to an electric field between two electrodes (or plates) when a voltage is applied across them. In this case, charge of equal magnitude but opposite sign accumulates on the opposing capacitor plates. The capacitance depends on the area of the plates, separation, and dielectric material between them. The basic structure of a chip capacitor shown in Fig. 1.13(*a*) consists of two parallel plates, each of area $A = W \times l$, which are separated by a dielectric material of thickness d and permittivity $\varepsilon_0\varepsilon_r$, where ε_0 is the free-space permittivity (8.85×10^{-12} farads/m) and ε_r is the relative dielectric constant.

Chip capacitors are usually used in hybrid integrated circuits when relatively high capacitance values are required. In the parallel-plate configuration, equation to calculate the capacitance is commonly written as

$$C(\text{pF}) = 8.85 \times 10^{-3} \varepsilon_r \frac{Wl}{d} \tag{1.116}$$

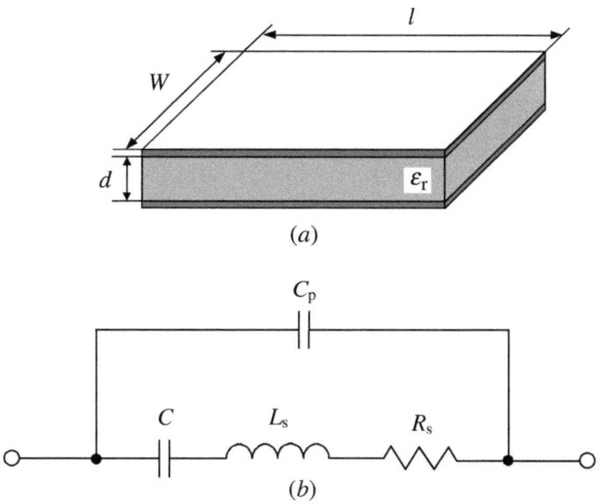

FIGURE 1.13 Parallel capacitor topology and its equivalent circuit.

where W, l, and d are dimensions in millimeters. Generally, the low-frequency bypass capacitor values are expressed in microfarads and nanofarads, high-frequency blocking and tuning capacitors are expressed in picofarads, and parasitic or fringing capacitances are written in femtofarads. This basic formula given by Eq. (1.116) can also be applied to capacitors based on a multilayer technique [8]. The lumped-element equivalent circuit of a capacitor is shown in Fig. 1.13(*b*), where L_s is the series plate inductance, R_s is the series contact and plate resistance, and C_p is the parasitic parallel capacitance. When $C >> C_p$, the frequency ω_{SRF}, at which the reactances of series elements C and L_s become equal, is called the *capacitor self-resonant frequency*, and the capacitor impedance is equal to the resistance R_s.

For monolithic applications where relatively low capacitances (typically < 0.5 pF) are required, planar series capacitances in the form of microstrip or interdigital configurations can be used. These capacitors are simply formed by gaps in the center conductor of the microstrip lines, and they do not require any dielectric films. The gap capacitor shown in Fig. 1.14(*a*) can be equivalently represented by a series coupling capacitance and two parallel fringing capacitances [17]. The interdigital capacitor is a multifinger periodic structure, as shown in Fig. 1.14(*b*), where the capacitance occurs across a narrow gap between thin-film transmission-line conductors [18]. These gaps are essentially very long and folded to use a small amount of area. In this case, it is important to keep the size of the capacitor very small relative to a wavelength so that it can be treated as a lumped element. A larger total width-to-length ratio results in the desired higher shunt capacitance and lower series inductance. An approximate expression for the total

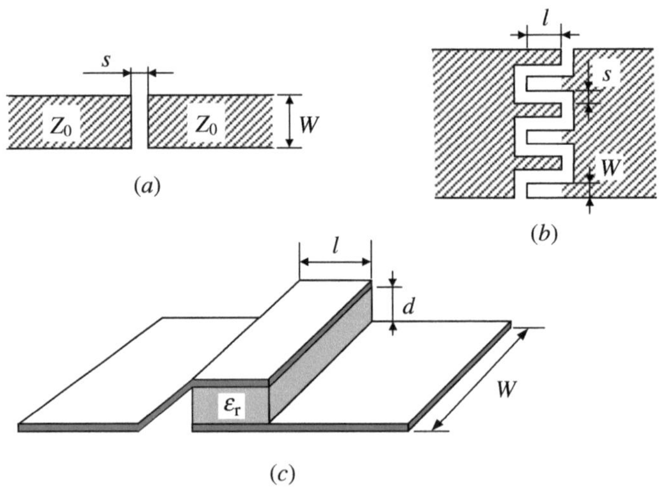

Figure 1.14 Different series capacitor topologies.

capacitance of interdigital structure with $s = W$ and length l less than a quarter wavelength can be given by

$$C\ (\text{pF}) = (\varepsilon_r + 1)l[(N-3)A_1 + A_2] \tag{1.117}$$

where N is the number of fingers and

$$A_1(\text{pF}/\mu\text{m}) = 4.409\ \tanh\left[0.55\left(\frac{h}{W}\right)^{0.45}\right]\times 10^{-6} \tag{1.118}$$

$$A_2(\text{pF}/\mu\text{m}) = 9.92\ \tanh\left[0.52\left(\frac{h}{W}\right)^{0.5}\right]\times 10^{-6} \tag{1.119}$$

where h is the spacing from the ground plane.

Series planar capacitors with larger values, which are called the metal-insulator-metal (MIM) capacitors, can be realized by using an additional thin dielectric layer (typically < 0.5 μm) between two metal plates, as shown in Fig. 1.14(*c*) [8]. The bottom plate of the capacitor uses a thin unplated metal, and typically the dielectric material is silicon nitride (Si_3N_4) for integrated circuits on GaAs and silicon dioxide (SiO_2) for integrated circuits on Si. The top plate uses a thick plated conductor to reduce the loss in the capacitor. These capacitors are used to achieve higher capacitance values in small areas (10 pF and greater), with typical tolerances from 10 to 15%. The capacitance can be calculated according to Eq. (1.116).

1.7 Transmission Line

Transmission lines are widely used in the matching circuits of the power amplifiers, in the resonant and feedback circuits of the oscillators, as elements of the directional couplers, power combiners, and dividers. When the propagated signal wavelength is compared to its physical dimension, the transmission line can be considered as a two-port network with distributed parameters, where the voltages and currents vary in magnitude and phase over length.

Schematically, a transmission line is often represented as a two-wire line, as shown in Fig. 1.15(*a*), where its electrical parameters are distributed along its length. The physical properties of a transmission line are determined by four basic parameters:

- The series inductance L due to the self-inductive phenomena of two conductors.
- The shunt capacitance C in view of the close proximity between two conductors.

- The series resistance R due to the finite conductivity of the conductors.
- The shunt conductance G that is related to the dielectric losses in the material.

As a result, a transmission line of length Δx represents a lumped-element circuit shown in Fig. 1.15(*b*), where ΔL, ΔC, ΔR, and ΔG are the series inductance, the shunt capacitance, the series resistance, and the shunt conductance per unit length, respectively. If all these elements are distributed uniformly along the transmission line and their values do not depend on the chosen position of Δx, this transmission line is called the *uniform transmission line* [19]. Any finite length of the uniform transmission line can be viewed as a cascade of section length Δx.

To define the distribution of the voltages and currents along the uniform transmission line, it is necessary to write the differential equations using Kirchhoff's voltage law for instantaneous values of the voltages and currents in the line section of length Δx distant x from its beginning. For the sinusoidal steady-state condition, the telegrapher equations for $V(x)$ and $I(x)$ are given by

$$\frac{d^2V(x)}{dx^2} - \gamma^2 V(x) = 0 \tag{1.120}$$

$$\frac{d^2I(x)}{dx^2} - \gamma^2 I(x) = 0 \tag{1.121}$$

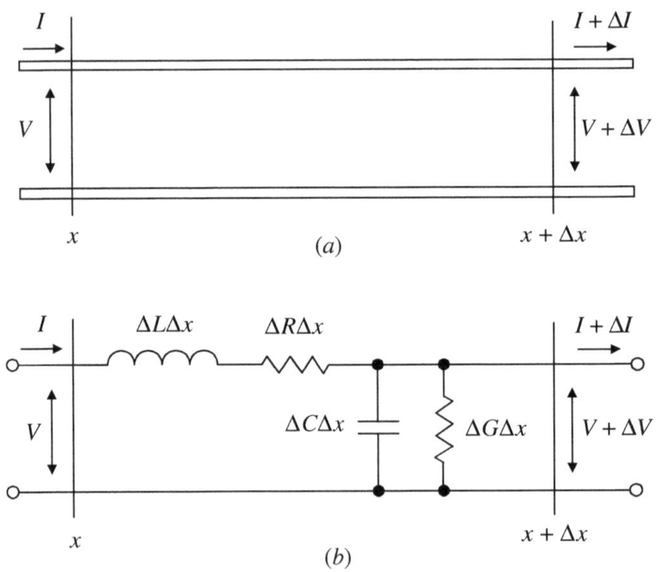

FIGURE 1.15 Transmission line schematics.

where $\gamma = \alpha + j\beta = \sqrt{(\Delta R + j\omega\Delta L)\,(\Delta G + j\omega\Delta C)}$ is the complex propagation constant (which is a function of frequency), α is the attenuation constant, and β is the phase constant. The general solutions of Eqs. (1.120) and (1.121) for voltage and current of the traveling wave in the transmission line can be written as

$$V(x) = A_1 \exp(-\gamma x) + A_2 \exp(\gamma x) \tag{1.122}$$

$$I(x) = \frac{A_1}{Z_0} \exp(-\gamma x) - \frac{A_2}{Z_0} \exp(\gamma x) \tag{1.123}$$

where $Z_0 = \sqrt{(\Delta R + j\omega\Delta L)/(\Delta G + j\omega\Delta C)}$ is the characteristic impedance of the transmission line, $V_i = A_1\exp(-\gamma x)$ and $V_r = A_2\exp(\gamma x)$ represent the incident voltage and the reflected voltage, and $I_i = A_1\exp(-\gamma x)/Z_0$ and $I_r = A_2\exp(\gamma x)/Z_0$ are the incident current and the reflected current, respectively. From Eqs. (1.122) and (1.123), it follows that the characteristic impedance of the transmission line Z_0 represents the ratio of the incident (reflected) voltage to the incident (reflected) current at any position on the line as

$$Z_0 = \frac{V_i(x)}{I_i(x)} = \frac{V_r(x)}{I_r(x)} \tag{1.124}$$

For a lossless transmission line with the attenuation constant $\alpha = 0$, the propagation constant $\gamma = j\beta = j\omega\sqrt{\Delta L \Delta C}$, and the phase constant $\beta = \omega\sqrt{\Delta L \Delta C}$ when $R = G = 0$, the voltage and current do not change with position. Consequently, the characteristic impedance is reduced to $Z_0 = \sqrt{L/C}$ and represents a real number. The wavelength is defined as $\lambda = 2\pi/\beta = 2\pi/\omega\sqrt{\Delta L \Delta C}$ and the phase velocity as $v_p = \omega/\beta = 1/\sqrt{\Delta L \Delta C}$.

Figure 1.16 represents a transmission line of the characteristic impedance Z_0 terminated with a load Z_L. In this case, the constants A_1 and A_2 are determined at the position $x = l$ by

$$V(l) = A_1 \exp(-\gamma\, l) + A_2 \exp(\gamma\, l) \tag{1.125}$$

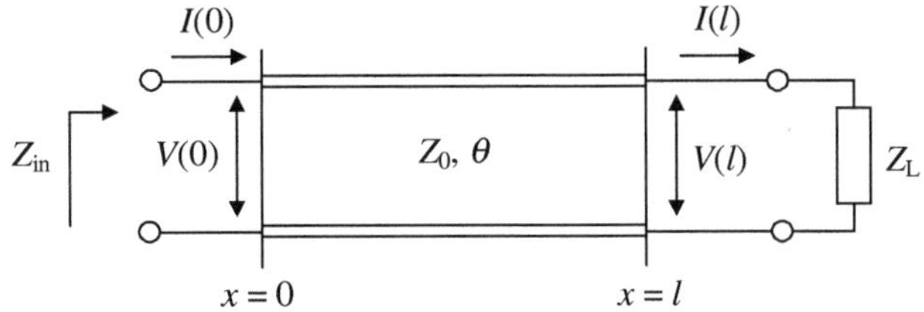

FIGURE 1.16 Loaded transmission line.

$$I(l) = \frac{A_1}{Z_0}\exp(-\gamma\, l) - \frac{A_2}{Z_0}\exp(\gamma\, l) \qquad (1.126)$$

and equal to

$$A_1 = \frac{V(l) + Z_0 I(l)}{2}\ \exp(\gamma\, l) \qquad (1.127)$$

$$A_2 = \frac{V(l) - Z_0 I(l)}{2}\exp(-\gamma\, l) \qquad (1.128)$$

As a result, the wave equations for voltage $V(x)$ and current $I(x)$ can be rewritten as

$$V(x) = \frac{V(l) + Z_0 I(l)}{2}\exp[\gamma(l-x)] + \frac{V(l) - Z_0 I(l)}{2}\exp[-\gamma(l-x)] \qquad (1.129)$$

$$I(x) = \frac{V(l) + Z_0 I(l)}{2Z_0}\exp[\gamma(l-x)] - \frac{V(l) - Z_0 I(l)}{2Z_0}\exp[-\gamma(l-x)] \qquad (1.130)$$

which allows their determination at any position on the transmission line.

The voltage and current amplitudes at $x = 0$ as functions of the voltage and current amplitudes at $x = l$ can be determined from Eqs. (1.129) and (1.130) as

$$V(0) = \frac{V(l) + Z_0 I(l)}{2}\exp(\gamma\, l) + \frac{V(l) - Z_0 I(l)}{2}\exp(-\gamma\, l) \qquad (1.131)$$

$$I(0) = \frac{V(l) + Z_0 I(l)}{2Z_0}\exp(\gamma\, l) - \frac{V(l) - Z_0 I(l)}{2Z_0}\exp(-\gamma\, l) \qquad (1.132)$$

By using the ratios $\cosh x = [\exp(x) + \exp(-x)]/2$ and $\sinh x = [\exp(x) - \exp(-x)]/2$, Eqs. (1.131) and (1.132) can be rewritten in the form

$$V(0) = V(l)\cosh(\gamma\, l) + Z_0 I(l)\sinh(\gamma\, l) \qquad (1.133)$$

$$I(0) = \frac{V(l)}{Z_0}\sinh(\gamma\, l) + I(l)\cosh(\gamma\, l) \qquad (1.134)$$

which represents the transmission equations of the symmetrical reciprocal two-port network expressed through the *ABCD*-parameters when $AD - BC = 1$ and $A = D$. Consequently, the transmission *ABCD*-matrix of the lossless transmission line with $\alpha = 0$ can be given by

$$[ABCD] = \begin{bmatrix} \cos\theta & jZ_0 \sin\theta \\ \dfrac{j\sin\theta}{Z_0} & \cos\theta \end{bmatrix} \tag{1.135}$$

Using the formulas to transform *ABCD*-parameters into *S*-parameters yields

$$[S] = \begin{bmatrix} 0 & \exp(-j\theta) \\ \exp(-j\theta) & 0 \end{bmatrix} \tag{1.136}$$

where $\theta = \beta l$ is the electrical length of the transmission line.

In the case of the loaded lossless transmission line, the reflection coefficient Γ is defined as the ratio between the reflected voltage wave and the incident voltage wave given at x as

$$\Gamma(x) = \frac{V_r}{V_i} = \frac{A_2}{A_1}\exp(2j\beta x) \tag{1.137}$$

By taking into account Eqs. (1.127) and (1.128), the reflection coefficient for $x = l$ can be defined as

$$\Gamma = \frac{Z - Z_0}{Z + Z_0} \tag{1.138}$$

where Γ represents the load reflection coefficient and $Z = Z_L = V(l)/I(l)$. If the load is mismatched, only part of the available power from the source is delivered to the load. This power loss is called the *return loss* (*RL*) and is calculated in decibels as

$$RL = -20 \log_{10}|\Gamma| \tag{1.139}$$

For a matched load when $\Gamma = 0$, a return loss is of ∞ dB. A total reflection with $\Gamma = 1$ means a return loss of 0 dB when all incident power is reflected.

According to the general solution for voltage at a position x in the transmission line,

$$V(x) = V_i(x) + V_r(x) = V_i[1 + \Gamma(x)] \tag{1.140}$$

Hence, the maximum amplitude (when the incident and reflected waves are in phase) is

$$V_{max}(x) = |V_i|\,[1 + |\Gamma(x)|] \tag{1.141}$$

and the minimum amplitude (when these two waves are 180° out of phase) is

$$V_{\text{min}}(x) = |V_{\text{i}}|[1 - |\Gamma(x)|] \tag{1.142}$$

The ratio of V_{max} to V_{min}, which is a function of the reflection coefficient Γ, represents the *voltage standing wave ratio* (*VSWR*). The *VSWR* is a measure of mismatch and can be written as

$$VSWR = \frac{V_{\text{max}}}{V_{\text{min}}} = \frac{1 + |\Gamma|}{1 - |\Gamma|} \tag{1.143}$$

which can change from 1 to ∞ (where $VSWR = 1$ implies a matched load). For a load impedance with zero imaginary part when $Z_{\text{L}} = R_{\text{L}}$, the *VSWR* can be calculated using $VSWR = R_{\text{L}}/Z_0$ when $R_{\text{L}} \geq Z_0$ and $VSWR = Z_0/R_{\text{L}}$ when $Z_0 \geq R_{\text{L}}$.

From Eqs. (1.133) and (1.134), it follows that the input impedance of the loaded lossless transmission line can be obtained by

$$Z_{\text{in}} = \frac{V(0)}{I(0)} = Z_0 \frac{Z_{\text{L}} + jZ_0 \tan(\theta)}{Z_0 + jZ_{\text{L}} \tan(\theta)} \tag{1.144}$$

which gives an important dependence between the input impedance, the transmission-line parameters (electrical length and characteristic impedance), and the arbitrary load impedance.

References

1. L. O. Chua, C. A. Desoer, and E. S. Kuh, *Linear and Nonlinear Circuits*, New York: McGraw-Hill, 1987.
2. D. R. Cunningham and J. A. Stuller, *Circuit Analysis*, New York: John Wiley & Sons, 1995.
3. G. D. Vendelin, A. M. Pavio, and U. L. Rohde, *Microwave Circuit Design Using Linear and Nonlinear Techniques*, New York: John Wiley & Sons, 2005.
4. D. M. Pozar, *Microwave Engineering*, New York: John Wiley & Sons, 2004.
5. D. A. Frickey, "Conversions between S, Z, Y, h, ABCD, and T Parameters Which Are Valid for Complex Source and Load Impedances," *IEEE Trans. Microwave Theory Tech.*, vol. MTT-42, pp. 205–211, Feb. 1994.
6. J. Shekel, "Matrix Representation of Transistor Circuits," *Proc. IRE*, vol. 40, pp. 1493–1497, Nov. 1952.
7. E. F. Terman, *Radio Engineer's Handbook*, New York: McGraw-Hill, 1945.
8. I. J. Bahl, *Lumped Elements for RF and Microwave Circuits*, Boston: Artech House, 2003.
9. H. A. Wheeler, "Simple Inductance Formulas for Radio Coils," *Proc. IRE*, vol. 16, pp. 1398–1400, Oct. 1928.
10. S. S. Mohan, M. del Mar Hershenson, S. P. Boyd, and T. H. Lee, "Simple Accurate Expressions for Planar Spiral Inductances," *IEEE J. Solid-State Circuits*, vol. SC-34, pp. 1419–1424, Oct. 1999.
11. J. M. Lopez-Villegas, J. Samitier, C. Cane, P. Losantos, and J. Bausells, "Improvement of the Quality Factor of RF Integrated Inductors by Layout

Optimization," *IEEE Trans. Microwave Theory Tech.*, vol. MTT-48, pp. 76–83, Jan. 2000.
12. J. R. Long and M. A. Copeland, "The Modeling, Characterization, and Design of Monolithic Inductors for Silicon RF IC's," *IEEE J. Solid-State Circuits*, vol. SC-32, pp. 357–369, Mar. 1997.
13. N. A. Talwalkar, C. P. Yue, and S. S. Wong, "Analysis and Synthesis of On-Chip Spiral Inductors," *IEEE Trans. Electron Devices*, vol. ED-52, pp. 176–182, Feb. 2005.
14. N. M. Nguyen and R. G. Meyer, "Si IC-Compatible Inductors and *LC* Passive Filters," *IEEE J. Solid-State Circuits*, vol. SC-25, pp. 1028–1031, Aug. 1990.
15. J. Craninckx and M. S. J. Steyaert, "A 1.8-GHz Low-Phase-Noise CMOS VCO Using Optimized Hollow Spiral Inductors," *IEEE J. Solid-State Circuits*, vol. SC-32, pp. 736–744, May 1997.
16. S. L. March, "Simple Equations Characterize Bond Wires," *Microwaves & RF*, vol. 30, pp. 105–110, Nov. 1991.
17. P. Benedek and P. Silvester, "Equivalent Capacitances of Microstrip Gaps and Steps," *IEEE Trans. Microwave Theory Tech.*, vol. MTT-20, pp. 729–733, Nov. 1972.
18. G. D. Alley, "Interdigital Capacitors and Their Applications in Lumped Element Microwave Integrated Circuits," *IEEE Trans. Microwave Theory Tech.*, vol. MTT-18, pp. 1028–1033, Dec. 1970.
19. S. A. Schelkunoff, "The Impedance Concept and Its Application to Problems of Reflection, Refraction, Shielding and Power Absorption," *Bell Syst. Tech. J.*, vol. 17, pp. 17–48, Jan. 1938.

CHAPTER 2

Nonlinear Circuit Design Methods

This chapter presents the most commonly used design techniques for analyzing nonlinear circuits, in particular, transistor power amplifiers. There are several approaches to analyze and design nonlinear circuits, depending on their main specifications. Among them, an analysis both in the time domain to determine transient circuit behavior and in the frequency domain to improve power and spectral performances when parasitic effects such as instability and spurious emission must be eliminated or minimized. By using the time-domain technique, it is relatively easy to describe a nonlinear circuit with differential equations, which can be solved either analytically for simple cases or numerically in a common case. The time-domain analysis is limited by its inability to operate with the circuit immittance (impedance or admittance) parameters, and it can be practically applied only for the circuits with lumped parameters or ideal transmission lines. The frequency-domain analysis is less ambiguous because a relatively complex circuit often can be reduced to one or more sets of immittances at each harmonic component. For example, by using a quasilinear approach, the nonlinear circuit parameters averaged over the fundamental component allow one to apply a linear circuit analysis. Advanced modern computer-aided design (CAD) simulators incorporate both frequency-domain and time-domain methods, as well as optimization techniques to provide all necessary design cycles. Some practical equations, such as the Taylor and Fourier series expansions, Bessel functions, trigonometric identities, and concept of the conduction angle, which simplify the circuit design procedure, are given. The dynamic X-parameters are introduced as a novel way to build behavioral models for power amplifiers that include long-term memory effects.

2.1 Frequency-Domain Analysis

The best way to understand the electrical behavior of the power amplifier and the fastest way to calculate its basic characteristics, such as output power, efficiency, power gain, or harmonic suppression, is to use a frequency-domain analysis. Generally, such an analysis is based on the determination of the output response of the nonlinear active device when the multiharmonic signal is applied to its input port, which analytically can be written as

$$i(t) = f\,[v(t)] \tag{2.1}$$

where $i(t)$ is the output current, $v(t)$ is the input voltage, and $f(v)$ is the nonlinear transfer function of the active device. Unlike the frequency-domain analysis, the time-domain analysis establishes the relationships between voltage and current in each circuit element in the time domain when a system of nonlinear integro-differential equations is obtained by applying Kirchhoff's law to the circuit to be analyzed.

The voltage $v(t)$ in the frequency domain generally represents a multiple frequency signal at the device input by

$$v(t) = V_0 + \sum_{k=1}^{N} V_k \cos(\omega_k t + \phi_k) \tag{2.2}$$

where V_0 is the dc voltage, V_k is the voltage amplitude and ϕ_k is the phase of the k-order harmonic component ω_k, $k = 1, 2, \ldots, N$, and N is the number of harmonics.

The frequency-domain analysis based on substituting Eq. (2.2) into Eq. (2.1) for a particular nonlinear transfer function of the active device determines an output spectrum as a sum of the fundamental-frequency and higher-order harmonic components, the amplitudes and phases of which will determine the output signal spectrum. Generally, this is a complicated procedure that requires a harmonic-balance technique to numerically calculate an accurate nonlinear circuit response. However, the solution can be found analytically in a simple way when it is necessary to estimate only the basic performance in the form of the output power and efficiency. In this case, a technique based on a piecewise-linear approximation of the device transfer function can provide a clear insight to the basic behavior of the power amplifier and its operation modes. It can also serve as a good starting point for a final CAD and optimization procedure.

The result of the frequency-domain analysis is shown as a summation of the harmonic components, the amplitudes and phases of which will determine the output signal spectrum. This problem can be solved analytically by using the trigonometric identities, piecewise-linear approximation, or Bessel functions.

2.1.1 Trigonometric Identities

The use of the trigonometric identities is very convenient when the transfer characteristic of the nonlinear element can be represented by the power series

$$i(v) = a_0 + a_1 v + a_2 v^2 + \ldots + a_n v^n \tag{2.3}$$

If the simplest case is assumed, in which the input signal consists of the fundamental harmonic component only written as

$$v = V\cos(\omega t + \phi) \tag{2.4}$$

then, by substituting Eq. (2.4) into Eq. (2.3), the power series can be rewritten as

$$i = a_0 + a_1 V\cos(\omega t + \phi) + a_2 V^2 \cos^2(\omega t + \phi) + \ldots + a_n V^n \cos^n(\omega t + \phi) \tag{2.5}$$

To represent the right-hand side of Eq. (2.5) as a sum of the first-order cosine components, the following trigonometric identities, which replace the *n*th-order cosine components, can be used:

$$\cos^2\psi = \frac{1}{2}(1 + \cos 2\psi) \tag{2.6}$$

$$\cos^3\psi = \frac{1}{4}(3\cos\psi + \cos 3\psi) \tag{2.7}$$

$$\cos^4\psi = \frac{1}{8}(3 + 4\cos 2\psi + \cos 4\psi) \tag{2.8}$$

$$\cos^5\psi = \frac{1}{16}(10\cos\psi + 5\cos 3\psi + \cos 5\psi) \tag{2.9}$$

where $\psi = \omega t + \phi$.

By using the appropriate substitutions from Eqs. (2.6) through (2.9) and equating the signal frequency component terms, Eq. (2.5) can be rewritten as

$$\begin{aligned} i = I_0 + I_1\cos(\omega t + \phi) + I_2 \cos 2(\omega t + \phi) \\ + I_3 \cos 3(\omega t + \phi) + \ldots + I_n \cos n(\omega t + \phi) \end{aligned} \tag{2.10}$$

where

$$I_0 = a_0 + \frac{1}{2} a_2 V^2 + \frac{3}{8} a_4 V^4 + \ldots$$

$$I_1 = a_1 V + \frac{3}{4} a_3 V^3 + \frac{5}{8} a_5 V^5 + \ldots$$

$$I_2 = \frac{1}{2} a_2 V^2 + \frac{1}{2} a_4 V^4 + \ldots$$

$$I_3 = \frac{1}{4} a_3 V^3 + \frac{5}{16} a_5 V^5 + \ldots$$

Comparing Eqs. (2.3) and (2.10) results in the following conclusions:

- For nonlinear elements, the output spectrum contains frequency components, which are multiples of the input signal frequency. The number of the highest-frequency component is equal to the maximum degree of the power series. Therefore, to find the amplitude of nth-harmonic response, the current-voltage characteristic of nonlinear elements should be approximated by the order of not less than an n-order power series.
- The output dc and even-order harmonic components are determined only by the even voltage degrees in the device transfer characteristic given by Eq. (2.3). The odd-order harmonic components are defined only by the odd voltage degrees for the single-harmonic input signal given by Eq. (2.4).
- The current phase ψ_k of the kth-order harmonic component $\omega_k = k\omega$ is k times greater than the input signal current phase ψ,

$$\psi_k = \omega_k t + \phi_k = k(\omega t + \phi) \tag{2.11}$$

that is also applied to their initial phases defined as

$$\phi_k = k\phi \tag{2.12}$$

2.1.2 Piecewise-Linear Approximation

The piecewise-linear approximation of the active device current-voltage transfer characteristic is a result of replacing the actual nonlinear dependence $i = f(v_{in})$, where v_{in} is the voltage applied to the device input, by an approximated one that consists of straight lines tangential to the actual dependence at the specified points. Such a piecewise-linear approximation for the case of two straight lines is shown in Fig. 2.1(*a*).

The output current waveforms for the actual current-voltage dependence (dashed curves) and its piecewise-linear approximation by two straight lines (solid curves) are plotted in Fig. 2.1(*b*). Under large-signal operation mode, the waveforms corresponding to these two dependences are practically the same with negligible deviation for small values of the output current close to the pinch-off region of the device operation and significant deviation close to the saturation region of the device operation. However, the latter case results in a significant nonlinear distortion and is used only for high-efficiency operation modes when the active period of the device operation is minimized. Hence, at least two first output current components, dc and fundamental, can be calculated through a Fourier-series expansion with sufficient accuracy. Therefore, such a piecewise-linear approximation with two straight lines can be effective for a very

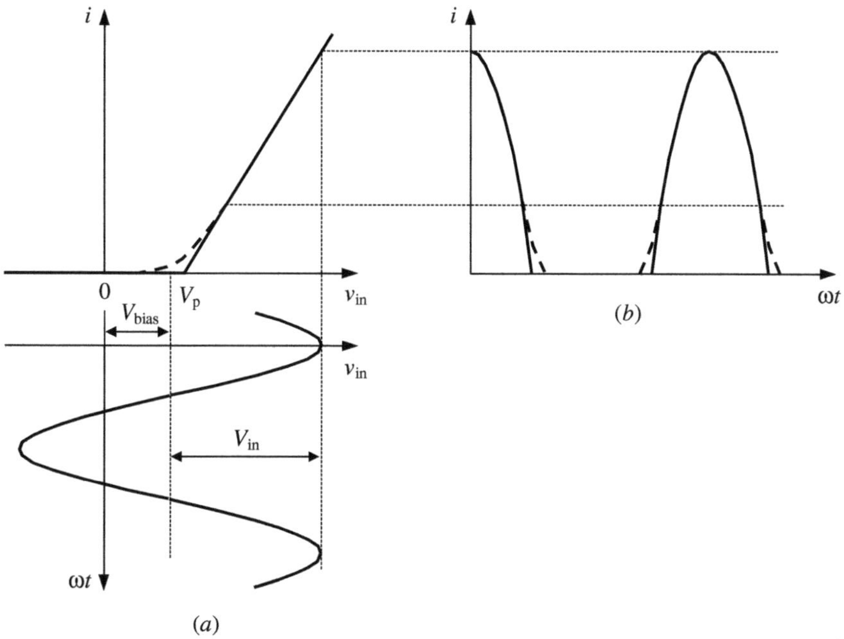

FIGURE 2.1 Piecewise-linear approximation.

quick estimate of the output power and collector (or drain) efficiency of the power amplifier.

In this case, the piecewise-linear transfer current-voltage characteristic of the transistor is defined by

$$i = \begin{cases} 0 & \text{when } v_{in} \le V_p \\ g_m(v_{in} - V_p) & \text{when } v_{in} \ge V_p \end{cases} \tag{2.13}$$

where g_m is the device transconductance and V_p is the pinch-off voltage.

Consider the effect of the input cosinusoidal signal

$$v_{in} = V_{bias} + V_{in} \cos\omega t \tag{2.14}$$

on the nonlinear element, the current-voltage characteristic of which is approximated by two straight lines, as shown in Fig. 2.2, where V_{bias} is the dc bias voltage.

At the point on the plot when the voltage $v_{in}(\omega t)$ becomes equal to the pinch-off voltage V_p and where $\omega t = \theta$, the output current $i(\theta)$ takes a zero value. At this moment,

$$V_p = V_{bias} + V_{in} \cos\theta \tag{2.15}$$

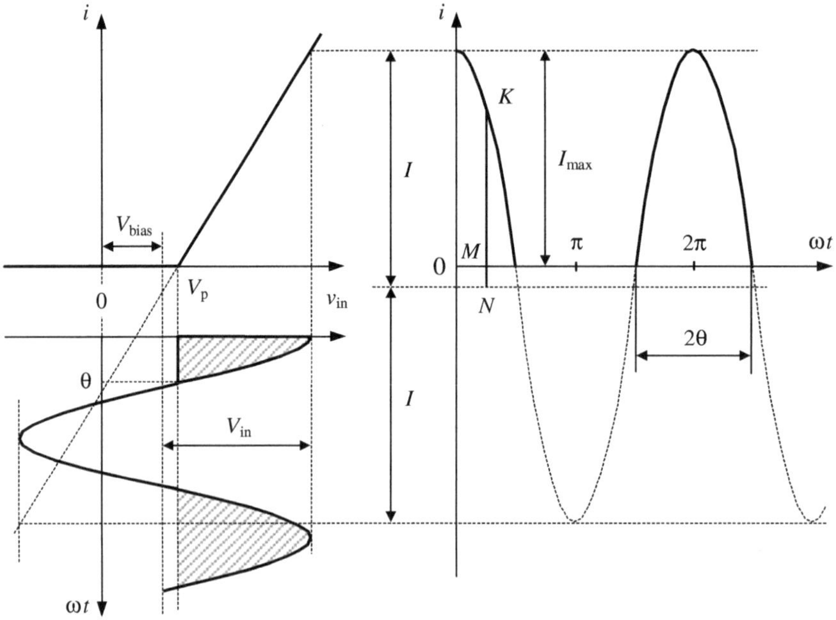

FIGURE 2.2 Schematic definition of conduction angle.

and θ can be calculated from

$$\cos\theta = -\frac{V_{bias} - V_p}{V_{in}} \tag{2.16}$$

As a result, the output current represents a sequence of the cosinusoidal pulses with the maximum height of I_{max} and width of 2θ, and can be calculated within the interval $-\theta \le \omega t \le \theta$ as

$$i = KN - MN = I\cos\omega t - I\cos\theta = I(\cos\omega t - \cos\theta) \tag{2.17}$$

When $\omega t = 0$, then $i = I_{max}$, and

$$I_{max} = g_m V_{in}(1 - \cos\theta) \tag{2.18}$$

where $I = g_m V_{in}$.

Generally, the angle θ characterizes the class of the active device operation. If $\theta = \pi$ or 180°, the device operates in the active region during the entire period (Class-A operation). When $\theta = \pi/2$ or 90°, the device operates half a wave period in the active region and half a wave period in the pinch-off region (Class-B operation). The values of $\theta > 90°$ correspond to Class-AB operation with a certain value of the quiescent current. Therefore, the double angle 2θ is called the *conduction angle*, the value of which directly indicates a class of the active device operation.

The Fourier-series expansion of the even function when $i(\omega t) = i(-\omega t)$ contains only even component functions and can be written as

$$i(\omega t) = I_0 + I_1 \cos\omega t + I_2 \cos 2\omega t + I_3 \cos 3\omega t + \dots \tag{2.19}$$

with the dc, fundamental-frequency, and nth-order harmonic components calculated by

$$I_0 = \frac{1}{2\pi}\int_{-\theta}^{\theta} g_{\mathrm{m}} V_{\mathrm{in}} \ (\cos\omega t - \cos\theta)\ d\omega t = \gamma_0(\theta)\, I \tag{2.20}$$

$$I_1 = \frac{1}{\pi}\int_{-\theta}^{\theta} g_{\mathrm{m}} V_{\mathrm{in}} (\cos\omega t - \cos\theta)\ \cos\omega t\ d\omega t = \gamma_1(\theta)\, I \tag{2.21}$$

$$I_{\mathrm{n}} = \frac{1}{\pi}\int_{-\theta}^{\theta} g_{\mathrm{m}} V_{\mathrm{in}} (\cos\omega t - \cos\theta)\ \cos n\omega t\ d\omega t = \gamma_{\mathrm{n}}(\theta)\, I \tag{2.22}$$

where $\gamma_{\mathrm{n}}(\theta)$ are called the coefficients of expansion of the output current cosinusoidal pulse or the current coefficients [1, 2]. They can be analytically defined as

$$\gamma_0(\theta) = \frac{1}{\pi}(\sin\theta - \theta\ \cos\theta) \tag{2.23}$$

$$\gamma_1(\theta) = \frac{1}{\pi}(\theta - \sin\theta\ \cos\theta) \tag{2.24}$$

$$\gamma_{\mathrm{n}}(\theta) = \frac{2}{\pi}\,\frac{\sin n\theta\ \cos\theta - n\ \cos n\theta\ \sin\theta}{n(n^2-1)} \tag{2.25}$$

where $n = 2, 3, \dots$.

The coefficients $\gamma_{\mathrm{n}}(\theta)$ for the dc, fundamental-frequency, and second- and higher-order current harmonic components are shown in Fig. 2.3. The maximum value of $\gamma_{\mathrm{n}}(\theta)$ is achieved when $\theta = 180°/n$. A special case is $\theta = 90°$, when odd current coefficients are equal to zero, that is, $\gamma_3(\theta) = \gamma_5(\theta) = \dots = 0$. The ratio between the fundamental-frequency and dc components $\gamma_1(\theta)/\gamma_0(\theta)$ varies from 1 to 2 for any values of the conduction angle, with a minimum value of 1 for $\theta = 180°$ and a maximum value of 2 for $\theta = 0°$. It is necessary to focus on the fact that, for example, the third-harmonic current coefficient $\gamma_3(\theta)$ becomes negative within the interval of $90° < \theta < 180°$. This implies appropriate phase changes of the third-harmonic current component when its values are negative. Consequently, if the harmonic components for which $\gamma_{\mathrm{n}}(\theta) > 0$ achieve positive maximum values at the times corresponding to the midpoints of the current waveform, the harmonic components for which $\gamma_{\mathrm{n}}(\theta) < 0$ can achieve negative maximum values at these times. As a result, combination of different harmonic components with proper loading will result to flattening of the

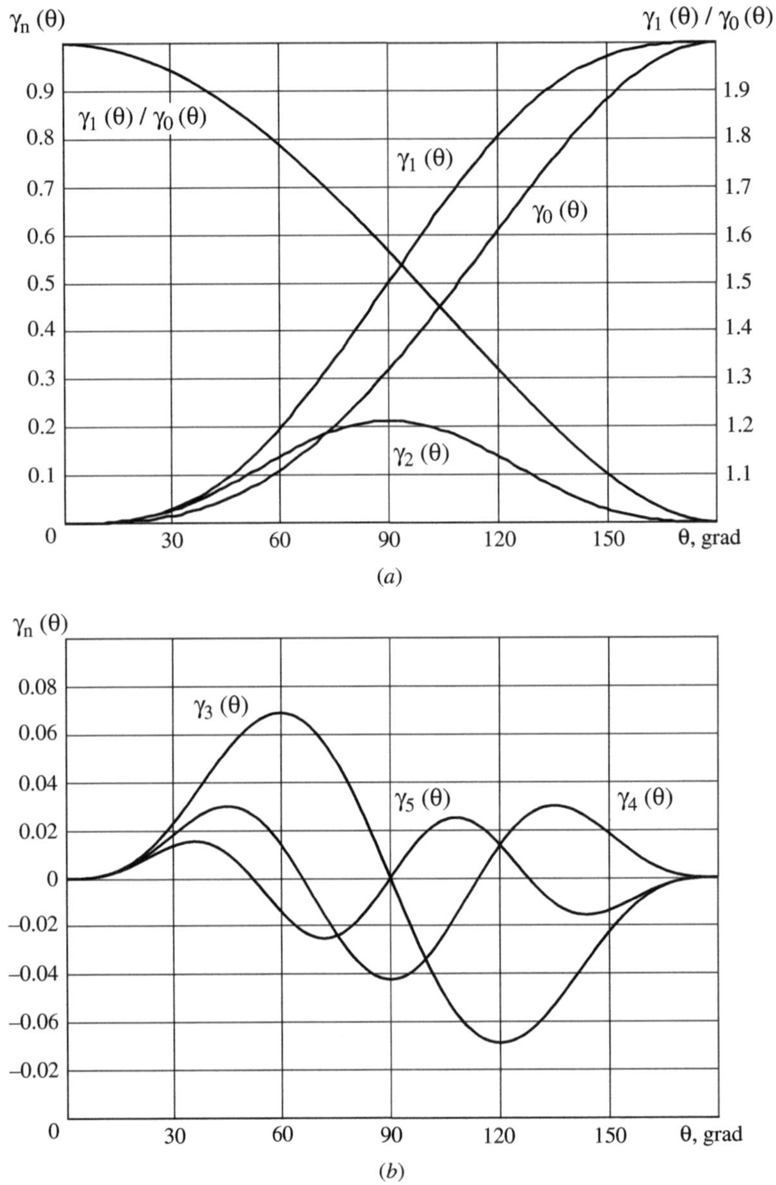

FIGURE 2.3 Coefficients $\gamma_n(\theta)$ for dc, fundamental, and higher-order current components.

current or voltage waveforms, thus improving efficiency of the power amplifier. The amplitude of the corresponding current harmonic component can be obtained by

$$I_n = \gamma_n(\theta)\, g_m V_{in} = \gamma_n(\theta)\, I \tag{2.26}$$

A plot of the Fourier amplitudes along the frequency axis, given by Eqs. (2.20) to (2.22), is called the *single-sideband amplitude spectrum* of $i(t)$. The normalized power of $i(t)$, according to Parseval's theorem for real periodic waveforms, is equal to the sum of the normalized powers of the individual components written as

$$\frac{1}{2\pi}\int_0^{2\pi} i^2(\omega t)d\omega t = I_0^2 + \frac{1}{2}\sum_{n=1}^{\infty} I_n^2 \tag{2.27}$$

A plot of the normalized powers corresponding to the Fourier components in Eq. (2.19) versus frequency is called the *single-sideband power spectrum* of $i(t)$. The height of each spectral component is numerically equal to the average power that would be dissipated if the nth-harmonic current were applied across the terminals of a 1-Ω resistance. Consequently, the sum of the heights of all components in the power spectrum is numerically equal to the average power that would be dissipated if $i(t)$ were applied across the terminals of a 1-Ω resistance.

Sometimes it is necessary for an active device to provide a constant value of I_{max} at any value of θ. This requires an appropriate change of the input amplitude V_{in}. In this case, it is more convenient to use the coefficients α_n when the nth-order current harmonic amplitude I_n is related to the maximum current waveform amplitude I_{max} as

$$\alpha_n = I_n / I_{max} \tag{2.28}$$

From Eqs. (2.18), (2.22), and (2.28), it follows that

$$\alpha_n = \frac{\gamma_n(\theta)}{1 - \cos\theta} \tag{2.29}$$

and the maximum value of $\alpha_n(\theta)$ is achieved when $\theta = 120°/n$.

2.1.3 Bessel Functions

The Bessel functions are used to analyze the power amplifier operation mode when a nonlinear behavior of the active device is described by exponential functions. For example, the transfer current-voltage characteristic of the bipolar transistor is approximated by the simplified exponential dependence (neglecting reverse base-emitter current) written as

$$i(v) = I_{sat}\left[\exp\left(\frac{v_{in}}{V_T}\right) - 1\right] \tag{2.30}$$

where I_{sat} is the minority carrier saturation current and V_T is the temperature voltage. If the input signal is given in a simple form of Eq. (2.14), Eq. (2.30) can be rewritten as

$$i(\omega t) = I_{sat}\left[\exp\left(\frac{V_{bias}}{V_T}\right)\exp\left(\frac{V_{in}\cos\omega t}{V_T}\right) - 1\right] \tag{2.31}$$

The current $i(\omega t)$ in Eq. (2.31) is the even function of ωt, and consequently it can be represented by the Fourier-series expansion given by Eq. (2.19). The Fourier harmonic components can be determined from

$$\exp\left(\frac{V_{\text{in}}\cos\omega t}{V_{\text{T}}}\right) = I_0\left(\frac{V_{\text{in}}}{V_{\text{T}}}\right) + 2\sum_{k=1}^{\infty} I_k\left(\frac{V_{\text{in}}}{V_{\text{T}}}\right)\cos(k\omega t) \tag{2.32}$$

where $I_k(V_{\text{in}}/V_{\text{T}})$ are the kth-order modified Bessel functions of the first kind for an argument $V_{\text{in}}/V_{\text{T}}$, which are shown in Fig. 2.4 for zero-order and first-order components. It should be noted that $I_0(0) = 1$ and $I_k(0) = 0$, and with an increase in the harmonic component number, its amplitude appropriately decreases.

According to Eq. (2.32), the current $i(\omega t)$ defined by Eq. (2.30) can be rewritten as

$$i(\omega t) = I_{\text{sat}}\left[\exp\left(\frac{V_{\text{bias}}}{V_{\text{T}}}\right) I_0\left(\frac{V_{\text{in}}}{V_{\text{T}}}\right) - 1\right] + 2I_{\text{sat}}\exp\left(\frac{V_{\text{bias}}}{V_{\text{T}}}\right) I_1\left(\frac{V_{\text{in}}}{V_{\text{T}}}\right)\cos\omega t$$

$$+\ 2I_{\text{sat}}\exp\left(\frac{V_{\text{bias}}}{V_{\text{T}}}\right) I_2\left(\frac{V_{\text{in}}}{V_{\text{T}}}\right)\cos 2\omega t + 2I_{\text{sat}}\exp\left(\frac{V_{\text{bias}}}{V_{\text{T}}}\right) I_3\left(\frac{V_{\text{in}}}{V_{\text{T}}}\right)\cos 3\omega t + \ldots \tag{2.33}$$

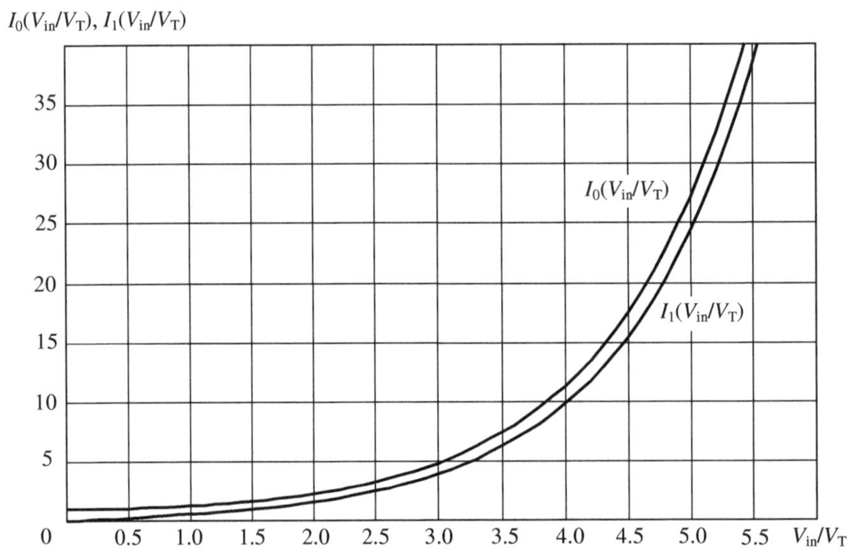

FIGURE 2.4 Zero-order and first-order modified Bessel functions of the first kind.

As a result, comparing Eq. (2.33) with Eq. (2.19) allows the dc, fundamental-frequency, and *n*th-order Fourier current components to be determined as

$$I_0 = I_{sat}\left[\exp\left(\frac{V_{bias}}{V_T}\right) I_0\left(\frac{V_{in}}{V_T}\right) - 1\right] \tag{2.34}$$

$$I_1 = 2I_{sat}\exp\left(\frac{V_{bias}}{V_T}\right) I_1\left(\frac{V_{in}}{V_T}\right) \tag{2.35}$$

$$I_n = 2I_{sat}\exp\left(\frac{V_{bias}}{V_T}\right) I_n\left(\frac{V_{in}}{V_T}\right) \tag{2.36}$$

where $n = 2, 3, \ldots$.

The following relationships can be helpful when using the Bessel functions:

$$2\frac{dI_n(V_{in}/V_T)}{d(V_{in}/V_T)} = I_{n+1}\left(\frac{V_{in}}{V_T}\right) + I_{n-1}\left(\frac{V_{in}}{V_T}\right) \tag{2.37}$$

$$\frac{dI_0(V_{in}/V_T)}{d(V_{in}/V_T)} = I_1\left(\frac{V_{in}}{V_T}\right) \tag{2.38}$$

$$\frac{2n}{(V_{in}/V_T)} I_n\left(\frac{V_{in}}{V_T}\right) = I_{n-1}\left(\frac{V_{in}}{V_T}\right) - I_{n+1}\left(\frac{V_{in}}{V_T}\right) \tag{2.39}$$

$$I_n\left(-\frac{V_{in}}{V_T}\right) = (-1)^n I_n\left(\frac{V_{in}}{V_T}\right) \tag{2.40}$$

2.2 Time-Domain Analysis

A time-domain analysis establishes the relationships between voltage and current in each circuit element in the time domain when a system of equations is obtained by applying Kirchhoff's law to the circuit to be analyzed. The transmission line in the time domain can be represented as an element with finite delay time depending on its electrical length. Generally, in a nonlinear circuit, such a system will be composed of nonlinear integro-differential equations. The solution to this system can be found by applying numerical integration methods. Therefore, the choice of the time interval and the initial point is very important to provide a compromise between the speed and accuracy of calculation. In this case, the smaller the interval used, the smaller the error achieved. However, the number of points to be calculated for each period will be greater, which makes the calculation slower.

To analyze a nonlinear system in the time domain, it is necessary to know the voltage-current relationships for all elements used in a particular RF or microwave circuit. For example, for linear resistance R, with the sinusoidal voltage applied and the sinusoidal current flowing through it, the voltage-current relationship in the time domain is given by

$$V = RI \tag{2.41}$$

where V is the voltage amplitude and I is the current amplitude.

For linear capacitance C,

$$i(t) = \frac{dq(t)}{dt} = \frac{dq}{dv}\frac{dv}{dt} = C\frac{dv}{dt} \tag{2.42}$$

For linear inductance L,

$$v(t) = \frac{d\varphi(t)}{dt} = \frac{d\varphi}{di}\frac{di}{dt} = L\frac{di}{dt} \tag{2.43}$$

where φ is the magnetic flux across the inductance.

In order to obtain the expressions for appropriate incremental capacitance and inductance, nonlinear dependences, such as $q(v)$ or $\varphi(i)$, should each be expanded in a Taylor series, subtracting the dc components and substituting into Eqs. (2.42) and (2.43). Then, for the quasilinear case, the capacitance and inductance can be determined, respectively, by

$$C(V_{\text{dc}}) = \left.\frac{dq(v)}{dv}\right|_{v = V_{\text{dc}}} \tag{2.44}$$

$$L(I_{\text{dc}}) = \left.\frac{d\phi(i)}{di}\right|_{i = I_{\text{dc}}} \tag{2.45}$$

where V_{dc} is the dc voltage across the capacitor and I_{dc} is dc current flowing through the inductance.

Figure 2.5 shows the electrical schematic of the first-order RL circuit connected to the independent voltage source V_{dc} at the moment $t = 0$. When the voltage source first is connected to the circuit containing an inductor as an energy-storage element, the current in the RL circuit does not immediately reach its steady-state value. This means that there is a transient response that characterizes the step-driven circuit with an energy-storage element. Let us assume zero initial conditions of the RL circuit and now we wish to solve for current $i(t)$ at all $t > 0$. According to Kirchhoff's voltage law, the algebraic sum of the voltages across the circuit elements ($v_{\text{L}} = Ldi/dt$, $v_{\text{R}} = Ri$, and $v = V_{\text{dc}}$) should be equal to zero, that is, $v_{\text{L}} + v_{\text{R}} - v = 0$, resulting in the linear nonhomogeneous first-order differential equation written as

$$L\frac{di}{dt} + Ri(t) = V_{\text{dc}} \tag{2.46}$$

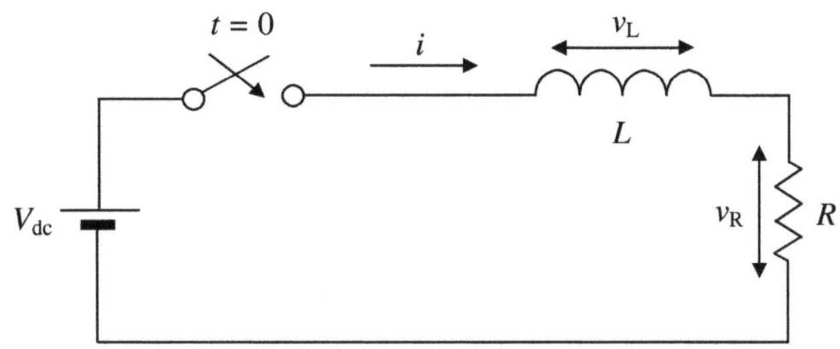

FIGURE 2.5 Schematic of *RL* circuit with connected source for $t > 0$.

whose general solution can be obtained as

$$i(t) = \frac{V_{dc}}{R} + A\,\exp\left(-\frac{R}{L}\,t\right) \tag{2.47}$$

The unknown coefficient A can be found from the initial conditions when the circuit current $i(0) = 0$ for $t = 0$. Consequently,

$$i(t) = \frac{V_{dc}}{R}\left[1 - \exp\left(-\frac{t}{\tau}\right)\right] \tag{2.48}$$

and

$$v_L(t) = L\frac{di}{dt} = \frac{V_{dc}}{R}\exp\left(-\frac{t}{\tau}\right) \tag{2.49}$$

where $\tau = L/R$ is the time constant of the *RL* circuit.

Thus, the current $i(t)$ in the *RL* circuit starts to grow from its zero value of $i(0) = 0$ at $t = 0$ asymptotically (by means of exponential functions) approaching the final dc value of $i(\infty) = I = V_{cc}/R$ at $t \to \infty$ when $v_L(\infty) = 0$ and $v_R(\infty) = V_{dc}$, according to Eqs. (2.48) and (2.49).

Figure 2.6 shows the electrical schematic of the nonlinear parallel circuit, which contains a capacitor as a nonlinear element depending

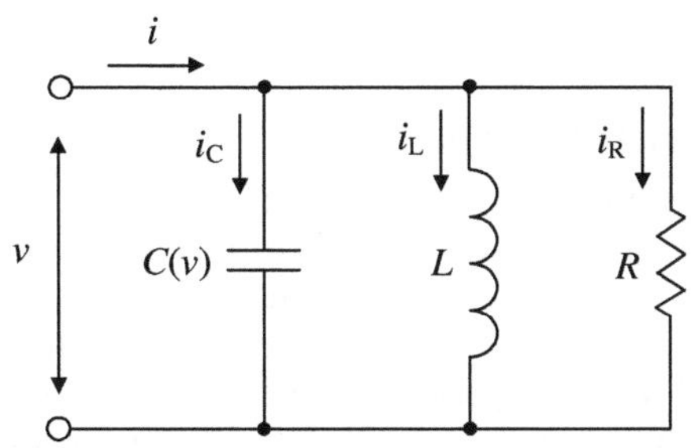

FIGURE 2.6 Schematic of nonlinear parallel circuit.

on the applied voltage. Such a capacitor can be represented by the nonlinear capacitance of *p-n* junction of the bipolar transistor or internal gate-source and gate-drain capacitances of the MOSFET or MESFET devices. Let us assume that the loaded quality factor of the parallel circuit is high enough to expect that the voltage across the circuit is sinusoidal, that is, $v = V \sin\omega t$, even if the current i contains the higher-order harmonic components. Then, according to Eq. (2.42) where $C = C(v)$, the current flowing through the nonlinear capacitance C can be written as

$$i_C = C(v)\, \omega V \cos\omega t \tag{2.50}$$

In the case of the second-order polynomial approximation of the nonlinear capacitance given by

$$C(v) = C_0 + A_1 v + A_2 v^2 \tag{2.51}$$

where C_0, A_1, and A_2 are the coefficients with positive values. Equation (2.50) can be rewritten by using the trigonometric identity of Eq. (2.6) as

$$i_C = \left(C_0 + \frac{A_2}{2} V^2 + A_1 V \sin\omega t - \frac{A_2}{2} V^2 \cos 2\omega t \right) \omega V \cos\omega t \tag{2.52}$$

From Eq. (2.52), it follows that the fundamental harmonic of the current i_{C1} through the nonlinear capacitance C can be represented by

$$i_{C1} = \left(C_0 + \frac{A_2}{4} V^2 \right) \omega V \cos\omega t \tag{2.53}$$

The same result for current i_{C1} can be obtained by including the equivalent (or fundamentally averaged) capacitance C_{avr} instead of the nonlinear capacitance according to

$$C_{avr} = C_0 + \frac{A_2}{4} V^2 \tag{2.54}$$

whose value increases for larger values of the voltage amplitude V.

2.3 Newton-Raphson Algorithm

To describe the nonlinear circuit behavior, it is necessary to solve the nonlinear algebraic equation or system of equations, which do not generally admit a closed form solution analytically. One of the most common numerical methods to solve such equations is a method based on the Newton-Raphson algorithm [3]. The initial guess for this method is chosen using a Taylor-series expansion of the nonlinear function. Consider a practical case when the device is represented by a two-port network where all nonlinear elements are functions of the two unknown voltages, input voltage v_{in} and output voltage v_{out}. As a result, after combining linear and nonlinear circuit elements, a system of two equations can be written as

$$f_1(v_{in}, v_{out}) = 0 \tag{2.55}$$

$$f_2(v_{in}, v_{out}) = 0 \tag{2.56}$$

By assuming that the variables v_{in0} and v_{out0} are the initial approximate solution of a system of Eqs. (2.55) and (2.56), these variables can be rewritten as $v_{in} = v_{in0} + \Delta v_{in}$ and $v_{out} = v_{out0} + \Delta v_{out}$, where Δv_{in} and Δv_{out} are the linear increments of the variables. Applying a Taylor-series expansion to Eqs. (2.55) and (2.56) yields

$$f_1(v_{in0} + \Delta v_{in}, v_{out0} + \Delta v_{out}) = f_1(v_{in0}, v_{out0}) + \left.\frac{\partial f_1}{\partial v_{in}}\right|_{\substack{v_{in}=v_{in0}\\ v_{out}=v_{out0}}} \Delta v_{in}$$

$$+ \left.\frac{\partial f_1}{\partial v_{out}}\right|_{\substack{v_{in}=v_{in0}\\ v_{out}=v_{out0}}} \Delta v_{out} + o\,(\Delta v_{in}^2 + \Delta v_{out}^2 + \dots) = 0 \tag{2.57}$$

$$f_2(v_{in0} + \Delta v_{in}, v_{out0} + \Delta v_{out}) = f_2(v_{in0}, v_{out0}) + \left.\frac{\partial f_2}{\partial v_{in}}\right|_{\substack{v_{in}=v_{in0}\\ v_{out}=v_{out0}}} \Delta v_{in}$$

$$+ \left.\frac{\partial f_2}{\partial v_{out}}\right|_{\substack{v_{in}=v_{in0}\\ v_{out}=v_{out0}}} \Delta v_{out} + o(\Delta v_{in}^2 + \Delta v_{out}^2 + \dots) = 0 \tag{2.58}$$

where $o(\Delta v_{in}^2 + \Delta v_{out}^2 + \dots)$ denotes the second- and higher-order components.

By neglecting the second- and higher-order components, Eqs. (2.57) and (2.58) can be rewritten in the matrix form as

$$-\begin{bmatrix} f_1 \\ f_2 \end{bmatrix} = \begin{bmatrix} \dfrac{\partial f_1}{\partial v_{in}} & \dfrac{\partial f_1}{\partial v_{out}} \\ \dfrac{\partial f_2}{\partial v_{in}} & \dfrac{\partial f_2}{\partial v_{out}} \end{bmatrix} \begin{bmatrix} \Delta v_{in} \\ \Delta v_{out} \end{bmatrix} \tag{2.59}$$

Equation (2.59) can be rewritten in the phasor form as

$$-\mathbf{F} = \mathbf{J}\,\Delta\mathbf{v} \tag{2.60}$$

where **J** is the Jacobian matrix of a system of Eqs. (2.55) and (2.56).

The solution of Eq. (2.60) for a nonsingular matrix **J** can be obtained by

$$\Delta\mathbf{v} = -\mathbf{J}^{-1}\mathbf{F} \tag{2.61}$$

This means that if

$$\mathbf{v}_0 = \begin{bmatrix} v_{\text{in0}} \\ v_{\text{out0}} \end{bmatrix} \tag{2.62}$$

is the initial guess voltage of this system of equation, then the next (more precise) solution can be written as

$$\mathbf{v}_1 = \mathbf{v}_0 - \mathbf{J}^{-1}\mathbf{F} \tag{2.63}$$

where

$$\mathbf{v}_1 = \begin{bmatrix} v_{\text{in1}} \\ v_{\text{out1}} \end{bmatrix} \tag{2.64}$$

Then, starting with the initial guess $\mathbf{v}_0$, we compute $\mathbf{v}_1$ at the first iteration. For the iteration $n + 1$, one can write

$$\mathbf{v}_{\text{n+1}} = \mathbf{v}_{\text{n}} - \mathbf{J}^{-1}\mathbf{F}(\mathbf{v}_{\text{n}}) \tag{2.65}$$

The iterative Eq. (2.65) is given for a system of two equations. However, it can be directly extended to a system of k nonlinear equations with k unknown parameters. This iterative procedure is terminated after $(n + 1)$ iterations whenever

$$|\mathbf{x}_{\text{n+1}} - \mathbf{x}_{\text{n}}| = \sqrt{\sum_{\text{k=1}}^{\text{K}} (x_{\text{n+1}}^{\text{k}} - x_{\text{n}}^{\text{k}})^2} < \varepsilon \tag{2.66}$$

where ε is a small positive number depending on the desired accuracy. For practical purposes, it is desirable that the Newton-Raphson algorithm should converge in a few steps. Therefore, the choice of an appropriate initial guess is crucial to the success of the algorithm.

Consider the simple circuit with the diode and resistor shown in Fig. 2.7. According to Kirchhoff's voltage law,

$$v = v_{\text{R}} + v_{\text{D}} \tag{2.67}$$

where $v_{\text{R}} = iR$.

The electrical behavior of the diode is described by

$$i(v_{\text{D}}) = I_{\text{sat}}\left[\exp\left(\frac{v_{\text{D}}}{V_{\text{T}}}\right) - 1\right] \tag{2.68}$$

Rearranging Eq. (2.68) gives the equation for v_{D} in the form

$$v_{\text{D}} = V_{\text{T}} \ln\left(\frac{i}{I_{\text{sat}}} + 1\right) \tag{2.69}$$

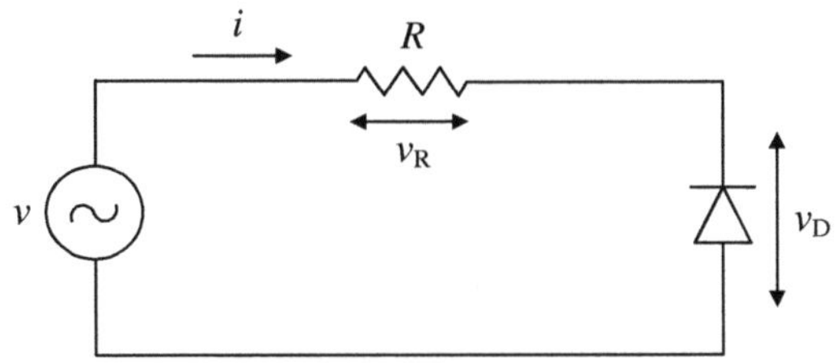

FIGURE 2.7 Circuit schematic with resistor, diode, and voltage source.

Thus, from Eqs. (2.68) and (2.69), it follows that

$$v = iR + V_T \ln\left(\frac{i}{I_{sat}} + 1\right) \tag{2.70}$$

This allows current i to be determined for a specified voltage v. However, because it is impossible to analytically solve this equation for current i in an explicit form, the solution must be found numerically.

Consider a dc voltage source V with a dc current I. For the sinusoidal voltage source, it is necessary to calculate the Bessel functions for the dc, fundamental-frequency, and higher-order current harmonic components. In this case, it is convenient to rewrite Eq. (2.70) as

$$f(I) = IR + V_T \ln\left(\frac{I}{I_{sat}} + 1\right) - V = 0 \tag{2.71}$$

from which

$$f'(I) = R + V_T \frac{1}{I + I_{sat}} \tag{2.72}$$

Then, applying the iterative algorithm for a single variable results in

$$I_n = I_{n-1} - \frac{f(I_{n-1})}{f'(I_{n-1})} \tag{2.73}$$

Using Eqs. (2.71) and (2.72) finally yields

$$I_n = I_{n-1} - \frac{I_{n-1}R + V_T \ln\left(\frac{I_{n-1}}{I_{sat}} + 1\right) - V}{R + V_T \frac{1}{I_{n-1} + I_{sat}}} \tag{2.74}$$

The results of the numerical calculation of the currents I_n for each iteration for $V_T = 25.9$ mA/V, $R = 5\ \Omega$, $V = 5$ V, $I_{sat} = 10\ \mu$A, and initial current $I_0 = 50$ mA are given in Table 2.1. The calculation error $\varepsilon_n = I_N - I_n$,

n	I_n, A	ε_n
0	0.05	0.899371786
1	0.878469005	0.070902781
2	0.948955229	0.000416557
3	0.949371786	

Table 2.1 Three-Step Iteration Procedure

where $n = 0, 1, \ldots, N$, illustrates the fast convergence to the solution for each iteration step. The error at each subsequent iteration step is approximately proportional to the square one of error at the previous step. If the required accuracy of $\varepsilon < 0.1\%$ is set in advance, the iteration procedure will be stopped at the third iteration step.

2.4 Quasilinear Method

To simplify the analysis and design procedure of the power amplifier, in some cases it is enough to apply a quasilinear method, which is based on the use of the ratios between the fundamental-frequency components of currents and voltages and the replacement of the transistor nonlinear elements by equivalent fundamentally averaged linear ones. The derivation of equations for the equivalent averaged linear elements in terms of the signal voltages is based on the static current-voltage and voltage-capacitance transistor characteristics.

For example, for a bipolar transistor, whose simplified equivalent circuit is shown in Fig. 2.8, all elements of its equivalent circuit are nonlinear, depending significantly on the operation mode, especially the transconductance g_m, base-emitter capacitance C_π, and collector capacitance C_c. The base-emitter capacitance C_π consists of the diffusion and junctions components and, at high frequencies, its

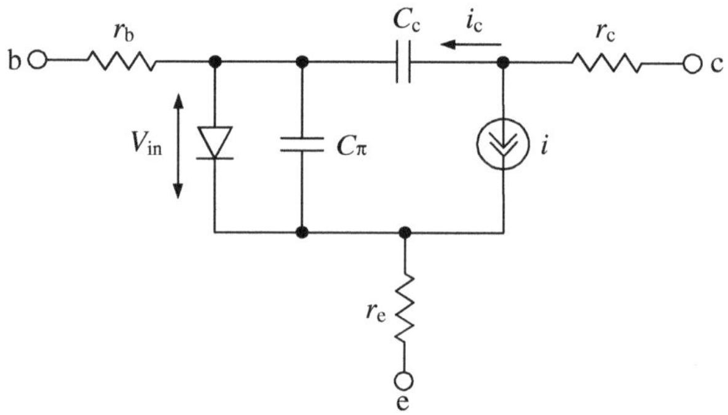

Figure 2.8 Bipolar transistor simplified equivalent circuit.

reactance is sufficiently high to shunt the base-emitter forward-biased diode. By taking into account that the device transition frequency is obtained by $f_T = g_m/2\pi C_\pi$, it is sufficient to consider the only nonlinear elements g_m, $\omega_T = 2\pi f_T$, and C_c, as the base resistance r_b poorly depends on a bias mode. The fundamentally averaged large-signal transconductance (or average transconductance) can be easily determined from Eq. (2.35) as

$$g_{m1} = \frac{I_1}{V} = \frac{2I_{sat}}{V} \exp\left(\frac{V_{cc}}{V_T}\right) I_1\left(\frac{V_{in}}{V_T}\right) \tag{2.75}$$

where V_{cc} is the collector dc bias voltage.

The collector capacitance represents a junction capacitance and can be approximated by

$$C_c = \frac{C_{c0}}{\left(1+\dfrac{v_c}{\varphi}\right)^\gamma} \tag{2.76}$$

where φ is the built-in junction potential, γ is the junction sensitivity, and C_{c0} is the initial capacitance when $v_c = 0$.

If our interest is restricted to the fundamental frequency, and $v_c = V_{cc} + V_c \sin\omega t$, where V_c is the voltage fundamental-frequency amplitude across the capacitance C_c, then the current flowing through the collector capacitance is defined for the quasilinear case as

$$i_c = C_c(v_c)\frac{dv_c}{dt} = \frac{\omega C_{c0} V_c \cos\omega t}{\left(1+\dfrac{V_{cc}}{\varphi}+\dfrac{V_c}{\varphi}\sin\omega t\right)^\gamma} = \frac{\omega\, C_c(V_{cc})\, V_c \cos\omega t}{(1+\xi \sin\omega t)^\gamma} \tag{2.77}$$

where $C_c(V_{cc})$ is the small-signal capacitance at the operating bias point and $\xi = V_c/(V_{cc} + \varphi)$.

As a result, the averaged large-signal collector capacitance C_{c1} can be calculated through the fundamental-frequency Fourier-series component as

$$C_{c1}(V_c) = \frac{I_{c1}}{\omega V_c} = \frac{C_c(V_{cc})}{\pi}\int_0^{2\pi} \frac{\cos^2\omega t}{(1+\xi \sin\omega t)^\gamma} d\omega t \tag{2.78}$$

Figure 2.9 shows the voltage dependences of the fundamentally averaged collector capacitance in a large-signal mode. Within the values of $\xi < 1$, the maximum large-signal value of $C_{c1}(V_c)$ deviates from the small-signal value of $C_c(V_{cc})$ by not more than 20% for an abrupt junction with $\gamma = 0.5$.

For a MESFET device with the simplified equivalent circuit shown in Fig. 2.10, the drain current i_d is a function of the gate-source

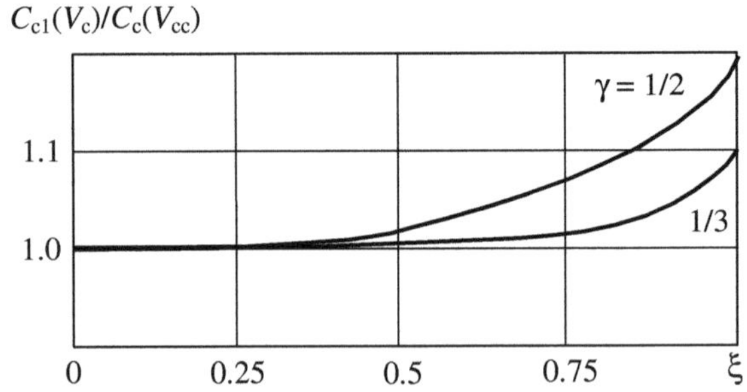

FIGURE 2.9 Large-signal behavior of device junction capacitance.

voltage v_{gs} and the drain-source voltage v_{ds}, which can be expanded in a two-dimensional Taylor series as

$$i_d(v_{gs}, v_{ds}) = I_0 + \left.\frac{\partial f}{\partial v_{gs}}\right|_{\substack{v_{gs}=V_s \\ v_{ds}=V_{dd}}} (v_{gs} - V_g) + \left.\frac{\partial f}{\partial v_{ds}}\right|_{\substack{v_{gs}=V_g \\ v_{ds}=V_{dd}}} (v_{ds} - V_{dd})$$

$$+\frac{1}{2}\left[\left.\frac{\partial^2 f}{\partial v_{gs}^2}\right|_{\substack{v_{gs}=V_g \\ v_{ds}=V_{dd}}} (v_{gs} - V_g)^2 + 2\left.\frac{\partial^2 f}{\partial v_{gs}\,\partial v_{ds}}\right|_{\substack{v_{gs}=V_g \\ v_{ds}=V_{dd}}} (v_{gs} - V_g)(v_{ds} - V_{dd}) \right.$$

$$\left. + \left.\frac{\partial^2 f}{\partial v_{ds}^2}\right|_{\substack{v_{gs}=V_g \\ v_{ds}=V_{dd}}} (v_{ds} - V_{dd})^2 + \ldots \right] \tag{2.79}$$

where V_g is the gate dc bias voltage and V_{dd} is the drain dc supply voltage.

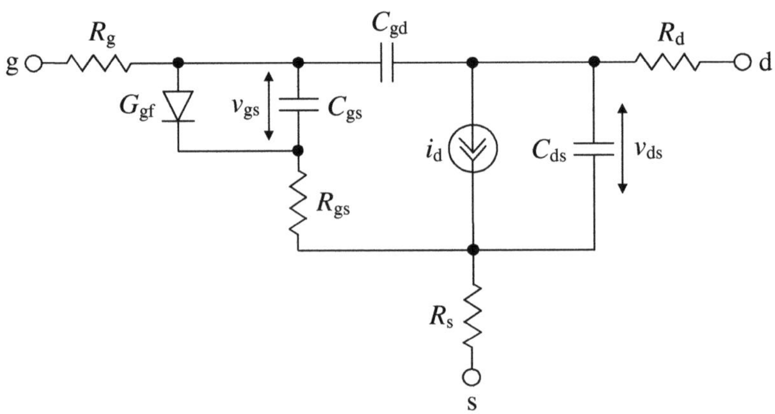

FIGURE 2.10 MESFET simplified equivalent circuit.

In the small-signal quasilinear case, the high-degree terms are neglected and

$$i_{\rm d}(v_{\rm gs}, v_{\rm ds}) = I_0 + \left.\frac{\partial f}{\partial v_{\rm gs}}\right|_{\substack{v_{\rm gs}=V_{\rm g} \\ v_{\rm ds}=V_{\rm dd}}} (v_{\rm gs} - V_{\rm g}) + \left.\frac{\partial f}{\partial v_{\rm ds}}\right|_{\substack{v_{\rm gs}=V_{\rm g} \\ v_{\rm ds}=V_{\rm dd}}} (v_{\rm ds} - V_{\rm dd}) \quad (2.80)$$

The gate-source and drain-source instantaneous voltages can respectively be written as

$$v_{\rm gs} = V_{\rm g} + V_{\rm gs}\cos(\omega t + \phi) \quad (2.81)$$

$$v_{\rm ds} = V_{\rm dd} + V_{\rm ds}\cos\omega t \quad (2.82)$$

where $V_{\rm gs}$ and $V_{\rm ds}$ are the gate-source and drain-source voltage amplitudes, and ϕ is the phase difference between these voltages.

Consequently, the instantaneous drain current given by Eq. (2.79) can be rewritten as

$$i_{\rm d}(\omega t) = I_0 + g_{\rm m1} V_{\rm gs}\cos(\omega t + \phi) + G_{\rm ds1} V_{\rm ds}\cos\omega t \quad (2.83)$$

where

$$g_{\rm m1} = \left.\frac{I_{\rm d}}{V_{\rm gs}}\right|_{V_{\rm ds}=0} \quad (2.84)$$

is the linearized large-signal transconductance,

$$G_{\rm ds1} = \left.\frac{I_{\rm d}}{V_{\rm ds}}\right|_{V_{\rm gs}=0} \quad (2.85)$$

is the linearized differential output conductance, I_0 is the dc drain current, $I_{\rm d}$ is the fundamental drain current amplitude, and $G_{\rm ds1} = 1/R_{\rm ds1}$ [4].

Multiplying the right- and left-hand sides of Eq. (2.83) by $\sin\omega t$ and integrating over the entire signal period result in the fundamentally averaged transconductance $g_{\rm m1}$ obtained by

$$g_{\rm m1} = -\frac{1}{\pi V_{\rm gs}\sin\phi}\int_0^{2\pi} i_{\rm d}(\omega t)\sin\omega t\, d\omega t \quad (2.86)$$

Similarly, multiplying by $\sin(\omega t + \phi)$ results in the fundamentally averaged drain-source conductance $G_{\rm ds1}$ obtained by

$$G_{\rm ds1} = \frac{1}{\pi V_{\rm ds}\sin\phi}\int_0^{2\pi} i_{\rm d}(\omega t)\sin(\omega t + \phi)\, d\omega t \quad (2.87)$$

The average large-signal gate-source capacitance C_{gs1} can be calculated similarly to that of for the abrupt collector capacitance C_{c1} of the bipolar transistor with $\gamma = 0.5$. The average gate forward conductance G_{gf1} is defined by

$$G_{gf1} = \frac{2I_{sat}}{V_{gs}} \exp\left(\frac{V_g}{V_T}\right) I_1\left(\frac{V_{gs}}{V_T}\right) \tag{2.88}$$

where I_{sat} is the saturation current of the Schottky barrier and $I_1(V_{gs}/V_T)$ is the first-order modified Bessel function of the first kind.

The gate charging resistance R_{gs} varies with the gate-source capacitance C_{gs} in such a way that the charging time constant $\tau_g = R_{gs}C_{gs}$ varies insignificantly and it can be treated as a constant in a quasilinear approximation.

2.5 Harmonic Balance Method

Harmonic balance analysis determines the periodic steady-state circuit responses because the basis signal set chosen to represent the physical signals in the circuit consists of sinusoidal functions [5, 6]. The harmonic balance method requires that the circuit be divided into two subcircuits connected by wires forming multiports. One subcircuit contains the linear components of the circuit, and another contains the only nonlinear device model parameters, as shown in Fig. 2.11. The linear parasitic elements of the device are taken into account by the linear subcircuit. Sources and loads are concentrated in a separate multiport network. The N wires at the linear-nonlinear interface connect the two subcircuits and define corresponding nodes. Current flowing out of one subcircuit must equal to that flowing into another. Matching the frequency components in each wire satisfies the continuity equation for current. An iterative process calculates the current at each wire so that the obtained dependences are satisfied for both linear and nonlinear sides of the entire circuit.

The nonlinear subcircuit is represented by a set of nonlinear equations

$$i_k(t) = f[v_1(t), v_2(t), \ldots, v_N(t)] \tag{2.89}$$

where f is the arbitrary nonlinear function and can include differentiation and integration, i_k and v_k are the kth-port current and voltage,

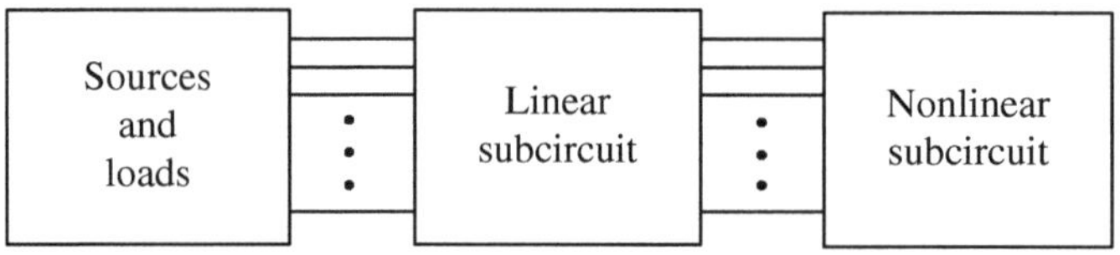

FIGURE 2.11 Harmonic balance circuit representation.

respectively, and $k = 1, \ldots, N$. The linear subcircuit response is calculated in the frequency domain at each harmonic component by linear analysis and is represented by an $N \times (N + M)$ matrix. The M additional variables are the additional external nodes, to which voltage or current sources are connected. In the case of an applied input signal containing a sum of harmonics of $\omega, 2\omega, \ldots, K\omega$, there will be $(K + 1)$ matrices so that the matrix relationship between the port voltages and currents can be written as

$$\begin{bmatrix} v_1(k\omega) \\ \vdots \\ v_N(k\omega) \end{bmatrix} = \begin{bmatrix} H_{11}(k\omega) & H_{12}(k\omega) & \cdots & H_{1(N+M)}(k\omega) \\ H_{21}(k\omega) & H_{22}(k\omega) & \cdots & H_{2(N+M)}(k\omega) \\ \vdots & \vdots & \ddots & \vdots \\ H_{N1}(k\omega) & H_{N2}(k\omega) & \cdots & H_{N(N+M)}(k\omega) \end{bmatrix} \begin{bmatrix} i_1(k\omega) \\ \vdots \\ i_N(k\omega) \\ i_{N+1}(k\omega) \\ \vdots \\ i_{N+M}(k\omega) \end{bmatrix} \tag{2.90}$$

where $H_{ij}(k\omega)$ are the impedance or transfer ratios, depending on which of the variables are voltages and which are currents, and $k = 0, 1, \ldots, K$. The harmonic balance program finds a simultaneous solution of Eqs. (2.89) and (2.90) for $v_1, v_2, \ldots, v_N$, so that $i_1, i_2, \ldots, i_N$ can be determined.

Figure 2.12 illustrates the application of the harmonic balance method to a three-terminal MESFET device. The MESFET device is represented by only nonlinear elements, whereas all its parasitics, matching and output networks are incorporated into a linear subcircuit. The source terminal is chosen as a reference, so that $N = 2$. Here, the voltages v_1 and v_2 are independent variables, as are the currents i_1

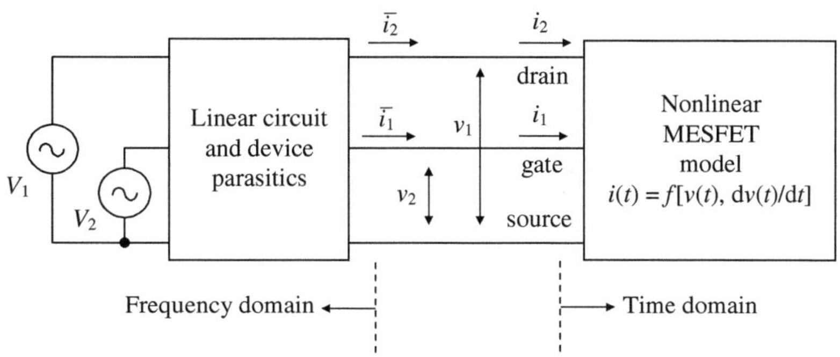

FIGURE 2.12 Application of harmonic balance method to three-terminal MESFET device.

and i_2. Additional applied sources are the external voltage sources V_1 and V_2. As a result, the output current flowing into the load or voltage across the load can be readily found once i_1 and i_2 are determined. Because Eq. (2.89) is stated in the time domain and Eq. (2.90) is stated in the frequency domain, time-to-frequency conversion is achieved using a discrete Fourier transform. An initial estimate must be made for $i_j(t)$ and $v_j(t)$, $j = 1, \ldots, N$, because their values are unknown at the beginning of calculation. Iteration between Eqs. (2.89) and (2.90) is performed using a discrete Fourier transform to obtain the frequency components calculated in the time domain using Eq. (2.89) until a self-consistent set of variables that satisfy the current continuity equations is attained. The continuity equation for current states that the "nonlinear" currents i_j must be equal to the "linear" currents $\bar{i}_j$, that corresponds to zero error function as a solution.

Figure 2.13 shows the equivalent circuit of a power amplifier, including a nonlinear MESFET circuit model, the input and output matching circuits, and the source and load impedances [6]. The device model includes both linear and nonlinear elements, which can be characterized by the appropriate measurement or modeling. The largest contribution to the nonlinear device behavior is made by the nonlinear drain current source I_d, forward-bias gate current source I_g, and gate-to-drain current source I_b, which are assumed to be functions of the instantaneous internal gate voltage V_g and drain voltage V_d. An analysis of the circuit shown in Fig. 2.13 using Kirchhoff's voltage and current laws in the frequency domain results in two complex algebraic equations with V_g and V_d as the independent variables for each kth Fourier component:

$$AV_{gk} + BV_{dk} = C \tag{2.91}$$

$$DV_{gk} + EV_{dk} = F \tag{2.92}$$

where

$$A = 1 + jk\omega[(Z_2 + R_{gs})C_{gs} + (Z_1 + Z_S)(C_{gs} + C_{gd} + jk\omega R_{gs}C_{gs}C_{gd})]$$

$$B = jk\omega[-(Z_1 + Z_S)C_{gd} + Z_2C_{ds}]$$

$$C = -(Z_1 + Z_2 + Z_S)I_g + (Z_1 + Z_S)I_b - Z_2I_d + V_S$$

$$D = jk\omega[Z_2C_{gs} - (Z_3 + Z_L)(C_{ds} + jk\omega R_{gs}C_{gs}C_{ds})]$$

$$E = 1 + jk\omega[Z_2C_{gs} + (Z_3 + Z_L)(C_{gd} + C_{gs})]$$

$$F = -Z_2I_g - (Z_3 + Z_L)I_b + (Z_2 + Z_3 + Z_L)I_d + V_{dd}$$

where $Z_1 = R_g + jk\omega L_g$, $Z_2 = R_s + jk\omega L_s$, $Z_3 = R_d + jk\omega L_d$, and $k = 1, \ldots, N$.

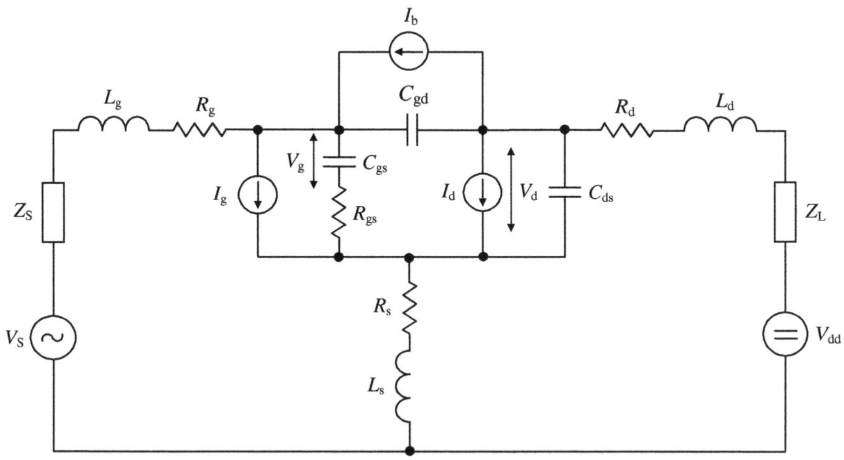

FIGURE 2.13 MESFET power amplifier equivalent circuit.

Here, the matrices *A*, *B*, *D*, and *E* describe the linear network, and the matrices *C* and *F* represent the independent driving sources. The nonlinearity in these equations is contained in the currents I_g, I_b, and I_d, which commonly depend on both V_g and V_d. Consequently, Eqs. (2.91) and (2.92) determine the voltages V_g and V_d only implicitly. The system of these equations is best solved numerically using an iterative technique.

The iterative solution process begins with an initial approximation to the solution, which can be made by neglecting the gate forward-bias current I_g and the gate-to-drain current I_b and assuming a linear device output conductance and transconductance. Setting and substituting the initial values of the bias current and the fundamental drain current amplitude and phase into Eqs. (2.91) and (2.92), which are rearranged with the real and imaginary parts, allows the appropriate bias gate and drain voltages and fundamental gate and drain voltages to be calculated in the frequency domain. Rewriting the voltages $v_g(t)$ and $v_d(t)$ in the time domain as

$$v_g(\omega t) = V_g + V_{g1}\cos(\omega t + \phi_{g1}) \tag{2.93}$$

$$v_d(\omega t) = V_{dd} + V_{d1}\cos(\omega t + \phi_{d1}) \tag{2.94}$$

and substituting them into the nonlinear current expressions $i_g(v_g, v_d)$ and $i_d(v_g, v_d)$ give the Fourier components for each current harmonic with the appropriate amplitude and phase. Then, at each iteration step, inserting every gate and drain current written in a complex form into Eqs. (2.91) and (2.92) allows the complex gate and drain harmonic voltages to be obtained. Their representation in the time domain can be given by

$$v_{g}(\omega t)=\sum_{k=1}^{N} V_{gk}\cos(k\omega t+\psi_{gk}) \tag{2.95}$$

$$v_{d}(\omega t)=\sum_{k=1}^{N} V_{dk}\cos(k\omega t+\psi_{dk}) \tag{2.96}$$

which, in turn, allows the complex gate and drain currents to be defined by inserting into their nonlinear expressions. Finally, when the accuracy of the solution is satisfied, the iterative process ends. In this case, an error function, which defines the duration of this iterative procedure, can be written in a mean square form.

2.6 X-Parameters

The scattering *S*-parameters have been used to represent linear electrical networks for simulation and design since the 1960s, when measuring *S*-parameters was made possible with the introduction of the network analyzer [7]. In this case, the *S*-parameters allow designers to accurately describe the properties of very complicated linear networks as simple "black boxes" and then integrate them with other electrical components using linear frequency-domain simulators. Furthermore, they have proved to be an important initial step in determining the non-linear transistor models by measuring the small-signal scattering parameters characterizing the electrical behavior of nonlinear active devices in the entire frequency operating range under different bias voltages. However, new transistor technologies and demand for greater accuracy lead to an increase in device and model complexity, driving the need for enhanced characterization techniques and specialized test equipment. Note that the annotation "black box" refers to the fact that no knowledge is used or required concerning the internal circuitry of the device under test (DUT).

As an alternative, a large-signal network analysis (LSNA) can be provided which is empowered by a unique combination of two mathematical transformations [8]. The first one is the transformation between the traveling voltage-wave formalism and voltage-current representation, whereas the second one is the transformation between the time domain and the frequency domain. With an LSNA technology, the incident and reflected waves are measured at the terminals of the device under large-signal operation conditions, including harmonics and intermodulation products. In this case, a new nonlinear model called the *polyharmonic distortion (PHD) model* can be identified with advanced nonlinear measurements similar to *S*-parameters for the linear case. The PHD modeling is a black-box, frequency-domain modeling technique, and this approach, also referred to as *X*-parameters, can be used as a natural extension of *S*-parameters under large-signal conditions to create the device large-signal model [9, 10]. In this case, a DUT is connected to a large-signal network analyzer, and model is automatically extracted that accurately describes all kinds of nonlinear behavior such as amplitude and phase of harmonics, compression characteristics, spectral regrowth, and amplitude-dependent input and output impedances.

The model theory is derived from a multiharmonic linearization around a periodic steady state under a large-signal drive with several small tones one at time at each port and at each harmonic of the fundamental up to the maximum number needed for model. For a given DUT, it is necessary to determine the set of the describing functions F_{pm} that correlate all the relevant input spectral components A_{qn} and output spectral components B_{pm}, where q and p range from one to the number of signal ports, whereas m and n range from zero to the highest harmonic index [10]. This is mathematically expressed as

$$B_{pm} = F_{pm}(A_{11}, A_{12}, \dots, A_{21}, A_{22}, \dots) \tag{2.97}$$

from which practical models can be developed in the frequency domain. In this case, it is assumed that the fundamental frequency is a known constant. The PHD model is a particular approximation of Eq. (2.97) as a result of its linearization.

The concept of describing functions is shown in Fig. 2.14(*a*). The basic feature of the describing functions is related to the fact that F_{pm} describes a time-invariant system. This implies that applying an arbitrary delay to the input signals, in our case the incident *A*-waves, always results in exactly the same time delay for the output signals, the reflected *B*-waves. In the frequency domain, applying a time delay is equivalent to the application of a linear phase shift (proportional to frequency), and as such this fact can mathematically be expressed as

$$\begin{aligned} B_{pm}\exp(jm\theta) = F_{pm}[\, & A_{11}\exp(j\theta),\ A_{12}\exp(j2\theta),\ \dots, \\ & A_{21}\exp(j\theta), A_{22}\exp(j2\theta),\ \dots\,] \end{aligned} \tag{2.98}$$

where θ can be made equal to the inverted phase of A_{11} because Eq. (2.98) is valid for all values of θ.

By introducing the phasor P set by the phase of the large-signal component A_{11} and defined as $P = \exp(j\varphi A_{11})$, the spectral mapping of Eq. (2.98) can be rewritten substituting $\exp(j\theta)$ by P^{-1} as

$$B_{pm} = F_{pm}(|A_{11}|, A_{12}P^{-2}, A_{13}P^{-3}, \dots, A_{21}P^{-1}, A_{22}P^{-2}, \dots)\, P^{m} \tag{2.99}$$

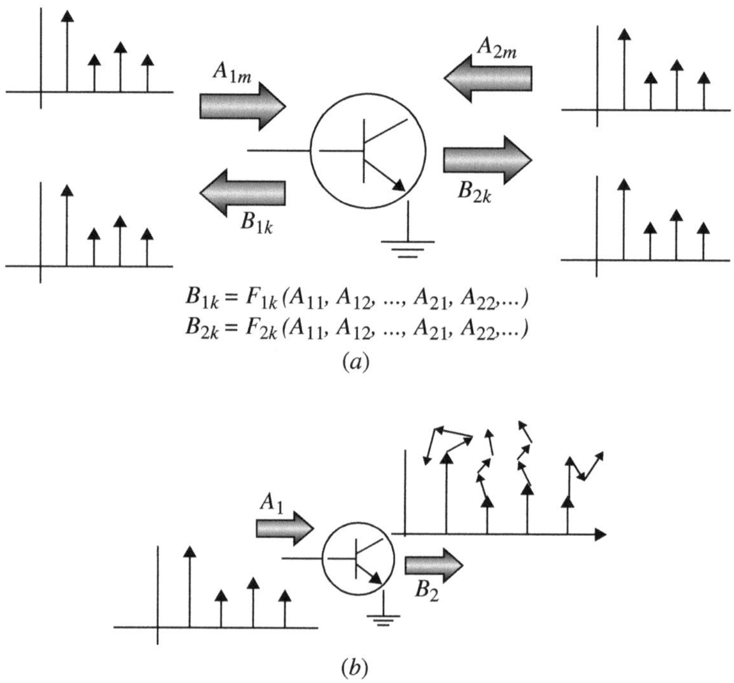

FIGURE 2.14 Concept of describing functions and harmonic superposition principle.

where the first input argument is a positive real number, which is the amplitude of the fundamental component at the input port 1, rather than a complex number.

Generally, the superposition principle is not valid under large-signal nonlinear operation conditions. However, in many practical cases such as in power amplifiers driven by a narrowband input signal, there is only one dominant large-signal input component (A_{11}) present, whereas all other input components (the harmonic frequencies) are relatively small. In this case, it is possible to use the superposition principle for the relatively small input components, which can be called the *harmonic superposition principle*. The harmonic superposition principle is graphically illustrated in Fig. 2.14(*b*), where, for simplicity, only A_{1m} and B_{2n} are present and components A_{2m} and B_{1n} are neglected. As an example, consider the case when the input spectral component A_{11} creates four output spectral components B_{2n}. Then, leaving the A_{11} excitation the same and adding a relatively small A_{12} component will result in a first deviation of all B_{2n} components in the output spectrum. The same results happen for the subsequent second and third deviations for all output components simultaneously when the relatively small third (A_{13}) and fourth (A_{14}) spectral components are added at the input. The harmonic superposition principle holds when the overall deviation of the output spectrum B_2 is the superposition of all individual deviations. The harmonic superposition is a key assumption to the PHD model. A linearization of Eq. (2.99) versus all components except the large-signal A_{11} leads to

$$B_{pm} = K_{pm}(|A_{11}|)P^{m} + \sum_{qn} G_{pq,mn}(|A_{11}|)P^{m}\,\mathrm{Re}(A_{qn}P^{-n}) + \sum_{qn} H_{pq,mn}(|A_{11}|)\,P^{m}\,\mathrm{Im}(A_{qn}P^{-n}) \tag{2.100}$$

where

$$K_{pm}(|A_{11}|) = F_{pm}(|A_{11}|, 0, \ldots, 0) \tag{2.101}$$

$$G_{pq,mn}(|A_{11}|) = \left.\frac{\partial F_{pm}}{\partial \mathrm{Re}(A_{qn}P^{-n})}\right|_{|A_{11}|,\,0,\,\ldots,\,0} \tag{2.102}$$

$$H_{pq,mn}(|A_{11}|) = \left.\frac{\partial F_{pm}}{\partial \mathrm{Im}(A_{qn}P^{-n})}\right|_{|A_{11}|,\,0,\,\ldots,\,0} \tag{2.103}$$

The PHD model equation is derived by substituting the real and imaginary parts of the input arguments in Eq. (2.100) by a linear combination of the input arguments and their corresponding conjugates. Because

$$\mathrm{Re}(A_{qn}P^{-n}) = \frac{A_{qn}P^{-n} + \mathrm{conj}(A_{qn}P^{-n})}{2} \tag{2.104}$$

$$\mathrm{Im}(A_{qn}P^{-n}) = \frac{A_{qn}P^{-n} - \mathrm{conj}(A_{qn}P^{-n})}{-2j} \tag{2.105}$$

Eq. (2.100) can be rewritten as

$$B_{pm} = K_{pm}(|A_{11}|)P^{m} + \sum_{qn} G_{pq,mn}(|A_{11}|)P^{m}\,\frac{A_{qn}P^{-n} + \mathrm{conj}(A_{qn}P^{-n})}{2} + \sum_{qn} H_{pq,mn}(|A_{11}|)P^{m}\,\frac{A_{qn}P^{-n} - \mathrm{conj}(A_{qn}P^{-n})}{-2j} \tag{2.106}$$

Rearranging the corresponding terms finally results in the relatively simple PHD model equation

$$B_{pm} = \sum_{qn} S_{pq,mn}(|A_{11}|)P^{m-n}A_{qn} + \sum_{qn} T_{pq,mn}(|A_{11}|)\,P^{m+n}\mathrm{conj}(A_{qn}) \tag{2.107}$$

where new functions $S_{pq,mn}$ and $T_{pq,mn}$ are defined through functions K_{pm}, $G_{pq,mn}$, and $H_{pq,nm}$ introduced in Eq. (2.100).

The basic PHD model given in Eq. (2.107) simply describes that the B-waves result from a linear mapping of the A-waves, similar to classic S-parameters. At the same time, there are some significant differences. For example, the right-hand side of Eq. (2.107) contains a contribution associated with not only the A-waves but also with the conjugate part of the A-waves. However, the conjugate part of the A-waves is not present at all with S-parameters because the contribution of an A-wave to a particular B-wave is not a function of the phase of that A-wave. Any phase shift in A-wave will just result in the same phase shift of the contribution to the particular B-wave. This is no longer the case when a large fundamental signal (A_{11} wave) is present at the input of the DUT. In this case, the large-signal A_{11} wave creates a phase reference point for the other entire incident A-waves, and the contribution to the B-waves of a particular A-wave depends on the phase relationship between this particular A-wave and the large-signal A_{11} wave. This relative phase dependence is expressed in Eq. (2.107) through the presence of the conjugate A-wave terms [10].

For a simple case of a B_{21} (fundamental at the output) depending on A_{21} (reflected fundamental at the output) and A_{11} (incident fundamental at the input), Eq. (2.107) reduces to

$$B_{\text{pm}} = S_{21,11}(|A_{11}|)\, A_{11} + S_{22,11}(|A_{11}|)\, A_{21} + T_{22,11}(|A_{11}|)\, P^2 \text{conj}(A_{11}) \tag{2.108}$$

which can be rewritten in a normalized form using the phasor definition $P = \exp(j\varphi A_{11})$ as

$$\begin{aligned} \frac{B_{21}}{A_{21}} &= S_{21,11}(|A_{11}|)\frac{A_{11}}{A_{21}} + S_{22,11}(|A_{11}|) \\ &\quad + T_{22,11}(|A_{11}|)\exp[j2\varphi(A_{11}) - j2\varphi(A_{21})] \end{aligned} \tag{2.109}$$

that demonstrates the phase difference between A_{11} and A_{21} through the complex exponential. Besides the relative phase dependence, the PHD model can provide another feature when compared to S-parameters, namely that it relates input and output spectral components that have different frequencies. It describes, for example, how A_{13}, which is the third harmonic of the incident wave, will contribute to a change in B_{22}, which is the second harmonic at port 2.

To describe different nonlinear DUT characteristics from the PHD model, a highly simplified model containing exclusively the $S_{21,11}$ term

$$B_{21} = S_{21,11}(|A_{11}|)\, A_{11} \tag{2.110}$$

can be considered. Dividing both sides of Eq. (2.110) by A_{11} reveals that the amplitude of the function $S_{21,11}$ corresponds to the compression

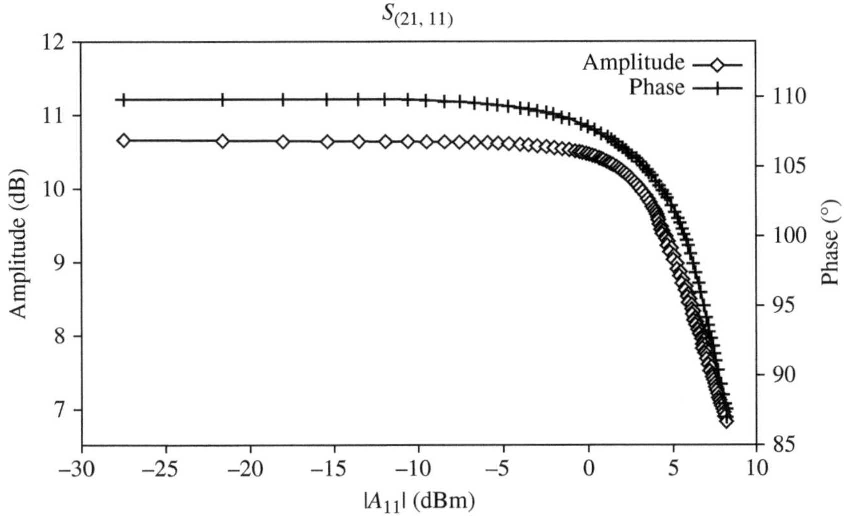

FIGURE 2.15 Amplitude and phase characteristics of DUT.

characteristic of the DUT, while the *AM-PM* (amplitude-phase) conversion characteristic is given by the phase of $S_{21,11}$ as

$$S_{21,11}(|A_{11}|) = \frac{B_{21}}{A_{11}} \tag{2.111}$$

Figure 2.15 shows the measured amplitude and phase of $S_{21,11}$ of a wideband microwave integrated circuit amplifier with a fundamental frequency of 9.9 GHz. The comparison of the amplifier gain and *AM-PM* characteristics as a function of the incident power over a decade frequency range from 600 MHz to 6 GHz from highly linear operation to over 2.5-dB gain compression showed the close agreement between the derived PHD behavioral model and the circuit-level model over the same range of operating conditions [9]. The PHD model can also predict the generation of harmonics, with simple equations up to the fourth harmonic written as

$$B_{21} = S_{21,21}(|A_{11}|)PA_{11} \tag{2.112}$$

$$B_{31} = S_{21,31}(|A_{11}|)\,P^2 A_{11} \tag{2.113}$$

$$B_{41} = S_{21,41}(|A_{11}|)\,P^3 A_{11} \tag{2.114}$$

The PHD nonlinear behavioral model, which is characterized by *X*-parameters, can be applied to the input signals that are represented as a modulated carrier when a complex envelope domain representation of the *A*-wave and *B*-wave signals is used and a quasistatic relationship between the *A*-waves and the *B*-waves is assumed [10]. In the envelope domain, a time-varying signal $x(t)$ can be expressed through its complex envelope representation by a series of the time-varying complex functions $X_h(t)$ by

$$x(t) = \mathrm{Re}\left[\sum_{h} X_h(t)\exp(j2\pi h f_c t)\right] \tag{2.115}$$

where f_c is the carrier frequency. When this envelope representation is applied to the *A*-waves and the *B*-waves, the PHD model of Eq. (2.107) can be rewritten as

$$B_{pm}(t) = \sum_{qn} S_{pq,mn}(|A_{11}(t)|)\, P^{m-n}(t) A_{qn}(t) + \sum_{qn} T_{pq,mn}(|A_{11}(t)|) P^{m+n}(t)\, \mathrm{conj}[A_{qn}(t)] \tag{2.116}$$

which then can be used to calculate the amplitude and phase of the *B*-wave complex envelopes as a function of the *A*-wave complex envelopes. The resulting time-dependent *B*-wave complex envelopes are converted into the frequency domain via Fourier transform, where the resulting spectra are used to calculate typical nonlinear parameters, for example, the adjacent and alternate channel leakage power ratios $ACLR_1$ and $ACLR_2$, respectively.

Because of the time-varying thermal and trapping effects in the device itself, the nonlinear characteristics of the active device, and hence its *X*-parameters, are dependent on these slow processes in the device, which are called the *memory effects*. Generally, the memory effects can be classified as either short-term or long-term. If the short-term memory effects are caused by physical dynamics that occur at the timescale of the carrier, for example, the frequency dispersion, then the long-term memory effects are caused by slowly varying processes, which can also include the self-heating or self-biasing effects. The PHD behavioral model can account for the memory effect modifications, which are valid if the rate of change of the effect is significantly less than the modulation speed. In this case, the modified PHD model equation is rewritten as a superposition of a static part and a dynamic part, where the static part behaves like a conventional PHD model and the dynamic part represents the long-term memory effects, which are described as the integral effect of a general nonlinear function of the instantaneous amplitude of the input signal $A(t)$, the past values of the input signal $A(t - \tau)$, and how long ago that past value occurred (variable τ) [11].

References

1. A. I. Berg, *Theory and Design of Vacuum-Tube Generators* (in Russian), Moscow: GEI, 1932.
2. P. H. Osborn, "A Study of Class B and C Amplifier Tank Circuits," *Proc. IRE*, vol. 20, pp. 813–834, May 1932.
3. J. K. Fidler and C. Nightingale, *Computer Aided Circuit Design*, New York: John Wiley & Sons, 1978.
4. Y. Tajima, B. Wrona, and K. Mishima, "GaAs FET Large-Signal Model and its Application to Circuit Designs," *IEEE Trans. Electron Devices*, vol. ED-28, pp. 171–175, Feb. 1981.
5. R. Gilmore, "Nonlinear Circuit Design Using the Modified Harmonic Balance Algorithm," *IEEE Trans. Microwave Theory Tech.*, vol. MTT-34, pp. 1294–1307, Dec. 1986.
6. D. L. Peterson, A. M. Pavio, and B. Kim, "A GaAs FET Model for Large-Signal Applications," *IEEE Trans. Microwave Theory Tech.*, vol. MTT-32, pp. 276–281, Mar. 1984.
7. D. Vye, "Fundamentally Changing Nonlinear Microwave Design," *Microwave J.*, vol. 53, pp. 22–40, Mar. 2010.
8. J. Verspecht, "Large-Signal Network Analysis," *IEEE Microwave Mag.*, vol. 6, pp. 82–92, Dec. 2005.
9. D. E. Root, J. Verspecht, D. Sharrit, J. Wood, and A. Cognata, "Broad-Band Poly-Harmonic Distortion (PHD) Behavioral Models from Fast Automated Simulations and Large-Signal Vectorial Network Measurements," *IEEE Trans. Microwave Theory Tech.*, vol. MTT-53, pp. 3656–3664, Nov. 2005.
10. J. Verspecht and D. E. Root, "Polyharmonic Distortion Modeling," *IEEE Microwave Mag.*, vol. 7, pp. 44–57, Jun. 2006.
11. P. Roblin, D. E. Root, J. Verspecht, Y. Ko, and J. P. Teyssier, "New Trends for the Nonlinear Measurement and Modeling of High-Power RF Transistors and Amplifiers with Memory Effects," *IEEE Trans. Microwave Theory Tech.*, vol. MTT-60, pp. 1964–1978, Jun. 2012.

CHAPTER 3

Nonlinear Active Device Modeling

Accurate device modeling is an extremely important procedure in circuit design, especially in monolithic integrated circuits. Better approximations of the final design can only be achieved if the nonlinear device behavior is described accurately. Once a device model has been incorporated into a circuit simulator, it requires the parameters to specify the device characteristics according to the model equations. The next step is to provide a procedure for adequate extraction of *S*-parameter data. This poses the problem of how to extract the device parameters from the measurement results accurately, rapidly, and without unnecessary measurement complexity. The best way is to minimize the device bias points under *S*-parameter measurements and to combine the analytical approach with a final optimization procedure to provide the best fitting of the experimental curves and empirical equation-based model curves. Numerical optimization is often used to fit the model *S*-parameters to the measured parameters as the resulting device element values depend on the starting values and are not unique. The analytical approach incorporates a derivation of the basic intrinsic device parameters from its equivalent circuit based on *S*- to *Y*- or *Z*-parameter transformations using sufficiently simple equations. However, it is crucial to choose an appropriately large-signal model that is most suitable for a specific active device, accurately describes the nonlinear behavior of its equivalent circuit parameters, and contains a reasonable number of model parameters to be determined.

This chapter describes all necessary steps for an accurate device modeling procedure, beginning with determining the small-signal equivalent circuit parameters. A variety of nonlinear models, including noise models, for MOSFETs, MESFETs, HEMTs, bipolar devices including HBTs, which are very prospective to design modern microwave hybrid and monolithic integrated circuits, are given and discussed.

3.1 Power MOSFETs

Personal wireless communication services have been driving the development of silicon MOSFET worldwide to provide reliable low-cost and high-performance technology. For example, the laterally diffused MOSFET (LDMOSFET) device structures have proven to be highly efficient, high gain, and linear for high-power RF and microwave base station applications. To develop low-cost silicon-integrated circuits using CMOS technology for higher speed and higher-frequency integrated circuits and subsystems within shorter design time, it is necessary to create accurate device models to allow efficient CAD simulation. Several well-known physically based MOSFET models were developed to describe the device electrical behavior [1, 2]. However, some of them such as Level 1, Level 2, or Level 3 large-signal models are very simple and cannot describe the current-voltage and voltage-capacitance characteristics with acceptable accuracy. Other popular models, as the BSIM3v3 or higher-version models, are too much complicated and quite formal in general so that, for better learning of the device basic properties, it is useful to consider its intrinsic nonlinear core circuit only. Besides, BSIM3v3 may not be as accurate for RFIC simulation due to their derivative discontinuity. Moreover, microwave parasitic effects in silicon MOSFETs are not easy physically predictable. Table-based models, such as the HP Root model, are only accurate for the characterized structures and measurement conditions. An empirical analytical modeling approach is an explicit and valid compromise between physical models and data-based models.

3.1.1 Small-Signal Equivalent Circuit

To describe accurately the nonlinear properties of the large MOSFET devices, it is necessary to take into account the distributed nature of the gate capacitance, because the channel resistance is not equal to zero. In this case, the device channel can be modeled as a bias-dependent *RC* distributed transmission line along the channel length, as shown in Fig. 3.1. This one-dimensional approach assumes a gradual channel approximation when the quantity of charge in the channel is controlled completely by the gate electrode, only fields in the vertical dimension influence the depletion region, and channel current is provided entirely by drift with a constant mobility. Despite some drawbacks related to short-channel devices, this approach allows a compromise between accuracy and simplicity of a model derivation.

The *ABCD*-matrix of the given *RC* transmission line can be written as

$$[ABCD] = \begin{bmatrix} \cosh \gamma L & Z_0 \sinh \gamma L \\ \dfrac{\sinh \gamma L}{Z_0} & \cosh \gamma L \end{bmatrix} \tag{3.1}$$

where $\gamma = \sqrt{j\omega R'_{\text{ch}} C'_{\text{g}}}$ is the propagation constant, $Z_0 = R'_{\text{ch}}/\gamma$ is the characteristic transmission line impedance, L is the channel length,

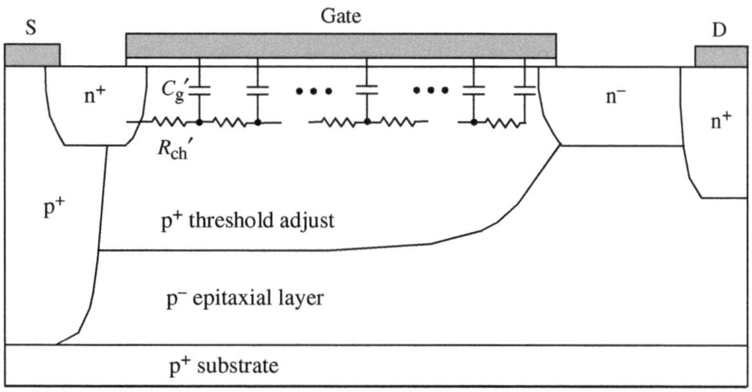

FIGURE 3.1 Schematic representation of MOSFET distributed channel structure.

$R'_{ch} = R_{ch}/L$, $C'_g = C_g/L$, R_{ch} is the channel charging resistance, which is a result of noninstantaneous respond to the changes of the gate-source voltage, and C_g is the total gate capacitance. In this case, the equivalent gate-source impedance Z_{gs} can be written using matrix Eq. (3.1) as

$$Z_{gs} = \frac{A}{C} = R_{ch}\frac{\coth\gamma L}{\gamma L} \tag{3.2}$$

The first-order approximation of Z_{gs} obtained from a power series expansion of Eq. (3.2) yields

$$Z_{gs} = R_{ch}\frac{\coth\gamma L}{\gamma L} \cong \frac{R_{ch}}{\gamma L}\left(\frac{1}{\gamma L} + \frac{\gamma L}{3}\right) = \frac{R_{ch}}{3} + \frac{1}{j\omega C_g} \tag{3.3}$$

From Eq. (3.3), it follows that the MOSFET intrinsic gate-source circuit can be modeled using a simple series circuit with the resistance $R_{gs} = R_{ch}/3$ and the capacitance $C_{gs} = C_g$. Figure 3.2(*a*) shows the intrinsic transistor equivalent circuit corresponding to the first-order channel approximation.

The second-order approximation of Z_{gs} is also derived from a series expansion of Eq. (3.2) as

$$Z_{gs} = R_{ch}\frac{\coth\gamma L}{\gamma L} \cong \frac{R_{ch}}{\gamma L}\left[\frac{1}{\gamma L} + \frac{\gamma L}{3} - \frac{(\gamma L)^2}{45}\right] = \frac{R_{ch}}{3}\left(1 - \frac{j\omega C_g R_{ch}}{15}\right)$$

$$+\frac{1}{j\omega C_g} \cong \frac{R_{ch}}{3} \Big/ \left(1 + j\omega\frac{R_{ch}}{3}\frac{C_g}{5}\right) + \frac{1}{j\omega C_g} \tag{3.4}$$

As a result, the second-order approximation of the device channel structure can be realized by series connection of the capacitance $C_{gs1} = C_g$ and the parallel *RC* circuit, which consists of the resistance $R_{gs} = R_{ch}/3$ and capacitance $C_{gs2} = C_g/5$. The intrinsic transistor equivalent circuit corresponding to the second-order approximation is shown in Fig. 3.2(*b*).

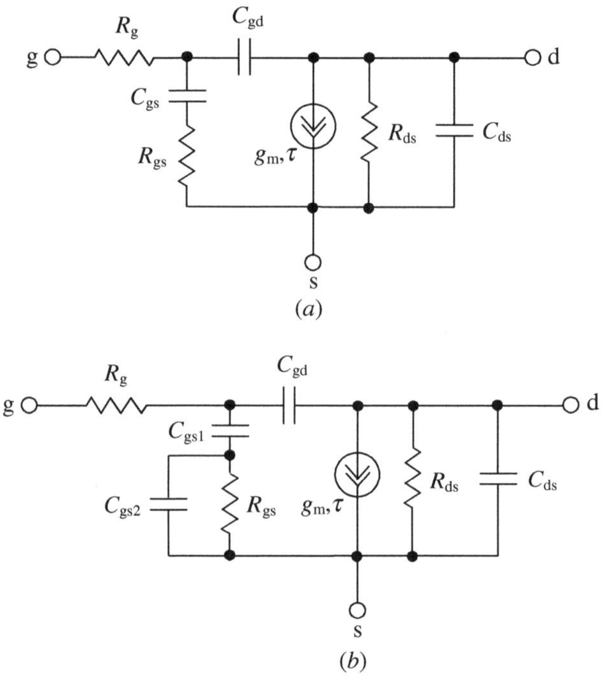

Figure 3.2 Intrinsic MOSFET equivalent circuit corresponding to (*a*) first- and (*b*) second-order channel approximation.

For a high-power MOSFET device whose channel width is significantly larger than its channel length, the distributed nature of the total gate resistance $R_{tot} = R_{sh}W/L$ across the width W (where R_{sh} is the sheet resistance of the gate material) has to be taken into consideration. In this case, silicon MOSFET can be decomposed into n devices, each with a width of W/n and a gate resistance of R_{tot}/n. For $n \to \infty$, it will be viewed as array of the small transistors distributed along the gate of the device. Commonly, it is necessary to consider a two-dimensional power MOSFET distributed model because it shows a distributed-gate nature along both the channel length and channel width. However, for a short-channel MOSFET, the distributed-gate effect along the channel length can be taken into account only in the frequency range close to the transition frequency f_T and higher. When $\omega R_{gs}C_{gs} << 1$, an analysis of the distributed-gate model along the channel width based on transmission line theory shows that all transistor Y-parameters should be modified by the term $\tanh(\gamma W)/(\gamma W)$ [3]. However, a linear power series expansion of this term as

$$\frac{\tanh(\gamma W)}{\gamma W} = 1 - j\omega C_g \frac{R_{tot}}{3} \cong \frac{1}{1 + j\omega C_g \frac{R_{tot}}{3}} \tag{3.5}$$

leads to only the additional use of a series lumped gate resistance $R_{tot}/3$ that does not alter the structure of the transistor equivalent circuit. Consequently, the overall gate resistance R_g can generally be divided in two consecutive series resistances as $R_g = R_{ge} + R_{gi}$, where R_{ge} is the extrinsic contact and ohmic gate electrode resistance and $R_{gi} = R_{tot}/3$ is the intrinsic gate resistance due to the distributed-gate structure of the power MOSFET.

The complete small-signal MOSFET equivalent circuit with the extrinsic parasitic elements is shown in Fig. 3.3. Here, R_{ds} is the differential channel resistance as a result of the channel length modulation by the drain voltage, C_{ds} is the drain-source capacitance, L_g is the gate lead inductance, R_s and L_s are the source bulk and ohmic resistance and lead inductance, R_d and L_d are the drain bulk and ohmic resistance and lead inductance, and C_{gp} and C_{dp} are the gate and source pad capacitances, respectively.

3.1.2 Determination of Equivalent Circuit Elements

To characterize the transistor electrical properties, it is sufficient to use the grounded-source intrinsic Y-parameters. Their two-port admittance matrix is written as

$$Y = \begin{bmatrix} \dfrac{j\omega C_{gs}}{1+j\omega\tau_g} + j\omega C_{gd} & -j\omega C_{gd} \\ \dfrac{g_m \exp(-j\omega\tau)}{1+j\omega\tau_g} - j\omega C_{gd} & G_{ds} + j\omega(C_{ds} + C_{gd}) \end{bmatrix} \tag{3.6}$$

where $G_{ds} = 1/R_{ds}$, $\tau_g = R_{ds}C_{gs}$, and τ is the effective channel carrier transit time. The intrinsic gate resistance R_{gi} can be considered as an external gate element. In this case, the MOSFET intrinsic Y-matrix is

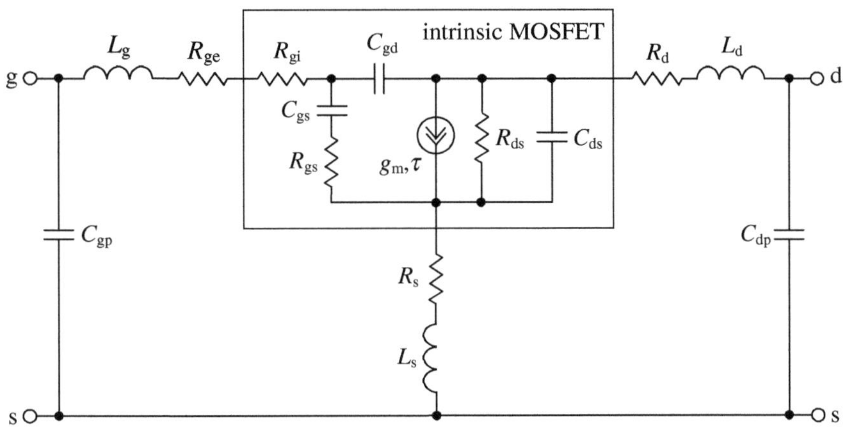

Figure 3.3 Small-signal MOSFET equivalent circuit with extrinsic linear elements.

the same as for the MESFET or HEMT devices. Consequently, to determine the elements of the intrinsic MOSFET small-signal equivalent circuit, it is possible to use the same analytical approach, which allows the determination of its elements through the real and imaginary parts of the device intrinsic admittance Y-parameters.

For known extrinsic parasitic elements, the determination of the intrinsic Y-parameters from experimental data can be obtained through the following procedure, which is shown in Fig. 3.4:

1. Measurement of the S-parameters of the extrinsic device.
2. Transformation of the S-parameters to the admittance Y-parameters with subtraction of the parallel capacitances C_{gp} and C_{dp}.

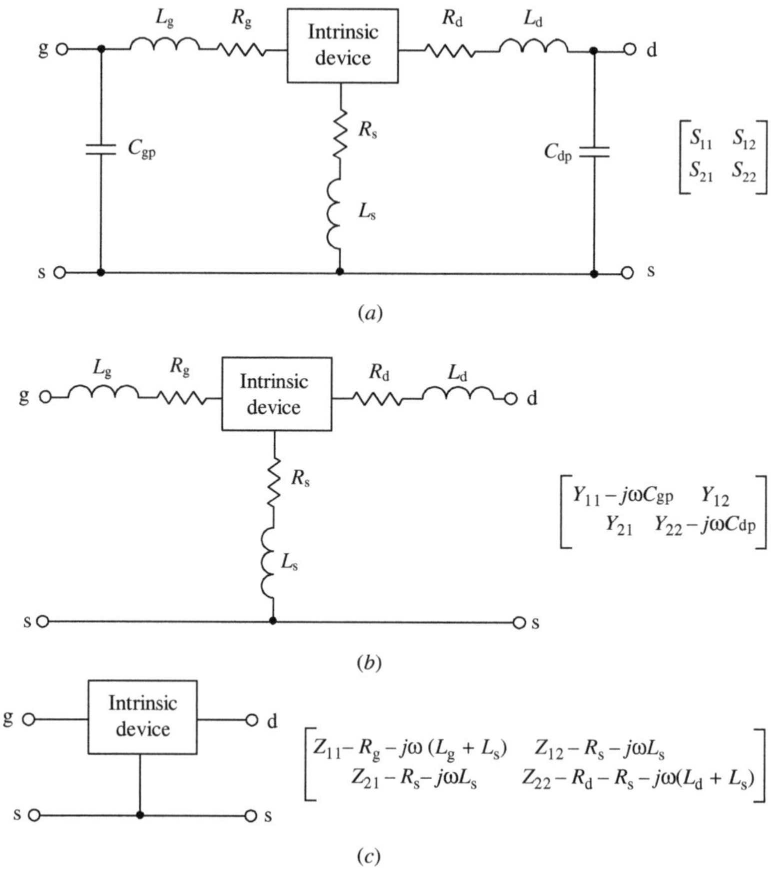

FIGURE 3.4 Method for extracting the device intrinsic Z-parameters.

3. Transformation of the admittance Y-parameters to the impedance Z-parameters with subtraction of series inductances L_g, L_s, L_d and resistances R_g, R_s, R_d.
4. Transformation of the impedance Z-parameters to the admittance Y-parameters of the intrinsic device two-port network.

The device extrinsic parasitic resistances and inductances can be directly determined from measurements performed at zero drain-source voltage. For known values of C_{gp} and C_{dp} and for forward gate biasing condition, when gate-source voltage is substantially larger than pinch-off voltage, the device equivalent circuit can be presented, as shown in Fig. 3.5. Then, the extrinsic impedance Z-parameters are written as

$$Z_{11} = R_s + R_g + j\left[\omega(L_s + L_g) - \frac{1}{\omega C_{gs}} \frac{1 + j\omega C_{gs} R_{gs}}{1 + \frac{C_{gd}}{C_{gs}}(1 + j\omega C_{gs} R_{gs})}\right] \tag{3.7}$$

$$Z_{12} = Z_{21} = R_s + j\omega L_s \tag{3.8}$$

$$Z_{22} = R_s + R_d + j\omega(L_s + L_d) \tag{3.9}$$

As a result, when the parasitic source inductance L_s is defined directly from measured ImZ_{12} as well as parasitic source resistance R_s from measured ReZ_{12}, the parasitic drain inductance L_d and resistance R_d can be calculated from measured ReZ_{22} and ImZ_{22}, respectively.

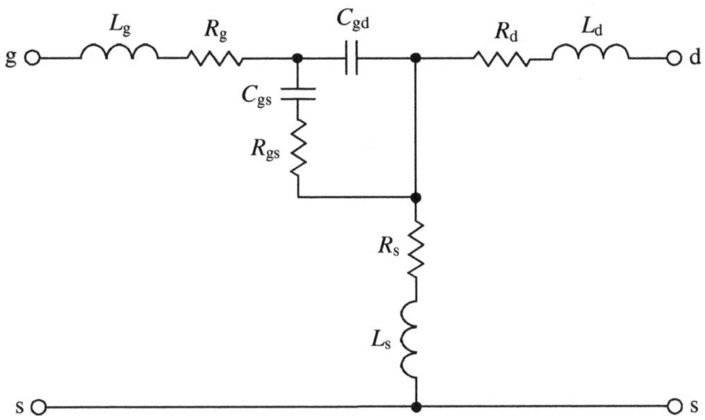

FIGURE 3.5 Device equivalent circuit corresponding to forward gate biasing condition.

The parasitic gate resistance R_g can be directly calculated from measured $\mathrm{Re}Z_{11}$ for a specified value of R_s.

The input, output and feedback MOSFET capacitances can be determined from the low-frequency measurements. In this case, at zero-drain bias and for the gate voltages lower than the pinch-off voltage V_p, the device small-signal equivalent circuit can be simplified to the one shown in Fig. 3.6. For low-frequency measurements of less than a 100 MHz when the extrinsic parasitic resistances and inductances have no influence on the device behavior, the imaginary parts of the Y-parameters can be written as

$$\mathrm{Im}\,Y_{11} = j\omega(C_{gs} + C_{gp} + C_{gd}) \tag{3.10}$$

$$\mathrm{Im}\,Y_{12} = \mathrm{Im}\,Y_{21} = -j\omega C_{gd} \tag{3.11}$$

$$\mathrm{Im}\,Y_{22} = j\omega(C_{ds} + C_{dp} + C_{gd}) \tag{3.12}$$

The small-signal equivalent circuit parameters can also be extracted using computer optimization. For example, the intrinsic device parameters are calculated analytically for each bias condition as the functions of the extrinsic parasitic elements, which are treated as global bias-independent data in the associated fitting process [4]. The measured and simulated S-parameters in a wide frequency range are compared to minimize the error function

$$\varepsilon_{ij} = \frac{1}{4}\sum_{i,j=1}^{2}\left(\frac{1}{N}\sum_{n=1}^{N}\frac{|\mathrm{meas}S_{ijn} - \mathrm{sim}S_{ijn}|}{\mathrm{meas}S_{ijn}}\right) \times 100\% \tag{3.13}$$

where $\mathrm{meas}S_{ijn}$ are the measured S-parameter data, $\mathrm{sim}S_{ijn}$ are the simulated S-parameter data, and N is the number of frequencies, where $i, j = 1, 2$. As a result, for the power lateral double-diffused MOSFET device with the gate length $L = 1.25$ μm and the gate width

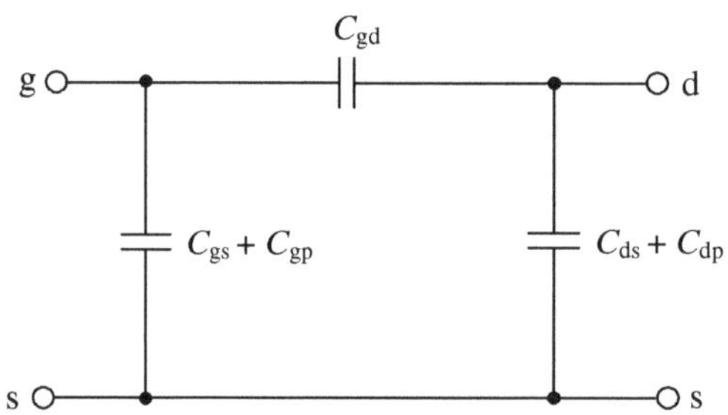

FIGURE 3.6 Device equivalent circuit corresponding to zero-drain bias condition.

$W = 1.44$ mm in a frequency range of 50 MHz up to 10 GHz, maximum error does not exceed 2.5%. This indicates a good accuracy of the MOSFET small-signal equivalent circuit model shown in Fig. 3.3.

3.1.3 Nonlinear I-V Models

An empirical approach to approximate the nonlinear behavior of the drain current source I_{ds} (V_{gs}, V_{ds}) of a JFET device is described in [5]. In this case, instead of using separate equations for the triode and pinch-off regions, irrespective of the device geometry and material parameters, a general expression based on hyperbolic functions was proposed:

$$I_{ds} = I_{dss}\left(1 - \frac{V_{gs}}{V_p}\right)^2 \tanh\alpha \left|\frac{V_{ds}}{V_p - V_{gs}}\right| \tag{3.14}$$

where I_{dss} is the saturation drain current for $V_{gs} = 0$, α is the saturation voltage parameter, and V_p is the pinch-off voltage. As a result, good agreement was obtained between the predicted and experimental results, which showed a promising prospect of such a simple empirical model.

A similar and sufficiently simple nonlinear model using a hyperbolic function and incorporating self-heating effect was later proposed to describe the *I-V* characteristics of a MOSFET device:

$$I_{ds} = \beta_{eff} V_{gst}^{VGexp}(1 + \lambda\, V_{ds})\tanh\left(\frac{\alpha\, V_{ds}}{V_{gst}^{SATexp}}\right) \tag{3.15}$$

where

$$\beta_{eff} = \beta / (\,1 + \mu_{crit} V_{gst}^{GMexp})$$

$$V_{gst} = V_{st} \ln\left[1 + \exp\left(\frac{V_{gst1}}{V_{st}}\right)\right]$$

$$V_{gst1} = V_{gs} - V_{th0} - \gamma\, V_{ds}$$

where *GMexp*, *SATexp*, and μ_{crit} are the channel current parameters [6].

An empirical nonlinear model, which is single-piece and continuously differentiable, developed for silicon LDMOS transistors is given by

$$I_{ds} = \beta V_{gst}^{VGexp}(1 + \lambda V_{ds})\tanh\left(\frac{\alpha V_{ds}}{V_{gst}}\right)[1 + K_1 \exp(V_{BReff1})]$$

$$+\, I_{ss} \exp\left(\frac{V_{ds} - V_{BR}}{V_T}\right) \tag{3.16}$$

where

$$V_{gst} = V_{st} \ln\left[1+\exp\left(\frac{V_{gst2}}{V_{st}}\right)\right]$$

$$V_{gst2} = V_{gst1} - \frac{1}{2}\left(V_{gst1} + \sqrt{(V_{gst1} - V_K)^2 + \Delta^2} - \sqrt{V_K^2 + \Delta^2}\right)$$

$$V_{gst1} = V_{gs} - V_{th0} - \gamma V_{ds}$$

$$V_{BReff1} = \frac{V_{ds} - V_{BReff}}{K_2} + M_3 \frac{V_{ds}}{V_{BReff}}$$

$$V_{BReff} = \frac{V_{BR}}{2}[1+\tanh(M_1 - V_{gst}M_2)]$$

where λ is the drain current slope parameter; β is the transconductance parameter; V_{th0} is the forward threshold voltage; V_{st} is the subthreshold slope coefficient; V_T is the temperature voltage; I_{ss} is the forward diode leakage current; V_{BR} is the breakdown voltage; K_1, K_2, M_1, M_2, and M_3 are the breakdown parameters; and V_K, $VGexp$, Δ, and γ are the gate-source voltage parameters [7].

In many applications, it is necessary to take into consideration the MOSFET operation in the weak-inversion region when the gate-source voltage V_{gs} is smaller than the threshold voltage V_{th}, for example, to improve the conversion gain of a mixer or reduce the intermodulation distortion of a Class AB power amplifier when device is biased around the onset of the strong-inversion region from the weak-inversion region for low drain quiescent current. The drain current in the weak-inversion region is mainly dominated by the diffusion component that increases exponentially with the gate voltage [2]. On the other hand, in the strong-inversion saturation region when the gate-source voltage is greater than the threshold voltage, the drain current is proportional to the square of ($V_{gs} - V_{th}$).

To obtain continuous behavior from weak-inversion region to strong-inversion region for the drain current and a compromise between accurate device modeling and ease of circuit analysis, the following analytical function can be used:

$$I_{ds}(V_{gs}) = A\{\ln[1+\exp(B(V_{gs} - V_{th}))]\}^2 \tag{3.17}$$

where A and B are the approximation parameters. In this case, the drain current is effectively proportional to the square of ($V_{gs} - V_{th}$) when V_{gs} is larger than V_{th} and exponentially decreases with the gate-source voltage when V_{gs} is smaller than V_{th}. The approximation parameters A and B are defined from the following conditions:

$$I_{ds}\Big|_{V_{gs}=V_{th}} = I_{th} \qquad \left.\frac{\partial I_{ds}}{\partial V_{gs}}\right|_{V_{gs}=V_{th}} = S_{th} \tag{3.18}$$

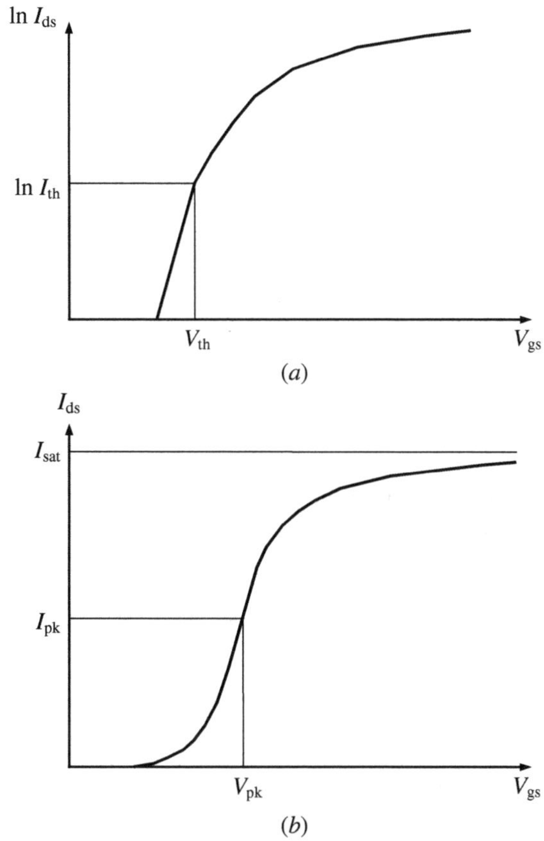

FIGURE 3.7 Drain current versus gate-source voltage.

where I_{th} is the threshold drain current, as shown in Fig. 3.7(*a*), and S_{th} is a slope of the current-voltage transfer characteristic in the threshold point. Then,

$$A = \frac{I_{th}}{(\ln 2)^2} \qquad B = \frac{S_{th}}{I_{th}} \ln 2 \tag{3.19}$$

Consequently, the transfer characteristic can be defined in terms of only two physical parameters I_{th} and S_{th}, which are easily calculated from the device measurements. By using similar analytical approach, the EKV MOST model has been successfully applied to low-voltage and low-current analog circuit design and simulation, referring voltage V_g, V_d, and V_s to the device local substrate [8].

In this case, the drain current I_d can be expressed as

$$I_d(V_g, V_s, V_d) = I_F(V_g, V_s) - I_R(V_g, V_d) \tag{3.20}$$

where $I_F(V_g, V_s)$ is the forward current component and $I_R(V_g, V_d)$ is the reverse current component. The current components I_F and I_R are

$$I_F(V_g,V_s) = I_s\left\{\ln\left[1+\exp\left(\frac{(V_g - V_{th0} - nV_s)}{2nV_T}\right)\right]\right\}^2 \tag{3.21}$$

$$I_R(V_g,V_d) = I_s\left\{\ln\left[1+\exp\left(\frac{(V_g - V_{th0} - nV_d)}{2nV_T}\right)\right]\right\}^2 \tag{3.22}$$

where $I_s = 2n\beta\, V_T^2$, β is the transfer parameter, V_T is the temperature voltage (26 mV at 300 K), V_{th0} is the gate-to-bulk threshold voltage defined with zero channel inversion charge, n is the slope factor defined from the device physics.

The function to describe the *I-V* curves of the HEMT devices was used to link the linear and saturation regions by a suitable continuous analytical dependence as

$$I_{ds}(V_{gs}, V_{ds}) = (I_{max}^{-1} + I_{dso}^{-1})^{-1} \tag{3.23}$$

where I_{dso} is the exponential function of V_{gs} and V_{ds}, and I_{max} is the maximum channel current written as

$$I_{max}(V_{ds}) = I_{pk}(1+\lambda\, V_{ds})\tanh(\alpha\, V_{ds}) \tag{3.24}$$

where I_{pk} is the drain current, which corresponds to the maximum slope of the I_{ds}–V_{gs} characteristic, as shown in Fig. 3.7(*b*) [9]. The saturation voltage parameter α affects a slope of the I_{ds}–V_{ds} characteristics in the linear region; parameter λ determines a slope of the same drain characteristics in the saturation region.

In view of the monotonous behavior of I_{ds}–V_{ds} curves, the entire drain current-voltage characteristics of a MOSFET device can be described as

$$I_{ds}(V_{gs}, V_{ds}) = I_0 \Big/ \left[1+\left(\frac{I_0}{I_{max}}\right)^{n}\right]^{\frac{1}{n}} \tag{3.25}$$

where

$$I_0 = \frac{I_{th}}{(\ln 2)^2(1-\beta V_{gs})}\left\{\ln\left[1+\exp\left(\frac{S_{th}\ln 2}{I_{th}}(V_{gs} - V_{th})\right)\right]\right\}^2$$

$$I_{max} = I_{sat}(1+\lambda V_{ds})\tanh(\alpha V_{ds})$$

where I_{sat} is the saturated drain current, α is the saturation voltage parameter (which affects a slope of the I_{ds}-V_{ds} characteristics in the linear region), λ is the parameter that determines a slope of the same drain characteristics in the saturation region, $V_{th} = V_{th0} - \sigma V_{ds}$, n and β are the fitting parameters that determine a slope of the transfer characteristics under large values of V_{gs}, and σ is the parameter that expresses empirically the dependence of the threshold voltage on V_{ds} [10]. Better accuracy can be achieved by using I_{sat} for I_{max} in Eq. (3.24) instead of I_{pk}.

To verify the new empirical *I-V* model, a high-power LDMOSFET with the gate width $W = 4$ cm and gate length $L = 1.1$ μm (LP801 from Polyfet) was used. The theoretical and experimental output $I_{ds}(V_{ds})$ curves are shown in Fig. 3.8. To define a deviation between the measured and modeled simulated data and express the error in percentage, the following current mean-square relative error function was chosen:

$$\varepsilon_{sq} = \frac{1}{M}\sum_{m=1}^{M}\left[\frac{1}{N}\sum_{n=1}^{N}\left(\frac{\text{meas}I_{dsnm} - \text{sim}I_{ds\,nm}}{\text{meas}I_{dsnm}}\right)^2\right] \times 100\% \tag{3.26}$$

where measI_{dsnm} are the measured drain current values, simI_{dsnm} are the simulated drain current values, N is the number of measured drain voltage bias points for the appropriate gate bias voltage V_{gs}, and M is the number of measured gate voltage bias points. The resulting current mean-square error of a family of the output I_{ds}–V_{ds} characteristics

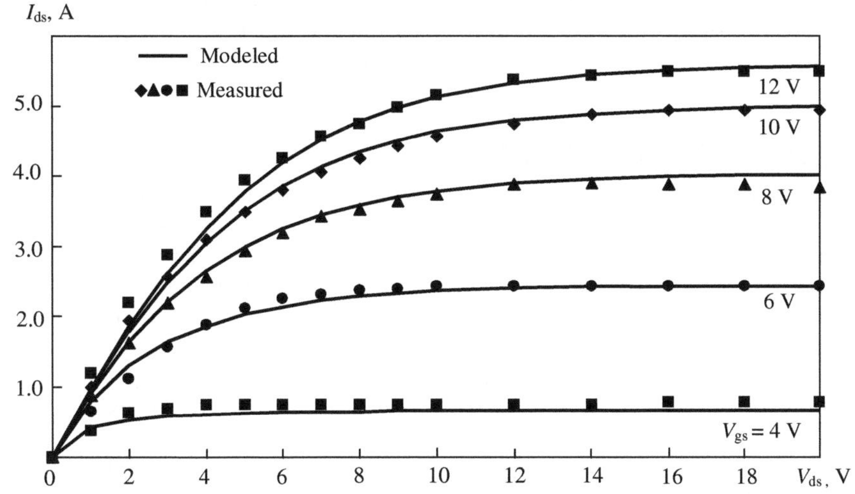

FIGURE 3.8 Measured and modeled $I_{ds}(V_{ds})$ curves of high-voltage LDMOSFET.

Parameters	α, 1/V	β, 1/V	λ, 1/V	V_{th}, V	I_{th}, V	S_{th}, A/V	I_{sat}, A	n	σ
Values	0.15	0	0.0005	2.7	0.115	0.2	6.8	1.0	0

TABLE 3.1 Simulated *I-V* Model Parameters of High-Voltage LDMOSFET

was 0.5%. The values of the simulated model parameters are indicated in Table 3.1.

The theoretical approximation of $I_{ds}(V_{ds})$ curves is sufficient and accurate for a high-power LDMOSFET and requires only six fitting parameters; four of which are easily determined by the experimental curves. An empirical model allows the description of I_{ds} and g_m as a function of V_{gs} in a weak-inversion region, as well as in the strong-inversion region of the LDMOSFET operation. Substantially better fitting to $I_{ds}(V_{gs})$ and $g_m(V_{gs})$ experimental curves is achieved compared with the results obtained from two-dimensional simulations as shown in Fig. 3.9 [11].

To verify that this model is applicable to low-voltage MOSFET devices, the appropriate low-voltage RF power MOSFET with a gate width $W = 2$ mm was chosen [12]. Figure 3.10 shows the transistor theoretical and experimental drain current $I_{ds}(V_{ds})$ curves. In this case, to improve the sensitivity of drain current I_{ds} under large values of V_{gs}, the term $(1 - \beta V_{gs})$ was introduced to the approximation function given in Eq. (3.25). The resulting current mean-square error of a family of the output I_{ds}–V_{ds} characteristics was 0.42% only. The values of simulated I_{ds}–V_{ds} model parameters are presented in Table 3.2. The transfer I_{ds}–V_{gs} characteristics have been compared with the same characteristics calculated by means of the BSIM3v3 model developed for modeling of deep submicron device. The results shown in Fig. 3.11 demonstrate a good agreement with the experimental curves and practically the same as in the case of the BSIM3v3 approximation.

3.1.4 Nonlinear C-V Models

The input capacitance C_{gs} normally influences the intermodulation distortion (IMD) level especially when the frequency increases in the microwave region [13]. Generally, the calculation of the gate-source capacitance C_{gs} or the gate-drain capacitance C_{gd} from the charges corresponding to the strong-inversion model only results in a mathematically complicated expression [2]. Therefore, in most cases when it is necessary to predict efficiency, gain, or output power of the power amplifier, the capacitances C_{gs} and C_{gd} can be modeled as the fixed capacitances measured at the quiescent bias voltage, and the *p-n* junction diode capacitance model can be applied to the capacitance C_{ds} [11]. The gate-drain capacitance C_{gd} can also be considered as the bias-dependent junction capacitance [14]. With the increase in the drain bias

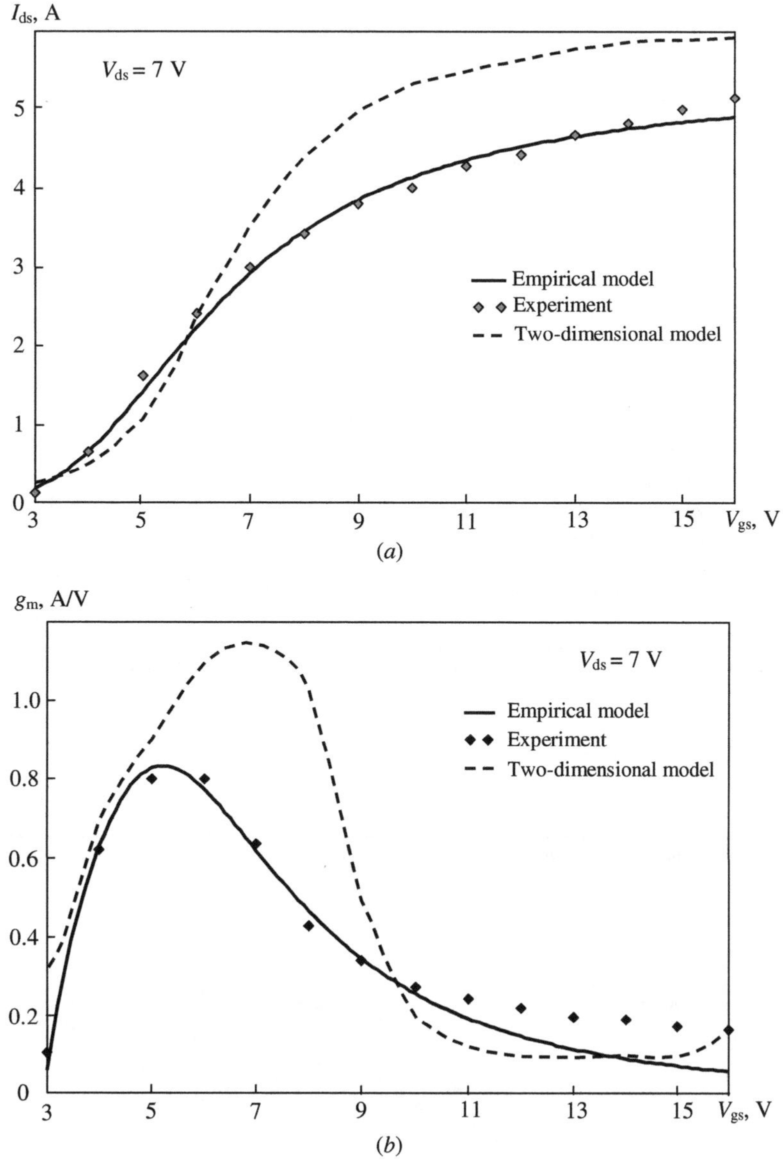

FIGURE 3.9 Measured and modeled $I_{ds}(V_{gs})$ and $g_m(V_{gs})$ curves of high-voltage LDMOSFET.

voltage, a depletion region is formed under the oxide in the lightly doped drain region. Therefore, the capacitance C_{gd} can be considered as a junction capacitance, which strongly depends on the drain-source bias voltage V_{ds}. According to the accurate charge model calculations,

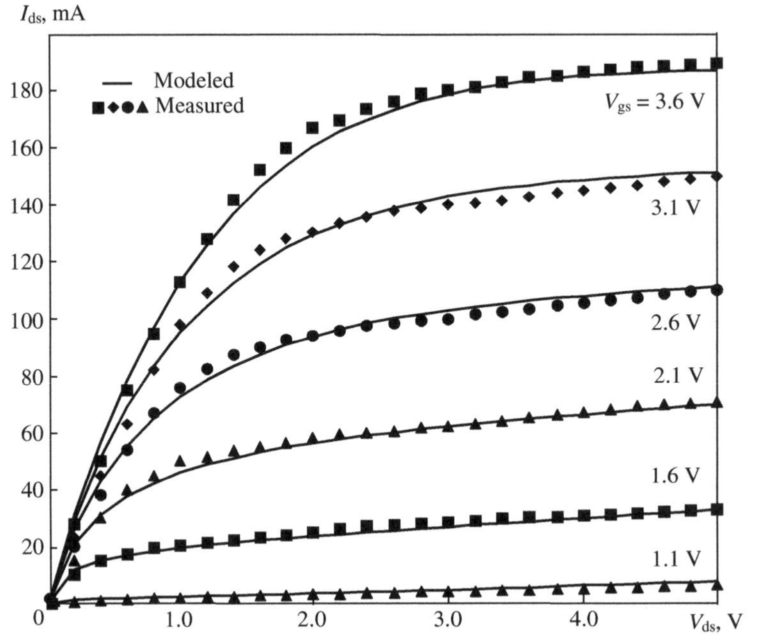

FIGURE 3.10 Measured and modeled $I_{ds}(V_{ds})$ curves of low-voltage MOSFET.

Parameters	α, 1/V	β, 1/V	λ, 1/V	V_{th}, V	I_{th}, mA	S_{th}, A/V	I_{sat}, A	n	σ
Values	0.58	0.17	0	0.9	0.029	1.05	0.31	0.78	0.05

TABLE 3.2 Simulated *I-V* Model Parameters of Low-Voltage MOSFET

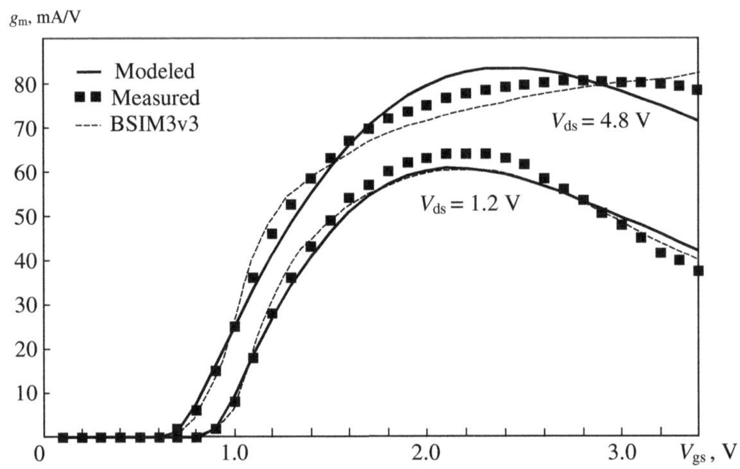

FIGURE 3.11 Measured and modeled $g_m(V_{gs})$ curves of low-voltage MOSFET.

the gate-drain capacitance C_{gd} has a strong dependence on V_{gs} only in the moderate-inversion region when $V_{gs} - V_{th} < 1$ V [2]. In this region, the behavior of C_{gd} is similar to C_{gs}, and can be evaluated using the same hyperbolic tangent functions. However, for high-voltage LDMOSFET devices, because the dependence of C_{gd} on V_{gs} is quite small, it seems sufficient to limit the dependence to V_{ds} only. The drain-source capacitance C_{ds} varies due to the change in the depletion region, which is mainly determined by the value of V_{ds}.

The gate-source capacitance C_{gs} can be described as a function of the gate-source bias voltage. First, we consider an appropriate behavior of each of its main composite part: the intrinsic gate-source capacitance C_{gsi}, including both the gate-source and the source-substrate charge fluctuations, and the gate-substrate capacitance C_{gbi}. The gate-source voltage dependence of these components is substantially different [2]. The intrinsic gate-substrate capacitance C_{gbi} is constant in the accumulation region where it is equal to the total intrinsic oxide capacitance C_{ox}, slightly decreases in the weak-inversion region, significantly reduces in the moderate-inversion region, and becomes practically constant in the strong-inversion or saturation region. The intrinsic gate-source capacitance C_{gsi} grows rapidly in the moderate-inversion region and equals $2C_{ox}/3$ in the saturation region. The voltage dependence of the total gate-source capacitance C_{gs} as a sum of its components C_{gsi} and C_{gbi} on V_{gs} is shown in Fig. 3.12.

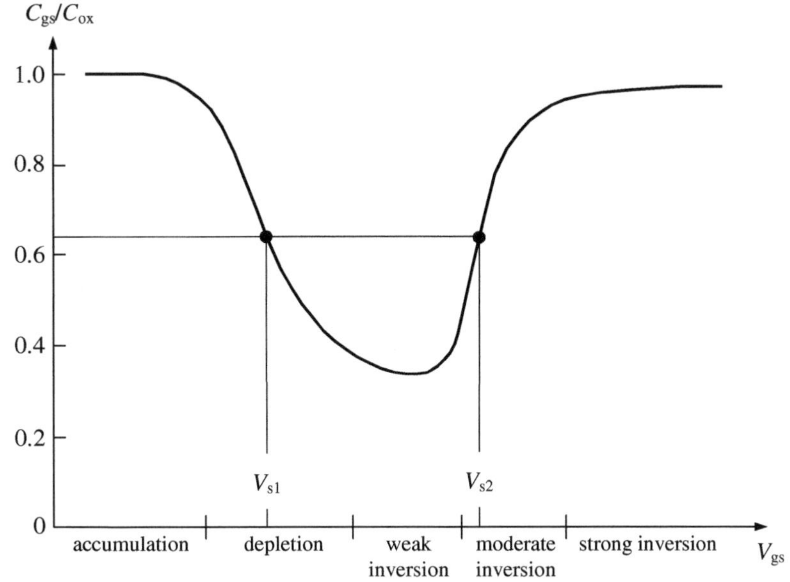

FIGURE 3.12 Gate-source capacitance versus gate-source voltage.

A hyperbolic tangent function can be used for each of two parts of the dependence $C_{gs}(V_{gs})$, where the gate-source capacitance can be approximated by

$$C_{gs} = C_{gsmin} + C_s \left\{1 + \tanh\left[\frac{S}{C_s}(V_{gs} - V_s)\right]\right\} \tag{3.27}$$

where $C_s = (C_{gsmax} - C_{gsmin})/2$, C_{gsmax} is the maximum gate-source capacitance, C_{gsmin} is the minimum gate-source capacitance, $S = (S_1, S_2)$ is the slope of $C_{gs}(V_{gs})$ at each bend point $V_{gs} = (V_{s1}, V_{s2})$, as shown in Fig. 3.12,

$$S_1 = \frac{\partial C_{gs}}{\partial V_{gs}}\bigg|_{V_{gs} = V_{s1}} \qquad S_2 = \frac{\partial C_{gs}}{\partial V_{gs}}\bigg|_{V_{gs} = V_{s2}} \tag{3.28}$$

The total gate-source capacitance C_{gs} as a function of V_{gs} can be described by

$$C_{gs} = C_{gsmax} - C_{gso}\left\{1 + \tanh\left[\frac{S_1}{C_s}(V_{gs} - V_{s1})\right]\right\} \times \left\{1 + \tanh\left[\frac{S_2}{C_s}(V_{gs} - V_{s2})\right]\right\} \tag{3.29}$$

where $C_{gsmax} = C_{ox}$ and C_{gso} is the model fitting parameter [10].

The approximation function for the gate-source capacitance C_{gs} as the dependence of V_{gs} can also be expressed by using the two components, both containing the hyperbolic functions:

$$C_{gs} = C_{gs1} + C_{gs2}\{1 + \tanh\,[C_{gs6}(V_{gs} + C_{gs3})]\} + C_{gs4}[1 - \tanh(C_{gs5}V_{gs})] \tag{3.30}$$

where C_{gs1}, C_{gs2}, C_{gs3}, C_{gs4}, C_{gs5}, and C_{gs6} are the approximation parameters [7].

If we consider the dependence of the gate-source capacitance C_{gs} on V_{ds} for submicron MOSFET devices when C_{gs} slightly increases with the increase in V_{ds}, the approximation expression for C_{gs} as a function of both V_{gs} and V_{ds} can be written as

$$C_{gs} = C_{gs0} + C_{gs1}\{A_s + B_s \tanh[C_s(V_{gs} - V_{th})]\} \times \{D_s + E_s[1 + \tanh(V_{gs} - V_{ds})]\tanh[F_s V_{ds} - G_s V_{gs}]\} \tag{3.31}$$

where C_{gs0} is the bias-dependent capacitance; V_{th} is the threshold voltage; C_{gs1} is the scaling factor; and A_s, B_s, C_s, D_s, E_s, F_s, and G_s are the model fitting parameters [15].

On the other hand, for high-voltage MOSFET devices, the dependencies of the gate-drain capacitance C_{gd} and drain-source capacitance C_{ds} on V_{ds} can be accurately evaluated by the junction capacitance model as

$$C_{\mathrm{gd(ds)}} = C_{\mathrm{gdo(dso)}} \left(\frac{\varphi + V_{\mathrm{dso}}}{\varphi + V_{\mathrm{ds}}} \right)^{\mathrm{m}} \tag{3.32}$$

where m (m_1 for C_{gd} or m_2 for C_{ds}) is the junction sensitivity depending on a doping profile ($m = 1/3$ for the linearly graded junction, $m = 1/2$ for the abrupt junction, and $m > 1/2$ for the hyperabrupt junction), φ is the contact potential, and C_{gdo} and C_{dso} are the junction capacitances when $V_{ds} = V_{dso}$. For practical profiles of the junction, which are neither exactly abrupt nor exactly linearly graded, one often chooses the parameters m and φ to obtain the best matching between the theoretical model and the measurements.

The results of the approximation of C_{ds} and C_{gd} as the junction capacitances for high-power LDMOSFET (LP801 from Polyfet) are shown in Figs. 3.13 and 3.14, respectively. The fitting parameters of the approximation curves $C_{gd}(V_{ds})$ and $C_{ds}(V_{ds})$ are given in Table 3.3.

The resulting drain-source capacitance average normalized difference error is only 4.3% according to

$$\varepsilon_{\mathrm{Cds}} = \frac{1}{N} \sum_{n=1}^{N} \frac{|\ \mathrm{meas}C_{\mathrm{ds\,n}} - \mathrm{sim}C_{\mathrm{ds\,n}}\ |}{\mathrm{meas}C_{\mathrm{ds\,n}}} \times 100\% \tag{3.33}$$

where measC_{dsn} are the measured capacitance values, simC_{dsn} are the simulated capacitance values, and N is the number of measured voltage bias points.

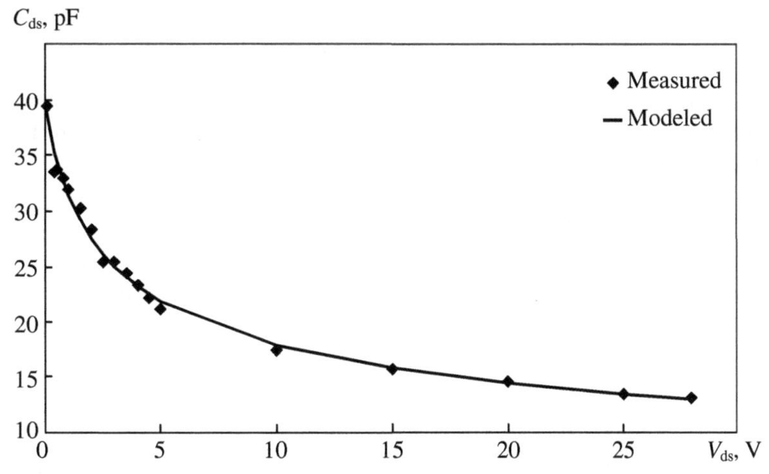

Figure 3.13 Measured and modeled $C_{ds}(V_{ds})$ dependencies of high-voltage LDMOSFET.

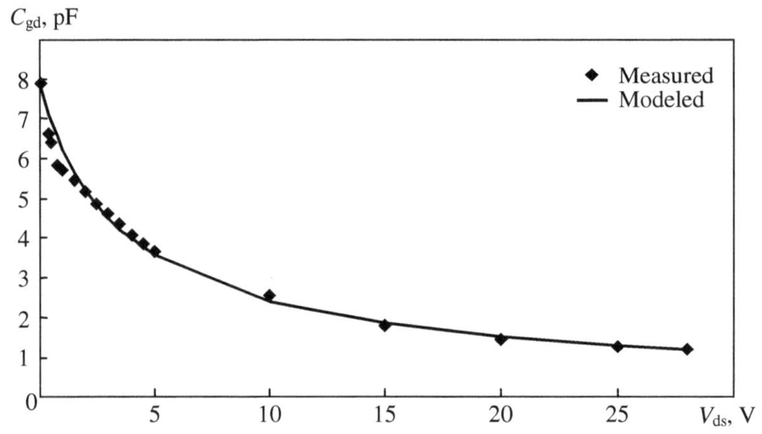

FIGURE 3.14 Simulated $C_{gd}(V_{ds})$ junction model parameters.

Capacitance	$C_{gd(ds)o}$, pF	m	φ, V
C_{gd}	7.88	0.8	2.94
C_{ds}	39.42	0.33	1.0

TABLE 3.3 Simulated *C-V* Junction Model Parameters

For submicron MOSFET devices, to take into account the dependence of C_{gd} both on V_{gs} and V_{ds}, the approximation expression for the C_{gd} is written as

$$C_{gd} = C_{gd0} + C_{gd1}\{A_d + B_d \tanh[C_d(D_d V_{gs} - V_{ds}) - V_{th}]\} \tag{3.34}$$

where C_{gd0} is the bias-dependent capacitance, V_{th} is the threshold voltage, C_{gd1} is the scaling factor, whereas A_d, B_d, C_d, and D_d are the model fitting parameters [15].

3.1.5 Charge Conservation

To describe the small- and large-signal device models, it is necessary to satisfy the charge conservation condition. For a three-terminal MOSFET device, the matrix equation for the small-signal charging circuit in frequency domain is given by

$$\begin{bmatrix} I_g \\ I_d \\ I_s \end{bmatrix} = j\omega \begin{bmatrix} C_{gg} & -C_{gd} & -C_{gs} \\ -C_{dg} & C_{dd} & -C_{ds} \\ -C_{sg} & -C_{sd} & C_{ss} \end{bmatrix} \begin{bmatrix} V_g \\ V_d \\ V_s \end{bmatrix} \tag{3.35}$$

where I_g, I_d, and I_s are the terminal current amplitudes; V_g, V_d, and V_s are the terminal voltage amplitudes; and the capacitance between any two device terminals (k, l) is described as $C_{kl} = \partial Q_k / \partial V_l$ [11]. To transform a three-terminal device into a two-port network with a common source terminal, the current and voltage terminal conditions of $I_g = I_{gs}$, $I_d = I_{ds}$, $I_s = -(I_{gs} + I_{ds})$, $V_g - V_s = V_{gs}$, and $V_d - V_s = V_{ds}$ should be taken into account. In addition, the following relationships between the terminal capacitances for three-terminal devices are valid:

$$\begin{aligned} C_{gg} &= C_{gd} + C_{gs} = C_{dg} + C_{sg} \\ C_{dd} &= C_{dg} + C_{ds} = C_{gd} + C_{sd} \\ C_{ss} &= C_{sg} + C_{sd} = C_{gs} + C_{ds} \end{aligned} \tag{3.36}$$

The admittance Y_c-matrix for such a capacitive two-port network is

$$Y_c = \begin{bmatrix} j\omega\,(C_{gs} + C_{gd}) & -j\omega\, C_{gd} \\ -j\omega\,(C_{gd} + C_m) & j\omega\,(C_{ds} + C_m + C_{gd}) \end{bmatrix} \tag{3.37}$$

where $C_m = C_{dg} - C_{gd}$ is the transcapacitance, C_{gd} represents the effect of the drain on the gate, and C_{dg} represents the effect of the gate on the drain, and these effects are different [2, 16]. Similarly to the *I-V* characteristics, there is no reason to expect that the effect of the drain voltage on the gate current, which is zero assuming no leakage current, is the same as the effect of the gate voltage on the drain current, which is significantly large.

Therefore, for power MOSFET devices, because the transcapacitance C_m is substantially less than C_{gs}, it can be translated to an additional delay time τ_c in a frequency range up to f_T by its combining with the transconductance g_m according to

$$\begin{aligned} g_m - j\omega C_m &= g_m \sqrt{1 + \left(\frac{\omega}{\omega_T}\frac{C_m}{C_{gs}}\right)^2} \exp\left[-j\tan^{-1}\left(\frac{\omega}{\omega_T}\frac{C_m}{C_{gs}}\right)\right] \\ &\cong g_m \exp(-j\omega\tau_c) \end{aligned} \tag{3.38}$$

where $\tau_c = C_m/(\omega_T C_{gs})$. To satisfy the charge conservation condition, the total delay time τ shown in the MOSFET equivalent circuit in Fig. 3.3 represents both the ideal transit time and delay time due to the transcapacitance. The transcapacitance C_m can be easily added to the drain-source capacitance C_{ds} under parameter extraction procedure.

3.1.6 Gate-Source Resistance

The gate-source resistance R_{gs} is determined by the effect of the channel inertia in responding to rapid changes of the time varying gate-source voltage, and it varies in such a manner that the charging time $\tau_g = R_{gs}C_{gs}$ remains approximately constant. Thus, the increase of R_{gs} in the velocity saturation region (when the channel conductivity decreases) is partially compensated by the decrease of C_{gs} due to nonuniform channel charge distribution [17]. The effect of R_{gs} becomes significant at higher frequencies close to the transition frequency f_T of the MOSFET and may not be taken into account when designing RF circuits operating below 2 GHz [12, 18]. For example, for the MOSFET with the depletion region doping concentration value N_A = 1700 μm^{-3}, the phase of the small-signal transconductance g_m near f_T reaches the value only of –15° [2].

3.1.7 Temperature Dependence

Silicon MOSFET devices are very sensitive to the operation temperature T and their characteristics are strongly temperature dependent [2]. The main parameters responsible for this are the effective carrier mobility μ and the threshold voltage V_{th}, resulting in the increase in the drain current through $V_{th}(T)$ and the decrease in the drain current through $\mu(T)$ with temperature. Increasing temperature decreases the slope of the $I_{ds}(V_{gs})$ curves. A certain value of V_{gs} can be found, at which the drain current becomes practically temperature-independent over a large temperature range. The variation of V_{th} with temperature in a wide range from –50° to +200°C represents a nonlinear function, which is slowly decreased with temperature and can be approximated by

$$V_{th}(T) = V_{th}(T_{nom}) + V_{T1}\Delta T + V_{T2}\Delta T^2 \tag{3.39}$$

where $\Delta T = T - T_{nom}$, T_{nom} = 300 K (27°C), and V_{T1} and V_{T2} are the linear and quadratic temperature coefficients for threshold voltage [11].

The variation of μ with temperature can be taken into account by introducing the appropriate temperature variation of I_{sat} in Eq. (3.25). The temperature variation of I_{sat} represents an almost straight line, which decreases with temperature [11]. The temperature dependence $I_{sat}(T)$ can be approximated by the linear function as

$$I_{sat}(T) = I_{sat}(T_{nom}) + I_T\Delta T \tag{3.40}$$

where I_T is the linear temperature coefficient for the saturation current.

The temperature dependencies of the MOSFET capacitances and series resistances can be described by the following linear equations:

$$C(T) = C(T_{nom}) + C_T\Delta T \tag{3.41}$$

$$R(T) = R(T_{nom}) + R_T\Delta T \tag{3.42}$$

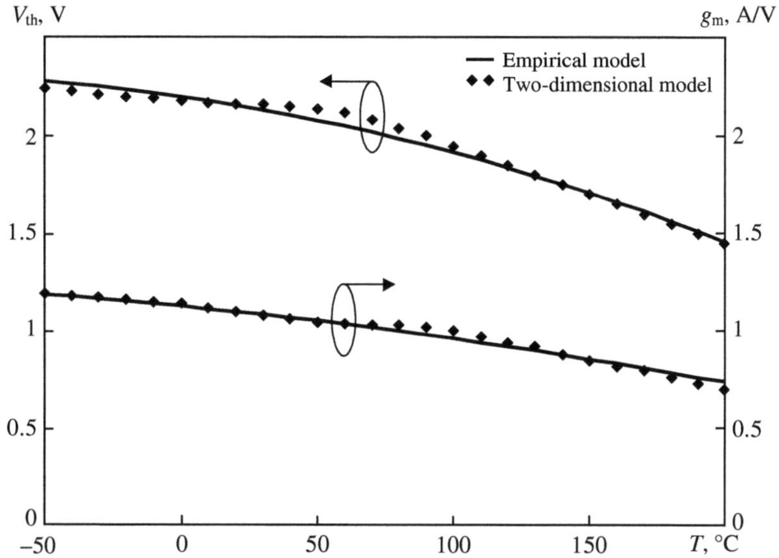

FIGURE 3.15 Modeled temperature dependencies of V_{th} and g_m.

where $C = (C_{gs}, C_{ds}, C_{gd})$, $R = (R_g, R_s, R_d)$, and R_T and C_T are the linear temperature coefficients for the capacitances and resistances, respectively [1, 19].

Figure 3.15 shows the modeled temperature dependencies of $V_{th}(T)$ and $g_m(T)$ for high-power LDMOSFET LP801. The results obtained by using the simple approximation Eqs. (3.39) and (3.40) with the values $V_{T1} = -2$ mV/°C, $V_{T2} = -8.55$ μV/(°C)^2 and $I_T = -4.5$ mA/°C show a good prediction of the I–V characteristics in a wide temperature range.

At high value of V_{gs} and V_{ds} under dc measurement, the slope of I_{ds}-V_{ds} curves can be negative that occurs due to the self-heating effect in a highly dissipated power region. For the drain current model given by Eq. (3.15), this effect can be taken into account by adding a linear component describing the temperature dependence as

$$\beta(T) = \beta(T_{nom}) + \beta_T \Delta T_j \tag{3.43}$$

$$\gamma(T) = \gamma(T_{nom}) + \gamma_T \Delta T_j \tag{3.44}$$

where $\Delta T_j = R_{th}P_{dis} + \Delta T$, R_{th}(°C/W) is the thermal resistance, P_{dis} is the dc power consumption in watts caused by dc biasing, and β_T and γ_T are the linear temperature coefficients with negative values, respectively [6].

Another way of taking into account the effect of the negative conductance at high biasing is to write the nonlinear I_{ds}-V_{ds} model as follows:

$$I_{ds}(T, p_T) = \frac{I_{ds}(T)}{1 + p_T V_d I_{ds}(T)} \tag{3.45}$$

where the drain current source $I_{ds}(T)$ is given by Eq. (3.25), $V_d = V_{ds}/\sqrt{1+(\omega\tau_{th})^2}$, p_T is the self-heating temperature coefficient, V_{ds} is the drain-source supply voltage, $\tau_{th} = R_{th}C_{th}$ is the thermal time constant, R_{th} is the thermal resistance, C_{th} is the thermal capacitance, and T_j is the function of ambient temperature T and p_T. A thermal equivalent circuit can be added to the large-signal MOSFET model as a parallel $R_{th}C_{th}$ circuit [7]. The thermal resistance R_{th} can be extracted from the temperature measurement of the dc characteristics. Because the slope of the dc measured $I_{ds}(V_{ds})$ curves changes its sign from positive to negative, the temperature coefficient p_T can be evaluated under the condition of

$$\frac{dI_{ds}(T, p_T)}{dV_{ds}} = 0 \tag{3.46}$$

As a result,

$$p_T = \frac{1}{I_{ds}^2(T)} \frac{dI_{ds}(T)}{dV_{ds}} \tag{3.47}$$

where $I_{ds}(V_{ds})$ curves are determined by measurement of the pulsed $I_{ds}(V_{ds})$ curves at ambient temperature T, and the value of $I_{ds}(T)$ is fixed the same as for zero slope of $I_{ds}(T, p_T)$.

The thermal time constant τ_{th} can be extracted by comparing pulsed $I_{ds}(V_{ds})$ curves calculated under different pulse widths and duty factors. A plot of I_{ds} as a function of pulse width under the fixed gate-source and drain-source bias voltages gives an appropriate value of τ_{th}.

Figure 3.16 shows the comparison between the measured (dc measurement) and modeled $I_{ds}(V_{ds})$ curves for a 12.5-V LDMOSFET cell with the gate geometry of $L = 1.25$ μm and $W = 1.44$ mm. The effect of self-heating for the model parameters $p_T = 0.035$ 1/AV, $\alpha = 0.2$ 1/V, $\beta = \sigma = 0$, $S_{th} = 37$ mA/V, $I_{th} = 1.5$ mA, $V_{th} = 2.25$ V, $I_{sat} = 0.48$ A, $n = 1$, and $\lambda = 0.0005$ 1/V is described by Eq. (3.45).

3.2 MESFETs and HEMTs

3.2.1 Small-Signal Equivalent Circuit

Adequate representation for MESFETs and HEMTs in a frequency range up to at least 25 GHz can be provided using a nonlinear small-signal model shown in Fig. 3.17(*a*), which is very similar to a nonlinear

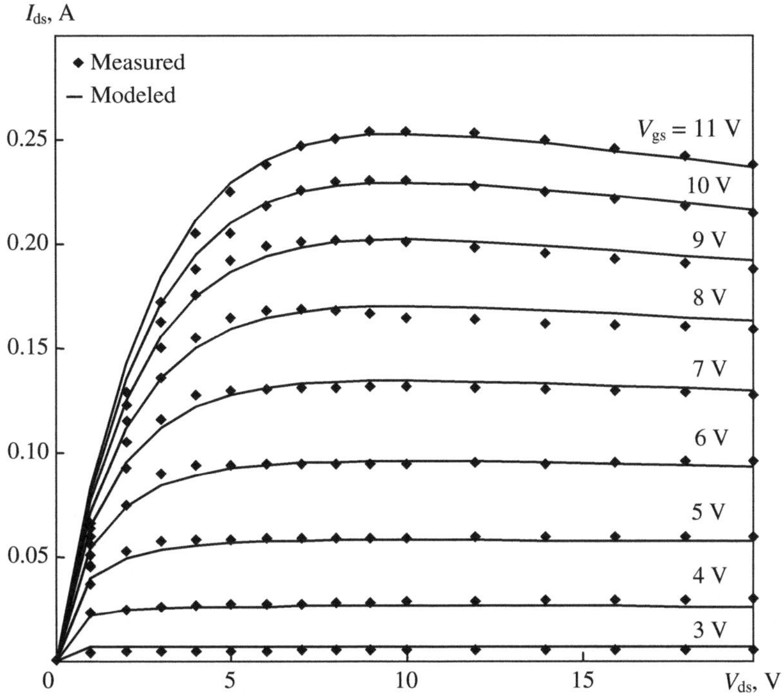

FIGURE 3.16 Measured and modeled $I_{ds}(V_{ds})$ curves of low-voltage LDMOSFET.

MOSFET small-signal model [20–22]. Here, the extrinsic elements R_g, L_g, R_d, L_d, R_s, and L_s are the bulk and ohmic resistances and lead inductances associated with the gate, drain, and source, whereas C_{gp} and C_{dp} are the gate and source pad capacitances, respectively. The capacitance C_{dsd} and resistance R_{dsd} model the dispersion of the MESFET or HEMT voltage-ampere (*I-V*) characteristics due to the trapping effect in the device channel, which leads to discrepancies between dc measurement and *S*-parameter measurement at high frequencies [23, 24].

The intrinsic model is described by the channel charging resistance R_{gs}, which represents the resistive path for charging of the gate-source capacitance C_{gs}, the feedback gate-drain capacitance C_{gd}, the output differential resistance R_{ds}, the drain-source capacitance C_{ds}, and the transconductance g_m. The gate-source capacitance C_{gs} and gate-drain capacitance C_{gd} represent the charge depletion region and are nonlinear functions of the gate-source and drain-source voltages. For negative gate-source voltage and small drain-source voltage, these capacitances are practically equal. However, when the drain-source voltage is increased beyond the current saturation point, the gate-drain capacitance C_{gd} is much more heavily back-biased than the gate-source capacitance C_{gs}. Therefore, the gate-source capacitance

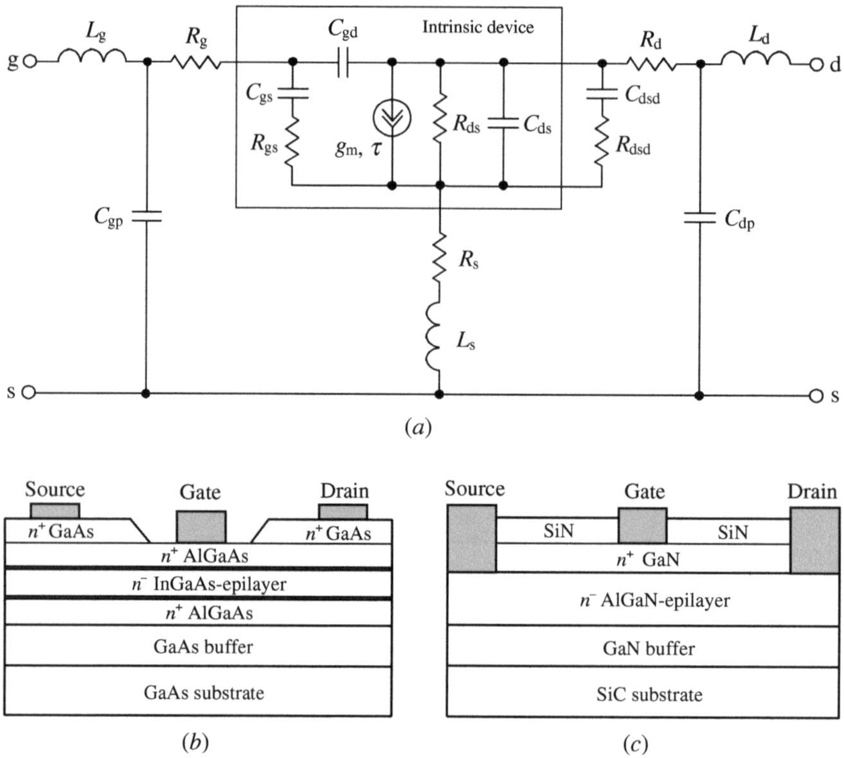

FIGURE 3.17 Nonlinear MESFET and HEMT model with HEMT physical structures.

C_{gs} is significantly more important and usually dominates the input impedance of the MESFET or HEMT device. The influence of the drain-source capacitance C_{ds} on the device behavior is insignificant and its value is bias independent. To model the transit time of electrons along the channel, the transconductance g_m usually includes the time constant τ.

Figure 3.17(*b*) shows the cross-section of a physical structure of an InGaAs/AlGaAs HEMT device, where an undoped InGaAs *n*-epilayer is used as a channel and two heavily *n*-doped AlGaAs layers with a high energetic barrier for holes are necessary to maximize high electron mobility in the channel. In this case, spacing between AlGaAs layer and InGaAs channel is optimized to achieve high breakdown voltage. An example of the physical structure of a AlGaN/GaN HEMT device is shown in Fig. 3.17(*c*), where an undoped AlGaN *n*-epilayer is used as a channel, *n*-type doped GaN layer can suppress dispersion in the device current-voltage characteristics, and SiN passivation layer with optimized parameters contributes to lower-trap device structure [25]. Thermal conductivity of a GaN HEMT device is improved by using a SiC substrate. Note that the

GaN-based technology can provide wider bandwidth and higher efficiency of the power amplifier due to high charge density and ability to operate at higher voltages for GaN HEMT devices, which are characterized by lower output capacitance and on-resistance [26–28]. Typical GaN HEMT MMIC process also includes the MIM capacitors, SiC substrate vias, and the thin-film and bulk resistors [29]. A 0.4-μm gate-length 28-V process provides 4.5 W/mm of gate periphery for circuits between dc and 8 GHz, whereas a 0.25-μm gate-length 40-V process provides 7 W/mm of gate periphery for circuits between dc and 18 GHz.

Figure 3.18 shows the enhanced nonlinear equivalent circuit of a high-power GaN HEMT device, where the delay network (C_{del1}, C_{del2}, and R_{del}) is used to describe the high-frequency delay effects, the source spreading resistance (R_{s1} and R_{s2}) are added to describe influence of the device channel to the increase in magnitude of S_{21} with frequency, and the electrothermal elements R_{therm} and C_{therm} are inserted to estimate channel temperature rise due to power dissipation [30].

To describe accurately the small-signal and large-signal device models, it is necessary to satisfy charge conservation condition. The models for the device gate-source capacitance C_{gs} and gate-drain capacitance C_{gd} should be derived from the charge model. There are commonly four partial derivatives of the two device terminal charges, the gate charge Q_g and the drain charge Q_d with regard to V_{gs} and V_{ds}, which appropriately represent totally four capacitances, as shown in Fig. 3.19(*a*) [23]. However, the intrinsic small-signal equivalent circuit contains only two capacitances. Consequently, in this case, they can be defined as

$$C_{gs} = \frac{\partial(Q_g + Q_d)}{\partial V_{gs}} \qquad C_{gd} = \frac{\partial\,(Q_g + Q_d)}{\partial V_{gd}} \tag{3.48}$$

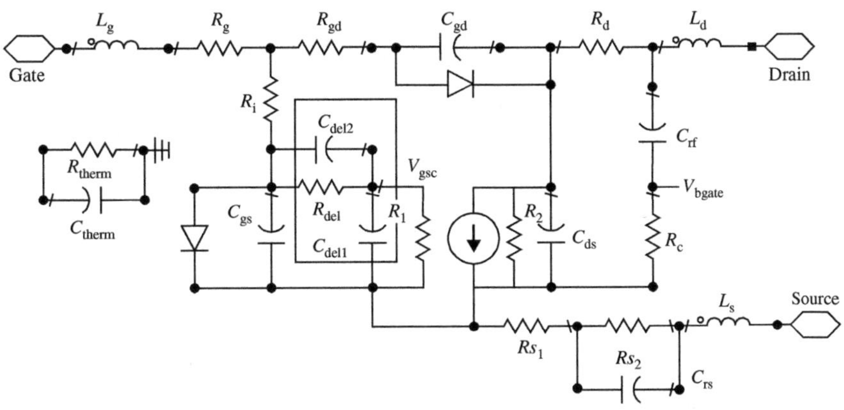

FIGURE 3.18 Nonlinear GaN HEMT model with electrothermal elements.

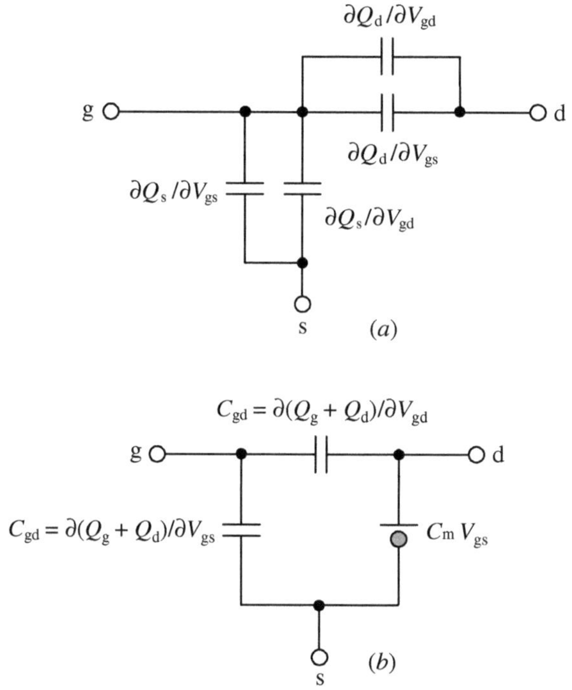

FIGURE 3.19 Capacitance equivalent circuits consistent with charge conservation.

The admittance Y_c-matrix for such a capacitive two-port network is written as

$$Y_c = \begin{bmatrix} j\omega(C_{gs} + C_{gd}) & -j\omega C_{gd} \\ -j\omega(C_{gd} - C_m) & j\omega C_{gd} \end{bmatrix} \tag{3.49}$$

where C_m is an additional transcapacitance, which is determined by

$$C_m = \frac{\partial Q_g}{\partial V_{gd}} - \frac{\partial Q_d}{\partial V_{gs}} \tag{3.50}$$

By adding the transcapacitance C_m, the capacitance equivalent circuit becomes consistent with the charge conservation condition, as shown in Fig. 3.19(*b*). Given that, for the MESFET devices, the transcapacitance C_m is substantially less than the gate-source capacitance C_{gs} in a frequency range where $\omega \le \omega_T$, it can be translated to an additional delay time τ_c by its combining with the small-signal transconductance g_m according to $g_m - j\omega C_m \approx g_m \exp(-j\omega\tau_c)$, where $\tau_c = C_m/(\omega_T C_{gs})$.

3.2.2 Determination of Equivalent Circuit Elements

To characterize the device electrical properties, first we consider the admittance Y-parameters derived from the intrinsic small-signal equivalent circuit as

$$Y_{11} = \frac{j\omega C_{gs}}{1 + j\omega C_{gs} R_{gs}} + j\omega C_{gd} \tag{3.51}$$

$$Y_{12} = -j\omega C_{gd} \tag{3.52}$$

$$Y_{21} = \frac{g_m \exp(-j\omega\tau)}{1 + j\omega C_{gs} R_{gs}} - j\omega C_{gd} \tag{3.53}$$

$$Y_{22} = \frac{1}{R_{ds}} + j\omega(C_{ds} + C_{gd}) \tag{3.54}$$

By dividing these equations into their real and imaginary parts, the parameters of the device small-signal equivalent circuit can be determined as follows [31]:

$$C_{gd} = -\frac{\text{Im } Y_{12}}{\omega} \tag{3.55}$$

$$C_{gs} = \frac{\text{Im}\,Y_{11} - \omega C_{gd}}{\omega}\left[1 + \left(\frac{\text{Re}\,Y_{11}}{\text{Im } Y_{11} - \omega C_{gd}}\right)^2\right] \tag{3.56}$$

$$R_{gs} = \frac{\text{Re}\,Y_{11}}{(\text{Im}\,Y_{11} - \omega C_{gd})^2 + (\text{Re}\,Y_{11})^2} \tag{3.57}$$

$$g_m = \sqrt{(\text{Re}\,Y_{21})^2 + (\text{Im } Y_{21} + \omega C_{gd})^2}\ \sqrt{1 + (\omega C_{gs} R_{gs})^2} \tag{3.58}$$

$$\tau = \frac{1}{\omega}\sin^{-1}\left(\frac{-\,\omega C_{gd} - \text{Im}\,Y_{21} - \omega C_{gs} R_{gs}\,\text{Re}\,Y_{21}}{g_m}\right) \tag{3.59}$$

$$C_{ds} = \frac{\text{Im}\,Y_{22} - \omega C_{gd}}{\omega} \tag{3.60}$$

$$R_{ds} = \frac{1}{\text{Re}\,Y_{22}} \tag{3.61}$$

Equations (3.55) through (3.61) are valid for the entire frequency range and for the drain voltages of $V_{ds} > 0$. If we assume that all extrinsic parasitic elements are already known, the only remaining problem is to determine the admittance Y-parameters of the intrinsic two-port network from experimental data. Consecutive stages shown in Fig. 3.20 can represent such a determination procedure in a following sequence [32]:

- Measurement of the *S*-parameters of the extrinsic device.
- Transformation of the *S*-parameters to the impedance *Z*-parameters with subtraction of the series inductances L_g and L_d.
- Transformation of the impedance *Z*-parameters to the admittance *Y*-parameters with subtraction of the parallel capacitances C_{gp} and C_{dp}.
- Transformation of the admittance *Y*-parameters to the impedance *Z*-parameters with subtraction of the series resistances R_g, R_s, R_d, and inductance L_s.
- Transformation of the impedance *Z*-parameters to the admittance *Y*-parameters of the intrinsic device two-port network.

The device extrinsic parasitic elements can be directly determined from measurements performed at $V_{ds} = 0$. Figure 3.21 shows the distributed *RC* channel network under the device gate for zero drain bias condition, where ΔC_g is the distributed gate capacitance, ΔR_{diode} is the distributed Schottky diode resistance, and ΔR_c is the distributed channel resistance. For any gate bias voltages, by taking into account the negligible influence of C_{gp} and C_{dp}, the extrinsic impedance *Z*-parameters are

$$Z_{11} = R_s + R_g + \frac{R_c}{3} + \frac{nkT}{qI_g} + j\omega(L_s + L_g) \tag{3.62}$$

$$Z_{12} = Z_{21} = R_s + \frac{R_c}{2} + j\omega L_s \tag{3.63}$$

$$Z_{22} = R_s + R_d + R_c + j\omega(L_s + L_d) \tag{3.64}$$

where R_c is the total channel resistance under the gate, nkT/qI_g is the differential resistance of the Schottky diode, n is the ideality factor, k is the Boltzmann constant, T is the Kelvin temperature, q is the electron charge, and I_g is the dc gate current. As a result, if the parasitic inductance L_s can be determined directly from measured $\mathrm{Im}Z_{12}$, the parasitic inductances L_g and L_d are calculated from measured $\mathrm{Im}Z_{11}$ and $\mathrm{Im}Z_{22}$, respectively. The resistance R_c is the channel technological parameter, which is usually known. The measured real parts of *Z*-parameters yield the values of R_s, R_g, and R_d.

At zero drain bias and for the gate voltages lower than the pinch-off voltage V_p, the small-signal equivalent circuit can be simplified to

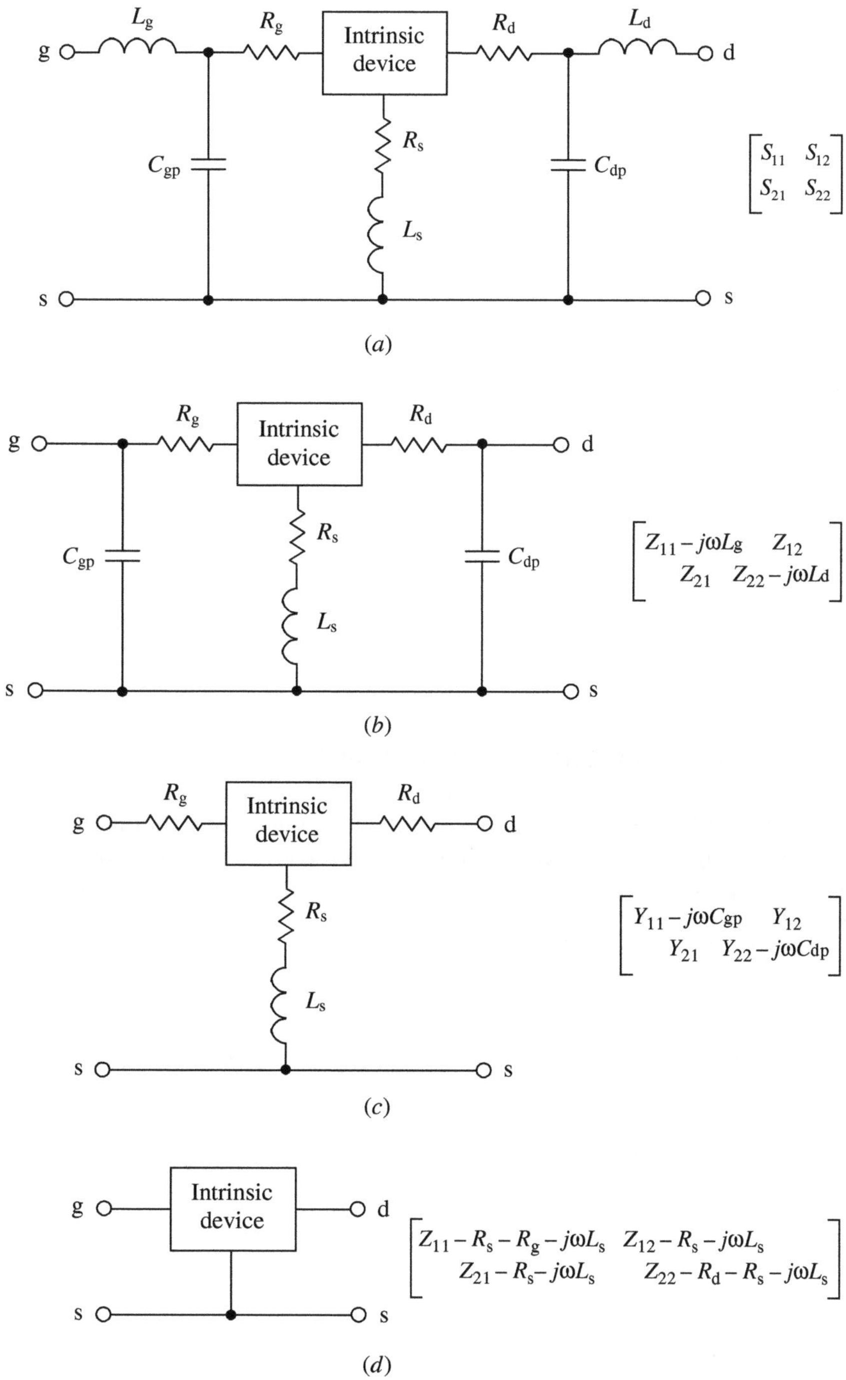

FIGURE 3.20 Method for extracting device intrinsic Z-parameters.

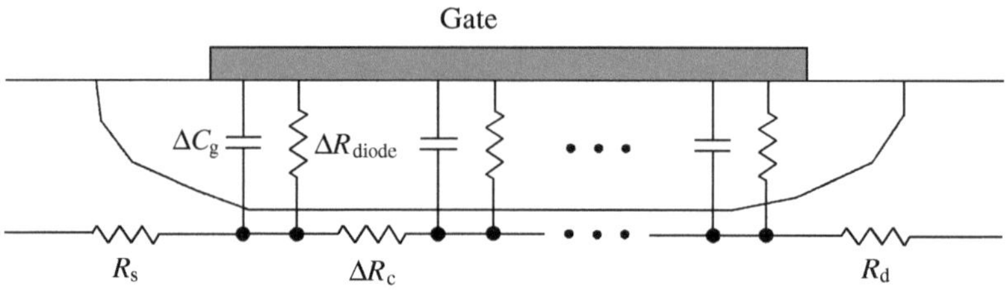

FIGURE 3.21 Distributed *RC* channel network schematic under device gate.

the one shown in Fig. 3.22. Here, the capacitance C_b represents the fringing capacitance due to the depleted layer extension at each side of the gate. For low-frequency measurements usually up to a few gigahertz, when the extrinsic parasitic resistances and inductances have no influence on the device behavior, the imaginary parts of the Y-parameters can be written as

$$\mathrm{Im}\,Y_{11} = j\omega(C_{gp} + 2C_b) \tag{3.65}$$

$$\mathrm{Im}\,Y_{12} = \mathrm{Im}\,Y_{21} = -j\omega C_b \tag{3.66}$$

$$\mathrm{Im}\,Y_{22} = j\omega(C_b + C_{dp}) \tag{3.67}$$

3.2.3 Curtice Quadratic Nonlinear Model

One of the first simple nonlinear intrinsic large-signal models for a MESFET device for use in the design of GaAs integrated circuits is shown in Fig. 3.23 [33]. It consists of a voltage-controlled source $I_{ds}(V_{gs}, V_{ds})$, the gate-source capacitance $C_{gs}(V_{gs})$, and a clamping diode between gate and source. The gate-drain capacitance C_{gd} is assumed to be constant.

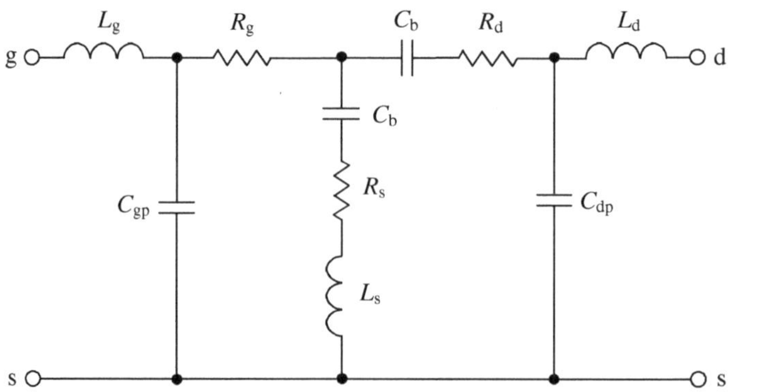

FIGURE 3.22 Small-signal MESFET circuit at zero drain bias voltage.

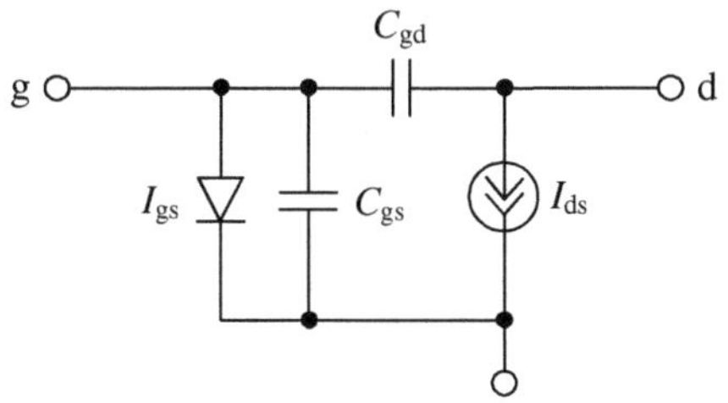

FIGURE 3.23 Curtice quadratic nonlinear intrinsic model.

An analytical function proposed to describe the nonlinear current source behavior is

$$I_{ds} = \beta(V_{gs} - V_p)^2(1 + \lambda\, V_{ds})\tanh(\alpha\, V_{ds}) \tag{3.68}$$

where β is the transconductance parameter determined from experimental data, V_p is the pinch-off voltage, and λ is the slope of the drain characteristic in the saturated region. Because of the finite value of maximum charge velocity of about 10^7 cm/s during transient operation, change in the gate voltage does not cause an instantaneous change in the drain current. For example, it takes the order of 10 ps to change the drain current after the gate voltage is changed in a 1-μm gate-length MESFET. Consequently, the most important result of this effect is a time delay between gate-source voltage and drain current. Therefore, the current source given by Eq. (3.68) as $I_{ds}[V_{gs}(t), V_{ds}(t)]$ should be altered to be $I_{ds}[V_{gs}(t - \tau), V_{ds}(t)]$, where τ is equal to the transit time under the gate.

The time delay effect is not easily added to most circuit analysis program. Therefore, a simple and sufficiently accurate way to solve this problem is to assume the current source to be of the form

$$I_{ds}(V_{gs}) - \tau \frac{dI_{ds}(V_{gs})}{dt} \tag{3.69}$$

where the second term is considered the first term of the Taylor series expansion of $I_{ds}(t - \tau)$ in time when for small τ the error is quite small,

$$\frac{dI_{ds}(V_{gs})}{dt} = \left.\frac{dI_{ds}(V)}{dV_{gs}}\right|_{V_{gs}} \frac{dV_{gs}}{dt} \tag{3.70}$$

The gate-source capacitance C_{gs} and gate-drain capacitance C_{gd} can be treated as the voltage-dependent Schottky-barrier diode capacitances. For the negative gate-source and small drain-source voltages, these capacitances are practically equal. However, when the drain-source voltage is increased beyond the current saturation point, the gate-drain capacitance C_{gd} is much more heavily back-biased than the gate-source capacitance C_{gs}. Therefore, the gate-source capacitance C_{gs} is significantly more important and usually dominates the input impedance of the MESFET device. In this case, an analytical expression to approximate the gate-source capacitance C_{gs} is

$$C_{gs} = C_{gs0} / \sqrt{1 - \frac{V_{gs}}{V_{gsi}}} \tag{3.71}$$

where C_{gs0} is the gate-source capacitance for $V_{gs} = 0$ and V_{gsi} is the built-in gate voltage. When V_{gs} approaches V_{gsi}, the denominator in Eq. (3.71) must not be allowed to approach zero because C_{gs} will continue to increase as the depletion width reduces, so that a forward bias condition occurs and the diffusion gate-source capacitance becomes of great importance. The built-in voltage V_{gsi} should be equal to the built-in voltage of the Schottky-barrier junction plus some part of the voltage drop along the channel under the gate.

3.2.4 Materka-Kacprzak Nonlinear Model

The large-signal MESFET model for computer calculation of GaAs MESFET amplifier characteristics with similar equivalent circuit is shown in Fig. 3.24, where the gate-source diode is connected in parallel to the gate-source capacitance C_{gs} [34, 35]. The main nonlinear elements include the equivalent gate-source capacitance C_{gs}, the diode in parallel to C_{gs}, which represents the current in the gate-channel junction, the drain current source I_{ds}, and the gate-drain current source I_{gd}, which represents the effect of the gate-drain breakdown. The remaining parameters of this model are assumed to be linear.

The voltage-controlled drain current source I_{ds} (V_{gs}, V_{ds}) is given by

$$I_{ds} = I_{dss}\left(1 - \frac{V_{gs}}{V_p}\right)^2 \tanh\left(\frac{\alpha V_{ds}}{V_{gs} - V_p}\right) \tag{3.72}$$

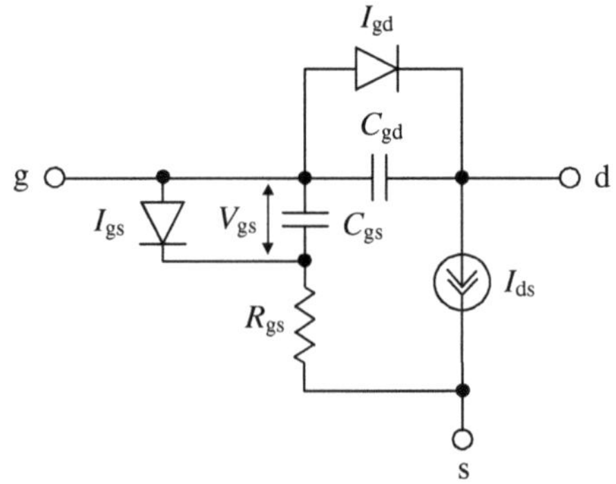

FIGURE 3.24 Materka-Kacprzak nonlinear intrinsic model.

where $V_p = V_{p0} + \gamma V_{ds}$ is the variable pinch-off voltage, I_{dss} is the saturation drain current for $V_{gs} = 0$, α is the saturation voltage parameter, γ is the voltage slop parameter of the pinch-off voltage, V_{p0} is the pinch-off voltage for $V_{ds} = 0$. To take into account the time delay between the drain current I_{ds} and the gate voltage V_{gs}, it is necessary to calculate the instantaneous drain current $I_{ds}(t)$ from Eq. (3.72) with $V_{gs} = V_{gs}(t - \tau)$ and $V_{ds} = V_{ds}(t)$, where τ is the model parameter.

The nonlinear gate-source capacitance C_{gs} can be described for $V_{gs} < 0.8V_{gsi}$ using Eq. (3.71) as a Schottky-barrier capacitance. For $V_{gs} \geq 0.8V_{gsi}$, C_{gs} is approximated by a straight line with the slope equal to the derivative dC_{gs}/dV_{gs} obtained from Eq. (3.71) at $V_{gs} = 0.8V_{gsi}$.

The current of the gate-source diode I_{gs} and gate-drain current I_{gd} can be obtained by

$$I_{gs} = I_{gss}[\exp(\alpha_s V_{gs}) - 1] \tag{3.73}$$

$$I_{gd} = I_{gdsr}[\exp(\alpha_{sr} V_{gd}) - 1] \tag{3.74}$$

where I_{gss}, I_{gdsr}, α_s, and α_{sr} are the model parameters. It should be noted that the current source I_{gd} does not represent any forward biased *p-n* or Schottky-barrier junction connected between the gate and drain terminals and only approximates the breakdown current. Therefore, the breakdown current obtained by Eq. (3.74) is negligibly small but can be increased considerably at large values of the gate-drain voltages.

3.2.5 Chalmers (Angelov) Nonlinear Model

A simple and accurate large-signal model for different submicron gate-length HEMT devices and commercially available MESFETs, which is capable of modeling the drain current-voltage characteristics and its derivatives, as well as the gate-source and gate-drain capacitances, is shown in Fig. 3.25 [36, 37]. This model can be used

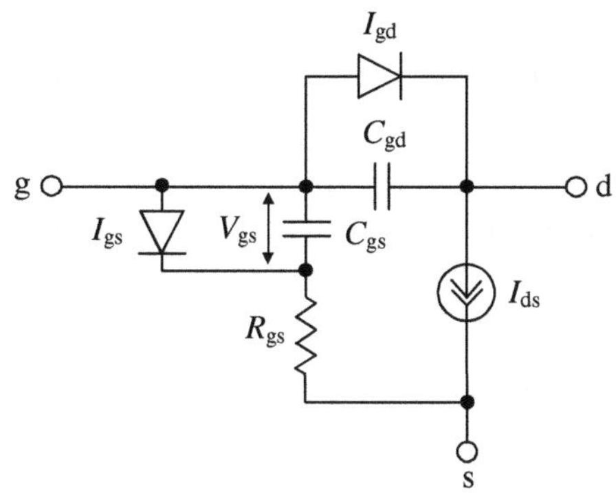

FIGURE 3.25 Angelov nonlinear intrinsic model.

not only for large-signal analysis of the power amplifier but also for predicting the performance of multipliers and mixers including intermodulation simulation.

The drain current source is described by using the hyperbolic functions as

$$I_{ds} = I_{pk}(1+\tanh\psi)(1+\lambda V_{ds})\tanh(\alpha V_{ds}) \tag{3.75}$$

where I_{pk} is the drain current at maximum transconductance with the contribution from the output conductance subtracted, λ is the channel length modulation parameter, and $\alpha = \alpha_0 + \alpha_1 \tanh\psi$ is the saturation voltage parameter, where α_0 is the saturation voltage parameter at pinch-off and α_1 is the saturation voltage parameter at $V_{gs} > 0$.

The parameter ψ is a power series function centered at V_{pk} with the bias voltage V_{gs} as a variable,

$$\psi = P_1(V_{gs} - V_{pk}) + P_2(V_{gs} - V_{pk})^2 + P_3(V_{gs} - V_{pk})^3 + \ldots \tag{3.76}$$

where V_{pk} is the gate voltage for maximum transconductance g_{mpk}. The model parameters as a first approximation can be easily obtained from the experimental $I_{ds}(V_{gs}, V_{ds})$ dependencies at a saturated channel condition when all higher terms in ψ are assumed to be zero, and λ is the slope of the $I_{ds}(V_{ds})$ curves.

The intrinsic maximum transconductance g_{mpk} is calculated from the measured maximum transconductance g_{mpkm}, by taking into account feedback effect due to the source resistance R_s, as

$$g_{mpk} = \frac{g_{mpkm}}{1 - g_{mpkm}R_s} \tag{3.77}$$

To evaluate the gate voltage V_{pk} and parameter P_1, it is necessary to define the derivatives of the drain current. If higher order terms of ψ are neglected, the transconductance g_m becomes equal to

$$g_m = \frac{\partial I_{ds}}{\partial V_{gs}} = I_{pk} P_1 \text{ sech } [P_1(V_{gs} - V_{pk})]^2(1+\lambda V_{ds})\tanh(\alpha V_{ds}) \tag{3.78}$$

The gate voltage V_{pk} that depends on the drain voltage can be extracted by finding the gate voltages for maximum transconductance, at which the second derivative of the drain current is equal to zero. In this case, it is advisable to use the simplified expression

$$V_{pk}(V_{ds}) = V_{pk0} + (V_{pks} - V_{pk0})(1+\lambda V_{ds})\tanh(\alpha V_{ds}) \tag{3.79}$$

where V_{pk0} is measured at V_{ds} closely to zero and V_{pks} is measured at V_{ds} in the saturation region.

A good fitting of P_1 and good results in harmonic balance simulations can be obtained using

$$P_1 = P_{1\mathrm{sat}}\left[1+\left(\frac{P_{10}}{P_{1\mathrm{sat}}}-1\right)\frac{1}{\cosh^2(BV_{\mathrm{ds}})}\right] \tag{3.80}$$

where $P_{10} = g_{m0}/I_{pk0}$ at V_{ds} close to zero and B is the fitting parameter ($B \approx 1.5\alpha$). The parameter P_2 makes the derivative of the drain current asymmetric, whereas the parameter P_3 changes the drain current values at voltages V_{gs} close to pinch-off voltage V_p. Three terms in Eq. (3.76) are usually enough to describe the behavior of the different MESFET or HEMT devices with acceptable accuracy. For example, it can be used to predict the large-signal behavior of the pHEMT devices using in high-power, high-efficiency 60-GHz MMICs [38]. By using three additional terms of a gate power-series function, the better accuracy can be achieved in large-signal modeling of AlGaN/GaN HEMT devices on SiC substrate [39]. This model can also be improved by incorporating two additional analytical expressions to model the device behavior in a saturation region, which is important to design high-efficiency switchmode power amplifiers, for example, in an inverse Class-F mode [40].

The same hyperbolic functions can be used to model the intrinsic device capacitances. When 5 to 10% accuracy is sufficient, the gate-source capacitance C_{gs} and gate-drain capacitance C_{gd} can be described by

$$C_{\mathrm{gs}} = C_{\mathrm{gs0}}[1+\tanh(P_{\mathrm{1gsg}}V_{\mathrm{gs}})][1+\tanh(P_{\mathrm{1gsd}}V_{\mathrm{ds}})] \tag{3.81}$$

$$C_{\mathrm{gd}} = C_{\mathrm{gd0}}[1+\tanh(P_{\mathrm{1gdg}}V_{\mathrm{gs}})]\,[1-\tanh(P_{\mathrm{1gdd}}V_{\mathrm{ds}}+P_{\mathrm{1cc}}V_{\mathrm{gs}}V_{\mathrm{ds}})] \tag{3.82}$$

where the product $P_{1cc}V_{gs}V_{ds}$ reflects the cross-coupling of V_{gs} and V_{ds} on C_{gd} and the coefficients P_{1gsg}, P_{1gsd}, P_{1gdg}, and P_{1gdd} are the fitting parameters. These dependences, unlike the commonly diode-like models, are suitable for HEMT devices with an undoped AlGaAs spacer-layer in view of the saturation effect of the gate-source capacitance C_{gs} for increasing V_{gs}. This is due to the absence of parasitic MESFET channel formation in the undoped AlGaAs layer, found in HEMTs with a doped AlGaAs layer. The approximate behavior of normalized gate-source capacitance C_{gs}/C_{gs0} (curve 1) and gate-drain capacitance C_{gd}/C_{gd0} (curve 2) as functions of V_{ds} for zero gate-source voltage is shown in Fig. 3.26, where C_{gs0} and C_{gd0} are the gate-source and gate-drain capacitances at $V_{ds} = 0$, respectively. The character of the curves is the same for both positive and negative V_{gs}, except that the capacitance range decreases for the same range of V_{ds} with the decrease in V_{gs}.

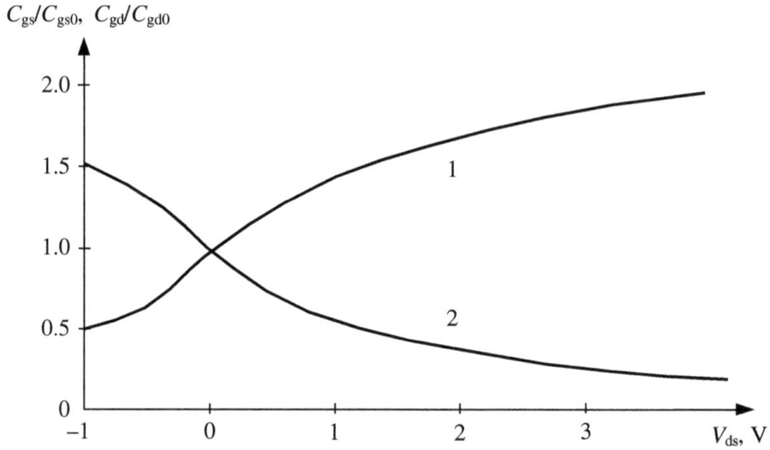

FIGURE 3.26 Gate-source and gate-drain capacitances versus drain-source voltage.

The drain-source dispersive resistance R_{dsd} as a nonlinear function of the gate-source voltage V_{gs} can be defined by

$$R_{dsd} = R_{dsd0} + \frac{R_{dsdp}}{1 + \tanh \psi} \tag{3.83}$$

where R_{dsd0} is the minimum value of R_{dsd} and R_{dsdp} determines the value of R_{dsd} at the pinch-off [22].

3.2.6 IAF (Berroth) Nonlinear Model

An analytical charge conservative large-signal model for HEMT devices, which is valid for a frequency range up to 60 GHz, is shown in Fig. 3.27 [41]. In this model, the current sources I_{gs}, I_{gd}, and I_{ds} and capacitances C_{gs} and C_{gd} are considered as the nonlinear elements.

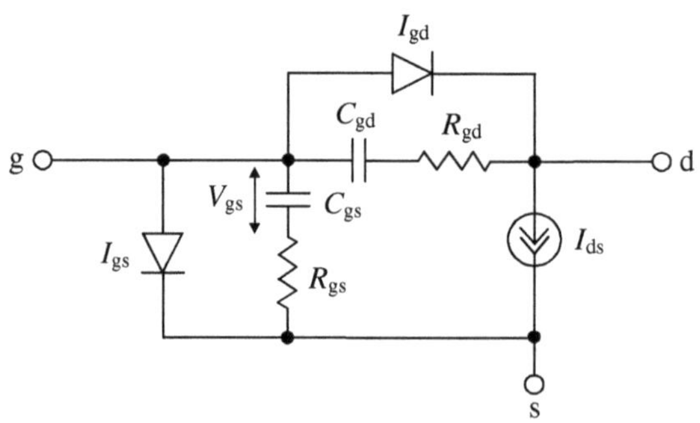

FIGURE 3.27 Berroth nonlinear intrinsic model.

The drain current source is represented by the following nonlinear equation with 10 fitting parameters:

$$I_{ds} = f(V_{gs})\left[1 + \frac{\lambda}{1 + \Delta_{\lambda}(V_{gs} - V_c + 2/\beta)} V_{ds}\right]\tanh(\alpha V_{ds}) \quad (3.84)$$

where

$$f(V_{gs}) = CD_{vc}\{1 + \tanh[\beta(V_{gs} - V_c) + \gamma(V_{gs} - V_c)^3]\}$$
$$+CD_{vsb}\{1 + \tanh[\delta(V_{gs} - V_{sb})]\}$$

α is the slope of drain current in the pinch-off region, β is the slope parameter of drain current, γ is the slope parameter of drain current in the pinch-off region, λ is the slope of drain current in the saturation region, Δ_λ is the gate voltage parameter for slope of drain current, δ is the drain current slope parameter correction term, V_c is the gate voltage for maximum transconductance, V_{sb} is the gate voltage for maximum transconductance correction term, CD_{vc} is the drain current multiplication factor, and CD_{vsb} is the drain current multiplication factor correction term.

To describe the I_{gs} and I_{gd} current sources, the diode model for both forward and reverse bias operation modes was used in the form

$$I_{diode} = I_{dsat}\left[\exp\left(\frac{V_{diode}}{nV_T}\right) - 1\right] \quad (3.85)$$

where I_{dsat} is the forward bias fitting parameter, V_T is the temperature voltage, and n is the diode ideality factor.

The nonlinear capacitances C_{gs} and C_{gd} are calculated by differentiating the voltage-depending charge function $Q_g(V_{gs}, V_{ds})$ with respect to V_{gs} and V_{ds}, which leads to the input capacitance $C_{11} = C_{gs} + C_{gd}$ and transcapacitance $C_{12} = -C_{gd}$, respectively,

$$Q_g(V_{gs}, V_{ds}) = A f_1 f_2 + E(V_{gs} - 0.5V_{ds}) \quad (3.86)$$

where

$$f_1 = \frac{1}{B}\ln\cosh\{B[(V_{gs} - V_1) - 0.5\tanh(CV_{ds})]\} + (V_{gs} - V_1) - 0.5\tanh(CV_{ds})$$

$$f_2 = 1 + D\ln\cosh(FV_{ds})$$

V_1 is the transition voltage, and A, B, C, D, E, and F are the model fitting parameters.

3.2.7 Model Selection

A large-signal device model should be sufficiently accurate for all operation conditions and as simple as possible. Let us first compare several large-signal models. Some models are not consistent with a common small-signal model; for example, the charging gate-source resistance R_{gs} is not described in the Curtice quadratic, and in most of the models the effect of the frequency dispersion of the transconductance and output conductance is omitted. Besides, another important problem of some large-signal models is the failure to carry out a charge conversation condition. The Materka model provides better description of the slopes of the voltage-current characteristics in the pinch-off region [21]. Although the Materka model does not fulfill charge conservation, it seems to be an acceptable compromise between accuracy and model simplicity for MESFETs but not for pseudomorphic HEMTs, where it is preferable to use the Angelov method [21]. A diode-like capacitance model does not approximate the measured *C-V* characteristics closely enough. For HEMT devices, an analytical charge conservative Berroth model can be used to model the large-signal parameters in the frequency range up to 60 GHz.

3.3 BJTs and HBTs

3.3.1 Small-Signal Equivalent Circuit

The complete bipolar transistor small-signal equivalent circuit with extrinsic parasitic elements is shown in Fig. 3.28(*a*). Based on this hybrid π-type representation, the electrical properties of the bipolar transistors, in particularly HBT devices, can be described with sufficient accuracy up to 30 GHz [42, 43]. Here, the extrinsic elements R_b, L_b, R_c, L_c, R_e, and L_e are the series resistances and lead inductances associated with the base, collector, and emitter, and C_{pbe}, C_{pbc}, and C_{pce} are the parasitic capacitances associated with the contact pads, respectively. The lateral resistance with the base semiconductor resistance underneath the base contact and the base semiconductor resistance underneath the emitter are combined into a base-spreading resistance r_b. The intrinsic model is described by the dynamic diode resistance r_π, the total base-emitter junction capacitance and base charging capacitance C_π, the transconductance g_m, and the output Early resistance r_{ce}, which model the effect of base-width modulation on the transistor characteristics due to variations in the collector-base depletion region [44]. To increase the usable operating frequency range of the device up to 50 GHz, it is necessary to properly include the collector-current delay time in the current source as $g_m \exp(-j\omega\tau_\pi)$, where τ_π is the transit time [45].

The cross-section of a physical structure of an AlGaAs/GaAs HBT device is shown in Fig. 3.28(*b*), with heavily *p*-doped base to reduce base resistance and lightly *n*-doped emitter to minimize the emitter capacitance [46]. The lightly *n*-doped collector region allows

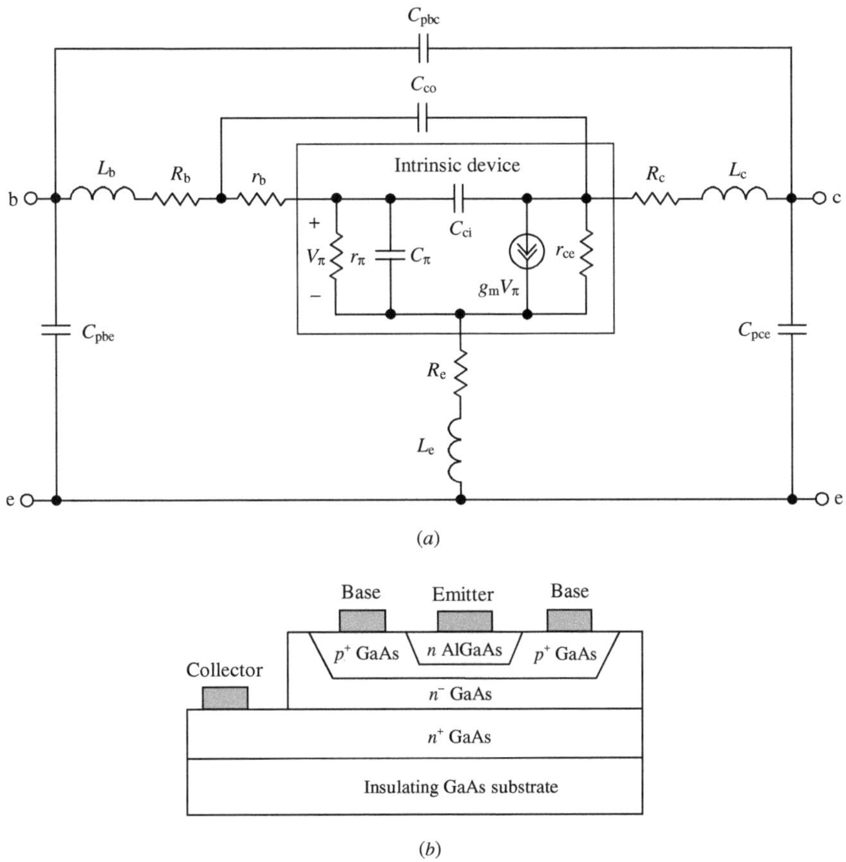

FIGURE 3.28 Nonlinear BJT and HBT model with HBT physical structure.

collector-base junction to sustain relatively high voltages without breaking down. The forward-bias emitter-injection efficiency is very high because the wider-bandgap AlGaAs emitter injects electrons into the GaAs *p*-base at lower energy level, but the holes are prevented from flowing into the emitter by a high-energy barrier, thus resulting in the ability to decrease base length, base-width modulation, and increase frequency response. By using a wide bandgap InGaP layer instead of an AlGaAs one, the device performance over temperature can be improved [47]. The high-linearity power performance in Class-AB condition at the backoff power level, the ruggedness under mismatch and overdrive condition, and the long lifetime of the InGaP/GaAs HBT technology make it very attractive for the 28-V power amplifier applications [48]. The growth process used for a high-voltage HBT device is identical to the process used for the conventional low-voltage HBT device, which is widely used in handset power amplifiers, except for changes to the collector because of the higher voltage operating requirements. The epitaxial growth process starts with a highly doped *n*-type collector layer and lightly *n*-doped collector drift region, then followed by a heavily doped *p*-type base layer and an InGaP emitter layer, and finishes with an InGaAs cap layer [49]. As a result, the high-voltage HBT devices exhibit collector-base breakdown voltages higher than 70 V.

Figure 3.29 shows the modified version of a bipolar transistor equivalent circuit, where $C_c = C_{co} + C_{ci}$, $r_{b1} = r_b C_{ci}/C_c$, and $r_{b2} = r_b C_{co}/C_c$ [50]. Such an equivalent circuit becomes possible due to an equivalent π- to T-transformation of the elements r_b, C_{co}, and C_{ci} and a condition $r_b << (C_{ci} + C_{co})/\omega C_{ci} C_{co}$, which is usually fulfilled over a frequency range close to the device maximum frequency f_{max}. Then, from a comparison of the transistor nonlinear models, for a bipolar transistor in Fig. 3.29, for a MOSFET device in Fig. 3.3, and for a MESFET device in Fig. 3.17(*a*), it is easy to detect the circuit similarity of all these equivalent circuits, which means that the basic circuit design procedure is very similar for any type of bipolar or field-effect transistors. The difference is in the device physics and values of the model parameters. However, techniques for the representation of the input and output impedances, stability analysis based on the feedback effect, or derivation of power gain and efficiency are very similar.

3.3.2 Determination of Equivalent Circuit Elements

If all extrinsic parasitic elements of the device equivalent circuit shown in Fig. 3.28(*a*) are known, the intrinsic two-port network

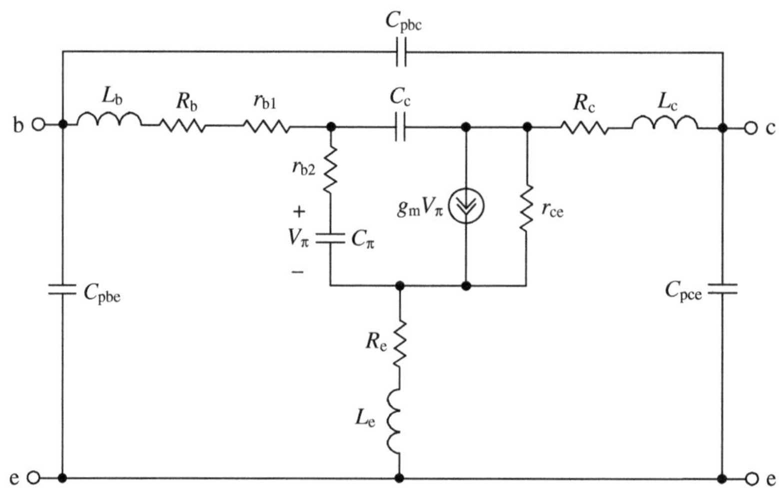

Figure 3.29 Nonlinear BJT and HBT model with HBT physical structure.

parameters can be embedded with the following determination procedure [42]:

- Measurement of the *S*-parameters of the transistor with extrinsic elements.
- Transformation of the *S*-parameters to the admittance *Y*-parameters with subtraction of the parasitic shunt capacitances C_{pbe}, C_{pbc}, and C_{pce}.
- Transformation of the new *Y*-parameters to the impedance *Z*-parameters with subtraction of the parasitic series elements L_b, R_b, L_e, R_e, L_c, and R_c.
- Transformation of the new *Z*-parameters to the *Y*-parameters with subtraction of the parasitic shunt capacitance C_{co}.
- Transformation of the new *Y*-parameters to the *Z*-parameters with subtraction of the parasitic series resistance r_b.
- Transformation of the new *Z*-parameters to the *Y*-parameters of the intrinsic device two-port network.

The bipolar transistor intrinsic *Y*-parameters can be written as

$$Y_{11} = \frac{1}{r_\pi} + j\omega(C_\pi + C_{ci}) \tag{3.87}$$

$$Y_{12} = -j\omega C_{ci} \tag{3.88}$$

$$Y_{21} = g_m \exp(-j\omega\tau_\pi) - j\omega C_{ci} \tag{3.89}$$

$$Y_{22} = \frac{1}{r_{ce}} + j\omega C_{ci} \tag{3.90}$$

After separating Eqs. (3.87) through (3.90) into their real and imaginary parts, the elements of the intrinsic small-signal equivalent circuit can be determined analytically as

$$C_\pi = \frac{\text{Im}(Y_{11} + Y_{12})}{\omega} \tag{3.91}$$

$$r_\pi = \frac{1}{\text{Re } Y_{11}} \tag{3.92}$$

$$C_{ci} = -\frac{\text{Im}\,Y_{12}}{\omega} \tag{3.93}$$

$$g_m = \sqrt{(\text{Re}\,Y_{21})^2 + (\text{Im } Y_{21} + \text{Im}\,Y_{12})^2} \tag{3.94}$$

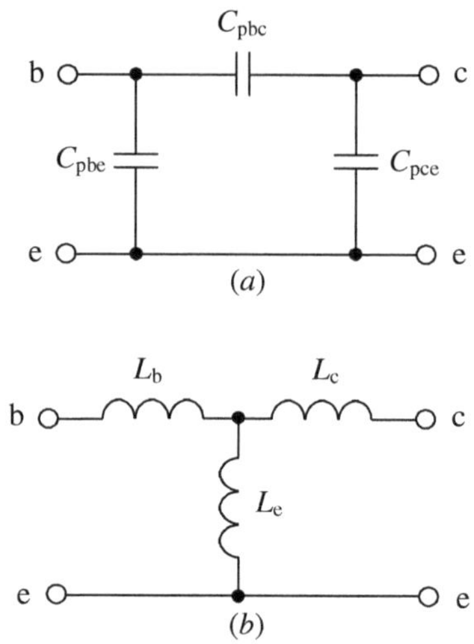

FIGURE 3.30 Models for parasitic pad capacitances and lead inductances.

$$\tau_\pi = \frac{1}{\omega}\cos^{-1}\frac{\mathrm{Re}\,Y_{21} + \mathrm{Re}\,Y_{12}}{\sqrt{(\mathrm{Re}\,Y_{21})^2 + (\mathrm{Im}\,Y_{21} + \mathrm{Im}\,Y_{12})^2}} \tag{3.95}$$

$$r_{ce} = \frac{1}{\mathrm{Re}\,Y_{22}} \tag{3.96}$$

The parasitic capacitances associated with the pads can be determined by the measurement of the open test structure with the corresponding circuit model shown in Fig. 3.30(*a*). When the values of these pad capacitances are known, it is easy to determine the values of the parasitic series inductances by measuring the shorted test structure with corresponding circuit model shown in Fig. 3.30(*b*).

The values of the series parasitic resistances can be calculated on the basis of the physical parameters of the device or by adding them to the intrinsic device parameters (with the appropriate numerical solution of a nonlinear system of eight equations with eight independent variables using iterative technique) [43]. In the latter case, it is assumed that influence of the transit time τ_π on the bipolar electrical properties in a frequency range up to 30 GHz is negligible.

The external parasitic parallel capacitance C_{co}, as well as the other device capacitances, can be estimated from the device behavior at low frequencies and cutoff operating conditions [51, 52]. For these conditions, the device small-signal equivalent circuit is reduced to that with the capacitive elements only, as shown in Fig. 3.31.

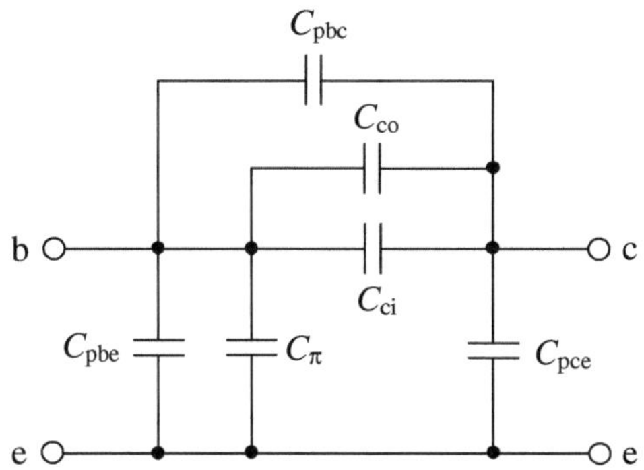

Figure 3.31 Small-signal model at low frequencies and cutoff operation mode.

The device capacitances can be directly calculated from measured Y-parameters by

$$C_{pbe} + C_{\pi} = \frac{\mathrm{Im}(Y_{11} + Y_{12})}{\omega} \tag{3.97}$$

$$C_{pce} = \frac{\mathrm{Im}(Y_{22} + Y_{12})}{\omega} \tag{3.98}$$

$$C_{pbc} + C_{co} + C_{ci} = -\frac{\mathrm{Im}\,Y_{12}}{\omega} \tag{3.99}$$

Because C_{pbc} and C_{co} are the bias-independent capacitances and C_{ci} is the base-collector junction capacitance, the extraction of $C_{pbc} + C_{co}$ can be carried out by fitting the sum $C_{pbc} + C_{co} + C_{ci}$ to the expression for junction capacitance at different base-collector voltages. If an approximation expression for C_{co} is given by

$$C_{co} = C_{jco} / \sqrt{1 + \frac{V_{bc}}{\varphi_c}} \tag{3.100}$$

then the extraction of the parameters C_{jco}, φ_c, and C_{ci} can be performed using the linear equation

$$\left(\frac{C_{jco}}{\frac{\mathrm{Im}\,Y_{12}}{\omega} + C_{pbc} + C_{ci}} \right)^2 = 1 + \frac{1}{\varphi_c} V_{bc} \tag{3.101}$$

As a result, linearizing this equation by choosing the proper value for C_{ci} with known value of C_{pbc} gives the values for the remaining two parameters C_{jco} and φ_c from the slope and intercept point of the final linearized dependence.

3.3.3 Equivalence of Intrinsic π- and T-Type Topologies

The small-signal equivalent circuit of a bipolar transistor can be represented by both π-type and T-type topologies. The T-type equivalent circuit representation is appealing because all the model parameters can be directly tied to the physics of the device and an excellent fit between measured and simulated S-parameters in the frequency range up to 30 to 40 GHz can be provided [51–53]. The small-signal HBT equivalent circuit with a T-type topology is shown in Fig. 3.32.

There is one-to-one correspondence between π-type and T-type device models. Comparing both small-signal equivalent circuits shown in Figs. 3.28(a) and 3.32 demonstrates the only difference in the representation of the intrinsic device models enclosed in boxes. From Fig. 3.33, it follows that the admittances $Y_e = I_e/V_{be}$ for π- and T-type models can be determined as

$$Y_e = \frac{1}{r_e} + j\omega C_e = \frac{1}{r_\pi} + j\omega C_\pi + g_m \exp(-j\omega\tau_\pi) \tag{3.102}$$

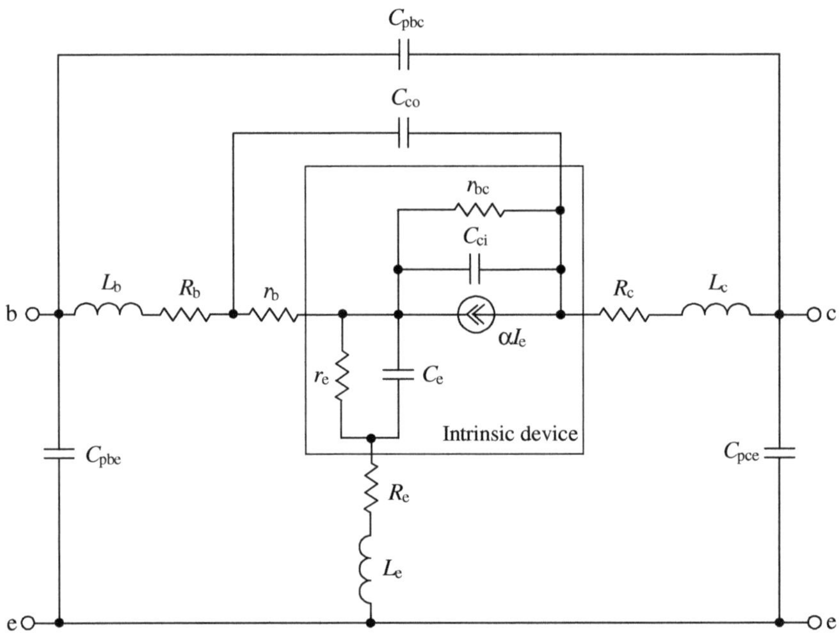

FIGURE 3.32 Small-signal T-model of bipolar device.

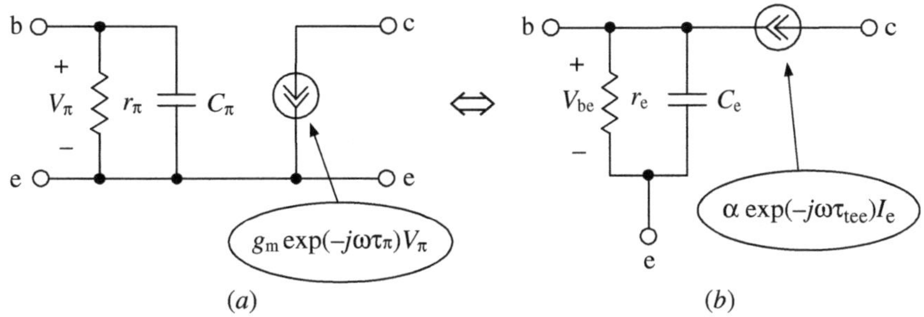

Figure 3.33 Intrinsic π-type and T-type bipolar device topologies.

In this case, the collector source currents for both models are the same,

$$\alpha \exp(-j\omega\tau_{tee})I_e = g_m \exp(-j\omega\tau_\pi)V_\pi \tag{3.103}$$

where τ_{tee} is the transit time for a T-type model, $\alpha = \alpha_0/(1 + j\omega\tau_\alpha)$, $\tau_\alpha = 1/(2\pi f_\alpha)$, f_α is the alpha cutoff frequency, and α_0 is the low-frequency collector-to-emitter current gain.

The expressions for intrinsic π-model parameters can be derived through the intrinsic T-model parameters as [45]

$$g_m = \frac{\alpha_0}{\sqrt{1+(\omega\tau_\alpha)^2}}\sqrt{\left(\frac{1}{r_e}\right)^2 + (\omega C_e)^2} \tag{3.104}$$

$$\tau_\pi = \tau_{tee} - \frac{1}{\omega}[\tan^{-1}(\omega C_e r_e) + \tan^{-1}(\omega\tau_\alpha)] \tag{3.105}$$

$$\frac{1}{r_\pi} = \frac{1}{r_e} - g_m \cos(\omega\tau_\pi) \tag{3.106}$$

$$C_\pi = C_e - g_m \frac{\sin(\omega\tau_\pi)}{\omega} \tag{3.107}$$

Both π-type and T-type bipolar device topologies can describe the transistor electrical properties in a very wide frequency range and, when optimized, up to 50 GHz.

3.3.4 Nonlinear Bipolar Device Modeling

Because the bipolar transistor can be considered to be an interacting pair of p-n junctions, the approach to model its nonlinear properties is the same as that used for the diode modeling. The simple large-signal Ebers-Moll model with a single current source between the collector

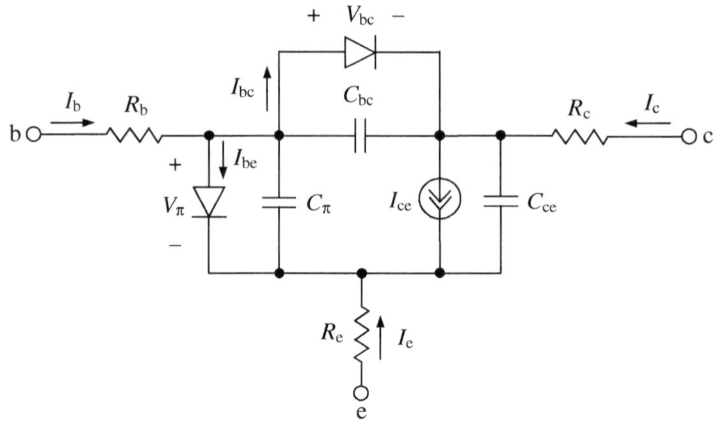

FIGURE 3.34 Large-signal Ebers-Moll model.

and the emitter is shown in Fig. 3.34 [1, 54]. The collector-emitter source current I_{ce} is defined by

$$I_{ce} = I_{sat}\left[\exp\left(\frac{V_\pi}{V_T}\right) - \exp\left(\frac{V_{bc}}{V_T}\right)\right] \tag{3.108}$$

where I_{sat} is the bipolar transistor reverse saturation current and V_T is the thermal voltage obtained by

$$V_T = \frac{kT}{q} \tag{3.109}$$

where q is the electron charge, k is Boltzmann's constant, and T is the temperature in Kelvin.

The device terminal currents are defined as $I_c = I_{ce} - I_{bc}$, $I_e = -I_{ce} - I_{be}$, and $I_b = I_{be} + I_{bc}$, where the diode currents are given by

$$I_{be} = \frac{I_{sat}}{\beta_F}\left[\exp\left(\frac{V_\pi}{V_T}\right) - 1\right] \tag{3.110}$$

$$I_{bc} = \frac{I_{sat}}{\beta_R}\left[\exp\left(\frac{V_{bc}}{V_T}\right) - 1\right] \tag{3.111}$$

where β_F and β_R are the large-signal forward current gain and reverse current gain of a common-emitter bipolar transistor, respectively.

The device capacitances C_π and C_{bc} each consist of the two components and are modeled by the diffusion capacitance and junction capacitance, respectively, as

$$C_\pi = \tau_F \frac{dI_{be}}{dV_\pi} + C_{jeo}\left(1 - \frac{V_\pi}{\varphi_e}\right)^{-m_e} \tag{3.112}$$

$$C_{bc} = \tau_R \frac{dI_{bc}}{dV_{bc}} + C_{jco}\left(1 - \frac{V_{bc}}{\varphi_c}\right)^{-m_c} \tag{3.113}$$

where τ_F and τ_R are the ideal total forward time and reverse transit time, C_{jeo} and C_{jco} are the base-emitter and base-collector zero-bias junction capacitances, and m_e and m_c are the base-emitter and base-collector junction grading factors, respectively.

The collector-emitter substrate capacitance C_{ce} should be taken into account when designing the integrated circuits. Its representation is adequate for many cases, because the epitaxial-layer-substrate junction is reversed-biased for isolation purposes, and usually it is modeled as a capacitance with constant value.

The Ebers-Moll model cannot describe the second-order effects that can influence the accuracy of Eq. (3.108). One deals with the width of the neutral base region when the widths of the depletion regions change with V_π and V_{bc}, thus providing a base-width modulation (Early effect [44]). Others involve the device behavior at low and high currents, resulting in a variation of the large-signal forward current gain β_F with collector current I_{ce}. For example, as the number of electrons injected into the base approaches the background doping concentration, the number of holes will increase to maintain space charge neutrality and the gain will drop (Webster effect [55]). At the same time, if the electron concentration overwhelms the doping in the collector, the base will push out increasing the neutral base region and again reduce the gain (Kirk effect [56]). In addition, to find a better approximation of the distributed structure of the base-collector junction at microwave frequencies, the junction capacitance C_{bc} should be divided into two separate capacitances: internal C_{ci} and external C_{co}. The lateral resistance and the base semiconductor resistance underneath the base contact and the base semiconductor resistance underneath the emitter are combined into a base-spreading resistance r_b. Figure 3.35 shows the modified π-type Gummel-Poon nonlinear model of the bipolar transistor, which can describe the nonlinear electrical behavior of bipolar transistors, in particularly HBT devices, with a sufficient accuracy up to 20 GHz [57, 58].

For the Gummel-Poon large-signal model, the collector source current I_{ce} is determined by

$$I_{ce} = \frac{I_{ss}}{q_b}\left[\exp\left(\frac{V_\pi}{n_F V_T}\right) - \exp\left(\frac{V_{bc}}{n_R V_T}\right)\right] \tag{3.114}$$

where I_{ss} is the BJT fundamental constant defined at zero-bias condition, n_F is the forward current emission coefficient, n_R is the reverse current emission coefficient, and q_b is the variable parameter defined by

$$q_b = \frac{q_1}{2} + \sqrt{\left(\frac{q_1}{2}\right)^2 + q_2} \tag{3.115}$$

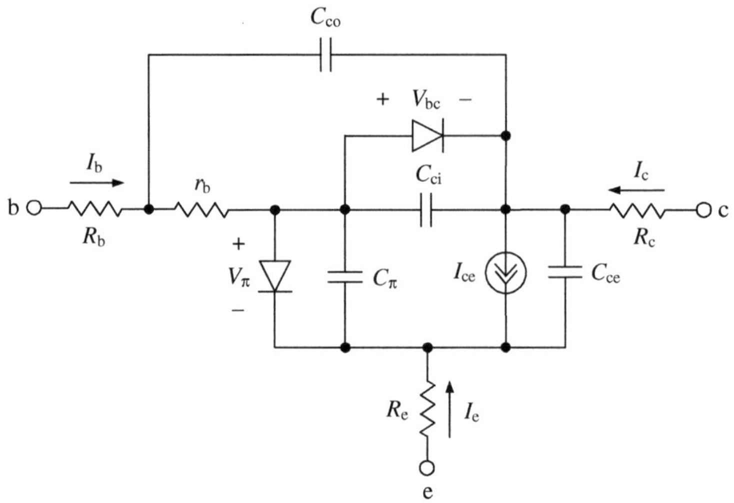

FIGURE 3.35 Large-signal Gummel-Poon model.

where

$$q_1 = 1 + \frac{V_\pi}{V_B} + \frac{V_{bc}}{V_A} \tag{3.116}$$

$$q_2 = \frac{I_{ss}}{I_{KF}}\left[\exp\left(\frac{V_\pi}{n_F V_T}\right) - 1\right] + \frac{I_{ss}}{I_{KR}}\left[\exp\left(\frac{V_{bc}}{n_R V_T}\right) - 1\right] \tag{3.117}$$

V_A is the forward Early voltage, V_B is the reverse Early voltage, I_{KF} is the forward knee current for high-level injection in the normal active region, and I_{KR} is the reverse knee current for low-level injection in inverse region [1, 59]. Although the physical mechanism of the collector current behavior given by Eqs. (3.115) through (3.117) is related to silicon BJTs, the highly doped base found in HBTs makes it unlikely that their base-collector potential will significantly change base width, or that the injected electron concentration will approach the background doping at typical current levels [60]. However, mainly for historical reasons, and so the model can still be used for silicon BJTs, they are usually retained for HBTs, although the physics is not exactly relevant.

The currents through model diodes are defined by

$$I_{be} = \frac{I_{sat}(0)}{\beta_{FM}(0)}\left[\exp\left(\frac{V_\pi}{n_F V_T}\right) - 1\right] + C_2 I_{sat}(0)\left[\exp\left(\frac{V_\pi}{n_{EL} V_T}\right) - 1\right] \tag{3.118}$$

$$I_{bc} = \frac{I_{sat}(0)}{\beta_{RM}(0)}\left[\exp\left(\frac{V_{bc}}{n_R V_T}\right) - 1\right] + C_4 I_{sat}(0)\left[\exp\left(\frac{V_{bc}}{n_{CL} V_T}\right) - 1\right] \quad (3.119)$$

where $I_{sat}(0)$ is the saturation current for $V_{bc} = 0$, $\beta_{FM}(0)$ and $\beta_{RM}(0)$ are the large-signal forward and reverse current gains of a common-emitter BJT in mid-current region for $V_{bc} = 0$, C_2 and C_4 are the forward and reverse low-current nonideal base current coefficients, respectively, and n_{EL} and n_{CL} are the nonideal low-current, base-emitter, and base-collector emission coefficients, which is usually close to 2.0, respectively.

The nonlinear behavior of the intrinsic base resistance r_b can be described by

$$r_b = r_{bm} + 3(r_{b0} - r_{bm})\frac{\tan z - z}{z \tan^2 z} \quad (3.120)$$

where

$$z = \frac{\sqrt{1 + \frac{144}{\pi^2}\frac{I_b}{I_{rb}}} - 1}{\frac{24}{\pi^2}\sqrt{\frac{I_b}{I_{rb}}}} \quad (3.121)$$

r_{bm} is the minimum base resistance that occurs at high current level, r_{b0} is the base resistance at zero bias with small base current level, and I_{rb} is the current where the base resistance falls halfway to its minimum value [1, 59].

The intrinsic device capacitances C_π, C_{ci}, and C_{co} are modeled by the diffusion capacitance and junction capacitance, respectively, as

$$C_\pi = \frac{d}{dV_\pi}\left(\tau_{FF}\frac{I_{cc}}{q_b}\right) + C_{jeo}\left(1 - \frac{V_\pi}{\varphi_e}\right)^{-m_e} \quad (3.122)$$

$$C_{ci} = \tau_R \frac{dI_{bc}}{dV_{bc}} + k_c C_{jco}\left(1 - \frac{V_{bc}}{\varphi_c}\right)^{-m_c} \quad (3.123)$$

$$C_{co} = C_{jco}(1 - k_c)\left(1 - \frac{V_{bco}}{\varphi_c}\right)^{-m_c} \quad (3.124)$$

where k_c is the fraction of the base-collector junction capacitance connected to the base resistance r_b, V_{bco} is the voltage through the capacitance C_{co}, and τ_{FF} is the modulated transit time defined by

$$\tau_{FF} = \tau_F\left[1 + X_{\tau F}\left(\frac{I_{cc}}{I_{cc} + I_{\tau F}}\right)^2 \exp\left(\frac{V_{bc}}{1.44 V_{\tau F}}\right)\right] \quad (3.125)$$

where $X_{\tau F}$ is the transit time bias dependence coefficient, $I_{\tau F}$ is the high-current parameter for effect on τ_F, $V_{\tau F}$ is the value of V_{bc} where the exponential equals to 0.5, and

$$I_{cc} = \frac{I_{ss}}{q_b}\left[\exp\left(\frac{V_\pi}{n_F V_T}\right) - 1\right] \tag{3.126}$$

As it follows from Eq. (3.122), the nonlinear behavior of capacitance C_π strongly depends on the effect of transit time modulation characterized by τ_{FF}. This transit charge variation results in significant changes of the transition frequency f_T at various operation conditions. For example, at medium currents, f_T reaches its peak value and is practically constant. Here, the ideal transit time is defined by $\tau_F = 1/(2\pi f_T)$ and the dominated base-emitter diffusion capacitance increases linearly with collector current. At low currents, f_T is dominated by the junction capacitance and increases with the increase in collector current. At high currents, the widening of the charge-neutral base region and pushing of the entire space-charge region toward the heavily doped collector region (Kirk effect) degrades the frequency response of the transistor by increasing the transit time and decreasing f_T. In this case, the transit time is modeled by τ_{FF}.

A simple nonlinear HBT model for computer-aided simulations can be based on the representation of the collector current source through the power series and diffusion capacitances through the hyperbolic functions [61]. To equivalently estimate the input impedance of a bipolar transistor, it needs to take into account that C_{ce} is usually much smaller than C_c. As a result, the equivalent output capacitance can be defined as $C_{out} \cong C_c$. The input equivalent R_{in} can approximately be represented by the base resistance r_b, whereas the input equivalent capacitance can be defined as $C_{in} \cong C_\pi + C_c$. The feedback effect of the collector capacitance C_c through C_{c0} and C_{ci} is sufficiently high when load variations are directly transferred to the device input with a significant extent.

The more complicated models, such as VBIC, HICUM, or MEXTRAM, include the effects of self-heating of a bipolar transistor, take into account the parasitic *p-n-p* transistor formed by the base, collector, and substrate regions, provide an improved description of depletion capacitances at large forward bias, and take into account avalanche and tunneling currents and other nonlinear effects corresponding to distributed high-frequency effects [60, 62]. The most common implementation of a lumped equivalent thermal network, describing the HBT self-heating behavior, represents a single- or two-section low-pass *RC* network with thermal resistances and capacitances.

References

1. G. Massobrio and P. Antognetti, *Semiconductor Device Modeling with SPICE*, New York: McGraw-Hill, 1993.
2. Y. P. Tsividis and K. Suyama, "MOSFET Modeling for Analog Circuit CAD: Problems and Prospects," *IEEE J. Solid-State Circuits*, vol. SC-29, pp. 210–216, Mar. 1994.
3. E. Abou-Allam and T. Manku, "A Small-Signal MOSFET Model for Radio Frequency IC Applications," *IEEE Trans. Computer-Aided Design*, vol. 16, pp. 437–447, May 1997.
4. F. Lin and G. Kompa, "FET Model Parameter Extraction Based on Optimization with Multiplane Data-Fitting and Bidirectional Search—A New Concept," *IEEE Trans. Microwave Theory Tech.*, vol. MTT-42, pp. 1114–1121, Jul. 1994.
5. T. Taki, "Approximation of Junction Field-Effect Transistor Characteristics by a Hyperbolic Function," *IEEE J. Solid-State Circuits*, vol. SC-13, pp. 724–726, Oct. 1978.
6. D. Heo, E. Chen, E. Gebara, S. Yoo, J. Lascar, and T. Anderson, "Temperature Dependent MOSFET RF Large Signal Model Incorporating Self Heating Effects," *1999 IEEE MTT-S Int. Microwave Symp. Dig.*, vol. 2, pp. 415–418.
7. W. R. Curtice, J. A. Pla, D. Bridges, T. Liang, and E. E. Shumate, "A New Dynamic Electro-Thermal Nonlinear Model for Silicon RF LDMOS FETs," *1999 IEEE MTT-S Int. Microwave Symp. Dig.*, vol. 2, pp. 419–422.
8. C. C. Enz and E. A. Vittoz, *Charge-Based MOS Transistor Modeling: The EKV Model for Low Power and RF IC Design*, Chichester: John Wiley & Sons, 2006.
9. Y. C. Chen, D. L. Ingram, H. C. Yen, R. Lai, and D. C. Streit, "A New Empirical *I–V* Model for HEMT Devices," *IEEE Microwave and Guided Wave Lett.*, vol. 8, pp. 342–344, Oct. 1998.
10. A. V. Grebennikov and F. Lin, "An Efficient CAD-Oriented Large-Signal MOSFET Model," *IEEE Trans. Microwave Theory Tech.*, vol. MTT-48, pp. 1732–1742, Oct. 2000.
11. P. Perugupalli, M. Trivedi, K. Shenai, and S. K. Leong, "Modeling and Characterization of an 80 V Silicon LDMOSFET for Emerging RFIC Applications," *IEEE Trans. Electron Devices*, vol. ED-45, pp. 1468–1478, Jul. 1998.
12. M. C. Ho, K. Green, R. Culbertson, J. Y. Yang, D. Ladwig, and P. Ehnis, "A Physical Large Signal Si MOSFET Model for RF Circuit Design," *1997 IEEE MTT-S Int. Microwave Symp. Dig.*, vol. 2, pp. 391–394.
13. J. Angel Garcia, A. Mediavilla, J. Carlos Pedro, N. B. Carvalho, A. Tazon, and J. L. Garcia, "Characterizing the Gate to Source Nonlinear Capacitor Role on FET IMD Performance," *1998 IEEE MTT-S Int. Microwave Symp. Dig.*, vol. 3, pp. 1635–1637.
14. D. Moncoqut, D. Farenc, P. Rossel, G. Charitat, H. Tranduc, J. Victory, and I. Pages, "LDMOS Transistor for Smart Power Circuits: Modeling and Design," *Proc. 1996 IEEE Bipolar/BiCMOS Circuits and Technology Meeting (BCTM)*, pp. 216–219.
15. C. E. Biber, M. L. Schmatz, T. Morf, U. Lott, and W. Bachtold, "A Nonlinear Microwave MOSFET Model for SPICE Simulators," *IEEE Trans. Microwave Theory Tech.*, vol. MTT-46, pp. 604–610, May 1998.
16. S. Lai, C. Fager, D. Kuylenstierna, and I. Angelov, "LDMOS Modeling," *IEEE Microwave Mag.*, vol. 14, pp. 108–116, Jan./Feb. 2013.
17. R. Sung, P. Bendix, and M. B. Das, "Extraction of High-Frequency Equivalent Circuit Parameters of Submicron Gate-Length MOSFET's," *IEEE Trans. Electron Devices*, vol. ED-45, pp. 1769–1775, Aug. 1998.
18. B. J. Cheu and P. K. Ko, "Measurement and Modeling of Short-Channel MOS Transistor Gate Capacitances," *IEEE J. Solid-State Circuits*, vol. SC-22, pp. 464–472, Jun. 1987.
19. R. E. Anholt and S. E. Swirhun, "Experimental Investigations of the Temperature Dependence of GaAs FET Equivalent Circuits," *IEEE Trans. Electron Devices*, vol. ED-39, pp. 2029–2036, Sep. 1992.

20. F. Ellinger, J. Kucera, and W. Baechtold, "Improvements on a GaAs MESFET Model for Nonlinear RF Simulations," *1998 IEEE MTT-S Int. Microwave Symp. Dig.*, vol. 3, pp. 1623–1626.
21. J.-M. Dortu, J.-E. Muller, M. Pirola, and G. Ghione, "Accurate Large-Signal GaAs MESFET and HEMT Modeling for Power MMIC Amplifier Design," *Int. J. Microwave and Millimeter-Wave Computer-Aided Eng.*, vol. 5, pp. 195–208, Sep. 1995.
22. I. Angelov, L. Bengtsson, and M. Garcia, "Extensions of the Chalmers Nonlinear HEMT and MESFET Model," *IEEE Trans. Microwave Theory Tech.*, vol. MTT-44, pp. 1664–1674, Oct. 1996.
23. C-J. Wei, Y. Tkachenko, and D. Bartle, "An Accurate Large-Signal Model of GaAs MESFET Which Accounts for Charge Conservation, Dispersion, and Self-Heating," *IEEE Trans. Microwave Theory Tech.*, vol. MTT-46, pp. 1638–1644, Nov. 1998.
24. A. Jarndal and G. Kompa, "Large-Signal Model for AlGaN/GaN HEMTs Accurately Predicts Trapping- and Self-Heating-Induced Dispersion and Intermodulation Distortion," *IEEE Trans. Electron Devices*, vol. ED-54, pp. 2830–2836, Nov. 2007.
25. K. Joshin and T. Kikkawa, "High-Power and High-Efficiency GaN HEMT Amplifiers," *2008 IEEE Radio and Wireless Symp. Dig.*, pp. 65–68.
26. T. Ishida, "GaN HEMT Technologies for Space and Radio Applications," *Microwave J.*, vol. 54, pp. 56–66, Aug. 2011.
27. U. K. Mishra, P. Parikh, and Y.-F. Wu, "AlGan/GaN HEMT—An Overview of Device Operation and Applications," *Proc. IEEE*, vol. 90, pp. 1022–1031, Jun. 2002.
28. D. W. Runton, B. Trabert, J. B. Shealy, and R. Vetury, "History of GaN," *IEEE Microwave Mag.*, vol. 14, pp. 82–93, May 2013.
29. R. S. Pengelly, S. M. Wood, J. W. Milligan, S. T. Sheppard, and W. L. Pribble, "A Review of GaN on SiC High Electron-Mobility Power Transistors and MMICs," *IEEE Trans. Microwave Theory Tech.*, vol. MTT-60, pp. 1764–1783, Jun. 2012.
30. I. Angelov, M. Thorsell, K. Andersson, N. Rorsman, E. Kuwata, H. Ohtsuka, and K. Yamanaka, "On the Large-Signal Modeling of High Power AlGaN/GaN HEMTs," *2012 IEEE MTT-S Int. Microwave Symp. Dig.*, pp. 1–3.
31. M. Berroth and R. Bosch, "Broad-Band Determination of the FET Small-Signal Equivalent Circuit," *IEEE Trans. Microwave Theory Tech.*, vol. MTT-38, pp. 891–895, Jul. 1990.
32. G. Dambrine, A. Cappy, F. Heliodore, and E. Playez, "A New Method for Determining the FET Small-Signal Equivalent Circuit," *IEEE Trans. Microwave Theory Tech.*, vol. MTT-36, pp. 1151–1159, Jul. 1988.
33. W. R. Curtice, "A MESFET Model for Use in the Design of GaAs Integrated Circuits," *IEEE Trans. Microwave Theory Tech.*, vol. MTT-25, pp. 448–456, May 1980.
34. T. Kacprzak and A. Materka, "Compact DC Model of GaAs FET's for Large-Signal Computer Calculation," *IEEE J. Solid-State Circuits*, vol. SC-18, pp. 211–213, Apr. 1983.
35. A. Materka and T. Kacprzak, "Computer Calculation of Large-Signal GaAs FET Amplifier Characteristics," *IEEE Trans. Microwave Theory Tech.*, vol. MTT-33, pp. 129–135, Feb. 1985.
36. I. Angelov, H. Zirath, and N. Rorsman, "A New Empirical Nonlinear Model for HEMT and MESFET Devices," *IEEE Trans. Microwave Theory Tech.*, vol. MTT-40, pp. 2258–2266, Dec. 1992.
37. I. Angelov, H. Zirath, and N. Rorsman, "Validation of a Nonlinear Transistor Model by Power Spectrum Characteristics of HEMT's and MESFET's," *IEEE Trans. Microwave Theory Tech.*, vol. MTT-43, pp. 1046–1051, May 1995.
38. O. S. A. Tang, K. H. G. Duh, S. M. J. Liu, P. M. Smith, W. F. Kopp, T. J. Rogers, and D. J. Pritchard, "Design of High-Power, High-Efficiency 60-GHz MMIC's Using an Improved Nonlinear PHEMT Model," *IEEE J. Solid-State Circuits*, vol. SC-32, pp. 1326–1333, Sep. 1997.
39. I. Angelov, V. Desmaris, K. Dynefors, P. A. Nilsson, N. Rorsman, and H. Zirath, "On the Large-Signal Modelling of AlGaN/GaN HEMTs and SiC MESFETs," *13th Europ. GAAS Symp. Dig.*, pp. 309–312, 2005.

40. A. Garcia-Osorio, J. R. Loo-Yau, J. A. Reynoso-Hernandez, S. Ortega, and J. L. del Valle-Padilla, "An Empirical *I-V* Nonlinear Model Suitable for GaN FET Class F PA Design," *Microwave and Optical Technology Lett.*, vol. 53, pp. 1256–1259, Jun. 2011.
41. R. Osorio, M. Berroth, W. Marsetz, L. Verweyen, M. Demmler, H. Massler, M. Neumann, and M. Schlechtweg, "Analytical Charge Conservative Large Signal Model for MODFETs Validated up to MM-Wave Range," *1998 IEEE MTT-S Int. Microwave Symp. Dig.*, vol. 2, pp. 595–598.
42. D. Costa, W. U. Liu, and J. S. Harris, "Direct Extraction of the AlGaAs/GaAs Heterojunction Bipolar Transistor Small-Signal Equivalent Circuit" *IEEE Trans. Electron Devices*, vol. ED-38, pp. 2018–2024, Sep. 1991.
43. H. Ghaddab, F. M. Ghannouchi, and F. Bouallegue, "Small-Signal Modeling of HBT's Using a Hybrid Optimization/Statistical Technique," *IEEE Trans. Microwave Theory Tech.*, vol. MTT-46, pp. 292–298, Mar. 1998.
44. J. M. Early, "Effects of Space-Charge Layer Widening in Junction Transistors," *Proc. IRE*, vol. 40, pp. 1401–1406, Nov. 1952.
45. D. A. Teeter and W. R. Curtice, "Comparison of Hybrid Pi and Tee HBT Circuit Topologies and Their Relationship to Large Signal Modeling," *1997 IEEE MTT-S Int. Microwave Symp. Dig.*, vol. 2, pp. 375–378.
46. P. M. Asbeck, M. F. Chang, K.-C. Wang, D. L. Miller, G. J. Sullivan, N. H. Sheng, E. A. Sovero, and J. A. Higgins, "Heterojunction Bipolar Transistors for Microwave and Millimeter-Wave Integrated Circuits," *IEEE Trans. Microwave Theory Tech.*, vol. MTT-35, pp. 1462–1468, Dec. 1987.
47. Y.-S. Lin and J.-J. Jiang, "Temperature Dependence of Current Gain, Ideality Factor, and Offset Voltage of AlGaAs/GaAs and InGaP/GaAs HBTs," *IEEE Trans. Electron Devices*, vol. ED-56, pp. 2945–2951, Dec. 2009.
48. N. L. Wang, W. Ma, S. Xu, E. Camargo, X. Sun, P. Hu, Z. Tang, H. F. Chau, A. Chen, and C. P Lee, "28-V High-Linearity and Rugged InGaP/GaAs HBT," *2006 IEEE MTT-S Int. Microwave Symp. Dig.*, pp. 881–884.
49. C. Steinberger, T. Landon, C. Suckling, J. Nelson, J. Delaney, J. Hitt, l. Witkowski, G. Burgin, R. Hajji, and O. Krutko, "250 W HVHBT Doherty with 57% WCDMA Efficiency Linearized to –55 dBc for 2c11 6.5 dB PAR," *IEEE J. Solid-State Circuits*, vol. SC-43, pp. 2218–2228, Oct. 2008.
50. V. M. Bogachev and V. V. Nikiforov, *Transistor Power Amplifiers* (in Russian), Moskva: Energiya, 1978.
51. A. Samelis and D. Pavlidis, "DC to High-Frequency HBT-Model Parameter Evaluation Using Impedance Block Conditioned Optimization," *IEEE Trans. Microwave Theory Tech.*, vol. MTT-45, pp. 886–897, Jun. 1997.
52. B. Li, F. S. Prasad, L.-W. Yang, and S. C. Wang, "A Semianalytical Parameter-Extraction Procedure for HBT Equivalent Circuit," *IEEE Trans. Microwave Theory Tech.*, vol. MTT-46, pp. 1427–1435, Oct. 1998.
53. U. Schaper and B. Holzapfl, "Analytical Parameter Extraction of the HBT Equivalent Circuit with *T*-like Topology from Measured *S*-parameters," *IEEE Trans. Microwave Theory Tech.*, vol. MTT-43, pp. 493–498, Mar. 1995.
54. J. J. Ebers and J. L. Moll, "Large-Signal Behavior of Junction Transistors," *Proc. IRE*, vol. 42, pp. 1761–1772, Dec. 1954.
55. W. M. Webster, "On the Variation of Junction-Transistor Current Amplification Factor with Emitter Current," *Proc. IRE*, vol. 42, pp. 914–920, Jun. 1954.
56. C. T. Kirk, "A Theory of Transistor Cutoff Frequency (fT) Falloff at High Current Densities," *IEEE Trans. Electron Devices*, vol. ED-9, pp. 164–174, Mar. 1962.
57. N. M. Rohringer and P. Kreuzgruber, "Parameter Extraction for Large-Signal Modeling of Bipolar Junction Transistors," *Int. J. Microwave and Millimeter-Wave Computer-Aided Eng.*, vol. 5, pp. 161–272, Sep. 1995.
58. J. P. Fraysee, D. Floriot, P. Auxemery, M. Campovecchio, R. Quere, and J. Obregon, "A Non-Quasi-Static Model of GaInP/AlGaAs HBT for Power Applications," *1997 IEEE MTT-S Int. Microwave Symp. Dig.*, vol. 2, pp. 377–382.
59. H. K. Gummel and H. C. Poon, "An Integral Charge Control Model of Bipolar Transistors," *Bell Syst. Tech. J.*, vol. 49, pp. 827–852, May/Jun. 1970.

60. J. McMacken, S. Nedeljkovich, J. Gering, and D. Halchin, "HBT Modeling," *IEEE Microwave Mag.*, vol. 9, pp. 48–72, Apr. 2008.
61. I. Angelov, K. Choumei, and A. Inoue, "An Empirical HBT Large-Signal Model for CAD," *Int. J. RF and Microwave Computer-Aided Eng.*, vol. 13, pp. 518–533, Nov. 2003.
62. M. Reisch, *High-Frequency Bipolar Transistors*, Berlin: Springer, 2003.

CHAPTER 4

Impedance Matching

This chapter begins by describing the main principles and impedance matching tools. Generally, an optimum matching solution depends on the circuit requirement, such as the simplicity in practical realization, the frequency bandwidth and minimum power ripple, design implementation and adjustability, stable operation conditions, and sufficient harmonic suppression. As a result, many types of the matching networks are available, including lumped elements and transmission lines. To simplify and visualize the matching design procedure, an analytical approach, which allows calculation of the parameters of the matching circuits using simple equations, and Smith chart traces are discussed. In addition, several examples of the narrow-band and broadband power amplifiers using bipolar or MOSFET devices are given, including detailed design considerations and explanations. Finally, design formulas and curves are given for several types of transmission lines, such as stripline, microstrip line, slotline and coplanar waveguide (CPW).

4.1 Main Principles

Impedance matching is necessary to provide maximum delivery to the load of the RF power available from the source. This means that, when the electrical signal propagates in the circuit, a portion of this signal will be reflected at the interface between the sections with different impedances. Therefore, it is necessary to establish the conditions, which allow to fully transmitting the entire electrical signal without any reflection. To determine an optimum value of the load impedance Z_L, at which the power delivered to the load is maximal, consider the equivalent circuit shown in Fig. 4.1(*a*).

The power delivered to the load can be defined as

$$P = \frac{1}{2} V_{\text{in}}^2 \,\text{Re}\left(\frac{1}{Z_L}\right) = \frac{1}{2} V_S^2 \left| \frac{Z_L}{Z_S + Z_L} \right|^2 \text{Re}\left(\frac{1}{Z_L}\right) \qquad (4.1)$$

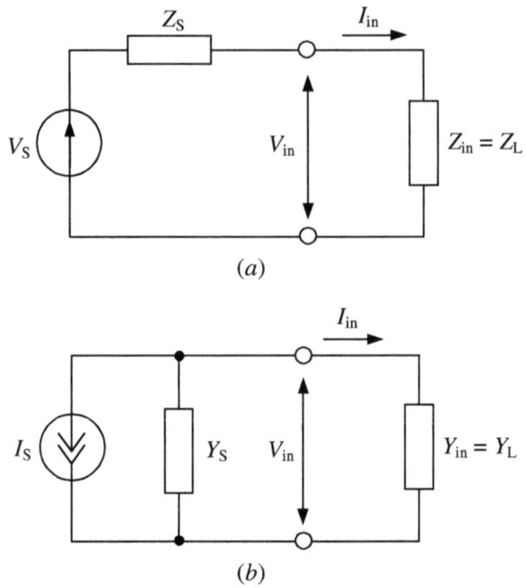

FIGURE 4.1 Equivalent circuits with (*a*) voltage and (*b*) current sources.

where $Z_S = R_S + jX_S$ is the source impedance, $Z_L = R_L + jX_L$ is the load impedance, V_S is the source voltage amplitude, and V_{in} is the load voltage amplitude. Substituting the real and imaginary parts of the source and load impedances Z_S and Z_L into Eq. (4.1) yields

$$P = \frac{1}{2} V_S^2 \frac{R_L}{(R_S + R_L)^2 + (X_S + X_L)^2} \tag{4.2}$$

Assume the source impedance Z_S is fixed and it is necessary to vary the real and imaginary parts of the load impedance Z_L until maximum power is delivered to the load. To maximize the output power, the following analytical conditions in the form of derivatives with respect to the output power are necessary to apply:

$$\frac{\partial P}{\partial R_L} = 0 \qquad \frac{\partial P}{\partial X_L} = 0 \tag{4.3}$$

Then, applying these conditions to Eq. (4.2), the following system of two equations can be obtained:

$$\frac{1}{(R_L + R_S)^2 + (X_L + X_S)^2} - \frac{2R_L(R_L + R_S)}{[(R_L + R_S)^2 + (X_L + X_S)^2]^2} = 0 \tag{4.4}$$

$$\frac{2X_L(X_L + X_S)}{[(R_L + R_S)^2 + (X_L + X_S)^2]^2} = 0 \tag{4.5}$$

Simplifying Eqs. (4.4) and (4.5) results in

$$R_S^2 - R_L^2 + (X_L + X_S)^2 = 0 \tag{4.6}$$

$$X_L(X_L + X_S) = 0 \tag{4.7}$$

By solving Eqs. (4.6) and (4.7) simultaneously for R_S and X_S, one can obtain

$$R_S = R_L \tag{4.8}$$

$$X_L = -X_S \tag{4.9}$$

or, in an impedance form,

$$Z_L = Z_S^* \tag{4.10}$$

where * denotes the complex-conjugate value [1, 2].

Equation (4.10) is called the *impedance conjugate matching condition*, and its fulfillment results in a maximum power delivered to the load for fixed source impedance. It should be noted that the term "impedance" was introduced by Oliver Lodge in 1889, and it meant the ratio V/I in the special circuit composed of a resistance and an inductance, where I and V are the amplitudes of an alternating current and the driving force which produced it [3].

The maximum power delivered to the load must be equal to

$$P = \frac{V_S^2}{8R_S} \tag{4.11}$$

The *admittance conjugate matching condition*, applied to the equivalent circuit shown in Fig. 4.1(*b*), is written as

$$Y_L = Y_S^* \tag{4.12}$$

which can be readily obtained in the same way. In this case, the maximum power delivered to the load can be defined by

$$P = \frac{I_S^2}{8G_S} \tag{4.13}$$

where $G_S = \mathrm{Re}Y_S$ is the source conductance and I_S is the source current amplitude.

Thus, the conjugate matching conditions in a common case can be determined through the immittance *W*-parameters, which represent any system of the impedance *Z*-parameters or admittance *Y*-parameters, in the form of

$$W_L = W_S^* \tag{4.14}$$

where W_S is the source immittance and W_L is the load immittance. The term "immittance" was introduced by Bode to refer to a complex number, which may be either the impedance or the admittance of a system [4].

For a single-stage power amplifier, the matching circuit is connected between the source and the input of an active device, as shown in Fig. 4.2(*a*), and between the output of an active device and the load, as shown in Fig. 4.2(*b*). For a multistage power amplifier, the load represents an input circuit of the next stage. Therefore, the matching circuit is connected between the output of the active device of the preceding amplifier stage and the input of the active device of the succeeding stage of the power amplifier, as shown in Fig. 4.2(*c*). The main objective is to properly transform the load immittance W_L to the optimum device output immittance W_{out}, the value of which is properly determined by the supply voltage, the output power, the saturation voltage of the active device, and the selected class of the active device operation to maximize the operating efficiency and output power of the power amplifier. It should be

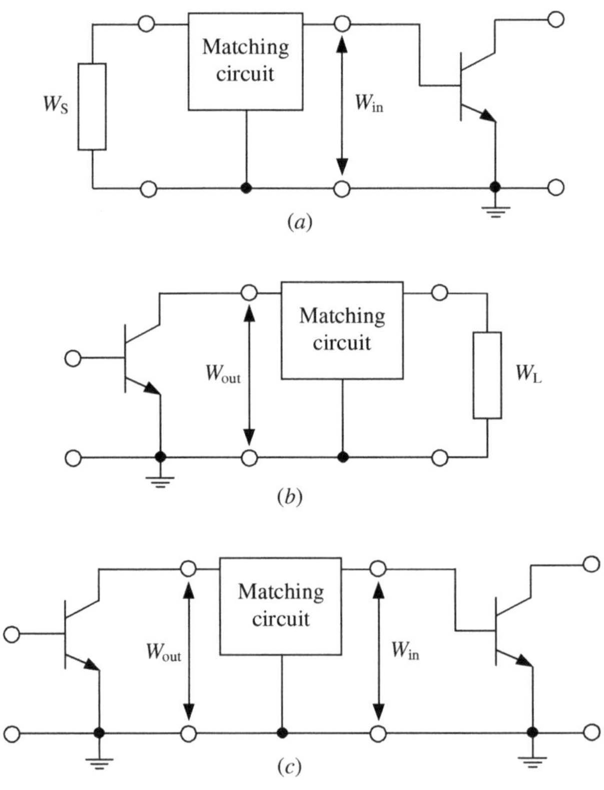

FIGURE 4.2 Matching circuit arrangements.

noted that Eq. (4.14) is given in a general immittance form without indication of whether it is used in a small-signal or large-signal application. In the latter case, this only means that the device immittance *W*-parameters are fundamentally averaged over large-signal swing across the device-equivalent circuit parameters and that the conjugate-matching principle is valid in both the small-signal application and the large-signal application where the optimum equivalent device output resistance (or conductance) at the fundamental is matched to the load resistance (or conductance) and the effect of the device output reactive elements is eliminated by the conjugate reactance of the load network. In addition, the matching circuits should be designed to realize the required voltage and current waveforms at the device output, to provide the stability of operation conditions, and to satisfy the requirements for the power amplifier amplitude and phase characteristics. The losses in the matching circuits must be as small as possible to deliver the output power to the load with maximum efficiency. Finally, it is desirable that the matching circuit be easy to tune.

4.2 Smith Chart

The hemisphere Smith chart is one of the tools most widely used to match circuit designs because it gives a clear and simple graphical representation of the consecutive matching design procedure [5, 6]. However, another hemisphere chart with a standing-wave indicator to measure the impedance ratio of a load on line where a grid of circular lines was marked with magnitude and angle of impedance, corresponding to latitude and longitude on a hemisphere, was simultaneously proposed by Carter [7, 8]. It should be mentioned that, more than one year prior to Smith's publication in 1939, a nomograph equivalent to the normalized impedance Smith chart was presented by a Japanese engineer Mizuhashi as a tool for graphically calculating impedances [9]. Generally, the Smith chart can be directly applied for matching procedure using both lumped elements and transmission lines. The Smith chart is particularly useful for matching circuit designs that use the transmission lines because analytical calculations in this case are very complicated. Also, when using the complete Smith chart, the circuit parameters such as voltage standing wave ratio (*VSWR*), reflection coefficient, return loss, or losses in the transmission line can be directly calculated. The Smith chart is a very important tool, being a part of modern computer-aided design software and test equipment and providing a useful visual way to understand the circuit behavior.

The Smith chart represents a relationship between the load impedance Z and the reflection coefficient Γ, which can be written in the normalized form of

$$\frac{Z}{Z_0} = \frac{1+\Gamma}{1-\Gamma} \tag{4.15}$$

where the normalized impedance Z/Z_0 and the reflection coefficient Γ can be written through their real and imaginary parts, respectively, as

$$\frac{Z}{Z_0} = \frac{R}{Z_0} + j\frac{X}{Z_0} \tag{4.16}$$

$$\Gamma = \Gamma_r + j\Gamma_i \tag{4.17}$$

Then, substituting Eqs. (4.16) and (4.17) into Eq. (4.15) results in

$$\frac{R}{Z_0} + j\frac{X}{Z_0} = \frac{1+\Gamma_r + j\Gamma_i}{1-\Gamma_r - j\Gamma_i} \tag{4.18}$$

By equating the real and imaginary parts of Eq. (4.18), one can obtain

$$\left(\Gamma_r - \frac{R}{R+Z_0}\right)^2 + \Gamma_i^2 = \left(\frac{Z_0}{R+Z_0}\right)^2 \tag{4.19}$$

$$(\Gamma_r - 1)^2 + \left(\Gamma_i - \frac{Z_0}{X}\right)^2 = \left(\frac{Z_0}{X}\right)^2 \tag{4.20}$$

As a result, in the (Γ_r, Γ_i) coordinate plane, Eq. (4.19) represents a family of circles centered at points $\Gamma_r = R/(R+Z_0)$ and $\Gamma_i = 0$ with radii of $Z_0/(R+Z_0)$, which are called the *constant-*(R/Z_0) *circles*. At the same time, Eq. (4.20) represents a family of circles at points $\Gamma_r = 1$ and $\Gamma_i = Z_0/X$ with radii of Z_0/X, which are called the *constant-*(X/Z_0) *circles*. These constant-(R/Z_0) and constant-(X/Z_0) circles with different normalized parameters are shown in Fig. 4.3(*a*), where the points $\Gamma_r = -1$ and $\Gamma_r = 1$ are also indicated. The plot of such circles is called the *impedance Smith chart* (or the *Z Smith chart*). The curve from the point *A* to the point *C* represents the impedance transformation from the pure resistance of 25 Ω to the inductive impedance of (25 + *j*25) Ω, which can be provided by using the inductance connected in series with the resistance.

Equations (4.19) and (4.20) can be rewritten in the admittance form with the real part *G* and imaginary part *B* when a relationship between the impedance *Y* and the reflection coefficient Γ can be written as

$$\frac{G}{Y_0} + j\frac{B}{Y_0} = \frac{1-\Gamma_r - j\Gamma_i}{1+\Gamma_r + j\Gamma_i} \tag{4.21}$$

where $Y_0 = 1/Z_0$.

Then,

$$\left(\Gamma_r + \frac{G}{G+Y_0}\right)^2 + \Gamma_i^2 = \left(\frac{Y_0}{G+Y_0}\right)^2 \tag{4.22}$$

$$(\Gamma_r + 1)^2 + \left(\Gamma_i + \frac{Y_0}{B}\right)^2 = \left(\frac{Y_0}{B}\right)^2 \tag{4.23}$$

From Eqs. (4.22) and (4.23), it follows that the constant-(G/Y_0) circles are centered at $\Gamma_r = -G/(G+Y_0)$ and $\Gamma_i = 0$ with radii of $Y_0/(G+Y_0)$. The constant-(B/Y_0) circles are centered at points $\Gamma_r = -1$ and $\Gamma_i = -Y_0/B$ with radii of Y_0/B, which is shown in Fig. 4.3(*b*). These circles are centered at anti-symmetric points in contrast to the impedance Smith chart. The *admittance Smith chart* (or the *Y Smith chart*), the admittances of which coincide with the appropriate impedances plotted at the *Z* Smith chart, is the mirror-reflected impedance Smith chart as a result of its rotation by 180°. The curve from the point *C* to the point *D* shows the admittance transformation from the inductive admittance of (20 – *j*20) mS to the pure conductance of 20 mS (or resistance of 50 Ω), which can be provided by using the capacitance connected in parallel with the initial admittance. The impedances and admittances can be determined by any impedance or admittance Smith chart where the normalized parameters are indicated. In a generalized case, this diagram is called *the immittance Smith chart* [10]. However, in this case, the impedance point and its corresponding admittance value are located one against another at the same distance from center (1, 0).

Therefore, in some cases it is advisable to use the combined impedance-admittance Smith chart shown in Fig. 4.3(*c*) when, for any point, there exists the normalized impedance from the *Z* Smith chart and normalized admittance from the *Y* Smith chart. This *Z*-*Y* Smith chart avoids the necessity of impedance rotating by 180° to find the corresponding admittance. A combined impedance-admittance Smith chart is very convenient for matching using lumped elements. For example, it is necessary to convert the source active impedance of 25 Ω (point *A*) into the load resistance of 50 Ω (point *D*). First, the series inductance plotted at the *Z* Smith chart changes the source impedance by moving along constant-(X/Z_0) circle from point *A* as far as point *C*. Then, it is converted to the *Y* Smith chart. As a result, the parallel capacitance starting at this point changes the given admittance by moving along constant-(B/Y_0) circle from point *C* as far as point *D*. Such a transformation between these two resistances for normalizing impedance $Z_0 = 50\ \Omega$ is shown in Fig. 4.3(*c*).

When designing RF and microwave power amplifiers, it is very important to determine such parameters as *VSWR*, reflection coefficient, or losses in the matching circuits based on the lumped elements or using the transmission lines. The linear reference scales, indicating these additional characteristics, are added underneath the Smith chart, as shown in Fig. 4.4. Scales around its periphery show the calibrated electrical wavelength and the angles of the reflection coefficients. The Smith chart can be easily implemented for the graphical design using the transmission lines. Expression for the input impedance of the lossless transmission line can be written in terms of the reflection coefficient as

$$\frac{Z_{in}}{Z_0} = \frac{1 + \Gamma \exp(-j2\theta)}{1 - \Gamma \exp(-j2\theta)} \tag{4.24}$$

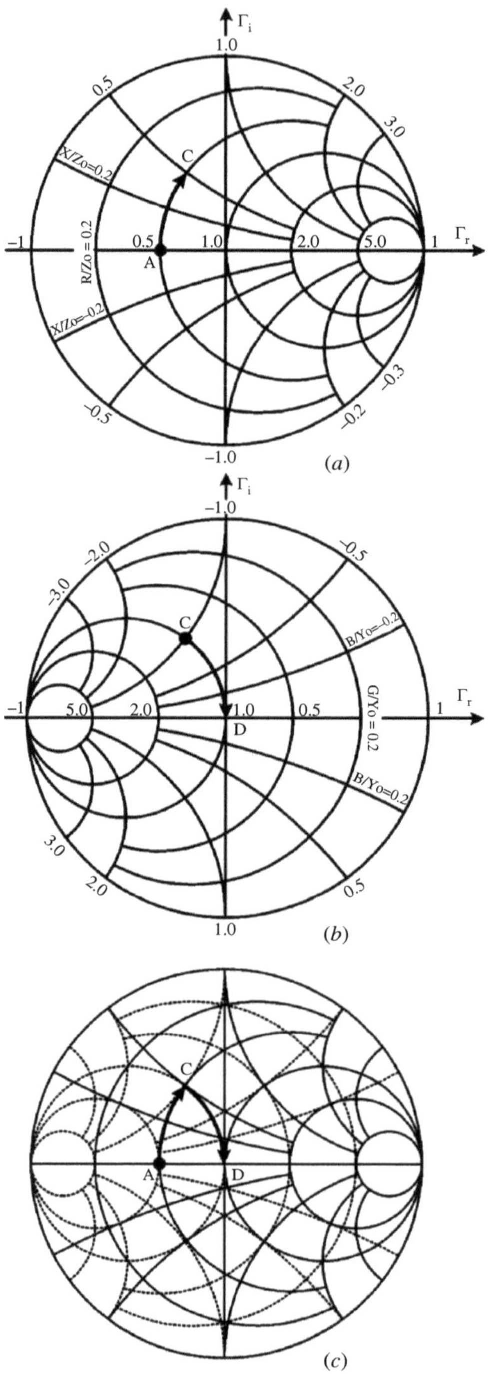

FIGURE 4.3 Simplified impedance and admittance Smith charts.

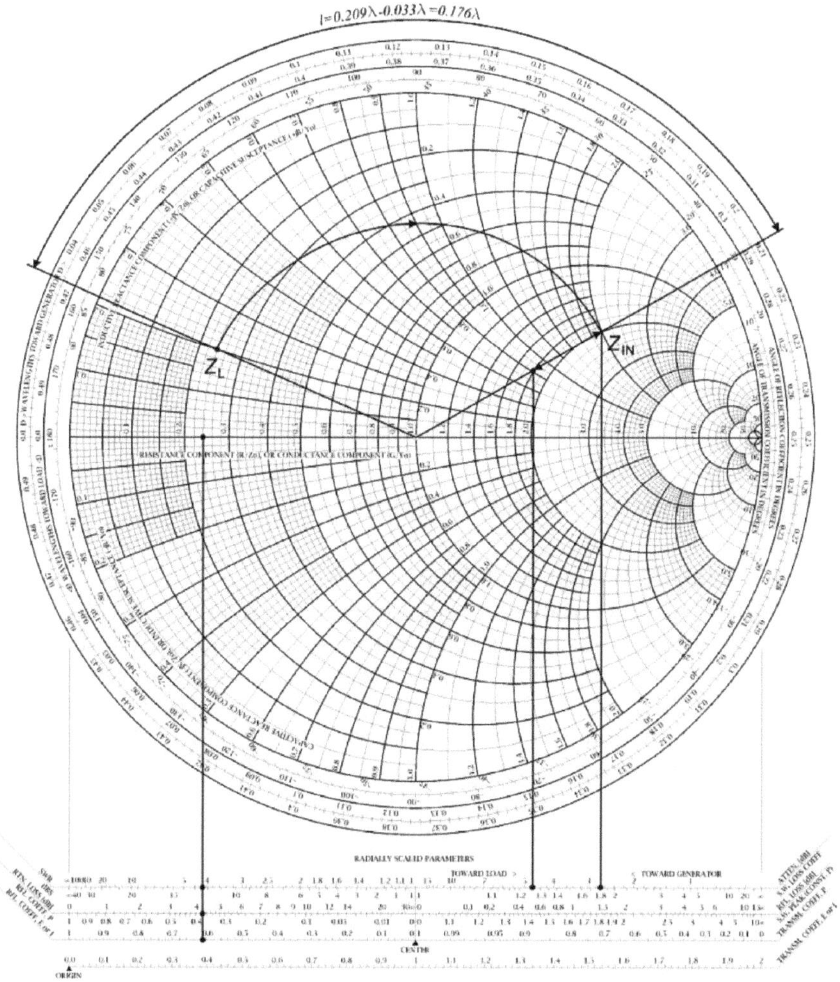

FIGURE 4.4 Smith chart.

This equation differs from Eq. (4.15) by the added phase angle only. This means that the normalized input impedance, seen looking into the transmission line with electrical length θ, can be found by rotating this impedance point clockwise by 2θ around the center of the Smith chart with the same radius $|\Gamma|$. By subtracting 2θ from the phase angle of the reflection coefficient, its value decreases in the clockwise direction according to the periphery scale. In this case, the half-wave transmission line provides a clockwise rotation of 2π or 360° around the center, thus returning the point to its original position.

For example, a typical 50-Ω transmission line provides a transformation of the load impedance from $Z_L = (12 + j10)\ \Omega$ to a new impedance of Z_{in} = (100 + j100) Ω. The normalized load impedance is Z_L/Z_0 = (0.24 + j0.2) Ω, which is plotted on the Smith chart, as shown in Fig. 4.4. By using a compass to draw the circle from this point to the intersection point with a real axis, we obtain $|\Gamma| = 0.61$ at the reflection coefficient scale, RL = 4.3 dB at the return loss scale, and $VSWR$ = 4.2 at the standing wave ratio (SWR) scale. To determine the angle of the reflection coefficient, it needs to draw a radial line through the load impedance point to the intersection of the periphery circle (an angle of 83° can be read). If a radial line is drawn through the point of the input impedance at the outer wavelength scale, the difference between points of 0.209λ and 0.033λ gives the length of the transmission line as 0.176λ. However, in the case of the transmission line with the insertion loss of 2 dB, the point obtained should be moved to the point of Z_{in} = (90 + j40) Ω, according to the attenuation scale.

4.3 Matching with Lumped Elements

4.3.1 Analytic Design Technique

Generally, there is a variety of configurations for matching networks, which can be used to connect a generating system efficiently to its useful load. However, in order to obtain high transmission efficiency, any of these networks should be properly designed. The lumped matching circuits in the form of (*a*) *L*-transformer, (*b*) π-transformer, or (*c*) *T*-transformer shown in Fig. 4.5 have proved for a long time to

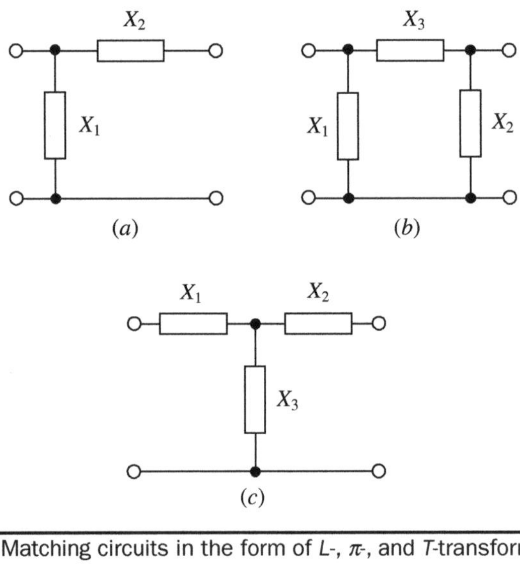

Figure 4.5 Matching circuits in the form of *L*-, π-, and *T*-transformers.

be effective for power amplifier design [11]. The simplest and most popular matching network is the matching circuit in the form of the *L*-transformer. The transforming properties of this matching circuit can be analyzed by using the equivalent transformation of a parallel into a series representation of *RX* circuit.

Consider the parallel *RX* circuit shown in Fig. 4.6(*a*), where R_1 is the real (resistive) and X_1 is the imaginary (reactive) parts of the circuit impedance $Z_1 = jX_1R_1/(R_1 + jX_1)$, and the series *RX* circuit shown in Fig. 4.6(*b*), where R_2 is the resistive and X_2 is the reactive parts of the circuit impedance $Z_2 = R_2 + jX_2$. These two circuits, series and parallel, can be considered equivalent at some frequency if $Z_1 = Z_2$, resulting in

$$R_2 + jX_2 = \frac{R_1X_1^2}{R_1^2 + X_1^2} + j\frac{R_1^2X_1}{R_1^2 + X_1^2} \tag{4.25}$$

Equation (4.25) can be rearranged by two separate equations for real and imaginary parts as

$$R_1 = R_2(1 + Q^2) \tag{4.26}$$

$$X_1 = X_2(1 + Q^{-2}) \tag{4.27}$$

where $Q = R_1/|X_1| = |X_2|/R_2$ is the circuit quality factor, which is equal for both the series and parallel *RX* circuits.

Consequently, if the reactive impedance $X_1 = -X_2(1 + Q^{-2})$ is connected in parallel to the series circuit R_2X_2, it allows the reactive impedance (or reactance) of the series circuit to be compensated. In this case, the input impedance of such a two-port network shown in Fig. 4.7 will be resistive only and equal to R_1. Consequently, to transform the resistance R_1 into the other resistance R_2 at the given frequency, it is sufficient to connect between them a two-port *L*-transformer with the opposite signs of the reactances X_1 and X_2, the parameters of which can be easily calculated from the following simple equations:

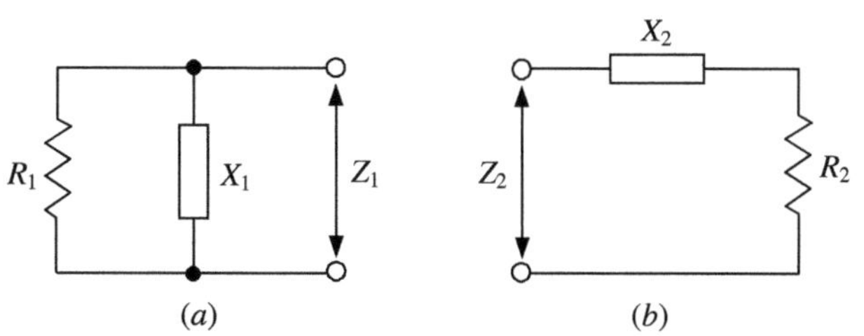

FIGURE 4.6 Impedance parallel and series equivalent circuits.

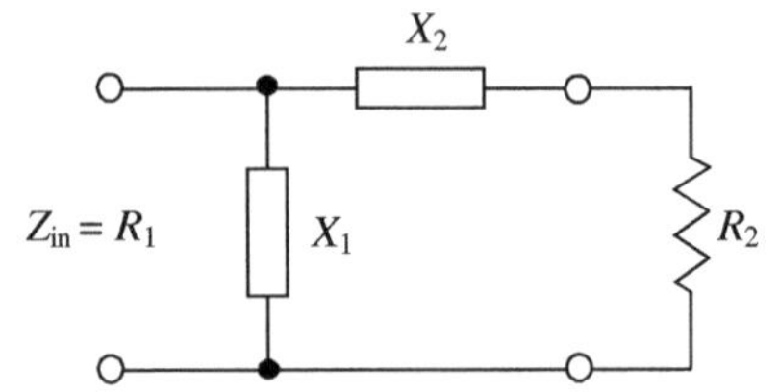

FIGURE 4.7 Input impedance of loaded two-port network.

$$|X_1| = \frac{R_1}{Q} \tag{4.28}$$

$$|X_2| = R_2 Q \tag{4.29}$$

where

$$Q = \sqrt{\frac{R_1}{R_2} - 1} \tag{4.30}$$

is the circuit (loaded) quality factor expressed through the resistances to be matched. Thus, to design a matching circuit with fixed resistances to be matched, first we need to calculate the circuit quality factor Q according to Eq. (4.30) and then to define the reactive elements according to Eqs. (4.28) and (4.29).

Because of the opposite signs of the reactances X_1 and X_2, the two possible circuit configurations (one in the form of a low-pass filter section and another in the form of a high-pass filter section) with the same transforming properties can be realized, as shown in Fig. 4.8 together with the design equations.

The lumped matching circuits in the form of the L-transformer loaded on the resistance R_2 can also be considered as some approximation of the parallel resonant circuit shown in Fig. 4.9. In this case, the series inductance L_2 and series resistance R_2 are transformed to

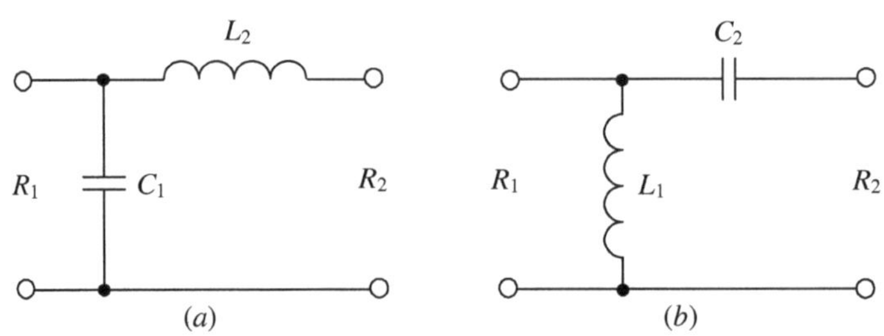

FIGURE 4.8 *L*-type matching circuits and relevant equations.

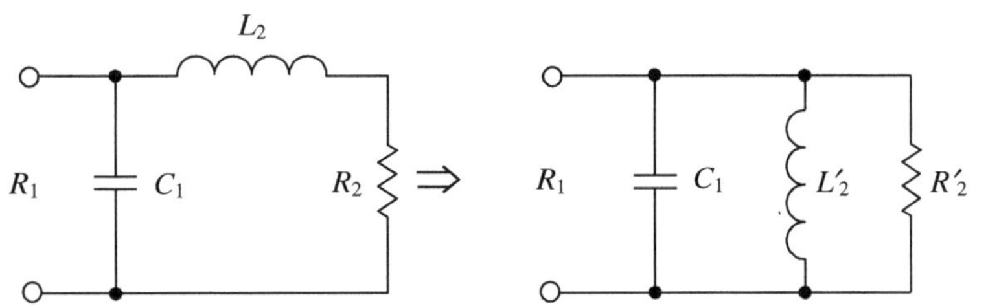

FIGURE 4.9 Parallel resonant circuit equivalent of loaded *L*-transformer.

the parallel inductance L_2' and parallel frequency-dependent resistance R_2', respectively, as

$$R_1 = R_2' = R_2(1+Q^2) \tag{4.31}$$

$$L_2' = L_2(1+Q^{-2}) \tag{4.32}$$

where $Q = \omega L_2/R_2$. The resonant frequency f_0 of such an equivalent parallel resonant circuit is determined from

$$\omega_0 = 2\pi f_0 = \sqrt{\frac{1}{L_2C_1} - \left(\frac{R_2}{L_2}\right)^2} \tag{4.33}$$

If this matching circuit has small values of Q, wider frequency bandwidth but poor out-of-band suppression can be achieved. However, with large values of Q, the frequency bandwidth is substantially reduced. For the case of $R_1/R_2 \geq 10$, which corresponds to the condition of $Q \geq 3$, the frequency bandwidth Δf and out-of-band suppression factor F_n of such an L-transformer can be approximately evaluated by the same formulas as for a parallel resonant circuit as

$$\Delta f = \frac{f_0}{Q} \tag{4.34}$$

$$F_n = Q(n^2 - 1) \tag{4.35}$$

where f_0 is the operating frequency and n is the harmonic number [12]. The out-of-band suppression factor F_n of the matching circuit represents the ratio of the selected harmonic component in the input (collector) current to the same harmonic component in the output (load) current. Figure 4.10 shows the frequency behavior of the input impedance magnitude $|Z_{in}|$ of the parallel resonant circuit.

The transformer efficiency η_T is determined by the ratio of P_L/P_{in}, where P_{in} is the power at the input of the transformer and P_L is the

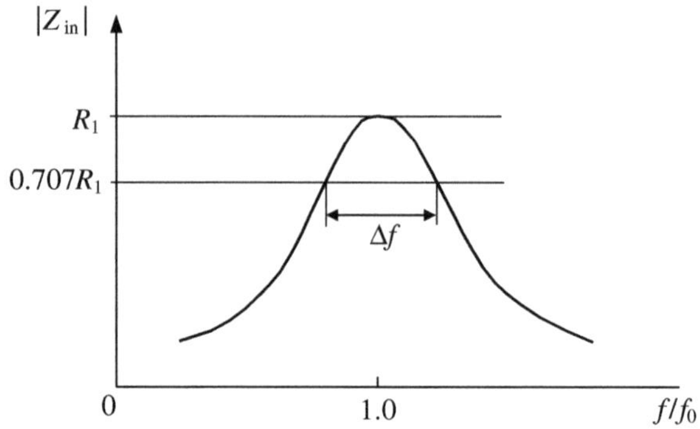

FIGURE 4.10 Frequency plot of input impedance for parallel resonant circuit.

transformer load power. The efficiency for the *L*-transformer with negligible losses in the capacitor can be calculated by

$$\eta_T = 1/\left(1+\frac{Q}{Q_{ind}}\right) \tag{4.36}$$

where Q_{ind} is the inductor quality factor [13]. From Eq. (4.36), it follows that the efficiency of the impedance transformer decreases with the increase in Q. This means that, for the same R_1 and series parasitic resistance of the circuit inductor, the lower resistance R_2 provides the higher current flowing through this inductor, which leads to an additional power dissipation. Similar result for the quality factor of the *L*-transformer with a lossy inductor derived through the insertion loss, and the more accurate expressions for the low-pass and high-pass *L*-type matching networks are given in [14]. An analysis of Eqs. (4.28) and (4.29) shows that, for the given resistances R_1 and R_2, each element of the *L*-transformer can have only one value at a fixed frequency. Consequently, it is difficult to satisfy simultaneously such contradictory requirements as efficiency, frequency bandwidth, and out-of-band suppression.

To avoid the parasitic low-frequency oscillations in the power amplifier and to increase the level of the harmonic suppression, it may be necessary to connect an additional L_0C_0 series circuit, the resonant frequency of which is equal to the operating frequency of the power amplifier, as shown in Fig. 4.11. In this case, the out-of-band suppression factor F_n defined by Eq. (4.35) is written as

$$F_n = Q_\Sigma(n^2 - 1) \tag{4.37}$$

where $Q_\Sigma = Q + Q_0$ and $Q_0 = \omega L_0/R_2$. Better harmonic suppression is achieved at the expense of the frequency bandwidth narrowing.

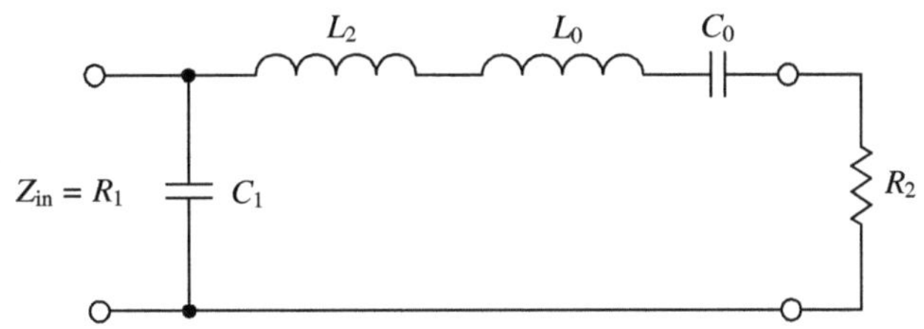

FIGURE 4.11 *L*-transformer with additional *LC* filter.

In practice, it makes sense to use the single two-port *L*-transformers in power amplifiers as the interstage matching circuits, where the requirements for out-of-band suppression and efficiency are not as high as for the output matching circuits. In this case, the main advantage of such an *L*-transformer is in its simplicity when only two reactive elements with fast tuning are needed. For larger values of $Q \geq 10$, it is possible to use a cascade connection of the *L*-transformers, which allows wider frequency bandwidth and transformer efficiency to be realized.

The matching circuits in the form of (*a*) π-transformer and (*b*) *T*-transformer can be realized by appropriate connection of two *L*-transformers, as shown in Fig. 4.12. For each *L*-transformer, the resistances R_1 and R_2 are transformed to some intermediate resistance R_0 with the value of $R_0 < (R_1, R_2)$ for a π-transformer and the value of $R_0 > (R_1, R_2)$ for a *T*-transformer. The value of R_0 is not fixed and can

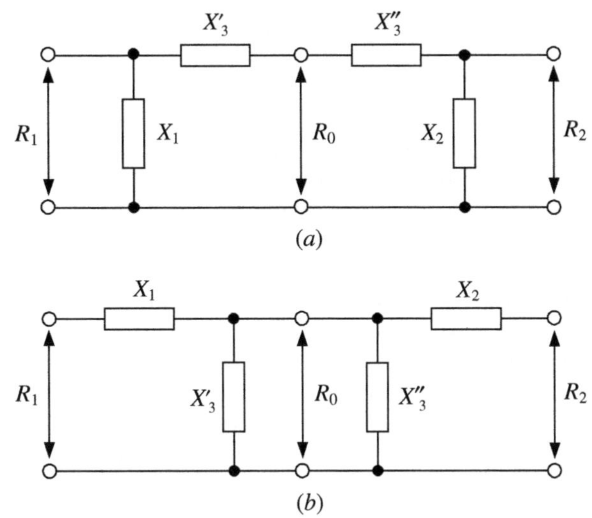

FIGUER 4.12 Matching circuits developed by connecting two *L*-transformers.

be chosen arbitrary depending on the frequency bandwidth. This means that, compared to the simple *L*-transformer with fixed parameters for the same ratio of R_2/R_1, the parameters of a π-transformer or a *T*-transformer can be different. However, they provide narrower frequency bandwidths due to higher quality factors because the intermediate resistance R_0 is either greater or smaller than each of the resistances R_1 and R_2. By taking into account the two possible circuit configurations of the *L*-transformer shown in Fig. 4.8, there is a possibility to design different circuit configurations of such two-port impedance transformers shown in Fig. 4.12(*a*), where $X_3 = X_3' + X_3''$, and in Fig. 4.12(*b*), where $X_3 = X_3'X_3''/(X_3' + X_3'')$.

Several of the most widely used two-port π-transformers, together with the design formulas, are shown in Fig. 4.13 [13]. The π-transformers are usually used as output matching circuits of high-power amplifiers in a Class-B operation mode when it is necessary to achieve a sinusoidal drain (or collector) voltage waveform by appropriate harmonic suppression. In addition, it is convenient to use some of them as interstage matching circuits in low-power and medium-power amplifiers when it is necessary to provide sinusoidal voltage waveforms both at the drain (or collector) of the preceding-stage transistor and at the gate (or base) of the succeeding-stage transistor. In this case, for the π-transformer with shunt capacitors, the input and output capacitances of these transistors can be easily included into the matching circuit elements C_1 and C_2, respectively. Finally, a π-transformer can be directly used as the load network for a high-efficiency Class-E mode with proper calculation of its design parameters.

Figure 4.14 shows the π-transformer with shunt capacitor and additional L_0C_0 series circuit, whose elements are defined by

$$\omega L_0 = \frac{1}{\omega C_0} = \frac{R_1 Q_0}{1 + Q_1^2} \tag{4.38}$$

where Q_0 is the arbitrary value depending on the specification requirements to the harmonic suppression. The other elements are defined by the design equations given in Fig. 4.13(*a*).

The π-transformer with two shunt capacitors shown in Fig. 4.13(*a*) represents a face-to-face connection of two simple low-pass *L*-transformers. As a result, there is no special requirement for the resistances R_1 and R_2, whose ratio R_1/R_2 can be greater or smaller than unity. As an example, the design equations correspond to the case of $R_1/R_2 >$ 1. However, as it will be further derived, the π-transformer with a series capacitor shown in Fig. 4.13(*b*) can only be used for impedance matching when $R_1/R_2 > 1$. Such a π-transformer represents a face-to-face connection of the high-pass and low-pass *L*-transformers, as shown in Fig. 4.15(*a*).

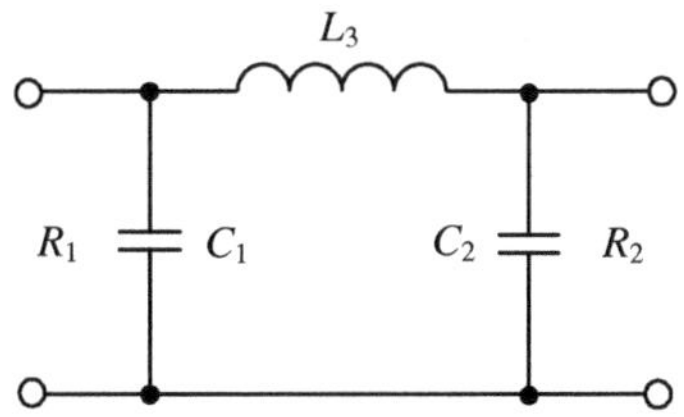

$$\omega C_1 = Q_1/R_1 \qquad \omega C_2 = Q_2/R_2 \qquad \omega L_3 = R_1(Q_1 + Q_2)/(1 + Q_1^2)$$

$$Q_2 = \sqrt{\frac{R_2}{R_1}(1 + Q_1^2) - 1} \qquad Q_1^2 > \frac{R_1}{R_2} - 1$$

(*a*)

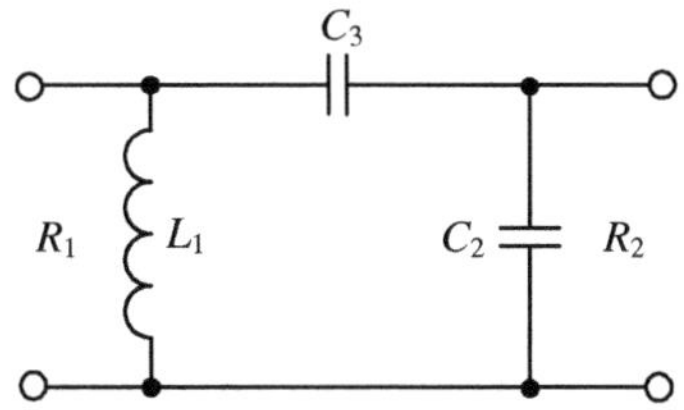

$$\omega L_1 = R_1/Q_1 \qquad \omega C_2 = Q_2/R_2 \qquad \omega C_3 = (1 + Q_2^2)/[R_2(Q_1 - Q_2)]$$

$$Q_2 = \sqrt{\frac{R_2}{R_1}(1 + Q_1^2) - 1} \qquad Q_1^2 > \frac{R_1}{R_2} - 1$$

(*b*)

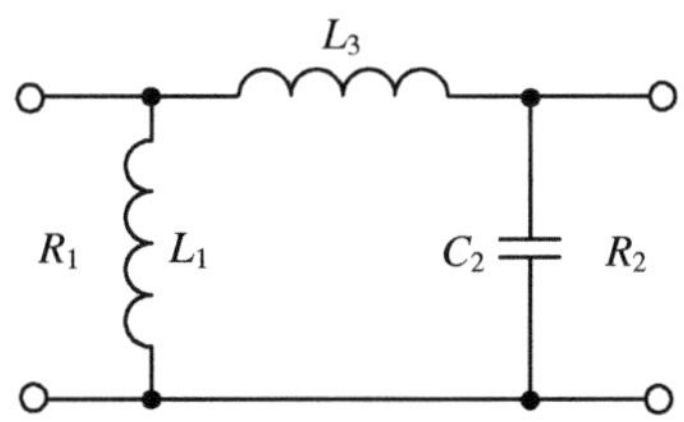

$$\omega L_1 = R_1/Q_1 \qquad \omega C_2 = Q_2/R_2 \qquad \omega L_3 = R_2(Q_2 - Q_1)/(1 + Q_2^2)$$

$$Q_1 = \sqrt{\frac{R_1}{R_2}(1 + Q_2^2) - 1} \qquad Q_2^2 > \frac{R_2}{R_1} - 1$$

(*c*)

FIGURE 4.13 π-transformers and relevant equations.

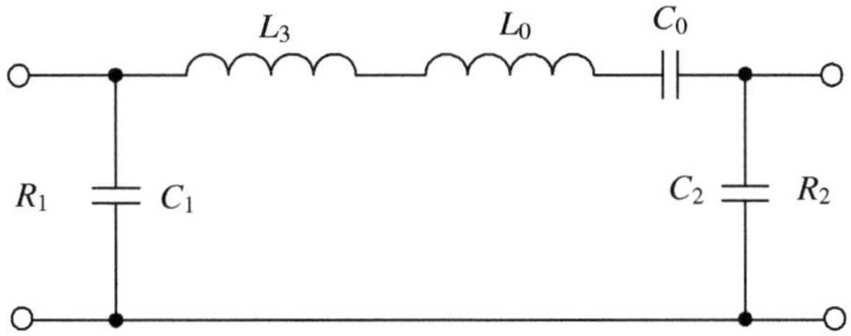

FIGURE 4.14 π-transformer with additional *LC*-filter.

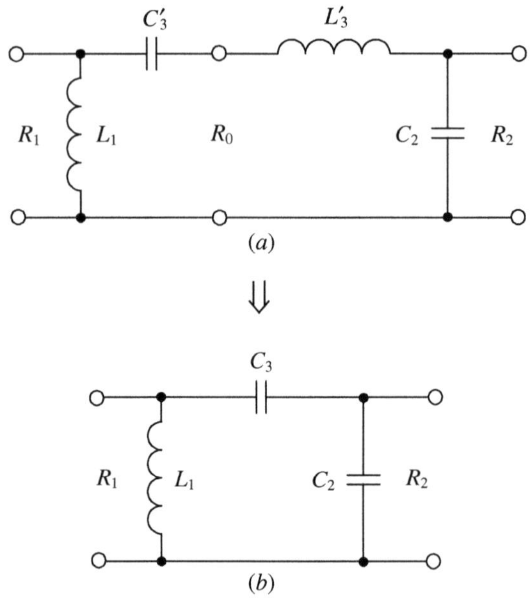

FIGURE 4.15 π-transformer with series capacitor.

The design equations for a high-pass section are written using Eqs. (4.28) through (4.30) as

$$\omega L_1 = \frac{R_1}{Q_1} \tag{4.39}$$

$$\omega C_3' = \frac{1}{Q_1 R_0} \tag{4.40}$$

$$Q_1^2 = \frac{R_1}{R_0} - 1 \tag{4.41}$$

where Q_1 is the circuit quality factor and R_0 is the intermediate resistance.

Similarly, for a low-pass section,

$$\omega C_2 = \frac{Q_2}{R_2} \tag{4.42}$$

$$\omega L_3' = Q_2 R_0 \tag{4.43}$$

$$Q_2^2 = \frac{R_2}{R_0} - 1 \tag{4.44}$$

Because it is assumed that $R_1 > R_2 > R_0$, from Eq. (4.41) it follows that the quality factor Q_1 of a high-pass L-transformer can be chosen from the condition of

$$Q_1^2 > \frac{R_1}{R_2} - 1 \tag{4.45}$$

Substituting Eq. (4.41) into Eq. (4.44) results in

$$Q_2 = \sqrt{\frac{R_2}{R_1}(1+Q_1^2) - 1} \tag{4.46}$$

Combining the reactances of two series elements (capacitor C_3' and inductor L_3') given by Eqs. (4.40) and (4.43) yields

$$\omega L_3' - \frac{1}{\omega C_3'} = R_0(Q_2 - Q_1) = \frac{R_2(Q_2 - Q_1)}{(1+Q_2^2)} \tag{4.47}$$

As a result, because $Q_1 > Q_2$, the total series reactance is negative, which can be provided by a series capacitance C_3, as shown in Fig. 4.15(*b*), with susceptance

$$\omega C_3 = \frac{1+Q_2^2}{R_2(Q_1 - Q_2)} \tag{4.48}$$

On the other hand, if $Q_2 > Q_1$ when $R_2 > R_1 > R_0$, then the total series reactance is positive that can be provided by a series inductance L_3, and all matching circuit parameters can be calculated according to the design equations given in Fig. 4.13(*c*). In this case, it needs first to choose the value of Q_2 for fixed resistances R_1 and R_2 to be matched, then to calculate the value of Q_1, and finally the values of the shunt inductance L_1, shunt capacitance C_2, and series inductance L_3.

Some of the matching circuit configurations of two-port T-transformers, together with the design formulas, are shown in Fig. 4.16 [13]. The T-transformers are usually used in the high-power amplifiers as the input, interstage, and output matching circuits, especially the matching circuit with shunt and series capacitors shown in Fig. 4.16(*b*). In this case, if a high value of the inductance L_2 is chosen, the current waveform at the input of the transistor with a small input resistive part will be close to sinusoidal. By using such a T-transformer for the output matching of a power amplifier, it is easy to realize a high-efficiency Class-F operation mode, because the series inductance connected to the drain (or collector) of the active device creates open-circuit harmonic impedance conditions.

The T-transformer with two series inductors, shown in Fig. 4.16(*a*), represents a back-to-back connection of two simple low-pass L-transformers. In this case, the resistance ratio R_1/R_2 can be greater or smaller than unity, similar to a π-transformer with two shunt capacitors shown

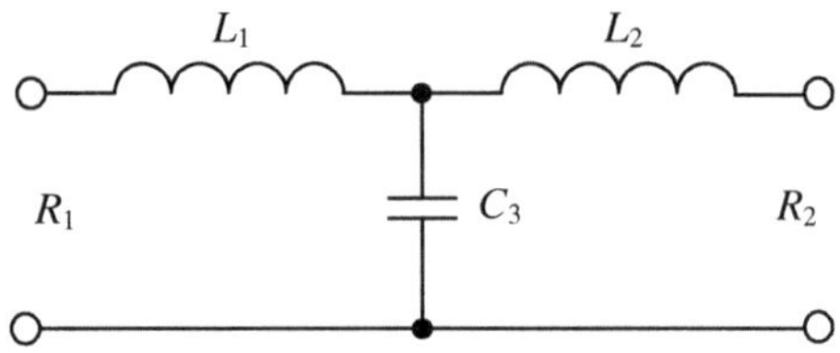

$$\omega L_1 = Q_1 R_1 \qquad \omega L_2 = Q_2 R_2 \qquad \omega C_3 = (Q_1 + Q_2)/[R_2(1+Q_2^2)]$$

$$Q_1 = \sqrt{\frac{R_2}{R_1}(1+Q_2^2) - 1} \qquad Q_2^2 > \frac{R_1}{R_2} - 1$$

(*a*)

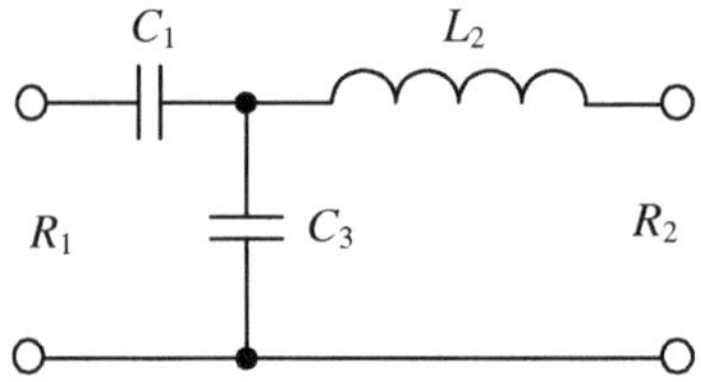

$$\omega C_1 = 1/R_1 Q_1 \qquad \omega L_2 = Q_2 R_2 \qquad \omega C_3 = (Q_2 + Q_1)/[R_2(1+Q_2^2)]$$

$$Q_1 = \sqrt{\frac{R_2}{R_1}(1+Q_2^2) - 1} \qquad Q_2^2 > \frac{R_1}{R_2} - 1$$

(*b*)

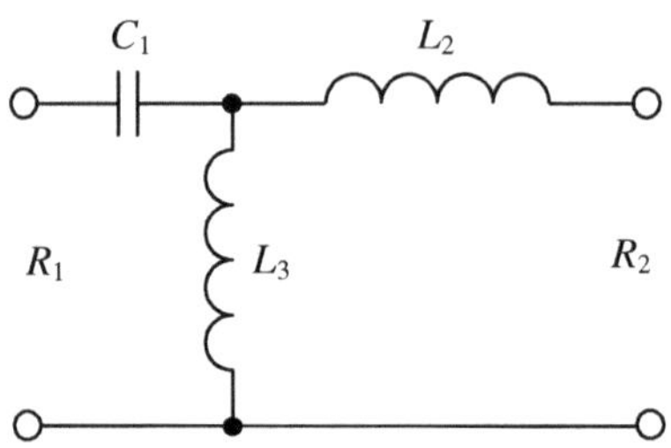

$$\omega C_1 = 1/R_1 Q_1 \qquad \omega L_2 = Q_2 R_2 \qquad \omega L_3 = R_2(1+Q_2^2)/(Q_1 - Q_2)$$

$$Q_2 = \sqrt{\frac{R_1}{R_2}(1+Q_1^2) - 1} \qquad Q_1^2 > \frac{R_2}{R_1} - 1$$

(*c*)

FIGURE 4.16 *T*-transformers and relevant equations.

in Fig. 4.13(*a*). As an example, the design equations correspond to the case of $R_1/R_2 > 1$. However, the *T*-transformer with series and shunt capacitors shown in Fig. 4.16(*b*) can only be used for impedance matching when $R_1/R_2 > 1$. Such a *T*-transformer represents a back-to-back

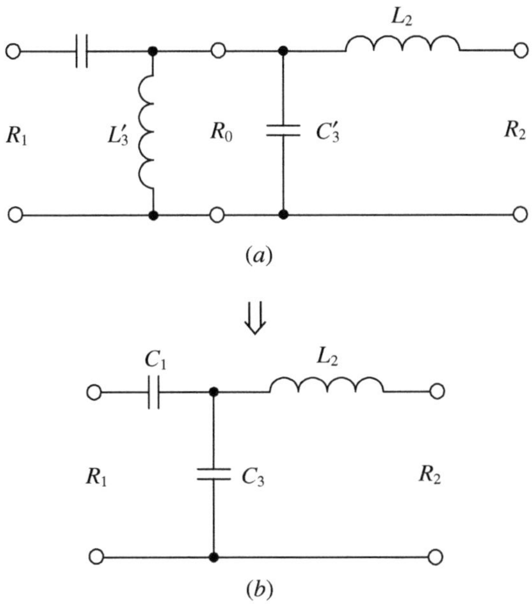

FIGURE 4.17 *T*-transformer with series and shunt capacitors.

connection of the high-pass and low-pass *L*-transformers, as shown in Fig. 4.17(*a*).

The design equations for a high-pass section of such a *T*-transformer are written using Eqs. (4.28) through (4.30) as

$$\omega C_1 = \frac{1}{R_1 Q_1} \tag{4.49}$$

$$\omega L_3' = \frac{R_0}{Q_1} \tag{4.50}$$

$$Q_1^2 = \frac{R_0}{R_1} - 1 \tag{4.51}$$

where Q_1 is the circuit quality factor and R_0 is the intermediate resistance.

Similarly, for a low-pass section,

$$\omega L_2 = Q_2 R_2 \tag{4.52}$$

$$\omega C_3' = Q_2 / R_0 \tag{4.53}$$

$$Q_2^2 = \frac{R_0}{R_2} - 1 \tag{4.54}$$

Because it is assumed that $R_0 > R_1 > R_2$, from Eq. (4.54) it follows that the quality factor Q_2 of a low-pass L-transformer can be chosen from the condition of

$$Q_2^2 > \frac{R_1}{R_2} - 1 \tag{4.55}$$

Substituting Eq. (4.54) into Eq. (4.51) results in

$$Q_1 = \sqrt{\frac{R_2}{R_1}(1+Q_2^2) - 1} \tag{4.56}$$

Combining the susceptances of two shunt elements (inductor L_3' and capacitor C_3') given by Eqs. (4.50) and (4.53) yields

$$\omega C_3' - \frac{1}{\omega L_3'} = \frac{Q_2 - Q_1}{R_0} = \frac{Q_2 - Q_1}{R_2(1+Q_2^2)} \tag{4.57}$$

As a result, because $Q_2 > Q_1$, the total shunt susceptance is positive, which can be provided by a shunt capacitance C_3, as shown in Fig. 4.17(*b*), with susceptance

$$\omega C_3 = \frac{Q_2 - Q_1}{R_2(1+Q_2^2)} \tag{4.58}$$

On the other hand, if $Q_1 > Q_2$ when $R_0 > R_2 > R_1$, then the total shunt susceptance is negative that can be provided by a shunt inductance L_3, and all matching circuit parameters can be calculated according to the design equations given in Fig. 4.16(*c*). In this case, it needs first to choose the value of Q_1 for fixed resistances R_1 and R_2 to be matched, then to calculate the value of Q_2, and finally the values of the series capacitance C_1, series inductance L_2, and shunt inductance L_3.

If the elements of π- and T-transformers are chosen according to the condition $X_1 = X_2 = -X_3$, then the input impedance Z_{in} of the transformer loaded by the resistance R_L (from any side) is equal to

$$Z_{in} = R_{in} = \frac{X^2}{R_L} \tag{4.59}$$

where $X = |X_i|$, $i = 1, 2, 3$. As a result, the input impedance Z_{in} will be resistive, regardless of the value of the load resistance R_L. For example, setting $R_L = R_2$ for the transformers shown in Figs. 4.13(*a*) and 4.16(*a*) yields

$$R_1 = \frac{X^2}{R_2} \tag{4.60}$$

When $X_1 \neq X_2 \neq -X_3$, the input impedances of π- and T-transformers will be resistive for only one particular value of R_L.

4.3.2 Bipolar UHF Power Amplifier

Consider a design example of a 10-W 300-MHz bipolar power amplifier with a supply voltage of 12.5 V, providing a power gain of at least 10 dB. The first design step is to select an appropriate active device that allows both simplifying the circuit design procedure and satisfying the specified requirements. Usually, the manufacturer states the values of the input and output impedances (or admittances) at the nominal operation point in the datasheet for the device. For example, the above requirements can be realized by an *n-p-n* silicon bipolar transistor operating at 300 MHz with the input impedance $Z_{in} = (1.3 + j0.9)\ \Omega$ and output admittance $Y_{out} = (150 - j70)$ mS, which is intended for power amplification in a Class AB with supply voltages up to 13.5 V.

In this case, the input impedance Z_{in} is expressed as a series combination of the transistor input resistance and inductive reactance, whereas the output admittance Y_{out} is represented by a parallel combination of the transistor output resistance and inductive reactance. This means that, at the required operating frequency, effect of the series parasitic collector lead inductance exceeds that of the shunt collector capacitance, which gives a net inductive reactance to the equivalent output circuit of an active device. To match the series input inductive impedance to the standard 50-Ω source impedance, it is convenient to use a matching circuit in the form of a *T*-transformer shown in Fig. 4.16(*b*), where the series capacitor C_1 can serve also as a blocking capacitor. Figure 4.18 shows the complete input network including the input device elements and matching circuit. From the reactive part of the input impedance Z_{in} it follows that, at the operating frequency of 300 MHz, the input inductance will be equal approximately to 0.5 nH.

It is necessary first to calculate the circuit quality factor Q_2, which is needed to determine the parameters of the matching circuit, resulting in

$$Q_2 > \sqrt{\frac{R_1}{R_{in}} - 1} = 6.1$$

Consequently, the value of Q_2 must be larger than 6.1. For example, $Q_2 = 6.5$ provides a 3-dB bandwidth of 300 MHz/6.5 = 46 MHz.

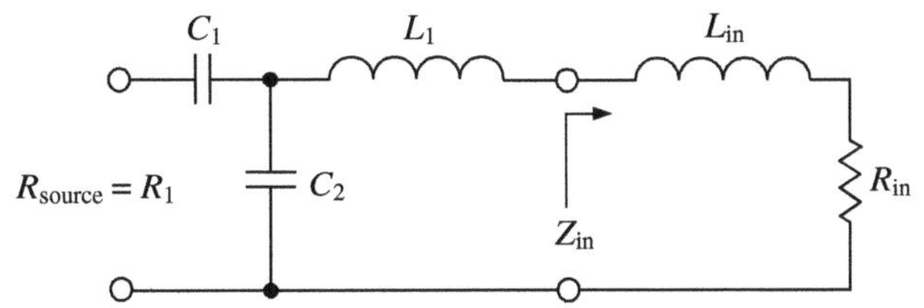

FIGURE 4.18 Complete input network circuit.

In this case, a value of Q_1 will be equal to 0.35. As a result, the values of the input matching circuit elements are

$$C_1 = \frac{1}{\omega Q_1 R_1} = 30 \text{ pF}$$

$$L_1 + L_{\text{in}} = \frac{Q_2 R_{\text{in}}}{\omega} = 4.5 \text{ nH} \Rightarrow L_1 = 4.0 \text{ nH}$$

$$C_2 = \frac{Q_2 - Q_1}{\omega R_{\text{in}}(1+Q_2^2)} = 59 \text{ pF}$$

This type of a T-transformer is widely used in practical matching circuit design because of its design simplicity and implementation convenience. In addition, the sufficiently small value of a series capacitance C_1 can contribute to elimination of the low-frequency parasitic oscillations in the case of a multistage power amplifier. The function of each element can be graphically traced on the Smith chart as shown in Fig. 4.19.

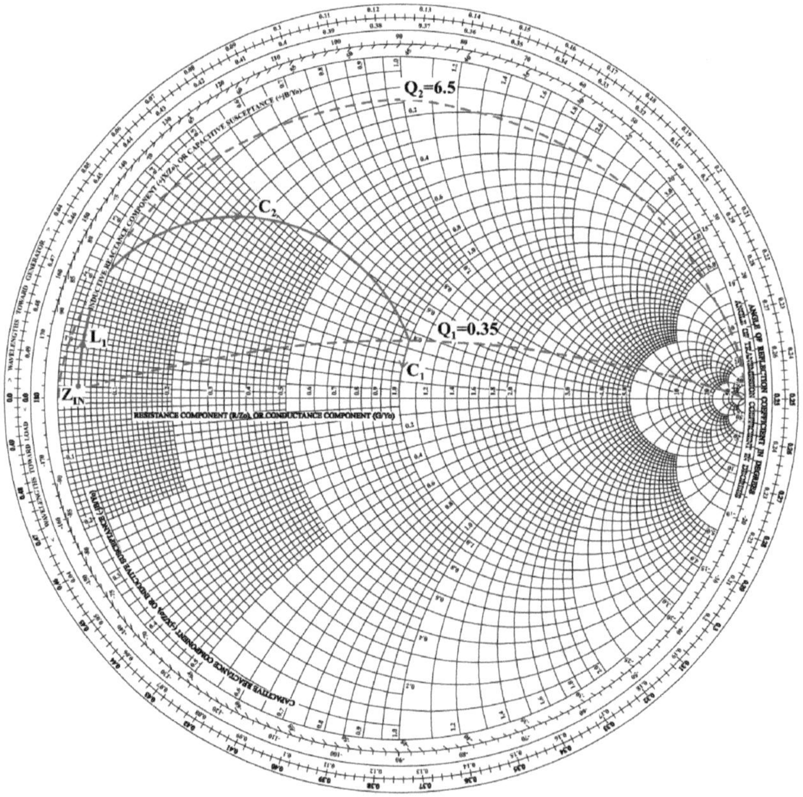

FIGURE 4.19 Smith chart with elements from Fig. 4.18.

The easiest and most convenient way to plot the traces of the matching circuit elements is by plotting initially the traces of $Q_2 = 6.5$ and $Q_1 = 0.35$. The circle of equal Q is plotted by taking into account that, for each point located at this circle, a ratio of X/R or B/G must be the same. The trace of the series inductance L_1 must be plotted as far as the intersection point with Q_2-circle. This means that, beginning at Z_{in}, a curve of increasing inductive reactance must be plotted as far as Q_2-circle. The value of L_1 is determined from the normalized inductive impedance at this intersection point. Then, because of a normalization to 50 Ω, the chart value must be multiplied by the factor of 50.

The trace of the parallel capacitance C_2 must be plotted using admittance circles. The previous impedance point located at Q_2-circle is converted to its appropriate admittance one. This point is symmetrical to the impedance point regarding the center point and is located on a straight line from the intersection point drawn through the center of the Smith chart into its lower half at the same distance from the center point. A curve from this point with constant conductance and increasing capacitive susceptance is plotted. These points are transformed to appropriate impedances using a line through the center point extended at equal distance on the other side and stop when the transformed curve reaches the $Q_1 = 0.35$ circle. In other words, it is necessary to transform mentally or to use a transparent admittance Smith chart (impedance Smith chart rotated on 180°) to plot a curve for C_2 on the upper half of the impedance Smith chart. The difference between the susceptances at the beginning and the end of this curve determines the value of C_2. Then, a curve of reducing inductive reactance is plotted down to the center point in order to determine a value of the series capacitance C_1.

A similar design philosophy can be applied to the design of the output matching circuit shown in Fig. 4.20. However, by taking into account the presence of the output shunt inductance, it makes sense to use a matching circuit in the form of a π-transformer shown in Fig. 4.13(*c*). The equivalent output resistance in a Class-B mode can be analytically evaluated by

$$R_{out} = \frac{(V_{cc} - V_{sat})^2}{2P_{out}} \cong \frac{(0.9\, V_{cc})^2}{2P_{out}} = 6.3\ \Omega$$

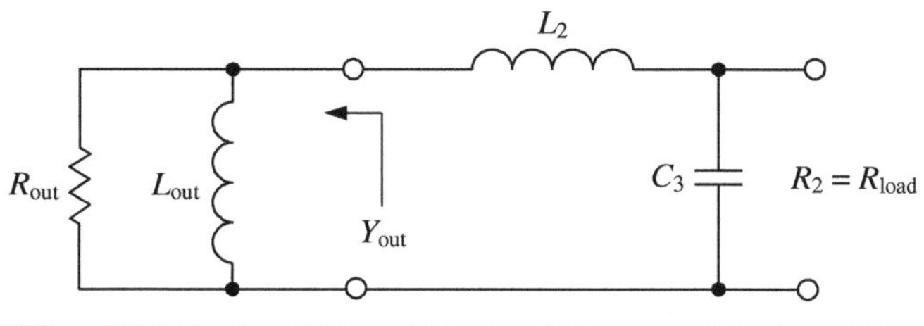

FIGURE 4.20 Complete output network circuit.

where V_{cc} is the supply voltage, V_{sat} is the saturation voltage, and P_{out} is the output power at the fundamental frequency. Its value is very close to the measurement value given by the real part of the device output admittance calculated as $R_{out} = 1/0.15 = 6.7\ \Omega$. A value of the inductance L_{out} is approximately equal to 7.6 nH.

The quality factor Q_2 can be chosen as

$$Q_2 > \sqrt{\frac{R_2}{R_{out}} - 1} = 2.5$$

The calculated quality factor Q_1 of the device output circuit is

$$Q_1 = \frac{R_{out}}{\omega L_{out}} = 0.47$$

This value of L_{out} allows the output device admittance to be matched to a 50-Ω load using such a π-transformer because

$$Q_2 = \sqrt{\frac{R_2}{R_{out}}(1 + Q_1^2) - 1} = 2.8 > 2.5$$

As a result, the values of the other two elements of the output matching circuit are

$$C_3 = \frac{Q_2}{\omega R_2} = 31\ \text{pF}$$

$$L_2 = \frac{R_2(Q_2 - Q_1)}{\omega(1 + Q_2^2)} = 6.8\ \text{nH}$$

A blocking capacitor that performs dc supply decoupling can be connected after the π-transformer with a sufficiently high value of its capacitance to avoid any negative influence on the matching circuit. Alternatively, it can be used in series with the inductor L_2, in order to form a series resonant circuit, as shown in Fig. 4.11. The design of the output circuit using the Smith chart is given in Fig. 4.21. Initially, it is necessary to transform the output admittance Y_{out} to the output impedance Z_{out} using the straight line of the Smith chart, putting the Z_{out} point at the same distance from the center point as for the Y_{out} point. Then, the effect of increasing series inductance L_2, by moving from the Z_{out} point along the curve of the constant R and increasing X until intersection with the $Q_2 = 2.8$ circle, is plotted. To determine the parallel capacitance C_3, it is necessary to transform this point to the corresponding admittance one and plot the curve of the constant G and increasing B, which must intersect the center point of the Smith chart.

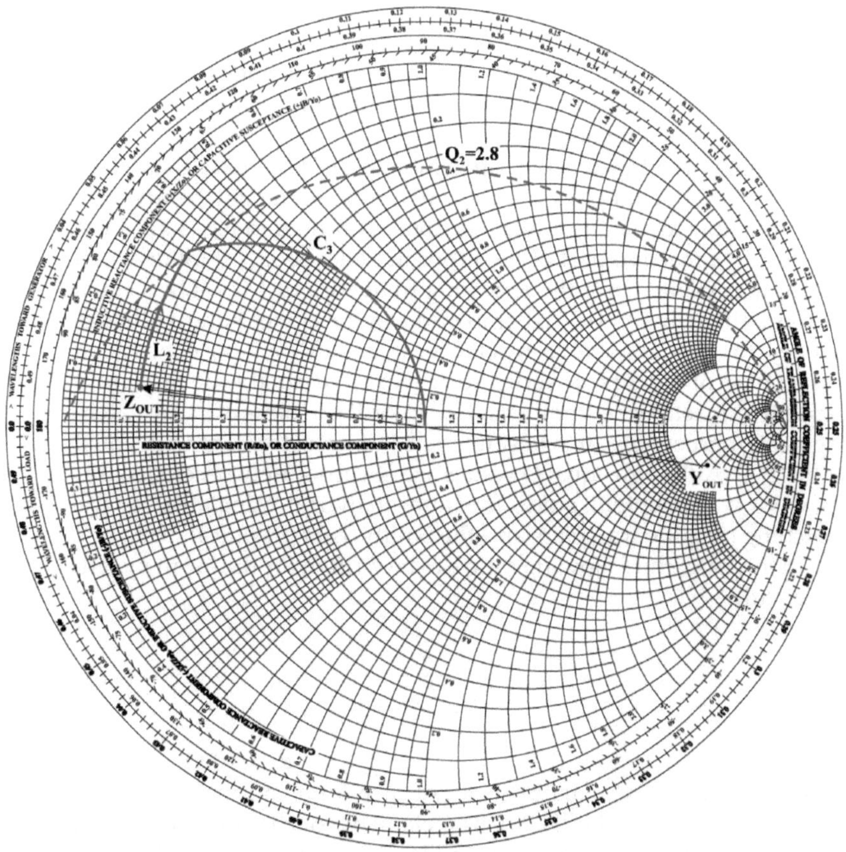

FIGURE 4.21 Smith chart with elements from Fig. 4.20.

4.3.3 MOSFET VHF High-Power Amplifier

Now let us demonstrate lumped matching circuit technique to design a 150-W MOSFET power amplifier with the supply voltage of 50 V, operating in a frequency bandwidth of 132 to 174 MHz and providing a power gain greater than 10 dB. These requirements can be satisfied using a silicon *n*-channel enhancement-mode VDMOSFET device designed for power amplification in the VHF frequency range. In this case, the center bandwidth frequency is equal to $f_c = \sqrt{132 \times 174} = 152$ MHz. For this frequency, the manufacturer states the following values of the input and output impedances: $Z_{in} = (0.9 - j1.2)\ \Omega$ and $Z_{out} = (1.8 + j2.1)\ \Omega$. Both Z_{in} and Z_{out} represent the series combination of an input or output resistance with a capacitive or inductive reactance, respectively. To cover the required frequency bandwidth, the low-Q matching circuits should be used that allows reduction of the in-band amplitude ripple and improvement of the input *VSWR*.

The value of a quality factor for 3-dB bandwidth level must be less than $Q = 152/(174 - 132) = 3.6$. As a result, it is very convenient to design the input and output matching circuits using the simple L-transformers in the form of low-pass and high-pass filter sections with a constant value of Q [15].

To match the input series capacitive impedance to the standard 50-Ω source impedance in a sufficiently wide frequency bandwidth, it is preferable to use three filter sections, as shown in Fig. 4.22. From the negative reactive part of the input impedance Z_{in}, it follows that the input capacitance at the operating frequency of 152 MHz is equal to approximately 873 pF. To compensate at the center bandwidth frequency for this capacitive reactance, it is sufficient to connect an inductance of 1.3 nH in series to it. Now, when the device input capacitive reactance is compensated, in order to simplify the matching design procedure, it is best to cascade L-transformers with equal value of Q. Although equal Q values are not absolutely necessary, this provides a convenient guide for both analytical calculation of the matching circuit parameters and the Smith chart graphical design.

In this case, the following ratio can be written for the input matching circuit:

$$\frac{R_1}{R_2} = \frac{R_2}{R_3} = \frac{R_3}{R_{in}} \tag{4.61}$$

resulting in $R_2 = 13$ Ω and $R_3 = 3.5$ Ω for $R_{source} = R_1 = 50$ Ω and $R_{in} = 0.9$ Ω. Consequently, a quality factor of each L-transformer is equal to $Q = 1.7$ according to Eq. (4.30). The elements of the input matching circuit using the formulas given in Fig. 4.8 can be calculated as $L_1 = 31$ nH, $C_1 = 47$ pF, $L_2 = 6.2$ nH, $C_2 = 137$ pF, $L_3 = 1.6$ nH, and $C_3 = 509$ pF.

This equal-Q approach significantly simplifies the matching circuit design using the Smith chart. When calculating a value of Q, it is necessary to plot a circle of equal Q values on the Smith chart. Then, each element of the input matching circuit can be readily determined, as shown in Fig. 4.23. Each trace for the series inductance must be plotted until the intersection point with Q-circle, whereas each trace for the parallel capacitance should be plotted until intersection with a horizontal real axis.

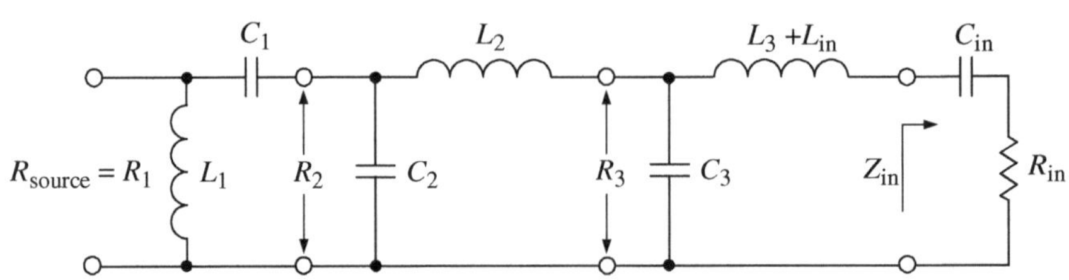

FIGURE 4.22 Complete broadband input matching circuit.

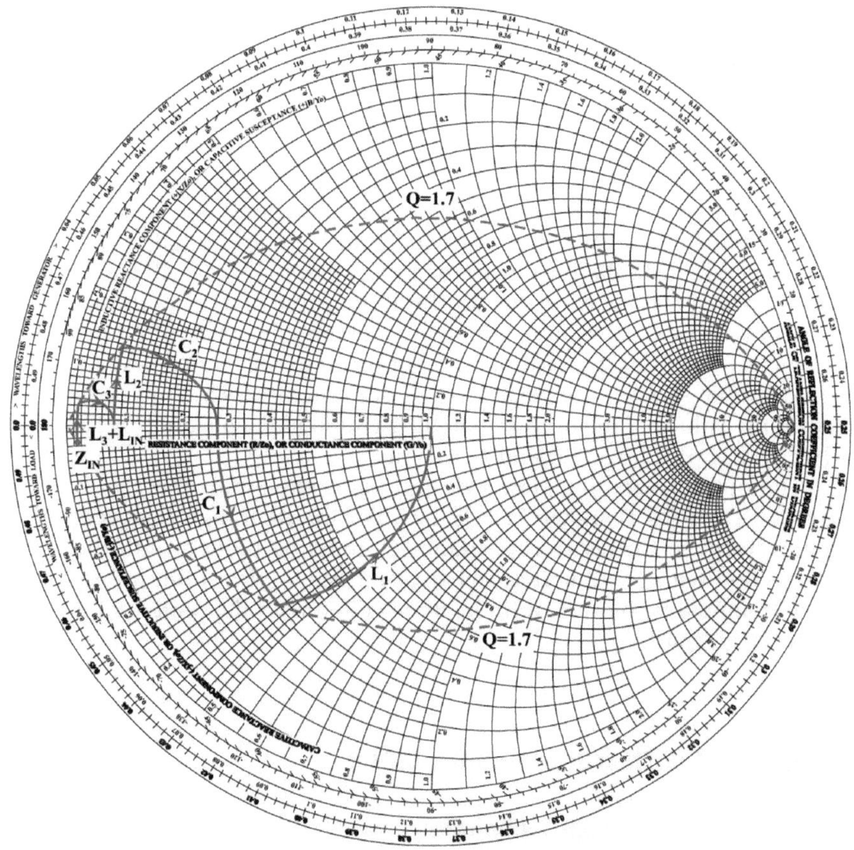

FIGURE 4.23 Smith chart with elements from Fig. 4.22.

To match the output series inductive impedance to the standard 50-Ω load impedance, it is sufficient to use only two filter sections, as shown in Fig. 4.24. At the operating frequency of 152 MHz, the transistor series output inductance is equal to approximately 2.2 nH. This inductance can be used as a part of the *L*-transformer in the form of a

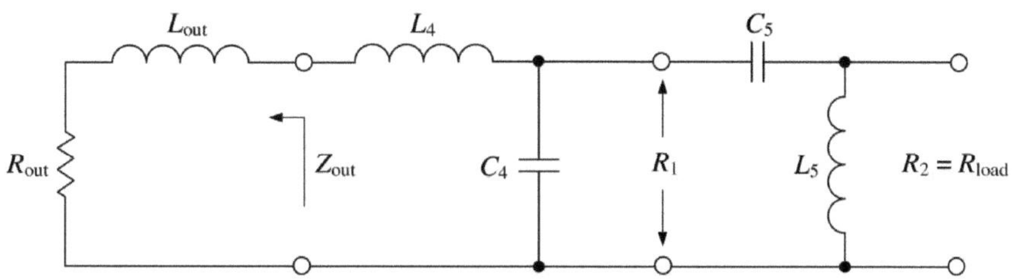

FIGURE 4.24 Complete broadband output network circuit.

low-pass filter section. For an output matching circuit, the condition of equal-Q values results in

$$\frac{R_2}{R_1} = \frac{R_1}{R_{out}} \tag{4.62}$$

with the value of $R_1 = 9.5\ \Omega$ for $R_{load} = R_2 = 50\ \Omega$ and $R_{out} = 1.8\ \Omega$. Consequently, a quality factor of each L-transformer is equal to $Q = 2.1$, which is substantially smaller than the value of Q for 3-dB bandwidth level. Now it is necessary to check the value of a series inductance of the low-pass section, which must exceed the value of 2.2 nH for correct matching procedure. The appropriate calculation gives a value of total series inductance $L_4 + L_{out}$ of approximately 4 nH. As a result, the values of the output matching circuit elements are $L_4 = 1.8$ nH, $C_4 = 231$ pF, $C_5 = 52$ pF, and $L_5 = 25$ nH.

The output matching circuit design using the Smith chart with constant Q-circles is shown in Fig. 4.25. For the final high-pass section, a trace for the series capacitance C_5 must be plotted until the intersection with $Q = 2.1$ circle, whereas a trace for the shunt inductance L_5 should be plotted until the intersection with the center point of the Smith chart.

4.4 Matching with Transmission Lines

4.4.1 Analytic Design Technique

At very high frequencies, it is very difficult to implement lumped elements with predefined accuracy in view of a significant effect of their parasitic parameters, for example, the parasitic interturn and direct-to-ground capacitances for lumped inductors and the stray inductance for lumped capacitors. However, these parasitic parameters become a part of a distributed LC structure such as a transmission line [3]. In this case, for a microstrip line, the series inductance is associated with the flow of current in the conductor and the shunt capacitance is associated with the strip separated from the ground by the dielectric substrate. If the line is wide, the inductance is reduced but the capacitance is large. However, for a narrow line, the inductance is increased, but the capacitance is small.

Figure 4.26 shows the impedance matching circuit in the form of a transmission-line transformer connected between the source impedance Z_S and load impedance Z_L. The input impedance as a function of the length of the transmission line with arbitrary load impedance is

$$Z_{in} = Z_0 \frac{Z_L + jZ_0 \tan\theta}{Z_0 + jZ_L \tan\theta} \tag{4.63}$$

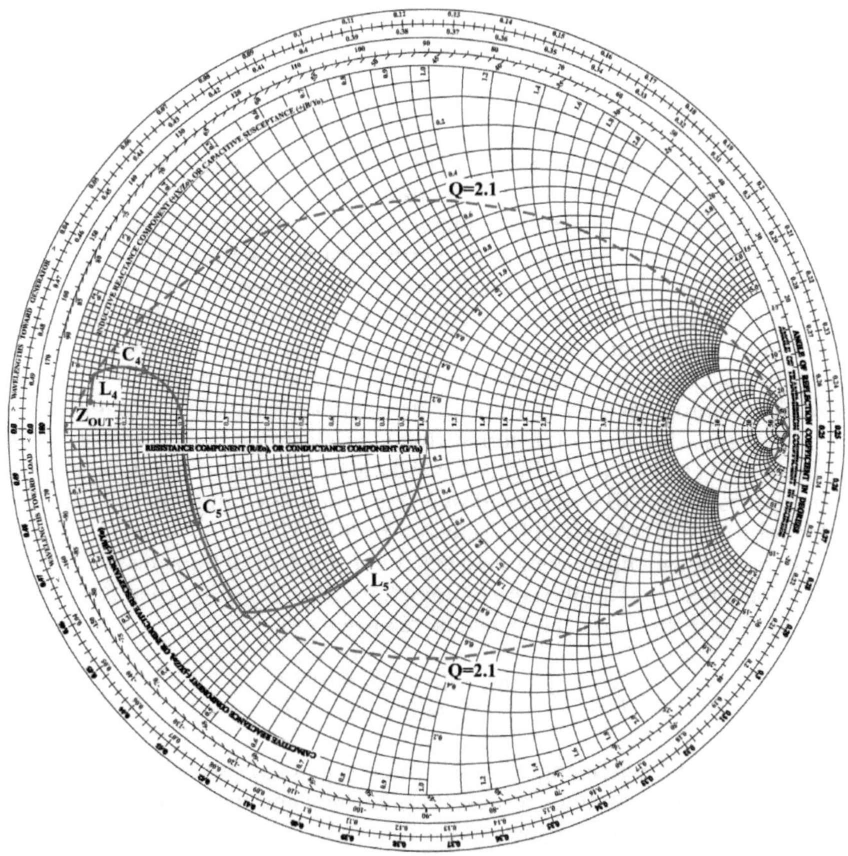

Figure 4.25 Smith chart with elements from Figure 4.24.

where Z_0 is the characteristic impedance, $\theta = \beta l$ is the electrical length of the transmission line, $\beta = \omega \sqrt{\mu_r \varepsilon_r}/c$ is the phase constant, c is the speed of light in free space, μ_r is the substrate permeability, ε_r is the substrate permittivity, ω is the angular frequency, and l is the geometrical length of the transmission line [6, 16].

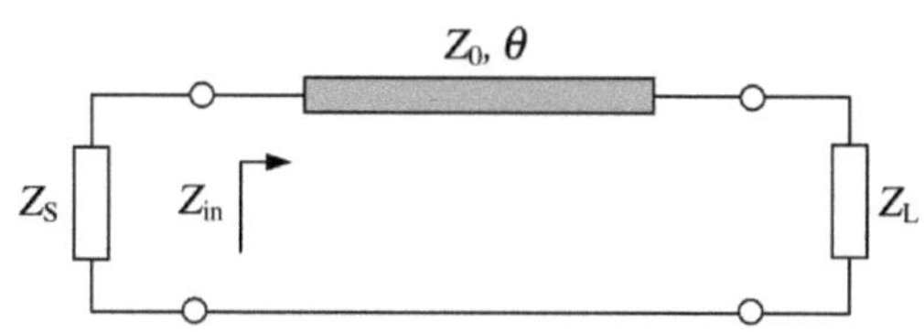

Figure 4.26 Transmission-line impedance transformer.

For a quarter-wavelength transmission line with $\theta = \pi/2$, the expression for Z_{in} is simplified to

$$Z_{in} = Z_0^2 / Z_L \tag{4.64}$$

from which it follows that, for example, a 50-Ω load is matched to a 12.5-Ω source with the characteristic impedance of 25 Ω.

Usually, such a quarter-wavelength impedance transformer is used for impedance matching in a narrow frequency bandwidth of 10 to 20%, and its length is chosen at the bandwidth center frequency. However, using a multisection quarterwave transformer widens the bandwidth and expands the choice of the substrate to include materials with high dielectric permittivity, which reduces the transformer's size. For example, by using a transformer composed of seven quarter-wave transmission lines of different characteristic impedances, whose lengths are selected at the highest bandwidth frequency, the power gain flatness of ±1 dB is achieved over the frequency range of 5 to 10 GHz for a 15-W GaAs MESFET power amplifier [17].

To provide a complex-conjugate matching of the input transmission-line impedance Z_{in} with the source impedance $Z_S = R_S + jX_S$ when $R_S = \mathrm{Re}Z_{in}$ and $X_S = -\mathrm{Im}Z_{in}$, Eq. (4.63) can be rewritten as

$$R_S - jX_S = Z_0 \frac{R_L + j(X_L + Z_0 \tan\theta)}{Z_0 - X_L \tan\theta + jR_L \tan\theta} \tag{4.65}$$

For a quarter-wavelength transformer, Eq. (4.65) can be divided into two separate equations representing the real and imaginary parts of the source impedance Z_S as

$$R_S = Z_0^2 \frac{R_L}{R_L^2 + X_L^2} \tag{4.66}$$

$$X_S = -Z_0^2 \frac{X_L}{R_L^2 + X_L^2} \tag{4.67}$$

For a purely real load impedance with $X_L = 0$, a quarterwave transmission line with the characteristic impedance Z_0 can provide impedance matching for a purely active source and load only when

$$Z_0 = \sqrt{R_S R_L} \tag{4.68}$$

Generally, Eq. (4.65) can be divided into two equations representing the real and imaginary parts,

$$R_S(Z_0 - X_L \tan\theta) - R_L(Z_0 - X_S \tan\theta) = 0 \tag{4.69}$$

$$X_S(X_L \tan\theta - Z_0) - Z_0(X_L + Z_0 \tan\theta) + R_S R_L \tan\theta = 0 \tag{4.70}$$

Solving Eqs. (4.69) and (4.70) for the two independent variables Z_0 and θ yields

$$Z_0 = \sqrt{\frac{R_S(R_L^2 + X_L^2) - R_L(R_S^2 + X_S^2)}{R_L - R_S}} \tag{4.71}$$

$$\theta = \tan^{-1}\left(Z_0 \frac{R_S - R_L}{R_S X_L - X_S R_L}\right) \tag{4.72}$$

As a result, the transmission line with the characteristic impedance Z_0, determined by Eq. (4.71), and the electrical length θ, determined by Eq. (4.72), can match any source and load impedances when the impedance ratio gives a positive value inside the square root in Eq. (4.71).

For a particular case of a purely active source when $Z_S = R_S$, the ratio between the load and transmission-line parameters can be expressed by

$$X_L Z_0 (1 - \tan^2\theta) + (Z_0^2 - X_L^2 - R_L^2)\tan\theta = 0 \tag{4.73}$$

Then, for the electrical length of the transmission line $\theta = \pi/4$, the expression for the characteristic impedance Z_0 given by Eq. (4.71) can be simplified to

$$Z_0 = |Z_L| = \sqrt{R_L^2 + X_L^2} \tag{4.74}$$

whereas the required real source impedance R_S should be equal to

$$R_S = R_L \frac{Z_0}{Z_0 - X_L} \tag{4.75}$$

Consequently, any load impedance can be transformed to a real source impedance defined by Eq. (4.75) using a $\lambda/8$ transformer, the characteristic impedance of which is equal to the magnitude of the load impedance [18].

Applying the same approach to match purely resistive load with the source impedance, the total matching circuit that includes two $\lambda/8$ transformers and a $\lambda/4$ transformer can provide impedance matching between any complex source impedance Z_S and load impedance Z_L, as shown in Fig. 4.27.

In practice, to simplify the power amplifier design at microwave frequencies, the simple matching circuits are very often used, including an *L*-transformer with a series transmission line as the basic matching section. It is convenient to analyze the transforming properties of this matching circuit by substituting the equivalent transformation of the parallel *RX* circuit into the series one. For example, R_1 is the resistance and $X_1 = -1/\omega C$ is the reactance of the impedance $Z_1 = jR_1X_1/(R_1 + jX_1)$ for a parallel *RC* circuit, and $R_{in} = \text{Re}Z_{in}$ is the resistance and

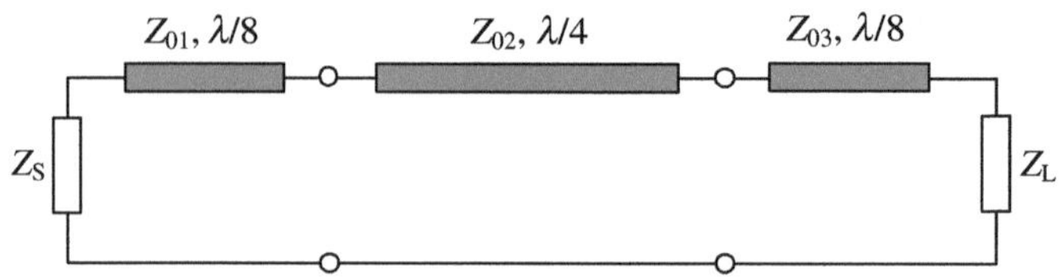

FIGURE 4.27 Transmission-line transformer for any source and load impedances.

$X_{in} = \mathrm{Im}Z_{in}$ is the reactance of the impedance $Z_{in} = R_{in} + jX_{in}$ for the series transmission-line circuit shown in Fig. 4.28.

For a complex-conjugate matching when $Z_1 = Z_{in}^*$, we obtain

$$\frac{R_1 X_1^2}{R_1^2 + X_1^2} + j\frac{R_1^2 X_1}{R_1^2 + X_1^2} = R_{in} - jX_{in} \tag{4.76}$$

The solution of Eq. (4.76) can be written in the form of two expressions for real and imaginary impedance parts as

$$R_1 = R_{in}(1 + Q^2) \tag{4.77}$$

$$X_1 = -X_{in}(1 + Q^{-2}) \tag{4.78}$$

where $Q = R_1/|X_1| = X_{in}/R_{in}$ is the quality factor equal for both parallel capacitive and series transmission-line circuits. By using Eq. (4.63), the real and imaginary parts of the input impedance Z_{in} can be written as

$$R_{in} = Z_0^2 R_2 \frac{1 + \tan^2\theta}{Z_0^2 + (R_2 \tan\theta)^2} \tag{4.79}$$

$$X_{in} = Z_0 \tan\theta \frac{Z_0^2 - R_2^2}{Z_0^2 + (R_2 \tan\theta)^2} \tag{4.80}$$

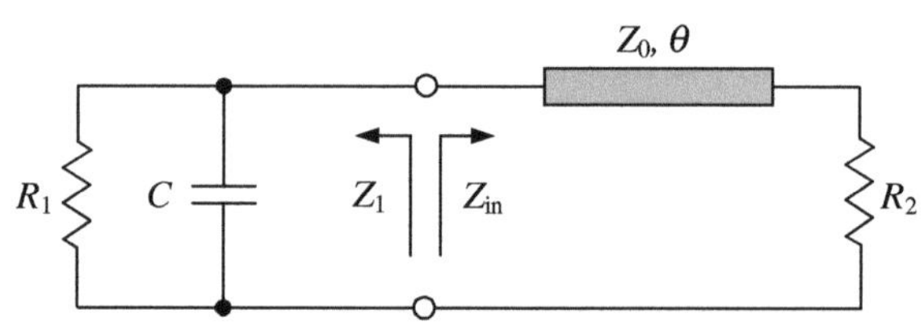

FIGURE 4.28 *L*-transformer with series transmission line.

From Eq. (4.80), it follows that an inductive input impedance (necessary to compensate for the capacitive parallel component) is provided when $Z_0 > R_2$ for $\theta < \pi/2$ and $Z_0 < R_2$ for $\pi/2 < \theta < \pi$. As a result, to transform the resistance R_1 into the other resistance R_2 at the given frequency, it is necessary to connect a two-port low-pass *L*-transformer (including a parallel capacitor and a series transmission line) between them. When one parameter (usually the characteristic impedance Z_0) is known, the matching circuit parameters can be calculated from the following two equations:

$$C = \frac{Q}{\omega R_1} \tag{4.81}$$

$$\sin 2\theta = \frac{2Q}{\dfrac{Z_0}{R_2} - \dfrac{R_2}{Z_0}} \tag{4.82}$$

where

$$Q = \sqrt{\frac{R_1}{R_2}\left[\cos^2\theta + \left(\frac{R_2}{Z_0}\right)^2 \sin^2\theta\right] - 1} \tag{4.83}$$

is the circuit (loaded) quality factor defined as a function of the resistances R_1 and R_2 and the parameters of the transmission line (characteristic impedance Z_0 and electrical length θ).

It follows from Eqs. (4.82) and (4.83) that the electrical length θ can be calculated as a result of the numerical solution of a transcendental equation with one unknown parameter. However, it is more convenient to combine these two equations and to rewrite them in the implicit form of

$$\frac{R_1}{R_2} = \frac{1 + \left(\dfrac{Z_0}{R_2} - \dfrac{R_2}{Z_0}\right)^2 \sin^2\theta \cos^2\theta}{\cos^2\theta + \left(\dfrac{R_2}{Z_0}\right)^2 \sin^2\theta} \tag{4.84}$$

Figure 4.29 shows the resistance ratio of R_1/R_2 as a function of the normalized parameter Z_0/R_2 and electrical length θ in the form of two nomographs: for the case of $Z_0/R_2 > 0$ shown in Fig. 4.29(*a*) and for the case of $Z_0/R_2 < 0$ shown in Fig. 4.29(*b*). When the input resistance R_1 and load resistance R_2 are known in advance, it is easy to evaluate the required value of θ for a fixed transmission-line characteristic impedance Z_0 using these nomographs. The graphical results show that, in contrast to a lumped *L*-transformer, a transmission-line *L*-transformer can match purely resistive source and load impedances with any ratio of R_1/R_2.

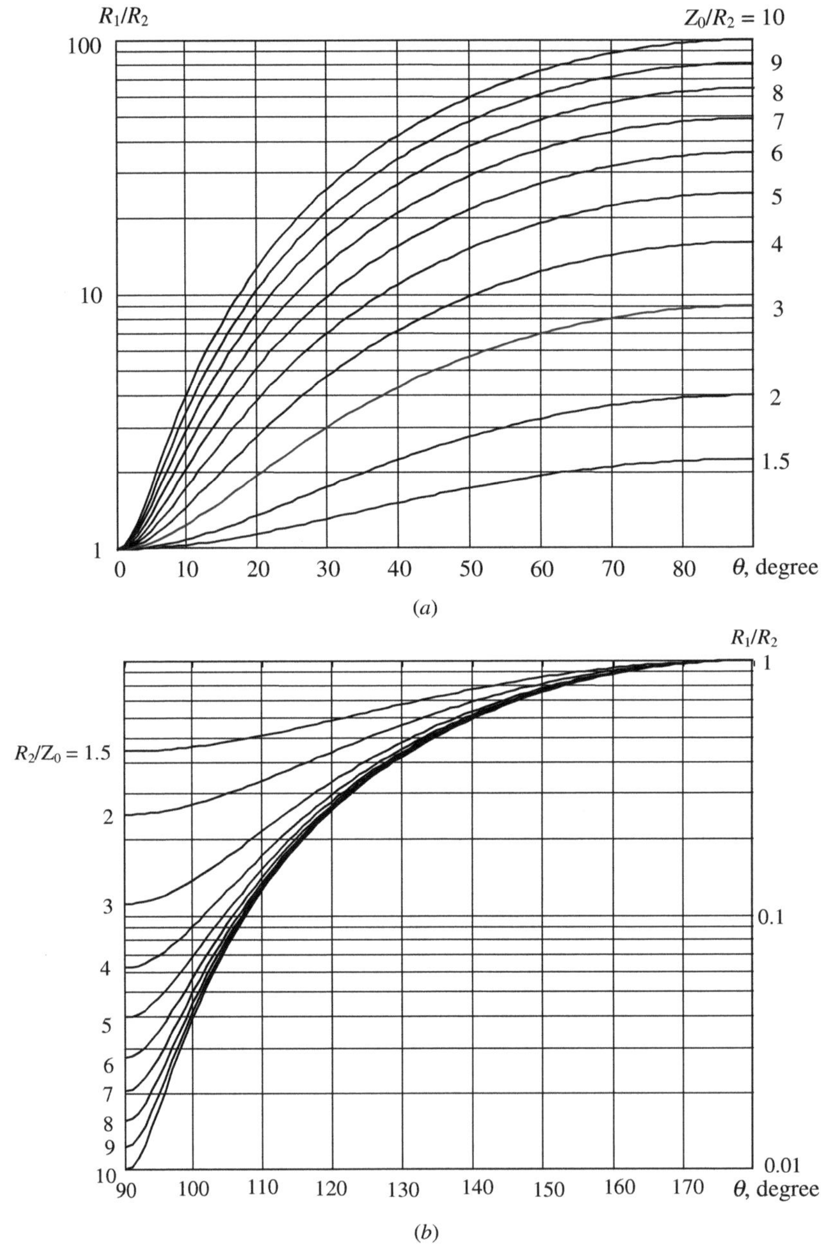

FIGURE 4.29 Nomographs for calculating transmission-line *L*-transformer.

A π-transformer can be realized by back-to-back connection of the two *L*-transformers, as shown in Fig. 4.30(*a*), where the resistances R_1 and R_2 are transformed into some intermediate resistance R_0. In this case, to minimize the length of a transmission line, the value of R_0

should be smaller than that of both R_1 and R_2, that is, $R_0 < (R_1, R_2)$. The same procedure for a T-transformer shown in Fig. 4.30(*b*) gives a value of R_0 that is larger than that of both R_1 and R_2, that is, $R_0 > (R_1, R_2)$. Then, for a T-transformer two shunt adjacent capacitors are combined, whereas, for a π-transformer two adjacent series transmission lines are combined into a single transmission line with total electrical length.

For a π-transformer, the lengths of each part of the combined transmission line can be calculated by equating the imaginary parts of the impedances from both sides at the reference plane A–A' to zero, which means that the intermediate impedance R_0 is real. This leads to two quadratic equations to calculate the electrical lengths θ_1 and θ_2 of the combined series transmission line written as

$$\tan^2\theta_1 - \frac{R_1}{Z_0 Q_1}\left[1-(1+Q_1^2)\left(\frac{Z_0}{R_1}\right)^2\right]\tan\theta_1 - 1 = 0 \tag{4.85}$$

$$\tan^2\theta_2 - \frac{R_2}{Z_0 Q_2}\left[1-(1+Q_2^2)\left(\frac{Z_0}{R_2}\right)^2\right]\tan\theta_2 - 1 = 0 \tag{4.86}$$

where $Q_1 = \omega C_1 R_1$ and $Q_2 = \omega C_2 R_2$.

However, to simplify the matching circuit design procedure, it is very helpful to use the nomographs shown in Fig. 4.29. If the values of R_1 and R_2 are selected in advance to set the intermediate resistance R_0 and the characteristic impedance of the transmission line Z_0 is known, the values of θ_1 and θ_2 can be easily determined from one of these nomographs.

A widely used two-port π-transformer with two shunt capacitors along with its design formulas is shown in Fig. 4.31 [19]. Such a

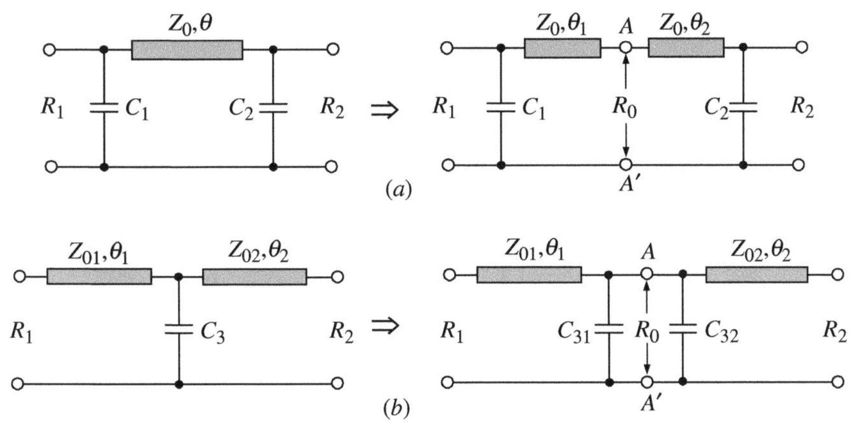

FIGURE 4.30 π- and T-transformers with transmission lines.

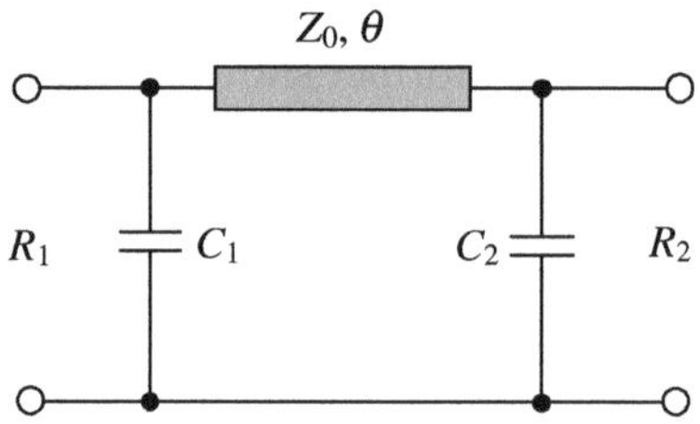

$$\omega C_1 = \frac{Q}{R_1} \qquad \omega C_2 = \frac{Q}{R_2} \qquad \theta = \theta_1 + \theta_2$$

$$Q_2 = \sqrt{\frac{R_2}{R_1}(1 + Q_1^2) - 1} \qquad Q_1 > \sqrt{\frac{R_1}{R_2} - 1}$$

FIGURE 4.31 Transmission-line π-transformer and relevant equations.

transformer can be conveniently used as an output matching circuit when the device collector (or drain-source) capacitance can be considered as a first shunt capacitance and the parasitic series lead inductance can easily be added to a series transmission line. Also, it is convenient to use this transformer as the matching circuits in balanced power amplifiers where the shunt capacitors can be connected between series transmission lines due to effect of virtual grounding. The schematic of a transmission-line two-port T-transformer with the series and shunt capacitors along with the design formulas is given in Fig. 4.32 [19].

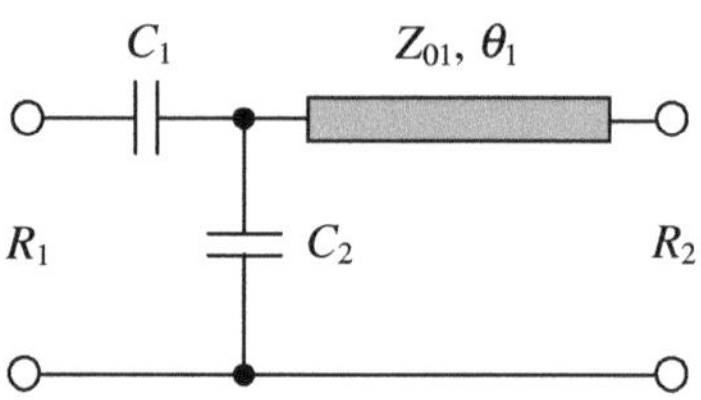

$$\omega C_1 = \frac{1}{R_1 Q_1} \qquad \sin 2\theta_1 = 2Q_2 / \left(\frac{Z_{01}}{R_2} - \frac{R_2}{Z_{01}} \right) \qquad \omega C_2 = \frac{Q_2 - Q_1}{R_1 (1 + Q_1^2)}$$

$$Q_1 = \sqrt{\frac{R_2}{R_1} \frac{1 + Q_2^2}{\cos^2\theta_1 + (R_2/Z_{01})\sin^2\theta_1} - 1} \qquad Q_2 > \sqrt{\frac{R_1}{R_2} - 1}$$

FIGURE 4.32 Transmission-line T-transformer and relevant equations.

4.4.2 Equivalence between Circuits with Lumped and Distributed Parameters

Generally, the input impedance of the transmission line at a particular frequency can be expressed as that of a lumped element, the equivalence of which for a shunt inductor L is shown in Fig. 4.33(*a*) and for a shunt capacitor C is shown in Fig. 4.33(*b*). If load represents a short when $Z_L = 0$, from Eq. (4.63) it follows that

$$Z_{in} = jZ_0 \tan\theta \tag{4.87}$$

which corresponds to the inductive input impedance for $\theta < \pi/2$. The equivalent inductance L at the required angular frequency ω is calculated from

$$L = \frac{X_{in}}{\omega} = \frac{Z_0 \tan\theta}{\omega} \tag{4.88}$$

where $X_{in} = \text{Im}Z_{in}$ is the input reactance. This means that the network shunt inductor can equivalently be replaced with an short-circuited transmission line of characteristic impedance Z_0 and electrical length θ.

Similarly, when $Z_L = \infty$,

$$Z_{in} = -jZ_0 \cot\theta \tag{4.89}$$

which corresponds to the capacitive input impedance for $\theta < \pi/2$. The equivalent capacitance C at the required angular frequency ω is determined from

$$C = -\frac{1}{\omega X_{in}} = \frac{\tan\theta}{\omega Z_0} \tag{4.90}$$

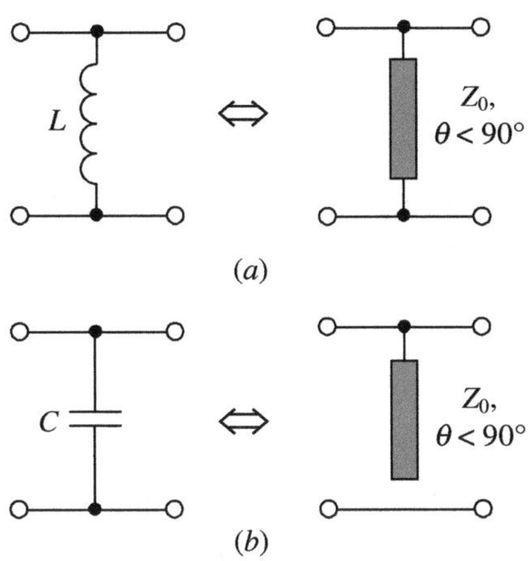

FIGURE 4.33 Equivalence between lumped element and transmission line.

These equivalences between lumped elements and transmission lines are exact at the required design frequency only. The reactance of the inductor increases linearly with increasing frequency, whereas the reactance of a shorted line increases as $\tan\theta$. For a short transmission line when $\theta << 90°$, the input impedance increases linearly with frequency as $\tan\theta \approx \theta$. Therefore, the short-circuited transmission line behaves like an inductor and open-circuited transmission line behaves like a capacitor over a range of frequencies where the electrical lengths of these transmission lines are much less than 90°.

To define the single-frequency equivalence between π-transformers using lumped and distributed elements, it is convenient to use two-port transmission *ABCD*-parameters. The transmission *ABCD*-matrices of these π-transformers shown in Fig. 4.34(*a*) can respectively be given by

$$[ABCD]_{\mathrm{L}} = \begin{bmatrix} 1 & 0 \\ j\omega C & 1 \end{bmatrix} \begin{bmatrix} 1 & j\omega L \\ 0 & 1 \end{bmatrix} \begin{bmatrix} 1 & 0 \\ j\omega C & 1 \end{bmatrix}$$

$$= \begin{bmatrix} 1-\omega^2 LC & j\omega L \\ j\omega C(2-\omega^2 LC) & 1-\omega^2 LC \end{bmatrix} \tag{4.91}$$

$$[ABCD]_{\mathrm{T}} = \begin{bmatrix} 1 & 0 \\ j\omega C & 1 \end{bmatrix} \begin{bmatrix} \cos\theta & jZ_0 \sin\theta \\ j\dfrac{\sin\theta}{Z_0} & \cos\theta \end{bmatrix} \begin{bmatrix} 1 & 0 \\ j\omega C & 1 \end{bmatrix}$$

$$= \begin{bmatrix} \cos\theta - \omega C_{\mathrm{T}} Z_0 \sin\theta & jZ_0 \sin\theta \\ \dfrac{j}{Z_0}(2\omega C_{\mathrm{T}} Z_0 \cos\theta + \sin\theta - \omega^2 C_{\mathrm{T}}^2 Z_0^2 \sin\theta) & \cos\theta - \omega C_{\mathrm{T}} Z_0 \sin\theta \end{bmatrix} \tag{4.92}$$

where θ is the electrical length of a transmission line at the required angular frequency ω.

Because equivalent circuits must have equal matrix elements, then, for $A_{\mathrm{L}} = A_{\mathrm{T}}$ and $B_{\mathrm{L}} = B_{\mathrm{T}}$, we can write

$$1-\omega^2 LC = \cos\theta - \omega C_{\mathrm{T}} Z_0 \sin\theta \tag{4.93}$$

$$j\omega L = jZ_0 \sin\theta \tag{4.94}$$

As a result, the equivalent series inductance can be expressed through the parameters of the transmission line as

$$L = \frac{Z_0 \sin\theta}{\omega} \tag{4.95}$$

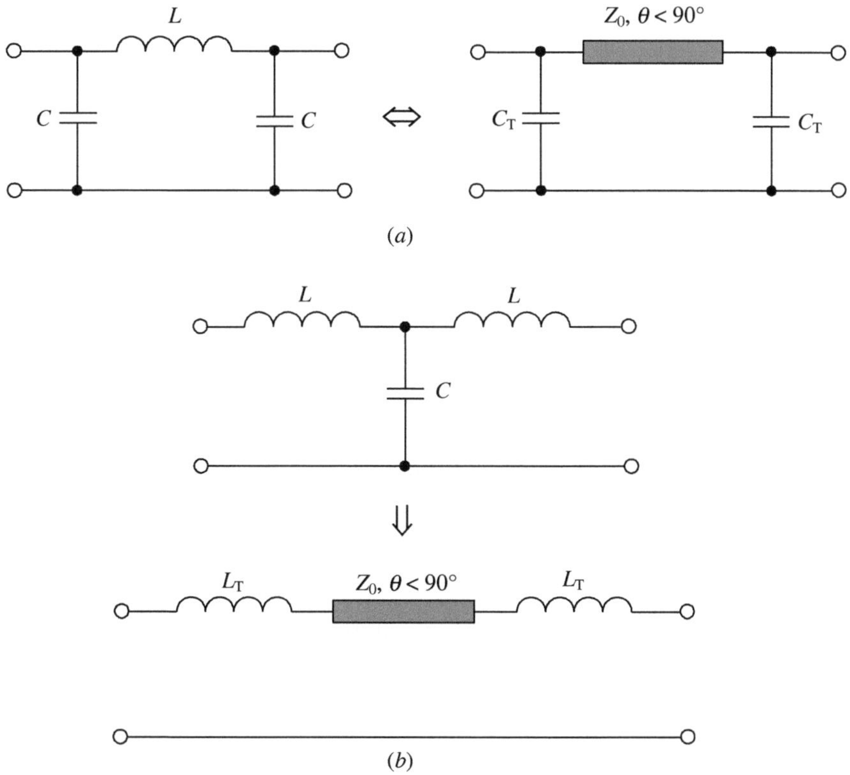

FIGURE 4.34 Lumped and transmission line matching-circuit equivalence.

whereas the relationship between the shunt capacitance C from a lumped π-transformer and the shunt capacitance C_T from a transmission-line π-transformer can be obtained by

$$C_T = \frac{\cos\theta + \omega C Z_0 \sin\theta - 1}{\omega Z_0 \sin\theta} \tag{4.96}$$

From Eq. (4.96), it follows that $C_T \approx C$ for high values of the characteristic impedance Z_0 when $\omega C Z_0 >> 1$ or small values of the electrical length θ when $\cos\theta \approx 1$. Consequently, the series lumped inductor L can be replaced by a short transmission line, the parameters of which can be calculated according to Eq. (4.95). The characteristic impedance of the series transmission line is chosen to be sufficiently high to provide better accuracy.

Similarly, to define the single-frequency equivalence between a lumped T-transformer and a transmission-line transformer shown in

Fig. 4.34(*b*), their transmission *ABCD*-matrices can respectively be given by

$$[ABCD]_L = \begin{bmatrix} 1 & j\omega L \\ 0 & 1 \end{bmatrix} \begin{bmatrix} 1 & 0 \\ j\omega C & 1 \end{bmatrix} \begin{bmatrix} 1 & j\omega L \\ 0 & 1 \end{bmatrix}$$

$$= \begin{bmatrix} 1-\omega^2 LC & j\omega L(2-\omega^2 LC) \\ j\omega C & 1-\omega^2 LC \end{bmatrix} \tag{4.97}$$

$$[ABCD]_T = \begin{bmatrix} 1 & j\omega L_T \\ 0 & 1 \end{bmatrix} \begin{bmatrix} \cos\theta & jZ_0 \sin\theta \\ j\dfrac{\sin\theta}{Z_0} & \cos\theta \end{bmatrix} \begin{bmatrix} 1 & j\omega L_T \\ 0 & 1 \end{bmatrix}$$

$$= \begin{bmatrix} \cos\theta - \dfrac{\omega L_T \sin\theta}{Z_0} & jZ_0 \left(2\dfrac{\omega L_T}{Z_0}\cos\theta + \sin\theta - \dfrac{\omega^2 L_T^2}{Z_0^2}\sin\theta \right) \\ j\dfrac{\sin\theta}{Z_0} & \cos\theta - \dfrac{\omega L_T \sin\theta}{Z_0} \end{bmatrix} \tag{4.98}$$

where θ is the electrical length of a transmission line at the required angular frequency ω.

For equivalent matrix elements $A_L = A_T$ and $C_L = C_T$,

$$1-\omega^2 LC = \cos\theta - \frac{\omega L_T}{Z_0}\sin\theta \tag{4.99}$$

$$j\omega C = j\frac{\sin\theta}{Z_0} \tag{4.100}$$

As a result, the equivalent shunt capacitance C can be expressed through the parameters of the transmission line as

$$C = \frac{\sin\theta}{\omega Z_0} \tag{4.101}$$

whereas the relationship between the series inductance L from a lumped *T*-transformer and the series inductance L_T from a transmission-line transformer can be obtained by

$$L_T = \frac{\cos\theta + \dfrac{\omega L}{Z_0}\sin\theta - 1}{\dfrac{\omega}{Z_0}\sin\theta} \tag{4.102}$$

From Eq. (4.102), it follows that $L_T \approx L$ for small values of the characteristic impedance Z_0 when $\omega L/Z_0 >> 1$ or small values of the electrical length θ when $\cos\theta \approx 1$. Consequently, the shunt lumped capacitor C can be replaced by a short transmission line, the parameters of which can be calculated according to Eq. (4.101). In this case, the characteristic impedance of the series transmission line is chosen to be sufficiently low to provide better accuracy.

4.4.3 Narrow-Band Microwave Power Amplifier

As an example, consider the design of a transmission-line output matching circuit for a 1.6-GHz bipolar power amplifier that operates at a supply voltage of 24 V and provides an output power of 5 W with a power gain of about 10 dB. These requirements can be satisfied by using an *n-p-n* silicon microwave transistor intended to operate in Class-AB power amplifiers for a frequency range of 1.5 to 1.7 GHz. At the operating frequency of 1.6 GHz, let the output device impedance be $Z_{out} = (5.5 - j6.5)\ \Omega$, which corresponds to a series combination of the transistor equivalent output resistance and capacitance. To match this capacitive impedance to the standard 50-Ω load, it is best to use a matching circuit in the form of a *T*-transformer shown in Fig. 4.32. Figure 4.35 shows the complete two-port network, including the output device impedance and matching circuit.

The circuit should compensate for the series capacitance inherent in the equivalent output impedance. For the small electrical length when $\tan\theta_{in} \approx \theta_{in}$ and characteristic impedance $Z_0 >> R_{out}$, Eqs. (4.79) and (4.80) in terms of the output resistance R_{out} and reactance X_{out} can respectively be simplified to

$$R_{out} \cong R_2 \tag{4.103}$$

$$\theta_2 \cong -\frac{X_{out}}{Z_0} = \frac{1}{\omega C_{out} Z_0} \tag{4.104}$$

where θ_2 is a part of the total transmission line that is required to compensate for the output capacitive reactance. If Z_0 is chosen to be 50 Ω, $\theta_2 = 6.5/50 = 0.13$ radians, which is equal to the electrical length of approximately 7.5°. Then, the value of a quality factor Q_2 is calculated as

$$Q_2 > \sqrt{\frac{R_1}{R_2} - 1} = 2.84$$

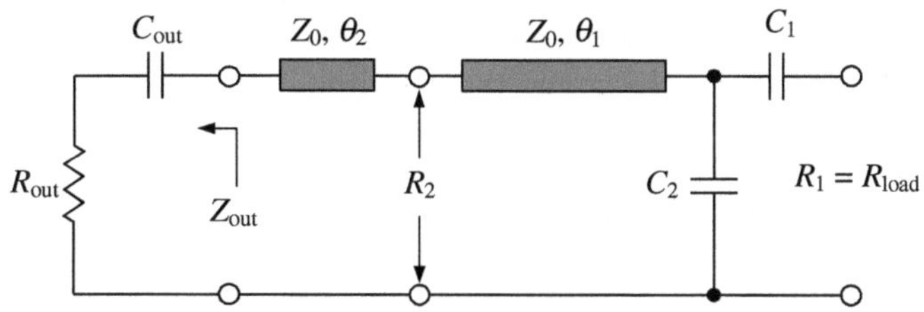

FIGURE 4.35 Complete output two-port network circuit.

Consequently, Q_2 must be larger than 2.84. For example, a value of $Q_2 = 3$ can be chosen to yield a 3-dB frequency bandwidth of 1.6 GHz/3 = 533 MHz. As a result, the values of the output matching circuit parameters are

$$\theta_1 = \frac{1}{2}\sin^{-1}\left[2Q_2 / \left(\frac{Z_0}{R_2} - \frac{R_2}{Z_0}\right)\right] = 21°$$

$$Q_1 = \sqrt{\frac{R_2}{R_1}\frac{1+Q_2^2}{\cos^2\theta_1 + (R_2/Z_0)^2\sin^2\theta_1} - 1} = 0.5$$

$$C_2 = \frac{Q_2 - Q_1}{\omega R_1(1+Q_1^2)} = 4 \text{ pF}$$

$$C_1 = \frac{1}{\omega Q_1 R_1} = 4 \text{ pF}$$

The function of each element for visual effect can be traced on the Smith chart, as shown in Fig. 4.36. The easiest and most convenient way to plot the traces of the matching circuit elements is to first plot the traces of Q_1 and Q_2, then plot the trace of the series transmission line as far as the intersection point with Q_2-circle, followed by the plotting the trace of the shunt capacitance C_2 as far as the intersection point with Q_1-circle, and finally plot the trace of the series capacitance C_1 to the center 50-Ω point. The plot of the series transmission line represents an arc of the circle with the center point at the center of the Smith chart.

4.4.4 Broadband UHF High-Power Amplifier

Now consider the design example of a broadband 150-W power amplifier that operates over a frequency bandwidth of 470 to 860 MHz, uses a 28-V supply voltage, and provides a power gain of more than 10 dB. In this case, it is convenient to use a high-power balanced LDMOSFET device specially designed for UHF TV transmitters. Let us assume that the manufacturer states the value of input impedance for each transistor-balanced part as $Z_{in} = (1.7 + j1.3)$ Ω at the center bandwidth frequency $f_c = \sqrt{470 \times 860} = 635$ MHz. The input impedance Z_{in} represents a series combination of the input resistance and inductive reactance. To cover the required frequency bandwidth, low-Q matching circuits should be used to reduce in-band amplitude ripple and improve input *VSWR*.

To achieve a 3-dB frequency bandwidth, the value of a quality factor must be less than $Q = 635/(860 - 470) = 1.63$. Based on this value of Q, the next step is to define a number of matching sections.

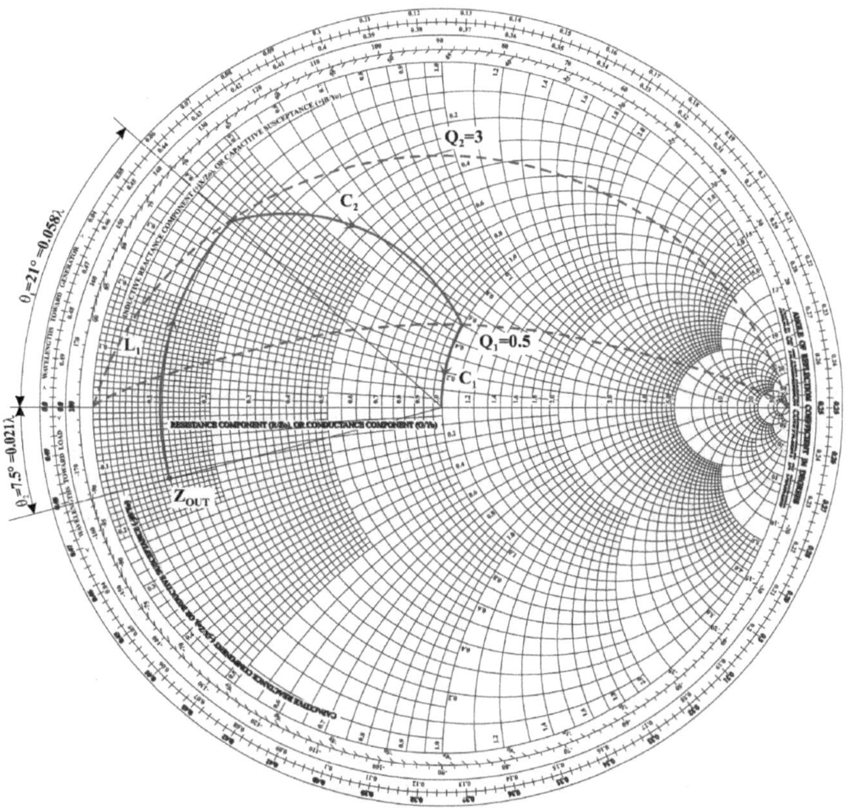

FIGURE 4.36 Smith chart with elements from Figure 4.35.

For example, for a single-stage input lumped matching circuit, the value of quality factor Q is

$$Q > \sqrt{\frac{50}{1.7} - 1} = 5.33$$

which means that the entire frequency range can be appropriately cover using a multistage matching circuit only.

In this case, it is important that the device input quality factor is smaller than 1.63, being equal to $Q_{in} = 1.3/1.7 = 0.76$. It is very convenient to design the input matching circuit (as well as the output matching circuit) by using simple low-pass L-transformers, composed of the series transmission line and shunt capacitor each, with a constant value of Q for each balanced part of the active device. Then, these two input matching circuits are combined by inserting shunt capacitors, the values of which are reduced twice, between the two series transmission lines.

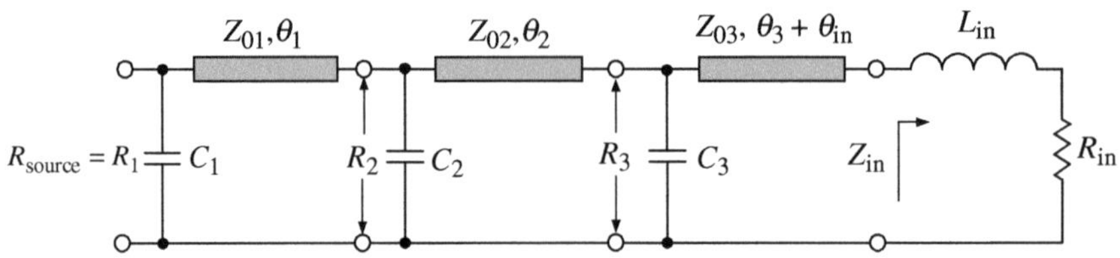

FIGURE 4.37 Complete broadband input two-port network circuit.

To match the series input inductive impedance to the standard 50-Ω source, we use three low-pass *L*-transformers, as shown in Fig. 4.37. In this case, the input resistance R_{in} can be assumed to be constant over entire frequency range. At the center bandwidth frequency of 635 MHz, the input inductance is equal approximately to 0.3 nH. By taking this inductance into account, it is necessary to subtract the appropriate value of the electrical length θ_{in} from the total electrical length θ_3. Because of the short size of this transmission line when $\tan\theta_{in} \approx \theta_{in}$, a value of θ_{in} can be easily calculated in accordance with

$$\theta_{in} \cong \frac{X_{in}}{Z_0} = \frac{\omega L_{in}}{Z_0} \tag{4.105}$$

According to Eq. (4.83), there are two simple possibilities to provide input matching using a technique with equal quality factors of *L*-transformers. One option is to use the same values of characteristic impedance for all transmission lines; the other one is to use the same electrical lengths for all transmission lines. Consider the first approach, which also allows direct use of the Smith chart, and choose the value of the characteristic impedance $Z_0 = Z_{01} = Z_{02} = Z_{03} = 50\ \Omega$. The ratio of input and output resistances can be written as

$$\frac{R_1}{R_2} = \frac{R_2}{R_3} = \frac{R_3}{R_{in}} \tag{4.106}$$

which gives the values $R_2 = 16.2\ \Omega$ and $R_3 = 5.25\ \Omega$ for $R_{source} = R_1 = 50\ \Omega$ and $R_{in} = 1.7\ \Omega$. The values of electrical lengths are determined from the nomograph shown in Fig. 4.29(*a*) as $\theta_1 = 30°$, $\theta_2 = 7.5°$, and $\theta_3 = 2.4°$.

To calculate the quality factor *Q*, from Eq. (4.83) which must be equal for each *L*-transformer, it is enough to know the electrical length θ_1 of the first *L*-transformer. The remaining two electrical lengths can be directly obtained from Eq. (4.82). As a result, the quality factor of each *L*-transformer is equal to a value of $Q = 1.2$. The values of the shunt capacitances using Eq. (4.81) are $C_1 = 6$ pF, $C_2 = 19$ pF, and $C_3 = 57$ pF.

For a constant *Q*, we can simplify significantly the design of the matching circuit by using the Smith chart. After calculating the value of *Q*, it is necessary to plot a constant *Q*-circle on the Smith chart. Figure 4.38 shows the input matching circuit design using the Smith

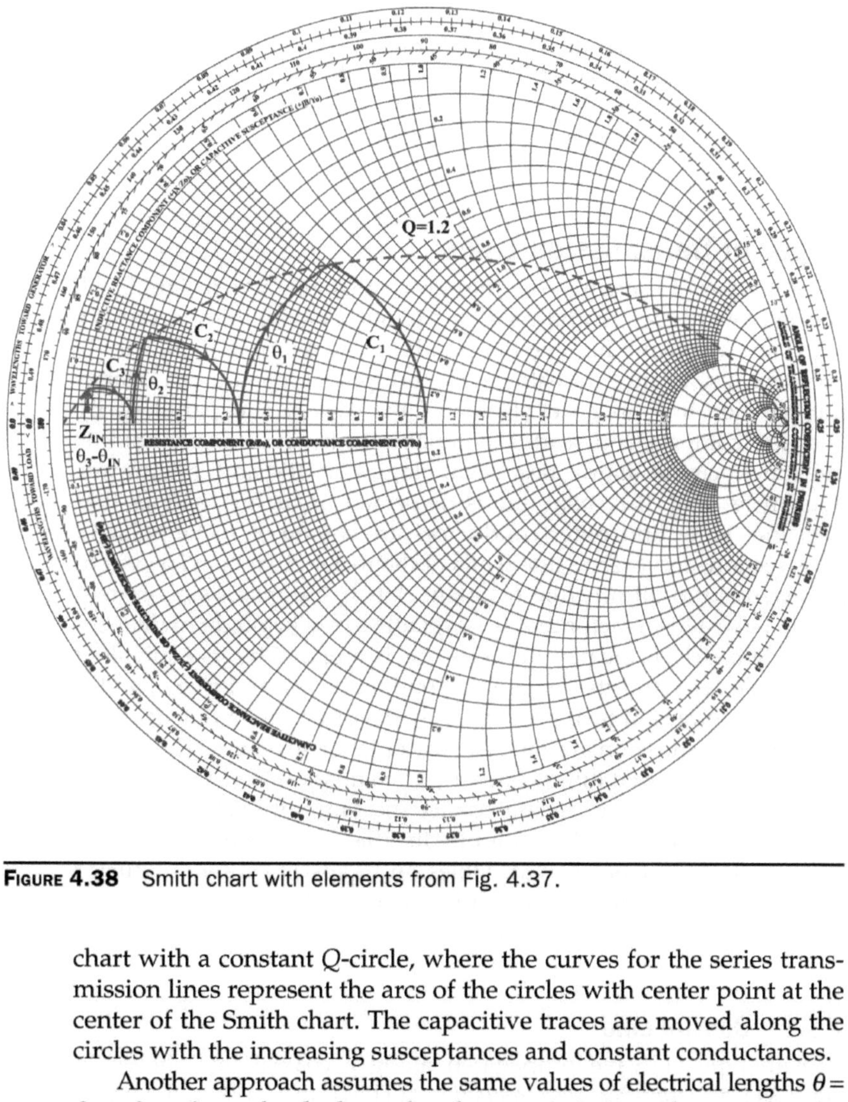

Figure 4.38 Smith chart with elements from Fig. 4.37.

chart with a constant Q-circle, where the curves for the series transmission lines represent the arcs of the circles with center point at the center of the Smith chart. The capacitive traces are moved along the circles with the increasing susceptances and constant conductances.

Another approach assumes the same values of electrical lengths $\theta = \theta_1 = \theta_2 = \theta_3$, and calculates the characteristic impedances of series transmission lines from Eq. (4.83) at equal ratios of the input and output resistances according to Eq. (4.106). Such an approach is more convenient in practical design, because, when using the microstrip lines with standard characteristic impedance $Z_0 = 50\ \Omega$, the electrical length of the transmission line adjacent to the active device input terminal is usually too short. In this case, it makes sense to set the characteristic impedance of the first transmission line to $Z_{01} = 50\ \Omega$. Then, the value of $\theta = 30°$ is determined directly from the nomograph shown in Fig. 4.29(*a*). Further calculation of Q from Eq. (4.81) for fixed θ and Z_0/R_2 yields $Q = 1.2$. The characteristic impedances of the remaining two transmission lines are then calculated easily from Eq. (4.81) or Eq. (4.83). Their values are $Z_{02} = 15.7\ \Omega$ and $Z_{03} = 5.1\ \Omega$ with the same values of the shunt capacitances. Generally, the characteristic impedances and electrical lengths of the transmission lines can be optimized for convenience of practical implementation.

4.5 Types of Transmission Lines

Several types of transmission lines are available when designing RF and microwave active and passive circuits. Coaxial lines have very high bandwidth and high power-handling capability, and are widely used for impedance transformers and power combiners. Planar transmission lines as an evolution of the coaxial and parallel-wire lines are compact and readily adaptable to hybrid and monolithic integrated circuit fabrication technologies at RF and microwave frequencies [20]. If coaxial line is deformed in such a manner that both the center and outer conductors are square or rectangular in cross section, and then if side walls of the rectangular coaxial system are extended to infinity, the resultant flat-strip transmission system would have a form factor that is adaptable to the printed-circuit technique. Similarly, if the parallel-wire line is replaced by its equivalent of a single wire and its image in a conducting ground plane, and if this single wire is in turn progressively distorted into a flat strip, the resulting transmission system is again a planar structure. There is an important aspect that differs flat-strip transmission lines from coaxial lines. In a coaxial line, an impedance discontinuity acts as a shunt capacitance, whereas a discontinuity in a flat strip has a series inductance in its equivalent circuit. Holes and gaps in center conductor strips also represent discontinuities that can be utilized in many applications to microwave circuitry.

4.5.1 Coaxial Line

A main type of wave propagated along a coaxial line shown in Fig. 4.39 is the *transverse electromagnetic* (TEM) wave. When the transverse fields of a TEM wave are the same as the static fields that can exist between the conductors, the electromagnetic properties of a coaxial line can be characterized by the following parameters:

the shunt capacitance per unit length

$$C = 2\pi\varepsilon_0\varepsilon_r / \ln\left(\frac{b}{a}\right) \tag{4.107}$$

where a is the radius of inner conductor and b is the inner radius of outer conductor,

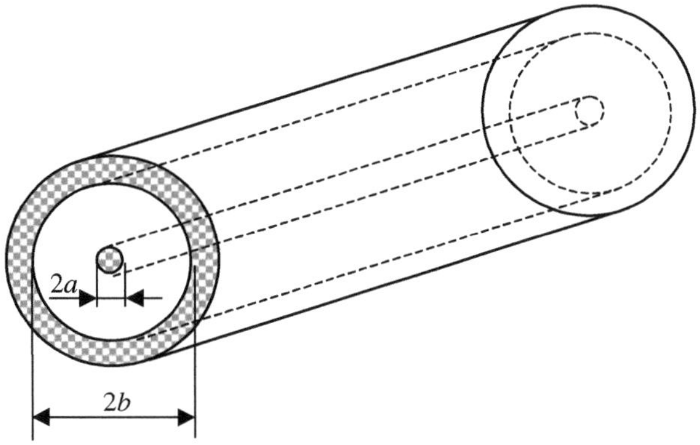

Figure 4.39 Coaxial line structure.

the series inductance per unit length

$$L = \frac{\mu_0 \mu_r}{2\pi} \ln\left(\frac{b}{a}\right) \tag{4.108}$$

where $\mu_0 = 4\pi \times 10^{-7}$ H/m is the permeability of free space and μ_r is the relative magnetic constant or substrate permeability,

the series resistance per unit length

$$R = \frac{R_s}{2\pi}\left(\frac{1}{b} + \frac{1}{a}\right) \tag{4.109}$$

where $R_s = \rho/\Delta(f) = \sqrt{\pi\mu_0 \rho f}$ is the surface resistivity, ρ is the metallization electrical resistivity, $\Delta(f)$ is the penetration depth, and f is the frequency,

the shunt conductance per unit length

$$G = 2\pi\sigma/\ln\left(\frac{b}{a}\right) = 2\pi\omega\varepsilon_0\varepsilon_r \tan\delta/\ln\left(\frac{b}{a}\right) \tag{4.110}$$

where σ is the dielectric conductivity and $\tan\delta$ is the dielectric loss tangent,

the characteristic impedance

$$Z_0 = \frac{\eta}{2\pi} \ln\left(\frac{b}{a}\right) \tag{4.111}$$

where $\eta = \sqrt{\mu/\varepsilon}$ is the wave impedance of the lossless coaxial line identical to the intrinsic impedance of the medium [21].

The conductor loss factor (in Np/m) can be written as

$$\alpha_{\mathrm{c}} = \frac{R}{2Z_0} \tag{4.112}$$

whereas the dielectric loss factor (in Np/m) can be defined by

$$\alpha_{\mathrm{d}} = \frac{GZ_0}{2} = \frac{\sigma\eta}{2} = \pi\sqrt{\varepsilon_{\mathrm{r}}}\,\frac{\tan\delta}{\lambda} \tag{4.113}$$

where λ is the free-space wavelength.

4.5.2 Stripline

The geometry of a commonly used stripline is shown in Fig. 4.40. The strip conductor of width W is placed between two flat dielectric substrates with the same dielectric constant. The outer surfaces of these substrates are metalized and serve as a ground conductor. In practice, the strip conductor is etched on one of the dielectric substrates by photolithography process. Because the stripline has two conductors and a homogeneous dielectric, it can support a pure TEM propagation mode, that is the usual mode of operation. The advantages of striplines are good electromagnetic shielding and low attenuation losses, which make them suitable for high-Q and low-interference applications. However, striplines require strong symmetry that makes their tuning complicated due to difficult access to center conductor. As a result, the stripline structure is not convenient for incorporating chip elements and associated bias circuitry.

The exact expression for the characteristic impedance of a lossless stripline of zero thickness is given by

$$Z_0 = \frac{30\pi}{\sqrt{\varepsilon_{\mathrm{r}}}}\frac{K(k)}{K(k')} \tag{4.114}$$

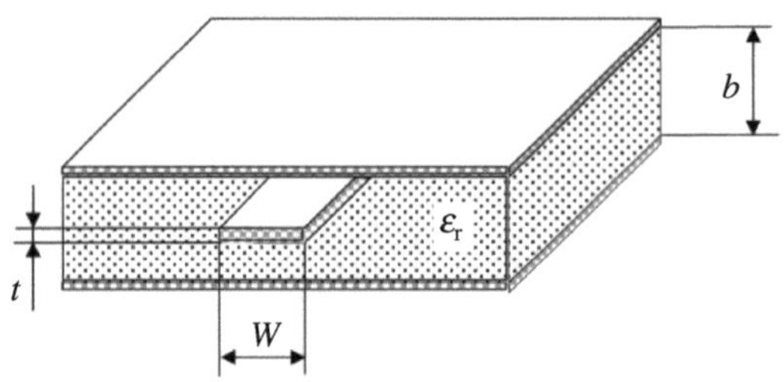

FIGURE 4.40 Stripline structure.

where

$$K(k) = \int_0^{\pi/2} \frac{d\varphi}{\sqrt{1-k^2\sin^2\varphi}} \tag{4.115}$$

is the complete elliptic integral of the first kind, $k = \text{sech}(\pi W/2b)$, and $k' = \sqrt{1-k^2}$ [22, 23].

An expression for the ratio $K(k)/K(k')$ can be simplified to

$$\frac{K(k)}{K(k')} = \begin{cases} \pi/\ln\left(2\,\dfrac{1+\sqrt{k'}}{1-\sqrt{k'}}\right) & \text{for } 0 \le k \le \dfrac{1}{\sqrt{2}} \\ \dfrac{1}{\pi}\ln\left(2\,\dfrac{1+\sqrt{k}}{1-\sqrt{k}}\right) & \text{for } \dfrac{1}{\sqrt{2}} \le k \le 1 \end{cases} \tag{4.116}$$

which provides the relative error lower than 3×10^{-6} [24].

In practice, it makes sense to use a sufficiently simple formula without complicated special functions [25]. In this case, the formula for Z_0 can be written within 1% of the exact results as

$$Z_0 = \frac{30\pi}{\sqrt{\varepsilon_r}} \frac{b}{W_e + 0.441b} \tag{4.117}$$

where W_e is the effective width of the center conductor defined by

$$\frac{W_e}{b} = \frac{W}{b} - \begin{cases} 0 & \text{for } \dfrac{W}{b} > 0.35b \\ \left(0.35 - \dfrac{W}{b}\right)^2 & \text{for } \dfrac{W}{b} < 0.35b \end{cases} \tag{4.118}$$

For a stripline with a TEM propagation mode, the dielectric loss factor α_d is the same as for coaxial line, which is determined by Eq. (4.113). An approximation result for the conductor loss factor α_c (in Np/m) can be obtained by

$$\alpha_c = \begin{cases} A\dfrac{2.7\times10^{-3}\, R_s\varepsilon_r Z_0}{30\pi(b-t)} & \text{for } Z_0\sqrt{\varepsilon_r} < 120 \\ B\dfrac{0.16\, R_s}{Z_0 b\pi} & \text{for } Z_0\sqrt{\varepsilon_r} > 120 \end{cases} \tag{4.119}$$

with

$$A = 1 + \frac{2W}{b - t} + \frac{1}{\pi}\frac{b+t}{b-t}\ln\frac{2b-t}{t} \tag{4.120}$$

$$B = 1 + \frac{b}{0.5W + 0.7t}\left(0.5 + \frac{0.414t}{W} + \frac{1}{2\pi}\ln\frac{4\pi W}{t}\right) \tag{4.121}$$

where t is the thickness of the strip [16].

Figure 4.41 shows the characteristic impedance Z_0 of a stripline as a function of the normalized strip width W/b for various ε_r according to Eqs. (4.117) and (4.118). Typical values of the main electrical and thermal properties of some substrate materials are listed in Table 4.1.

4.5.3 Microstrip Line

In a microstrip line, the grounded metallization surface covers only one side of dielectric substrate, as shown in Fig. 4.42. Such a configuration is equivalent to a pair-wire system for the image of the conductor in the ground plane that produces the required symmetry [26]. In this case, the electric and magnetic field lines are located in both the dielectric region between the strip conductor and the ground plane and in the air region above the substrate. As a result, the electromagnetic wave propagated along a microstrip line is not a pure TEM, because the phase velocities in these two regions are not the same. However, in a quasistatic approximation, which gives sufficiently accurate results as long as the height of the dielectric substrate is very

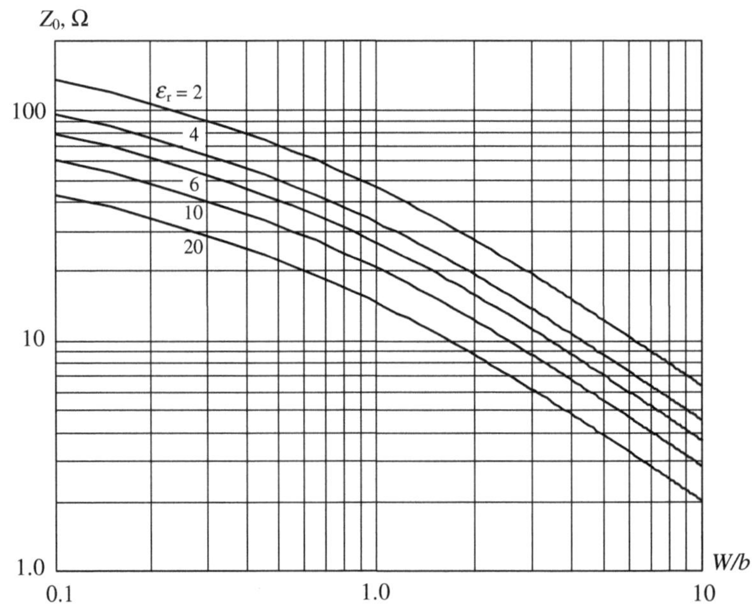

FIGURE 4.41 Stripline characteristic impedance versus W/b.

Typical Substrate	Dielectric Constant ε_r @ 10 GHz	Loss Tangent $\tan\delta$ @ 10 GHz	Coefficient of Thermal Expansion (CTE) ppm/°C
Alumina 99.5%	9.8	0.0003	6.7
Aluminum nitride	8.7	0.001	4.5
Barium tetratitanade	37	0.0002	8.3
Beryllium oxide	6.7	0.006	7.5
Epoxy glass FR-4	4.7	0.01	3.0
Fused quartz	3.78	0.0001	0.5
Gallium arsenide	13.1	0.0016	6.5
Silicon (1 kΩ·cm)	11.9	0.015	4.2
Teflon	2.5	0.0008	15

TABLE 4.1 Electrical and Thermal Properties of Substrate Materials

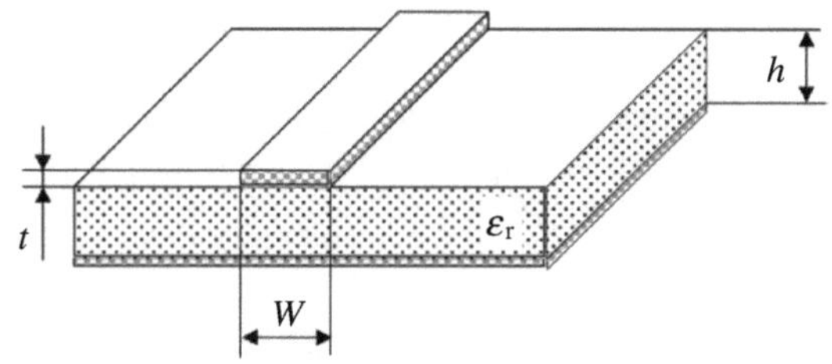

FIGURE 4.42 Microstrip line structure.

small compared with the wavelength, it is possible to obtain the explicit analytical expressions for the electrical characteristics. Because microstrip line is an open structure, it has a major fabrication advantage over the stripline due to simplicity of practical realization, interconnection, and adjustments.

The exact expression for the characteristic impedance of a lossless microstrip line with finite strip thickness is given by [27, 28]

$$Z_0 = \begin{cases} \dfrac{60}{\sqrt{\varepsilon_{re}}} \ln\left(\dfrac{8h}{W_e} + \dfrac{W_e}{4h}\right) & \text{for } \dfrac{W}{h} \le 1 \\ \dfrac{120\pi}{\sqrt{\varepsilon_{re}}}\left[\dfrac{W_e}{h} + 1.393 + 0.667 \ln\left(\dfrac{W_e}{h} + 1.444\right)\right]^{-1} & \text{for } \dfrac{W}{h} \ge 1 \end{cases} \tag{4.122}$$

where

$$\frac{W_e}{h} = \frac{W}{h} + \frac{\Delta W}{h} \tag{4.123}$$

$$\frac{\Delta W}{h} = \begin{cases} \dfrac{1.25}{\pi}\dfrac{t}{h}\left(1 + \ln\dfrac{4\pi W}{t}\right) & \text{for } \dfrac{W}{h} \le 1/2\pi \\ \dfrac{1.25}{\pi}\dfrac{t}{h}\left(1 + \ln\dfrac{2h}{t}\right) & \text{for } \dfrac{W}{h} \ge 1/2\pi \end{cases} \tag{4.124}$$

$$\varepsilon_{re} = \frac{\varepsilon_r + 1}{2} + \frac{\varepsilon_r - 1}{2}\frac{1}{\sqrt{1 + 12h/W}} - \frac{\varepsilon_r - 1}{4.6}\frac{t}{h}\sqrt{\frac{h}{W}} \tag{4.125}$$

Figure 4.43 shows the characteristic impedance Z_0 of a microstrip line with zero strip thickness as a function of the normalized strip width W/h for various ε_r according to Eqs. (4.122) through (4.125).

In practice, it is possible to use a sufficiently simple formula to estimate the characteristic impedance Z_0 of a microstrip line with zero strip thickness written as [29]

$$Z_0 = \frac{120\pi}{\sqrt{\varepsilon_r}}\frac{h}{W}\frac{1}{1 + 1.735\varepsilon_r^{-0.0724}(W/h)^{-0.836}} \tag{4.126}$$

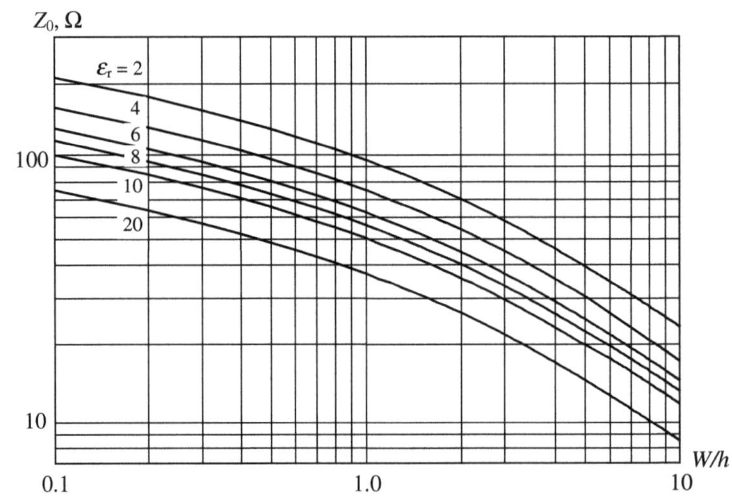

FIGURE 4.43 Microstrip characteristic impedance versus *W/h*.

For a microstrip line in a quasi-TEM approximation, the conductor loss factor α_c (in Np/m) as a function of the microstrip-line geometry can be obtained by

$$\alpha_c = \begin{cases} 1.38A \dfrac{R_s}{hZ_0} \dfrac{32-(W_e/h)^2}{32+(W_e/h)^2} & \text{for } \dfrac{W}{h} \le 1 \\ 6.1 \cdot 10^{-5} A \dfrac{R_s Z_0 \varepsilon_{re}}{h} \left(\dfrac{W_e}{h} + \dfrac{0.667\ W_e/h}{1.444 + W_e/h} \right) & \text{for } \dfrac{W}{h} \ge 1 \end{cases} \tag{4.127}$$

with

$$A = 1 + \frac{h}{W_e}\left(1 + \frac{1}{\pi}\ln\frac{2B}{t}\right) \tag{4.128}$$

$$B = \begin{cases} 2\pi W & \text{for } \dfrac{W}{h} \le 1/2\pi \\ h & \text{for } \dfrac{W}{h} \ge 1/2\pi \end{cases} \tag{4.129}$$

where W_e/h is given by Eqs. (4.123) and (4.124) [30].

The dielectric loss factor α_d (in Np/m) can be calculated by

$$\alpha_d = 27.3 \frac{\varepsilon_r}{\varepsilon_r - 1} \frac{\varepsilon_{re} - 1}{\sqrt{\varepsilon_{re}}} \frac{\tan\delta}{\lambda} \tag{4.130}$$

Conductor loss is a result of several factors related to the metallic material composing the ground plane and walls, among which are conductivity, skin effect, and surface ruggedness. For most microstrip lines (except some kinds of semiconductor substrate such as silicon), the conductor loss is much more significant than the dielectric loss. The conductor losses increase with increasing characteristic impedance due to greater resistance of narrow strips. The electrical resistivity of some conductor materials is given in Table 4.2.

4.5.4 Slotline

Slotlines are usually used when it is necessary to realize a high value of the characteristic impedance Z_0 [31, 32]. A slotline is dual to a microstrip line and represents a narrow slot between two conductive surfaces, one of which is grounded. Changing the width of the slot can easily change the characteristic impedance of the slotline. The transverse electric *H*-mode wave propagates along the slotline. The three basic types of slotlines include unilateral, antipodal, and bilateral. The geometry of a

Material	Symbol	Electrical Resistivity, μΩ·cm	Material	Symbol	Electrical Resistivity, μΩ·cm
Aluminum	Al	2.65	Palladium	Pd	10.69
Coppar	Cu	1.67	Platinum	Pt	10.62
Gold	Au	2.44	Silver	Ag	1.59
Indium	In	15.52	Tantalum	Ta	15.52
Iron	Fe	9.66	Tin	Sn	11.55
Lead	Pb	21.0	Titanium	Ti	55.0
Molybdenum	Mo	5.69	Tungsten	W	5.6
Nickel	Ni	8.71	Zink	Zn	5.68

TABLE 4.2 Electrical Resistivity of Conductor Materials.

unilateral slotline is shown in Fig. 4.44, with a narrow gap in the conductive coating on one side of the dielectric substrate and being bare on the other side of substrate. Slotline can be used either alone or with microstrip line on the opposite side of substrate.

It is difficult to provide exact analytical expressions to calculate the slotline parameters. However, equations for Z_0 can be obtained for a quasi-TEM approximation with zero conductor thickness and infinite width of the entire slotline system written as

for $0.02 \le W/h \le 0.2$

$$Z_0 = 72.62 - 15.283\ \ln\varepsilon_r + 50\left(1 - 0.02\frac{h}{W}\right)\left(\frac{W}{h} - 0.1\right)$$

$$+ (19.23 - 3.693\ \ln\varepsilon_r)\ \ln\left(10^2\frac{W}{h}\right) - \left(11.4 - 2.636\ \ln\varepsilon_r - 10^2\frac{h}{\lambda}\right)^2$$

$$\times\left[0.139\ \ln\varepsilon_r - 0.11 + \frac{W}{h}(0.465\ \ln\varepsilon_r + 1.44)\right] \qquad (4.131)$$

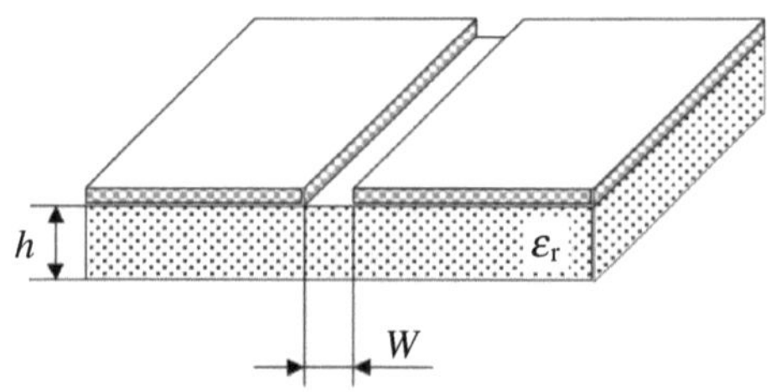

FIGURE 4.44 Slotline structure.

for $0.2 \le W/h \le 1.0$

$$Z_0 = 113.19 - 23.257\ln\varepsilon_r + 1.25\frac{W}{h}(114.59 - 22.531\ln\varepsilon_r)$$

$$+20\left(1-\frac{W}{h}\right)\left(\frac{W}{h}-0.2\right) - \left[0.15 + 0.1\ln\varepsilon_r + \frac{W}{h}(0.899\ln\varepsilon_r - 0.79)\right]$$

$$\times\left[10.25 - 2.171\ln\varepsilon_r + \frac{W}{h}(2.1 - 0.617\ln\varepsilon_r) - 10^2\frac{h}{\lambda}\right]^2 \qquad (4.132)$$

where $0.01 \le h/\lambda \le 0.25/\sqrt{\varepsilon_r - 1}$ and λ is the free-space wavelength [33].

Figure 4.45 shows the characteristic impedance Z_0 of a slotline within the error of 2% as a function of the normalized slot width W/h for $h/\lambda = 0.02$ and various $\varepsilon_r = 9.7, 11, 12, \ldots, 20$ calculated by Eqs. (4.131) and (4.132).

4.5.5 Coplanar Waveguide

A coplanar waveguide (CPW) is similar in structure to a slotline, the only difference being a third conductor centered in the slot region.

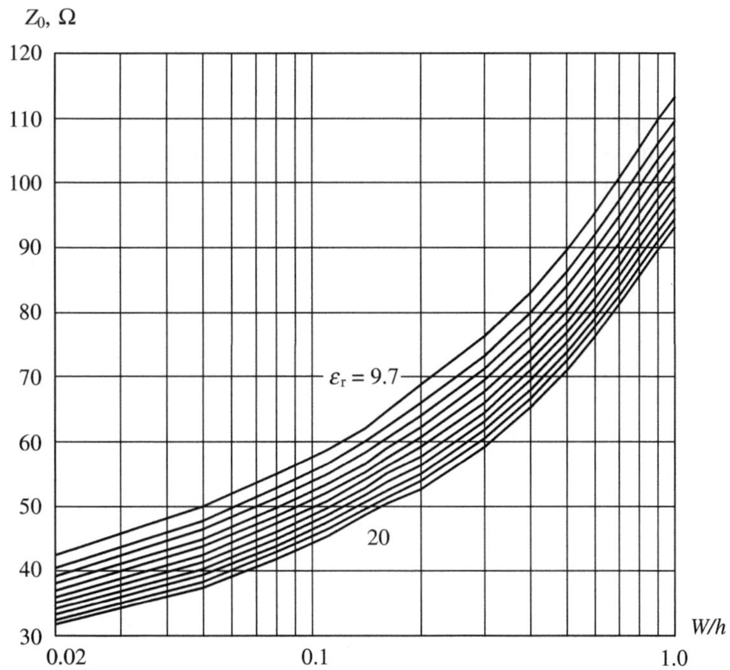

FIGURE 4.45 Slotline characteristic impedance versus W/h.

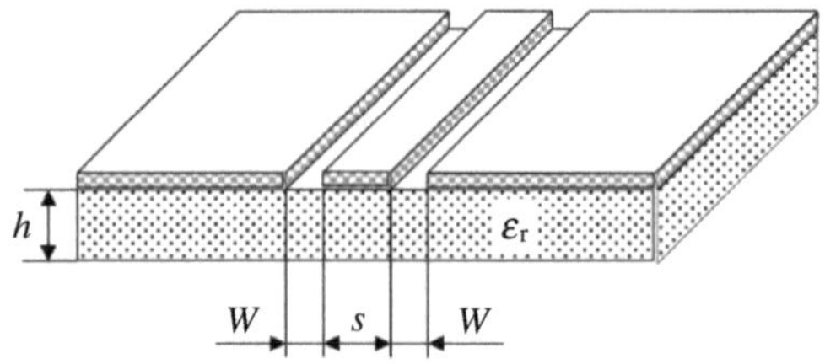

FIGURE 4.46 Coplanar waveguide structure.

The center strip conductor and two outer grounded conductors lie in the same plane on substrate surface, as shown in Fig. 4.46 [34, 35]. A coplanar configuration has some advantages such as low dispersion, ease of attaching shunt and series circuit components, no need for via holes, or simple realization of short-circuited ends, which makes a CPW suitable for hybrid and monolithic integrated circuits. In contrast to the microstrip and stripline, the CPW has shielding between adjacent lines that creates a better isolation between them. However, like microstrip and stripline, the CPW can be also described by a quasi-TEM approximation for both numerical and analytical calculations. Because of the high dielectric constant of the substrate, most of the RF energy is stored in the dielectric and the loading effect of the grounded cover is negligible if it is more than two slot widths away from the surface. Similarly, the thickness of the dielectric substrate with higher relative dielectric constants is not so critical, and practically it should be one or two times the width W of the slots.

The approximate expression of the characteristic impedance Z_0 for zero metal thickness that is satisfactorily accurate in a wide range of substrate thicknesses can be written as

$$Z_0 = \frac{30\pi}{\sqrt{\varepsilon_{re}}} \frac{K(k')}{K(k)} \tag{4.133}$$

where

$$\varepsilon_{re} = 1 + \frac{\varepsilon_r - 1}{2} \frac{K(k')}{K(k)} \frac{K(k_1)}{K(k_1')} \tag{4.134}$$

$k = \frac{s}{s+2W}$, $k_1 = \sinh\frac{\pi s}{4h} / \sinh\frac{\pi(s+2W)}{4h}$, $k' = \sqrt{1-k^2}$, $k_1' = \sqrt{1-k_1^2}$, and K is the complete elliptic integral of the first kind [36]. The values of ratios $K(k)/K(k')$ and $K(k_1)/K(k'_1)$ can be defined from Eq. (4.116). Figure 4.47 shows the characteristic impedance Z_0 of a CPW as a function of the parameter $s/(s + 2W)$ for various ε_r according to Eqs. (4.133) and (4.134).

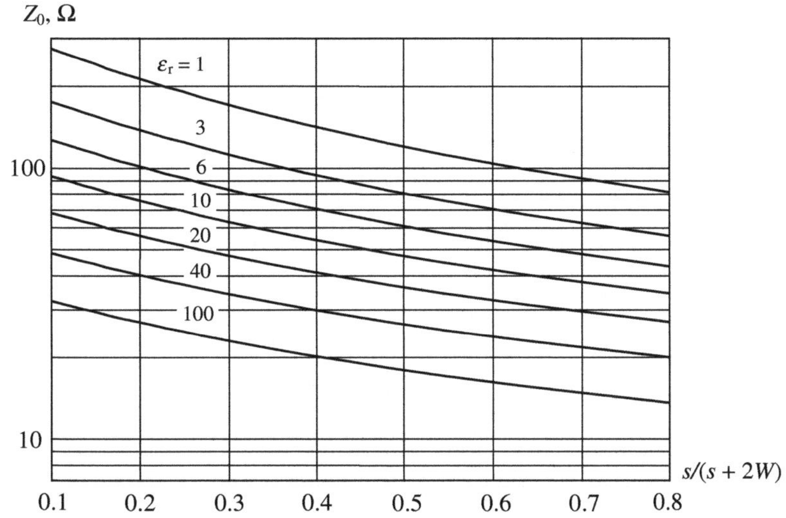

FIGURE 4.47 Coplanar waveguide characteristic impedance versus $s/(s + 2W)$.

References

1. H. W. Bode, "A Method of Impedance Correction," *Bell Syst. Tech. J.*, vol. 9, pp. 794–838, Oct. 1930.
2. S. Roberts, "Conjugate-Image Impedances," *Proc. IRE*, vol. 34, pp. 198–204, Apr. 1946.
3. S. A. Schelkunoff, "The Impedance Concept and Its Application to Problems of Reflection, Refraction, Shielding and Power Absorption," *Bell Syst. Tech. J.*, vol. 17, pp. 17–48, Jan. 1938.
4. H. W. Bode, *Network Analysis and Feedback Amplifier Design*, New York: Van Nostrand, 1945.
5. P. H. Smith, "Transmission Line Calculator," *Electronics*, vol. 12, pp. 29–31, Jan. 1939.
6. P. H. Smith, *Electronic Applications of the Smith Chart*, New York: Noble Publishing, 2000.
7. P. S. Carter, "Charts for Transmission-Line Measurements and Computations," *RCA Rev.*, vol. 3, pp. 355–368, Jan. 1939.
8. H. A. Wheeler, "Reflection Charts Relating to Impedance Matching," *IEEE Trans. Microwave Theory Tech.*, vol. MTT-32, pp. 1008–1021, Sep. 1984.
9. T. Mizubashi, "Theory of Four-Terminal Impedance Transformation Circuit and Matching Circuit (in Japanese)," *The Journal of the Institute of Electrical Communications Engineers in Japan*, vol. 21, pp. 1053–1058, Dec. 1937.
10. H. L. Krauss, C. W. Bostian, and F. H. Raab, *Solid State Radio Engineering*, New York: John Wiley & Sons, 1980.
11. W. L. Everitt, "Output Networks for Radio-Frequency Power Amplifiers," *Proc. IRE*, vol. 19, pp. 725–737, May 1931.
12. L. F. Gray and R. Graham, *Radio Transmitters*, New York: McGraw-Hill, 1961.
13. V. M. Bogachev and V. V. Nikiforov, *Transistor Power Amplifiers* (in Russian), Moskva: Energiya, 1978.
14. N. Nallam and S. Chatterjee, "Multi-Band Frequency Transformations, Matching Networks and Amplifiers," *IEEE Trans. Circuits Syst.–I: Regular Papers*, vol. CAS-I-60, pp. 1635–1647, Jun. 2013.

15. A. Tam, "Network Building Blocks Balance Power Amp Parameters," *Microwaves & RF*, vol. 23, pp. 81–87, Jul. 1984.
16. D. M. Pozar, *Microwave Engineering*, New York: John Wiley & Sons, 2004.
17. Y. Ito, M. Mochizuki, M. Kohno, H. Masuno, T. Takagi, and Y. Mitsui, "A 5-10 GHz 15-W GaAs MESFET Amplifier with Flat Gain and Power Responses," *IEEE Microwave and Guided Wave Lett.*, vol. 5, pp. 454–456, Dec. 1995.
18. D. H. Steinbrecher, "An Interesting Impedance Matching Network," *IEEE Trans. Microwave Theory Tech.*, vol. MTT-15, p. 382, Jun. 1967.
19. A. V. Grebennikov, "Create Transmission-Line Matching Circuits for Power Amplifiers," *Microwaves & RF*, vol. 39, pp. 113-122, Oct. 2000.
20. R. M. Barrett, "Microwave Printed Circuits—The Early Years," *IEEE Trans. Microwave Theory Tech.*, vol. MTT-32, pp. 983–990, Sep. 1984.
21. S. Ramo, J. R. Whinnery, and T. Van Duzer, *Fields and Waves in Communication Electronics*, New York: John Wiley & Sons, 1993.
22. S. B. Kohn, "Characteristic Impedance of the Shielded-Strip Transmission Line," *IRE Trans. Microwave Theory Tech.*, vol. MTT-2, pp. 52–55, Jul. 1954.
23. H. Howe, *Stripline Circuit Design*, Dedham: Artech House, 1974.
24. W. Hilberg, "From Approximations to Exact Relations for Characteristic Impedances," *IEEE Trans. Microwave Theory Tech.*, vol. MTT-17, pp. 259–265, May 1969.
25. I. J. Bahl and R. Garg, "A Designer's Guide to Stripline Circuits," *Microwaves*, vol. 17, pp. 90-96, Jan. 1978.
26. D. D. Grieg and H. F. Engelmann, "Microstrip—A New Transmission Technique for the Kilomegacycle Range," *Proc. IRE*, vol. 40, pp. 1644–1650, Dec. 1952.
27. E. O. Hammerstad, "Equations for Microstrip Circuit Design," *Proc. 5th Europ. Microwave Conf.*, pp. 268–272, 1975.
28. I. J. Bahl and R. Garg, "Simple and Accurate Formulas for Microstrip with Finite Strip Thickness," *Proc. IEEE*, vol. 65, pp. 1611–1612, Nov. 1977.
29. R. S. Carson, *High-Frequency Amplifiers*. New York: John Wiley & Sons, 1975.
30. K. C. Gupta, R. Garg, and R. Chadha, *Computer-Aided Design of Microwave Circuits*, Dedham: Artech House, 1981.
31. S. B. Cohn, "Slot Line on a Dielectric Substrate," *IEEE Trans. Microwave Theory Tech.*, vol. MTT-17, pp. 768–778, Oct. 1969.
32. E. A. Mariani, C. P. Heinzman, J. P. Agrios, and S. B. Cohn, "Slot Line Characteristics," *IEEE Trans. Microwave Theory Tech.*, vol. MTT-17, pp. 1091–1096, Dec. 1969.
33. R. Garg and K. C. Gupta, "Expression for Wavelength and Impedance of a Slotline," *IEEE Trans. Microwave Theory Tech.*, vol. MTT-24, p. 532, Aug. 1976.
34. C. P. Weng, "Coplanar Waveguide: A Surface Strip Transmission Line Suitable for Nonreciprocal Gyromagnetic Device Applications," *IEEE Trans. Microwave Theory Tech.*, vol. MTT-17, pp. 1087–1090, Dec. 1969.
35. R. N. Simons, *Coplanar Waveguide Circuits, Components, and Systems*. New York: John Wiley & Sons, 2001.
36. G. Ghione and C. Naldli, "Analytical Formulas for Coplanar Lines in Hybrid and Monolithic," *Electronics Lett.*, vol. 20, pp. 179–181, Feb. 1984.

CHAPTER 5

Power Transformers, Combiners, and Couplers

It is critical, particularly at higher frequencies, that the special types of combiners and dividers are used to avoid insufficient power performance of the individual active devices. The methods of configuration of the combiners or dividers differ depending on the operating frequency, frequency bandwidth, output power, and size requirements. Coaxial cable combiners with ferrite cores are used to combine the output powers of power amplifiers intended for wideband applications. The device output impedance is usually sufficiently small for high power levels, and the coaxial-line transformers are often used to match this impedance with a standard 50-Ω load, especially if it is necessary to do over wide frequency range. For narrowband applications, the *N*-way Wilkinson combiners are widely used due to their simple practical realization. At microwave frequencies, the size of combiners should be very small, and therefore the hybrid microstrip combiners (including different types of the microwaves hybrids and directional couplers) are commonly used to combine output powers of the power amplifiers. In this chapter, the basic properties of three- and four-port networks, as well as a variety of different combiners, transformers, and directional couplers, are described.

5.1 Basic Properties

Basic three- or four-port networks can be used to divide the output power of a single power source or to combine the output powers of two or more power amplifiers. Generally, the multiport network required to combine the output powers of N identical power sources is based on these basic networks. In this case, it is very important to match the output impedances of all power amplifiers with the load to provide the overall output power that is N times larger than the output power of a single power amplifier. Besides, changes in the operation condition of one power amplifier should not affect the operation conditions of the remaining power amplifiers. To satisfy this requirement, all input ports of the combiner should be decoupled to be mutually independent. When one of the power amplifiers is eliminated, the total output power must decrease by as little as possible to be within the limits of a maximum permissible level. In addition, the combiners can effectively be used for both narrow-band and broadband transmitters.

5.1.1 Three-Port Networks

The simplest devices used for power dividing and combining represent the three-port networks with one input and two outputs, as shown in Fig. 5.1(*a*), and two inputs and one output in the power combiner, as shown in Fig. 5.1(*b*). The scattering *S*-matrix of an arbitrary three-port network can be written as

$$[S] = \begin{bmatrix} S_{11} & S_{12} & S_{13} \\ S_{21} & S_{22} & S_{23} \\ S_{31} & S_{32} & S_{33} \end{bmatrix} \tag{5.1}$$

where $S_{ij} = S_{ji}$ for the symmetric scattering *S*-matrix when all components are passive and reciprocal. In this case, if all ports are appropriately matched (when $S_{ii} = 0$), the scattering *S*-matrix reduces to

$$[S] = \begin{bmatrix} 0 & S_{12} & S_{13} \\ S_{12} & 0 & S_{23} \\ S_{13} & S_{23} & 0 \end{bmatrix} \tag{5.2}$$

A lossless condition applied to a fully matched *S*-matrix given by Eq. (5.2) requires it to be a unitary matrix when

$$[S]^* [S] = 1 \tag{5.3}$$

where $[S]^*$ is the matrix, which is complex-conjugated to the original *S*-matrix [1, 2].

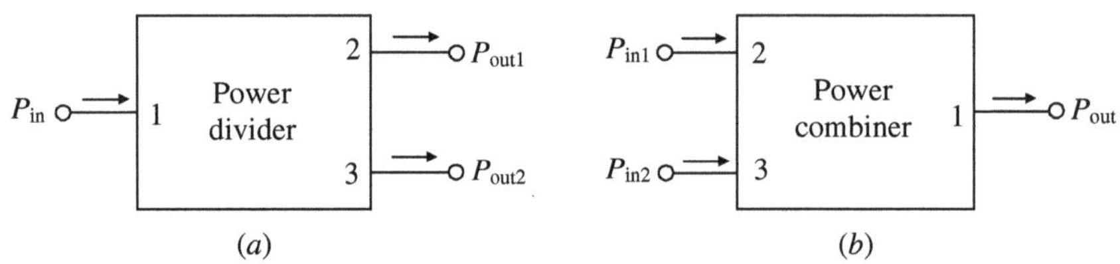

FIGURE 5.1 Schematic diagrams of power divider and power combiner.

As a result of multiplying two matrices,

$$|S_{12}|^2 + |S_{13}|^2 = |S_{12}|^2 + |S_{23}|^2 = |S_{13}|^2 + |S_{23}|^2 = 1 \tag{5.4}$$

$$S_{13}^* S_{23} = S_{23}^* S_{12} = S_{12}^* S_{13} = 0 \tag{5.5}$$

From Eq. (5.5), it follows that at least two of three available parameters S_{12}, S_{13}, and S_{23} should be zero, which is inconsistent with at least one condition given by Eq. (5.4). This means that a three-port network cannot be lossless, reciprocal, and matched at all ports. However, if any one of these three conditions is not fulfilled, a physically realizable device is possible for practical implementation. A lossless and reciprocal three-port network can be realized if only two of its ports are matched, or it is lossy being reciprocal and fully matched at all three ports, as in the case of the resistive divider.

If a reciprocal three-port network represents the 3-dB power divider when, for a given input power at the port 1, the output powers at the ports 2 and 3 are equal, then, according to Eq. (5.4),

$$|S_{12}| = |S_{13}| = \frac{1}{\sqrt{2}} \tag{5.6}$$

5.1.2 Four-Port Networks

The four-port networks are used for directional power coupling when, for a given input signal at the port 1, the output signals are delivered to the ports 2 and 3, and no power is delivered to the port 4 (ideal case), as shown in Fig. 5.2. The scattering S-matrix of a reciprocal four-port network matched at all its ports is given by

$$[S] = \begin{bmatrix} 0 & S_{12} & S_{13} & S_{14} \\ S_{12} & 0 & S_{23} & S_{24} \\ S_{13} & S_{23} & 0 & S_{34} \\ S_{14} & S_{24} & S_{34} & 0 \end{bmatrix} \tag{5.7}$$

where $S_{ij} = S_{ji}$ for the symmetric scattering S-matrix when all components are passive and reciprocal. In this case, the power supplied to

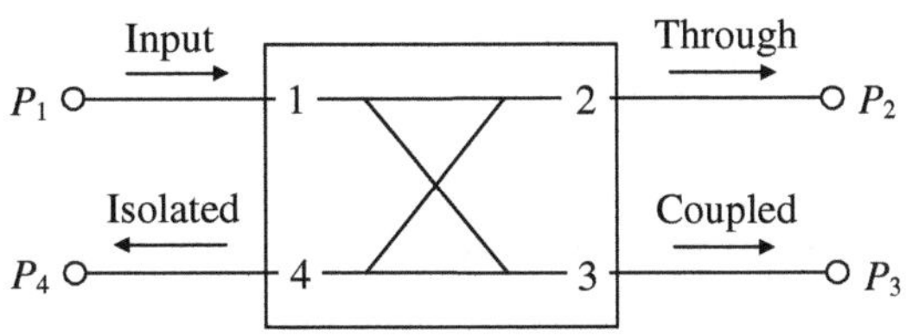

FIGURE 5.2 Schematic diagram of directional coupler.

the input port 1 is coupled to the coupled port 3 with a coupling factor $|S_{13}|^2$, whereas the remainder of the input power is delivered to the through port 2 with a coupling factor $|S_{12}|^2$.

For a lossless four-port network, the unitary condition of the fully matched S-matrix given by Eq. (5.7) results in

$$|S_{12}|^2 + |S_{13}|^2 = |S_{12}|^2 + |S_{24}|^2 = |S_{13}|^2 + |S_{34}|^2 = |S_{24}|^2 + |S_{34}|^2 = 1 \tag{5.8}$$

which implies a full isolation between ports 2 and 3, and ports 1 and 4 respectively when

$$S_{14} = S_{41} = S_{23} = S_{32} = 0 \tag{5.9}$$

and that

$$|S_{13}| = |S_{24}| \qquad |S_{12}| = |S_{34}| \tag{5.10}$$

The scattering S-matrix of such a directional coupler, matched at all its ports with two decoupled two-port networks, reduces to

$$[S] = \begin{bmatrix} 0 & S_{12} & S_{13} & 0 \\ S_{12} & 0 & 0 & S_{24} \\ S_{13} & 0 & 0 & S_{34} \\ 0 & S_{24} & S_{34} & 0 \end{bmatrix} \tag{5.11}$$

The directional coupler can be classified according to the phase shift ϕ between its two output ports 2 and 3 as the in-phase coupler with $\phi = 0$, quadrature coupler with $\phi = 90°$ or $\pi/2$, and out-of-phase coupler with $\phi = 180°$ or π. The following important quantities are used to characterize the directional coupler:

- The power-split ratio or power division ratio K^2, which is calculated as the ratio of powers at the output ports when all ports are nominally (reflectionless) terminated,

$$K^2 = \frac{P_2}{P_3} \tag{5.12}$$

- The insertion loss C_{12}, which is calculated as the ratio of powers at the input port 1 relative the output port 2,

$$C_{12} = 10 \log_{10} \frac{P_1}{P_2} = -20 \log_{10} |S_{12}| \tag{5.13}$$

- The coupling C_{13}, which is calculated as the ratio of powers at the input port 1 relative to the output port 3,

$$C_{13} = 10 \log_{10} \frac{P_1}{P_3} = -20 \log_{10} |S_{13}| \tag{5.14}$$

- The directivity C_{34}, which is calculated as the ratio of powers at the output port 3 relative to the isolated port 4,

$$C_{34} = 10\log_{10}\frac{P_3}{P_4} = 20\log_{10}\frac{|S_{13}|}{|S_{14}|} \tag{5.15}$$

- The isolation C_{14} and C_{23}, which are calculated as the ratios of powers at the input port 1 relative to the isolated port 4 and between the two output ports (output port 2 is considered an input port), respectively,

$$C_{14} = 10\log_{10}\frac{P_1}{P_4} = -20\log_{10}|S_{14}| \tag{5.16}$$

$$C_{23} = 10\log_{10}\frac{P_2}{P_3} = -20\log_{10}|S_{23}| \tag{5.17}$$

- The voltage standing wave ratio at each port or $VSWR_i$, where $i = 1, 2, 3, 4$, which is calculated as

$$VSWR_i = \frac{1 + |S_{ii}|}{1 - |S_{ii}|} \tag{5.18}$$

In an ideal case, the directional coupler would have $VSWR_i = 1$ at each port, an insertion loss $C_{12} = 3$ dB, a coupling $C_{13} = 3$ dB, an infinite isolation and a directivity $C_{14} = C_{23} = C_{34} = \infty$.

5.2 Transmission-Line Transformers and Combiners

The transmission-line transformers and combiners can provide very wide operating bandwidths and operate up to frequencies of 3 GHz and higher [3, 4]. They are widely used in matching networks for antennas and power amplifiers in the HF and VHF bands, as well as in mixer circuits, and their low losses make them especially useful in high-power circuits [5, 6]. Typical structures for transmission-line transformers consist of parallel wires, coaxial cables, or bifilar twisted wire pairs. In the latter case, the characteristic impedance can easily be determined by the wire diameter, the insulation thickness, and, to some extent, by the twisting pitch [7, 8]. For coaxial cable transformers with correctly chosen characteristic impedance, the theoretical high-frequency bandwidth limit is reached when the cable length comes in order of a half wavelength, with the overall achievable bandwidth being about a decade. By introducing the low-loss high-permeability ferrites alongside a good-quality semi-rigid coaxial or symmetrical strip cable, the low frequency limit can be significantly improved providing bandwidths of several or more decades.

The concept of a broadband impedance transformer consisted of a pair of interconnected transmission lines was first disclosed and described by Guanella [9, 10]. Figure 5.3(*a*) shows the Guanella transformer system with a transmission-line character achieved by an arrangement comprising one pair of cylindrical coils, which are wound in the same sense and are spaced a certain distance apart by an intervening dielectric. In this case, one cylindrical coil is located inside the insulating cylinder and the other coil is located on the outside of this cylinder. For the currents flowing through both windings in opposite directions, the corresponding flux in the coil axis is negligibly small. However, for the currents flowing in the same direction through both coils, the latter may be assumed to be connected in parallel, and a coil pair represents a considerable inductance for such currents and acts like a choke coil. With the terminal 4 being grounded, such a 1:1 transformer provides matching of a balanced source to an unbalanced load and is called the *balun* (*ba*lanced-to-*un*balanced transformer). In this case, if the terminal 2 is grounded, it represents simply a delay line. In a particular case, when the terminals 2 and 3 are grounded, the transformer performs as a phase inverter. The series-parallel connection of a plurality of coil pairs can produce a match between unequal source and load resistances. Figure 5.3(*b*) shows a 4:1 impedance (2:1 voltage) transmission-line transformer where the

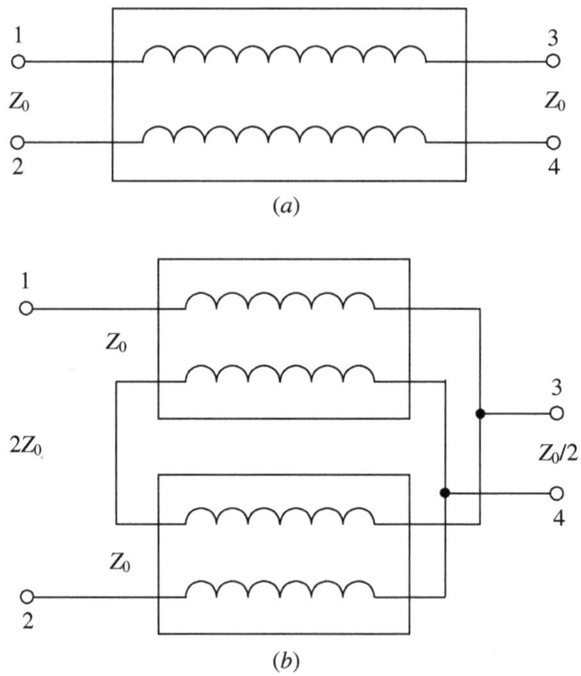

FIGURE 5.3 Schematic configurations of Guanella 1:1 and 4:1 transformers.

two pairs of cylindrical transmission-line coils are connected in series at the input and in parallel at the output. For the characteristic impedance Z_0 of each transmission line, this results in two times higher impedance $2Z_0$ at the input and two times lower impedance $Z_0/2$ at the output. By grounding the terminal 4, such a 4:1 impedance transformer provides impedance matching of the balanced source to the unbalanced load. In this case, when the terminal 2 is grounded, it performs as a 4:1 *unun* (*un*balanced-to-*un*balanced transformer). With a series-parallel connection of n coil pairs with the characteristic impedance Z_0 each, the input impedance is equal to nZ_0 and the output impedance is equal to Z_0/n. Because Guanella transformer adds voltages that have equal delays through the transmission lines, such a technique results in the so-called *equal-delay* transmission-line transformers.

The simplest transmission-line transformer represents a quarterwave transmission line whose characteristic impedance is chosen to give the correct impedance transformation. However, this transformer provides a narrow-band performance valid only around frequencies, for which the transmission line is odd multiples of a quarter-wavelength. If a ferrite sleeve is added to the transmission line, common-mode currents flowing in both transmission-line inner and outer conductors in phase and in the same direction are suppressed and the load may be balanced and floating above ground [11, 12]. If the characteristic impedance of the transmission line is equal to the terminating impedances, the transmission is inherently broadband. If not, there will be a dip in the response at the frequency, at which the transmission line is a quarter-wavelength long.

A coaxial cable transformer, whole the physical configuration and equivalent circuit representation are shown in Fig. 5.4(*a*) and 5.4(*b*), respectively, consists of the coaxial line arranged inside the ferrite core or wound around the ferrite core. Because of its practical configuration, the coaxial cable transformer takes a position between the lumped and distributed systems. Therefore, at lower frequencies its equivalent circuit represents a conventional low-frequency transformer shown in Fig. 5.4(*c*), whereas at higher frequency it is a transmission line with the characteristic impedance Z_0 shown in Fig. 5.4(*d*). The advantage of such a transformer is that the parasitic interturn capacitance determines its characteristic impedance, whereas in the conventional wire-wound transformer with discrete windings this parasitic capacitance negatively contributes to the transformer frequency performance.

When $R_S = R_L = Z_0$, the transmission line can be considered a transformer with a 1:1 impedance transformation. To avoid any resonant phenomena, especially for complex loads, which can contribute to the significant output power variations, as a general rule, the length

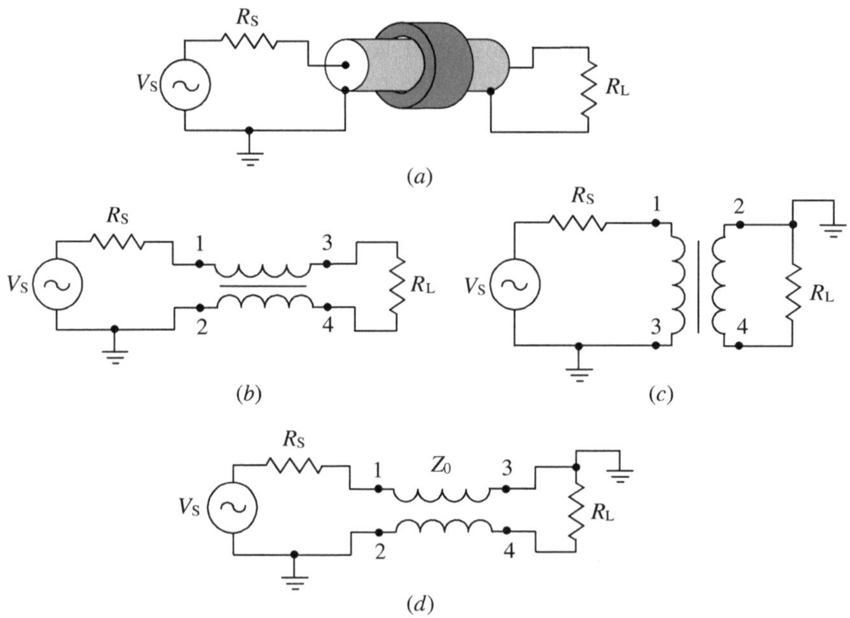

FIGURE 5.4 Schematic configurations of coaxial cable transformer.

l of the transmission line is kept to no more than an eight of a wavelength λ_{min} at the highest operating frequency,

$$l \le \frac{\lambda_{min}}{8} \tag{5.19}$$

where λ_{min} is the minimum wavelength in the transmission line corresponding to the high operating frequency f_{max}.

The low-frequency bandwidth limit of a coaxial cable transformer is determined by the effect of the magnetizing inductance L_m of the outer surface of the outer conductor according to the equivalent low-frequency transformer model shown in Fig. 5.5(*a*), where the transmission line is represented by the ideal 1:1 transformer [6]. The resistance R_0 represents the losses of the transmission line. An approximation to the magnetizing inductance can be made by considering the outer surface of the coaxial cable to be the same as that of a straight wire (or linear conductor) which, at higher frequencies where the skin effect causes the current to be concentrated on the outer surface, would have the self-inductance defined by

$$L_m = 2l\left[\ln\left(\frac{2l}{r}\right) - 1\right] \text{ nH} \tag{5.20}$$

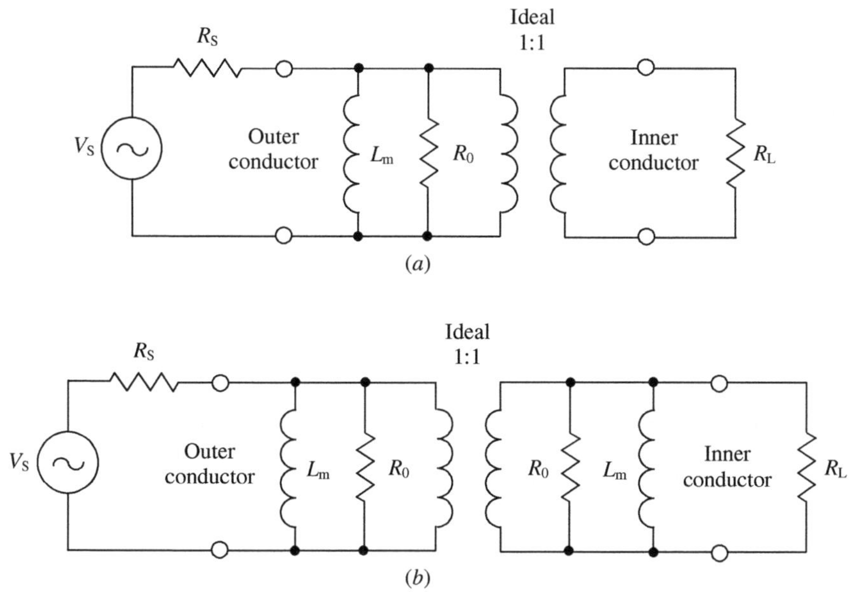

FIGURE 5.5 Low-frequency models of 1:1 coaxial cable transformer.

where l is the length of the coaxial cable in cm and r is the radius of the outer surface of the outer conductor in cm [6].

High permeability of core materials results in shorter transmission lines. If a toroid is used for the core, the magnetizing inductance L_m is obtained by

$$L_m = 4\pi\, n^2 \mu\, \frac{A_e}{L_e}\ \text{nH} \tag{5.21}$$

where n is the number of turns, μ is the core permeability, A_e is the effective cross-sectional area of the core in cm^2, and L_e is the average magnetic path length in cm [13].

By considering the transformer equivalent circuit shown in Fig. 5.5(*a*), the ratio between the power delivered to the load P_L and power available at the source $P_S = V_S^2 / 8R_S$, when $R_S = R_L$, can be obtained from

$$\frac{P_L}{P_S} = \frac{(2\omega L_m)^2}{R_S^2 + (2\omega L_m)^2} \tag{5.22}$$

which gives the minimum operating frequency f_{min} for a given magnetizing inductance L_m as

$$f_{min} \geq \frac{R_S}{4\pi\, L_m} \tag{5.23}$$

when taking into account the maximum decrease of the output power by 3 dB.

The similar low-frequency model for a coaxial cable transformer using twisted or parallel wires is shown in Fig. 5.5(*b*) [6]. Here, the model is symmetrical as both conductors are exposed to any magnetic material, and therefore contribute identically to the losses and low-frequency performance of the transformer.

An approach using the transmission line based on a single bifilar wound coil to realize a broadband 1:4 impedance transformation was introduced by Ruthroff [14, 15]. In this case, by using a core material of sufficiently high permeability, the number of turns can be significantly reduced. Figure 5.6(*a*) shows the circuit schematic of an unbalanced-to-unbalanced 1:4 transmission-line transformer, where the terminal 4 is connected to the input terminal 1. As a result, for $V = V_1 = V_2$, the output voltage is twice the input voltage, and the transformer has a 1:2 voltage step-up ratio. As the ratio of input voltage to input current is one-fourth the load voltage to load current, the transformer is fully matched for maximum power transfer when $R_L = 4R_S$, and the transmission-line characteristic impedance Z_0 is equal to the geometric mean of the source and load impedances,

$$Z_0 = \sqrt{R_S R_L} \tag{5.24}$$

where R_S is the source resistance and R_L is the load resistance. Figure 5.6(*b*) shows the impedance transformer acting as a phase inverter, where the load resistance is included between the terminals 1 and 4 to become a 1:4 balun. This technique is called the *bootstrap effect*, which doesnot have the same high-frequency response as Guanella equal-delay approach because it adds a delayed voltage to a direct one [16]. The delay becomes excessive when the transmission line reaches a significant fraction of a wavelength.

Figure 5.7(*a*) shows the physical implementation of a 4:1 impedance Ruthroff transformer using a coaxial cable arranged inside the ferrite core. At lower frequencies, such a transformer can be considered an ordinary 2:1 voltage autotransformer. The insertion loss for a broadband 4:1 impedance transformer, as a function of the transmission-line electrical length θ, can be calculated from

$$\frac{P_L}{P_S} = \frac{4R_S R_L (1+\cos\theta)^2}{\left[2R_S(1+\cos\theta) + R_L\cos\theta\right]^2 + \left(\dfrac{R_S R_L + Z_0^2}{Z_0}\right)^2 \sin^2\theta} \tag{5.25}$$

where P_S is the maximum available power from the source with internal resistance R_S and P_L is the power delivered to the load R_L [14, 17]. For a matched transformer when $R_L = 4R_S$ and $Z_0 = 2R_S$, Eq. (5.25) reduces to

$$\frac{P_L}{P_S} = \frac{4\,(1+\cos\theta)^2}{(1+3\,\cos\theta)^2 + 4\sin^2\theta} \tag{5.26}$$

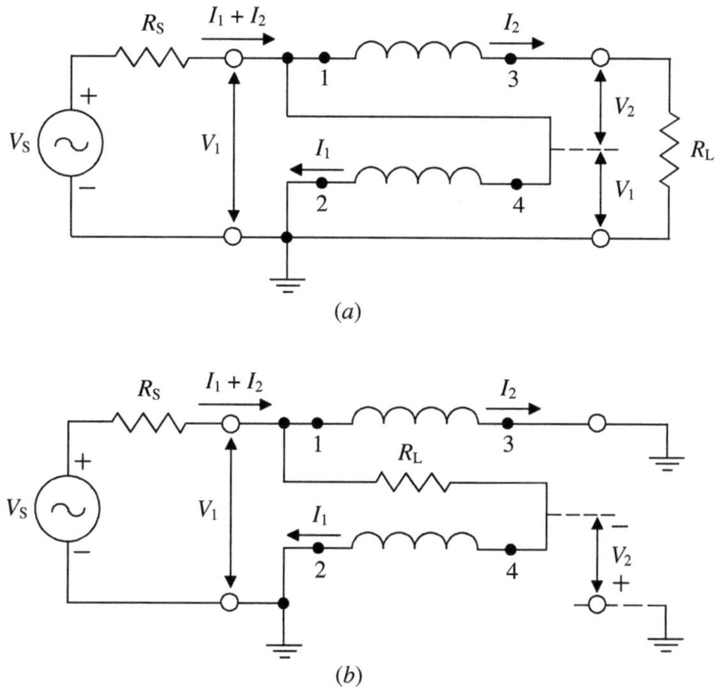

Figure 5.6 Schematic configurations of Ruthroff 1:4 impedance transformer.

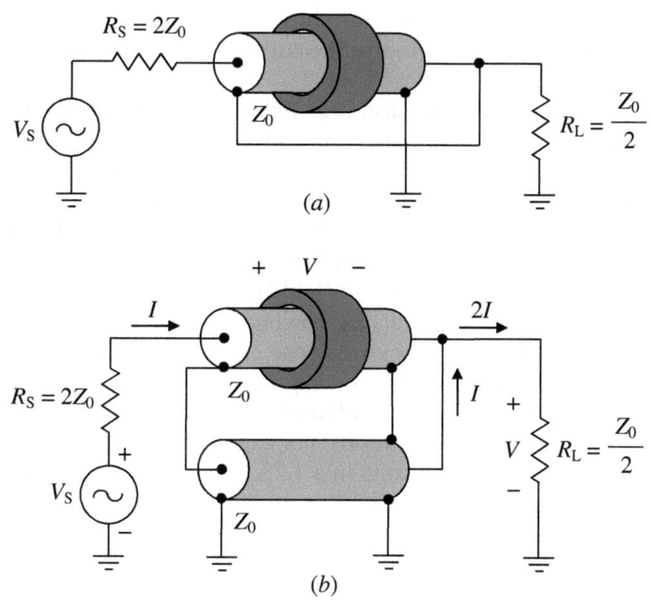

Figure 5.7 Schematic configurations of 4:1 coaxial cable transformer.

To improve the performance at higher frequencies, it is necessary to add an additional phase-compensating line of the same length, as shown in Fig. 5.7(*b*), resulting in a Guanella ferrite-based 4:1 impedance transformer. In this case, a ferrite core is necessary only for the upper line because the outer conductor of the lower line is grounded at both ends, and no current is flowing through it. A current I driven into the inner conductor of the upper line produces a current I that flows in the outer conductor of the upper line, resulting in a current $2I$ flowing into the load R_L. Because the voltage $2V$ from the transformer input is divided in two equal parts between the coaxial line and the load, such a transformer provides an impedance transformation from $R_S = 2Z_0$ into $R_L = Z_0/2$, where Z_0 is the characteristic impedance of each coaxial line. The bandwidth extension for the Ruthroff transformers can also be achieved by using the transmission lines with a step-function and exponential changes in their characteristic impedances [18, 19]. To adopt this transmission-line transformer for microwave planar applications, the coaxial line can be replaced by a pair of stacked strip conductors or coupled microstrip lines [20, 21].

Figure 5.8 shows similar arrangements for the 3:1 voltage coaxial cable transformers, which produce 9:1 impedance transformation [22]. A current I driven into the inner conductor of the upper line in Fig. 5.8(*a*) will cause a current I to flow in the outer conductor of the upper line. This current then produces a current I in the outer conductor of the lower line, resulting in a current $3I$ flowing into the load R_L. The lowest coaxial line can be removed, resulting in a 9:1 impedance coaxial cable transformer shown in Fig. 5.8(*b*). The characteristic impedance of each transmission line is specified by the voltage applied to the end of the line and the current flowing through the line and is equal to Z_0.

By using the transmission-line baluns with different integer-transformation ratios in certain connection, it is possible to obtain the fractional-ratio baluns and ununs [23, 24]. Figure 5.9 shows the transformer configuration for obtaining an impedance ratio of 2.25:1, which consists of a 1:1 Guanella balun on the top combined with a 1:4 Guanella balun where voltages on the left-hand side are in series and on the right-hand side are in parallel [25]. In this case, the left-hand side has the higher impedance. In a matched condition, this transformer should have a high-frequency response similar to a single transmission line. By grounding the corresponding terminals (shown by dashed lines), it becomes a broadband unun. Different ratios can be obtained with other configurations. For example, using a 1:9 Guanella balun below the 1:1 unit results in a 1.78:1 impedance ratio, whereas the impedance ratio becomes 1.56:1 with a 1:16 balun.

On the other hand, the overall 1:1.5 voltage transformer configuration, whose block schematic is shown in Fig. 5.10(*a*), can be achieved by using the cascade connection of a 1:3 voltage transformer to increase the impedance by 9 times, and a 2:1 voltage transformer to

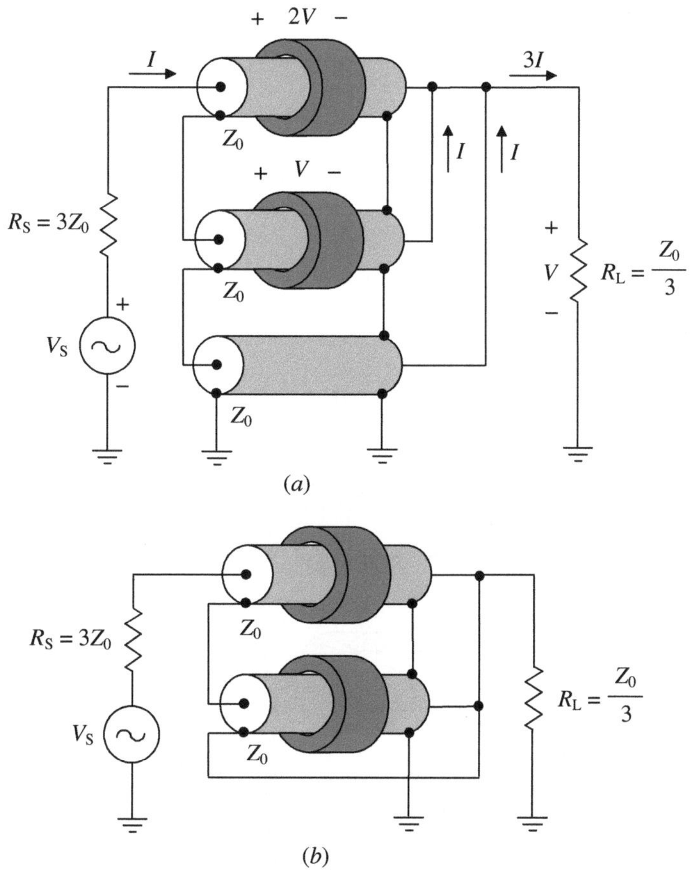

FIGURE 5.8 Schematic configurations of 9:1 coaxial cable transformer.

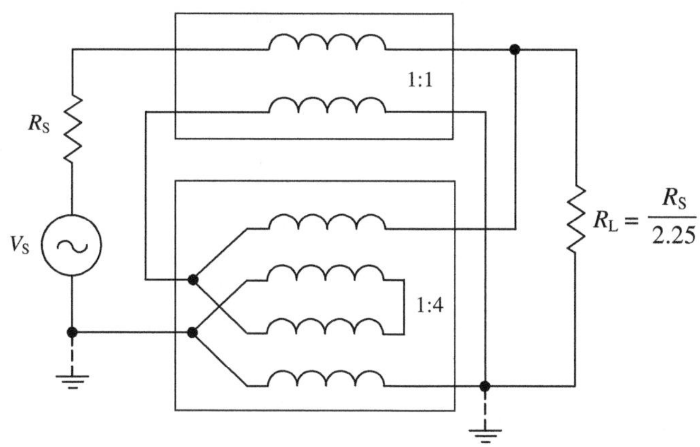

FIGURE 5.9 Schematic configuration of equal-delay 2.25:1 unun.

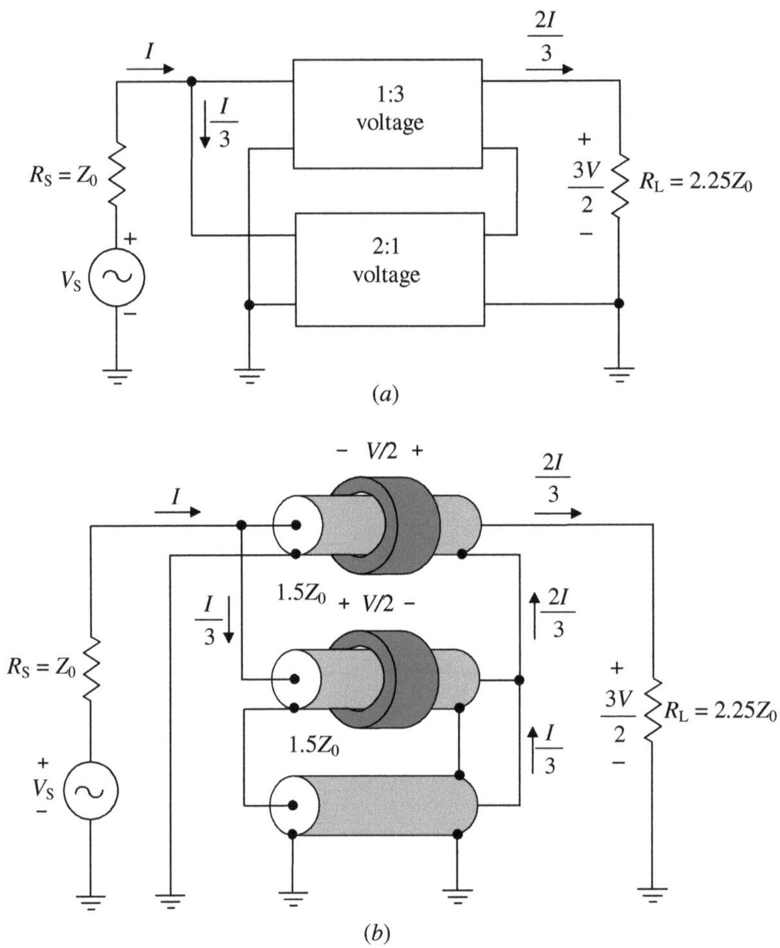

FIGURE 5.10 Schematic configurations of fractional 1:2.25 impedance transformer.

decrease the impedance by four times [23]. The transformer practical configuration using coaxial cables and ferrite cores is shown in Fig. 5.10(*b*). Here, the currents $I/3$ in the inner conductors of two lower lines cause an overall current $2I/3$ in the outer conductor of the upper line, resulting in a current $2I/3$ flowing into the load R_L. A load voltage $3V/2$ is 180° out of phase with a longitudinal voltage $V/2$ along the upper line, resulting in a voltage V at the transformer input. The lowest line also can be eliminated with direct connection of the points at both ends of its inner conductor, as in the case of the 2:1 and 3:1 Ruthroff voltage transformers shown in Figs. 5.7(*a*) and 5.8(*b*), respectively. If the source impedance is 50 Ω, the characteristic impedance of all

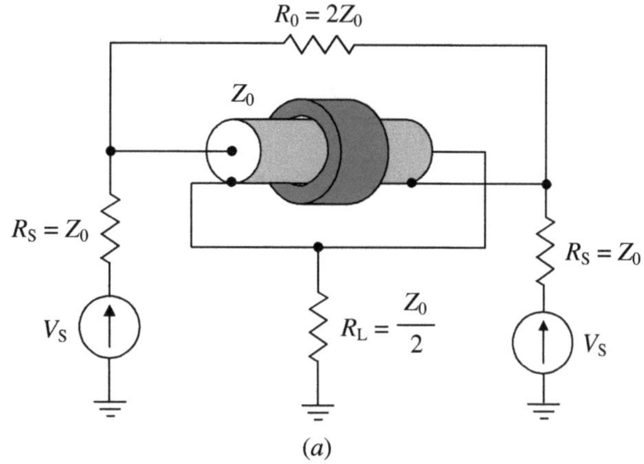

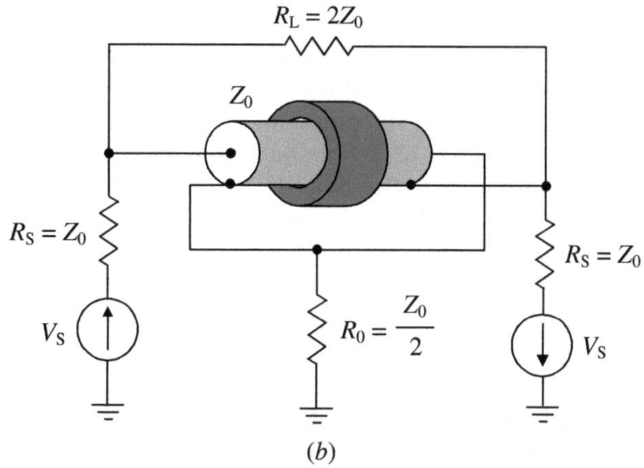

FIGURE 5.11 Hybrid combiners with one coaxial cable transformer.

three transmission lines should be 75 Ω. In this case, the matched condition corresponds to a load impedance of 112.5 Ω.

By using the coaxial cable transformers, various hybrid circuits can be developed, in which the input signals from the two power amplifiers are of exactly the same amplitude and properly phased [26]. Generally, a hybrid circuit can include one or two transformers interconnected so as to form a four-port device. Figure 5.11(*a*) shows the example of such a broadband hybrid circuit with one transformer, which combines the two in-phase signals when both signals are delivered to the load R_L and no signal will be dissipated in the ballast resistor R_0 [14]. The main advantage of this transformer is zero longitudinal voltage along the line for equal input powers, resulting in no losses occurred in the ferrite core. When one input signal source (e.g., power amplifier) defaults or disconnects, the longitudinal voltage

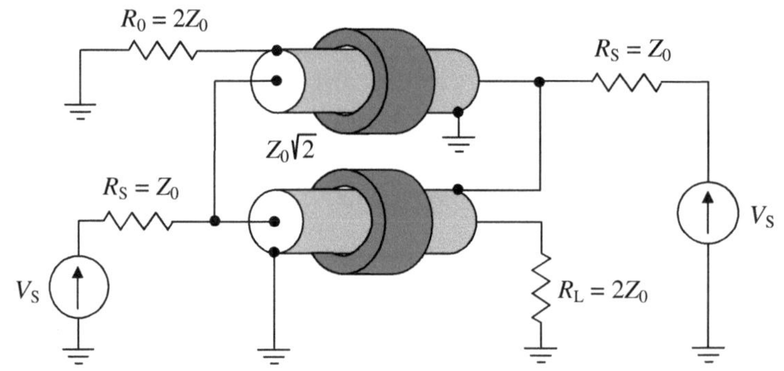

FIGURE 5.12 Two-cable hybrid combiner.

becomes equal to half a voltage of another input source. The ballast resistor should be of noninductive type and rated for 25% of the total power. The degree of isolation obtainable depends on the frequency and the overall design of the hybrid, with typical figures of 30 to 40 dB for 2-to 30-MHz operation [27]. For such a simple hybrid circuit, it is readily possible to combine the two 180° out-of-phase signals when the ballast resistor is considered the load, and the load resistor in turn is considered the ballast resistor, as shown in Fig. 5.11(*b*).

The schematic of a hybrid circuit with two coaxial cable transformers using as a power combiner is shown in Fig. 5.12 [28]. The advantage of this combiner is that both the load R_L and the ballast resistor R_0 are grounded. These hybrid transformer-based combiners can also be used for the power splitting when the output power from the single source is divided and delivered into the two independent loads. In this case, the original load and the two signal sources should be switched. The scattering *S*-matrix of a two-cable hybrid combiner, as a function of the electrical length θ of the coaxial cable, is given by

$$[S] = \frac{\sqrt{2}}{2 + j(1/\rho + \rho/2)\tan\theta} \begin{bmatrix} -\sec\theta & 1 \\ 1 & \sec\theta \end{bmatrix} \tag{5.27}$$

where ρ is the ratio of the coaxial-cable characteristic impedance to the source impedance. From a symmetric matrix given in Eq. (5.27), it follows that both combiner inputs are fully isolated from each other regardless of the transmission-line characteristic impedance and electrical length, which should be equal for both coaxial cables. In this case, the minimum insertion loss is achieved with the optimum value $\rho = \sqrt{2}$. However, varying ρ within the limits of 1 to 2 will only insignificantly affect the transformer performance.

It should be noted that the term "hybrid" came not from the fact that the transformer might be constructed of two different entities (e.g., cable and resistor), but just because it is being driven by two signals, as opposed to only one. This term initially was related to a hybrid coil (or bridge transformer) used in the switched telephone networks and represents a transformer that has three winding and is designed to be configured as a circuit having four branches. Consequently, the hybrid circuit represents a four-port device having two input ports, one sum port, and one difference port. The unique characteristic of the hybrid circuit is its ability to isolate the two input signal sources. At microwave frequencies, two classes of hybrid circuits can be distinguished: hybrid couplers, representing the ring, branch-line, or coupled-line structures, and hybrid junctions, in which the four transmission lines are joined at a common point.

Figure 5.13(*a*) shows the circuit structure of a coaxial cable two-way combiner where the input signals having the same amplitudes and

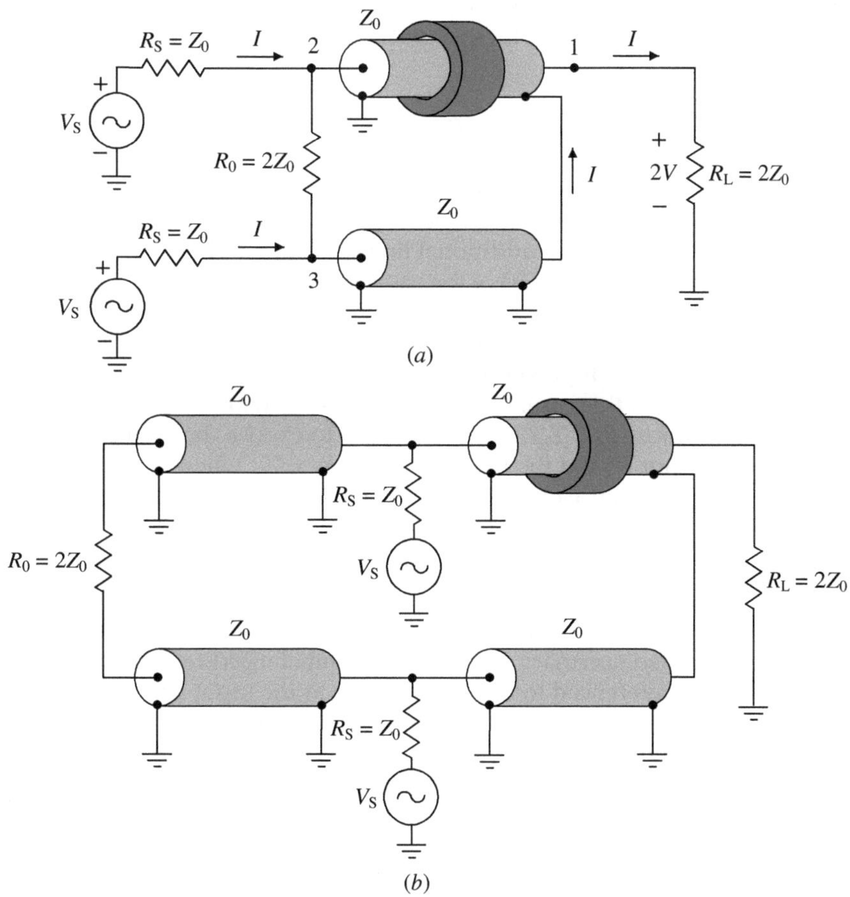

Figure 5.13 Coaxial cable combiners with increased isolation.

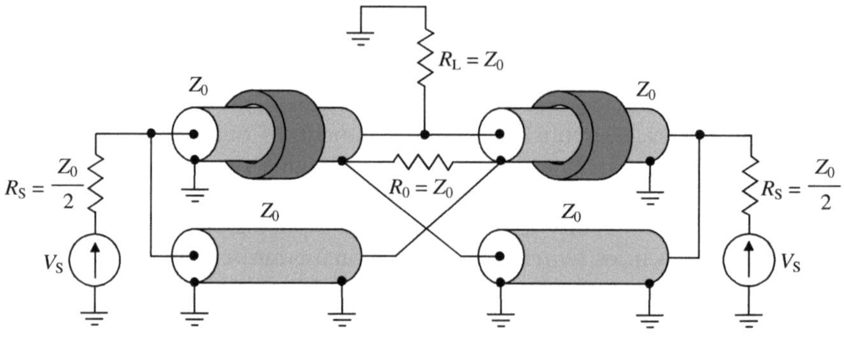

FIGURE 5.14 Fully matched and isolated coaxial cable combiner.

phases at ports 2 and 3 are matched at higher frequencies when all lines are of the same lengths and $R_S = Z_0 = R_L/2 = R_0/2$ [28, 29]. In this case, the isolation between these input ports can be calculated in decibels by

$$C_{12} = 10\log_{10}\left[4\ (1+4\cot^2\theta)\right] \tag{5.28}$$

where θ is the electrical length of each transmission line. In order to improve the isolation, the symmetrical ballast resistor R_0 should be connected through the two additional lines, as shown in Fig. 5.13(*b*), where all transmission lines have the same electrical lengths [4].

Figure 5.14 shows the coaxial cable two-way combiner, which is fully matched and isolated in pairs [4, 28]. Such combiners can be effectively used in high-power broadcasting VHF FM and VHF-UHF TV transmitters. In this case, for power amplifiers with the identical output impedances $R_S = Z_0/2$, it is necessary to choose the values of the ballast resistor R_0 and the load R_L of $R_0 = R_L = Z_0$, where Z_0 is the characteristic impedance of each transmission line of the same length.

5.3 Baluns

For a push-pull operation of the power amplifier with a balanced transistor, it is also necessary to provide the unbalanced-to-balanced transformation referenced to the ground both at the input and at the output of the power amplifier. The most suitable approach to solve this problem in the best possible manner at high frequencies and microwaves is to use the transmission-line baluns (balanced-to-unbalanced transmission-line transformers). The first transmission-line balun for coupling a single coaxial line having a quarter wavelength at the center bandwidth frequency to a push-pull coaxial line (or a pair of coaxial lines), which maintains perfect balance over a wide frequency range, was introduced and described by Lindenblad in 1939 [30, 31].

The main requirements to baluns are to provide an accurate 180° phase shift over the required frequency bandwidth with minimum loss and equal balanced impedances. Otherwise, a lack of symmetry will degrade the output power and efficiency. Besides, the symmetrical port must be well isolated from the ground to minimize an unwanted effect of the parasitic capacitances.

A wire-wound transformer, whose simplified equivalent schematic is shown in Fig. 5.15(*a*), provides an excellent broadband balun covering frequencies from low kHz to beyond 2 GHz in commercial applications. They are usually realized with a center-tapped winding that provides a short circuit to the even-mode (common-mode) signals, while having no effect on the differential (odd-mode) signal.

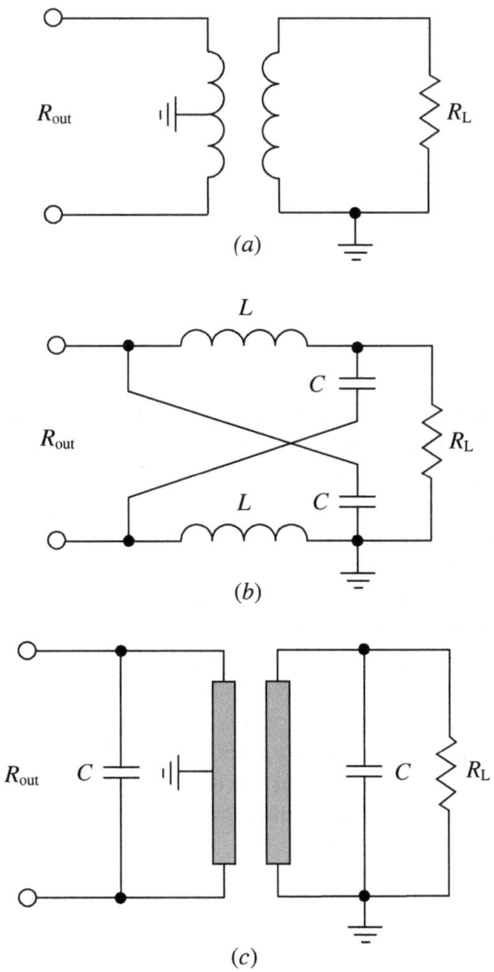

FIGURE 5.15 Different circuit configurations of 1:1 balun.

The wire-wound transformers are more expensive than the printed or lumped *LC* baluns, which are more suitable in practical mixer designs. However, unlike the wire-wound transformers, the lumped *LC* baluns are narrow-band as containing the resonant elements.

Figure 5.15(*b*) shows the circuit schematic of a lattice-type *LC* balun, which was proposed in the mid-1930s for combining powers in the push-pull amplifier to further deliver the overall power directly to antenna [32]. It consists of the two capacitors and two inductors, which produce the ±90° phase shifts at the output ports. The values of identical inductances *L* and capacitances *C* can be obtained by

$$L = \frac{\sqrt{R_{\text{out}} R_{\text{L}}}}{\omega_0} \tag{5.29}$$

$$C = \frac{1}{\omega_0 \sqrt{R_{\text{out}} R_{\text{L}}}} \tag{5.30}$$

where ω_0 is the center bandwidth radian frequency, R_{out} is the balanced output resistance, and R_{L} is the unbalanced load resistance. When designing this circuit, it needs to be confident that the operating frequency is well below the self-resonant frequencies of their components.

In monolithic microwave applications where the lumped inductances are usually replaced by the transmission lines, the designs with microstrip coupled lines, Lange couplers, or multilayer coupled structures are very popular. However, the electrical length of the transmission lines at the center bandwidth frequency is normally set to a quarter-wavelength, which is too large for applications in wireless communication systems. Therefore, it is very attractive to use the lumped-distributed balun structures, which can significantly reduce the balun size and, at the same time, can satisfy the required electrical characteristics. Figure 5.15(*c*) shows such a compact balun with a lumped-distributed structure consisting of the two coupled planar microstrip lines and two parallel capacitors, where the input transmission line is grounded at midpoint and the output transmission line is grounded at its one port [33]. Without these capacitors, it is necessary a very small spacing between the quarterwave microstrip lines to achieve a 3-dB coupling between them. However, by optimizing the balun elements around the center bandwidth frequency of 900 MHz, the planar structure of approximately 1/16th the size of the conventional quarter-wavelength structure was realized, with spacing $S = 8$ mils using an FR4 board with a substrate thickness of 30 mils.

Figure 5.16 shows the circuit arrangement with the two coaxial line transformers combined to provide a push-pull operation of the power amplifier by creating a balanced-to-unbalanced impedance transformation with higher spectral purity [28, 29]. Ideally, the 180° out-of-phase RF signals from both active devices biased in a Class-B

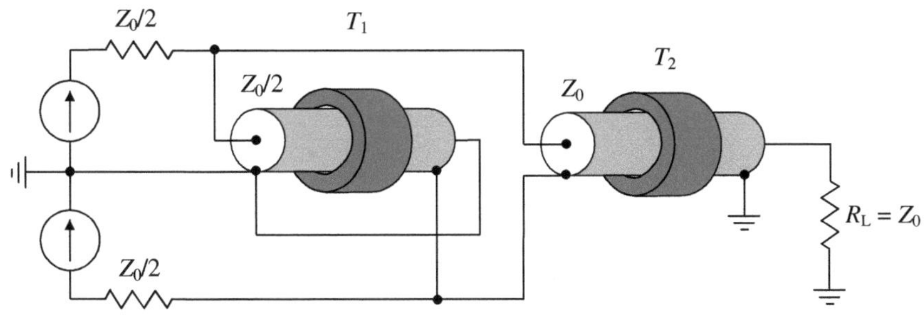

FIGURE 5.16 Circuit arrangement with two cable transformers for push-pull operation.

mode will have pure half-sinusoidal waveforms, which contain (according to the Fourier series expansion) only fundamental and even harmonic components. This implies a 180° shift between the fundamental components from both active devices and in-phase condition for the remaining even-harmonic components. In this case, the transformer T_1 representing a phase inverter is operated as a filter for even harmonics because currents flow through its inner and outer conductors in opposite directions. For each fundamental component flowing through its inner and outer conductors in the same directions, it works as an RF choke, the impedance of which depends on the core permeability. Consequently, because the transformer T_2 represents a 1:1 balun, in order to provide maximum power delivery to the load R_L, the output equivalent resistance of each active device should be two times smaller.

For a simple 1:1 transmission-line balun realized with a twisted wire pair or coaxial cable, the balanced end is isolated from the ground at the center bandwidth frequency. To compensate for the short-circuited line reactance over certain frequency bandwidth around center frequency, a series open-circuited transmission line was introduced by Marchand, resulting in a compensated balun, the simplified schematic of which is shown in Fig. 5.17(*a*) [34]. In this case, when the electrical length of the compensated line is a quarter-wavelength at the center bandwidth frequency, the load resistance R_L is seen unchanged. When this structure is realized with coaxial cables, to eliminate unwanted current existing in the outer conductor and corresponding radiation, it is necessary to additionally provide the certain coupling between the coaxial cables forming a transmission line with the two outer conductors, as shown in Fig. 5.17(*b*) [35]. Generally, the shunting reactance of this compensating line can reduce the overall balun reactance about center frequency or reverse it sign depending on the balanced load resistance, characteristic impedance of the compensating line, and coupling (characteristic impedance) between the outer conductors of these two lines. Hence, a compensating line can create a complementary reactance to a

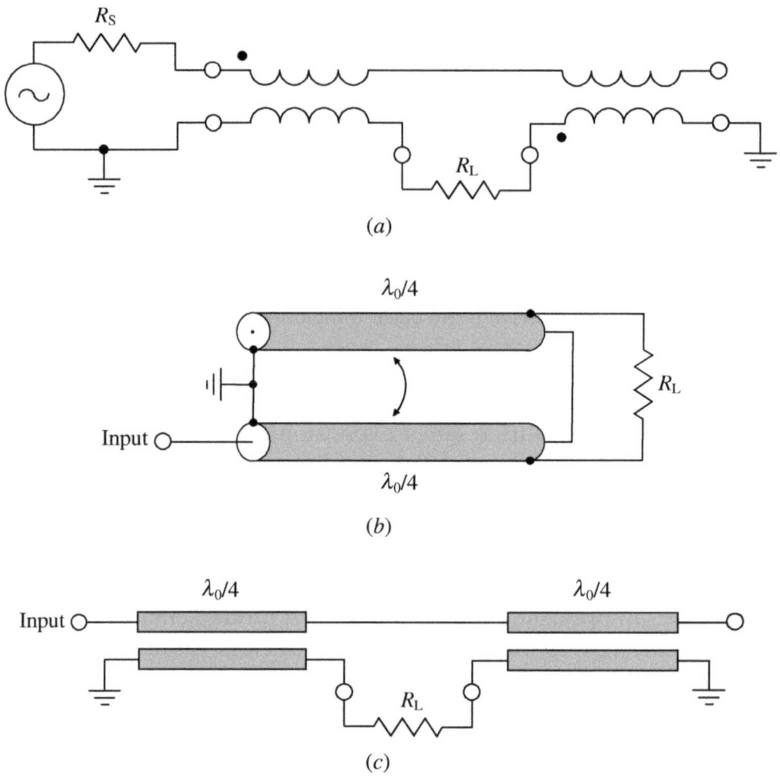

FIGURE 5.17 Schematic configurations of Marchand balun.

balanced load and provide an improved match over broader frequency range. At microwaves, the wire-wound transformers are usually replaced by a pair of the coupled transmission lines, as shown in Fig. 5.17(*c*), thus resulting in a compact planar structure. Note that generally the characteristic impedances of the coaxial or coupled transmission lines can be different to optimize the frequency bandwidth response.

Multilayer configurations make Marchand balun even more compact and can provide wide bandwidths due to the tight coupling between the coupled-line sections. Modeling and synthesis result of a two-layer monolithic Marchand balun configuration with the two-coupled lines, the basic structure of which is shown in Fig. 5.18(*a*), is discussed in [36]. In this configuration, the unbalanced terminal is connected to the microstrip line located at the upper metallization level, whereas the balanced load is connected to the microstrip lines located at the lower metallization level. The transmission-line sections in different layers are not isolated from each other. It should be

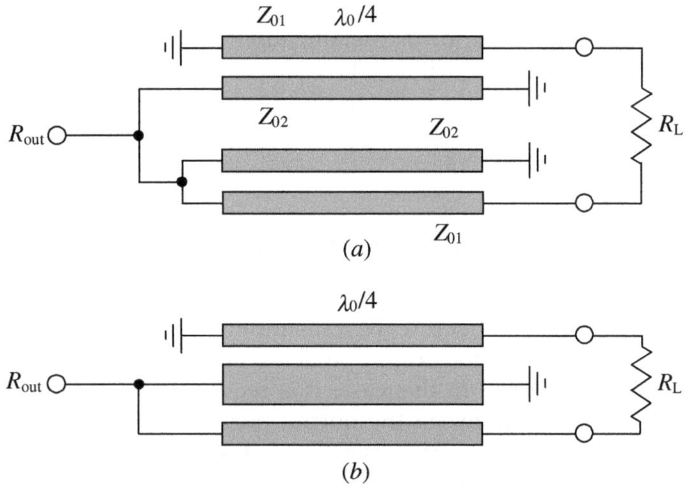

FIGURE 5.18 Schematic configurations of coupled-line Marchand balun.

noted that, for a given set of the output balanced and load unbalanced resistances R_{out} and R_L, the characteristic impedances of the outer and inner microstrip lines Z_{01} and Z_{02} are not unique, and they can be calculated from

$$C = \frac{1}{2}\sqrt{\frac{Z_{02}}{Z_{01}}} \tag{5.31}$$

$$Z_{02} = 4Z_{01} - \frac{R_{out}R_L}{Z_{01}} \tag{5.32}$$

where C is the coupling factor [37]. However, a different choice of Z_{01} and Z_{02} leads to a different frequency bandwidth. For example, for $R_{out} = 50\ \Omega$ and $R_L = 100\ \Omega$, it was found that using the symmetrical directional coupler with $Z_{01} = Z_{02} = 40.825\ \Omega$ results in a frequency bandwidth of 48.4% with $|S_{11}| < -10$ dB and amplitude imbalance within 0.91 dB, whereas the frequency bandwidth of 20.9% with an amplitude imbalance of less than 1.68 dB will be realized for the nonsymmetrical case when $Z_{01} = 38\ \Omega$ and $Z_{02} = 20.42\ \Omega$.

The design of a three-line microstrip balun, the basic schematic of which is shown in Fig. 5.18(*b*), is based on the equivalence between a six-port section of three coupled lines and a six-port combination of two couplers [37]. The results of circuit analysis and optimization show that the spacings between adjacent microstrip lines are so narrow that it is difficult to fabricate a single-layer three-line balun. For a two-layer three-line balun with the two coupled outer lines on the top metallization level, the spacing between these lines is significantly wider than in a single-layer case. However, wider frequency range can be achieved using a two-layer three-line balun with the two coupled outer lines at the lower metallization level. For example, the measurements results for this balun show that, being fabricated on the Duroid RT5880 substrate, it can provide a frequency range of 2.13 to 3.78 GHz with an amplitude imbalance within 2.12 dB and a phase error of less than 4.51°.

5.4 Wilkinson Power Dividers/Combiners

The in-phase power combiners and dividers are the important components of the RF and microwave transmitters when it is necessary to deliver a high level of the output power to antenna, especially in the phased-array systems. In this case, it is required to provide a high degree of isolation between the output ports over some frequency range for identical in-phase signals with equal amplitudes. Figure 5.19(*a*) shows a planar structure of the basic parallel beam *N*-way divider/combiner, which provides a combining of powers from the *N* signal sources. Here, the input impedance of the *N* transmission lines (connected in parallel) with the characteristic impedance of Z_0 each is

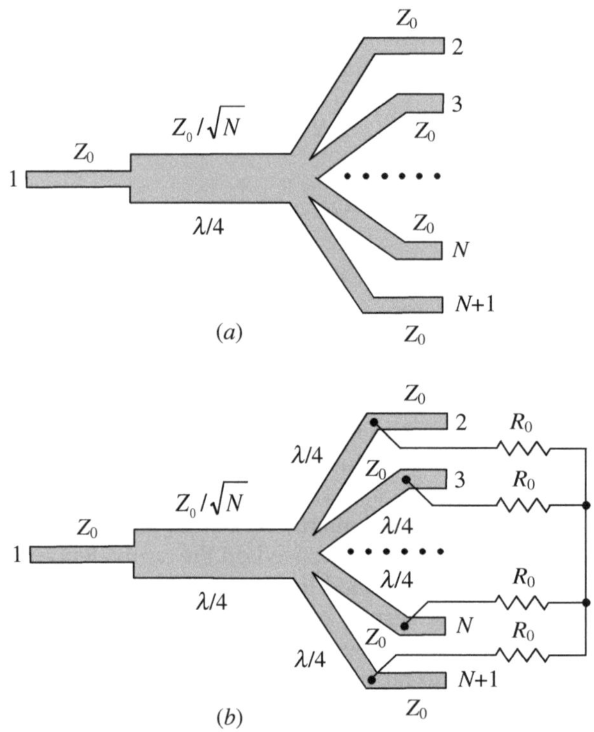

FIGURE 5.19 Circuit topologies of *N*-way in-phase combiners/dividers.

equal to Z_0/N. Consequently, an additional quarterwave transmission line with the characteristic impedance $Z_0/\sqrt{N}$ is required to convert the input impedance Z_0/N to the standard impedance Z_0. However, this N-way combiner cannot provide a sufficient isolation between the input ports. The impedances are matched only when all input signals have the same amplitudes and phases at any combiner input. The effect of any input on the remaining ones becomes smaller for combiners with greater number of inputs. For example, if the input signal is delivered into the input port 2 and all other $(N - 1)$ input ports and output port 1 are matched, then the power dissipated at any load connected to the matched input ports will be decreased by $(1 - 1/N^2)/(2N - 1)$ times and isolation between any two input ports expressed through S-parameters is obtained by

$$S_{ij} = -10 \log_{10}\left(\frac{1}{N^2}\frac{N^2 - 1}{2N - 1}\right) \tag{5.33}$$

where N is a number of the input ports and $i, j = 2, \ldots, N + 1$.

In most cases, better isolation is required than obtained by Eq. (5.33). The simplest way to provide full isolation between the input and output ports of the combiner is to connect the ferrite isolators (circulators) at the input ports $2, \ldots, N + 1$. In this case, the lengths of the transmission lines connected between each ferrite isolator and a quarterwave transmission line should be equal. Although the ferrite isolators increase the overall weight and dimensions of the combiner and contribute to additional insertion losses, nevertheless they provide a very simple combiner realization and protect the connected power amplifiers from the load variations. By using such a 12-way parallel beam combiner, the continuous output power of 1 kW for the L-band transmitter was obtained at the operating frequency of 1.25 GHz [38].

When one or more power amplifiers are destroyed for some reasons, the overall output power P_{out} and efficiency η_c of the combiner can be calculated, respectively, by

$$P_{out} = \frac{(N - M)^2}{N} P_1 \tag{5.34}$$

$$\eta_c = \frac{P_{out}}{P_{in}} = 1 - \frac{M}{N} \tag{5.35}$$

where $P_{in} = (N - M)P_1$, P_1 is the output power from a single power amplifier, N is the number of the input ports, and M is the number of the destroyed power amplifiers. Part of the output power of the remaining power amplifiers will be dissipated within the ferrite isolators (in ballast resistors of circulators). For each ferrite isolator connected to the operating power amplifier, the dissipated power P_{do} can be defined as

$$P_{do} = \left(\frac{M}{N}\right)^2 P_1 \tag{5.36}$$

whereas, for each isolator connected to the destroyed power amplifier, the dissipated power P_{dd} can be calculated from

$$P_{dd} = \left(\frac{N-M}{N}\right)^2 P_1 \tag{5.37}$$

In this case, by adding the ballast resistors $R_0 = Z_0$, the right-hand side terminals of which are combined together in a common junction, as shown in Fig. 5.19(*b*), matching of all ports, low loss, and high isolation between the input and output ports can be provided. Such kind of a simple N-way power divider is known as a *Wilkinson power divider* [39]. However, it should be mentioned that historically this divider/combiner was first reported at the end of the 1950s [40–42]. Originally, a Wilkinson power divider was composed of a coaxial line, in which the hollow inner conductor has been split into N splines of a quarter-wavelength each, with shorting plate connecting the splines at the input end and resistors connected in a radial manner between each spline at the output end and a common junction. The frequency response of the voltage standing wave ratio at the divider input port ($VSWR_{in}$), depending on the number of the output ports N, is shown in Fig. 5.20 [43].

The planar realization of the simplest two-way Wilkinson divider, which was originally implemented in a stripline ring structure [44], is shown in Fig. 5.21(*a*). It consists of the two quarterwave microstrip lines connected in parallel at the input end and the planar ballast resistor connected between the output ports of the microstrip lines. Despite its small dimensions and simple construction, such a divider provides a sufficient isolation between the output ports over a sufficiently wide frequency bandwidth when equal power division is

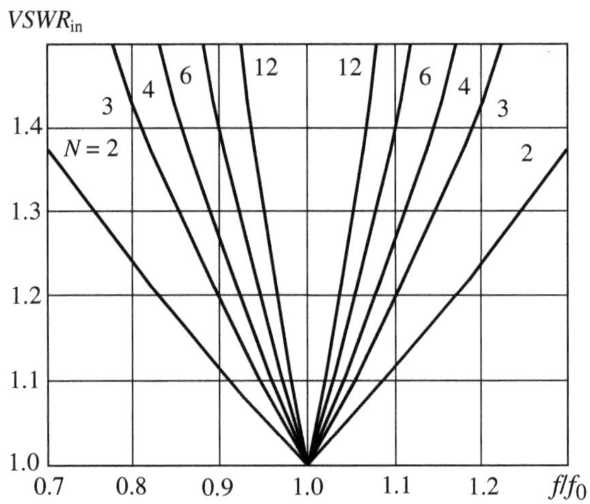

Figure 5.20 Frequency performance of N-way Wilkinson divider.

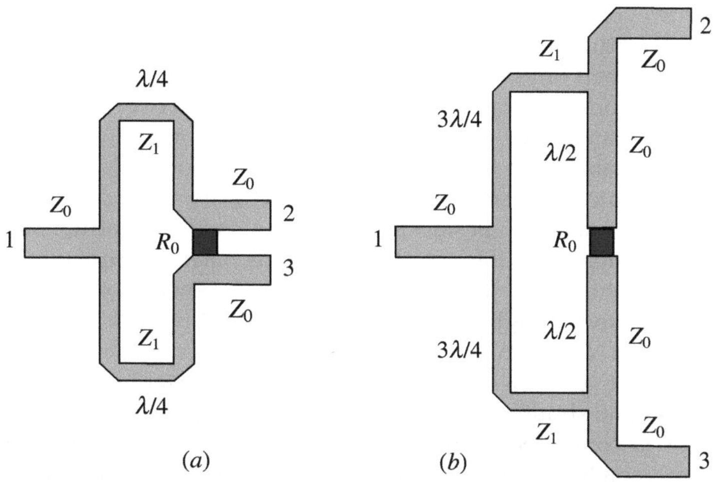

FIGURE 5.21 Microstrip realization of two-way Wilkinson dividers.

provided due to a symmetrical configuration with $R_0 = 2Z_0$ and $Z_1 = Z_0\sqrt{2}$. However, in practice, it is necessary to take into account the distributed RC structure of the ballast resistor when its size is sufficiently large, as well as manufacturing tolerances and discontinuities. As a result, in a frequency bandwidth of 30% with $VSWR_{in} \le 1.2$ at the input port 1 and $VSWR_{out} \le 1.03$ at the output ports 2 and 3, the isolation between the divider outputs can be better than 20 dB [45].

The scattering S-matrix of the ideally matched two-way Wilkinson divider at the center bandwidth frequency is obtained by

$$[S] = -\frac{j}{\sqrt{2}} \begin{bmatrix} 0 & 1 & 1 \\ 1 & 0 & 0 \\ 1 & 0 & 0 \end{bmatrix} \tag{5.38}$$

In microwave and millimeter-wave integrated circuits, in order to increase a self-resonant frequency of the ballast chip resistor, the overall MMIC dimensions must be very small. This means that the two branches of the power divider are very close to each other, which leads to strong mutual coupling between the output microstrip lines and, as a result, upsets the desired power-split ratio. A possible solution is to use the branches with the electrical lengths of $3\lambda/4$ instead of $\lambda/4$ and to include the two additional branches into a semicircle, as shown in Fig. 5.21(b) [46]. These additional branches should be of the half-wave electrical lengths with the characteristic impedances equal to Z_0. In this case, isolation can be better than 17 dB between all ports with the insertion loss of about 1.3 dB at the operating frequency of 30.4 GHz.

However, the ballast resistors of the conventional N-way Wilkinson combiners/dividers cannot be designed to represent a planar structure when their physical lengths and connecting wires are minimal to provide sufficient isolation among the output ports over the required frequency range. For example, the radial and fork N-way hybrids have reasonably wide frequency bandwidth, of about 20% and higher, but their match and isolation are not perfect even at the center bandwidth frequency [47]. Besides, because of a small size of the ballast resistor compared to the wavelength and its balanced structure, it is difficult to heat-sink it in the case of a high-power combining. In order to provide higher output power capability, it is possible to modify the N-way Wilkinson combiner/divider by replacing the ballast resistive star with a combination of the quarterwave transmission lines and shunt-connected resistors [48]. In this case, each ballast resistor is connected to the corresponding output port through a transmission line. At the same time, all ballast resistors are connected to a common floating starpoint by the transmission lines. Such a modification has an advantage of external isolation loads (high-power ballast resistors) and easy monitoring capability for imbalances at the output ports. For a two-way planar power combiner/divider, the circuit topology of which is shown in Fig. 5.22, the balanced 100-Ω ballast resistor is replaced by a transmission-line network and two 50-Ω resistors are connected to the ground acting as the 180° out-of-phase load, where $Z_1 = Z_0\sqrt{2}$, $Z_2 = Z_0 / \sqrt{2}$, and $Z_0 = R_0 =$ 50 Ω [49].

The cascade connection of the two-way Wilkinson power combiners/dividers can provide a multiway power dividing or power combining. The simplest practical realization is the binary power divider/combiner, which is composed of the n stages and each consecutive stage contains an increase by 2^N number of two-way dividers/

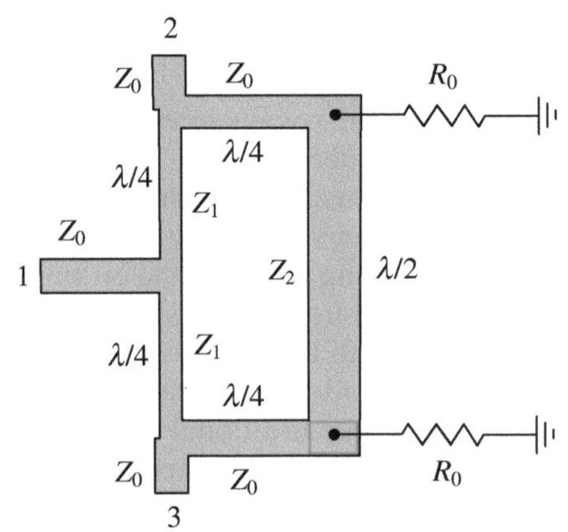

FIGURE 5.22 The Gysel high-power in-phase planar combiner/divider.

combiners [50]. For a single destroyed power amplifier, the power dissipated in the ballast resistors is equal to

$$P_{\mathrm{db}} = \left(1 - \frac{1}{N}\right) P_1 \tag{5.39}$$

The output power of $P_1/2$ is dissipated in the ballast resistor adjacent to the destroyed power amplifier; the output power of $P_1/4$ is dissipated in the ballast resistor of the next stage, and so on. It should be mentioned that the power divider with a number of outputs multiple to 4^N represents the convenient case when the characteristic impedances of the transmission lines are of the same value. Figure 5.23 shows the simple practical implementation of a four-way microstrip Wilkinson divider/combiner fabricated on alumina substrate with six 50-Ω quarterwave microstrip lines and two 100-Ω and one 50-Ω thin-film resistors [51]. This microstrip Wilkinson power divider/combiner can provide an insertion loss of less than 0.3 dB and an isolation between any outputs of about 20 dB in a frequency bandwidth of ±10% at microwave frequencies.

The frequency bandwidth performance of a Wilkinson power divider/combiner can be improved with an increasing number of its sections [45]. Generally, a broadband two-way Wilkinson power divider can contain N pairs of equal-length transmission lines and N bridging resistors distributed from the input port 1 to the output ports 2 and 3. For example, for $N = 2$, the theoretical minimum isolation in an octave band between the output ports 2 and 3 can achieve 27.3 dB with a *VSWR* at each port better than 1.1. In monolithic microwave integrated circuits, by using a two-metal layer GaAs HBT process when the bottom metal layer can realize a coplanar waveguide (CPW) transmission line and the top metal layer can realize a microstrip transmission line, the size of a two-section two-way power divider/combiner can be reduced. In this case, an isolation of 15 dB

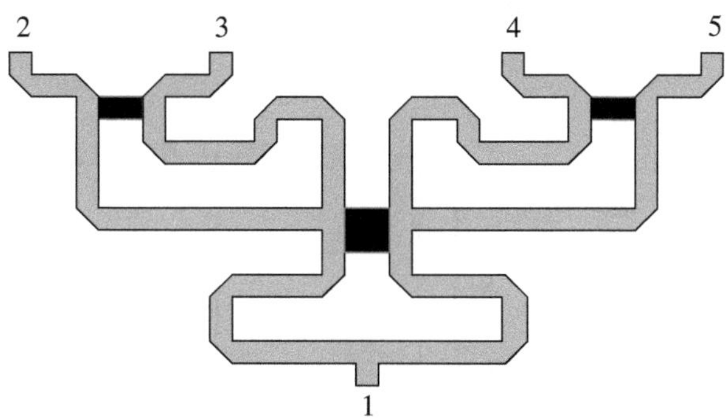

FIGURE 5.23 Practical four-way microstrip Wilkinson power combiner/divider.

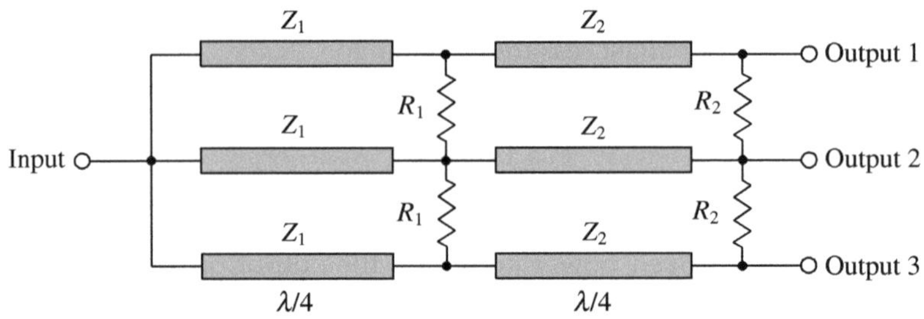

FIGURE 5.24 Microstrip three-way divider with improved isolation.

and a return loss of 15 dB can be achieved in a frequency bandwidth from 15 to 45 GHz [52].

Figure 5.24 shows the equivalent circuit representation of a three-way modified Wilkinson power divider/combiner [53]. By assuming that all the impedances of the input and three output ports are equal to 50 Ω, the characteristic impedances of the quarterwave transmission lines are selected for a maximally flat performance as $Z_1 = 114\ \Omega$ and $Z_2 = 65.8\ \Omega$. To match circuit at the center bandwidth frequency, the values of the ballast planar resistors should be equal to $R_1 = 64.95\ \Omega$ and $R_2 = 200\ \Omega$. In this case, the isolation between output ports of such a three-way divider demonstrates more than 20 dB in an octave frequency band.

Generally, high characteristic impedance values (usually higher than 100 Ω) for the transmission lines can create a problem in their practical microstrip implementation, because their narrow widths increase the insertion loss. In this case, using a recombinant power divider, the topology of which is shown in Fig. 5.25, provides an isolation of 20 dB in a frequency range of 72% for a maximum line impedance of 80 Ω and requires only three isolation resistors [54]. This three-way recombinant divider fabricated on a 25-mil thick 99.6% alumina substrate is characterized by an insertion loss of about 1 dB and a return loss of more than 12 dB in a frequency range of 6 to

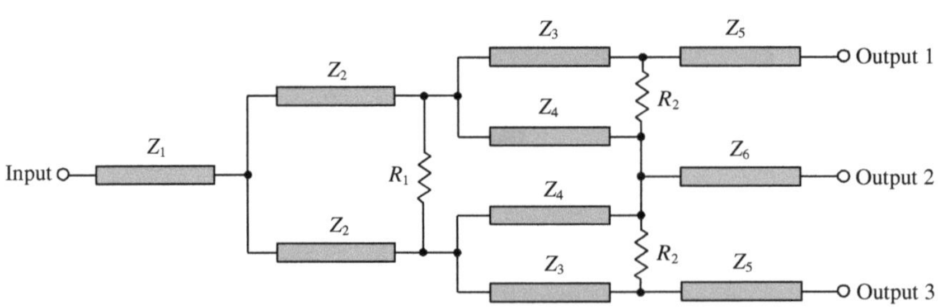

FIGURE 5.25 Microstrip three-way recombinant divider with improved isolation.

14 GHz. The design values for the quarterwave transmission lines were $Z_1 = 36\ \Omega$, $Z_2 = Z_3 = 40\ \Omega$, $Z_4 = 80\ \Omega$, and $Z_5 = Z_6 = 40\ \Omega$, with the ballast resistors $R_1 = 50\ \Omega$ and $R_2 = 100\ \Omega$. Over a 2:1 frequency bandwidth, the center-to-side and side-to-side isolations exceed 20 dB.

The power divider broadband properties can also be improved by using the more complicated phase-shifting circuit instead of a simple microstrip line. The phase shift between the two output ports 2 and 3 will be close to 90° in an octave frequency range if a Schiffman element based on the coupled microstrip lines is connected to one output port [55]. At the same time, an additional microstrip line with an electrical length of 270° at the center bandwidth frequency should be connected to the second output port.

In the design of a microwave distributed network, a power divider providing the two equal-phase outputs with unequal power splitting is often required. The split-tee power divider is a simple compact and broadband device. It provides two isolated equal-phase unequal-amplitude outputs with a good match at each port. Because a split-tee power divider is similar to the N-way equiphase equiamplitude power divider, it can be developed from this N-way divider by first connecting M of the output ports together to form one port and the remaining $N - M$ output ports together to form the other port, then connecting the quarterwave transformers to the resulting output ports to adjust their impedance level, and finally a power divider with two equiphase outputs and power ratio of $N/(N - M)$ is derived.

The basic schematic of a power divider with unequal output load impedances is shown in Fig. 5.26(a) [56]. This power divider is

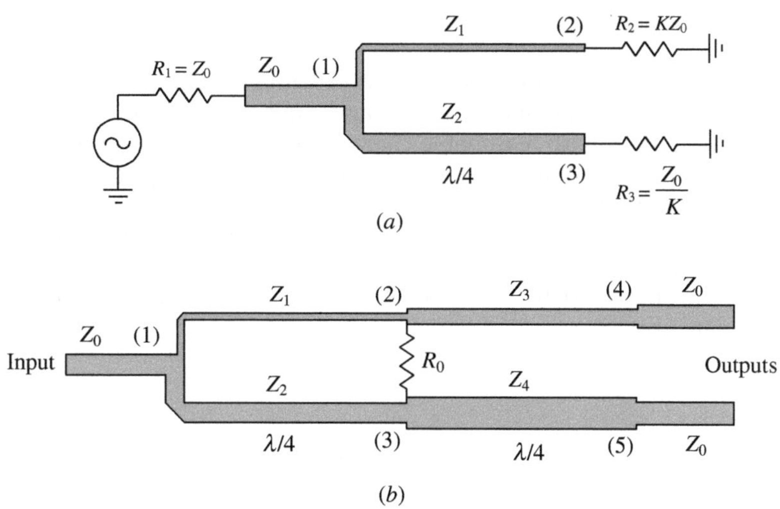

FIGURE 5.26 Split-tee power divider.

designed such that, when fed from the input port 1, the perfect match will be achieved at the center bandwidth frequency when the output power at port 3 is K^2 times the output power at port 2, and that between port 2 and ground is equal to the voltage between port 3 and ground when measured at equal distances from the input port 1. To satisfy these conditions, the characteristic impedances Z_1 and Z_2 for unequal loads $R_2 = KZ_0$ and $R_3 = Z_0/K$ are calculated from

$$Z_1 = KZ_0\sqrt{K + \frac{1}{K}} \tag{5.40}$$

$$Z_2 = \frac{Z_0}{K}\sqrt{K + \frac{1}{K}} \tag{5.41}$$

where both transmission lines are of a quarter wavelength at the center bandwidth frequency.

Because the voltages at ports 2 and 3 are equal with this design, a resistor may be placed between these two ports without causing any power dissipation. However, isolation between output ports and a good match seen looking in at any ports is obtainable because of this resistor. Finally, to transform the two unequal output impedances to the output impedance Z_0 equal for each output port, the characteristic impedances of additional quarterwave transformers Z_3 and Z_4 and ballast resistor R_0 shown in Fig. 5.26(*b*) are determined from

$$Z_3 = Z_0\sqrt{K} \tag{5.42}$$

$$Z_4 = \frac{Z_0}{\sqrt{K}} \tag{5.43}$$

$$R_0 = Z_0\left(K + \frac{1}{K}\right) \tag{5.44}$$

The three-way power divider with various output power ratios, which represents a planar structure and can be easily realized using microstrip lines with reasonable characteristic impedances, is shown in Fig. 5.27 [57]. When port 1 is an input port, the input power is divided by a ratio of M:N:K at the corresponding output ports 2, 4, and 6 with isolated ports 3 and 5. The electrical lengths of the transmission lines must be 90° except for the half-wave middle horizontal line. The characteristic impedances of the transmission lines can be calculated by

$$Z_1 = Z_0\sqrt{\frac{\Delta_1}{\Delta_2}} \tag{5.45}$$

$$Z_2 = Z_0\sqrt{\frac{\Delta_1}{M}} \tag{5.46}$$

$$Z_3 = Z_0 \tag{5.47}$$

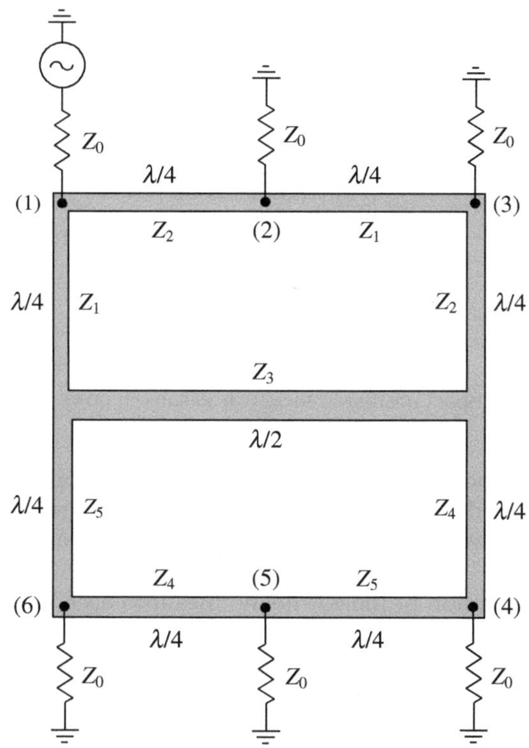

FIGURE 5.27 New type of three-way power divider.

$$Z_4 = Z_0\sqrt{\frac{\Delta_2}{N}} \tag{5.48}$$

$$Z_5 = Z_0\sqrt{\frac{\Delta_2}{K}} \tag{5.49}$$

where $\Delta_1 = M + N + K$ and $\Delta_2 = N + K$. For example, for a three-way divider with $M = 3$, $N = 2$, and $K = 1$, it follows that $Z_1 = Z_2 = 1.41Z_0$, $Z_4 = 1.22Z_0$, and $Z_5 = 1.73Z_0$. The same characteristic impedances are required for a 1:1:1 equal-power three-way divider; only the input port must be changed to port 4 in this case.

Figure 5.28 shows the compact microstrip three-way Wilkinson power divider designed to operate over a frequency range of 1.7 to 2.1 GHz with minimum combining efficiency of 93.8%, maximum amplitude imbalance of 0.35 dB, and isolation better than 15 dB [58]. To avoid any amplitude and phase imbalances between the divider 50-Ω output ports, the ballast resistor connected to its middle branch should be split into two equal parallel resistors. To obtain an ideal floating node, these two resistors are connected together with narrow microstrip lines that are as short as possible. Finally, to connect the

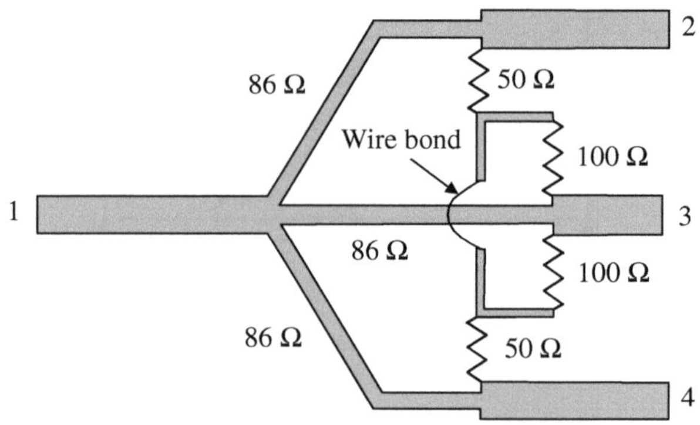

FIGURE 5.28 Compact microstrip three-way Wilkinson power divider.

resistors from both sides of the middle branch, a copper wire of 7-mil diameter is used. The most critical parameter is the isolation between ports 2 and 4, which can be improved by shortening the bondwire length.

Implementation of the 45° delay lines into each parallel path of the power amplifier and use of the in-phase Wilkinson combiners in pairs and a hybrid quadrature combiner in the output combining circuit, as shown in Fig. 5.29, can improve the overall power amplifier characteristics, which become more insensitive to variations of the load *VSWR* when the *ACLR* (adjacent channel leakage power ratio) reduces by more than 12 dB and efficiency increases by more than 10% for $VSWR = 3$ [59]. The basic idea is to spread the different impedances seen by the device outputs when phase delay of the reflected signals varies between 0 and 180° with a step of 45°, thus creating different impedances along the corresponding load *VSWR* circle on the Smith chart.

In this case, each from four active devices sees different load impedance, as seen from the Smith chart (normalized to 50 Ω) shown in Fig. 5.30, where the load variation circle is given for $VSWR = 2$, as

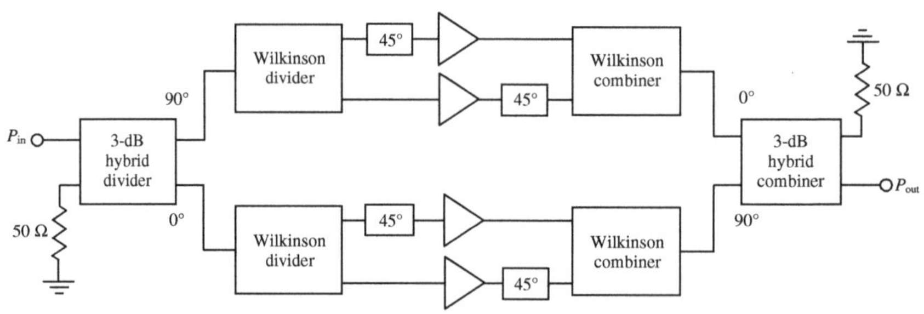

FIGURE 5.29 Balanced power amplifier topology with 45° delay lines.

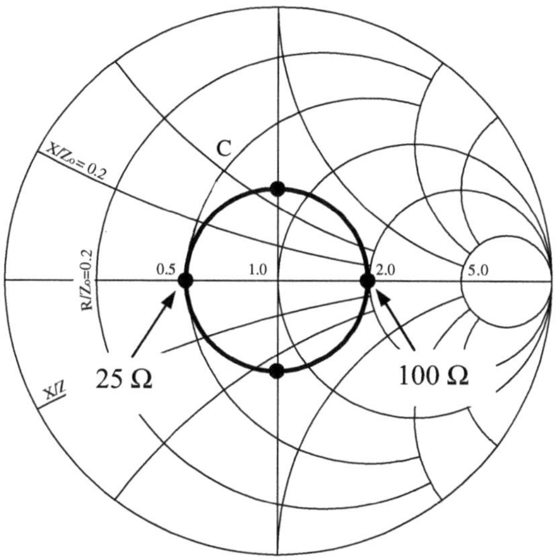

FIGURE 5.30 Smith-chart impedances for *VSWR* = 2.

an example. As a result, the only one device from entire four sees the high impedance of 100 Ω, while the impedances seen by the other three devices are spread along the Smith-chart circle, with the real parts being in between 100 and 25 Ω. This means that the output power is reduced only for one device and linearity is disturbed significantly only by one device, unlike a conventional parallel-operation power amplifier where it is distorted by all four devices. In a practical implementation, for a 3.5-V 29-dBm GaAs MESFET power amplifier designed to operate in a 900-MHz digital cellular phone system, the *ACLR* below –45 dBc with an efficiency of over 45% can be obtained for a load $VSWR \leq 3$ [59].

5.5 Branch-Line Hybrid Couplers

The branch-line hybrid couplers were first described in the 1940s, but the problem of their exact synthesis remained a puzzle for a number of years [60]. Initially, the branch-line hybrid was analyzed as a four-arm symmetrical network based on a superposition of the results obtained in the even and odd modes [61]. By writing the even- and odd-mode matrices together, the characteristic impedances of the branch lines and coupling into different ports can be obtained. A general synthesis procedure that can be applied to any structure of a multibranch hybrid, based on an invariance of the Richards's, variable $S = j\tan\theta$ to the transformation of $S \rightarrow 1/S$ apart from a 180° phase change, had become available a decade later [62]. As a result, with highly precise computer-design techniques available for branch-line hybrids, it became possible to generate any coupling value in the useful coupling range of 0 to 15 dB. Their waveguide designs that have been used in large complex feeds for phase-array radars are compact, highly predictable in amplitude and phase characteristics, and handle very high power. Coaxial, microstrip, or stripline implementations of the branch-line hybrids provide simple planar structures of moderate bandwidth capability, up to about 2/3 of an octave.

For a fully matched case with the standard 50-Ω source and load impedances when the characteristic impedances of its transverse branches are of 50 Ω and the characteristic impedances of its longitudinal main lines are of $50/\sqrt{2} = 35.4\ \Omega$, the microstrip branch-line hybrid shown in Fig. 5.31 represents a 3-dB directional coupler, for which the input power in the arm 1 divides evenly between the arms 2 and 3 with the phase shift of 90°. No power is delivered to the arm 4, because the signal flowing through different paths (lengths of $\lambda/4$ and $3\lambda/4$) have the same amplitude and opposite phases at this port. The branch-line hybrid does not depend on the load mismatch level for equal reflected coefficients from the outputs when all reflected power is dissipated in the 50-Ω ballast resistor. However, in practice, because of the quarter-wavelength transmission-line requirement, the bandwidth of such a single-stage quadrature branch-line hybrid is limited to 10 to 20%.

When all ports are matched, power entering the input port 1 is divided between the output ports 2 and 3 with a 90° phase shift between these outputs and no power is delivered to the isolated port 4. The scattering S-matrix of such a quadrature hybrid branch-line coupler is given by

$$[S] = -\frac{1}{\sqrt{2}}\begin{bmatrix} 0 & j & 1 & 0 \\ j & 0 & 0 & 1 \\ 1 & 0 & 0 & j \\ 0 & 1 & j & 0 \end{bmatrix} \tag{5.50}$$

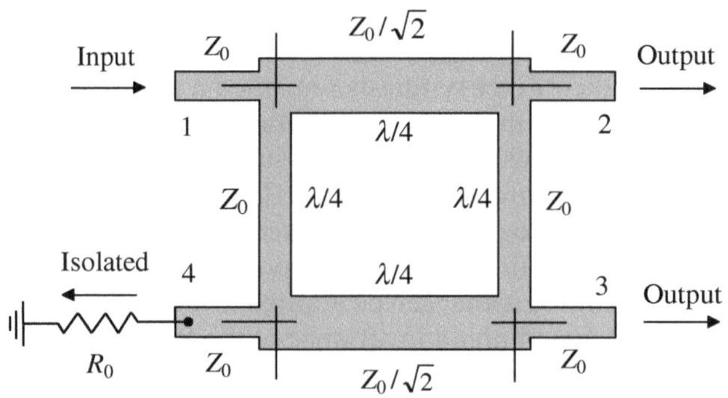

FIGURE 5.31 Microstrip branch-line quadrature hybrid coupler.

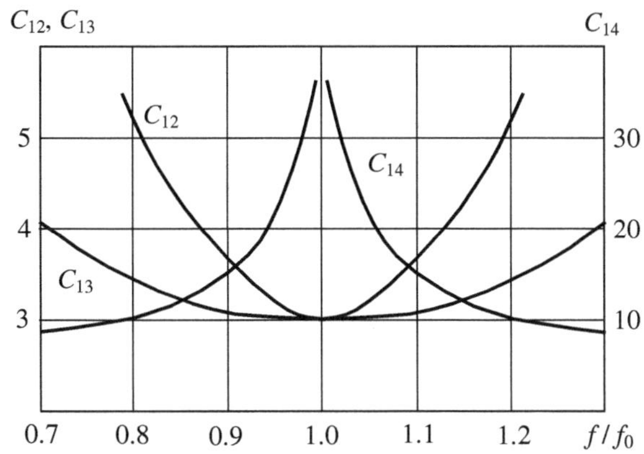

FIGURE 5.32 Bandwidth performance of single-section branch-line hybrid coupler.

Figure 5.32 shows the calculated frequency-bandwidth characteristics of a single-section branch-line hybrid coupler matched at the center bandwidth frequency with the load impedance $Z_L = Z_0 = 50\ \Omega$ [51]. At millimeter-wave frequencies, the lengths of the microstrip lines can actually get shorter than the widths and the mutual coupling between the input lines and discontinuities at the input increases significantly. This has a direct effect on the input/output matching conditions, frequency bandwidth, and isolation. To minimize the effect of these problems, the branch-line hybrid coupler can be designed as a two-section hybrid coupler using the three quarter-wavelength lines for the series main lines and the quarter-wavelengths lines for the shunt branch lines, with all inputs/outputs orthogonal to each other [63]. As a result, the return loss can achieve 10 dB or better over 90% of the band with the isolation of 10 dB or better over the whole band, and the difference in the coupling can be equal or less than 1 dB over about 75% of the frequency band from 26 to 40 GHz.

If one pair of terminating resistors has different values compared to the other pair, the resulting branch-line hybrid can operate as a directional coupler and an impedance transformer simultaneously [64]. Design values of the branch- and main-line characteristic impedances for a single-section branch-line hybrid coupler shown in Fig. 5.33 (related to the input source impedance Z_{0S} and output load impedance Z_{0L}) can be calculated from

$$Z_1 = \frac{Z_{0S}}{K} \tag{5.51}$$

$$Z_2 = \sqrt{\frac{Z_{0S}\,Z_{0L}}{1 + K^2}} \tag{5.52}$$

$$Z_3 = \frac{Z_1\,Z_{0L}}{Z_{0S}} \tag{5.53}$$

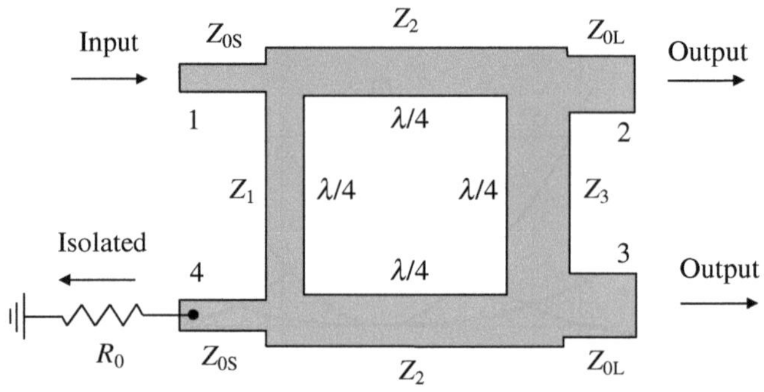

Figure 5.33 Microstrip branch-line quadrature impedance-transforming hybrid.

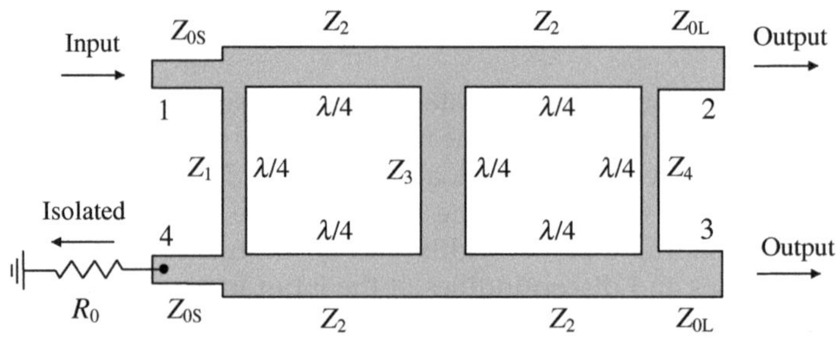

Figure 5.34 Broadband microwave branch-line quadrature hybrid.

where K is the voltage-split ratio between the output ports 2 and 3, and $R_0 = Z_{0S}$ [65]. Such a branch-line hybrid coupler with a 2:1 (50- to 25-Ω) impedance transformation ratio can provide an approximately 20% frequency bandwidth with a ±0.25-dB amplitude imbalance. However, for a fixed directivity, the frequency bandwidth of the branch-line impedance-transforming hybrid coupler increases as the output-to-input impedance ratio is reduced [64].

The operating bandwidth can be significantly increased by using multistage impedance-transforming hybrids. A two-section branch-line impedance-transforming quadrature hybrid coupler is shown in Fig. 5.34. To design this hybrid with the given impedance transformation ratio r and power-split ratio K^2, the branch- and main-line characteristic impedances should be chosen according to

$$Z_1 = Z_{0S}\sqrt{r\,\frac{t^2 - r}{t - r}} \tag{5.54}$$

$$\frac{Z_2^2}{Z_3} = Z_{0S}\sqrt{r - \left(\frac{r}{t}\right)^2} \tag{5.55}$$

$$Z_4 = Z_{0S}\frac{\sqrt{r\,(t^2 - r)}}{t - 1} \tag{5.56}$$

where $t = r\sqrt{1+K^2}$ [66]. The condition of $Z_2 = Z_3$ gives the maximum bandwidth when the best performance at the center bandwidth frequency is specified.

For an equal power division when $K = 1$, the condition of $t = r\sqrt{2}$ specifies a minimum value of r that is equal to 0.5. However, in practice, it is better to choose r in the range of 0.7 to 1.3, in order to provide the physically realizable branch-line characteristic impedances for a 50-Ω input impedance. For example, for the 50- to 35-Ω impedance transformation using a two-stage hybrid, the impedances are as follows: $Z_1 = 72.5$ Ω, $Z_2 = Z_3 = 29.6$ Ω, and $Z_4 = 191.25$ Ω. This gives the power balance between the output ports better than 0.5 dB with the return loss and isolation better than 20 dB over a frequency bandwidth of 25% for a 2-GHz quadrature hybrid.

For MMIC applications, the overall size of the quadrature branch-line hybrids with quarter-wavelength transmission lines is too large. Therefore, it is attractive to replace each quarterwave branch line with the combination of a short-length transmission line and two shunt capacitors providing similar bandwidth properties. In this case, consider the admittance matrix $[Y_a]$ for a quarterwave transmission line shown in Fig. 5.35(*a*) and the admittance matrix $[Y_b]$ for a circuit consisting of a short transmission line with the two shunt capacitors shown in Fig. 5.35(*b*), which are respectively given by

$$\left[Y_a\right] = \frac{1}{jZ_0}\begin{bmatrix} 0 & -1 \\ -1 & 0 \end{bmatrix} \tag{5.57}$$

$$\left[Y_b\right] = \frac{1}{jZ\sin\theta}\begin{bmatrix} \cos\theta - \omega CZ\sin\theta & -1 \\ -1 & \cos\theta - \omega CZ\sin\theta \end{bmatrix} \tag{5.58}$$

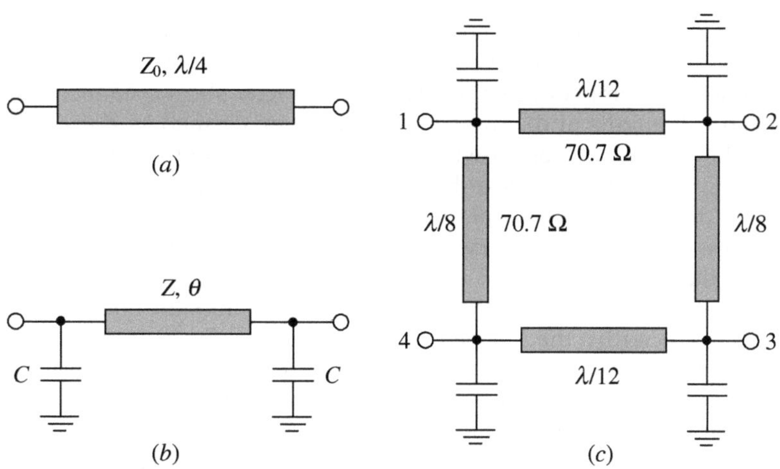

FIGURE 5.35 Reduced-size branch-line quadrature hybrid.

where Z_0 is the characteristic impedance of a quarterwave line, Z and θ are the characteristic impedance and electrical length of a shortened line, respectively, and C is the shunt capacitance. By equating the corresponding Y-parameters in Eqs. (5.57) and (5.58), the simple analytical ratios between the circuit parameters can be obtained as

$$Z = \frac{Z_0}{\sin\theta} \tag{5.59}$$

$$C = \frac{\cos\theta}{\omega Z_0} \tag{5.60}$$

from which it follows that the lengths of the hybrid transmission lines can be made much shorter by increasing their characteristic impedance Z. For example, when choosing the electrical length of $\theta = 45°$, the characteristic impedance of the transmission line increases by a factor of $\sqrt{2}$.

The circuit schematic of a reduced-size branch-line quadrature hybrid coupler is shown in Fig. 5.35(*c*) [67]. Compared to the conventional branch-line hybrid coupler with the characteristic impedances of its branch and mainlines of Z_0 and $Z_0/\sqrt{2}$, respectively, the circuit parameters of the reduced-size branch-line hybrid coupler are obtained by

$$\theta_1 = \sin^{-1}\left(\frac{Z_0}{Z}\right) \tag{5.61}$$

$$\theta_2 = \sin^{-1}\left(\frac{Z_0}{Z\sqrt{2}}\right) \tag{5.62}$$

$$\omega C Z_0 = \sqrt{1-\left(\frac{Z_0}{Z}\right)^2} + \sqrt{2-\left(\frac{Z_0}{Z}\right)^2} \tag{5.63}$$

where θ_1 and θ_2 are the electrical lengths of the shunt branch line and series main line, respectively. For a particular case of the standard characteristic impedance $Z_0 = 50\ \Omega$, the characteristic impedance and electrical lengths of the transmission lines are defined as $Z = Z_0\sqrt{2}$, $\theta_1 = 45°$, and $\theta_2 = 30°$, as shown in Fig. 5.35(*c*). The experimental results for a 25-GHz reduced-size branch-line quadrature hybrid coupler show that its bandwidth performance is slightly narrower than that of the conventional quarterwave hybrid coupler, but its overall size is more than 80% smaller.

To further reduce the MMIC size, the transmission lines can be fully replaced by the lumped planar inductors. Such an approach becomes possible because the symmetrical lumped LC-type π- or T-section is equivalent at a single frequency to the transmission-line section with the appropriate characteristic impedance and electrical length. The lumped-element equivalent circuit of a transmission-line branch-line hybrid is shown in Fig. 5.36(*a*) [68]. This circuit has also

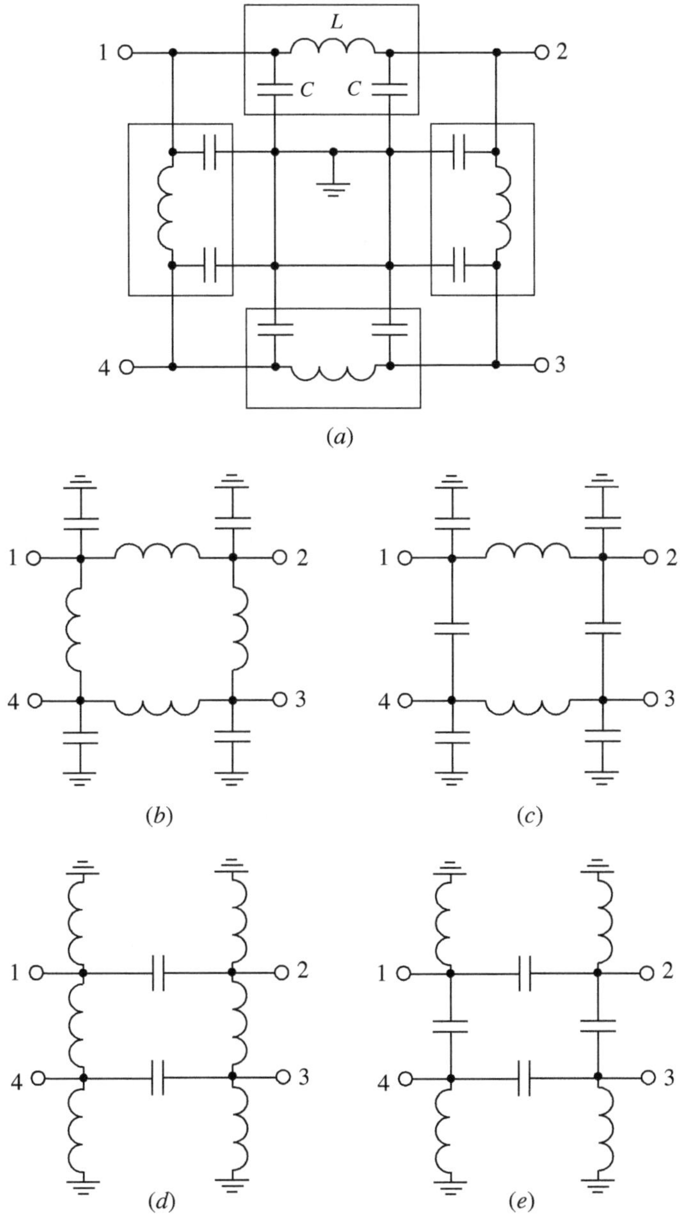

FIGURE 5.36 Equivalent circuits of lumped *LC*-type hybrid coupler.

some additional advantages when each its section can work as a separate impedance transformer, a low-pass filter, and a phase shifter. The circuit can be diced into four separate sections and cascaded for the desired transmission characteristics. The circuit analysis indicates that various types of networks fulfill the conditions required for an ideal hybrid. Therefore, greater design flexibility in the choice of the hybrid structure and performance is possible. Several possible single-section two-branch hybrid options are shown in Fig. 5.36 [69]. In this case, it should be noted that not only a low-pass section but also a high-pass section can be respectively used. In the latter case, the high-pass *LC* section is considered an equivalent single-frequency replacement for a 270° transmission line [70]. Figure 5.36(*c*) and 5.36(*d*) illustrates the use of both low-pass and high-pass sections simultaneously, whereas only high-pass sections compose the hybrid shown in Fig. 5.36(*e*). The performances of the hybrid couplers shown in Figs. 5.36(*b*) and 5.36(*e*) are very similar to those of the classical single-section branch-line hybrid coupler. The bandwidth performances of the hybrid couplers shown in Figs. 5.36(*c*) and 5.36(*d*) are narrower because the power balance between their output ports is much narrower. Broader bandwidth and lower output impedances can be provided with a two-section three-branch lumped-element hybrid coupler.

Figure 5.37(*a*) shows the equivalent circuit of a capacitively coupled lumped-element hybrid coupler, which is used for monolithic design of variable phase shifters [71]. However, the power and phase balance bandwidth at the output ports of this hybrid is very narrow,

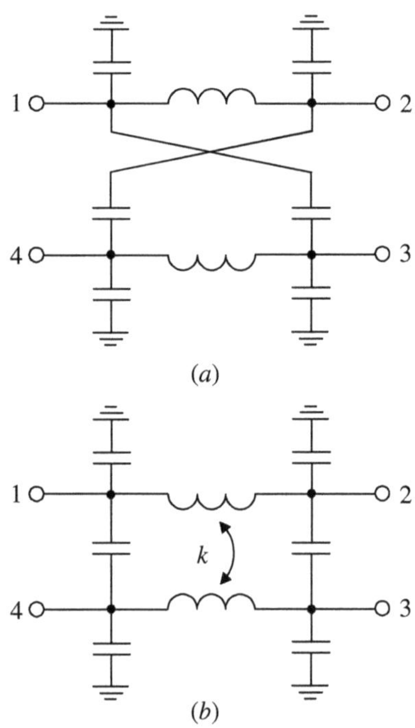

FIGURE 5.37 Equivalent circuits of lumped hybrid with capacitive and inductive coupling.

in the limits of a few percent. An alternative design of an inductively coupled lumped-element hybrid coupler is shown in Fig. 5.37(*b*) [72]. As a basic element, it includes lumped multiturn mutually coupled spiral inductors with the coupling coefficient, which can be realized using a bifilar (a sandwich of two multiturn spiral inductors with inner and outer windings) spiral transformer to achieve a coupling coefficient $k = 0.707$. In this case, this basic lumped-element configuration is completely equivalent to a transmission line in the vicinity of the center bandwidth frequency. The inductively coupled hybrid coupler can provide a power balance within 0.2 dB and phase balance within 1° in a frequency bandwidth of ±10% in 2-GHz wireless applications. However, during the design procedure, some parasitic effects should be taken into account. For example, the coupled inductor itself has a sufficient value of internal capacitance. Also, the finite value of the inductor quality factor results in a modest amplitude imbalance, but it leads to a significant phase deviation from ideal quadrature 90° difference. In this case, to compensate for the resulting performance degradation, the electromagnetic simulation of the structure and optimization of the values of the added shunt capacitors on both sides of the circuit are required. Two of these inductively coupled hybrids can be cascaded in tandem to significantly extend the frequency bandwidth. As a result, the phase shift of 93±6°, the insertion loss between 1 and 1.5 dB, the return loss better than 16 dB, and the isolation between the output ports better than 18 dB were measured over the frequency range from 2 to 6 GHz [72].

5.6 Coupled-Line Directional Couplers

The first directional couplers were composed of either a two-wire balanced line coupled to a second balanced line along a distance of quarter wavelength, or a pair of rods a quarter wavelength long between the ground planes [60]. Although the propagation of waves on systems of parallel conductors was investigated many decades ago in connection with the problem of crosstalk between open wire lines or cable pairs in order to eliminate the natural coupling rather than use it, the first exact design theory for TEM (transverse electromagnetic) transmission-line couplers was introduced by Oliver [73]. In terms of the even and odd electric-field modes describing a system of the coupled conductors, it can be stated that the coupling is backward with coupled wave on the secondary line propagating in the direction opposite to the direction of the wave on the primary line, the directivity will be perfect with *VSWR* equal to unity if $Z_0^2 = Z_{0e}Z_{0o}$ at all cross sections along the directional coupler, and the midband voltage coupling coefficient C of the directional coupler is defined as

$$C = \frac{Z_{0e} - Z_{0o}}{Z_{0e} + Z_{0o}} \tag{5.64}$$

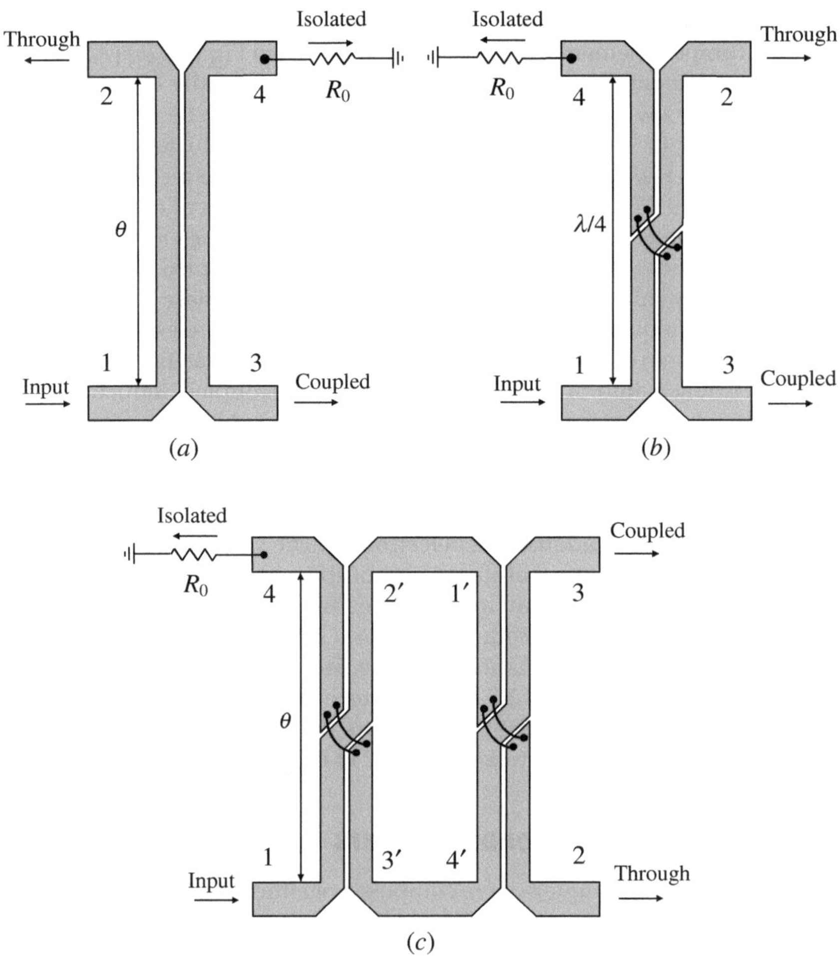

FIGURE 5.38 Coupled-line directional couplers.

where $C = 0$ for zero coupling and $C = 1$ for completely superposed transmission lines.

A coupled-line directional coupler, whose stripline single-section topology is shown in Fig. 5.38(*a*), can be used for broadband power dividing or combining. Its electrical properties are described using a concept of two types of excitations for the coupled lines in TEM approximation. In this case, for the even mode, the currents flowing in the strip conductors are equal in amplitude and flow in the same direction. The electric field has even symmetry about the center line, and no current flows between the two strip conductors. For the odd mode, the currents flowing in the strip conductors are equal in amplitude, but they flow in opposite directions. The electric field lines have an odd symmetry about the center line, and a voltage null exists between these two strip conductors. An arbitrary excitation of the coupled lines can always be treated as a superposition of appropriate amplitudes of even and odd modes. Therefore, the characteristic impedance for even excitation mode Z_{0e} and the characteristic impedance for the odd excitation mode Z_{0o} characterize the coupled lines. When the two coupled equal-strip lines are used in a standard system with characteristic impedance $Z_0 = \sqrt{Z_{0e}Z_{0o}}$, then

$$Z_{0e} = Z_0\sqrt{\frac{1+C}{1-C}} \tag{5.65}$$

$$Z_{0o} = Z_0\sqrt{\frac{1-C}{1+C}} \tag{5.66}$$

An analysis in terms of the scattering S-parameters gives $S_{11} = S_{14} = 0$ for any electrical lengths of the coupled lines and the output port 4 is isolated from the matched input port 1. Changing the coupling between the lines and their widths can change the characteristic impedances Z_{0e} and Z_{0o}. In this case,

$$S_{12} = \frac{\sqrt{1-C^2}}{\sqrt{1-C^2}\cos\theta + j\sin\theta} \tag{5.67}$$

$$S_{13} = \frac{jC\sin\theta}{\sqrt{1-C^2}\cos\theta + j\sin\theta} \tag{5.68}$$

where θ is the electrical length of the coupled-line section.

The voltage-split ratio K is defined as the ratio between voltages at ports 2 and 3 as

$$K = \left|\frac{S_{12}}{S_{13}}\right| = \frac{\sqrt{1-C^2}}{C\sin\theta} \tag{5.69}$$

where K can be controlled by changing the coupling coefficient C and electrical length θ.

For a quarter-wavelength-long coupler when $\theta = 90°$, Eqs. (5.67) and (5.68) reduce to

$$S_{12} = -j\sqrt{1-C^2} \tag{5.70}$$

$$S_{13} = C \tag{5.71}$$

from which it follows that equal voltage split between the output ports 2 and 3 can be provided with $C = 1/\sqrt{2}$.

If it is necessary to provide the output ports 2 and 3 at one side, it is best to use the construction of a microstrip directional coupler with crossed bondwires, as shown in Fig. 5.38(*b*). The strip crossover for a stripline directional coupler can be easily achieved with the three-layer sandwich. The microstrip 3-dB directional coupler fabricated on alumina substrate for idealized zero strip thickness should have the calculated strip spacing of less than 10 μm. Such a narrow value easily explains the great interest in the constructions of the directional couplers with larger spacing.

One of the effective solutions is to use a tandem connection of the two identical directional couplers, which alleviates the physical problem of tight coupling, as two individual couplers need only 8.34-dB coupling to achieve a 3-dB directional coupler [74, 75]. The tandem coupler shown in Fig. 5.38(*c*) has the electrical properties of the individual coupler when the output ports 1, 4 and 2, 3 are isolated in pairs, and the phase difference between the output ports 2 and 3 is of 90°.

From an analysis of the signal propagation from the input port 1 to the output ports 2 and 3 of the tandem coupler, when the signal from the input port 1 propagates to the output port 2 through the traces 1-2′-1′-2 and 1-3′-4′-2, whereas the signal flowing through the traces 1-2′-1′-3 and 1-3′-4′-3 is delivered to the output port 3, the ratio of the scattering parameters S_{12}^{T} and S_{13}^{T} of a tandem coupler can be expressed through the corresponding scattering parameters S_{12} and S_{13} of the individual coupler as

$$\frac{S_{12}^{\mathrm{T}}}{S_{13}^{\mathrm{T}}} = \frac{S_{12}^2 + S_{13}^2}{2S_{12}S_{13}} = -j\,\frac{1 - C^2(1+\sin^2\theta)}{2C\sqrt{1-C^2}\,\sin\theta} \tag{5.72}$$

As a result, the signal at the output port 2 overtakes that at the output port 3 by 90°. In this case, for a 3-dB tandem coupler with $\theta = 90°$, the magnitude of Eq. (5.72) must be equal to unity. Consequently, the required voltage coupling coefficient is calculated as

$$C = 0.5\sqrt{2-\sqrt{2}} = 0.3827$$

or

$$C_{12} = C_{13} = 8.34 \text{ dB}$$

As an example, a tandem 8.34-dB directional coupler has the dimensions of $W/h = 0.77$ and $S/h = 0.18$ for alumina substrate with $\varepsilon_r = 9.6$, where W is the strip width, S is the strip spacing, and h is the substrate thickness [51].

Another way to increase the coupling between the two edge-coupled microstrip lines is to use several parallel narrow microstrip lines interconnected with each other by the bondwires, as shown in Fig. 5.39. For a Lange coupler shown in Fig. 5.39(*a*), the four coupled

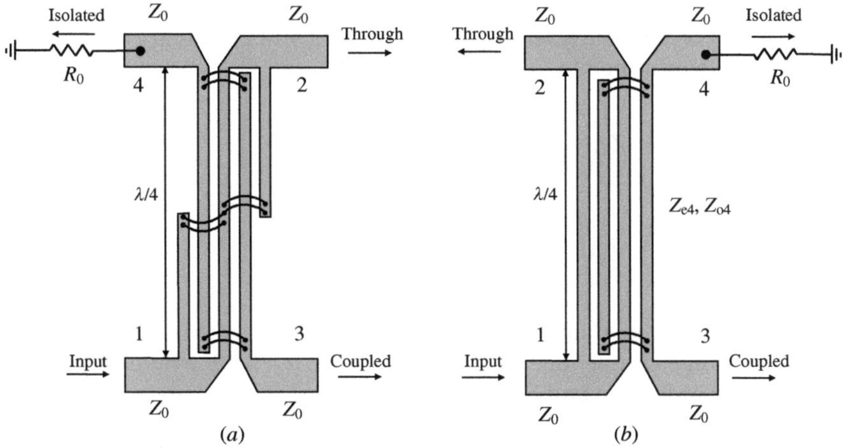

FIGURE 5.39 Lange directional couplers.

microstrip lines are used, achieving a 3-dB coupling over an octave or wider bandwidth [76]. In this case, the signal flowing to the input port 1 is distributed between the output ports 2 and 3 with the phase difference of 90°. However, this structure is quite complicated for practical implementation when, for alumina substrate with $\varepsilon_r = 9.6$, the dimensions of a 3-dB Lange coupler are $W/h = 0.107$ and $S/h = 0.071$, where W is the width of each strip and S is the spacing between adjacent strips.

Figure 5.39(*b*) shows the unfolded Lange coupler with four strips of equal length, which offers the same electrical performance but is easier for circuit modeling [77]. The even-mode characteristic impedance Z_{e4} and odd-mode characteristic impedance Z_{o4} of the Lange coupler with $Z_0^2 = Z_{e4}Z_{o4}$ in terms of the characteristic impedances of a two-conductor line (which is identical to any pair of adjacent lines in the coupler) can be obtained by

$$Z_{e4} = \frac{Z_{0o} + Z_{0e}}{3Z_{0o} + Z_{0e}} Z_{0e} \tag{5.73}$$

$$Z_{o4} = \frac{Z_{0e} + Z_{0o}}{3Z_{0e} + Z_{0o}} Z_{0o} \tag{5.74}$$

where Z_{0e} and Z_{0o} are the even- and odd-mode characteristic impedances of the two-conductor pair [78].

The midband voltage coupling coefficient C is given by

$$C = \frac{Z_{e4} - Z_{o4}}{Z_{e4} + Z_{o4}} = \frac{3\,(Z_{0e}^2 - Z_{0o}^2)}{3\,(Z_{0e}^2 + Z_{0o}^2) + 2Z_{0e}Z_{0o}} \tag{5.75}$$

The even- and odd-mode characteristic impedances Z_{0e} and Z_{0o}, as functions of the characteristic impedance Z_0 and coupling coefficient C, are determined by

$$Z_{0e} = Z_0\sqrt{\frac{1+C}{1-C}}\ \frac{4C-3+\sqrt{9-8C^2}}{2C} \tag{5.76}$$

$$Z_{0o} = Z_0\sqrt{\frac{1-C}{1+C}}\ \frac{4C+3-\sqrt{9-8C^2}}{2C} \tag{5.77}$$

For alumina substrate with $\varepsilon_r = 9.6$, the dimensions of such a 3-dB unfolded Lange coupler are $W/h = 0.112$ and $S/h = 0.08$, where W is the width of each strip and S is the spacing between the strips.

The design theory for TEM transmission-line couplers is based on an assumption of the same phase velocities of the even and odd propagation mode. However, this is not the case for coupled microstrip lines, because they have unequal even- and odd-mode phase velocities. In this case, the odd mode has more fringing electric field in the air region rather than the even mode with electric field concentrating mostly in the substrate underneath the microstrip lines. As a result, the effective dielectric permittivity in the latter case is higher, thus indicating a smaller phase velocity for the even mode. Consequently, it is required to apply the phase velocity compensation techniques in order to improve the coupler directivity, which decreases with increasing frequency. Figure 5.40(*a*) shows the topology of a typical wiggly-line coupler (with sawtooth shape of coupled lines), where wiggling the adjacent edges of the microstrip lines, which makes their physical lengths longer, slows the odd-mode wave without much affecting the even-mode wave [79]. High directivity can also be

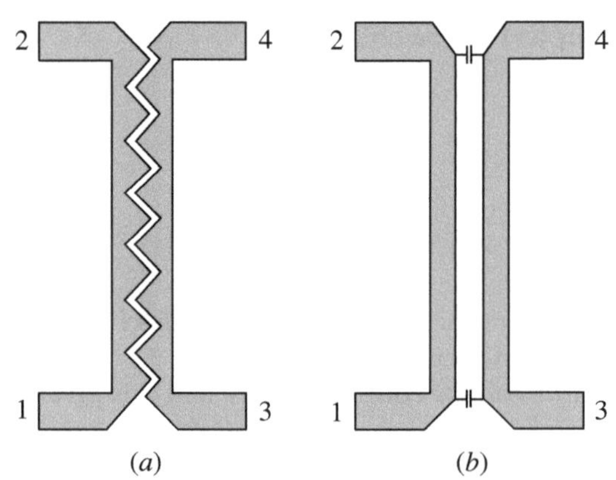

FIGURE 5.40 Coupled-line directional couplers.

achieved by using a capacitive compensation. Figure 5.40(*b*) shows the capacitively compensated microstrip directional coupler, where the two identical lumped capacitors are connected between coupled lines at their edges. Physically, these edge capacitors affect the odd mode by equivalent extension of the transmission-line electrical lengths, with almost no effect for even mode. For an ideal lossless operation condition at 12 GHz using standard alumina substrate, the compensated coupled-line microstrip directional coupler can improve directivity from 13.25 dB to infinity [80]. Capacitive compensation can be performed by a gap coupling of the open-circuit stub formed in a subcoupled line [81]. In this case, the coupler directivity can be improved by 23 dB in a frequency range from 1 to 2.5 GHz, compared to the directivity of the conventional uncompensated microstrip coupler.

At radio frequencies and low microwaves, the conventional quarter-wavelength directional coupler has very large dimensions that limit their practical application, especially in monolithic circuits. Figure 5.41 shows the circuit diagram of a reduced-size directional coupler consisting of the two coupled microstrip lines, whose electrical lengths are much smaller than a quarter-wavelength. The main problem of the coupler at frequencies, where the electrical length of its coupled lines is smaller than the quarter-wavelength, is that the degree of coupling linearly varies with frequency. To compensate this frequency behavior, the output port 3 can be connected to a series inductor L followed by a shunt resistor R [40, 82]. The inductance value depends on the coupling value, flatness, and midband frequency, whereas the resistance value depends on the impedance of the secondary line and inductance value. Such a microstrip reduced-size directional coupler with $L = 180$ nH and $R = 62\ \Omega$ can provide the coupling of about 30 dB with a flatness of ±0.1 dB, a directivity greater than 20 dB, an insertion loss less than 0.25 dB, and a *VSWR* less than 1.15 in a frequency bandwidth of 60% around

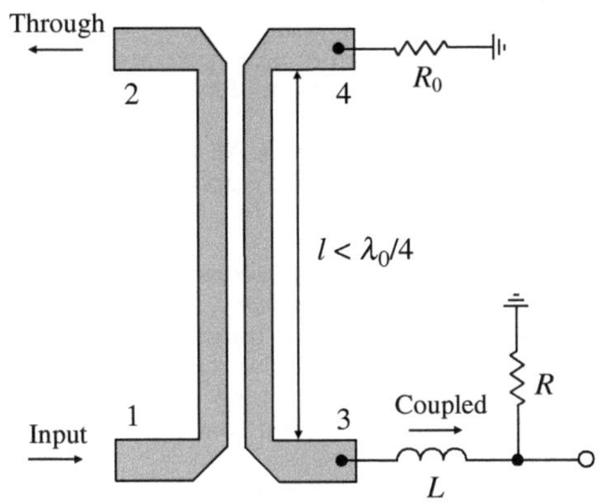

FIGURE 5.41 Coupled-line reduced-size directional coupler.

200 MHz. Tuning of the center bandwidth frequency and coupling can be simply realized by varying the inductance value.

References

1. D. M. Pozar, *Microwave Engineering*, New York: John Wiley & Sons, 2004.
2. R. E. Collin, *Foundation for Microwave Engineering*, New York: McGraw-Hill, 1992.
3. H. L. Krauss, C. W. Bostian, and F. H. Raab, *Solid State Radio Engineering*, New York: John Wiley & Sons, 1980.
4. Z. I. Model, *Networks for Combining and Distribution of High Frequency Power Sources* (in Russian), Moskva: Sov. Radio, 1980.
5. J. Sevick, *Transmission Line Transformers*, Norcross: Noble Publishing, 2001.
6. C. Trask, "Transmission Line Transformers: Theory, Design and Applications," *High Frequency Electronics*, vol. 4, pp. 46–53, Dec. 2005, vol. 5, pp. 26-33, Jan. 2006.
7. E. Rotholz, "Transmission-Line Transformers," *IEEE Trans. Microwave Theory Tech.*, vol. MTT-29, pp. 148–154, Apr. 1981.
8. J. Horn and G. Boeck, "Design and Modeling of Transmission Line Transformers," *Proc. 2003 IEEE SBMO/MTT-S Int. Microwave and Optoelectronics Conf.*, vol. 1, pp. 421–424.
9. G. Guanella, "New Method of Impedance Matching in Radio-Frequency Circuits," *The Brown Boveri Rev.*, vol. 31, pp. 327–329, Sep. 1944.
10. G. Guanella, "High-Frequency Matching Transformer," U.S. Patent 2,470,307, May 1949 (filed Apr. 1945).
11. R. K. Blocksome, "Practical Wideband RF Power Transformers, Combiners, and Splitters," *Proc. RF Technology Expo 86*, pp. 207–227, 1986.
12. J. L. B. Walker, D. P. Myer, F. H. Raab, and C. Trask, *Classic Works in RF Engineering: Combiners, Couplers, Transformers, and Magnetic Materials*, Norwood: Artech House, 2005.
13. J. Sevick, "Magnetic Materials for Broadband Transmission Line Transformers," *High Frequency Electronics*, vol. 4, pp. 46–52, Jan. 2005.
14. C. L. Ruthroff, "Some Broad-Band Transformers," *Proc. IRE*, vol. 47, pp. 1337–1342, Aug. 1959.
15. C. L. Ruthroff, "Broadband Transformers," U.S. Patent 3,037,175, May 1962 (filed May 1958).
16. J. Sevick, "A Simplified Analysis of the Broadband Transmission Line Transformer," *High Frequency Electronics*, vol. 3, pp. 48–53, Feb. 2004.
17. O. Pitzalis and T. P. M. Couse, "Practical Design Information for Broadband Transmission Line Transformer," *Proc. IEEE*, vol. 56, pp. 738–739, Apr. 1968.
18. R. T. Irish, "Method of Bandwidth Extension for the Ruthroff Transformer," *Electronics Lett.*, vol. 15, pp. 790–791, Nov. 1979.
19. S. C. Dutta Roy, "Optimum Design of an Exponential Line Transformer for Wide-Band Matching at Low Frequencies," *Proc. IEEE*, vol. 67, pp. 1563–1564, Nov. 1979.
20. M. Engels, M. R Jansen, W. Daumann, R. M. Bertenburg, and F. J. Tegude, "Design Methodology, Measurement and Application of MMIC Transmission Line Transformers," *1995 IEEE MTT-S Int. Microwave Symp. Dig.*, vol. 3, pp. 1635–1638.
21. S. P. Liu, "Planar Transmission Line Transformer Using Coupled Microstrip Lines," *1998 IEEE MTT-S Int. Microwave Symp. Dig.*, vol. 2, pp. 789–792.
22. O. Pitzalis and T. P. M. Couse, "Broadband Transformer Design for RF Transistor Power Amplifiers," *Proc. 1968 Electronic Components Conf.*, pp. 207–216.
23. S. E. London and S. V. Tomashevich, "Line Transformers with Fractional Transformation Factor," *Telecommunications and Radio Engineering*, vol. 28/29, pp. 129–130, Apr. 1974.
24. T. J. Gluszczak and J. D. Harmer, "Transmission Line Transformers with Integer-Ratio Voltage Transformations," *Proc. 1974 IEEE Int. Circuits and Systems Symp.*, pp. 368–370.

25. J. Sevick, "Design of Broadband Ununs with Impedance Ratios Less Than 1:4," *High Frequency Electronics*, vol. 3, pp. 44–50, Nov. 2004.
26. E. F. Sartori, "Hybrid Transformers," *IEEE Trans. Parts, Materials, and Packaging*, vol. PMP-4, pp. 59–66, Sep. 1968.
27. H. O. Granberg, "Broadband Transformers and Power Combining Techniques for RF," *Application Note AN749*, Motorola Semiconductor Product, 1975.
28. O. V. Alekseyev, *Broadband Transmitters* (in Russian), Moskva: Svyaz, 1978.
29. R. A. Gardenghi, "Solid State High Frequency Power Amplifier," U.S. Patent 3,406,352, Oct. 1968 (filed May 1965).
30. N. E. Lindenblad, "Television Transmitting Antenna for Empire State Building," *RCA Rev.*, vol. 3, pp. 387–408, Apr. 1939.
31. N. E. Lindenblad, "Junction between Single and Push-Pull Lines," U.S. Patent 2,231,839, Feb. 1941 (filed May 1939).
32. C. Lorenz, "Schaltungsanordnung zum Ubergang von Einer symmetrischen, Elektrischen Anordnung zu Einer Unsymmetrischen, Insbesondere bei Hochfrequenzanordnungen," German Patent DE603816, Sep. 1934.
33. B. P. Kumar and G. R. Branner, "Optimized Design of Unique Miniaturized Planar Baluns for Wireless Applications," *IEEE Microwave and Wireless Comp. Lett.*, vol. 13, pp. 134–136, Feb. 2003.
34. N. Marchand, "Transmission-Line Conversion Transformers," *Electronics*, vol. 17, pp. 142–145, Dec. 1944.
35. G. Oltman, "The Compensated Balun," *IEEE Trans. Microwave Theory Tech.*, vol. MTT-14, pp. 112–119, Mar. 1966.
36. C.-M. Tsai and K. C. Gupta, "A Generalized Model for Coupled Lines and its Applications to Two-Layer Planar Circuits," *IEEE Trans. Microwave Theory Tech.*, vol. MTT-40, pp. 2190–2199, Dec. 1992.
37. C. Cho and K. C. Gupta, "A New Design Procedure for Single-Layer and Two-Layer Three-Line Baluns," *IEEE Trans. Microwave Theory Tech.*, vol. MTT-46, pp. 2119–2514, Dec. 1998.
38. S. V. Bearse, "New Combining and Cooling Techniques Developed for 1 kW L-Band Transmitter," *Microwaves*, vol. 16, pp. 9–10, Sep. 1977.
39. E. J. Wilkinson, "An *N*-Way Hybrid Power Divider," *IRE Trans. Microwave Theory Tech.*, vol. MTT-8, pp. 116–118, Jan. 1960.
40. L. G. Maloratsky, *Passive RF & Microwave Integrated Circuits*, New York: Elsevier, 2004.
41. V. M. Katushkina and Z. I. Model, "*N*-Way Power Combiner of VHF-UHF Transmitters (in Russian)," *Elektrosvyaz*, vol. 7, pp. 17–25, Jul. 1959.
42. S. Y. London, "Independent Operation of High Power VHF Amplifiers on Common Load (in Russian)," *Voprosy Radioelektroniki, Ser. X, Tekhnika Svyazi*, vol. 6, pp. 87–96, Jun. 1959.
43. J. J. Taub and G. P. Kurpis, "A More General *N*-Way Hybrid Power Divider," *IEEE Trans. Microwave Theory Tech.*, vol. MTT-17, pp. 406–408, Jul. 1969.
44. J. R. Dent, "Strip Line Hybrid Ring," U.S. Patent 3,164,791, Jan. 1965 (filed May 1961).
45. S. B. Cohn, "A Class of Broad-Band Three-Port TEM-Mode Hybrids," *IEEE Trans. Microwave Theory Tech.*, vol. MTT-16, pp. 110–116, Feb. 1968.
46. D. Antsos, R. Crist, and L. Sukamto, "A Novel Wilkinson Power Divider with Predictable Performance at K and Ka-Band," *1994 IEEE MTT-S Int. Microwave Symp. Dig.*, vol. 2, pp. 907–910.
47. A. A. M. Saleh, "Planar Electrically Symmetric *n*-Way Hybrid Power Dividers/Combiners," *IEEE Trans. Microwave Theory Tech.*, vol. MTT-28, pp. 555–563, Jun. 1980.
48. U. H. Gysel, "A New N-Way Power Divider/Combiner Suitable for High-Power Applications," *1975 IEEE MTT-S Int. Microwave Symp. Dig.*, pp. 116–118.
49. R. Knochel and B. Mayer, "Broadband Printed Circuit 0°/180° Couplers and High Power Inphase Power Dividers," *1990 IEEE MTT-S Int. Microwave Symp. Dig.*, vol. 1, pp. 471–474.
50. H. F. Chapell, "Binary Power-Divider Design Approach," *IEEE Trans. Microwave Theory Tech.*, vol. MTT-22, pp. 580–581, May 1974.

51. O. A. Chelnokov, *Radio Transmitters* (in Russian), Moskva: Radio i Svyaz, 1982.
52. Y. Sun and A. P. Freundorfer, "Broadband Folded Wilkinson Power Combiner/Splitter," *IEEE Microwave and Wireless Components Lett.*, vol. 14, pp. 295–297, Jun. 2004.
53. N. Nagai, E. Maekawa, and K. Ono, "New *N*-way Hybrid Power Dividers," *IEEE Trans. Microwave Theory Tech.*, vol. MTT-25, pp. 1008–1012, Dec. 1977.
54. M. E. Goldfarb, "A Recombinant, In-Phase Power Divider," *IEEE Trans. Microwave Theory Tech.*, vol. MTT-39, pp. 1438–1440, Aug. 1991.
55. B. M. Schiffman, "A New Class of Broad-Band Microwave 90-Degree Phase Shifter," *IRE Trans. Microwave Theory Tech.*, vol. MTT-6, pp. 232–237, Apr. 1958.
56. L. I. Parad and R. L. Moynihan, "Split-Tee Power Divider," *IEEE Trans. Microwave Theory Tech.*, vol. MTT-13, pp. 91–95, Jan. 1965.
57. J.-S. Lim and S.-Y. Eom, "A New 3-Way Power Divider with Various Output Power Ratios," *1996 IEEE MTT-S Int. Microwave Symp. Dig.*, vol. 2, pp. 785–788.
58. D. Maurin and K. Wu, "A Compact 1.7-2.1 GHz Three-Way Power Combiner Using Microstrip Technology with Better Than 93.8% Combining Efficiency," *IEEE Microwave and Guided Wave Lett.*, vol. 6, pp. 106–108, Feb. 1996.
59. H. Ikeda, H. Kosugi, and T. Uwano, "A Low Distortion and High Efficiency Parallel-Operation Power Amplifier Combined in Different Phases in Wide Range of Load Impedances," *1996 IEEE MTT-S Int. Microwave Symp. Dig.*, pp. 535–538.
60. S. B. Cohn and R. Levy, "History of Microwave Passive Components with Particular Attention to Directional Couplers," *IEEE Trans. Microwave Theory Tech.*, vol. MTT-32, pp. 1046–1054, Sep. 1984.
61. J. Reed and G. J. Wheeler, "A Method of Analysis of Symmetrical Four-Port Networks," *IRE Trans. Microwave Theory Tech.*, vol. MTT-4, pp. 246–252, Oct. 1956.
62. R. Levy and L. Lind, "Synthesis of Symmetrical Branch-Guide Directional Couplers," *IEEE Trans. Microwave Theory Tech.*, vol. MTT-16, pp. 80–89, Feb. 1968.
63. P. Meaney, "A Novel Branch-Line Coupler Design for Millimeter-Wave Applications," *1990 IEEE MTT-S Int. Microwave Symp. Dig.*, vol. 1, pp. 585–588.
64. L. F. Lind, "Synthesis of Asymmetrical Branch-Guide Directional Coupler-Impedance Transformers," *IEEE Trans. Microwave Theory Tech.*, vol. MTT-17, pp. 45–48, Jan. 1969.
65. R. K. Gupta, S. E. Anderson, and W. J. Getsinger, "Impedance-Transforming 3-dB 90° Hybrids," *IEEE Trans. Microwave Theory Tech.*, vol. MTT-35, pp. 1303–1307, Dec. 1987.
66. S. Kumar, C. Tannous, and T. Danshin, "A Multisection Broadband Impedance Transforming Branch-Line Hybrid," *IEEE Trans. Microwave Theory Tech.*, vol. MTT-43, pp. 2517–2523, Nov. 1995.
67. T. Hirota, A. Minakawa, and M. Muraguchi, "Reduced-Size Branch-Line and Rat-Race Hybrid for Uniplanar MMIC's," *IEEE Trans. Microwave Theory Tech.*, vol. MTT-38, pp. 270–275, Mar. 1990.
68. M. Caulton, B. Hershenov, S.P. Knight, and R. E. DeBrecht, "Status of Lumped Elements in Microwave Integrated Circuits —Present and Future," *IEEE Trans. Microwave Theory Tech.*, vol. MTT-19, pp. 588–599, Jul. 1971.
69. R. W. Vogel, "Analysis and Design of Lumped- and Lumped-Distributed-Element Directional Couplers for MIC and MMIC Applications," *IEEE Trans. Microwave Theory Tech.*, vol. MTT-40, pp. 253–262, Feb. 1992.
70. S. J. Parisi, "180° Lumped Element Hybrid," *1989 IEEE MTT-S Int. Microwave Symp. Dig.*, vol. 3, pp. 1243–1246.
71. R. C. Frye, S. Kapur, and R. C. Melville, "A 2-GHz Quadrature Hybrid Implemented in CMOS Technology," *IEEE J. Solid-State Circuits*, vol. SC-38, pp. 550–555, Mar. 2003.
72. F. Ali and A. Podell, "A Wide-Band GaAs Monolithic Spiral Quadrature Hybrid and Its Circuit Applications," *IEEE J. Solid-State Circuits*, vol. SC-26, pp. 1394–1398, Oct. 1991.
73. B. M. Oliver, "Directional Electromagnetic Couplers," *Proc. IRE*, vol. 42, pp. 1686–1692, Nov. 1954.

74. G. D. Monteath, "Coupled Transmission Lines as Symmetrical Directional Couplers," *IEE Proc.*, vol. 102, part B, pp. 383–392, May 1955.
75. T. P. Shelton and J. A. Mosko, "Synthesis and Design of Wide-Band Equal-Ripple TEM Directional Couplers and Fixed Phase Shifters," *IEEE Trans. Microwave Theory Tech.*, vol. MTT-14, pp. 462–473, Oct. 1966.
76. J. Lange, "Interdigitated Stripline Quadrature Hybrid," *IEEE Trans. Microwave Theory Tech.*, vol. MTT-17, pp. 1150–1151, Dec. 1969.
77. R. Waugh and D. LaCombe, "Unfolding the Lange Coupler," *IEEE Trans. Microwave Theory Tech.*, Vol. MTT-20, pp. 777–779, Nov. 1972.
78. W. P. Ou, "Design Equations for an Interdigitated Directional Coupler," *IEEE Trans. Microwave Theory Tech.*, vol. MTT-23, pp. 253–255, Feb. 1975.
79. A. Podell, "A High Directivity Microstrip Coupler Technique," *1970 G-MTT Int. Microwave Symp. Dig.*, pp. 33–36.
80. M. Dydyk, "Accurate Design of Microstrip Directional Couplers with Capacitive Compensation," *1990 IEEE MTT-S Int. Microwave Symp. Dig.*, vol. 1, pp. 581–584.
81. C.-S. Kim, J.-S. Lim, D.-J. Kim, and D. Ahn, "A Design of Single and Multi-Section Microstrip Directional Coupler with the High Directivity," *2004 IEEE MTT-S Int. Microwave Symp. Dig.*, pp. 1895–1898.
82. L. Maloratsky, "Miniature Directional Coupler," U.S. Patent 5424694, Jun. 1995.

CHAPTER 6

Power Amplifier Design Fundamentals

Generally, the power amplifier design requires accurate active device modeling, impedance matching depending on the technical requirements and operation conditions, stability in operation, and simplicity in practical implementation. The quality of the power amplifier design is evaluated by realized maximum power gain under stable operation condition with minimum amplifier stages, and the requirement of linearity or high efficiency can be considered where it is needed. For a stable operation, it is necessary to evaluate the operating frequency domains where the active device may be potentially unstable. To avoid the parasitic oscillations, the stabilization circuit technique for different frequency domains (from low frequencies up to high frequencies close to the device transition frequency) is discussed. The key parameter of the power amplifier is its linearity, which is very important for many wireless communication applications. The relationships between the output power, 1-dB gain compression point, third-order intercept point, and intermodulation distortions of the third and higher orders are given and illustrated for different active devices. The device bias conditions that are generally different for linearity or efficiency improvement depend on the power amplifier operation class and the type of the active device. The bias circuits for the voltage-controlled MOSFET devices are simple due to its voltage control, because the gate dc current is equal to the gate leakage current. However, for the current-controlled bipolar devices, they differ substantially, depending on the device base current and class of operation. The basic Classes A, AB, B, and C of the power amplifier operation are introduced and analyzed, as well as classes of operation based on a finite number of harmonics are discussed and described. The concept and design of the push-pull amplifiers using balanced transistors are also presented.

6.1 Main Characteristics

Power amplifier design aims for maximum power gain and efficiency for a given value of output power with a predictable degree of stability. In order to extract the maximum power from a source generator, it is a well-known fact that the external load should have a vector value that is conjugate of the internal impedance of the source [1]. The power delivered from a generator to a load when matched on this basis will be called the available power of the generator [2]. In this case, the power gain of a four-terminal network is defined as the ratio of the power delivered to the load connected at the output terminals to the power available from the generator connected to the input terminals, usually measured in decibels, and this ratio is called the *power gain* irrespective of whether it is greater or less than one [3, 4].

Figure 6.1 shows the basic block schematic of a single-stage power amplifier circuit, which includes an active device, an input matching circuit to match with the source impedance, and an output matching circuit to match with the load impedance. Generally, the two-port active device is characterized by a system of the immittance W-parameters, that is, any system of the impedance Z-parameters, hybrid H-parameters, or admittance Y-parameters [5, 6]. The input and output matching circuits transform the source and load immittances W_S and W_L into specified values between ports 1 to 2 and 3 to 4, respectively, by means of which the optimal design operation mode of the power amplifier is realized.

The given technical requirements and the convenience of the design realization (using the appropriate types of the active devices and circuit schematics) determine the choice of system of the admittance Y-parameters or impedance Z-parameters. Let, for the given input and output voltages, the active device be characterized by a matrix of Y-parameters. In this case, the source of the input signal is represented by the current source with an internal admittance Y_S, and a load is characterized by the load admittance Y_L, as shown in Fig. 6.2(*a*). If a two-port active device is described by a system of Z-parameters, the source of the input signal is represented by the voltage source with an internal impedance Z_S, whereas the load is characterized by the load impedance Z_L, as shown in Fig. 6.2(*b*). In both cases, the admittances Y_S and Y_L and the impedances Z_S and Z_L are seen looking toward the source and load through the input and output matching circuits.

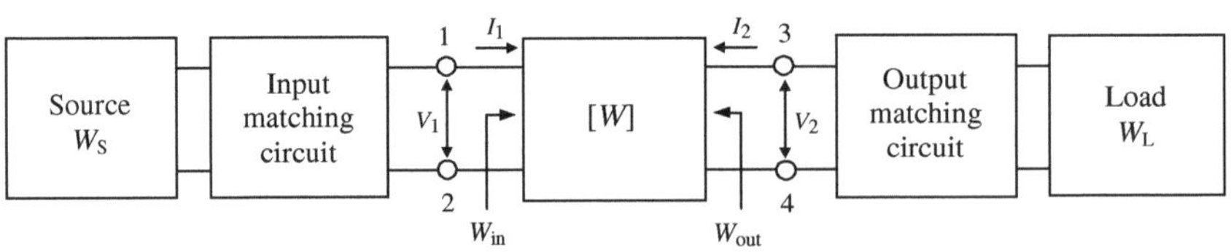

FIGURE 6.1 Block schematic of single-stage power amplifier.

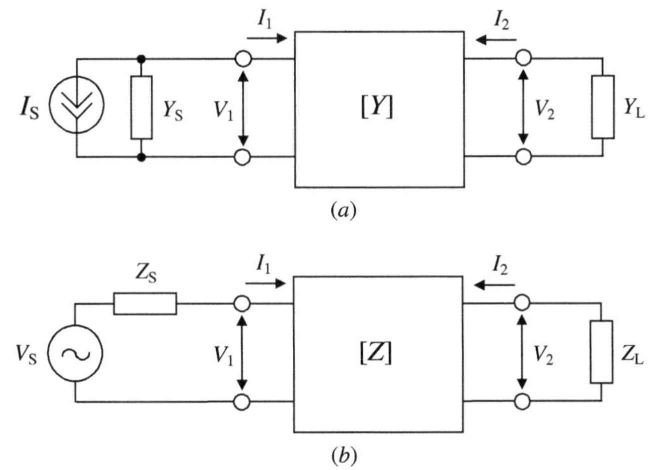

FIGURE 6.2 Two-port loaded amplifier networks.

To calculate the electrical characteristics of a power amplifier, first consider a system of the admittance Y-parameters. The active device in this case is described by the following system of two equations:

$$\begin{cases} I_1 = Y_{11}V_1 + Y_{12}V_2 \\ I_2 = Y_{21}V_1 + Y_{22}V_2 \end{cases} \tag{6.1}$$

Depending on impedance matching at the input and output device ports, several definitions of the amplifier power gain in terms of Y-parameters can be derived.

Operating power gain ($G_P = P_L/P_{in}$) is the ratio of power dissipated in the active load G_L to power delivered to the input port of the active device with admittance Y_{in}. This gain is independent of G_S but is strongly dependent on G_L.

Available power gain ($G_A = P_{out}/P_S$) is the ratio of power available at the output port of the active device with admittance Y_{out} to power available from the source G_S. This power gain depends on G_S but not G_L.

Transducer power gain ($G_T = P_L/P_S$) is the ratio of power dissipated in the active load G_L to power available from the source G_S. This power gain assumes complex-conjugate impedance matching at the input and output ports of the active device being dependent on both G_S and G_L.

Maximum available gain (MAG) is the theoretical power gain of the active device when its reverse transfer admittance Y_{12} is set equal to zero. It represents a theoretical limit on the gain that can be achieved with the given active device assuming of complex-conjugate impedance matching of the input and output ports of the active device with the source and load, respectively.

In practice, to characterize an amplifier circuit gain property, generally two types of power gain are used: operating power gain G_P and transducer power gain G_T. The operating power gain G_P is used when it is required to calculate the power at the input port of the device, which is necessary to provide the given power delivered to the load. However, to analyze the stability conditions, it is important to know both the value of the source impedance and the value of the load impedance. Therefore, in this case it is preferable to use the transducer power gain G_T, which is necessary to maximize within the restrictions imposed by the stability consideration.

First, consider the evaluation of G_P in terms of the two-port network Y-parameters. If V_1 is the amplitude at the input port of the active device, then

$$P_{in} = 0.5\, V_1^2\, \mathrm{Re}\, Y_{in} \tag{6.2}$$

where $Y_{in} = I_1 / V_1$ is the input admittance (between input ports 1 and 2) of a two-port network loaded on the admittance Y_L. Given that $I_2 = -Y_L V_2$, the expression for Y_{in} is defined from Eq. (6.1) as

$$Y_{in} = Y_{11} - \frac{Y_{12}\, Y_{21}}{Y_{22} + Y_L} \tag{6.3}$$

The output power dissipated in a load is obtained by

$$P_L = 0.5\, V_2^2\, \mathrm{Re}\, Y_L \tag{6.4}$$

As a result, the operating power gain G_P can be written as

$$G_P = \frac{P_L}{P_{in}} = \frac{|Y_{21}|^2\, \mathrm{Re}\, Y_L}{|Y_{22} + Y_L|^2\, \mathrm{Re}\, Y_{in}} \tag{6.5}$$

The operating power gain G_P does not depend on the source parameters and characterizes only the effectiveness of the power delivery from the input port of the active device to the load. This gain helps evaluate the gain property of a multistage amplifier when the overall operating power gain $G_{P(total)}$ is equal to the product of each stage G_P.

The transducer power gain G_T includes an assumption of the complex-conjugate matching of the load and the source. This means that $Y_{in} = Y_s^*$, where Y_s^* is the source admittance matched conjugately to the input port of the active device Y_{in}. As a result, the power available from the source is written as

$$P_S = \frac{I_S^2}{8\, \mathrm{Re}\, Y_S} \tag{6.6}$$

If $I_S = Y_S V_1 + I_1$, then the expression for the source current I_S using Eq. (6.1) can be defined by

$$I_S = \frac{(Y_{11} + Y_S)(Y_{22} + Y_L) - Y_{12} Y_{21}}{Y_{22} + Y_L}\, V_1 \tag{6.7}$$

From Eqs. (6.4) and (6.7) it follows that the transducer power gain G_T can be written as

$$G_T = \frac{P_L}{P_S} = \frac{4\left|Y_{21}\right|^2 \mathrm{Re}Y_S \ \mathrm{Re}Y_L}{\left|(Y_{11}+Y_S)(Y_{22}+Y_L)-Y_{12}Y_{21}\right|^2} \tag{6.8}$$

Similarly, the operating and transducer power gains G_P and G_T and the input and output impedances Z_{in} and Z_{out} can be expressed through the same analytical forms of a system of Z-parameters. Thus, by using the immittance W-parameters, they can be generalized as

$$W_{in} = W_{11} - \frac{W_{12}W_{21}}{W_{22} + W_L} \tag{6.9}$$

$$W_{out} = W_{22} - \frac{W_{12}W_{21}}{W_{11} + W_S} \tag{6.10}$$

$$G_P = \frac{\left|W_{21}\right|^2 \mathrm{Re}W_L}{\left|W_{22}+W_L\right|^2 \mathrm{Re}W_{in}} \tag{6.11}$$

$$G_T = \frac{4\left|W_{21}\right|^2 \mathrm{Re}W_S \,\mathrm{Re}W_L}{\left|(W_{11}+W_S)(W_{22}+W_L)-W_{12}W_{21}\right|^2} \tag{6.12}$$

where W_{ij} $(i, j = 1, 2)$ are the immittance two-port parameters of the active device equivalent circuit.

From Eqs. (6.9) and (6.10), it follows that if $W_{12} = 0$, then $W_{in} = W_{11}$ and $W_{out} = W_{22}$. As a result, in the case of the complex-conjugate matching at the input and output ports of the active device, the expression for MAG is obtained from Eq. (6.11) or (6.12) as

$$MAG = \frac{\left|W_{21}\right|^2}{4\ \mathrm{Re}W_{11}\,\mathrm{Re}W_{22}} \tag{6.13}$$

the magnitude of which depends only on the active device immittance parameters.

The bipolar transistor simplified small-signal π-hybrid equivalent circuit shown in Fig. 6.3 provides an example for a conjugately matched bipolar power amplifier. The impedance Z-parameters of the equivalent circuit of the bipolar transistor in a common-emitter configuration can be written as

$$Z_{11} = r_b + \frac{1}{g_m + j\omega C_\pi} \qquad Z_{12} = \frac{1}{g_m + j\omega C_\pi}$$

$$Z_{21} = -\frac{1}{j\omega C_c}\frac{g_m - j\omega C_c}{g_m + j\omega C_\pi} \qquad Z_{22} = \left(1 + \frac{C_\pi}{C_c}\right)\frac{1}{g_m + j\omega C_\pi} \tag{6.14}$$

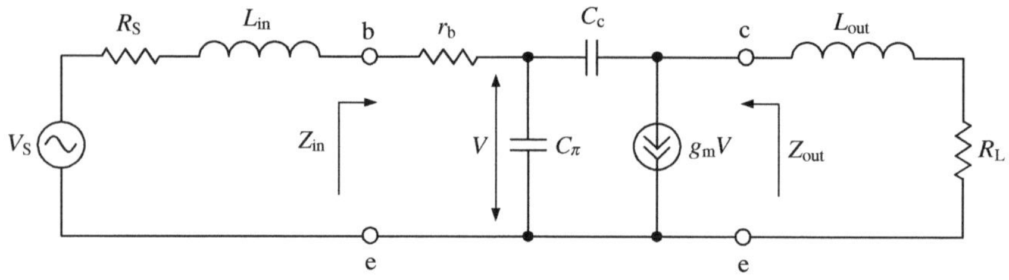

FIGURE 6.3 Simplified equivalent circuit of matched bipolar power amplifier.

where g_m is the device transconductance, r_b is the series base resistance, C_π is the base-emitter capacitance including both diffusion and junction components, and C_c is the feedback collector capacitance.

By setting the device feedback impedance Z_{12} to zero and complex-conjugate–matching conditions at the input of $R_S = \mathrm{Re}Z_{in}$ and $L_{in} = -\mathrm{Im}Z_{in}/\omega$ and at the output of $R_L = \mathrm{Re}Z_{out}$ and $L_{out} = -\mathrm{Im}Z_{out}/\omega$, the small-signal transducer power gain G_T can be calculated by

$$G_T = \left(\frac{f_T}{f}\right)^2 \frac{1}{8\pi f_T r_b C_c} \tag{6.15}$$

where $f_T = g_m/2\pi C_\pi$ is the device transition frequency. Equation (6.15) gives a well-known expression for maximum operating frequency of the BJT device when the maximum available gain is equal to unity,

$$f_{max} = \sqrt{\frac{f_T}{8\pi\, r_b C_c}} \tag{6.16}$$

Figure 6.4 shows the simplified circuit schematic for a conjugately matched FET (field-effect transistor) power amplifier. The admittance Y-parameters of the small-signal equivalent circuit of any FET device in a common-source configuration can be written as

$$Y_{11} = \frac{j\omega C_{gs}}{1 + j\omega C_{gs}R_{gs}} + j\omega C_{gd} \qquad Y_{12} = -j\omega C_{gd}$$
$$Y_{21} = \frac{g_m}{1 + j\omega C_{gs}R_{gs}} - j\omega C_{gd} \qquad Y_{22} = \frac{1}{R_{ds}} + j\omega(C_{ds} + C_{gd}) \tag{6.17}$$

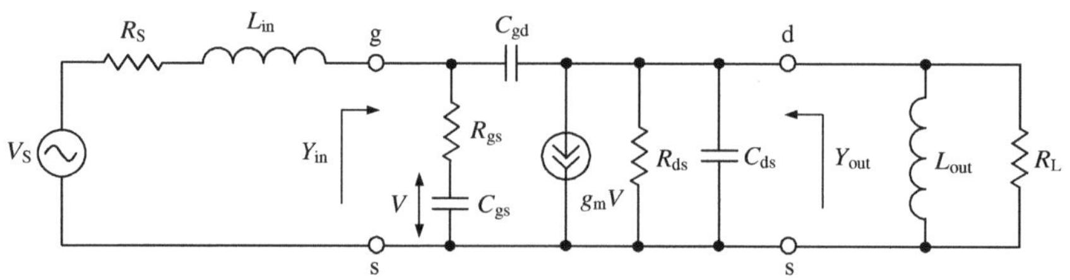

FIGURE 6.4 Simplified equivalent circuit of matched FET power amplifier.

where g_m is the device transconductance, R_{gs} is the gate-source resistance, C_{gs} is the gate-source capacitance, C_{gd} is the feedback gate-drain capacitance, C_{ds} is the drain-source capacitance, and R_{ds} is the differential drain-source resistance.

Because the value of the gate-drain capacitance C_{gd} is usually relatively small, effect of the feedback admittance Y_{12} can be neglected in a simplified case. Then, it is necessary to set $R_S = R_{gs}$ and $L_{in} = 1/\omega^2 C_{gs}$ for input matching, whereas $R_L = R_{ds}$ and $L_{out} = 1/\omega^2 C_{ds}$ for output matching. Hence, the small-signal transducer power gain G_T can approximately be calculated by

$$G_T(C_{gd} = 0) = MAG = \left(\frac{f_T}{f}\right)^2 \frac{R_{ds}}{4R_{gs}} \tag{6.18}$$

where $f_T = g_m/2\pi C_{gs}$ is the device transition frequency and MAG is the maximum available gain representing a theoretical limit for the power gain that can be achieved under complex-conjugate matching conditions. Equation (6.18) gives a well-known expression for maximum operating frequency of the FET device when the maximum available gain is equal to unity,

$$f_{max} = \frac{f_T}{2}\sqrt{\frac{R_{ds}}{R_{gs}}} \tag{6.19}$$

From Eqs. (6.15) and (6.18), it follows that the small-signal power gain of a conjugately matched power amplifier for any type of the active device drops off as $1/f^2$ or 6 dB per octave. Therefore, if a power gain G_T is known at the transition frequency f_T, $G_T(f)$ can readily be predicted at a certain frequency f by

$$G_T(f) = G_T(f_T)\left(\frac{f_T}{f}\right)^2 \tag{6.20}$$

It should be noted that previous analysis is based on the linear small-signal consideration when generally nonlinear device current source as a function of both input and output voltages can be characterized by the linear transconductance g_m as a function of the input voltage and the output differential resistance R_{ds} is a function of the output voltage. This is a result of a Taylor-series expansion of the output current as a function of the input and output voltages with maintaining only the dc and linear components. Such an approach helps understand and derive the maximum achievable power amplifier parameters in a linear approximation. In this case, the active device is operated in a Class-A mode when one-half of the dc power is dissipated in the device, whereas the other half is transformed to the fundamental-frequency output power flowing into the load, resulting in a maximum ideal collector efficiency of 50%. The device output resistance R_{out}

remains constant and can be calculated as a ratio of the dc supply voltage to the dc current flowing through the active device. In a nonlinear case, for a complex-conjugate–matching procedure, the device output immittance under large-signal consideration should be calculated using a Fourier-series analysis of the output current and voltage fundamental components. This means that, unlike a linear Class-A mode, the active device is operated in a device linear region only part of the entire period, and its output resistance is defined as a ratio of the fundamental-frequency output voltage to the fundamental-frequency output current. This is not a physical resistance resulting in a power loss inside the device, but an equivalent resistance required to use for a conjugate matching procedure. In this case, the complex conjugate matching is valid and necessary, first, to compensate for the reactive part of the device output impedance and second, to provide a proper load resistance, resulting in a maximum power gain for a given supply voltage and required output power delivered to the load. Note that this is not the maximum available small-signal power gain that can be achieved in a linear operation mode, but a maximum achievable large-signal power gain that can be achieved for a particular operation mode with a certain conduction angle. Of course, the maximum large-signal power gain is smaller than the small-signal power gain for the same input power, because the output power in a nonlinear operation mode also includes the powers at the harmonic components of the fundamental frequency.

Therefore, it makes more practical sense not to introduce separately the concepts of the gain match with respect to the linear power amplifiers and the power match in nonlinear power amplifiers because the maximum large-signal power gain as a function of the conduction angle corresponds to the maximum fundamental-frequency output power delivered to the load due to large-signal conjugate output matching. It is very important to provide a conjugate matching at both input and output device ports to achieve maximum power gain in a large-signal mode. In a Class-A mode, the maximum small-signal power gain ideally remains constant regardless of the output power level.

The transistor characterization in a large-signal mode can be done based on equivalent quasi-harmonic nonlinear approximation under the condition of sinusoidal port voltages [7]. In this case, the large-signal impedances are generally determined in the following manner. The designer tunes the load network (often by trial and errors) to maximize the output power to the required level using a particular transistor at a specified frequency and supply voltage. Then, the transistor is removed from the circuit and the impedance seen by the collector is measured at the carrier frequency. The complex-conjugate of the measured impedance then represents the equivalent large-signal output impedance of the transistor at that frequency, supply voltage, and output power. Similar design process is used to measure the input impedance of the transistor and to maximize power-added efficiency of the power amplifier. It is especially important with power transistors in the microwave region either to characterize the transistors in the packaging configuration, in which they will be utilized in the circuit application, or to accurately know the package parameters so that suitable corrections can be made if the transistors are mounted in a different environment [8].

To deliver maximum power to the load, it is necessary to use some impedance matching network that can modify the load as viewed from the generator [1]. The design of reactance networks to connect a resistive load efficiently to a source of power can be carried out most conveniently by the theory of image impedances. In this case, if the image impedances of such a network are pure resistances and it is connected between the generator and the load whose impedances are equal to these image impedances, the impedances will match at both junctions. Under these conditions with an assumption of pure reactances of the arms of the connecting network, no power will dissipate during transmission and so this maximum power will be delivered to the load. If the terminating impedances are not pure resistances, they can be made so at any single frequency by additional reactance in series with them. Not only can such reactance networks provide high operation efficiency but can also attenuate undesired harmonics. A variety of configurations can be designed to accomplish the desired result. Generally, the various L-type, T-type, and π-type configurations of reactances described in Chap. 4 are used for matching networks, depending on convenience of the circuit implementation and particular application technology. They can be implemented in both low- and high-pass topologies.

6.2 Power Gain and Stability

In early radiofrequency vacuum-tube transmitters, it was observed that the tubes and associated circuits may have damped or undamped oscillations depending on the circuit losses, the feedback coupling, the grid and anode potentials, and the reactance or tuning of the parasitic circuits [9, 10]. Various parasitic oscillator circuits such as the tuned-grid–tuned-anode circuit with capacitive feedback, Hartley, Colpitts, or Meissner oscillators can be realized at high frequencies, which potentially can be eliminated by adding a small resistor close to the grid or anode connections of the tubes for damping the circuits. Inductively coupled rather than capacitively coupled input and output circuits should be used wherever possible.

According to the immittance approach to the stability analysis of the active nonreciprocal two-port network, it is necessary and sufficient for its unconditional stability if the following system of equations can be satisfied for the given active device:

$$\mathrm{Re}\,[W_S(\omega) + W_{in}(\omega)] > 0 \tag{6.21}$$

$$\text{Im}\,[W_S(\omega)+W_{in}(\omega)]=0 \tag{6.22}$$

or

$$\text{Re}\,[W_L(\omega)+W_{out}(\omega)]\;>0 \tag{6.23}$$

$$\text{Im}\,[W_L(\omega)+W_{out}(\omega)]=0 \tag{6.24}$$

where ReW_S and ReW_L are considered to be greater than zero [11, 12]. The active two-port network can be treated as unstable or potentially unstable in the case of the opposite signs in Eqs. (6.21) and (6.23).

When Re$[W_S(\omega)] > 0$ and Re$[W_L(\omega)] > 0$, the requirements of amplifier unconditional stability can be simplified to

$$\text{Re}\,[W_{in}(\omega)]\;>0 \qquad \text{Re}\,[W_{out}(\omega)]>0 \tag{6.25}$$

According to Eqs. (6.9) and (6.10), the value of ReW_{in} depends on the load immittance W_L, whereas the variation of the source immittance W_S leads to the change of ReW_{out}. Therefore, if there is a negative value of ReW_{out}, the active two-port network will be potentially unstable in certain limits of the values of immittance W_S. Consequently, for unconditionally stable amplifier operation mode, it is necessary to satisfy the following condition:

$$\text{Re}\,[W_{out}(\omega)]\big|_{min}\;>0 \tag{6.26}$$

Analyzing Eq. (6.10) on extremum, the minimum positive value of ReW_{out} for a given constant value of ReW_S can be derived by solving equation $\partial\text{Re}W_{out}/\partial\text{Im}W_S=0$ as

$$\text{Re}\,W_{out}=\text{Re}\,W_{22}\;-\frac{|W_{12}\,W_{21}|+\text{Re}(W_{12}\,W_{21})}{2\text{Re}\;(W_{11}+W_S)} \tag{6.27}$$

The second term in the right-hand part of Eq. (6.27) as a function of ReW_S can take a maximum negative value when Re$W_S = 0$, resulting in

$$\text{Re}\,W_{out}=\text{Re}\,W_{22}\;-\frac{|W_{12}\,W_{21}|+\text{Re}(W_{12}\,W_{21})}{2\text{Re}\;W_{11}} \tag{6.28}$$

Consequently, the requirement of a positive minimum value of ReW_{out} given by Eq. (6.26) can be written as

$$2\text{Re}W_{11}\;\;\text{Re}W_{22}\;-|W_{12}\,W_{21}|-\text{Re}(W_{12}\,W_{21})\;>0 \tag{6.29}$$

Similar result can be obtained by optimizing the immittance W_{in} as a function of W_L. From Eq. (6.29) it follows that under such a condition the active device is unconditionally stable at any frequencies where this condition is fulfilled, regardless of any values of the source and load immittances W_S and W_L. By normalizing Eq. (6.29), a special relationship between the device immittance parameters called the *device stability factor* can be derived as

$$K = \frac{2\ \mathrm{Re}W_{11}\ \mathrm{Re}W_{22} - \mathrm{Re}\,(W_{12}W_{21})}{|W_{12}W_{21}|} \tag{6.30}$$

which shows a stability margin indicating how far from zero value are the real parts in Eqs. (6.9) and (6.11) being positive [12].

It should be noted that the applicability of Eqs. (6.29) and (6.30) is restricted to the following requirements:

$$\mathrm{Re}\,[W_{11}(\omega)] > 0 \qquad \mathrm{Re}\,[W_{22}(\omega)] > 0 \tag{6.31}$$

A comparison of Eqs. (6.29) and (6.30) shows that the active device is unconditionally stable if $K > 1$ and potentially unstable if $K < 1$. From Eq. (6.31) and the condition

$$|\mathrm{Re}\,(W_{12}W_{21})| \le |W_{12}W_{21}| \tag{6.32}$$

it follows that the smallest possible value of the device stability factor can be defined as $K = -1$, which means that the values of K can be arranged only in the interval $[-1, \infty)$.

When the active device is potentially unstable, an improvement of the power amplifier stability can be provided with the appropriate choice of the source and load immittances W_S and W_L. In this case, the circuit stability factor K_T is defined in the same way as the device stability factor K, but by taking into account $\mathrm{Re}W_S$ and $\mathrm{Re}W_L$ along with the device W-parameters. In this case, the circuit stability factor is given by

$$K_T = \frac{2\ \mathrm{Re}(W_{11} + W_S)\mathrm{Re}\,(W_{22} + W_L) - \mathrm{Re}(W_{12}W_{21})}{|W_{12}W_{21}|} \tag{6.33}$$

If the circuit stability factor $K_T \ge 1$, the power amplifier is unconditionally stable. However, the power amplifier becomes potentially unstable if $K_T < 1$. The value of $K_T = 1$ corresponds to the border of the circuit unconditional stability. The values of the circuit stability factor K_T and device stability factor K become equal if $\mathrm{Re}W_S = \mathrm{Re}W_L = 0$.

For the device stability factor $K > 1$, the operating power gain G_P has to be maximized [13, 14]. By analyzing Eq. (6.11) on extremum, it is possible to derive the optimum values $\mathrm{Re}\,W_L^o$ and $\mathrm{Im}\,W_L^o$, at which the operating power gain G_P is maximal, by solving the following system of two equations:

$$\frac{\partial G_P}{\partial\ \mathrm{Re}W_L} = 0 \qquad \frac{\partial G_P}{\partial\ \mathrm{Im}W_L} = 0 \tag{6.34}$$

As a result, the optimum values $\mathrm{Re}\,W_L^o$ and $\mathrm{Im}\,W_L^o$ depend on the immittance parameters of the active device and the device stability factor according to

$$\mathrm{Re}\,W_L^o = \frac{|W_{12}W_{21}|}{2\,\mathrm{Re}\,W_{11}}\sqrt{K^2-1} \tag{6.35}$$

$$\mathrm{Im}\,W_L^o = \frac{\mathrm{Im}(W_{12}W_{21})}{2\,\mathrm{Re}\,W_{11}} - \mathrm{Im}\,W_{22} \tag{6.36}$$

Substituting the obtained values of $\mathrm{Re}\,W_L^o$ and $\mathrm{Im}\,W_L^o$ into Eq. (6.11) yields an expression for calculating the maximum value of G_{Pmax} written as

$$G_{\mathrm{Pmax}} = \left|\frac{W_{21}}{W_{12}}\right| / (K + \sqrt{K^2-1}\,) \tag{6.37}$$

from which it follows that G_{Pmax} can be achieved only if $K > 1$, whereas $G_{\mathrm{Pmax}}^{\mathrm{A}} = |W_{21}/W_{12}|$ when $K = 1$.

If the source is conjugately matched with the input of the active device, the following conditions must be satisfied:

$$\mathrm{Re}\,W_S = \mathrm{Re}\,W_{\mathrm{in}} \qquad \mathrm{Im}\,W_S + \mathrm{Im}\,W_{\mathrm{in}} = 0 \tag{6.38}$$

Then, by substituting these expressions into Eq. (6.9), the optimum values $\mathrm{Re}\,W_S^o$ and $\mathrm{Im}\,W_S^o$ as functions of the immittance parameters of the active device and the device stability factor can be derived as

$$\mathrm{Re}\,W_S^o = \frac{|W_{12}W_{21}|}{2\,\mathrm{Re}\,W_{22}}\sqrt{K^2-1} \tag{6.39}$$

$$\mathrm{Im}\,W_S^o = \frac{\mathrm{Im}(W_{12}W_{21})}{2\,\mathrm{Re}\,W_{22}} - \mathrm{Im}\,W_{11} \tag{6.40}$$

A comparison of Eqs. (6.39) and (6.40) with Eqs. (6.35) and (6.36), respectively, shows that these expressions are identical. Consequently, the power amplifier with an unconditionally stable active device provides a maximum power gain operation only if the input and output of the active device are conjugately matched with the source and load impedances, respectively. For the lossless input matching circuit when the power available at the source is equal to the power delivered to the input port of the active device, that is, $P_S = P_{\mathrm{in}}$, the maximum operating power gain is equal to the maximum transducer power gain, that is, $G_{\mathrm{Pmax}} = G_{\mathrm{Tmax}}$.

6.3 Stabilization Circuit Technique

6.3.1 Frequency Domains of BJT Potential Instability

Domains of the device's potential instability include the operating frequency ranges where the device stability factor is $K < 1$. Within the bandwidth of these frequency domains, parasitic oscillations can occur, defined by internal positive feedback and operating conditions of the active device. Therefore, it is very important to determine effect of the device feedback parameters on the origin of the parasitic self-oscillations and to establish possible circuit configurations of the parasitic oscillators.

The instabilities may not be self-sustaining, induced by the RF drive power but remaining on its removal. One of the most serious cases of the power amplifier instability can occur with a variation of the load impedance. Under these conditions, the transistor may be destroyed almost instantaneously. However, even it is not destroyed, the instability can result in an increased level of the spurious emissions in the output spectrum of the power amplifier. Generally, the following classification for linear instabilities can be used [15]:

- Low-frequency oscillations produced by thermal feedback effects.
- Oscillations due to internal feedback.
- Negative resistance- or conductance-induced instabilities due to transit-time effects, avalanche multiplication, etc.
- Oscillations due to external feedback as a result of insufficient decoupling of the dc supply, etc.

The BJT stability factor expressed through the device impedance Z-parameters can be rewritten from Eq. (6.30) by

$$K = \frac{2\,R_{11}\,R_{22} - \mathrm{Re}\,(Z_{12}\,Z_{21})}{|Z_{12}\,Z_{21}|} \tag{6.41}$$

where $R_{11} = \mathrm{Re}Z_{11}$, $R_{22} = \mathrm{Re}Z_{22}$.

In a simplified case shown in Fig. 6.3, the only feedback element that influences the transistor stability factor is the collector capacitance C_c. Such a simplification of the equivalent circuit leads to the negligible errors across the frequency range of $f \le 0.3f_T$. At higher frequencies, an influence of the packaging parasitic reactive elements such as the bondwire and lead inductances and pad capacitances must be taken into account. The device impedance Z-parameters given by Eq. (6.14) can be rearranged as

$$\begin{aligned}
Z_{11} &= r_b + \frac{1}{g_m} / \left(1 + j\frac{\omega}{\omega_T}\right) \\
Z_{12} &= \frac{1}{g_m} / \left(1 + j\frac{\omega}{\omega_T}\right) \\
Z_{21} &= \left(\frac{1}{g_m} - \frac{1}{j\omega C_c}\right) / \left(1 + j\frac{\omega}{\omega_T}\right) \\
Z_{22} &= \left(\frac{1}{g_m} + \frac{1}{\omega_T C_c}\right) / \left(1 + j\frac{\omega}{\omega_T}\right)
\end{aligned} \tag{6.42}$$

where $\omega_T = 2\pi f_T$.

By using the ratios for Z-parameters given in Eq. (6.42), the device stability factor can be expressed through the parameters of the transistor equivalent circuit as

$$K = 2r_{\mathrm{b}}g_{\mathrm{m}}\frac{1+\dfrac{g_{\mathrm{m}}}{\omega_{\mathrm{T}}C_{\mathrm{c}}}}{\sqrt{1+\left(\dfrac{g_{\mathrm{m}}}{\omega C_{\mathrm{c}}}\right)^2}} \tag{6.43}$$

The device stability factor increases almost proportionally with the frequency at low frequencies and achieves its maximum value of

$$K = 2r_{\mathrm{b}}g_{\mathrm{m}}\left(1+\frac{g_{\mathrm{m}}}{\omega_{\mathrm{T}}C_{\mathrm{c}}}\right) \tag{6.44}$$

at higher frequencies. Equation (6.44) shows a ratio between the maximum device stability factor K and feedback capacitance C_{c}, and analytically demonstrates effect of the increase in the BJT stability factor with the decrease in the collector feedback capacitance.

At very low frequencies, the bipolar transistors are potentially stable and the fact that $K \to 0$ when $f \to 0$ can be explained by simplifying the bipolar equivalent circuit. In practice, at low frequencies, it is necessary to take into account the dynamic base-emitter resistance r_π and Early collector-emitter resistance r_{ce}, whose presence substantially increase the value of the device stability factor. This means that initially the magnitude of K decreases with the increase in the frequency and then begins to increase according to Eq. (6.43), which gives the only one unstable frequency domain with $K < 1$ and low boundary frequency f_{p1}. However, an additional region of possible low-frequency oscillations can occur due to thermal feedback where the collector junction temperature becomes frequently dependent, and the common-base configuration is especially affected by this [16].

Equating the device stability factor K with unity allows us to determine the high boundary frequency of a frequency domain of the bipolar transistor potential instability as

$$f_{\mathrm{p2}} = \frac{g_{\mathrm{m}}}{2\pi C_{\mathrm{c}}} / \sqrt{(2r_{\mathrm{b}}g_{\mathrm{m}})^2\left(1+\frac{g_{\mathrm{m}}}{\omega_{\mathrm{T}}C_{\mathrm{c}}}\right)^2 - 1} \tag{6.45}$$

When $r_{\mathrm{b}}g_{\mathrm{m}} > 1$ and $g_{\mathrm{m}} >> \omega_{\mathrm{T}}C_{\mathrm{c}}$, Eq. (6.45) reduces to

$$f_{\mathrm{p2}} \approx \frac{1}{4\pi\, r_{\mathrm{b}}C_\pi} \tag{6.46}$$

At higher frequencies, a presence of the parasitic reactive intrinsic transistor elements and package parasitics can be of great importance to the power amplifier stability. The series emitter lead inductance L_{e} shown in Fig. 6.5 provides a major effect upon the device stability factor.

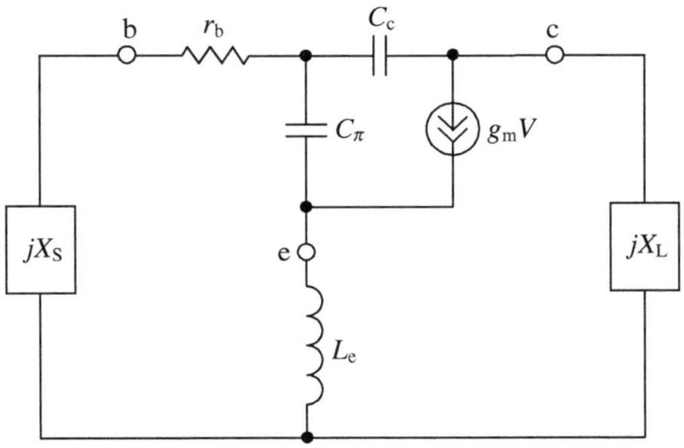

FIGURE 6.5 Simplified bipolar π-hybrid equivalent circuit with emitter lead inductance.

The presence of L_e leads to the appearance of the second frequency domain of potential instability at higher frequencies. The circuit analysis shows that the second frequency domain of potential instability is characterized by the low and high boundary frequencies f_{p3} and f_{p4} obtained by

$$f_{p3,4} = f_T\sqrt{\frac{1-4\omega_T r_b C_c}{8\omega_T r_b C_c} \mp \sqrt{\left(\frac{1-4\omega_T r_b C_c}{8\omega_T r_b C_c}\right)^2 - \frac{1+\kappa}{\omega_T r_b C_c \kappa^2}}} \tag{6.47}$$

where $\kappa = \omega_T L_e / r_b$, and can be realized only under the particular ratios between the normalized parameters $\omega_T L_e / r_b$ and $\omega_T r_b C_c$ [17].

For example, for sufficiently small values of $\omega_T r_b C_c$, the second domain of the potential instability cannot be realized under large values of L_e. With the decrease of L_e, the second frequency domain of potential instability narrows and disappears at some certain value of κ. The further decrease in L_e leads only to widening of the first frequency domain of potential instability and to the corresponding increase of the magnitude of the device stability factor K.

A condition to calculate the optimum value of κ_o (when the second frequency domain of the device potential instability disappears) is derived by equating the expression under the internal square root in Eq. (6.47) with zero as

$$\kappa_o = \frac{32\omega_T r_b C_c}{(1-4\omega_T r_b C_c)^2}\left[1+\sqrt{1+\frac{(1-4\omega_T r_b C_c)^2}{16\omega_T r_b C_c}}\right] \tag{6.48}$$

Figure 6.6 shows the generalized dependence of κ_o as a function of the normalized parameter $\omega_T r_b C_c$, according to Eq. (6.48). By using the curve plotted in Fig. 6.6, it is very convenient to define the presence or absence of the second domain of the device potential instability

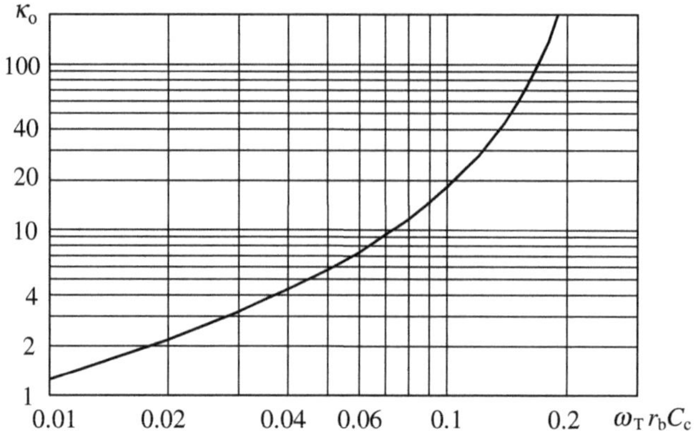

FIGURE 6.6 Generalized dependence of κ_0 versus normalized parameter $\omega_T r_b C_c$.

when the value of $\omega_T r_b C_c$ is known. As a result, the second domain disappears at all values of L_e when $\omega_T r_b C_c \geq 0.25$. The same situation occurs when $\omega_T r_b C_c < 0.25$ but for values of L_e at which $\kappa < \kappa_0$. It should be noted that the value of L_e when $\kappa = \kappa_0$ represents the best case scenario from the viewpoint of the frequency stability conditions. In this case, the first domain of the device instability is narrower than when $L_e = 0$, and the second domain of the device instability is not realized yet because of the sufficiently small value of L_e.

An appearance of the second frequency domain of the device potential instability is the result of the corresponding changes in the device feedback phase conditions and takes place only under a simultaneous effect of the collector capacitance C_c and emitter lead inductance L_e. If the effect of one of these factors is lacking, the active device is characterized by only the first domain of its potential instability.

Figure 6.7 shows the potentially realizable equivalent circuits of the parasitic oscillators. If the value of a series-emitter inductance L_e is negligible, the parasitic oscillations can occur only when the values of the source and load reactances are positive, that is, $X_S > 0$ and $X_L > 0$. In this case, the parasitic oscillator shown in Fig. 6.7(*a*) represents the inductive three-point circuit, where the inductive elements L_S and L_L in combination with the collector capacitance C_c form a Hartley oscillator. From a practical point of view, the more the value of the collector dc-feed inductance exceeds the value of the base-bias inductance, the more likely low-frequency parasitic oscillators can be created. It was observed that a very low inductance, even a short between the emitter and the base, can produce very strong and dangerous oscillations, which may easily destroy a transistor [15]. Therefore, it is recommended to increase the value of the base choke inductance and to decrease the value of the collector choke inductance because the ratio of the boundary values of the source and load inductances L_S and L_L,

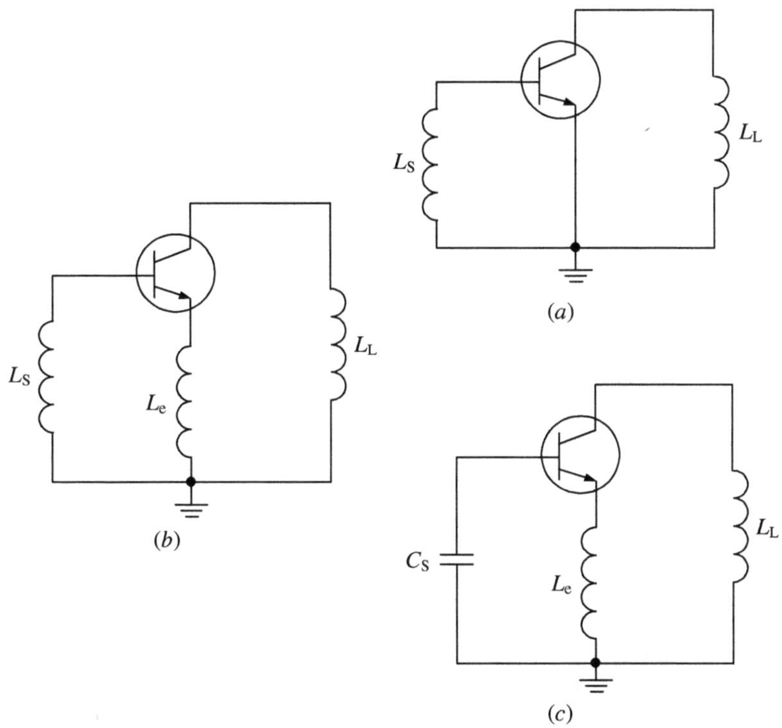

Figure 6.7 Equivalent circuits of parasitic bipolar oscillators.

which corresponds to the high boundary frequency of first potential instability domain, can be approximately estimated as [17]

$$\frac{L_L}{L_S} \approx \frac{1}{\omega_T r_b C_c} \tag{6.49}$$

The presence of L_e leads to the narrowing of the first frequency domain of the potential instability, which is limited to the high boundary frequency f_{p2}, and can contribute to appearance of the second frequency domain of the potential instability at higher frequencies. The parasitic oscillator that corresponds to the first frequency domain of the device potential instability can be realized only if the source and load reactances are inductive, that is, $X_S > 0$ and $X_L > 0$, with the equivalent circuit of such a parasitic oscillator shown in Fig. 6.7(*b*). The parasitic oscillator corresponding to the second frequency domain of the device potential instability can be realized only if the source reactance is capacitive and the load reactance is inductive, that is, $X_S < 0$ and $X_L > 0$, with the equivalent circuit shown in Fig. 6.7(*c*). The series emitter inductance L_e is an element of fundamental importance for the parasitic oscillator that corresponds to the second frequency domain of the device potential instability. It changes the circuit phase conditions so that it becomes possible to establish the oscillation phase-balance condition at high frequencies. Therefore, the decrease in the value of L_e is the most effective way to prevent the parasitic self-oscillation in the frequency range of $f > f_{p3}$. However, if it is possible to eliminate the parasitic oscillations at high frequencies by other means, increasing of L_e will result to narrowing of a low-frequency domain of potential instability, thus making the power amplifier potentially more stable, although at the expense of reduced power gain.

6.3.2 Frequency Domains of MOSFET Potential Instability

Based on the simplified device equivalent circuit shown in Fig. 6.4, it is convenient to represent the MOSFET stability factor through Y-parameters, as

$$K = \frac{2G_{11}G_{22} - \text{Re}\,(G_{12}G_{21})}{|Y_{12}Y_{21}|} \tag{6.50}$$

where $G_{11} = \text{Re}Y_{11}$, $G_{22} = \text{Re}Y_{22}$.

In this case, the only feedback element is the gate-drain capacitance C_{gd}, and such a simplification is sufficiently accurate across the frequency range of $f \le 0.3f_T$. At higher frequencies, an influence of the parasitic reactive parameters including the source inductance L_s as a most important one should be taken into account. By using the ratios for the admittance Y-parameters given in Eq. (6.17), the MOSFET stability factor can be expressed through the parameters of the transistor equivalent circuit as

$$K = \left[1 + \frac{2}{g_m R_{ds}}\left(1 + \frac{C_{gs}}{C_{gd}}\right)\right] \frac{\omega R_{gs} C_{gs}}{\sqrt{1 + (\omega R_{gs} C_{gs})^2}} \tag{6.51}$$

The device stability factor increases linearly from zero frequency and achieves its maximum value of

$$K = \left[1 + \frac{2}{g_m R_{ds}}\left(1 + \frac{C_{gs}}{C_{gd}}\right)\right] \tag{6.52}$$

at higher frequencies. Equation (6.52) determines the influence of the feedback gate-drain capacitance C_{gd} and transconductance g_m on maximum device stability factor K, which increases with the decrease in both these parameters.

For MOSFET transistors, the fact that $K \to 0$ when $f \to 0$ is the result of the simplification of the device equivalent circuit. However, at very low frequencies, it is necessary to take into account the gate leakage resistance R_{leak}. This means that initially the magnitude of K decreases with the increase in the frequency and then increases according to Eq. (6.51). Consequently, this gives the only one unstable frequency domain, with $K < 1$ and low boundary frequency f_{p1}, whose bandwidth depends on the parameters of the MOSFET equivalent circuit. Because of the very small ratio of R_{gs}/R_{leak}, typically of 10^{-6}, the value of the low boundary frequency f_{p1} is sufficiently small,

$$f_{p1} \approx 10^{-6}/2\pi R_{gs}C_{gs} \tag{6.53}$$

By equating the stability factor K with unity in Eq. (6.51), it is possible to determine the high boundary frequency of a frequency domain of the MOSFET potential instability as [18]

$$f_{p2} = \frac{1}{4\pi R_{gs}C_{gs}} \frac{g_m R_{ds}}{\sqrt{1+\dfrac{C_{gs}}{C_{gd}}}} \frac{1}{\sqrt{1+\dfrac{C_{gs}}{C_{gd}}+g_m R_{ds}}} \tag{6.54}$$

For usually available conditions (for power MOSFET devices) when $g_m R_{ds} = 10 \div 30$ and $C_{gd}/C_{gs} = 0.1 \div 0.2$, the expression to evaluate the high boundary frequency f_{p2} can be simplified to approximately

$$f_{p2} \approx \frac{1}{4\pi R_{gs}C_{gs}} \tag{6.55}$$

It should be noted that the power MOSFET devices have a substantially higher value of $g_m R_{ds}$ at small values of the drain current than at its high values. Consequently, at small values of the drain current, the MOSFET device is characterized by a wider domain of potential instability. This domain is significantly wider than the same first domain of the potential instability of the bipolar transistor.

The source inductance L_s shown in Fig. 6.8 creates a second frequency domain of potential instability at higher frequencies. The low

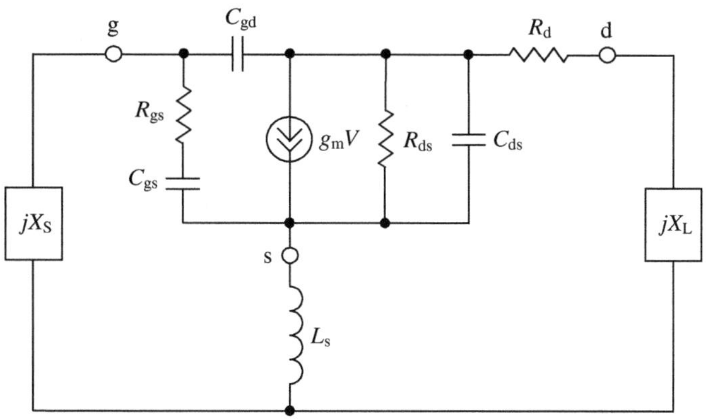

FIGURE 6.8 Simplified MOSFET equivalent circuit with source lead inductance.

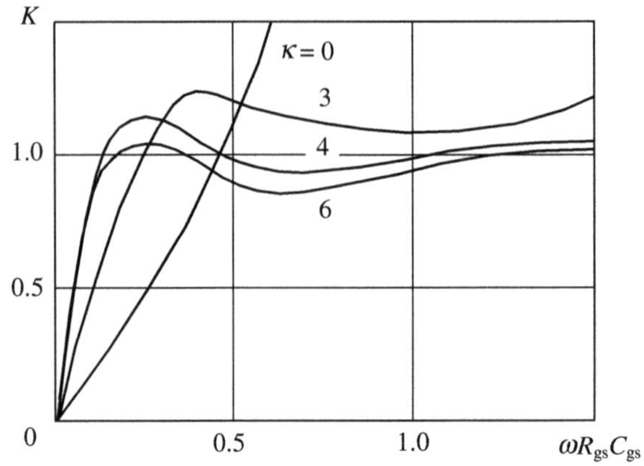

FIGURE 6.9 Device stability factor K versus normalized parameter $\omega R_{gs}C_{gs}$.

and high boundary frequencies f_{p3} and f_{p4} of the second frequency domain of the MOSFET potential instability have been numerically estimated for typical values of the MOSFET equivalent circuit parameters as $C_{ds}/C_{gs} = 0.5$, $R_d/R_{gs} = 0.75$, $g_m R_{gs} = 2$, and $C_{gd}/C_{gs} = 0.1$. Figure 6.9 shows the dependences of the device stability factor K on the normalized parameter $\omega R_{gs}C_{gs}$ for different values of $\kappa = \omega_T L_s/R_{gs}$ [18].

From Fig. 6.9 it follows that the second frequency domain of potential instability can be realized only under the certain values of κ. For example, the second frequency domain of potential instability becomes narrow with the decrease of L_s and disappears at $\kappa \cong 3.5$. The further decrease in L_s leads to a widening of the first frequency domain of potential instability and provides an increased magnitude of the device stability factor K.

In addition, an analysis of the influence of the source and load conductances G_S and G_L on the amplifier stability shows that the first frequency domain of potential instability disappears when $1/G_S R_{gs} = 2 \div 5$ and $g_m/G_L = 5 \div 10$. The second frequency domain of potential instability can disappear when $1/(G_S R_{gs}) \approx 2$ and $g_m/G_L \approx 2$.

Figure 6.10 shows the potentially realizable equivalent circuits of the parasitic oscillators. If the value of a series source inductance L_s is negligible, the parasitic oscillations can occur only when the values of the source and load reactances are positive, that is, $X_S > 0$ and $X_L > 0$. Consequently, the equivalent circuit of the parasitic oscillator shown in Fig. 6.10(*a*) represents the inductive three-point circuit, where the inductive elements L_S and L_L in combination with the gate-drain capacitance C_{gd} form a Hartley oscillator. The presence of the source inductance L_s leads to an appearance of the second frequency domain of potential instability at higher frequencies. As a result, the parasitic oscillator corresponding to the first frequency domain of the device

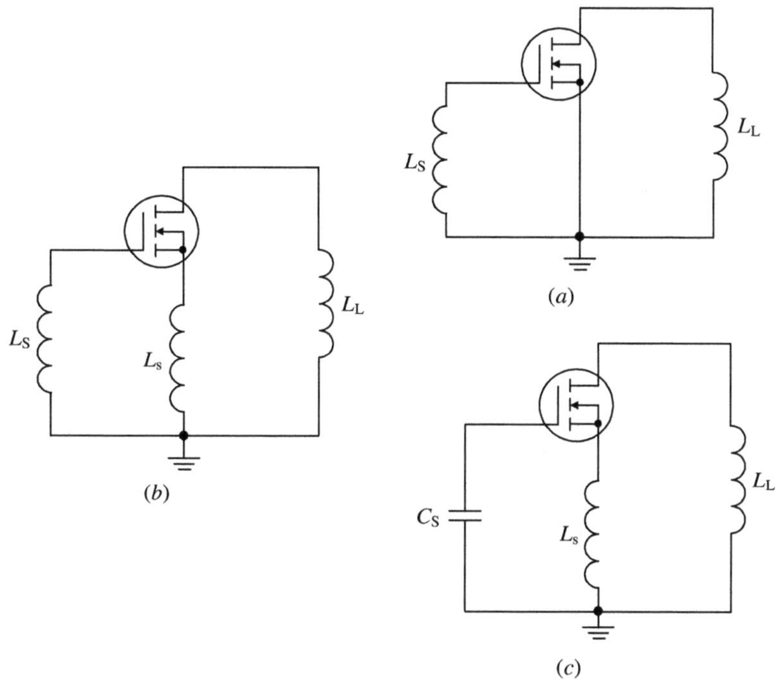

Figure 6.10 Equivalent circuits of parasitic MOSFET oscillators.

potential instability can be realized only if the source and load reactances are inductive, that is, ImX_S > 0 and ImX_L > 0, with the equivalent circuit of such a parasitic oscillator shown in Fig. 6.10(*b*). The parasitic oscillator corresponding to the second frequency domain of the device potential instability can be realized only if the source reactance is capacitive, that is ImX_S < 0, and the load reactance is inductive, that is, ImX_L < 0, with the equivalent circuit shown in Fig. 6.10(*c*).

Thus, to prevent the parasitic oscillations and to provide a stable operation mode of any power amplifier, it is necessary to take into consideration the following common requirements:

- Use an active device with stability factor $K > 1$.
- If it is impossible to choose an active device with $K > 1$, it is necessary to provide the circuit stability factor $K_T > 1$ by the appropriate choice of the real parts of the source and load immittances.
- Disrupt the equivalent circuits of the possible parasitic oscillators.
- Choose proper reactive parameters of the matching circuit elements adjacent to the input and output ports of the active device, which are necessary to avoid the self-oscillation conditions.

6.3.3 Some Examples of Stabilization Circuits

Generally, the parasitic oscillations can arise on any frequency within the potential instability domains for particular values of the source and load immittances W_S and W_L. The frequency dependences of W_S and W_L are very complicated and very often cannot be predicted exactly, especially in multistage power amplifiers. Therefore, it is very difficult to propose a unified approach to provide a stable operation mode of the power amplifiers with different circuit configurations and operation frequencies. In practice, the parasitic oscillations can arise close to the operating frequencies due to the internal positive feedback inside the transistor and at the frequencies sufficiently far from the operating frequencies due to the external positive feedback created by the surface mounted elements. As a result, the stability analysis of the power amplifier must include the methods to prevent the parasitic oscillations in different frequency ranges:

- At lower frequencies when the frequency of the parasitic oscillation f_p is significantly less than the operating frequency f_0, that is, $f_p << f_0$
- At higher frequencies when $f_p >> f_0$
- At the operating frequencies when $f_p \approx f_0$

Examples of the bias circuit configurations that can prevent the parasitic oscillations at lower frequencies are given in Fig. 6.11. The main idea in this case is to use a stabilizing resistor R_1 connected either in parallel to an RF choke L_1, as shown in Fig. 6.11(*a*), or together with a series bypass capacitor C_2 in parallel to the power supply, as shown in Fig. 6.11(*c*). To avoid worsening of the power amplifier performance when the value of R_1 is sufficiently small and comparable with the output immittance of the transistor, it is advisable to use an additional series inductance L_2, as shown in Fig. 6.11(*b*). The value of L_2 in this case should be approximately equal to $L_2 = (5 \div 10)R_1/\pi f_0$, and the value of C_2 can be chosen from the following condition: $1/2\pi R_1 f_0 << C_2 << 1/2\pi R_1 f_p$, where f_p corresponds to the high boundary frequency f_{p2} of the first frequency domain of the device potential instability if $f_0 >> f_{p2}$. When the impedance of the series circuit R_1C_2 is sufficiently high compared with the power supply impedance, it is advisable to use an additional RF choke L_2 with the following value: $L_2 >> R_1/2\pi f_0$, as shown in Fig. 6.11(*d*).

The equivalent circuit of the parasitic oscillator at higher frequencies is realized by means of the parasitic reactive parameters of the transistor and power amplifier circuit. The only possible equivalent circuit of such a parasitic oscillator at these frequencies is shown in Fig. 6.7(*c*), which can only be realized if the emitter lead inductance is present. Consequently, the electrical length of the emitter lead should be minimized, or, alternatively, the appropriate reactive impedances at the input and output terminals of the active device should be provided.

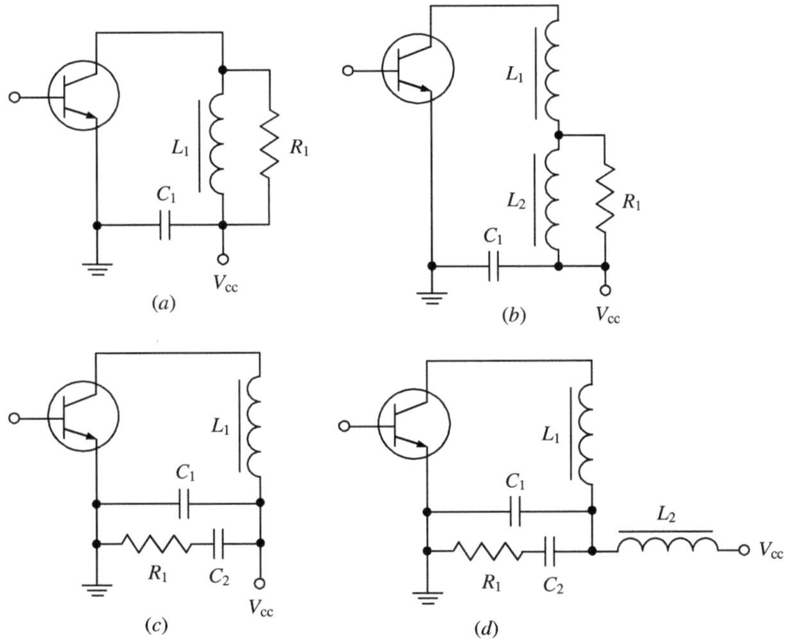

FIGURE 6.11 Bias circuit configurations preventing low-frequency oscillations.

For example, it is possible to avoid the parasitic oscillations at these frequencies if the inductive impedance is provided at the input of the transistor. The equivalent circuit of such a power amplifier is shown in Fig. 6.12, where the appropriate choice of the input and output matching circuits allows the inductive impedance at the input of the

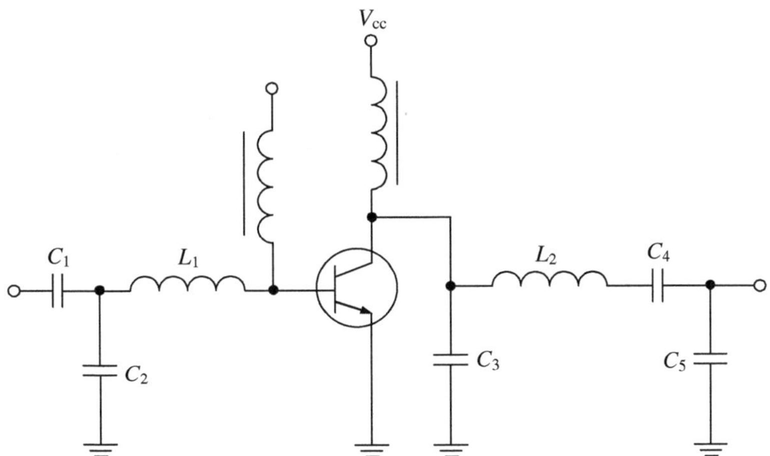

FIGURE 6.12 Circuit schematic preventing high-frequency oscillations.

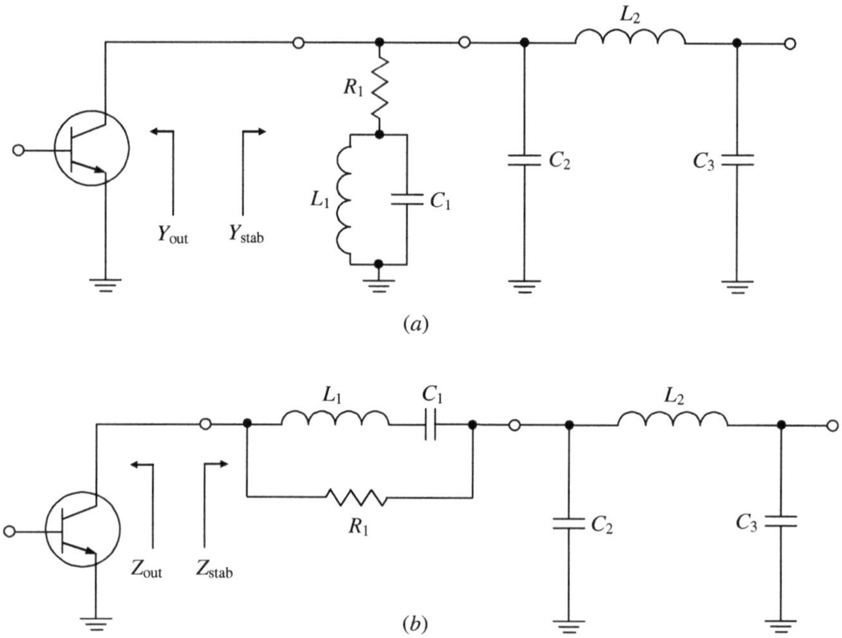

FIGURE 6.13 Circuit schematics with output stabilized *RLC* circuits.

transistor and the capacitive impedance at the output of the transistor to be provided at higher frequencies.

To increase the power amplifier stability, it is possible to use special stabilizing *RLC* circuits, as shown in Fig. 6.13. These stabilizing circuits are usually located between the transistor and the matching circuit, where they can be connected either in parallel, as shown in Fig. 6.13(*a*), or in series, as shown in Fig. 6.13(*b*), to the transistor output terminal, and the resonant circuit L_1C_1 is tuned to the operating frequency f_0 in both cases. As a result, the resistor R_1 provides a stabilizing effect beyond the operating frequency bandwidth. The selected minimum value of R_1 must satisfy the amplifier stability condition of $\text{Re}(W_{out} + W_{stab}) > 0$. Such stabilizing circuits can be connected at the input of the transistor as well. However, the use of these circuits leads to some changes in the frequency bandwidth: narrowing in the case of the parallel circuit configuration and widening in the case of the series one.

It should be noted that expressions in Eqs. (6.21) through (6.40) are given in terms of the device immittance *W*-parameters that allows the power gain and stability to be calculated using the impedance *Z*- or admittance *Y*-parameters of the device equivalent circuit and to physically understand the corresponding effect of each circuit parameter, but not through the scattering *S*-parameters, which are very convenient during the measurement procedure required for device modeling. Moreover, using modern simulation tools, it is even no need to draw stability circles on the Smith chart or analyze stability factor across the wide frequency range as *K*-factor is just a derivation from the basic stability conditions and usually is a function of linear parameters, which can only reveal linear instabilities. Besides, it is difficult to predict an unconditional stability for a multistage power amplifier because parasitic oscillations can be caused by the interstage circuits. In this case, the easiest and most effective way to provide a stable operation of the multistage power amplifier (or single-stage power amplifier) is to simulate the real part of the device input impedance $Z_{in} = V_{in}/I_{in}$ at the input terminal of each transistor at fundamental frequency as a ratio between the input voltage and current by placing a voltage node and a current meter, as shown in Fig. 6.14(*a*). If ReZ_{in} < 0, then either a small series resistor must be added to the device base terminal as a part of the input matching circuit or a load-network configuration can be properly selected to provide the resulting positive value of ReZ_{in}. In this case, not only linear instabilities with small-signal soft start-up oscillation conditions but also nonlinear instabilities with large-signal hard start-up oscillation conditions or parametric oscillations can be identified around operating region. Figure 6.14(*b*) shows the parallel *RC* stabilizing circuit with a bypass capacitor C_{bypass} connected in series to the input port of a GaN HEMT device [19]. In this case, using a stabilizing resistor R_{gate} and a low-value gate-bias resistor R_{bias} improves the stability factor considerably at low frequencies without affecting the device performance at higher frequencies.

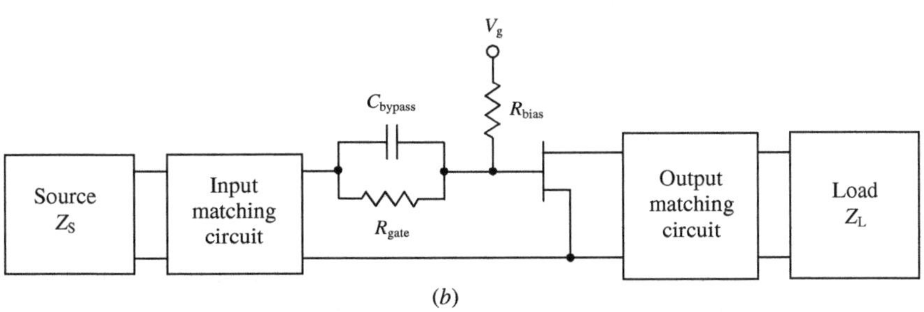

FIGURE 6.14 Single-stage power amplifiers with measured device input impedance.

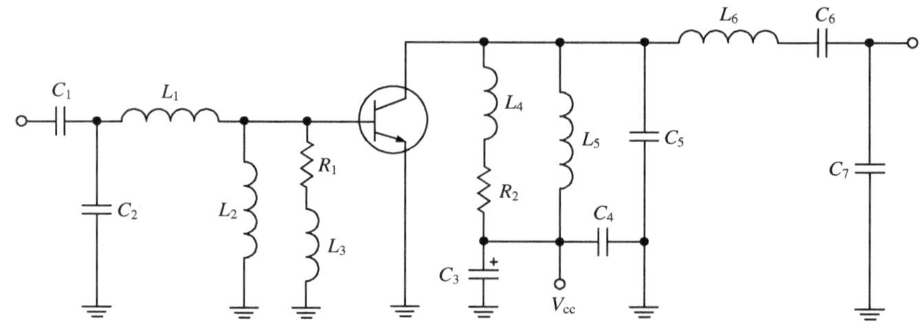

FIGURE 6.15 Stabilized bipolar Class C VHF power amplifier.

Figure 6.15 shows the example of a stabilized bipolar VHF power amplifier configured to operate in a zero-bias Class-C mode. Conductive input and output loadings due to resistances R_1 and R_2 eliminate a low-frequency instability domain. The series inductors L_3 and L_4 contribute to higher power gain if the resistance values are too small, and can compensate for the capacitive input and output device impedances. To provide a negative-bias Class-C mode, the shunt inductor L_2 can be removed. In this case, the only possible equivalent circuit of a parasitic oscillator at these frequencies is shown in Fig. 6.7(*c*). To avoid the parasitic oscillations at these frequencies, it is necessary to provide the inductive impedance at the input of the transistor and capacitive reactance at the output of the transistor. This is realized by the input series inductance L_1 and output shunt capacitance C_5.

The collector efficiency of the power amplifier can be increased by removing the shunt capacitor and series *RL*-circuit in the load network. The remaining series *LC* circuit provides high impedances at the second- and higher-order harmonic components of the output current, which flows now through the device collector capacitance unlike being grounded by the shunt capacitance. As a result, the bipolar Class-C power amplifier, whose circuit schematic is shown in Fig. 6.16, achieves a collector efficiency of 73% and a power gain of 9 dB with an output power of 13.8 W at an operating frequency of 160 MHz [20]. However, special care must be taken to eliminate parasitic spurious oscillations. In this case, the most important element in preventing the potential instability is the base bias resistor R_b. For example, for a relatively large-base choke inductance L_b and $R_b = 1$ kΩ, spurious oscillations exist at any tuning. Tuning becomes possible with no parasitic oscillations for the output *VSWR* less than 1.3 or supply voltage more than 22 V when R_b is reduced to 470 Ω. However, a very small reduction in input drive power causes spurious oscillations. Further reduction in R_b to 47 Ω provides a stable operation for output $VSWR \leq 7$ and supply voltages down to 7 V. Finally, no spurious oscillations occur at any load, supply voltage, and drive power level for $R_b = 26$ Ω.

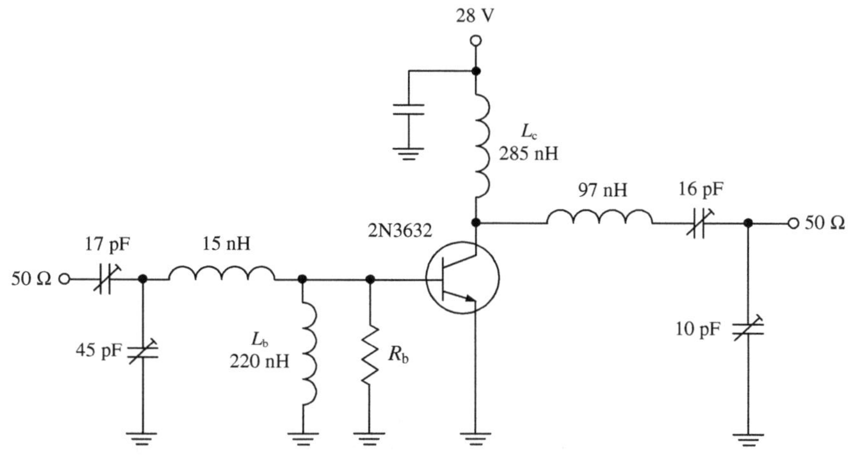

Figure 6.16 High-efficiency VHF bipolar Class-C power amplifier.

Because the transistor using as an active device in power amplifiers is characterized by a substantially nonlinear behavior, this can result in nonlinear instabilities, which provide generally the parametric generation of both harmonic and subharmonic components. The presence of subharmonics can be explained by the parametric varactor junction action of the collector-base voltage-dependent capacitance when the large-signal driving acts like pumping a varactor diode, as in a parametric amplifier [21, 22]. Such an amplifier exhibits negative resistance under certain conditions when a circuit starts oscillating at subharmonics or rational fractions of the operating frequency [23]. Generally, the parametric oscillations are the result of the external force impact on the element of the oscillation system by varying its parameter. Understanding of the physical origin of this parametric effect is very important in order to disrupt any potentially realizable parametric oscillator circuits. Especially, it is a serious concern for high-efficiency power amplifiers with very high voltage peak factor and voltage swing across the device nonlinear output capacitance, as the transistor is operated in pinch-off, active, and saturation regions.

6.4 Basic Classes of Operation: A, AB, B, and C

As established at the end of the 1910s, the amplifier efficiency may reach quite high values when suitable adjustments of the grid and anode voltages are made [24]. With resistive load, the anode current is in phase with the grid voltage, whereas it leads with the capacitive load and it lags with the inductive load. On the assumption that the anode current and anode voltage both have sinusoidal variations, the maximum possible output of the amplifying device would be just a half the dc supply power, resulting in an anode efficiency of 50%. However, by using a pulsed-shaped anode current, it is possible to achieve anode efficiency considerably in excess of 50%, potentially as high as 90%, by choosing the proper operation conditions. By applying the proper negative bias voltage to the grid terminal to provide the pulsed anode current of different width with the angle θ, the anode current becomes equal to zero, where the double angle 2θ represents a conduction angle of the amplifying device [25]. In this case, a theoretical anode efficiency approaches 100% when the conduction angle, during which the anode current flows, reduces to zero, with starting efficiency of 50% corresponding to the conduction angle of 360° or 100% duty ratio.

Generally, power amplifiers can be classified in three classes according to their mode of operation: *linear mode* when its operation is confined to the substantially linear portion of the active device characteristic curve; *critical mode* when the anode current ceases to flow, but operation extends beyond the linear portion up to the saturation and cutoff regions; and *nonlinear mode* when the anode current ceases to flow during a portion of each cycle, with a duration that depends on the grid bias [26]. When high efficiency is required, power amplifiers of the third class are used because the presence of harmonics contributes to the attainment of high efficiencies. In order to suppress harmonics of the fundamental frequency to deliver a sinusoidal signal to the load, a parallel resonant circuit can be used in the load network, which bypasses harmonics through a low-impedance path and, by virtue of its resonance to the fundamental, receives energy at that frequency. At the very beginning of the 1930s, power amplifiers operating in the first two classes with 100% duty ratio were called the Class-A power amplifiers, whereas those operating in the third class with 50% duty ratio were assigned to Class-B power amplifiers [27].

To analytically determine the operation classes of the power amplifier, consider a simple resistive stage shown in Fig. 6.17, where L_{ch} is the ideal choke inductor with zero series resistance and infinite reactance at the operating frequency, C_b is the dc-blocking capacitor with infinite value having zero reactance at the operating frequency, and R_L is the load resistor. The dc supply voltage V_{cc} is applied to both plates of the dc-blocking capacitor, being constant during entire signal period. The active device behaves as an ideal voltage- or current-controlled current source having zero saturation resistance.

Let us assume the input signal to be in a cosine form of

$$v_{in} = V_b + V_{in}\cos\omega t \tag{6.56}$$

where V_b is the input dc bias voltage. The operating point must be fixed at the middle point of the linear part of the device transfer characteristic with $V_{in} \le V_b - V_p$, where V_p is the device pinch-off voltage. Usually, to simplify an analysis of the power amplifier operation, the

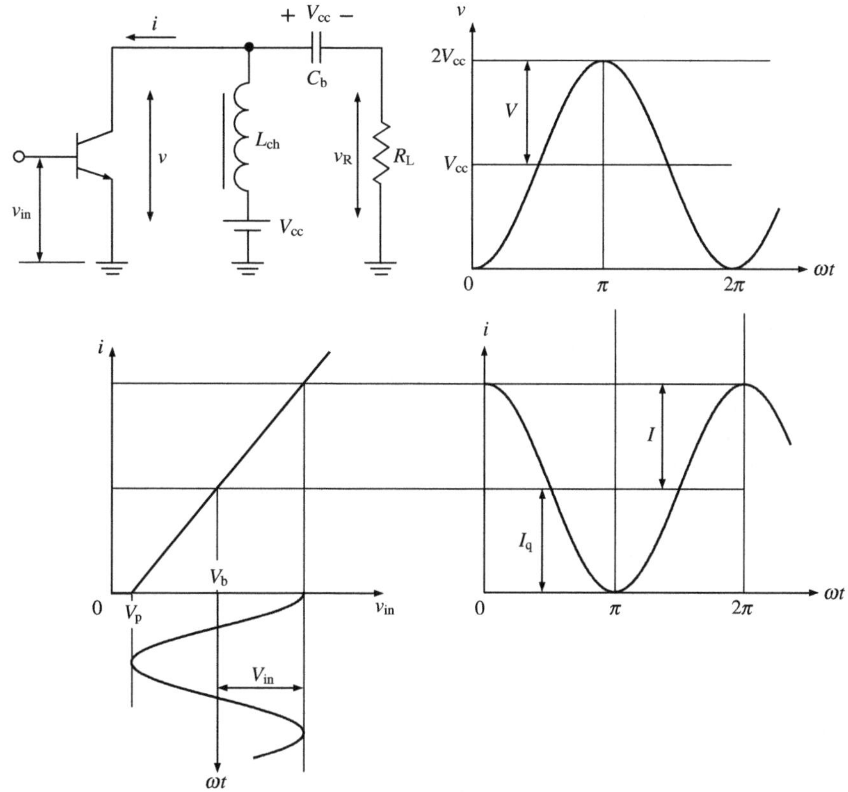

FIGURE 6.17 Voltage and current waveforms in Class-A operation.

device transfer characteristic is represented by a piecewise-linear approximation. As a result, the output current is cosinusoidal,

$$i = I_q + I \cos\omega t \tag{6.57}$$

with the quiescent current I_q greater or equal to the collector current amplitude I. In this case, the output collector current contains only two components—dc and cosine—and the averaged current magnitude is equal to a quiescent current or dc component I_q.

The output voltage v across the device collector represents a sum of the dc supply voltage V_{cc} and cosine voltage v_R across the load resistor R_L. Consequently, greater output current i results in greater voltage v_R across the load resistor R_L and smaller output voltage v across the device output. Thus, for a purely real load impedance $Z_L = R_L$, the collector voltage v is shifted by 180° relatively the input voltage v_{in} and can be written as

$$v = V_{cc} + V \cos(\omega t + 180°) = V_{cc} - V \cos\omega t \tag{6.58}$$

where V is the output voltage amplitude.

Substituting Eq. (6.57) into Eq. (6.58) yields

$$v = V_{cc} - (i - I_q)\, R_L \tag{6.59}$$

where $R_L = V/I$, and Eq. (6.59) can be rewritten as

$$i = \left(I_q + \frac{V_{cc}}{R_L} \right) - \frac{v}{R_L} \tag{6.60}$$

which determines a linear dependence of the collector current versus collector voltage. Such a combination of the cosine collector voltage and current waveforms is known as a Class-A operation mode. In practice, because of the inherent device nonlinearities, it is necessary to connect a parallel *LC* circuit with resonant frequency equal to the operating frequency to suppress any possible harmonic components.

Circuit theory prescribes that the collector efficiency η can be written as

$$\eta = \frac{P}{P_0} = \frac{1}{2}\frac{I}{I_q}\frac{V}{V_{cc}} = \frac{1}{2}\frac{I}{I_q}\xi \tag{6.61}$$

where

$$P_0 = I_q V_{cc} \tag{6.62}$$

is the dc output power,

$$P = \frac{IV}{2} \tag{6.63}$$

is the power delivered to the load resistance R_L at the fundamental frequency f_0, and

$$\xi = \frac{V}{V_{cc}} \tag{6.64}$$

is the collector voltage peak factor.

Then, by assuming the ideal conditions of zero saturation voltage when $\xi = 1$ and maximum output current amplitude when $I/I_q = 1$, from Eq. (6.61) it follows that the maximum collector efficiency in a Class-A operation mode is equal to

$$\eta = 50\% \tag{6.65}$$

However, as it also follows from Eq. (6.61), increasing the value of I/I_q can further increase the collector efficiency. This leads to a step-by-step nonlinear transformation of the current cosine waveform to its pulsed waveform when the amplitude of the collector current exceeds zero value during only a part of the entire signal period. In this case, an active device is operated in the active region followed

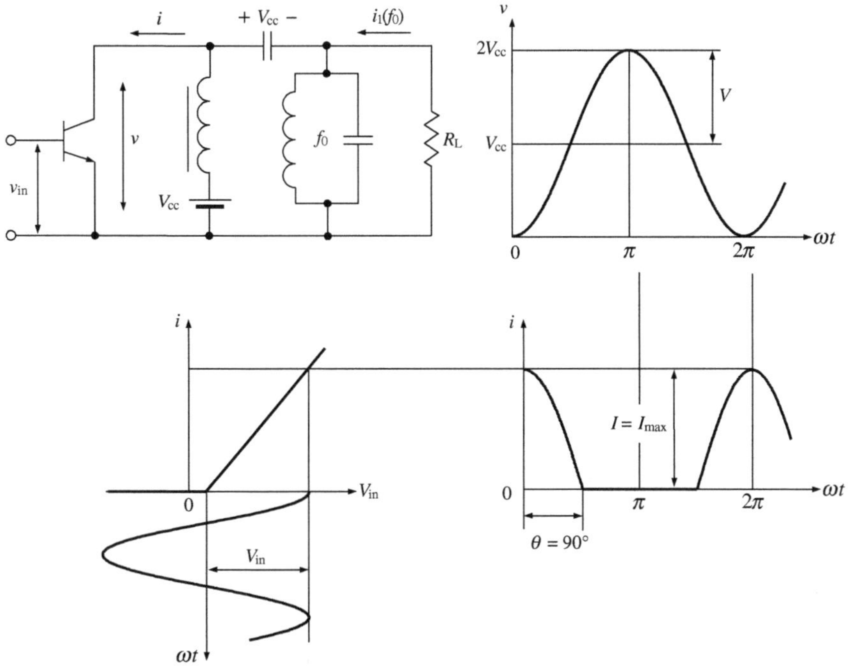

Figure 6.18 Voltage and current waveforms in Class-B operation.

by the operation in the pinch-off region when the collector current is zero, as shown in Fig. 6.18. As a result, the frequency spectrum at the device output will generally contain the second-, third-, and higher-order harmonics of the fundamental frequency. However, because of high-quality factor of the parallel resonant *LC* circuit, only the fundamental-frequency signal flows to the load, while the short-circuit conditions are fulfilled for higher-order harmonic components. Therefore, ideally the collector voltage represents a purely sinusoidal waveform with the voltage amplitude $V \le V_{cc}$.

Analytically such an operation can be written as

$$i(\omega t) = \begin{cases} I_q + I\cos\omega t & -\theta \le \omega t < \theta \\ 0 & \theta \le \omega t < 2\pi - \theta \end{cases} \qquad (6.66)$$

where the conduction angle 2θ is the angle of a current flow indicating the part of the RF current cycle, for which device conduction occurs and determines the moment when output current $i(\omega t)$ takes a zero value [28]. At this moment

$$i(\theta) = I_q + I\cos\theta = 0 \qquad (6.67)$$

and half the conduction angle θ can be calculated by

$$\cos\theta = -\frac{I_q}{I} \tag{6.68}$$

As a result, the basic definitions for nonlinear operation modes of a power amplifier through half the conduction angle θ can be introduced as

- When $\theta > 90°$, then $\cos\theta < 0$ and $I_q > 0$, corresponding to Class-AB operation
- When $\theta = 90°$, then $\cos\theta = 0$ and $I_q = 0$, corresponding to Class-B operation
- When $\theta < 90°$, then $\cos\theta > 0$ and $I_q < 0$, corresponding to Class-C operation.

The periodic pulsed output current $i(\omega t)$ can be written as a Fourier-series expansion

$$i(\omega t) = I_0 + I_1 \cos\omega t + I_2 \cos 2\omega t + I_3 \cos 3\omega t + \cdots \tag{6.69}$$

where the dc and fundamental-frequency components can be obtained, respectively, by

$$I_0 = \frac{1}{2\pi}\int_{-\theta}^{\theta} I(\cos\omega t - \cos\theta)\, d\omega t = I\gamma_0 \tag{6.70}$$

$$I_1 = \frac{1}{\pi}\int_{-\theta}^{\theta} I(\cos\omega t - \cos\theta)\cos\omega t\; d\omega t = I\gamma_1 \tag{6.71}$$

where

$$\gamma_0 = \frac{1}{\pi}(\sin\theta - \theta\cos\theta) \tag{6.72}$$

$$\gamma_1 = \frac{1}{\pi}(\theta - \sin\theta\cos\theta) \tag{6.73}$$

are the current coefficients for the dc and fundamental-frequency components, respectively [29, 30].

From Eq. (6.70) it follows that the dc current component is a function of θ in the operation modes with $\theta < 180°$, in contrast to a Class-A operation mode where $\theta = 180°$ and the dc current is equal to the quiescent current during the entire period.

The collector efficiency of a power amplifier with shunt resonant circuit, biased to operate in the nonlinear modes, can be obtained by

$$\eta = \frac{P_1}{P_0} = \frac{1}{2}\frac{I_1}{I_0}\xi = \frac{1}{2}\frac{\gamma_1}{\gamma_0}\xi \tag{6.74}$$

which is a function of θ only, where

$$\frac{\gamma_1}{\gamma_0} = \frac{\theta - \sin\theta\cos\theta}{\sin\theta - \theta\cos\theta} \tag{6.75}$$

The Class-B power amplifiers had been defined as those operating with a negative grid bias such that the anode current is practically zero with no excitation grid voltage, and in which the output power is proportional to the square of the excitation voltage [31]. If $\xi = 1$ and $\theta = 90°$, then from Eqs. (6.74) and (6.75) it follows that the maximum collector efficiency in a Class-B operation mode is equal to

$$\eta = \frac{\pi}{4} \cong 78.5\% \tag{6.76}$$

The fundamental-frequency power delivered to the load $P_L = P_1$ is defined as

$$P_1 = \frac{VI_1}{2} = \frac{VI\gamma_1(\theta)}{2} \tag{6.77}$$

showing its direct dependence on the conduction angle 2θ. This means that reduction in θ results in lower γ_1, and, to increase the fundamental-frequency power P_1, it is necessary to increase the current amplitude I. Because the current amplitude I is determined by the input voltage amplitude V_{in}, the input power P_{in} must be increased. The collector efficiency also increases with reduced value of θ and becomes maximum when $\theta = 0°$, where the ratio γ_1/γ_0 is maximal, as follows from Eq. 6.75. For example, the collector efficiency η increases from 78.5 to 92% when θ reduces from 90 to 60°. However, it requires the input voltage amplitude V_{in} to be increased by 2.5 times, resulting in lower values of the power-added efficiency (*PAE*), which is defined as

$$PAE = \frac{P_1 - P_{in}}{P_0} = \frac{P_1}{P_0}\left(1 - \frac{1}{G_P}\right) \tag{6.78}$$

where $G_P = P_1/P_{in}$ is the operating power gain.

The Class-C power amplifiers had been defined as those operating with a negative grid bias more than sufficient to reduce the anode current to zero with no excitation grid voltage, and in which the output power varies as the square of the anode voltage between limits [31]. The main distinction between Class B and Class C is in the duration of the output current pulses, which are shorter for Class C when the active device is biased beyond the pinch-off point. To achieve the maximum anode efficiency in Class C, the active device should be biased (negative) considerably lower the pinch-off point to provide the sufficiently low conduction angles [32].

In order to obtain an acceptable tradeoff between a high power gain and a high power-added efficiency in different situations, the conduction angle should be chosen within the range of $120° \leq 2\theta \leq 190°$. If it is necessary to provide high collector efficiency of the active device having a high gain capability, it is necessary to choose a Class-C operation mode with θ close to 60°. However, when the input power is limited and power gain is not sufficient, a Class-AB operation mode with small quiescent current when θ is slightly greater than 90° is recommended. In the latter case, the linearity of the power amplifier can be significantly improved. From Eq. (6.75) it follows that the ratio of the fundamental-frequency component of the anode current to the dc current is a function of θ only, which means that, if the operating angle is maintained constant, the fundamental component of the anode current will replicate linearly to the variation of the dc current, thus providing the linear operation of the Class-C power amplifier when dc current is directly proportional to the grid voltage [33].

6.5 Load Line and Output Impedance

The graphical method of laying down a load line on the family of the static curves representing anode current against anode voltage for various grid potentials was already well known in the 1920s [34]. If an active device is connected in a circuit, in which the anode load is a pure resistance, the performance may be analyzed by drawing the load line, where the lower end of the line represents the anode supply voltage and the slope of the line is defined by the load resistance; that is, the load resistance is equal to the value of the intercept on the voltage axis divided by the value of the intercept on the current axis.

In a Class A, the output voltage v across the device anode (collector or drain) represents a sum of the dc supply voltage V_{cc} and cosine voltage across the load resistance R_L, and can be defined by Eq. (6.58). In this case, the power dissipated in the load and the power dissipated in the device are equal when $V_{cc} = V$, and the load resistance $R_L = V/I$ is equal to the device output resistance R_{out} [30]. In a pulsed operation mode (Class AB, B, or C), because the parallel LC circuit is tuned to the fundamental frequency, ideally the voltage across the load resistor R_L represents a cosine waveform. By using Eqs. (6.58), (6.66), and (6.71), the relationship between the collector current i and voltage v during a time period of $-\theta \leq \omega t < \theta$ can be written as

$$i = \left(I_q + \frac{V_{cc}}{\gamma_1 R_L} \right) - \frac{v}{\gamma_1 R_L} \tag{6.79}$$

where the fundamental current coefficient γ_1 as a function of θ is determined by Eq. (6.73), and the load resistance is defined as $R_L = V/I_1$, where I_1 is the fundamental current amplitude. Equation (6.79) determining the dependence of the collector current on the collector voltage for any values of the conduction angle in the form of a straight-line function is called the *load line* of the active device. For a Class-A operation mode with $\theta = 180°$ when $\gamma_1 = 1$, the load line defined by Eq. (6.79) is identical to the load line defined by Eq. (6.60).

Figure 6.19 shows the idealized active device output I-V curves and load lines for different conduction angles according to Eq. (6.79)

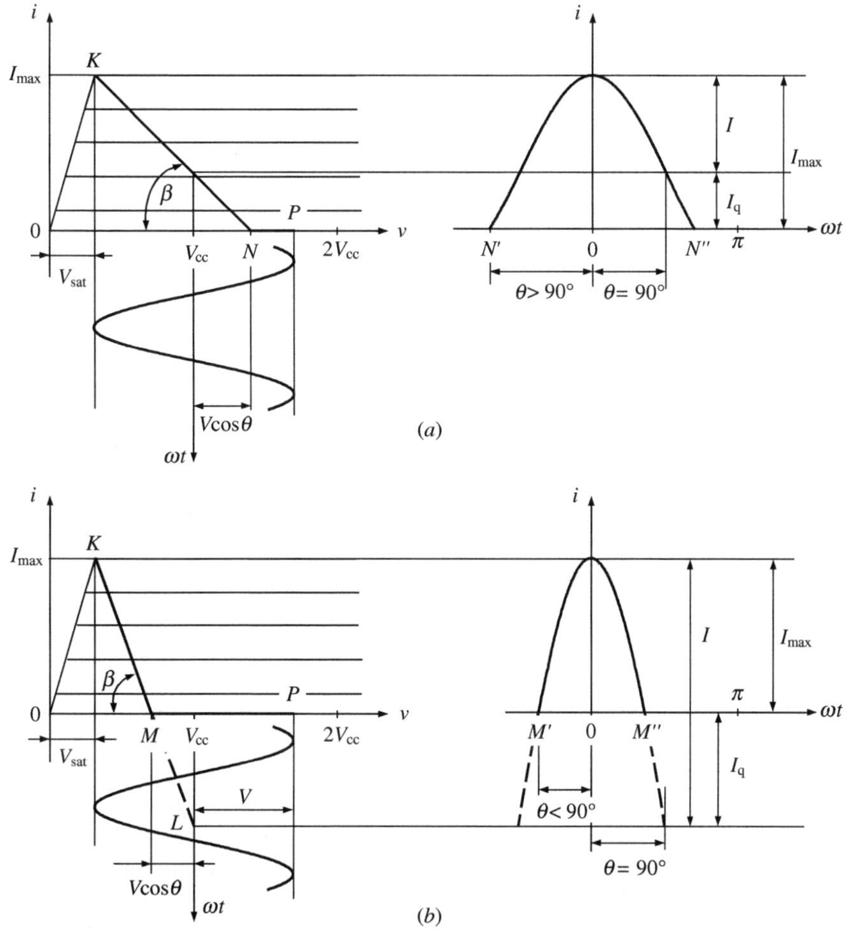

FIGURE 6.19 Collector current waveforms in Class-AB and Class-C operations.

with the corresponding collector and current waveforms. From Fig. 6.19 it follows that the maximum collector current amplitude I_{max} corresponds to the minimum collector voltage V_{sat} when $\omega t = 0$, and is the same for any conduction angle. The slope of the load line defined by its slope angle β is different for different conduction angles and values of the load resistance, and can be obtained by

$$\tan\beta = \frac{I}{V(1-\cos\theta)} = \frac{1}{\gamma_1 R_L} \tag{6.80}$$

from which it follows that greater slope angle β of the load line results in smaller value of the load resistance R_L for the same θ.

The load resistance R_L for the active device as a function of θ, which is required to terminate the device output to deliver the maximum output power to the load, can be written in a general form as

$$R_L(\theta) = \frac{V}{\gamma_1(\theta)\, I} \tag{6.81}$$

which is equal to the device equivalent output resistance R_{out} at the fundamental frequency [31]. The term "equivalent" means that this is not a real physical device resistance as in a Class-A mode, but its equivalent output resistance, whose value determines the optimum load, which should terminate the device output to deliver maximum fundamental-frequency output power. The equivalent output resistance is calculated as a ratio between the amplitudes of the collector cosine voltage and fundamental-frequency collector current component, whose value depends on half the conduction angle θ.

In a Class-B mode when $\theta = 90°$ and $\gamma_1 = 0.5$, the load resistance R_L^B is defined as $R_L^B = 2V/I_{max}$. Alternatively, by taking into account that $V_{cc} = V$ and $P_{out} = I_1 V$ for the fundamental-frequency output power, the load resistance R_L^B can be derived in a simple idealized analytical form with zero saturation voltage V_{sat} as

$$R_L^B = \frac{V_{cc}^2}{2P_{out}} \tag{6.82}$$

In general, the entire load line represents a broken line *PK* including a horizontal part, as shown in Fig. 6.19. Figure 6.19(*a*) represents a load line *PNK* corresponding to a Class-AB mode with $\theta > 90°$, $I_q > 0$, and $I < I_{max}$. Such a load line moves from point *K*, corresponding to the maximum output current amplitude I_{max} at $\omega t = 0$ and determining the device saturation voltage V_{sat}, through the point *N* located at the horizontal axis v, where $i = 0$ and $\omega t = \theta$. For a Class-AB operation, the conduction angle for the output current pulse between points *N′* and *N″* is greater than 180°. Figure 6.19(*b*) represents a load line *PMK*, corresponding to a Class-C mode with $\theta < 90°$, $I_q < 0$, and $I > I_{max}$. For a Class-C operation, the load line intersects a horizontal axis v at point *M*, and the conduction angle for the output-current pulse between points *M′* and *M″* is smaller than 180°. Hence, generally the load line represents a broken line with the first section having a slope angle β and the other horizontal section with zero current i. In a Class-B mode, the collector current represents half-cosine pulses with the conduction angle of $2\theta = 180°$ and $I_q = 0$.

Now let us consider a Class-B operation with increased amplitude of the cosine collector voltage. In this case, as shown in Fig. 6.20, an active device is operated in the saturation, active, and pinch-off regions, and the load line represents a broken line *LKMP* with three linear sections (*LK*, *KM*, and *MP*). The new section *KL* corresponds to the saturation region, resulting in the half-cosine output-current waveform with a depression in the top part. With further increase in

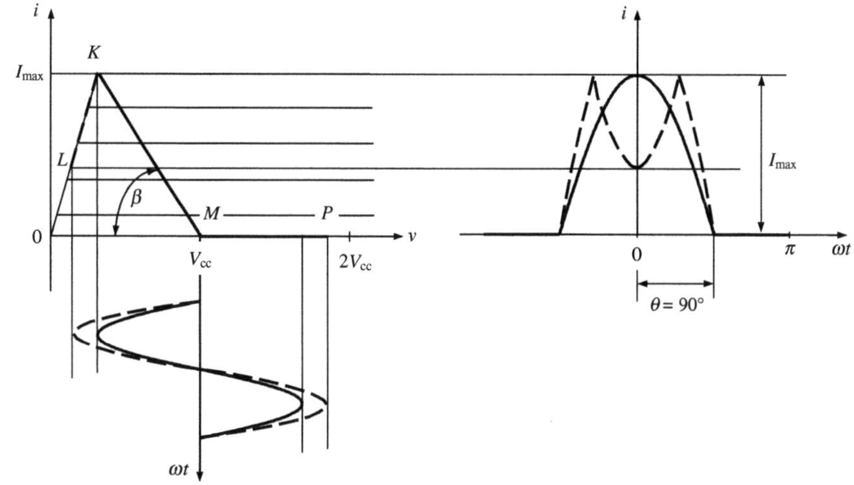

FIGURE 6.20 Collector current waveforms for the device operating in saturation, active, and pinch-off regions.

the output voltage amplitude, the output current pulse can be split into two symmetrical pulses, containing a significant level of the higher-order harmonic components. The same result can be achieved by increasing a value of the load resistance R_L when the load line is characterized by a smaller slope angle β.

The collector current waveform becomes asymmetrical for the complex load, whose impedance represents the load resistance and capacitive or inductive reactances. In this case, the Fourier-series expansion of the output current given by Eq. (6.69) includes a particular phase for each harmonic component. Then, the output voltage at the device collector is written as

$$v = V_{cc} - \sum_{n=1}^{\infty} I_n |Z_n| \cos(n\omega t + \phi_n) \tag{6.83}$$

where I_n is the amplitude of nth output current harmonic component, $|Z_n|$ is the magnitude of the load-network impedance at nth output current harmonic component, and ϕ_n is the phase of nth output current harmonic component. Assuming that Z_n is zero for $n = 2, 3, \ldots$, which is possible for the load network with a shunt parallel resonant circuit having negligible impedance at any harmonic component except the fundamental, Eq. (6.83) can be rewritten as

$$v = V_{cc} - I_1 |Z_1| \cos(\omega t + \phi_1) \tag{6.84}$$

As a result, for the inductive load impedance, the depression in the collector current waveform reduces and moves to the left-hand side of the waveform, whereas the capacitive load impedance causes

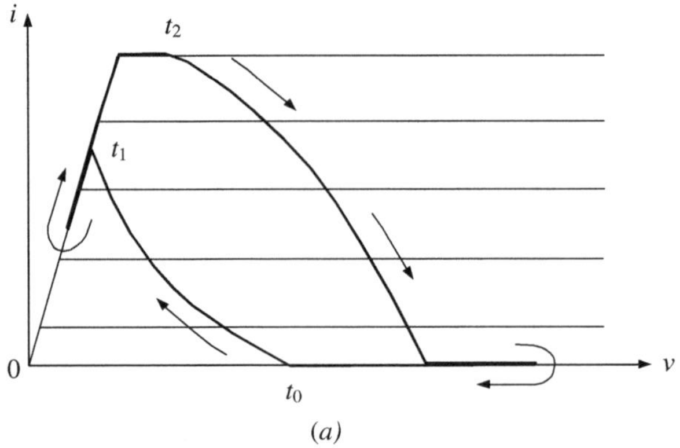

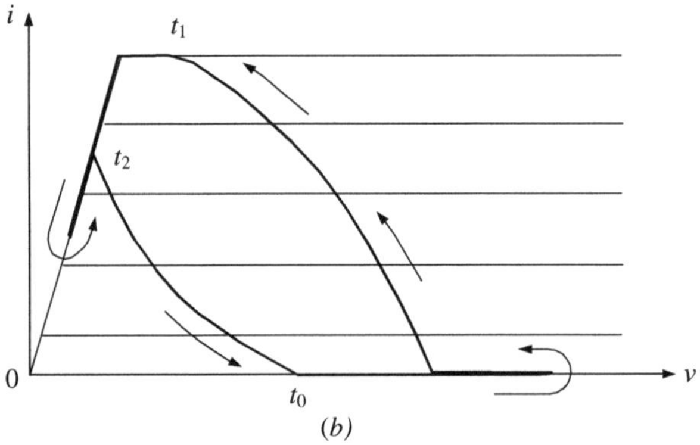

FIGURE 6.21 Load lines for (*a*) inductive and (*b*) capacitive load impedances.

the depression to deepen and shift to the right-hand side of the collector current waveform [35]. This effect can simply be explained by the different phase conditions for the fundamental (and generally for higher-order harmonic components) and is illustrated by the different load lines for (*a*) inductive and (*b*) capacitive load impedances shown in Fig. 6.21. Note that now the load line represents a two-dimensional curve with a complicated behavior.

6.6 Classes of Operation Based on Finite Number of Harmonics

Figure 6.22(*a*) shows the block diagram of a generic power amplifier, where the active device (which is shown as a MOSFET device but can be a bipolar transistor or any other suitable device) is controlled by its drive and bias to operate as a multiharmonic current source or switch, V_{dd} is the supply voltage, and I_0 is the dc current flowing through the RF choke [36]. The load-network bandpass filter is assumed linear and lossless, and it provides the drain load impedance $R_1 + jX_1$ at the

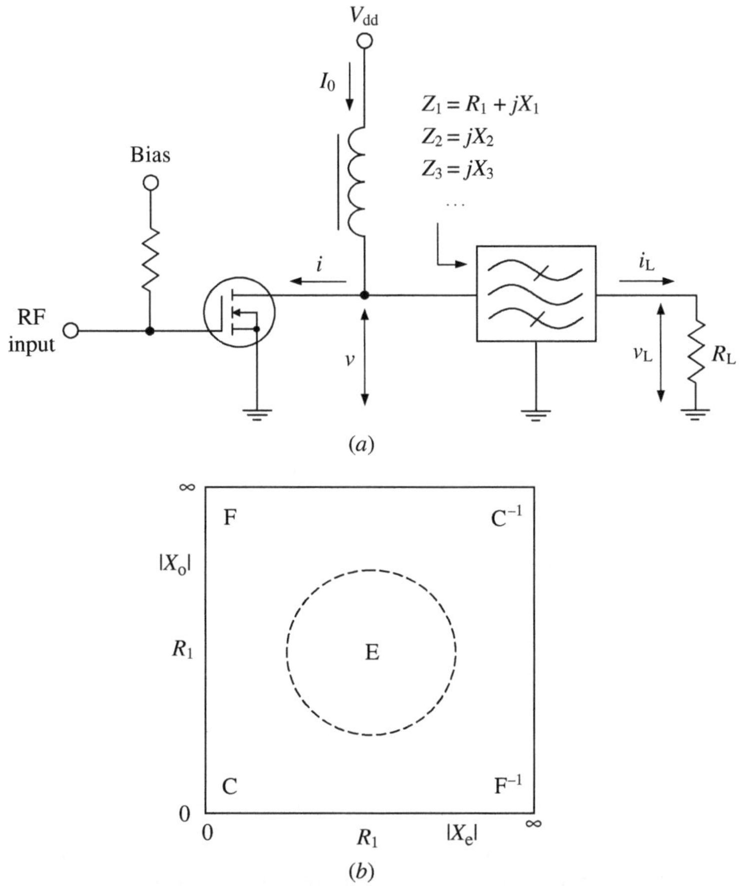

FIGURE 6.22 Basic power-amplifier structure and classes of amplification.

fundamental frequency and pure reactances X_k at each kth-harmonic component. For analysis simplicity, the load-network filter can incorporate the reactances of the RF choke and device drain-source capacitance, which is considered voltage independent.

Because such a basic power amplifier is assumed to generate power at only the fundamental frequency, harmonic components can be present generally in the voltage and current waveforms depending on class of operation. In a Class AB, B, or C, harmonics are present only in the drain current. However, in a Class-F mode, a given harmonic component is present in either drain voltage or drain current, but not both, and all or most harmonics are present in both the drain voltage and current waveforms in a Class-E mode. The required harmonics with optimum or near-optimum amplitudes can be produced by driving the power amplifier to saturation. The analysis based on a

Fourier-series expansion of the drain voltage and current waveforms shows that maximum achievable efficiency depends not on the class of operation but on the number of harmonics employed [36, 37]. For any set of harmonic reactances, the same maximum efficiency can be achieved by proper adjustment of the waveforms and the fundamental-frequency load reactance.

A mechanism for differentiating the various classes of power amplifier operation implemented with small numbers of harmonic components is shown in Fig. 6.22(*b*) [36]. It is based on the relative magnitudes of the even-(X_e) and odd-(X_o) harmonic impedances relative to the fundamental-frequency load resistance R_1. In this case, the classes of operation can be characterized in terms of a small number of harmonics as follows:

- Class F: Even-harmonic reactances are low and odd-harmonic reactances are high so that the drain voltage is shaped toward a square wave and drain current is shaped toward a half-sine wave.
- Inverse Class F (Class F^{-1}): Even-harmonic reactances are high and odd-harmonic reactances are low so that the drain voltage is shaped toward a half-sine wave and drain current is shaped toward a square wave.
- Class C: All harmonic reactances are low so that the drain current is shaped toward a narrow pulse.
- Inverse Class C (Class C^{-1}): All harmonic reactances are high so that the drain voltage is shaped toward a narrow pulse.
- Class E: All harmonic reactances are negative and comparable in magnitude to the fundamental-frequency load resistance.

The transition from "low" to "comparable" occurs in the range from $R_1/3$ to $R_1/2$, whereas that from "comparable" to "high" similarly occurs in the range from $2R_1$ to $3R_1$. In this case, the circular boundary is for illustration only, and the point at which an amplifier transitions from one class to another is somewhat judgmental and arbitrary, as there is no abrupt change in the mode of operation. All power amplifier degenerate to a Class-A mode when there is only a single (fundamental) frequency component. Class B is the special case of a pulsed operation with a conduction angle of 180°, which is represented by a half-sine current waveform based on even harmonics. Class D can be considered as a push-pull Class-F power amplifier, in which two active devices provide each other with paths for the even harmonics.

The transition from Class F to Class E and then to Class F^{-1} moves diagonally in Fig. 6.22(*b*) by progressively increasing X_2 from zero to ∞, while decreasing X_3 from ∞ to zero so that $X_3 = 1/X_2$. In a Class F with $X_2 = 0$ and $X_3 = \infty$, the voltage is a third-harmonic maximum-power waveform, whereas the current is a second-harmonic maximum-power waveform. For $X_2 = X_3 = -1$, the voltage waveform leans leftward and the current waveform leans rightward, thus approximating the all-harmonic Class-E waveforms. Finally, when $X_2 = \infty$ and $X_3 = 0$, the power amplifier operates in an inverse Class F (Class F^{-1}). The transition from Class F to Class C moves down to the left-hand side of Fig. 6.22(*b*) by setting X_2 at zero and progressively decreasing X_3 from ∞ to zero, and the waveforms remain almost unchanged for $X_3 \leq -3$. The explicit analytical expression for maximum achievable efficiency of finite-harmonic Class C with conduction angle $2\theta \to 0$ can be written as

$$\eta = \cos\left(\frac{\pi}{n+2}\right) \tag{6.85}$$

where n is the number of harmonics [38].

6.7 Mixed-Mode Class B and Nonlinear Effect of Collector Capacitance

In contrast to the conventional Class-C power amplifiers with a parallel resonant circuit resulting in a sinusoidal collector voltage waveform, the so-called mixed-mode Class-C configuration with a series resonant circuit was widely although somewhat accidentally adopted for most VHF and UHF transistor power amplifiers, which could provide better efficiency performance and where it is easier to provide the drive and bias [39, 40]. For low saturation resistance and significant nonlinear collector capacitance, it is difficult to maintain a sinusoidal collector voltage waveform. Instead, a nonlinear collector capacitance produces a voltage waveform containing harmonics in response to a sinusoidal current. As a result, the saturated bipolar transistor usually dominates the parallel-tuned circuit, flattening the collector voltage waveform [41]. Besides, it is also enough difficult in practice to implement the parallel-resonant circuit required for true Class-C operation in power amplifiers using either FET or BJT devices, especially with a high-quality factor. There are several additional difficulties in implementing true Class-C operation in solid-state power amplifiers, especially at VHF and UHF in view of the device lead lengths and stray reactances, causing a significant effect at these frequencies.

Figure 6.23 shows the simplified schematic of a mixed-mode Class-C power amplifier with a series resonant circuit in a load network, which provides the near-sinusoidal collector current and pulsed collector voltage with pulse duration less than one-half the period, depending on the value of the collector capacitance. The level of a Class-C operation with corresponding conduction angle is defined by the value of the resistor in a base bias circuit, where the inductor value is chosen to maximize the operating power gain.

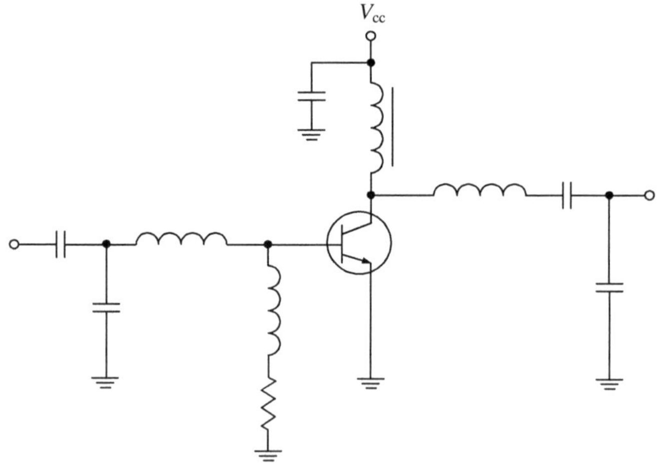

Figure 6.23 High-efficiency bipolar mixed-mode Class-C power amplifier.

As an example, by using a 28-V MRF373A LDMOSFET device in a series-tuned Class-C power amplifier, whose simplified circuit schematic is shown in Fig. 6.24(*a*), an output power of 50 W and a drain efficiency of 58% (by 10% lower than obtained with the idealized simulations) were achieved at 435 MHz [42]. Here, for the theoretical analysis, it was assumed that the transistor is driven so hard that its operation can be described by a switch, and the switch is turned on (closed) when the gate-source voltage is above the threshold voltage and turned off (open) when it is below. When the transistor is turned off, the total drain current i_d, which is a sum of the sinusoidal load current and dc supply current, charges or discharges the transistor drain-source capacitance C_{ds}. In this case, there are two basic power loss mechanisms: the transistor series loss, due to finite value of its saturation voltage, and the switching loss that accompanies the switch turning on. The switching loss is determined by the value of the capacitor voltage V_{csw} obtained prior to the start of switching on. Figure 6.24(*b*) shows the drain voltage and current waveforms, where the switch starts turning off at zero time instant and the drain voltage $v_d(\omega t)$ achieves maximum value V_{dmax} when the drain current $i_d(\omega t)$ turns negative through zero-crossing point. Significantly higher efficiency was achieved using the Cree CGH40120F transistor when the output power of 115 W was achieved under pulsed conditions with the drain efficiency between 78.4 and 82.7% at the same operating frequency for the duty ratio varying from 36 to 25% [43].

Generally, the dependence of the collector capacitance on the output voltage represents a nonlinear function. To evaluate the influence of the nonlinear collector capacitance on electrical behavior of the power amplifier, let us consider the load network including a series

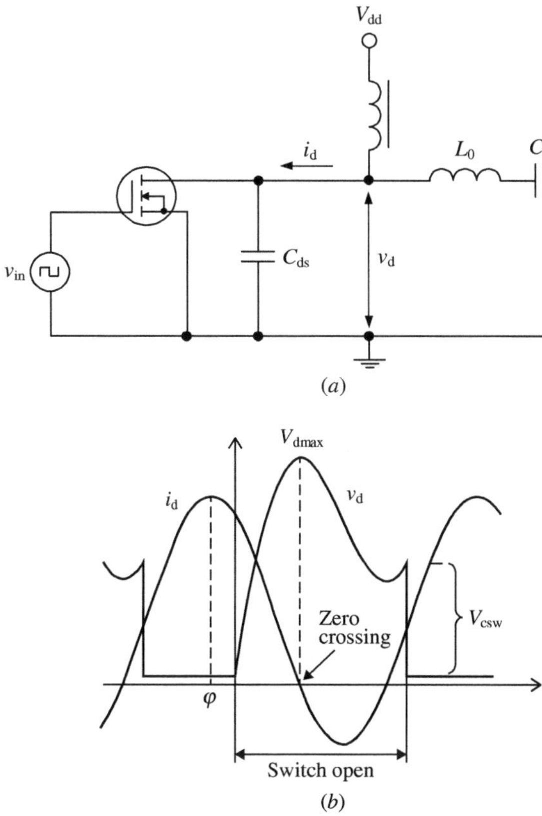

Figure 6.24 High-efficiency bipolar mixed-mode Class-C power amplifier.

resonant L_0C_0 circuit tuned to the fundamental frequency that provides open-circuit conditions for the second- and higher-order harmonic components of the output current and a low-pass L-type matching circuit with the series inductor L and shunt capacitor C, as shown in Fig. 6.25(*a*). The matching circuit is needed to match the equivalent output resistance R, corresponding to the required output power at the fundamental frequency, with the standard load resistance R_L. Figure 6.25(*b*) shows the simplified output equivalent circuit of the bipolar power amplifier.

The total output current flowing through the device collector can be written as

$$i = I_0 + \sum_{n=1}^{\infty} I_n \cos(n\omega t + \phi_n) \tag{6.86}$$

where I_n and ϕ_n are the amplitude and phase of the nth-harmonic component, respectively.

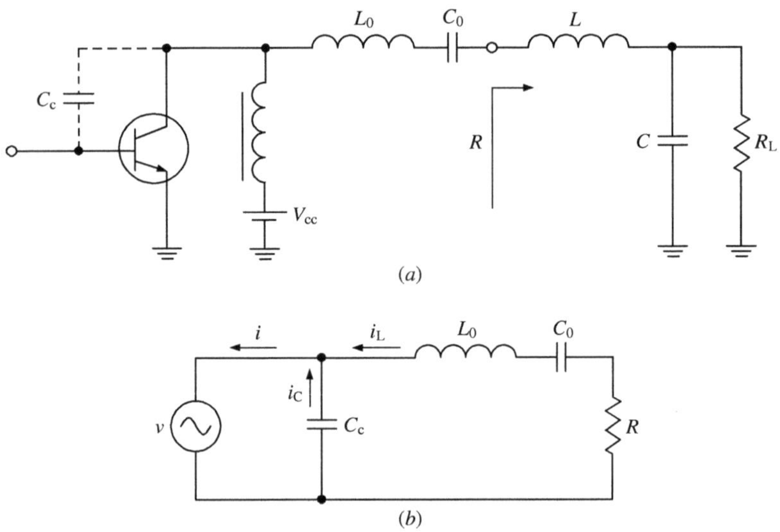

Figure 6.25 Circuit schematics of bipolar tuned power amplifier.

An assumption of a high-quality factor of the series resonant circuit allows the only fundamental-frequency current component to flow into the load. The current flowing through the nonlinear collector capacitance consists of the fundamental-frequency and higher-order harmonic components, which is written as

$$i_c = I_c \cos(\omega t + \phi_1) + \sum_{n=2}^{\infty} I_n \cos(n\omega t + \phi_n) \tag{6.87}$$

where I_C is the fundamental-frequency capacitor current amplitude.

The nonlinear behavior of the collector junction capacitance is described by

$$C_c = C_0 \left(\frac{\varphi + V_{cc}}{\varphi + v} \right)^{\gamma} \tag{6.88}$$

where C_0 is the collector capacitance at $v = V_{cc}$, V_{cc} is the supply voltage, φ is the contact potential, and γ is the junction sensitivity equal to 0.5 for abrupt junction.

As a result, the expression for charge flowing through collector capacitance can be obtained by

$$q = \int_0^v C(v)dv = \int_0^v \frac{C_0(\varphi + V_{cc})^{\gamma}}{(\varphi + v)^{\gamma}} dv \qquad 6.89)$$

When $v = V_{cc}$, then

$$q_0 = \frac{C_0(\varphi + V_{cc})}{1-\gamma}\left[1 - \left(\frac{\varphi}{\varphi + V_{cc}}\right)^{1-\gamma}\right] \tag{6.90}$$

Although the dc charge component q_0 is a function of the voltage amplitude, its variations at maximum voltage amplitude normally do not exceed 20% for $\gamma = 0.5$ [17]. Then, assuming q_0 is determined by Eq. (6.90) as a constant component, the total charge q of the nonlinear capacitance can be represented by the dc component q_0 and ac component Δq as

$$q = q_0 + \Delta q = q_0\left(1+\frac{\Delta q}{q_0}\right) = q_0\,\frac{(\varphi+v)^{1-\gamma}-\varphi^{1-\gamma}}{(\varphi+V_{cc})^{1-\gamma}-\varphi^{1-\gamma}} \tag{6.91}$$

Because $V_{cc} >> \varphi$ in the normal case, from Eq. (6.91) it follows that

$$\frac{v}{V_{cc}} = \left(1+\frac{\Delta q}{q_0}\right)^{\frac{1}{1-\gamma}} \tag{6.92}$$

where $q_0 \cong C_0 V_{cc}/(1-\gamma)$.

On the other hand, the charge component Δq can be written using Eq. (6.87) as

$$\Delta q = \int i_c(t)\,dt = \frac{I_C}{\omega}\sin(\omega t + \phi_1) + \sum_{n=2}^{\infty}\frac{I_n}{n\omega}\sin(n\omega t + \phi_n) \tag{6.93}$$

As a result, substituting Eq. (6.93) into Eq. (6.92) yields

$$\frac{v}{V_{cc}} = \left[1+\frac{I_C(1-\gamma)}{\omega C_0 V_{cc}}\sin(\omega t+\phi_1) + \sum_{n=2}^{\infty}\frac{I_n(1-\gamma)}{n\omega C_0 V_{cc}}\sin(n\omega t+\phi_n)\right]^{\frac{1}{1-\gamma}} \tag{6.94}$$

After applying a Taylor-series expansion to Eq. (6.94), it is sufficient to be limited to its first three terms to reveal the parametric effect. Then, equating the fundamental-frequency collector voltage components gives

$$\frac{v_1}{V_{cc}} = \frac{I_C}{\omega C_0 V_{cc}}\sin(\omega t+\phi_1) + \frac{I_C I_2 \gamma}{(2\omega\, C_0 V_{cc})^2}\cos(\omega t+\phi_2-\phi_1)$$
$$+\frac{I_2 I_3 \gamma}{12(\omega C_0 V_{cc})^2}\cos(\omega t+\phi_3-\phi_2) \tag{6.95}$$

Consequently, by taking into account that $v_1 = V_1\sin(\omega t + \phi_1)$, the fundamental voltage amplitude V_1 can be obtained from Eq. (6.95) as

$$\frac{V_1}{V_{cc}} = \frac{I_C}{\omega C_0 V_{cc}}\left[1+\frac{I_2\gamma}{4\omega C_0 V_{cc}}\cos(90^\circ+\phi_2-2\phi_1)\right.$$
$$\left.+\frac{I_2 I_3 \gamma}{12\omega C_0 V_{cc} I_C}\cos(90^\circ+\phi_3-\phi_2-\phi_1)\right] \tag{6.96}$$

Because a large-signal value of the abrupt-junction collector capacitance usually does not exceed 20%, the fundamental-frequency capacitor current amplitude I_C can be written in a first-order approximation as

$$I_C \cong \omega C_0 V_1 \tag{6.97}$$

As a result, from Eq. (6.96) it follows that, because of the parametric transformation due to the collector capacitance nonlinearity, the fundamental-frequency collector voltage amplitude increases by σ_p times according to

$$\sigma_p = 1 + \frac{I_2 \gamma}{4\omega C_0 V_{cc}} \cos(90° + \phi_2 - 2\phi_1) + \frac{I_2 I_3 \gamma}{12\,(\omega C_0)^2 V_1 V_{cc}} \cos(90° + \phi_3 - \phi_2 - \phi_1) \tag{6.98}$$

where $\sigma_p = \xi_p / \xi$ and ξ_p is the collector voltage peak factor with parametric effect [35].

From Eq. (6.98) it follows that to maximize the collector voltage peak factor and consequently the collector efficiency for a given value of the supply voltage V_{cc}, it is necessary to provide the following phase conditions:

$$\phi_2 = 2\phi_1 - 90° \tag{6.99}$$

$$\phi_3 = 3\phi_1 - 180° \tag{6.100}$$

Then, for $\gamma = 0.5$,

$$\sigma_p = 1 + \frac{I_2}{8\,\omega C_0 V_{cc}} + \frac{I_2 I_3}{24\,(\omega C_0)^2 V_1 V_{cc}} \tag{6.101}$$

Equation (6.101) shows the theoretical possibility to increase the collector voltage peak factor by 1.1 to 1.2 times, thus achieving collector efficiency of 85 to 90%. Physically, the improved efficiency can be explained by the transformation of powers corresponding to the second- and higher-order harmonic components into the fundamental-frequency output power due to the collector capacitance nonlinearity. However, this becomes effective only in the case of the load network with a series resonant circuit (mixed-mode Class C), because it ideally provides infinite impedance at the second- and higher-order harmonics (corresponding to inverse Class F with zero shunt capacitance), unlike the load network with a parallel resonant circuit (true Class C) having ideally zero impedance at these harmonics.

6.8 Load-Pull Characterization

In designing power amplifiers, it is important to know the transistor input impedance and load characteristics at a high input-driving power level and to optimize the output matching circuit on the basis of these large-signal characteristics. A computer-controlled technique for large-signal characterization of microwave power transistors used to map contours of constant power and efficiency on a Smith chart for dynamic matching of both input and output circuits was originally developed for the interstage matching between a varactor multiplier and a transistor power stage and then successfully applied to the broadband optimization of Class-A and Class-C transistor power amplifiers [44, 45]. The power-load contours consist of a series of curves on a Smith chart representing constant-output power, approaching the point of maximum output power, as shown in Fig. 6.26(*a*). When drawn at several frequencies over the band of interest, they represent loci of required output impedance for various

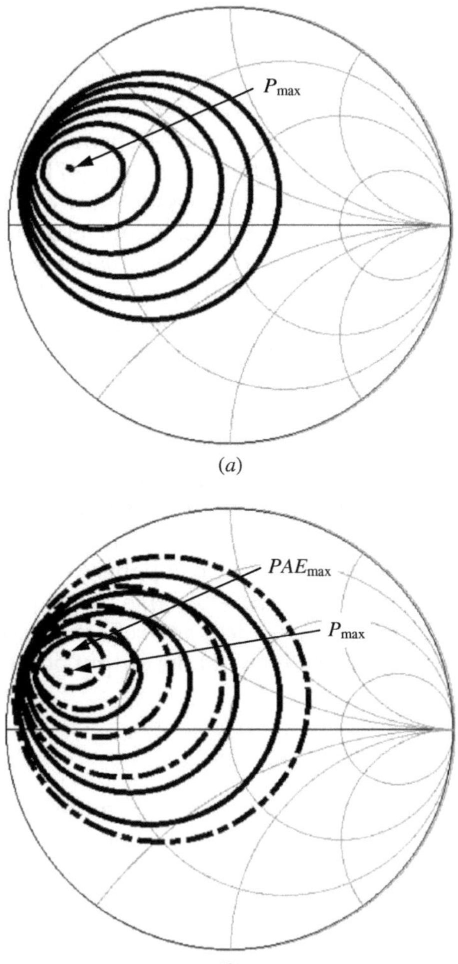

Figure 6.26 Constant-power and constant-efficiency load-pull contours.

output power levels. If constant-efficiency contours are overlaid on the constant-power contours, the efficiency is also known at each of the load impedances. Generally, the points for maximum output power and maximum efficiency can be located at different positions, each corresponding to its optimum impedance, as shown in Fig. 6.26(*b*), because the maximum power is achieved at the fundamental frequency, whereas the maximum efficiency significantly depends on the effect of the second- and higher-order harmonic components.

A power transistor operated in a Class-C mode is characterized by the output equivalent circuit, which is represented by a nonlinear multiharmonic current source and a nonlinear reactance. To obtain maximum output power, the load impedance must produce the maximum current and voltage swing. The relationship between the required load impedance and nonlinear output impedance determines the shape of the power-load contour, which can vary with input drive level from elliptical to circle contours [46]. With the conventional passive load-pull technique, external tuners are adjusted, demounted, and measured. In this case, the dc bias-dependent transistor small-signal impedances can be used as references [47]. To get a variable load, in addition to the traditional passive-network technique with variable elements, an active load-pull characterization can be used where the reflection coefficient is obtained using an auxiliary signal derived from the same test generator to inject into the output port [48, 49]. One of the advantages of this technique lies in the inherent simplicity of the calibration procedure, which is necessary to correct the transmission-line losses in the measurement system. However, its accuracy critically depends on the effective directivities of the directional couplers in the system [50]. The accuracy of the load-pull measurements can be improved by using ultralow-loss broadband tuners based on nonuniform, nonsymmetrical rectangular coaxial-to-microstrip directional couplers [51]. Independent tuning of the fundamental frequency and its second-harmonic component had become possible by using a scheme with frequency-selective tuners [52].

The load impedance Z_L, incident and reflected traveling waves a_2 and b_2, and reflection coefficient Γ_L at the output port of the transistor are related as

$$\Gamma_L = \frac{a_2}{b_2} = \frac{Z_L - Z_0}{Z_L + Z_0} \tag{6.102}$$

where Z_0 is the characteristic impedance of the system, in which the device under test (DUT) is used. The function of the tuners is to vary the magnitude and phase of the reflected signal so as to synthesize an appropriate Γ_L. This functionality can be obtained by altering the tuner setting by way of moving the slug or stub up and down and back and forth in a passive tuner, as shown in Fig. 6.27, or by actively injecting a magnitude- and phase-controlled signal in an active load-pull system [53]. In this case, as the stub (probe or slug) is inserted into the tuner

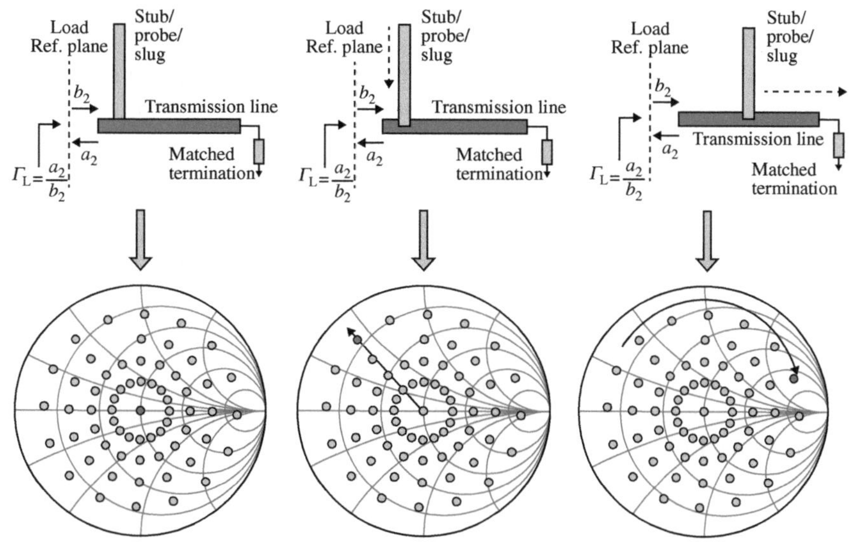

FIGURE 6.27 Passive tuner and reflection coefficient position.

transmission line, it introduces mismatch by adding parallel susceptance, and the parallel susceptance increases as the stub approaches the line and aids in the synthesis of the desired reflection coefficient. As a result, the magnitude of the impedance mismatch is determined by the stub position (depth) and the phase of the impedance mismatch is determined by the carriage position (length).

Generally, tuning with a load-pull system (mechanical or fully automatic) is a very complicated procedure, especially if necessary to take into account several harmonics of the fundamental frequency. Besides, it can only give the optimum input and output impedances, which are usually incorporated in datasheet for power transistors for specified output power, supply voltage, and frequency range. In a monolithic implementation, however, this is difficult to physically realize and this can be done based on the load-pull setup incorporated into the simulation tool. However, in all cases, this measurement should be followed by the subsequent load-network design with real elements. Therefore, in most cases where the device nonlinear model is available, the device equivalent output circuit can be represented by the output admittance where the equivalent large-signal output resistance can be estimated from Eq. (6.82) for fixed conduction angle 2θ and supply voltage V_{cc}, which is equal to the cosine voltage amplitude in an ideal case (assuming zero saturation voltage). For example, in a Class-B mode with $\theta = 90°$, it is written as $R_L = R_{out} = V/I_1 = V_{cc}^2/2P_{out}$, where P_{out} is the maximum fundamental-frequency power delivered to the load, and the output shunt reactance is represented by the sum of the output and feedback capacitances.

6.9 Linearity

To evaluate the nonlinear properties of the power amplifier, first it is necessary to consider the transfer function of the active device in the form of

$$i(\omega t) = f\left[v(\omega t)\right] \tag{6.103}$$

where $i(\omega t)$ is the output collector or drain current and $v(\omega t)$ is the input gate-source or base-emitter voltage. It is convenient to apply a power-series analysis, which is relatively easy to use and which gives a good intuitive sense of the nonlinear behavior of the active device. Let us assume that the nonlinearity is weak enough so that the power series converges. Then, the transfer function $f(v)$ can be approximated by its expanding around the dc voltage V_0 in a Taylor series as

$$f(v) = f(V_0) + \sum_{n=1}^{\infty} \frac{1}{n!} \left.\frac{\partial^{(n)} f(v)}{\partial v^n}\right|_{v=V_0} (v - V_0)^n \tag{6.104}$$

The nonlinear properties are usually determined by a two-tone excitation test signal with individual components separated slightly in frequency, which can be represented in a common case of unequal amplitudes as

$$v = V_0 + V_1 \cos\omega_1 t + V_2 \cos\omega_2 t \tag{6.105}$$

For first three derivatives, the output signal can be represented by a Taylor-series expansion with the appropriate equating of the frequency component terms as

$$
\begin{aligned}
i = {} & f(V_0) + \frac{1}{4} \left.\frac{\partial^2 f(v)}{\partial v^2}\right|_{v=V_0} (V_1^2 + V_2^2) \\
& + \left[\left.\frac{\partial f(v)}{\partial v}\right|_{v=V_0} + \frac{1}{4} \left.\frac{\partial^3 f(v)}{\partial v^3}\right|_{v=V_0} \left(\frac{1}{2} V_1^2 + V_2^2 \right) \right] V_1 \cos\omega_1 t \\
& + \left[\left.\frac{\partial f(v)}{\partial v}\right|_{v=V_0} + \frac{1}{4} \frac{\partial^3 f(v)}{\partial v^3} \left(V_1^2 + \frac{1}{2} V_2^2 \right) \right] V_2 \cos\omega_2 t \\
& + \frac{1}{4} \left.\frac{\partial^2 f(v)}{\partial v^2}\right|_{v=V_0} (V_1^2 \cos 2\omega_1 t + V_2^2 \cos 2\omega_2 t) \\
& + \frac{1}{24} \left.\frac{\partial^3 f(v)}{\partial v^3}\right|_{v=V_0} (V_1^3 \cos 3\omega_1 t + V_2^3 \cos 3\omega_2 t)
\end{aligned}
$$

$$+\frac{1}{2}\frac{\partial^2 f(v)}{\partial v^2}\bigg|_{v=V_o} V_1V_2\cos(\omega_1 \pm \omega_2)t + \frac{1}{8}\frac{\partial^3 f(v)}{\partial v^3}\bigg|_{v=V_o}$$
$$\left[V_1^2V_2\cos(2\omega_1 \pm \omega_2)t + V_1V_2^2\cos(\omega_1 \pm 2\omega_2)t\right] \dots \tag{6.106}$$

The following conclusions can be drawn from the above Taylor-series expansion of the active device transfer function:

- Variation of the device bias point is directly proportional to the second derivative (in a common case, even derivatives) of the transfer function.
- The device transfer function will be linear only if the third derivative (in a common case, odd derivatives) is equal to zero.
- Even-harmonic components result from even derivatives of the device transfer function, whereas odd-harmonic components result from odd derivatives of the device transfer function.
- First-order mixing products (total and differential) are provided by even derivatives of the device transfer function.
- Mixing products of the third and higher order are mainly determined by the odd derivatives of the device transfer function.
- Distortions, which are determined by the second derivative (second amplitude degree) or by the third derivative (third amplitude degree) of the device transfer function, are called the *second-order intermodulation distortions* (IMD_2) or the *third-order intermodulation distortions* (IMD_3), respectively.

From Eq. (6.106), it follows that the output current amplitude of the fundamental, second, and third harmonic or intermodulation components depends on the first, second, and third degree of the input voltage amplitude, respectively. Consequently, the output power levels of the linear, second-order, and third-order frequency components show a straight-line behavior and vary by 1 dB, 2 dB, and 3 dB, respectively, with an input power level of 1-dB variation. Further analysis of Eq. (6.106) would show that n-order components also vary by n dB with an input power level of 1-dB variation. As a result, these straight lines in terms of dBm intersect at the proper intercept points, and each this point is different for each order of intermodulation products. Consequently, if the intercept point is determined, for example, experimentally for a given type of the transistor, it is easy to evaluate the harmonic and n-order intermodulation output power levels at the arbitrary level of input power. Figure 6.28 shows the straight-line dependences for the fundamental, second-harmonic, and third-order intermodulation (IM_3) components with the corresponding intercept points IP_2 and IP_3, respectively.

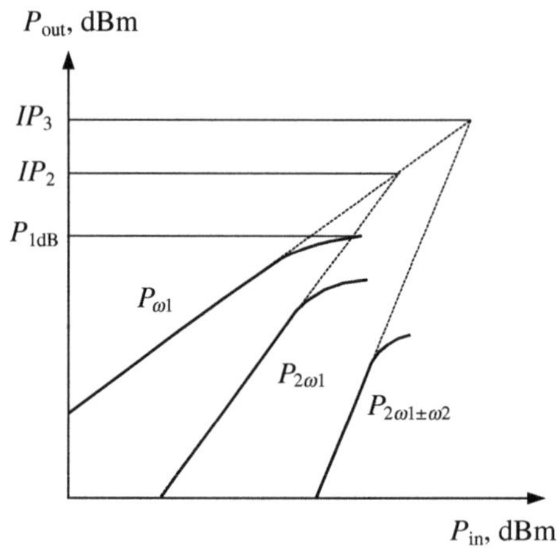

FIGURE 6.28 Straight-line dependences for harmonic and intermodulation components.

For any straight line, we can write the following equation:

$$P_{IM_n} = nP_{in} + P_{n0} \text{ (dBm)} \tag{6.107}$$

where P_{n0} is a constant that will be evaluated.

The linear fundamental-frequency output power is equal to

$$P_{\omega_1} = 10 \log_{10}(G_p P_{in}) = P_{in} + G_p \text{ (dBm)} \tag{6.108}$$

Equation (6.107) can be rewritten as

$$P_{IM_n} = nP_{\omega_1} + P_{n0} - nG_p \text{ (dBm)} \tag{6.109}$$

At the intercept point IP_n,

$$P_{IM_n} = P_{\omega_1} = IP_n \tag{6.110}$$

which yields a ratio

$$(1-n)\, IP_n = P_{no} - nG_p \text{ (dBm)} \tag{6.111}$$

Then,

$$P_{IM_n} = nP_{\omega_1} - (n-1)\, IP_n \text{ (dBm)} \tag{6.112}$$

Equation (6.112) can be used to evaluate the relationship between the fundamental output power P_{ω_1}, the output power corresponding to the n-order intermodulation component P_{IMn}, and the n-order intercept point IP_n at the input level below the device voltage saturation.

For example, the second harmonic component $P_{2\omega_1}$ and the third-order intermodulation component $P_{2\omega_1-\omega_2}$ can be easily evaluated as

$$P_{2\omega_1} = 2P_{\omega_1} - IP_2 \text{ (dBm)} \tag{6.113}$$

$$P_{2\omega_1-\omega_2} = 3P_{\omega_1} - 2IP_3 \text{ (dBm)} \tag{6.114}$$

The 1-dB compression level of the output power, at which a value of the power gain decreases by 1 dB compared to its small-signal value in a linear operation region, can be estimated by comparing the first- and third-order terms in Eq. (6.106) as

$$P_{1\text{dB}} = IP_3 - 9 \text{ dB (dBm)} \tag{6.115}$$

although, depending on the different devices with different types of their nonlinear transfer function, the difference $IP_3 - P_{1\text{dB}}$ may vary between 8 and 15 dB.

Equations (6.114) and (6.115) give a convenient and simple qualitative estimate of the basic nonlinear performance of the power amplifier. For example, if $IP_3 = 50$ dBm, then $P_{1\text{dB}} = 41$ dBm with the third-order nonlinear component $P_{2\omega_1-\omega_2} = 23$ dBm. To improve the linearity of the power amplifier, it is necessary to reduce the output power level for a given value of the intercept point. The level of –30 dB (relative to the fundamental-frequency output power P_{ω_1}) for the third-order intermodulation component $P_{2\omega_1-\omega_1}$ can be achieved only when $P_{\omega_1} = 35$ dBm, which means that the output power has to be reduced by 6 dB (or four times).

It is well known that setting the dc drain current of a GaAs MESFET device to approximately $0.5I_{\text{dss}}$, where I_{dss} is the saturated drain-source current, maximizes not only its gain but also intermodulation intercept points. First, the transfer $I_{\text{ds}}(V_{\text{gs}})$ curve is clearly more linear near sweet spot of $0.5I_{\text{dss}}$, where its slope, determining the device transconductance, is maximal, and thus mainly influences the first-degree coefficient of its Taylor-series expansion. Secondly, the nonlinearity of the $C_{\text{gs}}(V_{\text{gs}})$ curve significantly decreases with the bias point shifted toward higher values. To provide the high-efficiency operation mode of the MESFET power amplifier, it is necessary to use a value of the gate-source bias voltage quite close to the pinch-off voltage, with the appropriate worsening of its linear properties. However, in this case it is possible to choose the bias point with the drain quiescent current I_q in limits of $(0.1 \div 0.15)I_{\text{dss}}$, when the third-order intermodulation component IM_3 can be minimized at sufficiently high output power, as shown in Fig. 6.29. First of all, this is a result of the quadratic dependence of the drain current I_{ds} on the gate voltage V_{gs} near the pinch-off point. Besides, given a certain value of the gate-source bias voltage, the third-order intermodulation components and the intermodulation components, which are the results of an interaction

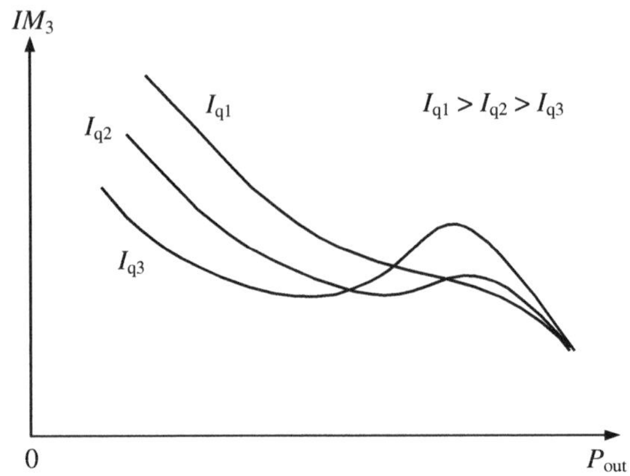

FIGURE 6.29 MESFET power amplifier intermodulation distortions for different quiescent currents.

of the second harmonics $2\omega_1$ and $2\omega_2$ and differential component $\omega_2 - \omega_1$ with the fundamental components ω_1 and ω_2, cancel each other. Because this cancellation depends on the load and source impedances at the frequencies far from the operating frequency bandwidth, it is necessary to provide an additional tuning of the input and output matching circuits in order to minimize intermodulation distortion.

Figure 6.30 shows the output power spectrum containing only n-order intermodulation components, which are the result of the effect of a two-tone input excitation. The amplitude of the higher-order intermodulation components decreases significantly when frequency increases. To evaluate the linear behavior of the transistor, it is sufficient to measure the amplitudes of the largest intermodulation

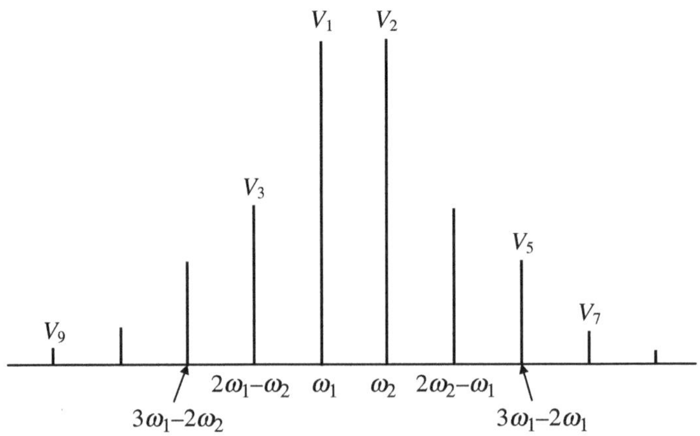

FIGURE 6.30 Typical output power spectrum for two-tone excitation.

components, that is, their third-order (IM_3) and the fifth-order (IM_5) components, which are defined as

$$IM_3 = 10\log_{10}\left(\frac{P_{2\omega_1-\omega_2}}{P}\right) = P_{2\omega_1-\omega_2} - P \text{ (dBc)} \tag{6.116}$$

$$IM_5 = 10\log_{10}\left(\frac{P_{3\omega_1-2\omega_2}}{P}\right) = P_{3\omega_1-2\omega_2} - P \text{ (dBc)} \tag{6.117}$$

where $P = P_{\omega_1} = P_{\omega_2}$ for equal two-tone signal amplitudes. On the other hand, the third-order intermodulation component IM_3 can be directly calculated from Eq. (6.114) as

$$IM_3 = P_{2\omega_1-\omega_2} - P_{\omega_1} = 2P_{\omega_1} - 2IP_3 \text{ (dBc)} \tag{6.118}$$

and, for example, for the power amplifier with IP_3 = 50 dBm and $P_{1\text{dB}}$ = 41 dBm, the third-order intermodulation component IM_3 is equal to −18 dBc.

The linearity of the MOSFET power amplifiers is strongly sensitive to the bias conditions when a choice of the optimum bias voltage in Class-AB operation mode improves the third-order intermodulation distortion by more than 10 dB [54]. This is a result of a quadratic character of the sufficiently long section of the device transfer dependence $I_{ds}(V_{gs})$. Minimum power gain flatness over the dynamic power range (linear P_{out} versus P_{in}) corresponds to the best linearity condition.

For bipolar transistors, a similar approach can be used when an improved intermodulation distortion is achieved by optimizing the collector quiescent current with proper base bias condition. In this case, in a Class-AB operation mode, the minimum level of the third-order intermodulation components is a function of the values of both the output power P_{out} and the collector quiescent current I_q. Low values of I_q give better linearity at higher power levels, whereas higher values of I_q give better linearity at lower power levels. Because the "sweet" point moves when different I_q are used, the minimum gain variations over the total dynamic range corresponds to the best linearity that can be achieved by adding the series resistor R with optimum value to the base bias circuit [55]. This series resistor is connected between the base bias voltage supply V_b and the bypass capacitor C_{bypass}, followed by a quarterwave microstrip line to provide better isolation between RF and dc paths, as shown in Fig. 6.31.

The use of a series resistor can minimize the gain variations and stabilize the level of the third-order intermodulation components over the dynamic range. For example, an increase in RF output power causes the appropriate increase in the dc collector current. This leads, in turn, to an increase in the dc base current with the corresponding increase of the voltage drop across the resistor R and corresponding decrease in a value of the base bias voltage. To determine the value of

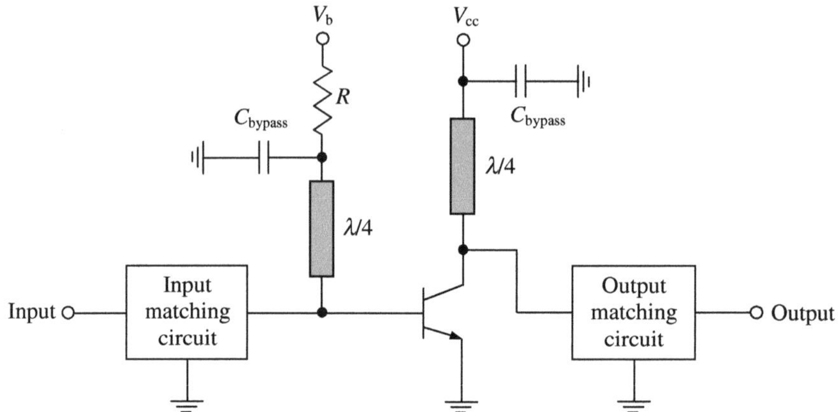

FIGURE 6.31 Bipolar power amplifier with linearizing bias resistor.

R, it needs to set a goal for the level of the intermodulation components at both high and low output power levels. Then, the value of R can be calculated according to

$$R = \frac{V_{b1} - V_{b2}}{I_{b2} - I_{b1}} \tag{6.119}$$

where V_{b1} is the dc voltage at low power level, V_{b2} is the dc voltage at high power level, I_{b1} is the dc base current at low power level, and I_{b2} is the dc base current at high power level.

In practice, the level of n-order intermodulation component is usually given relative to the peak envelope power P_{PEP} calculated from two-tone excitation signal measurements. In this case, the waveform of the signal dissipated in the load will be significantly different from the sine wave and the signal voltage can be written for a two-tone signal of equal amplitudes $V_1 = V_2 = V$ as

$$v_L = V_1 \sin\omega_1 t + V_2 \sin\omega_2 t = 2V \cos\Omega t \sin\omega t \tag{6.120}$$

where $\omega = (\omega_1 + \omega_2)/2$ is the center RF signal frequency and $\Omega = (\omega_2 - \omega_1)/2$ is the low intercarrier frequency or pulse envelope.

The waveform of such a two-tone driving signal with equal amplitudes is shown in Fig. 6.32, where $T = 2\pi/\Omega$ is the period of envelope. The peak envelope power P_{PEP} that corresponds to the output power with maximum amplitude $2V$ is equal to

$$P_{PEP} = \frac{(2V)^2}{2R_L} \tag{6.121}$$

where R_L is the load resistance. The total output power provided by each sinusoidal tone of a two-tone excitation signal with equal amplitudes is

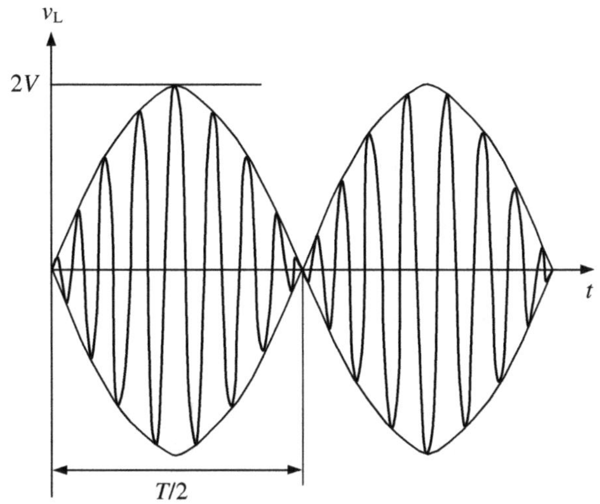

FIGURE 6.32 Two-tone driving signal.

$$P_{out} = P_{\omega 1} + P_{\omega 2} = \frac{V^2}{R_L} \qquad (6.122)$$

Comparing Eqs. (6.121) and (6.122) yields

$$P_{PEP} = 2P_{out} = 4P \qquad (6.123)$$

where $P = P_{\omega 1} = P_{\omega 2}$.

If the output amplitude and/or phase of the two- or multitone test signal are affected by the tone difference, the power amplifier exhibits so-called memory effect. Memory is caused by the storage of energy due to storage elements (capacitors and inductors) that has to be charged or discharged. Smooth memory effects (usually at low frequencies) do not usually affect the linearity of the power amplifier itself. A phase rotation of 10 to 20°, or an amplitude change of less than 0.5 dB, as a function of modulation frequency, has no dramatic effect on the linearity of the device. There are two memory effects: electrical and thermal. Electrical memory effects are produced by nonconstant node impedances within different frequency ranges corresponding to the dc, fundamental, and its higher-order harmonic components. Most of these effects are generated by frequency-dependent envelope impedance, and those close to the dc are the most harmful. Also, the greater difference between the tones, the more effect on higher-frequency tone can be caused by the device transit time. Thermal memory effects are generated by the junction temperature, which is modulated by the applied signal, and are much more prominent in a slow envelope sweep.

6.10 Push-Pull and Balanced Power Amplifiers

Generally, if it is necessary to increase an overall output power of the power amplifier, several active devices can be used in parallel, balanced, or push-pull configurations. In a parallel configuration, the active devices are not isolated from each other that requires a very good circuit symmetry, and the output impedance becomes too small in the case of high output power. The latter drawback can be eliminated by using a push-pull configuration, which provides increased values of the input and output impedances. For the same output power level, the input impedance Z_{in} and output impedance Z_{out} under a push-pull operation mode are approximately four times as high as that of in a parallel connection of the active devices. At the same time, the loaded quality factors of the input and output matching circuits remain unchanged because both the real and reactive parts of these impedances are increased by the factor of four. Very good circuit symmetry can be provided using balanced active devices with common emitters in a single package.

6.10.1 Basic Push-Pull Configurations

The push-pull configuration is widely used in many power amplifiers designed for different frequency ranges and with different output power levels. For the first time, a pair of vacuum tubes connected in a push-pull mode was described by Colpitts yet in the mid-1910th [56]. Such a connection balances out the even harmonics if the tube characteristics are identical, and is therefore a possible means of reducing distortions [57]. However, to achieve this linearity improvement, the push-pull circuits require transformers at their inputs and outputs. In a push-pull power amplifier, it is easy to eliminate or significantly reduce the feedback current, if the neutralizing identical coupling networks (capacitors) are connected crosswise, between the output of one vacuum tube and the input of the other vacuum tube [58].

The basic concept of a push-pull operation can be analyzed by using the corresponding circuit schematic shown in Fig. 6.33 [40]. It is most convenient to consider an ideal Class-B operation, which means that each transistor conducts exactly half a cycle (or 180°) with zero quiescent current. Let us also assume that the number of turns of both primary and secondary windings of the output transformer T_2 is equal ($n_1 = n_2$), and that the collector current of each transistor can be represented in the following half-sinusoidal form:

for the first transistor

$$i_{c1} = \begin{cases} +I_c \sin\omega t & 0 \le \omega t < \pi \\ 0 & \pi \le \omega t < 2\pi \end{cases} \tag{6.124}$$

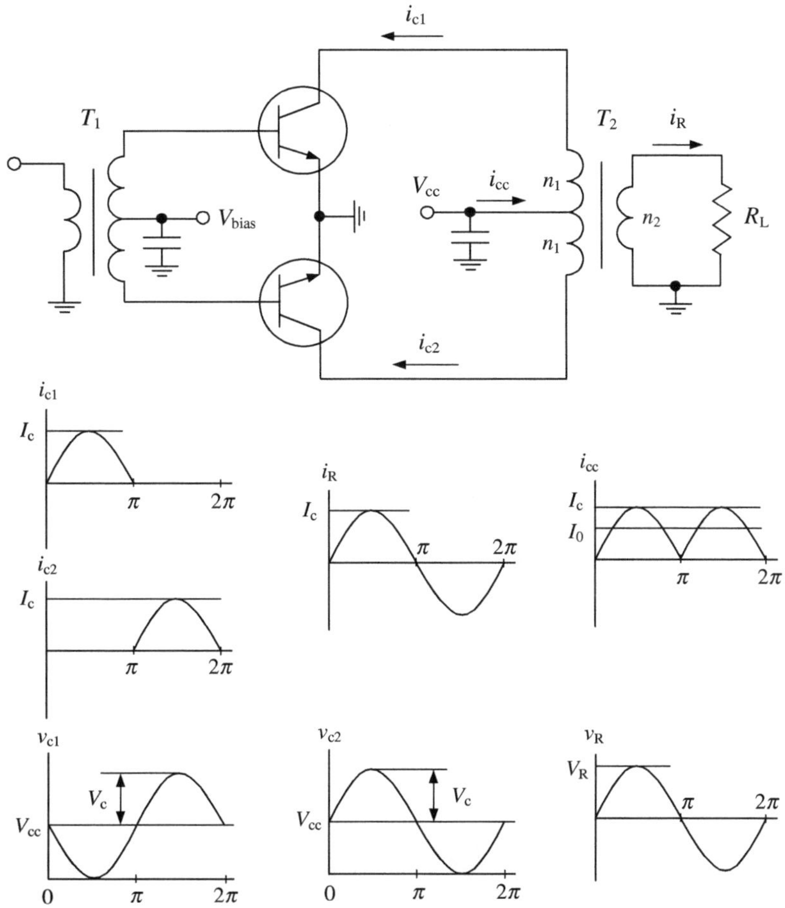

FIGURE 6.33 Basic concept of push-pull operation.

for the second transistor

$$i_{c2} = \begin{cases} 0 & 0 \le \omega t < \pi \\ -I_c \sin\omega t & \pi \le \omega t < 2\pi \end{cases} \tag{6.125}$$

where I_c is the output current amplitude.

Being transformed through the output transformer T_2 with the appropriate out-of-phase conditions, the total current flowing through the load R_L is obtained as

$$i_R(\omega t) = i_{c1}(\omega t) - i_{c2}(\omega t) = I_c \sin\omega t \tag{6.126}$$

The current flowing into the center tap of the primary windings of the output transformer T_2 is the sum of the collector currents, resulting in

$$i_{cc}(\omega t) = i_{c1}(\omega t) + i_{c2}(\omega t) = I_c\,|\sin\omega t| \tag{6.127}$$

Ideally, even-order harmonics being in phase are canceled out and should not appear at the load. In practice, a level of the second-harmonic component of 30 to 40 dB below the fundamental is allowable. However, it is necessary to connect a bypass capacitor to the center tap of the primary winding to exclude power losses due to even-order harmonics. The current $i_R(\omega t)$ produces the load voltage $v_R(\omega t)$ onto the load R_L as

$$v_R(\omega t) = I_c R_L \sin(\omega t) = V_R \sin(\omega t) \tag{6.128}$$

where V_R is the load voltage amplitude.

The total dc collector current is defined as the average value of $i_{cc}(\omega t)$, which yields

$$I_0 = \frac{1}{2\pi}\int_0^{2\pi} i_{cc}(\omega t)\, d\omega t = \frac{2}{\pi} I_c \tag{6.129}$$

For the ideal case of zero saturation voltage of both transistors when $V_c = V_{cc}$ and taking into account that $V_R = V_c$ for equal turns of windings when $n_1 = n_2$, the total dc power P_0 and fundamental-frequency output power P_{out} are obtained by

$$P_0 = \frac{2}{\pi} I_c V_{cc} \tag{6.130}$$

$$P_{out} = \frac{I_c V_{cc}}{2} \tag{6.131}$$

Consequently, the maximum theoretical collector efficiency that can be achieved in a push-pull Class-B operation mode is equal to

$$\eta = \frac{P_{out}}{P_0} = \frac{\pi}{4} \cong 78.5\% \tag{6.132}$$

In a balanced circuit, identical sides carry 180° out-of-phase signals of equal amplitude. If perfect balance is maintained on both sides of the circuit, the difference between signal amplitudes becomes equal to zero in each midpoint of the circuit, as shown in Fig. 6.34. This effect is called the *virtual grounding*, and this midpoint line is referred to as the *virtual ground*. The virtual ground, being actually inside the device package, reduces a common-mode inductance and results in better stability and usually higher power gain.

When a balanced transistor is used, new possibilities for both internal and external impedance matching procedure emerge. For instance, for a push-pull operation mode of two single-ended transistors, it is necessary to provide reliable grounding for input and output matching circuits for each device, as shown in Fig. 6.35(*a*). Using the balanced transistors simplifies significantly the matching circuit topologies with the series inductors and parallel capacitors connected between amplifying paths, as shown in Fig. 6.35(*b*), and dc-blocking capacitors are not needed.

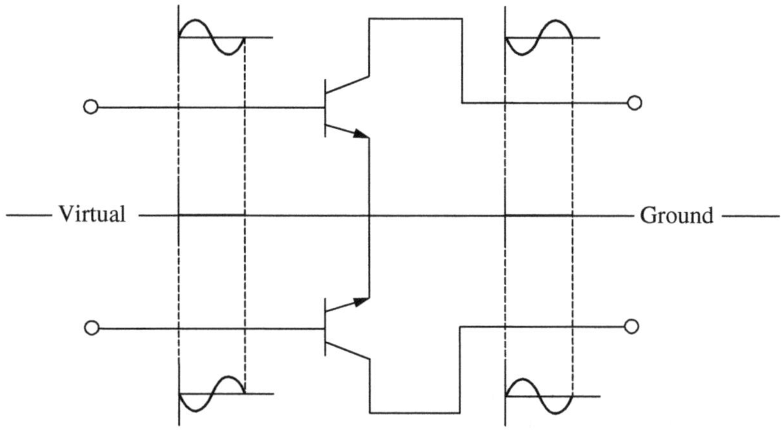

Figure 6.34 Basic concept of balanced transistor.

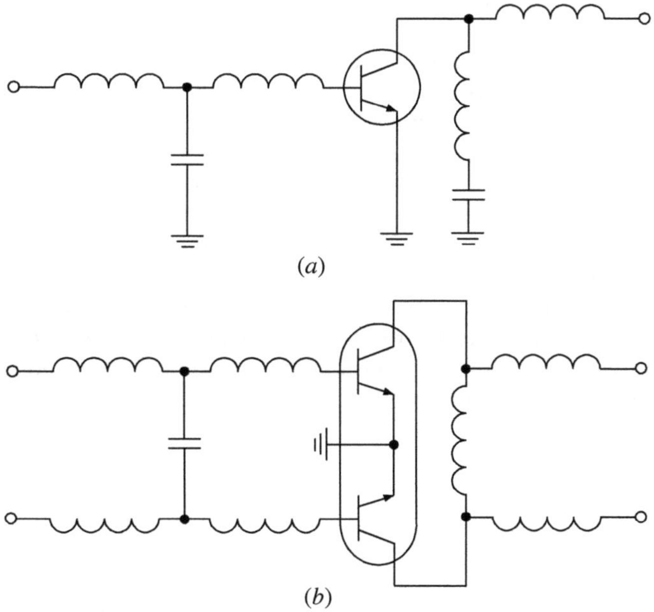

Figure 6.35 Matching techniques for single-ended and balanced transistors.

For a push-pull operation of the power amplifier with a balanced transistor, it is also necessary to provide the unbalanced-to-balanced transformation referenced to the ground both at the input and at the output of the power amplifier. The most suitable approach to solve this problem in the best possible manner at high frequencies and microwaves is to use the transmission-line transformers, as shown in Fig. 6.36. If the characteristic impedance Z_0 of the coaxial transmission

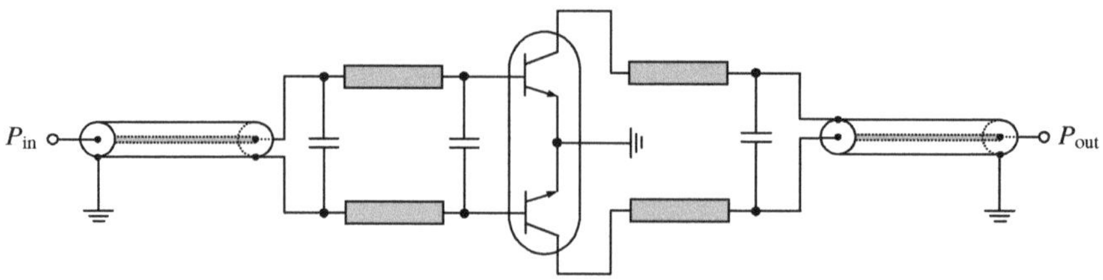

FIGURE 6.36 Push-pull power amplifier with balanced-to-unbalanced transformers.

line is equal to the input impedance at the unbalanced end of the transformer, the total impedance from both devices seen at the balanced end of the transformer will be equal to the input impedance. Hence, such a transmission-line transformer can be used as a 1:1 balanced-to-unbalanced transformer (balun). If $Z_0 = 50\ \Omega$, for the standard input impedance of 50 Ω, the impedance seen at each balanced part is equal to 25 Ω, which then is necessary to match with the appropriate input impedance of each part of a balanced transistor. The input and output matching circuits can easily be realized by using the series microstrip lines with parallel capacitors.

The miniaturized compact input unbalanced-to-balanced transformer shown in Fig. 6.37 covers the frequency bandwidth up to an octave with well-defined rejection-mode impedances [59]. To avoid the parasitic capacitance between the outer conductor and the ground, the coaxial semirigid transformer T_1 is mounted atop microstrip shorted stub l_1 and soldered continuously along its length. The electrical length of this stub is usually chosen from the condition of $\theta \leq \pi/2$ on the high bandwidth frequency depending on the matching requirements. To maintain circuit symmetry on the balanced side of the transformer network, another semirigid coaxial section T_2 with an

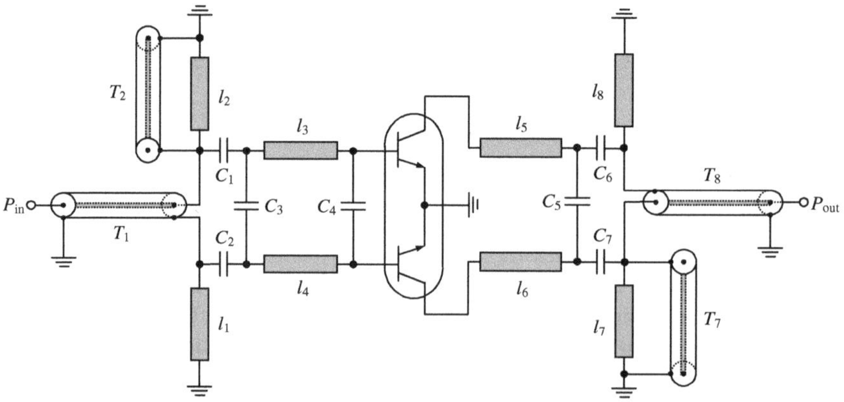

FIGURE 6.37 Push-pull power amplifier with compact unbalanced-to-balanced transformers.

unconnected center conductor is soldered continuously along microstrip shorted stub l_2. The lengths of T_2 and l_2 are equal to the lengths of T_1 and l_1, respectively. Because the input short-circuited microstrip stubs provide inductive impedances, the two series capacitors C_1 and C_2 of the same value are used for matching purposes, thereby forming the first high-pass matching section and providing dc blocking at the same time. The practical circuit realization of the output matching circuit and balanced-to-unbalanced transformer can be the same as for the input matching circuit.

6.10.2 Balanced Power Amplifiers

Balanced amplifier technique using the quadrature 3-dB couplers for power dividing and combining represents an alternative approach to the push-pull operation. Figure 6.38(*a*) shows the basic circuit schematic of a balanced amplifier, where two power amplifier units of the same performance are arranged between the input splitter and output combiner, each having a 90° phase difference between coupled and through ports. The fourth port of each quadrature coupler must be terminated with a ballast resistor R_{bal}, which is equal to 50 Ω for a 50-Ω system impedance. The input signal is split into two equal-amplitude components by the first 90° hybrid coupler with 0 and 90° paths, then amplified, and finally recombined by the second 90° hybrid coupler.

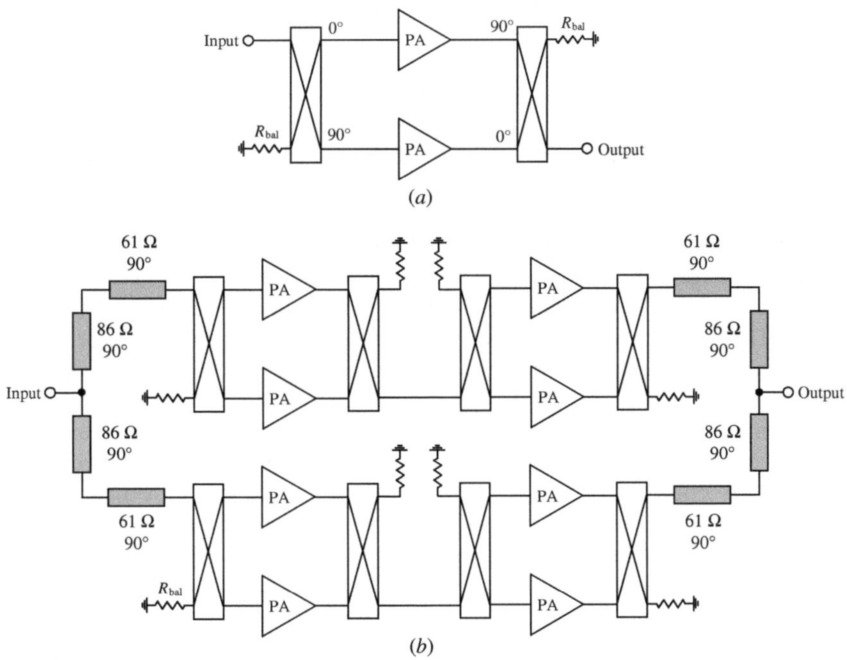

FIGURE 6.38 Schematics of balanced power amplifiers with quadrature hybrid couplers.

Because of proper phase shifting, both signals in the load of the isolated port of the combiner are cancelled out, and the load connected at the output port of the combiner sees the sum of these two signals. The theory of balanced amplifiers has been given by Kurokawa when the operating frequency bandwidth over 1.2 octaves can be obtained with single-section distributed quarterwave 3-dB directional couplers [60].

For a wide frequency range, the main advantages of the balanced design include the improved input and output impedance matching, gain flatness, intermodulation distortion, and potential design simultaneously for minimum noise figure and good input match. As an example, a four-stage balanced bipolar amplifier achieved a power gain of 20 ± 0.5 dB and an input *VSWR* less than 1.2 across the octave frequency bandwidth from 0.8 to 1.6 GHz [61]. By extending the wide operating frequency range to higher frequencies, an output power of around 23 dBm with gain variations close to 1 dB over 4.5 to 6.5 GHz and 8 to 12 GHz was achieved for the balanced microstrip GaAs MESFET power amplifiers [62, 63]. The balanced configuration has the distinct advantage that a low *VSWR* can be maintained, even though the individual amplifier stages may be mismatched. Besides, it provides a desirable feature that the signal level, at which saturation is achieved, will be increased by 3 dB, thus improving the linearity performance. Furthermore, the balanced circuit technique can also drastically reduce the third-order intermodulation products when the interfering signal enters from the output of a balanced amplifier for collocated transmitters [64].

If the individual amplifiers with equal performance in the balanced pair are not perfectly matched at certain frequencies, a signal in the 0° path of the coupler will be reflected from the corresponding amplifier and a signal in the 90° path of the coupler will be similarly reflected from the other amplifier. The reflected signals will again be phased with 90 and 0°, respectively, and the total reflected power as a sum of the in-phase reflected signals flows into the isolated port and dissipates on the ballast resistor R_{bal}. As a result, an input *VSWR* of the quadrature coupler does not depend on the equal load mismatch level. This gives a constant well-defined load to the driver stage, improving amplifier stability and driver power flatness across the operating frequency range. Generally, the stability factor of a balanced stage can be an order of magnitude higher than its single-ended equivalent, depending on the *VSWR* and isolation of the quadrature couplers. If one of the amplifiers fails or turned off, the balanced configuration provides a gain reduction of –6 dB only. Besides, the balanced structure provides ideally the cancellation in the load of the third-order products such as $2f_1 + f_2$, $2f_2 + f_1$, $3f_1$, $3f_2$, ..., and attenuation by 3 dB of the second-order products such as $f_1 \pm f_2$, $2f_1$, $2f_2$, ... In a microstrip implementation for octave-band power amplifiers, one of the most popular couplers for power dividing and combining is a 3-dB Lange hybrid coupler.

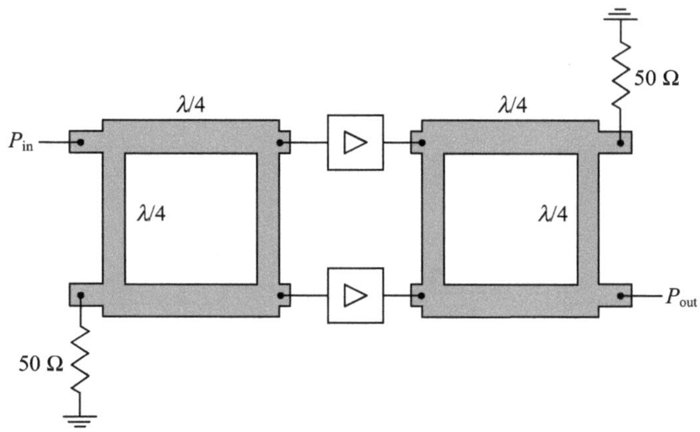

FIGURE 6.39 Power amplifier topology with branch-line hybrids.

Figure 6.38(*b*) shows the circuit schematic of a two-way balanced module consisting of two pairs of cascaded balanced amplifier stages, where the respective output powers are combined using simple two-element power combiners, which are composed of two quarterwave transmission lines with different characteristic impedances [65]. Based on this architecture and GaAs MESFET devices with the gate periphery of $1 \times 1000\ \mu m^2$, an output power of 1 W across 7.25-12 GHz was achieved.

To combine the output power from two or more transistors at microwaves, as shown in Fig. 6.39 for two power amplifiers with standard 50-Ω input and output impedances, the branch-line 90° microstrip-line hybrid combiners are the most popular, In this case, the characteristic impedances of the transverse branches should be of 50 Ω, whereas the longitudinal branches must have the characteristic impedance of $50/\sqrt{2} = 35.4\,\Omega$. In practice, because of the quarter-wavelength transmission line requirements, the bandwidth of such a balanced amplifier based on two quadrature branch-line hybrids is limited to 10 to 20%.

To simplify the matching requirements for balanced high-power amplifiers with small values of the device impedances, the quadrature branch-line 90° hybrids with impedance transforming property can be used [66]. Figure 6.40 shows an example of such a balanced GaAs MESFET power amplifier with an output power of 18 W and a power gain of 8.5 dB at 1.7 GHz. The characteristic impedances of all quarter-wavelength branch lines can be properly calculated by

$$Z_1 = Z_{in} \qquad Z_2 = \sqrt{\frac{Z_{in} Z_{out}}{2}} \qquad Z_3 = Z_{out} \tag{6.133}$$

Consequently, for a 50-to-20 Ω coupler, the characteristic impedances of the corresponding branch line are $Z_1 = 50\ \Omega$, $Z_2 = 22.4\ \Omega$, and

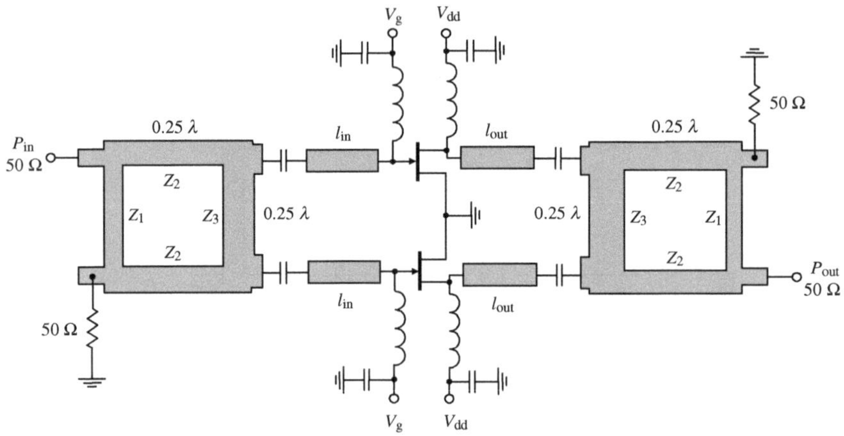

FIGURE 6.40 Balanced high-power GaAs MESFET amplifier with branch-line impedance-transforming hybrids.

$Z_3 = 20\ \Omega$. In this case, the input and output matching circuits for each transistor can be simplified to the series microstrip line with characteristic impedance of 20 Ω only, where $l_{in} = 7.47$ mm and $l_{out} = 6.67$ mm for alumina substrate in this particular case, with a series capacitor for dc blocking.

6.11 Bias Circuits

The simplest way to provide a proper dc biasing condition for a power MOSFET device in Class-A or Class-AB operation is to use the potentiometer-type voltage divider for the gate bias and choke inductor in the drain circuit, as shown in Fig. 6.41(*a*). However, in this case, any variations of the ambient temperature or bias voltage will lead to variations of quiescent current and, as a result, to appropriate variations of the output power, linearity, drain efficiency, and gain of the power amplifier. The threshold voltage V_{th} of the MOSFET transistor varies with temperature T linearly with the approximate velocity of $\Delta V_{th}/\Delta T \cong -2$ mV/°C. However, simple adding of a diode (or diode-connected MOSFET) in series to the variable resistor allows the quiescent current variation to be reduced substantially over temperature. A bias circuit corresponding to this stabilizing condition is shown in Fig. 6.41(*b*). For a high value of V_{th}, several diodes can be connected in series. The reason to use such a simple bias circuit for power MOSFET biasing is that its dc gate current is being equal to the gate leakage current only.

In contrast to MOSFET devices, where it can be possible to choose the optimum operating point with a practically zero temperature coefficient or to be limited to just an additional diode only, the bipolar

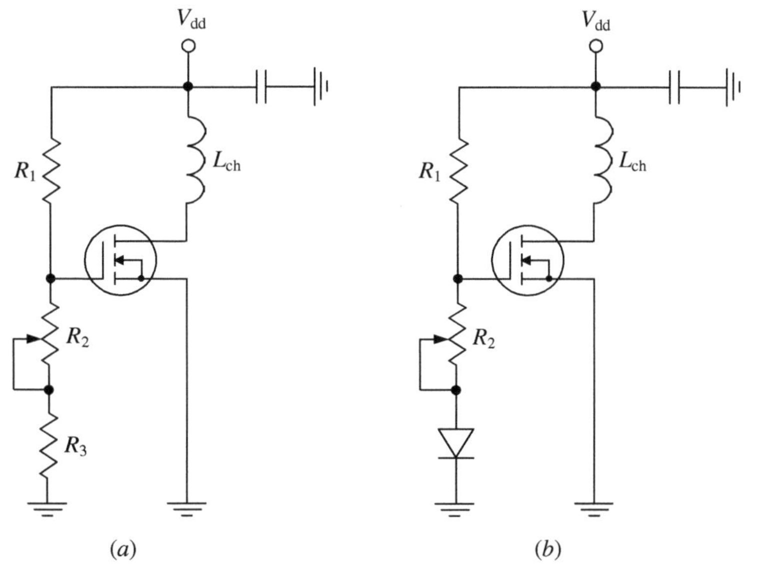

FIGURE 6.41 MOSFETs with simple bias circuits.

transistors require the more complicated approach of dc biasing depending on a class of operation. For example, in a Class-AB operation, the bias circuit has to deliver a dc voltage, which is slightly adjustable approximately within limits of 0.7 to 0.8 V with a wide range of the current values to stabilize the base current of the RF bipolar transistor. Besides, it is necessary to provide an operation mode of the power amplifier with temperature compensation (collector current stabilization over temperature) and minimum possible reference current (dc current from the reference dc voltage supply). One of the simplest versions of such a bias circuit with silicon diode temperature compensation is shown in Fig. 6.42(*a*). In this bias circuit, each silicon diode can be replaced by the *n-p-n* diode-connected transistor, the collector and the base of which are directly connected between each other.

A better temperature-compensating result can be achieved using the same transistors for RF and dc paths, only with reduced area sizes for bias circuit devices. Such an approach is usually used in monolithic integrated circuit design when transistor cells with different area sizes are used for both RF power device and bias circuit transistors. Figure 6.42(*b*) shows the temperature and supply-independent bias circuit, which is composed of the current driving transistor Q_4 and the compensation circuit, including the diode-connected transistors Q_1 and Q_2, compensating transistor Q_3, and resistors R_1 and R_2 [67]. By providing the same values for R_1 and R_2, the InGaP/GaAs MMIC power amplifier for WCDMA applications provides the quiescent

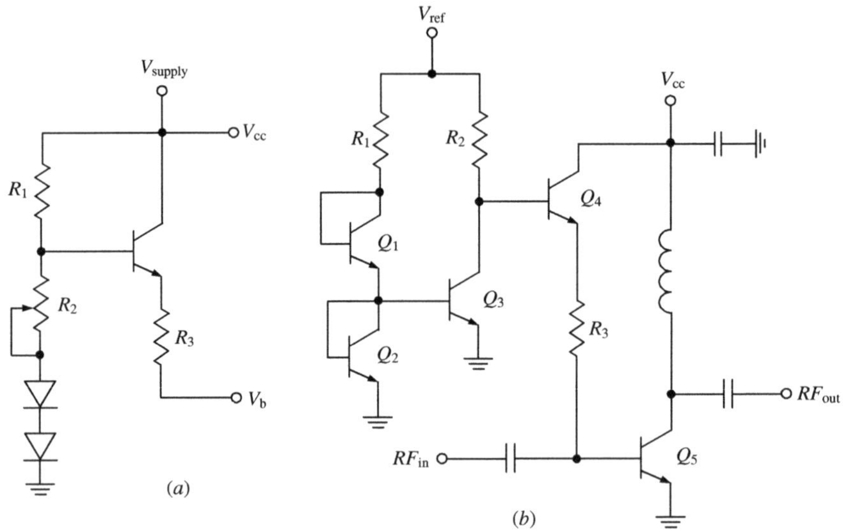

FIGURE 6.42 Typical bipolar Class-AB bias circuits.

current variations of only 6% for the temperature range of –30 to 90°C and 8.5% for the supply voltage range of 2.9 to 3.1 V, with the variations of the power gain of less than ±0.8 dB at the output power of 28 dBm.

Figure 6.43 shows a more complicated bias circuit commonly used for biasing the high-power bipolar transistors to provide their temperature-stable and reliable operation mode [68]. The temperature stabilization is provided with the parallel connection of the base-emitter

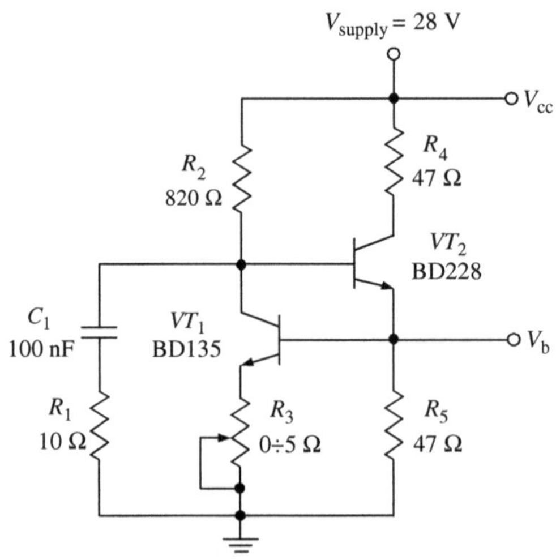

FIGURE 6.43 Typical bipolar Class-AB bias circuit for high-power amplifier.

diode junction of the transistor VT_1, whereas high value of the bias drive current for the RF power transistor is delivered by the transistor VT_2. If the dc collector current of an RF power transistor is 5 A and a value of its forward current gain β_F is approximately equal to 10, the maximum base current of the RF power transistor can be 0.5 A. In this bias circuit, the resistor R_5 is used to reduce the base current variations. At $V_b = 0.7$ V, the value of R_5 should be equal to 0.7 V/15 mA = 47 Ω for a current of 15 mA. Suppose that a value of the collector current of VT_1 is 30 mA. As a result, if the base-emitter junction voltage of VT_2 is equal to 0.8 V with a voltage across the resistor R_2 of 28 V − 1.5 V = 26.5 V, its value is 26.5 V/30 mA ≅ 820 Ω. The variable resistor R_3 serves to adjust the output voltage in limits of 0.1 V. To limit the maximum collector current of VT_2 by a value of 0.5 A, it is best to use the resistor R_4, whose maximum value is 26.5 V/0.5 A = 53 Ω, assuming the saturation voltage of 0.8 V for VT_2. It is sufficient to use its value of 47 Ω with a power dissipation of $(0.5\text{ A})^2 \times 47\ \Omega = 11.75$ W. Such a bias circuit can produce the parasitic oscillations near 1 MHz with highly capacitive loads. Therefore, to prevent these oscillations, it is necessary to connect the *RC* circuit between the collector of VT_1 and ground.

In most wireless communication systems, it is preferred that the power amplifier operates with high efficiency, maintaining an acceptable linearity over the desired supply voltage and output power ranges. However, there is a tradeoff between efficiency and linearity, with improvement in one coming at the expense of another. This means that it is necessary to provide an optimum stable fixed or adaptive bias point over wide temperature range and process variations. As a current-controlled device, the bipolar transistor in RF operation requires the dc base driving current, whose value depends on the output power and device parameters. Because technologically the bipolar device represents a parallel connection of the basic cells, it is important to use the ballast series resistors to avoid current imbalance and possible device collapse at higher current-density levels. Another important aspect is to keep the dc base-emitter bias point constant (or properly variable) over any RF input power variations in order to prevent the linearity degradation at maximum output power for power amplifiers with a variable envelope signal (such as EDGE, CDMA2000, WCDMA, or LTE).

The typical temperature-compensation current mirror bias circuit with one reference transistor Q_1 and one driving transistor Q_2 is shown in Fig. 6.44(*a*). This circuit keeps the quiescent current for the RF device Q_0 sufficiently constant over temperature variations, and the current flowing through resistor R_2 is sufficiently small. It is very important to provide the proper ratio between ballast resistors R_1 and R_0, equal to the reverse ratio of the device areas Q_0/Q_1. This can minimize the overall performance variation with temperature, as well as stabilize the dc bias point. The latter case is very important for the variable-envelope signals, as the dc bias voltage V_{be0} establishes the

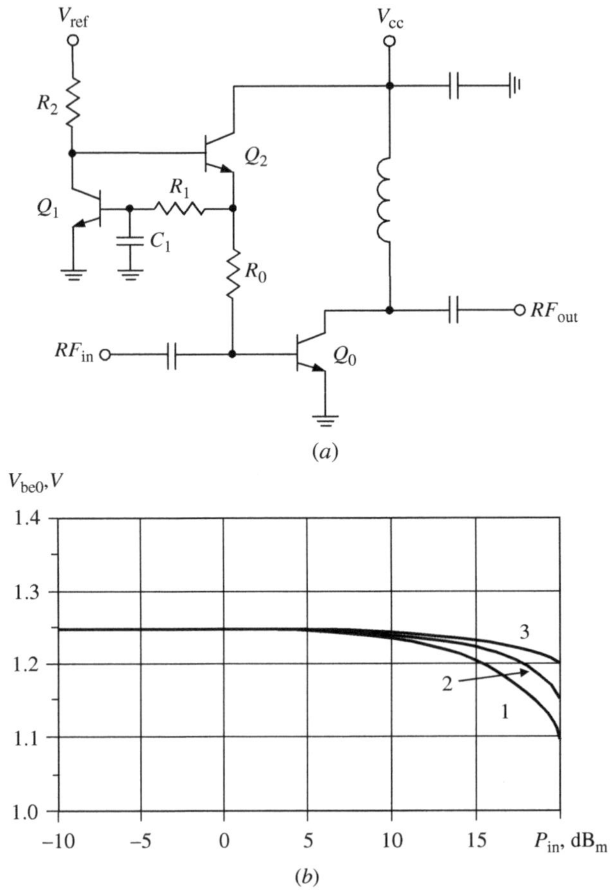

FIGURE 6.44 Bipolar power amplifier stage with current mirror bias circuit and its performance.

conduction angle and operation class for the RF device. If the dc base-emitter bias voltage reduces with the increase of RF input power, the Class-AB operation mode required for linear operation changes to a Class-C mode corresponding to nonlinear operation with zero quiescent current.

Figure 6.44(*b*) shows the dependence of the dc base-emitter bias voltage V_{be0} versus input power P_{in} for the second stage of a WCDMA InGaP/GaAs HBT power amplifier for three different cases, including two of them with ballast resistor $R_1 = 0$ (curve 1) and optimum ballast resistor R_1 (curve 2). From Fig. 6.44(*b*) it follows that including the ballast resistor with optimum value results in a more constant base-emitter dc bias voltage over a wider range of input powers, thus improving the linearity performance of the power amplifier at high

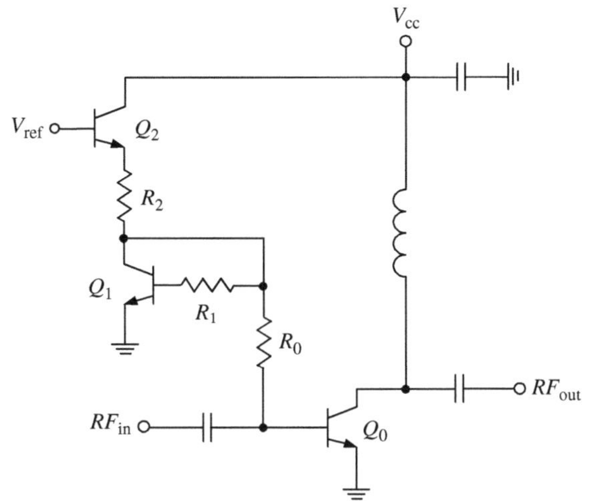

FIGURE 6.45 Bipolar power amplifier stage with emitter follower bias circuit.

power levels. In addition, it is best to use a shunt capacitor C_1 connected to the base terminal of the device Q_1 to form a low-pass *RC* filter, which provides better isolation of the bias circuit from RF signal, with a yet more constant base-emitter dc bias voltage (curve 3).

Figure 6.45 shows the emitter-follower bias circuit that provides temperature compensation and minimizing reference current requirements [69]. The emitter follower bias circuit requires only several tens of microamperes of reference current, whereas the current-mirror bias circuit requires a few milliamperes of reference current. Both the current-mirror and emitter-follower bias circuits have similar current-voltage behavior, but, for the same circuit parameters (R_0, R_1, and R_2) and device areas for Q_0, Q_1, and Q_2, the emitter follower bias circuit is the less sensitive to the reference voltage variations compared with a current mirror bias circuit. Variations of the collector supply voltage V_{cc} in limits of 3.0 to 5.0 V have no effect on the quiescent current set by the reference voltage V_{ref}.

To provide an efficient linear power amplifier operation over power dynamic range, it is necessary to minimize the quiescent current at backoff output powers, because maximum of the power density function for CDMA2000 or WCDMA standards with nonconstant envelope typically occurs at the output powers of about 25 to 30 dB below the saturation output power. As a current-controlled device, the bipolar transistor at RF operation requires the dc base driving current, whose value depends on the output power and device parameters. Unlike the current mirror and emitter follower bias circuits, the adaptive bias circuit can control both the temperature performance and dc power consumption as the output power varies by

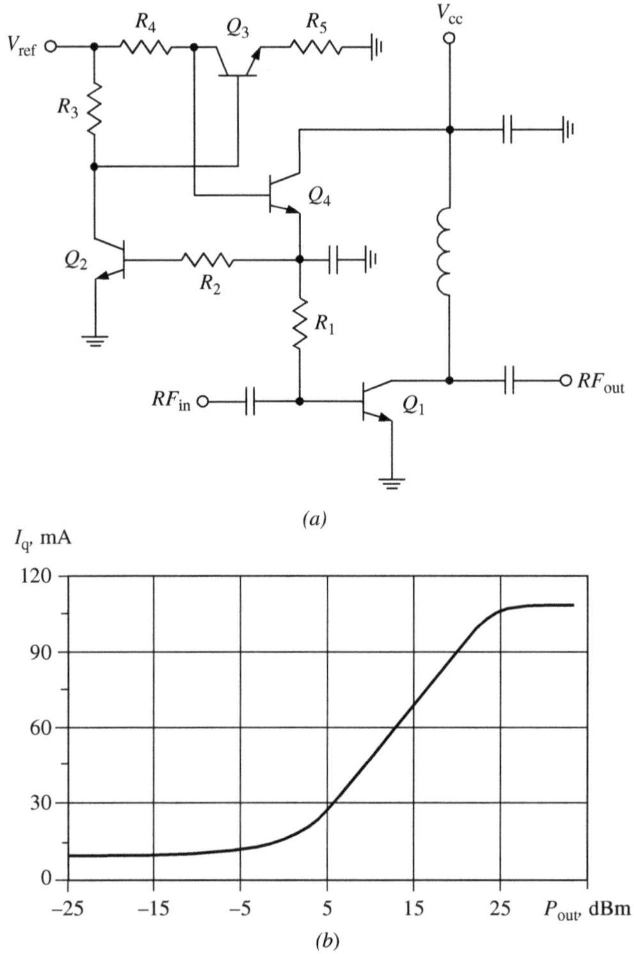

FIGURE 6.46 Adaptive bias circuit and its performance.

greatly improving a *PAE* when the output power is low and maintaining a high linearity of the power amplifier when its output power is high [70].

Figure 6.46(*a*) shows the schematic of the adaptive bias circuit, where the value of the ballast resistance R_1 is properly chosen to minimize the sensitivity to the base bias current variations and its value is scaled with ballasting resistor R_2 to the ratio of reverse device areas Q_2/Q_1 [71]. The bypass capacitor is needed to isolate dc bias circuit from the RF path. Because the RF transistor Q_1 is biased to Class AB with a small quiescent current and its collector current is a function of the input power, the collector current of the current mirror device Q_2 increases with input power. This increased current decreases the base voltage of the device Q_3, thus decreasing its collector current. Then, this decreased collector current increases the base voltage of the device Q_4 due to a smaller voltage drop across the resistor R_4, forcing its emitter current and the collector current of the RF device Q_1 to increase. Thus, the quiescent current of the adaptive bias circuit is an increasing function of the input power. This provides high quiescent current at high output powers when high linearity with tradeoff efficiency is achieved and low quiescent current at low output powers when high linearity and increased efficiency are obtained. As an example shown in Fig. 6.46(*b*), the quiescent current I_q varies from 110 mA at a maximum P_{out} = 33 dBm to 12 mA for low output powers. The minimum value of a quiescent current is defined by the values of resistors R_4 and R_5.

6.12 Practical Aspect of RF and Microwave Power Amplifiers

The typical hybrid topology of a microwave bipolar or GaAs MESFET power amplifier with the input and output matching circuit substrates and packaged transistor, usually designed for a Class-A operation mode, is shown in Fig. 6.47. Here, the input and output matching circuits are implemented as low-pass transmission-line *L*-transformers, where electrical lengths of the microstrip lines (series short-length lines and open-circuit stubs) depend on the transistor input and output impedances. The microstrip open-circuit stubs l_1 and l_4 represent the capacitive matching impedances, the series microstrip lines l_2 and l_3 provide the required inductive matching impedances, and C_1 and C_2 are the blocking capacitors. The isolation of the bias circuits from signal path at microwave frequency is usually performed using the quarterwave microstrip open-circuit and short-circuit stubs of different characteristic impedances.

The circuit schematic of a microwave linear GaAs MESFET power amplifier designed for the frequency bandwidth of 2.5 to 2.7 GHz in

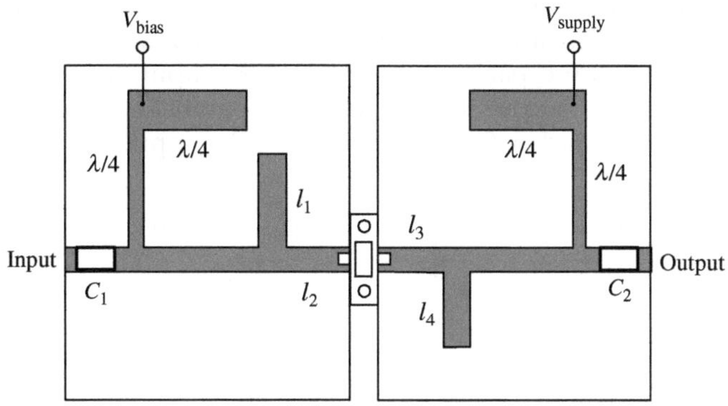

FIGURE 6.47 Typical topology of microwave power amplifier.

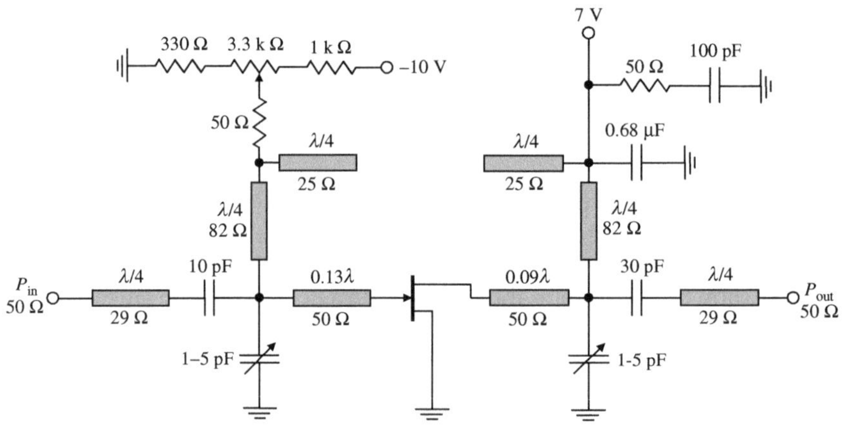

FIGURE 6.48 Circuit schematic of 2.5 to 2.7 GHz GaAs MESFET power amplifier.

a Class-AB operation mode is shown in Fig. 6.48. To match the device input and output inductive impedances $Z_{in} = (1.2 + j20)\ \Omega$ and $Z_{out} = (4.2 + j25)\ \Omega$ measured at a 5-W output power level with the source and load 50-Ω impedances, respectively, a combination of the microstrip quarterwave transformers and low-pass lumped *L*-transformers was used. Instead of the open-circuit microstrip stubs, the variable capacitances in limits of 1 to 5 pF were used for accurate matching tuning. A series *RC* circuit connected in parallel to the drain supply improves stability of operation by preventing parasitic oscillations at higher frequencies. To avoid the low-frequency oscillations, it is necessary to use bypass capacitor with a sufficiently large capacitance in parallel to the power supply.

Figure 6.49 shows the equivalent matching circuit topology of a microwave GaAs MESFET power amplifier designed for *S*-band application. A 38.4-mm gate width device was designed to achieve an output power of 20 W in a frequency bandwidth of 3.0 to 3.5 GHz with a 7-dB power gain and a *PAE* of 34% [72]. At the center bandwidth frequency, resistance of approximately 0.2 Ω connected in series with a 50-pF capacitance represents the input device impedance, whereas the output impedance is the parallel connection of a 2.5-Ω resistance and a 9-pF capacitance. The input quarterwave

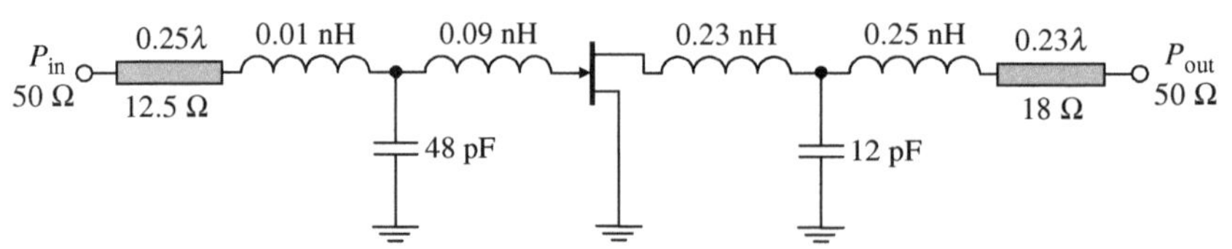

FIGURE 6.49 Circuit schematic of 3.0 to 3.5 GHz GaAs MESFET power amplifier.

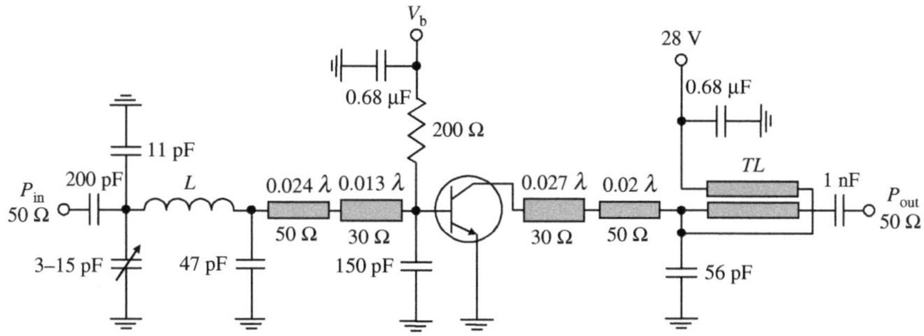

Figure 6.50 Bipolar VHF power amplifier for TV applications.

microstrip line with a characteristic impedance of 12.5 Ω provides an impedance transformation from 50 Ω to a sufficiently small impedance of (12.5 × 12.5/50) Ω = 3.125 Ω, and approximately the same impedance transformation is provided by the output impedance matching. For the corresponding impedance matching with input and output device impedances, the low-pass *T*-type lumped transformers are used with shunt chip capacitors and series lumped inductors realized by gold bondwires.

Figure 6.50 shows an example of the circuit schematic of a VHF linear bipolar power amplifier using both lumped elements and transmission lines for impedance matching. This power amplifier was developed to transmit the composite TV video and audio signal in a Class-A operation mode in a frequency range of 174 to 230 MHz with at least 10-dB power gain and a 30-W peak output power. The input matching circuit consists of both lumped elements composing a low-pass π-transformer and microstrip lines implemented on FR-4 substrate with a dielectric permittivity of $\varepsilon_r = 4.7$. A three-turn air-core inductor realizes the lumped inductor *L*, which is a part of the first lumped low-pass π-type matching section with shunt variable capacitor. The output matching circuit includes a 1:2 coaxial transformer *TL* with the characteristic impedance of 25 Ω and wavelength of $\lambda/8$ to provide a final 1:4 impedance matching, from 12.5 to 50 Ω.

In monolithic microwave applications, it is possible to increase the output power of the amplifier by connecting several transistors in parallel. Figure 6.51 shows the circuit schematic of a monolithic 5.5-GHz power amplifier, where two 4-mm-gate-width GaAs MESFET devices are used to provide maximum output power of 34 dBm with a peak *PAE* of 40% in a Class-B operation mode [73]. The input matching circuits are *T*-transformers with series microstrip lines and shunt capacitor, whereas two microstrip lines (parallel and series) provide output matching for each transistor to compensate for the device output reactance and to match real part of its output impedance with a load resistance, respectively.

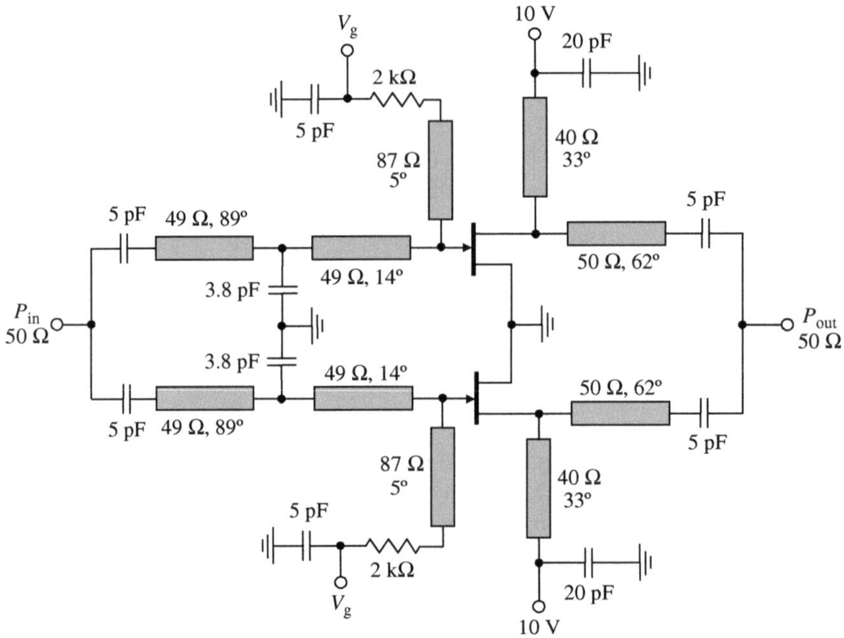

FIGURE 6.51 Circuit schematic of monolithic microwave GaAs MESFET power amplifier.

High-power RF amplifiers with operating frequencies of up to several GHz are very often designed using a push-pull operation mode. For example, a high-power UHF amplifier shown in Fig. 6.52 using PTF10120, which is an enhancement mode Ericsson's balanced

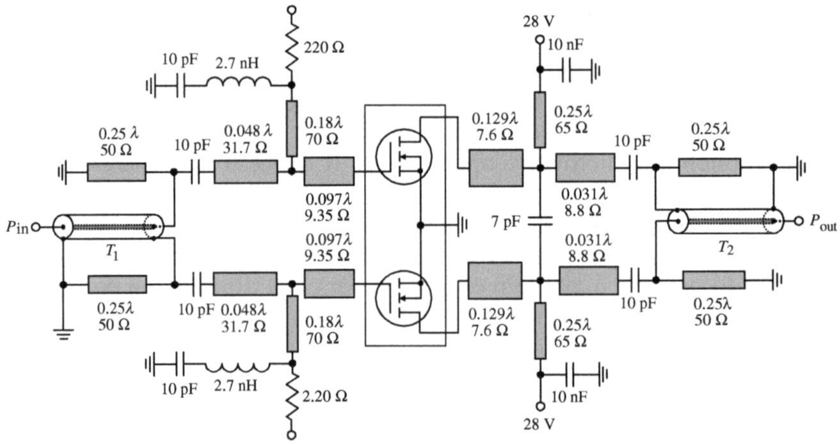

FIGURE 6.52 Circuit schematic of high-power UHF LDMOSFET push-pull amplifier.

LDMOSFET device, provides a 120-W output power at 1-dB small-signal gain compression in a frequency range of 1.93 to 1.99 GHz [74]. In this case, a power gain of 11 dB with a drain efficiency of about 40% for a quiescent current of 600 mA in a Class-AB operation mode was achieved. For a push-pull operation mode, the 1:1 input coaxial transformer T_1 and 1:1 output coaxial transformer T_2 are used to provide the corresponding unbalanced-to-balanced transformations. These baluns transform the single-ended input into two signals of equal amplitudes and 180° out of phase, and perform the corresponding opposite function at the output. The input and output matching networks combine distributed and lumped elements in a low-pass configuration.

A high-power RF push-pull amplifier can be designed by using a symmetrical mirror-image connected pair of MOSFETs with the same electrical characteristics. The circuit schematic of such an amplifier with a 300-W output power and 75% efficiency at the operating frequency of 81.36 MHz is shown in Fig. 6.53 [75]. The input ferrite transformer provides the 2:1 unbalanced-to-balanced transformation. Consequently, the gate impedance of each transistor, which was determined to be (0.3 + j2.75) Ω, should be matched to the 6.25-Ω transformer secondary winding impedance. The input matching circuit for each device represents a low-pass π-type section with shunt capacitors and series microstrip line; both TL_1 and TL_2 are 0.2 in wide (or 35 Ω) and 1.80 in long. The gates of both MOSFETs are connected to ground through the resistors on each side of the transformer secondary winding, although a single resistor at the secondary center

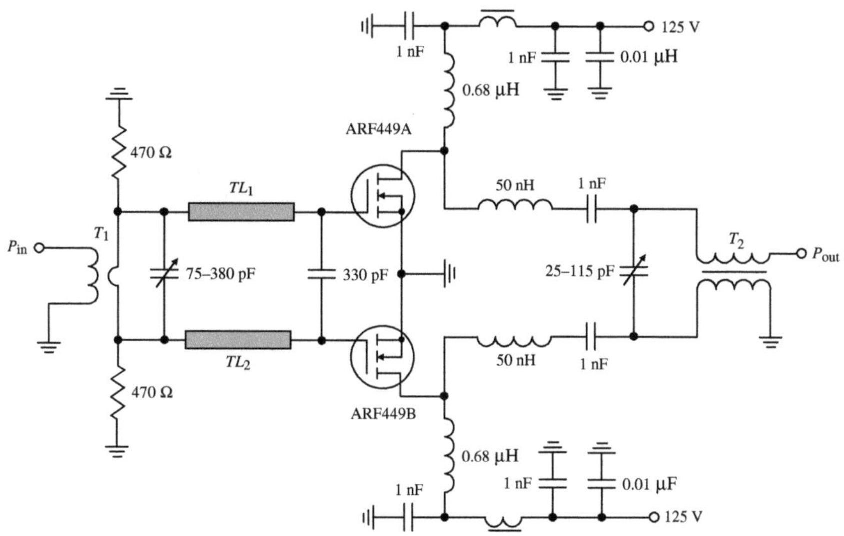

FIGURE 6.53 Circuit schematic of 300-W VHF MOSFET power amplifier.

tap would work as well. To match the output device impedance of (9.14 – *j*12.6) Ω with output coaxial transformer impedance of 25 Ω, it is sufficient to use a simple *L*-type matching section with a series inductor and a shunt tuning capacitor.

The MMIC power amplifiers are designed using foundry design kits for both active and passive components because they are located on the same substrate. In conjunction with circuit simulations, electromagnetic simulations can also be done for some specific sections of the input, interstage, and output matching networks. Figure 6.54(*a*) shows the circuit schematic of an MMIC driver amplifier implemented in a 2-μm GaInP/GaAs HBT technology [76]. Here, to provide

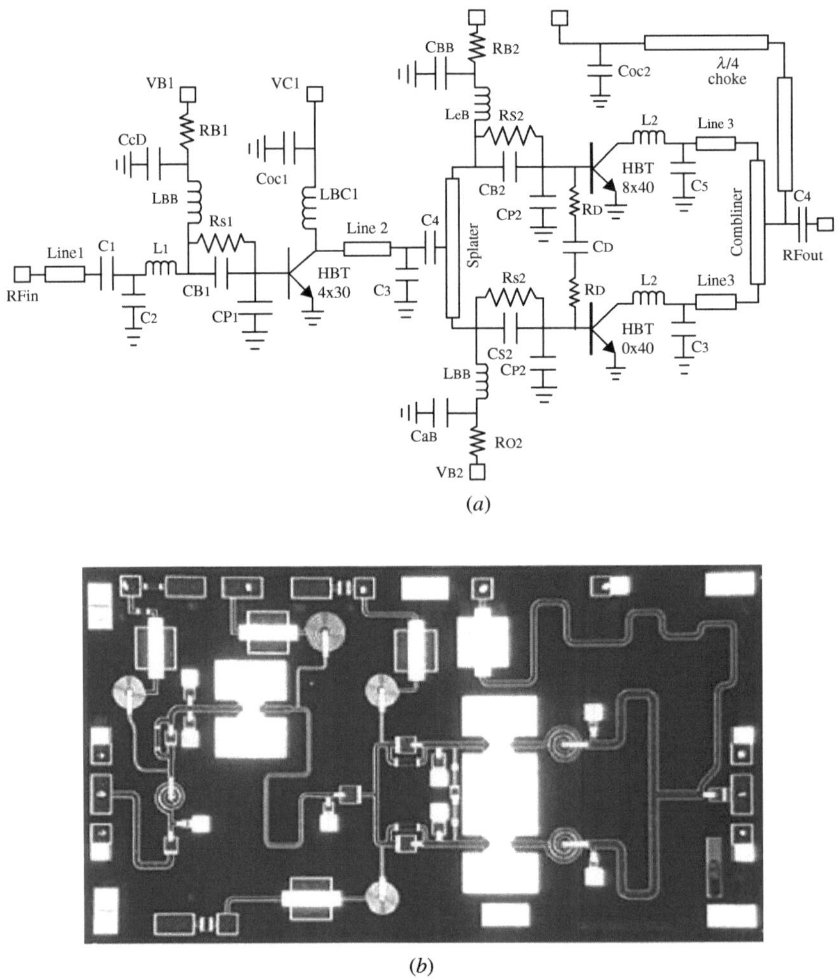

FIGURE 6.54 Schematic and photo of two-stage *X*-band MMIC driver amplifier. (*Courtesy of DEI-University of Bologna and CNR-IEIIT.*)

an amplifier unconditional stability from 300 MHz upward, parallel *RC* networks are used, each connected in series with the device base. Matching networks were implemented with a combination of the lumped *LC* sections and distributed microstrip structures. The final-stage two device cells with stabilization ($R_{S2}//C_{S2}$) and prematching (C_{P2}) networks at their inputs and matching circuit sections at their outputs are combined both by the corresponding microstrip *T*-junction, operating as a splitter at the input and as a combiner at the output, respectively. The collector bias network of the final stage is represented by a high-impedance quarterwave microstrip line, which provides the collector current symmetrically to both HBTs. In each base bias network, an integrated series resistor (R_{S1} or R_{S2}) is included, which is required to obtain (in conjunction with the integrated emitter ballast resistors) the stabilization of the device bias point in the specified temperature variation range. Moreover, the optimum choice of the base bias resistor is important also for the high-frequency gain compression characteristic of the device. All devices are biased in a Class AB with optimized quiescent currents for each stage. As a result, the driver amplifier with a die size of 1.8×3.4 mm^2 shown in Fig. 6.54(*b*) exhibits a power gain of about 20 dB, a *PAE* higher than 50%, and an output power of about 29 dBm at 1-dB compression point in a frequency bandwidth of 9.3 to 10.5 GHz. At higher gain compression, the driver amplifier can deliver more than 1-W output power with a *PAE* of about 57%.

References

1. W. L. Everitt, "Output Networks for Radio-Frequency Power Amplifiers," *Proc. IRE*, vol. 19, pp. 725–737, May 1931.
2. H. T. Friis, "Noise Figure of Radio Receivers," *Proc. IRE*, vol. 32, pp. 419–422, July 1944.
3. S. Roberts, "Conjugate-Image Impedances," *Proc. IRE*, vol. 34, pp. 198–204, Apr. 1946.
4. S. J. Haefner, "Amplifier-Gain Formulas and Measurements," *Proc. IRE*, vol. 34, pp. 500–505, Jul. 1946.
5. R. L. Pritchard, "High-Frequency Power Gain of Junction Transistors," *Proc. IRE*, vol. 43, pp. 1075–1085, Sep. 1955.
6. A. R. Stern, "Stability and Power Gain of Tuned Power Amplifiers," *Proc. IRE*, vol. 45, pp. 335–343, Mar. 1957.
7. L. S. Houselander, H. Y. Chow, and R. Spense, "Transistor Characterization by Effective Large-Signal Two-Port Parameters," *IEEE J. Solid-State Circuits*, vol. SC-5, pp. 77–79, Apr. 1970.
8. B. T. Vincent, "Large Signal Operation of Microwave Transistors," *IEEE Trans. Microwave Theory Tech.* vol. MTT-49, pp. 865–866, Nov. 1965.
9. B. J. Thompson, "Oscillations in Tuned Radio-Frequency Amplifiers," *Proc. IRE*, vol. 19, pp. 421–437, Mar. 1931.
10. G. W. Fyler, "Parasites and Instability in Radio Transmitters," *Proc. IRE*, vol. 23, pp. 985–1012, Sep. 1935.
11. F. B. Llewellyn, "Some Fundamental Properties of Transmission Systems," *Proc. IRE*, vol. 40, pp. 271–283, Mar. 1952.
12. D. F. Page and A. R. Boothroyd, "Instability in Two-Port Active Networks," *IRE Trans. Circuit Theory*, vol. CT-5, pp. 133–139, Jun. 1958.

13. J. M. Rollett, "Stability and Power Gain Invariants of Linear Two-Ports," *IRE Trans. Circuit Theory Appl.*, vol. CT-9, pp. 29–32, Jan. 1962.
14. J. G. Linvill and L. G. Schimpf, "The Design of Tetrode Transistor Amplifiers," *Bell Syst. Tech. J.*, vol. 35, pp. 813–840, Apr. 1956.
15. O. Muller and W. G. Figel, "Stability Problems in Transistor Power Amplifiers," *Proc. IEEE*, vol. 55, pp. 1458–1466, Aug. 1967.
16. O. Muller, "Internal Thermal Feedback in Fourpoles, Especially in Transistors," *Proc. IEEE*, vol. 52, pp. 924–930, Aug. 1964.
17. V. M. Bogachev and V. V. Nikiforov, *Transistor Power Amplifiers* (in Russian), Moskva: Energiya, 1978.
18. V. V. Nikiforov, E. P. Stroganova, and I. V. Shevnin, "An Influence of Parasitic Lead Inductances on Stability of Power MOSFET Amplifiers (in Russian)," *Radiotekhnika*, vol. 45, pp. 100–102, May 1990.
19. Application Note AN-010, *GaN for LDMOS Users*, Nitronex Corp., 2008.
20. J. Vidkjaer, "Instabilities in RF-Power Amplifiers Caused by a Self-Oscillation in the Transistor Bias Network," *IEEE J. Solid-State Circuits*, vol. SC-11, pp. 703–712, Oct. 1976.
21. D. R. Lohrmann, "Parametric Oscillations in VHF Transistor Power Amplifiers," *Proc. IEEE*, vol. 54, pp. 409–410, Mar. 1966.
22. D. R. Lohrmann, "Amplifiers Has 85% Efficiency while Providing up to 10 Watts Power over a Wide Frequency Band," *Electronic Design*, vol. 14, pp. 38–43, Mar. 1966.
23. P. Penfield and R. P. Rafuse, *Varactor Applications*, Cambridge: The M.I.T Press, 1962.
24. J. H. Morecroft and H. T. Friis, "The Vacuum Tubes as a Generator of Alternating-Current Power," *Trans. AIEE*, vol. 38, pp. 1415–1444, Oct. 1919.
25. D. C. Prince, "Vacuum Tubes as Power Oscillators, Part I," *Proc. IRE*, vol. 11, pp. 275–313, Jun. 1923.
26. A. A. Oswald, "Power Amplifiers in Trans-Atlantic Radio Telephony," *Proc. IRE*, vol. 13, pp. 313–324, Jun. 1925.
27. L. E. Barton, "High Audio Power from Relatively Small Tubes," *Proc. IRE*, vol. 19, pp. 1131–1149, Jul. 1931.
28. F. E. Terman and J. H. Ferns, "The Calculation of Class C Amplifier and Harmonic Generator Performance of Screen-Grid and Similar Tubes," *Proc. IRE*, vol. 22, pp. 359–373, Mar. 1934.
29. A. I. Berg, *Theory and Design of Vacuum-Tube Generators* (in Russian), Moskva: GEI, 1932.
30. P. H. Osborn, "A Study of Class B and C Amplifier Tank Circuits," *Proc. IRE*, vol. 20, pp. 813–834, May 1932.
31. C. E. Fay, "The Operation of Vacuum Tubes as Class B and Class C Amplifiers," *Proc. IRE*, vol. 20, pp. 548–568, Mar. 1932.
32. L. B. Hallman, "A Fourier Analysis of Radio-Frequency Power Amplifier Wave Forms," *Proc. IRE*, vol. 20, pp. 1640–1659, Oct. 1932.
33. W. L. Everitt, "Optimum Operating Conditions for Class C Amplifiers," *Proc. IRE*, vol. 22, pp. 152–176, Feb. 1934.
34. C. E. Kilgour, "Graphical Analysis of Output Tube Performance," *Proc. IRE*, vol. 19, pp. 42–50, Jan. 1931.
35. V. I. Kaganov, *Transistor Radio Transmitters* (in Russian), Moskva: Energiya, 1976.
36. F. H. Raab, "Class-E, Class-C, and Class-F Power Amplifiers based upon a Finite Number of Harmonics," *IEEE Trans. Microwave Theory Tech.*, vol. MTT-49, pp. 1462–1468, Aug. 2001.
37. F. H. Raab, "Maximum Efficiency and Output of Class-F Power Amplifiers," *IEEE Trans. Microwave Theory Tech.*, vol. MTT-49, pp. 1162–1166, Jun. 2001.
38. A. Juhas and L. A. Novak, "Comments on "Class-E, Class-C, and Class-F Power Amplifiers based Upon a Finite Number of Harmonics," *IEEE Trans. Microwave Theory Tech.*, vol. MTT-57, pp. 1623–1625, Jun. 2009.
39. J. Vidkjaer, "A Computerized Study of the Class-C-Biased RF-Power Amplifier," *IEEE J. Solid-State Circuits*, vol. SC-13, pp. 247–258, Apr. 1978.
40. H. L. Krauss, C. W. Bostian, and F. H.Raab, *Solid State Radio Engineering*, New York: John Wiley & Sons, 1980.

41. B. E. Rose, "Notes on Class-D Transistor Amplifiers," *IEEE J. Solid-State Circuits*, vol. SC-4, pp. 178–179, Jun. 1969.
42. J. Vidkjaer, "Series-Tuned High Efficiency RF-Power Amplifiers," *2008 IEEE MTT-S Int. Microwave Symp. Dig.*, pp. 73–76.
43. N. Le Gallou, J. Vidkjaer, and C. Poivey, "Suitability of GaN and LDMOS for 70-82% Efficiency 120-200W HPA Addressing Spaceborne P-Band Radar Applications," *Proc. 42nd Europ. Microwave Conf.*, pp. 691–694, 2012.
44. E. F. Belohoubek, A. Rosen, D. M. Stevenson, and A. Pressser, "Hybrid Integrated 10-Watt CW Broad-Band Power Source at *S* Band," *IEEE J. Solid-State Circuits*, vol. SC-4, pp. 360–366, Dec. 1969.
45. A. Presser and E. F. Belohoubek, "1–2 GHz High Power Linear Transistor Amplifier," *RCA Rev.*, vol. 33, pp. 737–751, Dec. 1972.
46. J. M. Cusack, S. M. Perlow, and B. S. Perlman, "Automatic Load Contour Mapping for Microwave Power Transistors," *IEEE Trans. Microwave Theory Tech.*, vol. MTT-12, pp. 1146–1152, Dec. 1974.
47. H. Abe and Y. Aono, "11-GHz GaAs Power MESFET Load-Pull Measurements Utilizing a New Method of Determining Tuner *Y* Parameters," *IEEE Trans. Microwave Theory Tech.*, vol. MTT-27, pp. 394–399, May 1979.
48. Y. Takayama, "A New Load-Pull Characterization Method for Microwave Power Transistors," *1976 IEEE MTT-S Int. Microwave Symp. Dig.*, pp. 218–220.
49. G. P. Bava, U. Pisani, and V. Pozzolo, "Active Load Technique for Load-Pull Characterization at Microwave Frequencies," *Electronics Lett.*, vol. 18, pp. 178–180, Feb. 1982.
50. C. Rauscher and H. A. Willing, "Simulation of Nonlinear Microwave FET Performance Using a Quasi-Static Model," *IEEE Trans. Microwave Theory Tech.*, vol. MTT-27, pp. 834–840, Oct. 1979.
51. V. Teppati, A. Ferrero, and U. Pisani, "Recent Advances in Real-Time Load-Pull Systems," *IEEE Trans. Instrum. Meas.*, vol. IM-57, pp. 2640–2646, Nov. 2008.
52. R. B. Stancliff and D. D. Poulin, "Harmonic Load-Pull," *1979 IEEE MTT-S Int. Microwave Symp. Dig.*, pp. 185–187.
53. M. S. Hashmi, F. M. Ghannouchi, P. J. Tasker, and K. Rawat, "Highly Reflective Load-Pull," *IEEE Microwave Mag.*, vol. 12, pp. 96–107, Jun. 2011.
54. A. Wood, "LDMOS Transistor Powers PCS Base-Station Amplifiers," *Microwaves & RF*, vol. 37, pp. 69–80, Mar. 1998.
55. K. Vennema, "Two-Tone Linearity in a 900 MHz Silicon Bipolar Class AB Amplifier," *Microwave J.*, vol. 39, pp. 88–93, Feb. 1996.
56. E. H. Colpitts, "System for the Transmission of Intelligence," U.S. Patent 1,137,384, Apr. 1915.
57. E. W. Kellogg, "Design of Non-distorting Power Amplifiers," *Trans. AIEE*, vol. 44, pp. 302–315, Feb. 1925.
58. S. Ballantine, "Method of and Means for Reducing Retroactive Currents in Push-Pull Amplifiers," U.S. Patent 1,560,332, Apr. 1923.
59. L. B. Lee, "Apply Wideband Techniques to Balanced Amplifiers," *Microwaves*, vol. 19, pp. 83–88, Apr. 1980.
60. K. Kurokawa, "Design Theory of Balanced Transistor Amplifiers," *Bell Syst. Tech. J.*, vol. 44, pp. 1675–1798, Oct. 1965.
61. R. S. Engelbrecht and K. Kurokawa, "A Wideband Low Noise L-band Balanced Transistor Amplifier," *Proc. IEEE*, vol. 53, pp. 237–247, Mar. 1965.
62. R. E. Neidert and H. A. Willing, "Wide-Band Gallium Arsenide Power MESFET Amplifiers," *IEEE Trans. Microwave Theory Tech.*, vol. MTT-24, pp. 342–350, Jun. 1976.
63. K. B. Niklas, R. B. Gold, W. T. Wilser, and W. R. Hitchens, "A 12-18 GHz Medium-Power GaAs MESFET Amplifier," *IEEE J. Solid-State Circuits*, vol. SC-13, pp. 520–527, Aug. 1978.
64. W. H. Ku, J. E. Erickson, R. E. Rabe, and G. L. Seasholtz, "Design Techniques and Intermodulation Analysis of Broad-Band Solid-State Power Amplifiers," *IEEE Trans. Electromagnetic Compatibility*, vol. EMC-19, pp. 57–65, May 1977.
65. K. B. Niklas, W. T. Wilser, R. B. Gold, and W. R. Hitchens, "Application of the Two-Way Balanced Amplifier Concept to Wide-Band Power Amplification Using GaAs MESFET's," *IEEE Trans. Microwave Theory Tech.*, vol. MTT-28, pp. 172–179, Mar. 1980.

66. I. D. Robertson, R. Herath, M. Gillick, and J. Bhari, "Solid State Power Amplifier Using Impedance-Transforming Branch-Line Couplers for *L*-band Satellite Systems," *Proc. 23rd Europ. Microwave Conf.*, pp. 448–449, 1993.
67. Y. S. Noh, J. H. Park, and C. S. Park, "A Temperature and Supply Independent Bias Circuit and MMIC Power Amplifier Implementation for WCDMA Applications," *IEICE Trans. Electron.*, vol. E88-C, pp. 725–728, Apr. 2005.
68. *RF Transmitting Transistor and Power Amplifier Fundamentals*, Philips Semiconductors, 1998.
69. T. Sato and C. Grigorean, "Design Advantages of CDMA Power Amplifiers Built with MOSFET Technology," *Microwave J.*, vol. 45, pp. 64–78, Oct. 2002.
70. Y. S. Noh and C. S. Park, "An Intelligent Power Amplifier MMIC Using a New Adaptive Bias Control Circuit for W-CDMA Applications," *IEEE J. Solid-State Circuits*, vol. SC-39, pp. 967–970, Jun. 2004.
71. A. Grebennikov, B. Zogl, H. Hermann., C. Roth, and W. Thomann, "High-Efficiency Balanced Switch-Path Monolithic SiGe HBT Power Amplifiers for Wireless Applications," *Proc. 37th Europ. Microwave Conf.*, pp. 1189–1192, 2007.
72. H. M. Macksey, H. Q. Tserng, and G. H. Westphal, "*S*-Band GaAs Power FET," *1982 IEEE MTT-S Int. Microwave Symp. Dig.*, pp. 150–152.
73. R. J. Trew, "MESFET Models for Microwave Computer-Aided Design," *Microwave J.*, vol. 33, pp. 115–130, May 1990.
74. C. Blair, "LDMOS Devices Provide High Power for Digital PCS," *Applied Microwave & Wireless*, vol. 10, pp. 84–88, Oct. 1998.
75. R. Frey, "A Push-Pull 300-Watt Amplifier for 81.36 MHz," *Applied Microwave & Wireless*, vol. 10, pp. 36–45, Apr. 1998.
76. C. Florian, R. P. Paganelli, and J. A. Lonac, "12-W *X*-Band MMIC HPA and Driver Amplifiers in InGaP-GaAs HBT Technology for Space SAR T/R Modules," *IEEE Trans. Microwave Theory Tech.*, vol. MTT-60, pp. 1805–1816, Jun. 2012.

CHAPTER 7

High-Efficiency Power Amplifiers

High efficiency of the power amplifier can be obtained by using overdriven Class-B, Class-F, or Class-E operation modes and their subclasses, depending on the technical requirements. In all cases, an efficiency improvement in practical implementation is achieved by providing the nonlinear operation conditions when an active device can simultaneously operate in pinch-off, active, and saturation regions, resulting in the nonsinusoidal collector current and voltage waveforms, symmetrical for Class-F and asymmetrical for Class-E operation modes. In Class-F power amplifiers analyzed in a frequency domain, the fundamental-frequency and harmonic load impedances are optimized by short-circuit termination and open-circuit peaking to control the voltage and current waveforms at the device output in order to obtain maximum efficiency. In Class-E power amplifiers analyzed in a time domain, an efficiency improvement is achieved by realizing the on/off active device switching operation (the pinch-off and saturation modes) with special current and voltage waveforms so that high voltage and high current do not concur at the same time.

7.1 Overdriven Class B

An improvement in both collector efficiency and output power described by conventional Class-B considerations can be realized if the load impedance at the fundamental and harmonically related frequencies presented to the device output of a tuned power amplifier stage are appropriately selected. In this case, a theoretical collector efficiency of 100% at 1.27 times the conventional value of output power designed in a Class-B mode is possible. Furthermore, if the power amplifier is overdriven when the RF drive is increased by 5.2 dB, the different load impedance can be derived so that 1.46 times the conventional value of output power can be achieved with 88% collector efficiency [1].

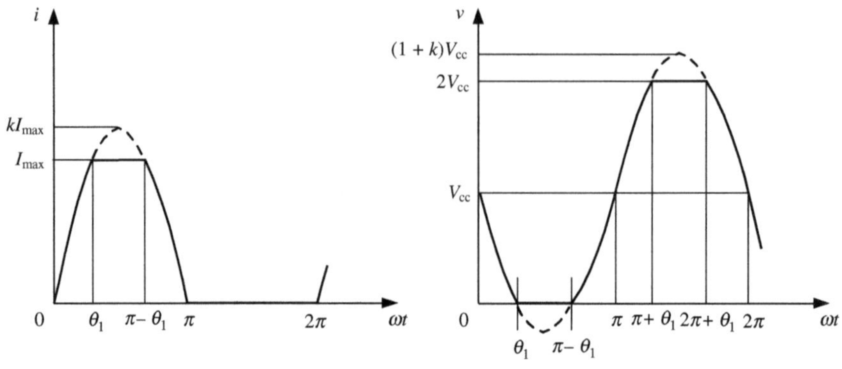

FIGURE 7.1 Overdriven Class-B collector current and voltage waveforms.

Figure 7.1 shows the idealized overdriven Class-B collector current and voltage waveforms. In this case, both voltage and current waveform amplitudes are increased, but their truncated peak values remain the same as in a conventional Class B and are equal to the value of the supply voltage V_{cc} and current peak value I_{max}, respectively. A Fourier analysis of the current and voltage waveforms as functions of the angular parameter θ_1 gives the following values of the voltage and current components:

for the dc voltage component,

$$V_0 = V_{cc} \tag{7.1}$$

for the fundamental voltage component,

$$V_1 = \frac{2\,V_{cc}}{\pi}\left(\frac{\theta_1}{\sin\theta_1} + \cos\theta_1\right) \tag{7.2}$$

for the odd voltage components ($n = 3, 5, \ldots$),

$$V_n = \frac{2\,V_{cc}}{\pi}\left[\frac{\sin(\theta_1 - n\theta_1)}{(1-n)\sin\theta_1} - \frac{\sin(\theta_1 + n\theta_1)}{(1+n)\sin\theta_1} + \frac{2\cos n\theta_1}{n}\right] \tag{7.3}$$

for the dc current component,

$$I_0 = \frac{I_{max}}{\pi}\left(\frac{\pi}{2} - \theta_1 + \tan\frac{\theta_1}{2}\right) \tag{7.4}$$

for the fundamental-current component,

$$I_1 = \frac{I_{max}}{\pi}\left(\frac{\theta_1}{\sin\theta_1} + \cos\theta_1\right) \tag{7.5}$$

for the odd-current components,

$$I_n = \frac{I_{max}}{\pi}\left[\frac{\sin(\theta_1 - n\theta_1)}{(1-n)\sin\theta_1} - \frac{\sin(\theta_1 + n\theta_1)}{(1+n)\sin\theta_1} + \frac{2\cos n\theta_1}{n}\right] \tag{7.6}$$

with zero voltage and current components for even n ($n = 2, 4, \ldots$).

The output power at the fundamental frequency $P_1 = V_1I_1/2$ can be calculated as a function of the angular parameter θ_1 by

$$P_1 = \frac{V_{cc}I_{max}}{\pi^2}\left(\frac{\theta_1}{\sin\theta_1} + \cos\theta_1\right)^2 \tag{7.7}$$

The dc power $P_0 = V_0I_0$ is also a function of the angular parameter θ_1, so that

$$P_0 = \frac{V_{cc}I_{max}}{\pi}\left(\frac{\pi}{2} - \theta_1 + \tan\frac{\theta_1}{2}\right) \tag{7.8}$$

and the out-of-band impedances are defined as

$$Z_n = \frac{2\,V_{cc}}{I_{max}} = R_L \quad \text{for odd } n \tag{7.9}$$

$$Z_n = 0 \quad \text{for even } n \tag{7.10}$$

where R_L is the load resistance. Consequently, the total power is delivered to the load not only at the fundamental (as in a conventional Class-B operation) but also at the odd-harmonic frequency components.

As a result, from Eq. (7.7) for the fundamental output power and from Eq. (7.8) for the dc power, the collector efficiency η can be written as

$$\eta = \frac{1}{\pi}\,\frac{\left(\dfrac{\theta_1}{\sin\theta_1} + \cos\theta_1\right)^2}{\dfrac{\pi}{2} - \theta_1 + \tan\dfrac{\theta_1}{2}} \tag{7.11}$$

For the extreme case of the rectangular voltage and current waveforms when θ_1 approaches zero, the collector efficiency η approaches

$$\eta = \frac{8}{\pi^2} = 81\% \tag{7.12}$$

This value is higher than in a conventional Class-B operation with a maximum collector efficiency η = 78.5%, as follows from Eq. (7.11) when $\theta_1 = 90°$. However, by analyzing Eq. (7.11) as a function of θ_1 on extremum, the maximum value of the collector efficiency in an overdriven Class-B operation mode can be increased up to 88.6% when $\theta_1 = 32.4°$.

To evaluate the power-added efficiency (*PAE*), it is necessary to calculate the effective operating power gain G_{Peff} in an overdriven operation. If it is assumed that the overdriven effect is absent, from Fig. 7.1 (dashed curves) it follows that in a conventional Class-B mode both the current and voltage amplitudes are k times larger than these in an overdriven Class-B mode, where $k = 1/\sin\theta_1$.

Consequently, for the same input power P_{in}, the fundamental output power in the conventional Class-B mode (P_1) will be larger than that in the overdriven mode (P_{1e}) by k^2 times. Besides, the effective operating power gain G_{Peff} will be smaller than that without overdriven effect (G_P), which is defined as

$$G_{Peff} = \frac{G_P}{k^2}\frac{P_{1e}}{P_1} = G_P\left(\frac{2\sin\theta_1}{\pi}\right)^2\left(\frac{\theta_1}{\sin\theta_1} + \cos\theta_1\right)^2 \tag{7.13}$$

Then, the *PAE* as a function of the angular parameter θ_1 can be calculated by

$$PAE = \frac{P_1 - P_{in}}{P_0} = \frac{P_1}{P_0}\left(1 - \frac{1}{G_{Peff}}\right) \tag{7.14}$$

Figure 7.2 shows the dependences of the power-added efficiencies versus θ_1 for different values of G_P, where a *PAE* of 77.2% in an overdriven mode is achieved with an optimum angular parameter θ_1 of 51.4° for a typical value of G_P = 12 dB in a conventional Class-B operation mode. However, in this case it is necessary to consider an excess of the voltage collector amplitude V_1 by $k = 1.28$ times over the dc supply voltage V_{cc}. The efficiency of a power amplifier can be maximized if the active device is operated ideally as a switch to provide nearly zero voltage and high current conditions when the transistor is turned on.

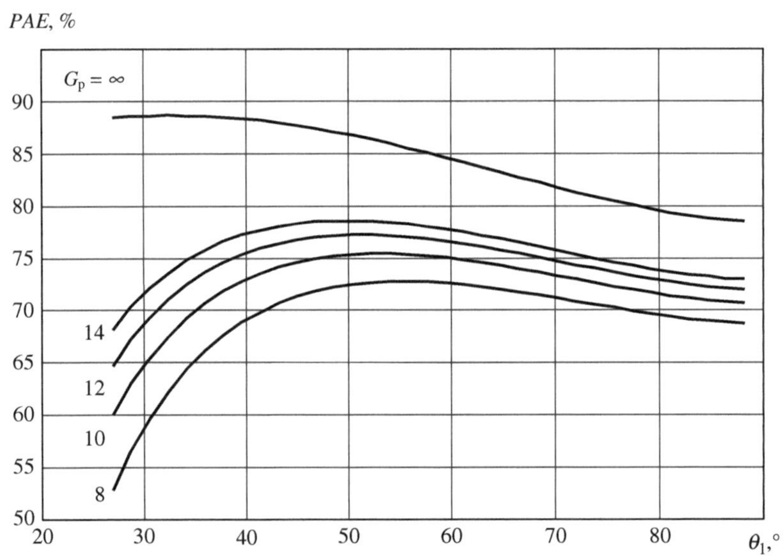

FIGURE 7.2 Power-added efficiencies for different values of power gain.

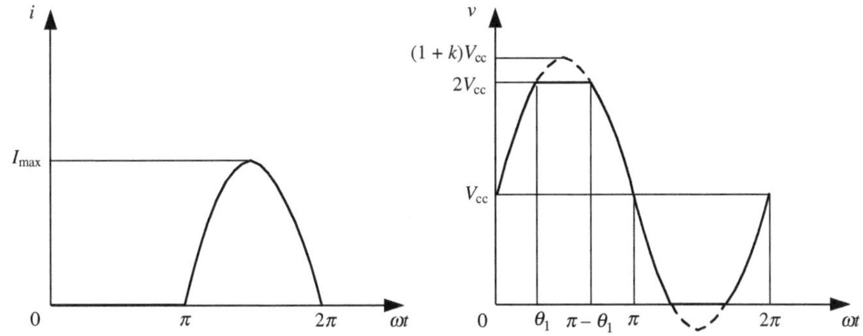

FIGURE 7.3 Optimum-efficiency Class-B collector current and voltage waveforms.

To further increase the efficiency in an overdriven Class-B operation mode, it is advisable to set a value of the angular parameter θ_1 to 90°, which provides a half-sinusoidal collector current waveform, and to approach θ_1 to zero for the collector voltage waveform, as shown in Fig. 7.3. In the extreme case of $\theta_1 = 0$, the collector voltage approaches a rectangular waveform.

7.2 Class-F Circuit Design

The practical possibility of improving efficiency by approximating the anode voltage waveform to square wave to minimize the value of the saturation voltage compared to the supply voltage over half an entire interval of $0 \leq \omega t \leq 2\pi$ was discussed in the early 1920s. As a potential solution, it was proposed to use the load network with a third-harmonic trap in series to the anode, as shown in Fig. 7.4(*a*) [2, 3]. However, the effect of the inclusion of a third-harmonic resonator was described and analyzed in detail only 1.5 decades later [4, 5]. It was shown that the symmetrical anode voltage waveform and level of its depression can be provided with opposite phase conditions at the waveform midpoints between the fundamental and third harmonic and optimum value of the ratio between their voltage amplitudes. In addition, it was noted that high operation efficiency can be achieved even when impedance of the third-harmonic resonator is equal or slightly greater than that of the fundamental tank. To maximize efficiency of the vacuum-tube amplifier with better approximating a square voltage anode waveform, it was also suggested to use an additional resonator tuned to the fifth harmonic, as shown in Fig. 7.4(*b*) [6].

Figure 7.5 shows that the shapes of the voltage and current waveforms can be significantly changed with increasing fundamental voltage amplitude by adding even one additional harmonic component being properly phased. For example, the combination of the fundamental-frequency and third-harmonic components being

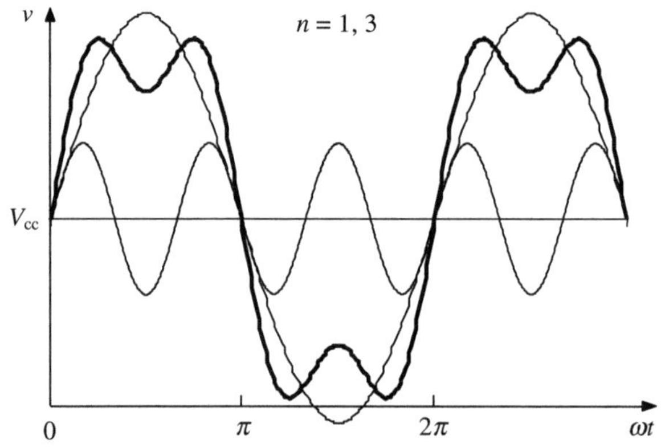

(a)

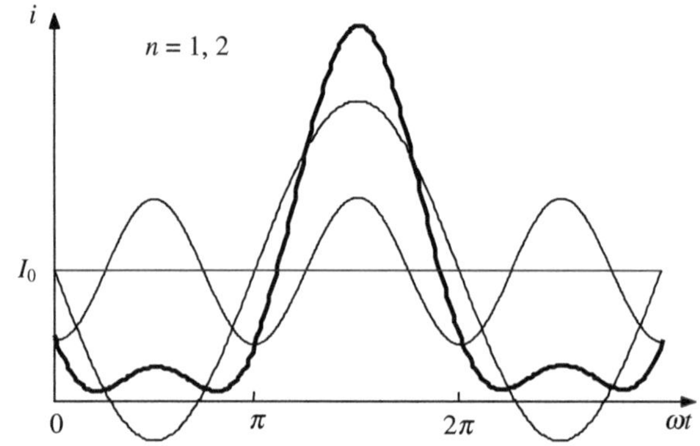

(b)

Figure 7.4 Fourier voltage and current waveforms with third and second harmonics.

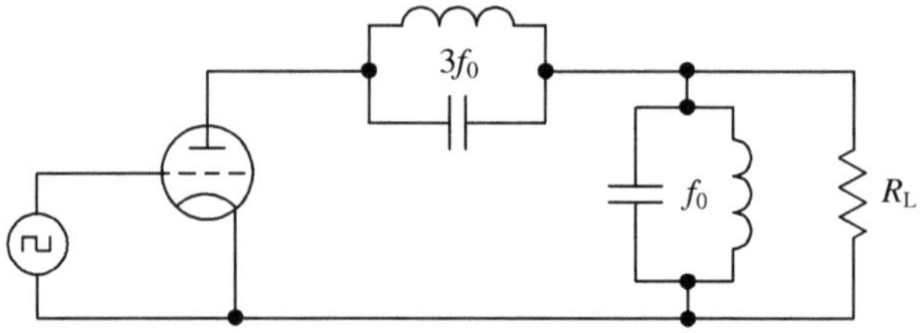

(a)

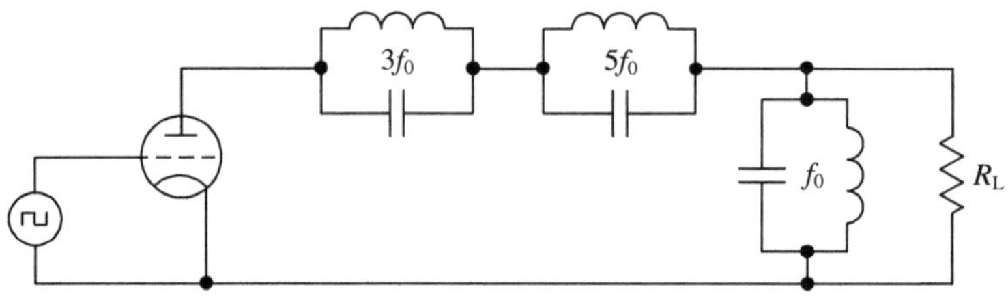

(b)

Figure 7.5 Biharmonic and polyharmonic power amplifiers.

180° out-of-phase at center points results in a flattened voltage waveform with depression in its center. It is clearly seen from Fig. 7.5(*a*) that the proper ratio between the amplitudes of the fundamental and third-harmonic components can provide the flattened voltage waveform with minimum depression and maximum difference between its peak amplitude and amplitude of the fundamental component. Similarly, the combination of the fundamental-frequency and second-harmonic components, being in phase at the center points, flattens the current waveform corresponding to the maximum values of the voltage waveform and sharpens the current waveform corresponding to the minimum values of the voltage waveform, as shown in Fig. 7.5(*b*). The optimum ratio between the amplitudes of the current fundamental-frequency and second-harmonic components can maximize a peak value of the current waveform, with its minimized value determined by the device saturation resistance in a practical circuit. Thus, power loss due to the active device can be minimized because the results of the integration over period when the minimum voltage corresponds to the maximum current will give a small value compared with the power delivered to the load.

7.2.1 Idealized Class-F Mode

Generally, an infinite number of the odd-harmonic tank resonators can maintain a square collector voltage waveform, also providing a half-sinusoidal current waveform. Figure 7.6(*a*) shows such a Class-F power amplifier with a multiple-resonator output filter to control the harmonic content of its collector (anode or drain) voltage and current waveforms, thereby shaping them to reduce dissipation and to increase efficiency [7].

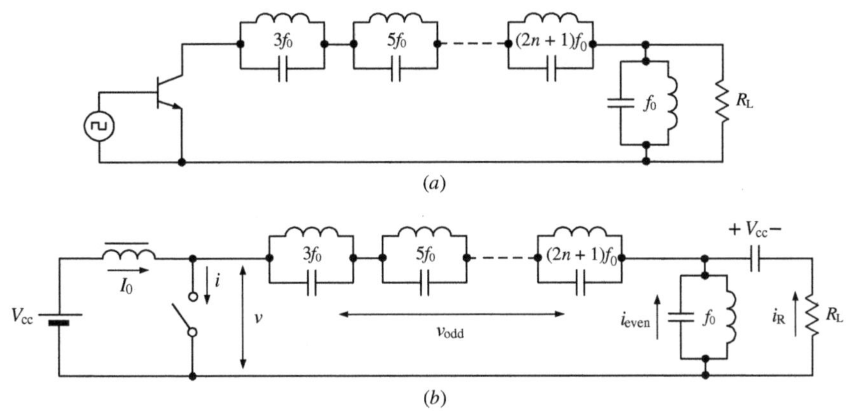

FIGURE 7.6 Basic circuits of Class-F power amplifier with parallel resonant circuits.

To simplify an analysis of a Class-F power amplifier, whose simple equivalent circuit is shown in Fig. 7.6(*b*), the following several assumptions are introduced:

- Transistor has zero saturation voltage, zero saturation resistance, and infinite off-resistance, and its switching action is instantaneous and lossless.
- RF choke allows only a dc current and has no resistance.
- Quality factors of all parallel resonant circuits have infinite impedance at the corresponding harmonic and zero impedance at other harmonics.
- There are no losses in the circuit except only into the load R_L.
- Operation mode with a 50% duty ratio.

To determine the idealized collector voltage and current waveforms, let us consider the distribution of voltages and currents in the load network, assuming the sinusoidal fundamental current flowing into the load as $i_R(\omega t) = I_R \sin(\omega t)$, where I_R is its amplitude. The voltage $v(\omega t)$ across the switch can be represented as a sum of the dc voltage V_{cc}, the fundamental voltage $v_R = i_R R_L$ across the load resistor, and the voltage v_{odd} across the odd-harmonic resonators,

$$v(\omega t) = V_{cc} + v_{odd}[(2n + 1)\omega t] + v_R(\omega t) \tag{7.15}$$

Because the time moment t was chosen arbitrarily, by introducing a phase shift of π, Eq. (7.15) can be rewritten for periodical sinusoidal functions as

$$v(\omega t + \pi) = V_{cc} - v_{odd}[(2n + 1)\omega t] - v_R(\omega t) \tag{7.16}$$

Then, the summation of Eqs. (7.15) and (7.16) yields

$$v(\omega t) = 2V_{cc} - v(\omega t + \pi) \tag{7.17}$$

From Eq. (7.17), it follows that the maximum value of the collector voltage cannot exceed a value of $2V_{cc}$, and the time duration with a maximum voltage of $v = 2V_{cc}$ coincides with the time duration with a minimum voltage of $v = 0$. Because the collector voltage is zero when the switch is turned off, the only possible waveform for the collector voltage is a square wave, composing of only dc, fundamental-frequency, and odd-harmonic components.

During the interval $0 < \omega t \leq \pi$ when the switch is turned off, the current $i(\omega t)$ flowing through the switch can be written as

$$i(\omega t) = I_0 + i_{even}(2n\omega t) + i_R(\omega t) \tag{7.18}$$

whereas during the interval $\pi < \omega t \leq 2\pi$ when the switch is turned on, the current $i(\omega t + \pi)$ is equal to zero, resulting in

$$0 = I_0 + i_{\text{even}}(2n\omega t) - i_{\text{R}}(\omega t) \tag{7.19}$$

Then, by substituting Eq. (7.19) into Eq. (7.18), we can rewrite Eq. (7.18) as

$$i(\omega t) = 2i_{\text{R}}(\omega t) = 2I_{\text{R}}\sin(\omega t) \tag{7.20}$$

from which it follows that the amplitude of the current flowing through the switch during the interval $0 < \omega t \leq \pi$ is two times greater than the amplitude of the fundamental current. Thus, in a general case of entire interval, Eq. (7.18) can be rewritten as

$$i(\omega t) = I_{\text{R}}\left(\sin\omega t + |\sin\omega t|\right) \tag{7.21}$$

which means that the switch current represents half-sinusoidal pulses with the amplitude equal to double-load current amplitude.

Consequently, for a purely sinusoidal current flowing into the load, which is shown in Fig. 7.7(*a*), the ideal collector voltage and current waveforms can be represented by the appropriate normalized waveforms shown in Fig. 7.7(*b*) and 7.7(*c*), respectively. Here, a sum of the fundamental and odd harmonics approximates a square voltage waveform and a sum of the fundamental and even harmonics approximates a half-sinusoidal collector current waveform. As a result, the shapes of the collector current and voltage waveforms provide a condition when the current and voltage do not overlap simultaneously. Such a condition, with symmetrical collector voltage and current waveforms, corresponds to an idealized Class-F operation mode with 100% collector efficiency.

A Fourier analysis of the current and voltage waveforms allows us to obtain the following equations for the dc current and the fundamental voltage and current components in the collector voltage and current waveforms:

The dc current I_0 can be calculated from Eq. (7.21) as

$$I_0 = \frac{1}{2\pi}\int_0^{\pi} 2I_{\text{R}}\sin\omega t\; d\omega t = \frac{2I_{\text{R}}}{\pi} \tag{7.22}$$

The fundamental current component can be calculated from Eq. (7.21) as

$$I_1 = \frac{1}{\pi}\int_0^{\pi} 2I_{\text{R}}\sin^2\omega t\; d\omega t = I_{\text{R}} \tag{7.23}$$

The fundamental voltage component can be calculated using Eq. (7.17) as

$$V_1 = V_{\text{R}} = \frac{1}{\pi}\int_{\pi}^{2\pi} 2V_{\text{cc}}\sin(\omega t + \pi)\; d\omega t = \frac{4\,V_{\text{cc}}}{\pi} \tag{7.24}$$

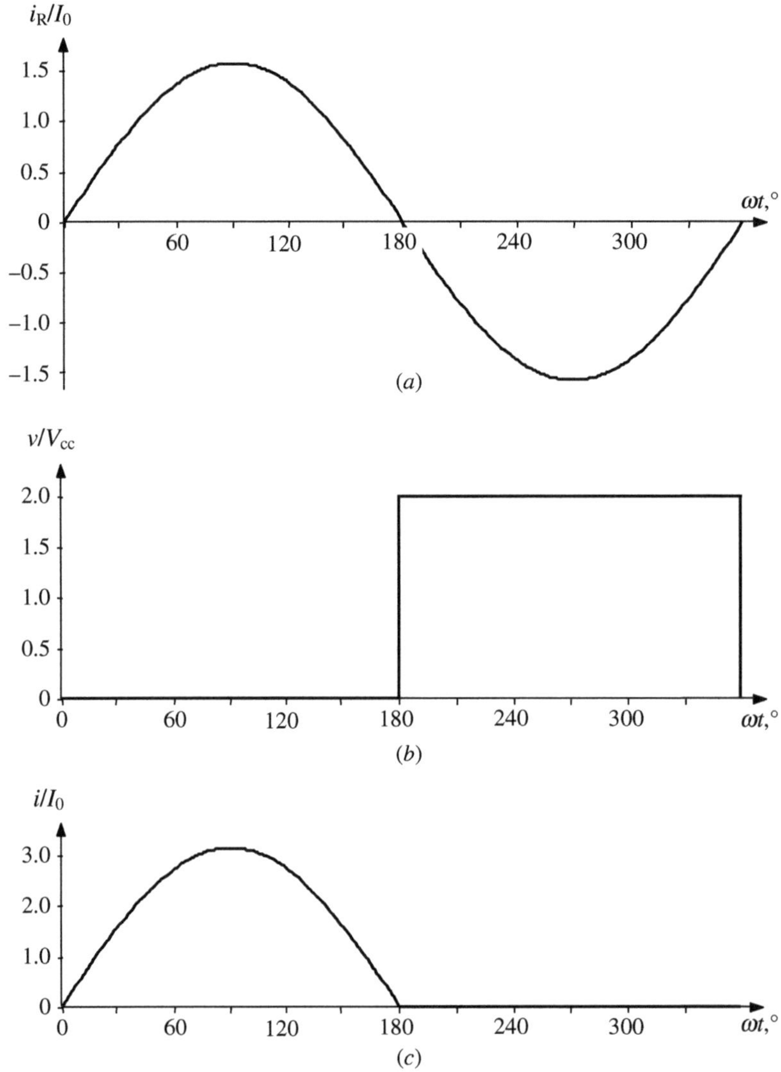

FIGURE 7.7 Ideal waveforms of Class-F power amplifier.

where $V_R = I_R R_L$ is the fundamental voltage amplitude across the load resistor R_L.

Then, the dc power and output power at the fundamental frequency are calculated by

$$P_0 = V_{cc} I_0 = \frac{2}{\pi} V_{cc} I_R \tag{7.25}$$

$$P_1 = \frac{V_1 I_1}{2} = \frac{2}{\pi} V_{cc} I_R \tag{7.26}$$

respectively, resulting in a theoretical collector efficiency with maximum value of

$$\eta = \frac{P_1}{P_0} = 100\% \tag{7.27}$$

In this case, the impedance conditions seen by the device collector for an idealized Class-F mode must be equal to

$$Z_1 = R_1 = \frac{8}{\pi^2}\frac{V_{cc}}{I_0} \tag{7.28}$$

$$Z_{2n} = 0 \text{ for even harmonics} \tag{7.29}$$

$$Z_{2n+1} = \infty \text{ for odd harmonics} \tag{7.30}$$

which are similar to that derived from the limiting case of the optimum efficiency Class-B mode [1].

7.2.2 Class F with Maximally Flat Waveforms

Although it is impossible to realize the ideal harmonic impedance conditions in practical implementation, the peaking of at least several current and voltage harmonic components can be provided to achieve a high operation efficiency of the power amplifier. The more the voltage waveform provided by higher-order harmonic components can be flattened, the less power dissipation due to flowing of the output current (when the output voltage is extremely small) occurs. To understand the basic design principles and to numerically calculate the power amplifier efficiency according to the contribution of an appropriate number of the harmonic components of voltage and current waveforms, it is convenient to use a design technique based on a Class-F approximation with maximally flat waveforms [8]. In this case, the load network is assumed ideal to deliver only the fundamental-frequency power to the load without loss. The active device represents an ideal multiharmonic current source with zero saturation voltage and output capacitance for providing instant switching between saturation and pinch-off operation regions. Flattening of the voltage and current waveforms to realize a Class-F operation can be accomplished by using odd-harmonic components to approximate a rectangular voltage waveform and even-harmonic components to approximate a half-sinusoidal current waveform given by

$$v(\omega t) = V_{cc} + V_1 \sin\omega t + \sum_{n=3,5,7,\ldots}^{\infty} V_n \sin n\omega t \tag{7.31}$$

$$i(\omega t) = I_0 - I_1 \sin\omega t - \sum_{n=2,4,6,\ldots}^{\infty} I_n \cos n\omega t \tag{7.32}$$

For the symmetrical flattened voltage waveforms shown in Fig. 7.8, the medium points where the voltage waveform reaches its maximum

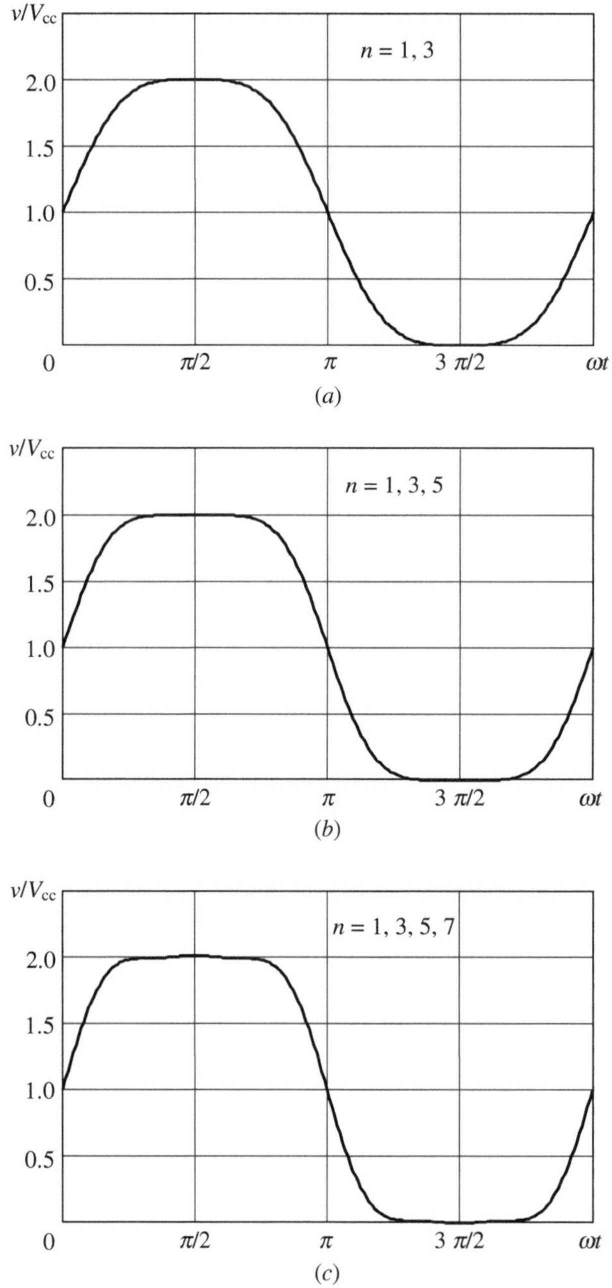

Figure 7.8 Voltage waveforms for *n*th-harmonic peaking.

and minimum values are at $\omega t = \pi/2$ and $\omega t = 3\pi/2$, respectively. Maximum flatness at minimum voltage requires the even-order derivatives to be zero at $\omega t = 3\pi/2$. As the odd-order derivatives are equal to zero because $\cos(n\pi/2) = 0$ for odd n, it is necessary to define the even-order derivatives of the voltage waveform given by Eq. (7.31).

For the third-harmonic peaking when only the third-harmonic component together with the fundamental one is present, their optimum amplitudes are defined as

$$V_1 = \frac{9}{8} V_{cc} \qquad V_3 = \frac{1}{8} V_{cc} \tag{7.33}$$

The voltage waveforms for the third-harmonic peaking ($n = 1, 3$), fifth-harmonic peaking ($n = 1, 3, 5$), and seventh-harmonic peaking ($n = 1, 3, 5, 7$) are shown in Fig. 7.8.

For the symmetrical current waveforms shown in Fig. 7.9, the medium points where the current waveform reaches its minimum and maximum values are at $\omega t = \pi/2$ and $\omega t = 3\pi/2$, respectively. As the odd-order derivatives are equal to zero because $\cos(\pi/2) = 0$ and $\sin(n\pi/2) = 0$ for even n, it is sufficient to determine the even-order derivatives of the current waveform given by Eq. (7.32). Maximum flatness at minimum current requires the even-order derivatives to be zero at $\omega t = \pi/2$.

For the second-harmonic peaking when only the second-harmonic component together with the fundamental one is present, their optimum amplitudes are defined by

$$I_1 = \frac{4}{3} I_0 \qquad I_2 = \frac{1}{3} I_0 \tag{7.34}$$

The current waveforms for the second-harmonic peaking ($n = 1, 2$), fourth-harmonic peaking ($n = 1, 2, 4$), and sixth-harmonic peaking ($n = 1, 2, 4, 6$) are shown in Fig. 7.9.

The effectiveness of the operations modes with different voltage and current harmonic peaking can be compared by calculating the collector (drain) efficiency η of each operation mode according to

$$\eta = \frac{P_1}{P_0} = \frac{1}{2} \frac{V_1}{V_{cc}} \frac{I_1}{I_0} \tag{7.35}$$

The resultant efficiencies for various combinations of the voltage and current harmonic components are given in Table 7.1, which shows that the efficiency increases with an increase in the number of voltage and current harmonic components. To increase efficiency, it is more desirable to provide harmonic peaking in consecutive numerical order (both for voltage and current harmonic components) than to increase the number of the harmonic components into only voltage or current waveforms. Class-F operation becomes mostly effective in

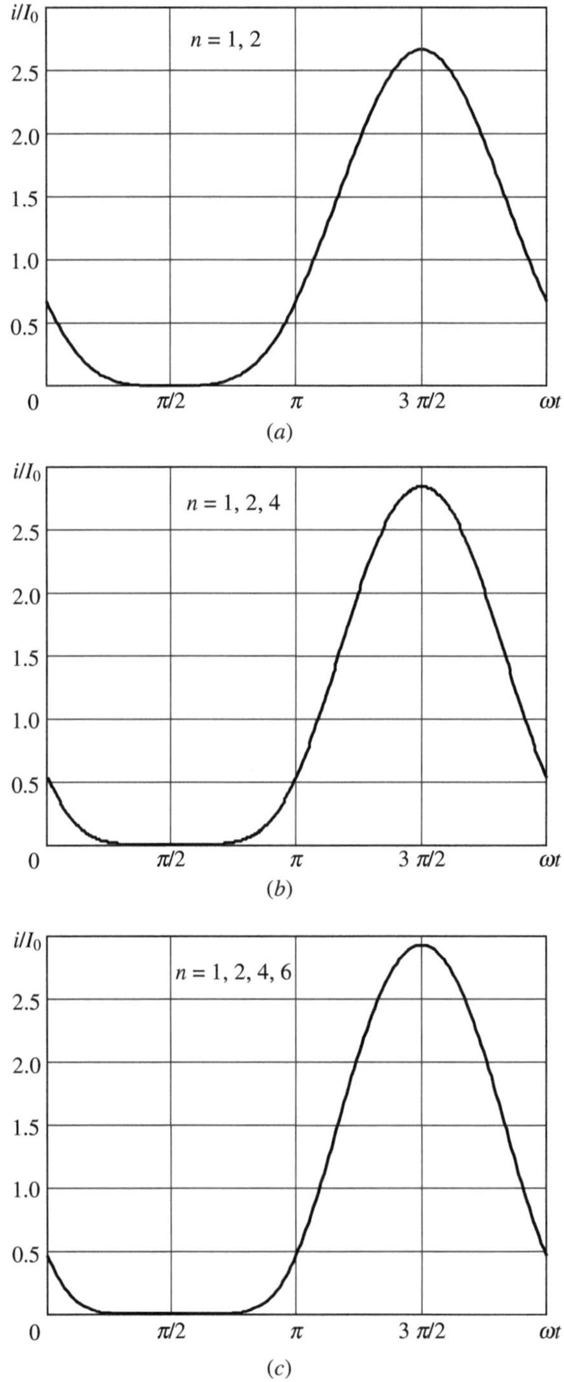

Figure 7.9 Current waveforms for *n*th-harmonic peaking.

Current Harmonic Components	**Voltage Harmonic Components**				
	1	1, 3	1, 3, 5	1, 3, 5, 7	1, 3, 5, …, ∞
1	1/2 = 0.500	9/16 = 0.563	75/128 = 0.586	1225/2048 = 0.598	$2/\pi = 0.637$
1, 2	2/3 = 0.667	3/4 = 0.750	25/32 = 0.781	1225/1536 = 0.798	$8/3\pi = 0.849$
1, 2, 4	32/45 = 0.711	4/5 = 0.800	5/6 = 0.833	245/288 = 0.851	$128/45\pi = 0.905$
1, 2, 4, 6	128/175 = 0.731	144/175 = 0.823	6/7 = 0.857	7/8 = 0.875	$512/175\pi = 0.931$
1, 2, 4,…, ∞	$\pi/4 = 0.785$	$9\pi/32 = 0.884$	$75\pi/256 = 0.920$	$1225\pi/4096 = 0.940$	1 = 1.000

Table 7.1 Resultant Efficiencies for Various Combinations of Voltage and Current Harmonic Components

comparison with Class-B operation if at least third-voltage harmonic peaking and fourth-current harmonic peaking are realized. An inclusion of fifth-voltage harmonic component increases the efficiency to 83.3%. An additional inclusion of sixth-current harmonic component into the current waveform and a seventh-voltage harmonic component into the voltage waveform leads to efficiencies up to 94%.

7.2.3 Class F with Quarterwave Transmission Line

Ideally, a control of an infinite number of the harmonics maintaining a square voltage waveform and a half-sinusoidal current waveform at the device output can be provided by using a serious quarterwave transmission line and a parallel-tuned resonant circuit, as shown in Fig. 7.10. This type of a Class-F power amplifier was initially proposed to be used at higher frequencies, where implementation of the load networks with only lumped elements is difficult and the parasitic device output (lead or package) inductor is sufficiently small [9]. In this case, the quarterwave transmission line transforms the load impedance according to

$$R = \frac{Z_0^2}{R_L} \tag{7.36}$$

where Z_0 is the characteristic impedance of a transmission line. For even harmonics, the short circuit on the load side of the transmission line is repeated, thus producing a short circuit at the drain. However, the short circuit at the load produces an open circuit at the drain for odd harmonics with resistive load at the fundamental.

Generally, at low drive level, the active device acts as a current source (voltage-controlled in the case of the MOSFETs or MESFETs and current-controlled in the case of bipolar transistors). As input drive increases, the active device enters saturation resulting in a harmonic-generation process. Because the quarterwave transmission line presents the high impedance conditions to all odd harmonics, all odd

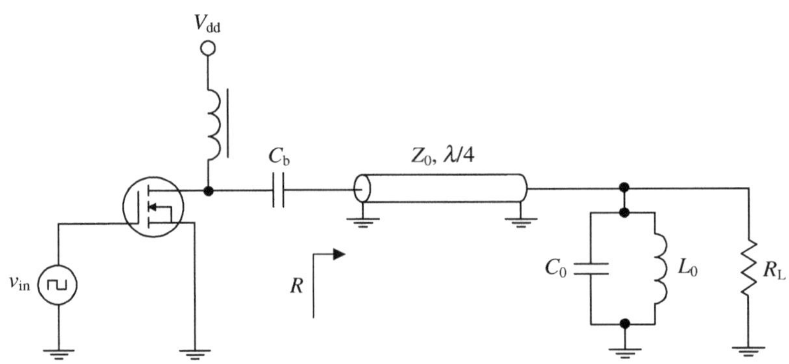

FIGURE 7.10 Class-F power amplifier with series quarterwave transmission line.

harmonics provide a proper contribution to the output voltage waveform. As a result, at high drive level, the output voltage waveform becomes a complete square wave and the active device is saturated for a full half-cycle. In this case, the transistor acts as a switch rather than a saturating current source.

An alternative configuration of the Class-F power amplifier with a shunt transmission line located in between the dc power supply and the device collector is shown in Fig. 7.11(*a*). In this case, there is no need to use an RF choke and a series-blocking capacitor because a series-fundamentally tuned resonant L_0C_0 circuit is used instead of a parallel fundamentally tuned resonant circuit. However, unlike the case with a series quarterwave transmission line, such a Class-F load-network configuration with a shunt quarterwave transmission line does not provide an impedance transformation. Therefore, the load resistance R, which is equal to the equivalent active device output

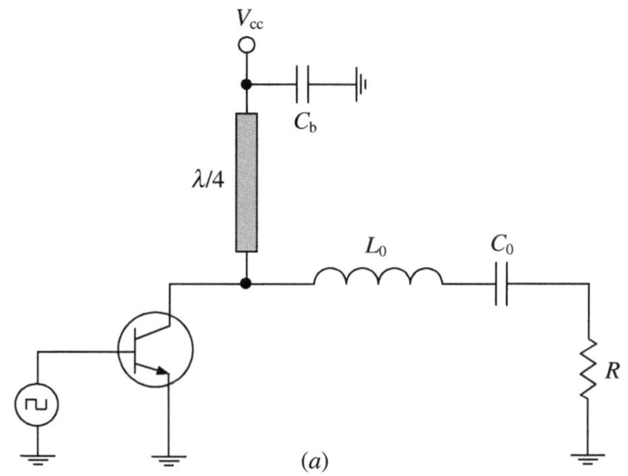

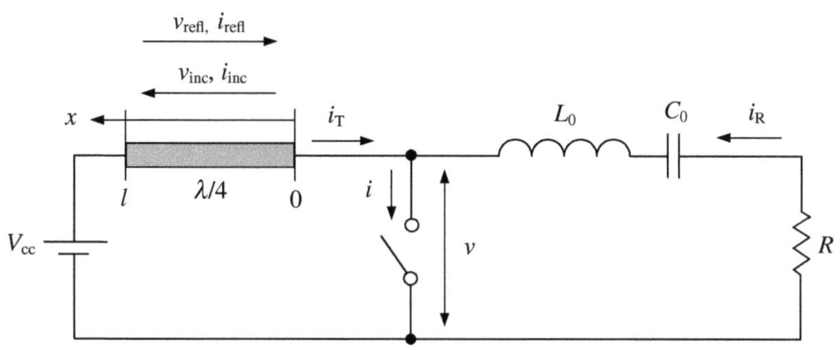

FIGURE 7.11 Class-F power amplifier with shunt quarterwave transmission line.

resistance at the fundamental frequency, must then be transformed to the standard load resistance R_L. Let us now derive analytically some basic fundamental properties of a quarterwave transmission line. The transmission line in the time domain can be represented as an element with finite delay time depending on its electrical length. Consider a simplified load network of the Class-F power amplifier shown in Fig. 7.11(*b*), which consists of a parallel quarterwave transmission line grounded at the end through power supply, a series fundamentally tuned L_0C_0 circuit, and a load resistance R. In an idealized case, the intrinsic device output capacitance is assumed to be negligible to affect the power amplifier RF performance. The loaded quality factor Q_L of the series resonant L_0C_0 circuit is high enough to provide the sinusoidal output current i_R flowing into the load R.

To define the collector voltage and current waveforms, consider the electrical behavior of a homogeneous lossless quarterwave transmission line connected to the dc voltage supply with RF grounding [10]. In this case, the voltage $v(t, x)$ in any cross section of such a transmission line can be represented as a sum of the incident voltage $v_{inc}(\omega t - 2\pi x/\lambda)$ and the reflected voltage $v_{refl}(\omega t + 2\pi x/\lambda)$, generally with an arbitrary waveform. When $x = 0$, the voltage $v(t, x)$ is equal to the collector voltage,

$$v(\omega t) = v(t, 0) = v_{inc}(\omega t) + v_{refl}(\omega t) \tag{7.37}$$

At the same time, at another end of the transmission line when $x = \lambda/4$, the voltage is constant and equal to

$$V_{cc} = v(t, \pi/2) = v_{inc}(\omega t - \pi/2) + v_{refl}(\omega t + \pi/2) \tag{7.38}$$

Because the time moment t was chosen arbitrarily, let us rewrite Eq. (7.38) using a phase shift of $\pi/2$ for each voltage by

$$v_{inc}(\omega t) = V_{cc} - v_{refl}(\omega t + \pi) \tag{7.39}$$

Substituting Eq. (7.39) into Eq. (7.37) yields

$$v(\omega t) = v_{refl}(\omega t) - v_{refl}(\omega t + \pi) + V_{cc} \tag{7.40}$$

Consequently, for the phase shift of π, the collector voltage can be obtained by

$$v(\omega t + \pi) = v_{refl}(\omega t + \pi) - v_{refl}(\omega t + 2\pi) + V_{cc} \tag{7.41}$$

For an idealized operation condition with a 50% duty ratio (or cycle) when during half a period the transistor is turned on and during another half a period the transistor is turned off with overall period of 2π, the voltage $v_{refl}(\omega t)$ can be considered the periodical function with a period of 2π,

$$v_{refl}(\omega t) = v_{refl}(\omega t + 2\pi) \tag{7.42}$$

As a result, the summation of Eqs. (7.40) and (7.41) results in the basic expression for collector voltage in the form

$$v(\omega t) = 2V_{cc} - v(\omega t + \pi) \tag{7.43}$$

From Eq. (7.43), which is similar to Eq. (7.17), it follows that the maximum value of the collector voltage cannot exceed a value of $2V_{cc}$ and the time duration with maximum voltage of $v = 2V_{cc}$ coincides with the time duration with minimum voltage of $v = 0$.

Similarly, the equation for the current i_T flowing into the quarter-wave transmission line can be obtained by

$$i_T(\omega t) = i_T(\omega t + \pi) \tag{7.44}$$

which means that the period of a signal flowing into the quarterwave transmission line is equal to π because it contains only even harmonics, because a shorted quarterwave transmission line has an infinite impedance at odd harmonics at its input.

Let the transistor operate as an ideal switch when it is turned on during the interval $0 < \omega t \le \pi$ where $v = 0$ and turned off during the interval $\pi < \omega t \le 2\pi$ where $v = 2V_{cc}$, according to Eq. (7.43). During the interval $\pi < \omega t \le 2\pi$ when the switch is turned off, the load is directly connected to the transmission line and $i_T = -i_R = -I_R \sin\omega t$. Consequently, during the interval $0 < \omega t \le \pi$ when the switch is turned on, $i_T = I_R \sin\omega t$ according to Eq. (7.44). Hence, the current flowing into the quarterwave transmission line at any ωt can be represented by

$$i_T(\omega t) = I_R |\sin\omega t| \tag{7.45}$$

where I_R is the amplitude of current flowing into the load.

Because the collector current is defined as $i = i_T + i_R$, then

$$i(\omega t) = I_R (\sin\omega t + |\sin\omega t|) \tag{7.46}$$

which means that the collector current represents half-sinusoidal pulses with the amplitude equal to double-load current amplitude.

Consequently, for a purely sinusoidal current flowing into the load due to the infinite loaded quality factor of the series fundamentally tuned L_0C_0 circuit shown in Fig. 7.7(*a*), the ideal collector voltage and current waveforms can be represented by the corresponding normalized square and half-sinusoidal waveforms shown in Figs. 7.7(*b*) and 7.7(*c*), respectively, where I_0 is the dc current. Here, a sum of odd harmonics approximates a square voltage waveform, and a sum of the fundamental and even harmonics approximates a half-sinusoidal collector current waveform. The waveform corresponding to the normalized current flowing into the quarterwave transmission line shown in Fig. 7.12 represents a sum of even harmonics. As a result, the shapes of

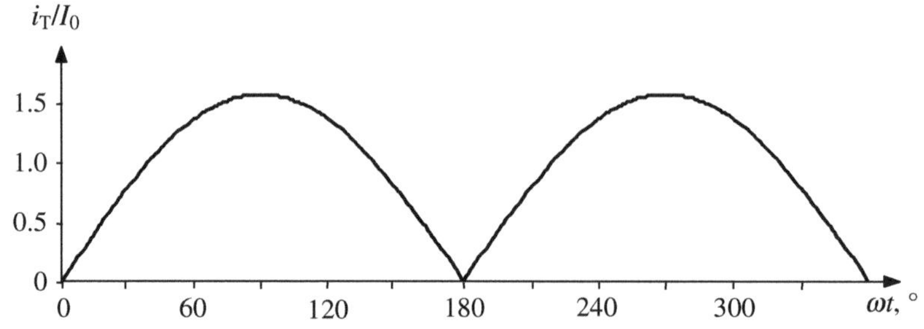

FIGURE 7.12 Ideal current waveform in quarterwave transmission line.

the collector current and voltage waveforms provide a condition where the current and voltage do not overlap simultaneously.

7.2.4 Effect of Saturation Resistance

It is useful to analytically estimate the effect of a saturation (or on-resistance) r_{sat} that is not equal to zero in a real transistor, and transistor therefore dissipates some amount of power due to the collector current flowing through this resistance when the transistor is turned on. The simplified equivalent circuit of a Class-F power amplifier with a quarterwave transmission line where the transistor is represented by a nonideal switch with the saturation resistance r_{sat} and parasitic output capacitance C_{out} is shown in Fig. 7.13. During the interval $0 < \omega t \leq \pi$ when the switch is turned on, the saturation voltage v_{sat} due to the current $i(\omega t)$ flowing through the switch can be written as

$$v_{sat}(\omega t) = V_{sat} \sin\omega t = 2 I_R r_{sat} \sin\omega t \tag{7.47}$$

where, by per Eq. (7.24), the saturation voltage amplitude V_{sat} can be obtained by

$$V_{sat} = 2\, V_R \frac{r_{sat}}{R} = \frac{8\, V_{cc}}{\pi} \frac{r_{sat}}{R} \tag{7.48}$$

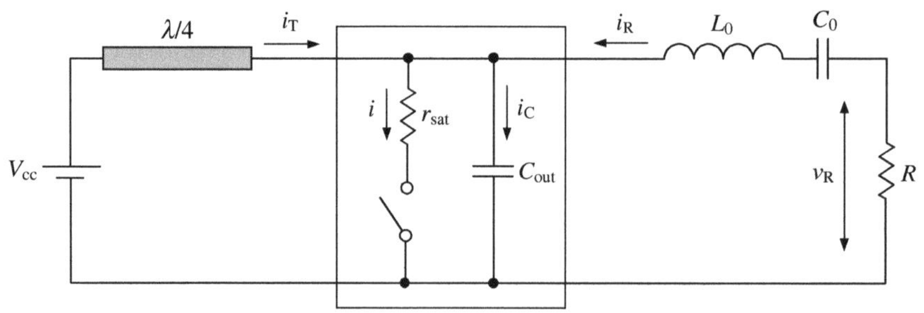

FIGURE 7.13 Effect of parasitic on-resistance and shunt capacitance.

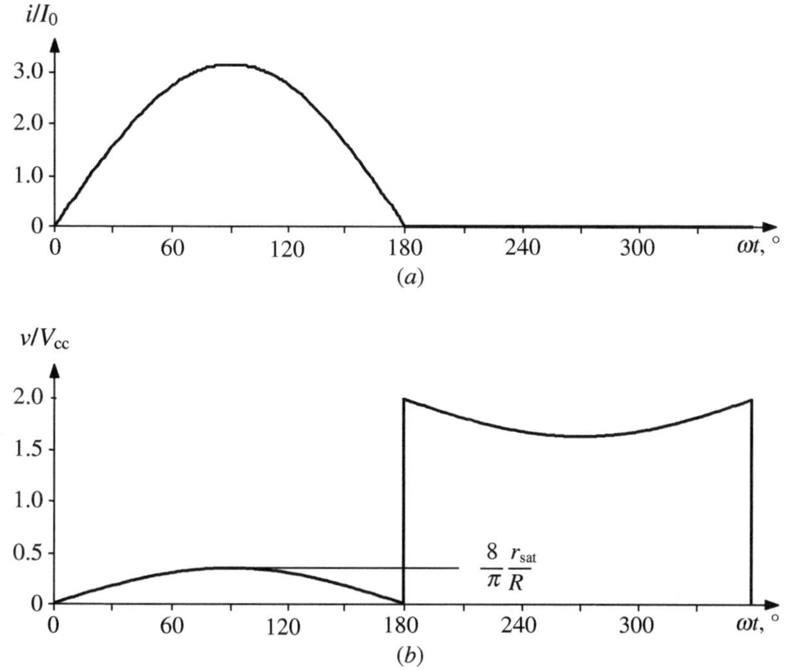

Figure 7.14 Idealized collector current and voltage waveforms with nonzero on-resistance.

The corresponding collector current and voltage waveforms are shown in Fig. 7.14, where the half-sinusoidal current flowing through the saturation resistance r_{sat} causes the deviation of the voltage waveform from the ideal square waveform. In this case, the bottom part of the voltage waveform becomes sinusoidal with the amplitude V_{sat} during the interval $0 < \omega t \le \pi$. From Eq. (7.43), it follows that the same sinusoidal behavior will correspond to the top part of the voltage waveform during the interval $\pi < \omega t \le 2\pi$.

The power losses and collector efficiency due to presence of the saturation resistance r_{sat} can be evaluated using Eqs. (7.20), (7.22), and (7.24) as

$$\frac{P_{sat}}{P_0} = \frac{1}{2\pi}\int_0^{2\pi} \frac{i^2(\omega t)\, r_{sat}}{I_0 V_{cc}}\, d\omega t = \frac{r_{sat}}{2\pi\, I_0 V_{cc}} \int_0^{2\pi} (2I_R)^2 \sin^2 \omega t \; d\omega t$$

$$= \frac{r_{sat} I_R}{V_{cc}} \frac{I_R}{I_0} = \frac{r_{sat}}{R}\frac{I_R}{I_0}\frac{V_R}{V_{cc}} = \frac{2r_{sat}}{R} \tag{7.49}$$

Hence, the collector efficiency can be calculated from

$$\eta = 1 - \frac{P_{sat}}{P_0} = 1 - \frac{2r_{sat}}{R} \tag{7.50}$$

In practice, the idealized collector voltage and current waveforms can be realized at low frequencies when the effect of the device output capacitance is negligible. At higher frequencies, the effect of the output capacitance contributes to a nonzero switching time, resulting in time periods when the collector voltage and collector current exist at the same time when simultaneously $v > 0$ and $i > 0$. Consequently, such a load network with shunt capacitance cannot provide the switchmode operation with an instantaneous transition from the device pinch-off to saturation mode. Therefore, during a nonzero time interval, the device operates in the active region as a nonlinear current source.

7.2.5 Load Networks with Lumped and Distributed Parameters

Theoretical results show that the proper control of only second and third harmonics can significantly increase the collector efficiency of the power amplifier by flattening the output voltage waveform. Because practical realization of a multielement high-order *LC* resonant circuit can cause a serious implementation problem, especially at higher frequencies, it is sufficient to be confined to a three- or four-element resonant circuit composing the load network of the power amplifier. In addition, it is necessary to take into account that, in practice, the combined extrinsic and intrinsic transistor output capacitance has a substantial effect on the efficiency. The device output capacitance C_{out} can represent the collector capacitance C_c in the case of the bipolar transistor or the sum of the drain-source capacitance and gate-drain capacitance, $C_{ds} + C_{gd}$, in the case of the FET device.

For a lumped-circuit power amplifier, a special three-element load network can be used to approximate the ideal Class-F mode by providing both high impedance at the fundamental and third harmonics and zero impedance at the second harmonic at the collector (or drain) by compensating for the influence of C_{out}. Examples of such load networks with additional parallel and series resonant circuits located between the dc power supply and device output are shown in Fig. 7.15 [12, 13]. Here, the output circuit of the active device is represented by a multiharmonic current source, and R_{out} is the equivalent output resistance at the fundamental frequency defined as a ratio of the fundamental voltage at the device output to the fundamental current flowing into the device.

The reactive part of the output admittance (or susceptance) $B_{net} = \mathrm{Im}(Y_{net})$ of the load network with a parallel resonant tank shown in Fig. 7.15(*b*), including the device output capacitance C_{out}, can be written as

$$B_{net} = \omega C_{out} - \frac{1-\omega^2 L_2 C_2}{\omega L_1 (1-\omega^2 L_2 C_2) + \omega L_2} \tag{7.51}$$

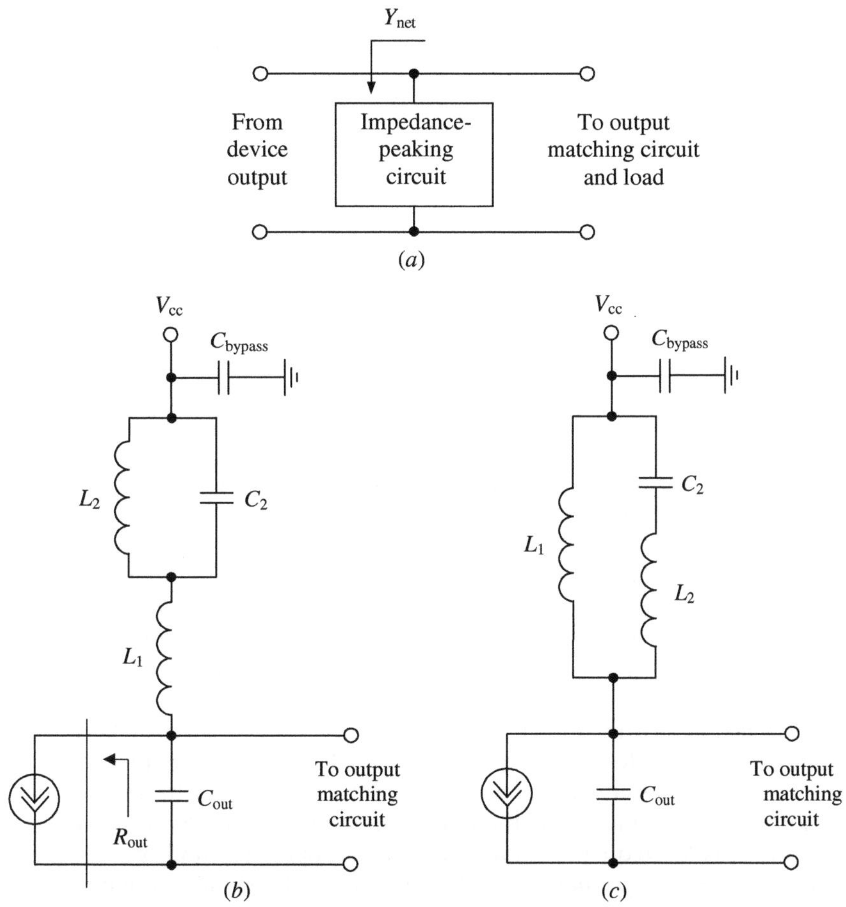

FIGURE 7.15 Load networks with parallel and series resonant circuits.

By applying three-harmonic impedance conditions at the device collector (or drain), open-circuited for the fundamental and third harmonic when $B_{net}(\omega_0) = B_{net}(3\omega_0) = 0$ and short-circuited for the second harmonic when $B_{net}(2\omega_0) = \infty$, the parameters of this impedance-peaking load network can be derived as

$$L_1 = \frac{1}{6\omega_o^2 C_{out}} \qquad L_2 = \frac{5}{3}L_1 \qquad C_2 = \frac{12}{5}C_{out} \tag{7.52}$$

where the sum of the reactance of the parallel resonant tank, consisting of an inductor L_2 and a capacitor C_2, and an inductor L_1 create resonances at the fundamental and third-harmonics, whereas the series capacitive reactance of the tank circuit in series with the inductance L_1 creates a short-circuit series resonance condition at the second-harmonic component [12, 13].

Applying the same conditions for the load network with a series resonant circuit shown in Fig. 7.15(*c*) results in the ratios between elements given by

$$L_1 = \frac{4}{9\omega_0^2 C_{out}} \qquad L_2 = \frac{9}{15} L_1 \qquad C_2 = \frac{15}{16} C_{out} \tag{7.53}$$

where an inductance L_2 and a capacitance C_2 create a short-circuit condition at the second harmonic, and all elements create the parallel-resonant tanks for the fundamental and third harmonics [14].

To determine the transfer performance of the impedance-peaking load network in the frequency domain, it is best to represent such a load network as shown in Fig. 7.15(*a*), then to simulate the small-signal *S*-parameters, and finally plot a magnitude of S_{21} in decibels over wide frequency range. As an example, the frequency-response characteristic of the load network with a parallel resonant circuit, whose parameters are calculated based on the fundamental frequency f_0 = 500 MHz, is shown in Fig. 7.16. In this case, the load-network parameters are C_{out} = 2.2 pF, R_{out} = 200 Ω, C_2 = 5.3 pF, L_1 = 7.7 nH, and L_2 = 12.8 nH with an inductor quality factor Q_{ind} = 20. It should be noted that the power amplifier efficiency can be even higher if the first element of the output matching circuit adjacent to the transistor output is in series and inductive to provide high-impedance conditions at higher-order harmonics.

As a first approximation for comparison between different operation modes, the output device resistance R_{out} at the fundamental frequency required to realize a Class-F operation mode with third-harmonic peaking can be estimated as the equivalent resistance determined at the fundamental frequency for an ideal Class-F operation and written as $R_{out} = R_1^{(F)} = V_1/I_1$, where V_1 and I_1 are the fundamental-frequency

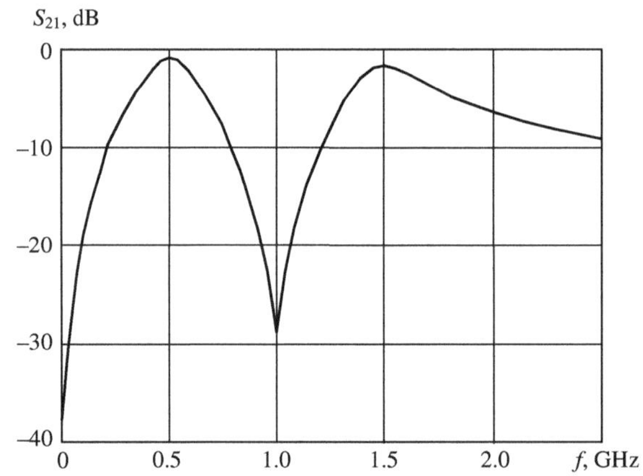

Figure 7.16 Frequency response of load network with parallel resonant circuit.

voltage and current amplitudes at the device output, respectively. For the same supply voltage V_{cc} and output power P_1 at the fundamental, assuming zero saturation voltage and using Eq. (7.28) yield

$$R_1^{(F)} = \frac{8}{\pi^2}\frac{V_{cc}}{I_0} = \frac{8}{\pi^2}\frac{V_{cc}^2}{P_1} = \left(\frac{4}{\pi}\right)^2 R_1^{(B)} \tag{7.54}$$

where $R_1^{(B)} = V_{cc}^2/2P_1$ is the output resistance at the fundamental in an ideal Class-B mode.

The ideal Class-F power amplifier with all even-harmonic short-circuit termination and third-harmonic peaking achieves a maximum drain efficiency of 88.4% [8]. Such an operation mode can be very conveniently realized by using the transmission lines in the load network. The impedance-peaking load-network topology of such a transmission-line power amplifier is shown in Fig. 7.17 [12, 13].

In this case, a quarterwave transmission line TL_1 located between the dc power supply and the drain terminal provides short-circuit termination for even harmonics. The electrical length θ_3 of an open-circuit stub TL_3 is chosen to have a quarter wavelength at the third-harmonic component to realize short-circuit condition at the right-hand side of the series transmission line TL_2, whose electrical length θ_2 should provide an inductive reactance to resonate with the device output capacitance C_{out} at the third harmonic. As a result, the electrical lengths of the transmission lines at the fundamental frequency can be obtained as

$$\theta_1 = \frac{\pi}{2} \quad \theta_2 = \frac{1}{3}\tan^{-1}\left(\frac{1}{3Z_0\,\omega_0 C_{out}}\right) \quad \theta_3 = \frac{\pi}{6} \tag{7.55}$$

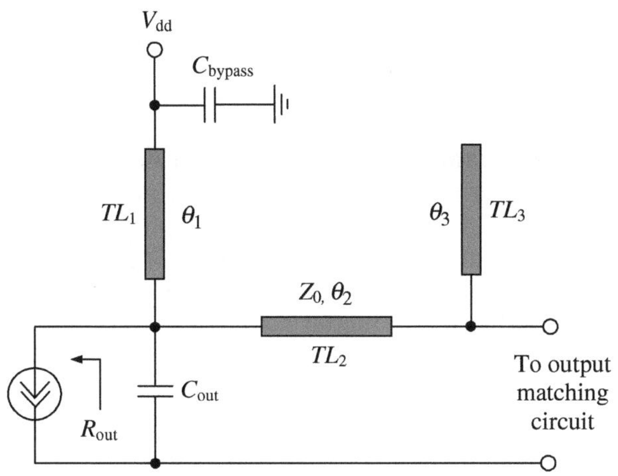

Figure 7.17 Transmission-line impedance-peaking circuit.

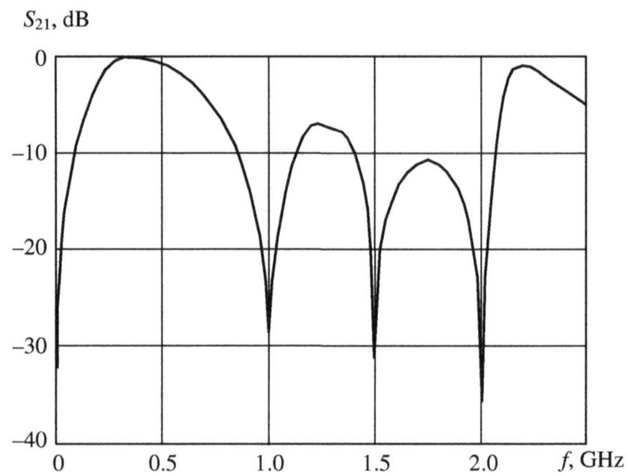

FIGURE 7.18 Frequency response of microstrip impedance-peaking load network.

where Z_0 is the characteristic impedance of the series transmission line TL_2 and ω_0 is the fundamental radian frequency. Figure 7.18 shows an example of the frequency-response characteristic of the microstrip impedance-peaking load network, assuming an alumina substrate for microstrip lines, with the device output resistance $R_{out} = 50\ \Omega$ and output capacitance $C_{out} = 2.2$ pF and the characteristic impedance of microstrip lines $Z_0 = 50\ \Omega$ and its electrical length $\theta_2 = 15°$. From Fig. 7.18, it follows that, for the corresponding short-circuit conditions for all even harmonics and third-harmonic peaking, an additional output impedance matching circuit to compensate for the reactive part and to match the real part of the device output impedance at the fundamental with the standard load impedance of 50 Ω at the fundamental frequency $f_0 = 500$ MHz, or this matching can be achieved by optimizing the characteristic impedances of the series microstrip line and open-circuit stub.

7.2.6 Design Examples of Class-F Power Amplifiers

The effectiveness of the Class-F load-network design technique can be demonstrated based on the example of high-power LDMOSFET amplifiers. The small-signal equivalent circuit of the LDMOSFET cell with a gate length of 1.25 μm and a gate width of 1.44 mm is shown in Fig. 7.19(*a*). The device model parameters were extracted from pulsed current-voltage (*I*–*V*) and small-signal *S*-parameter measurements. The parameters of the small-signal device equivalent circuit are given for Class AB with a quiescent current $I_q = 15$ mA at a supply voltage $V_{dd} = 28$ V. The measured and modeled output I_{ds}-V_{ds} characteristics of the high-power device with a total gate width of 28 × 1.44 mm are shown in Fig. 7.19(*b*) [15]. Based on these characteristics, it is easy

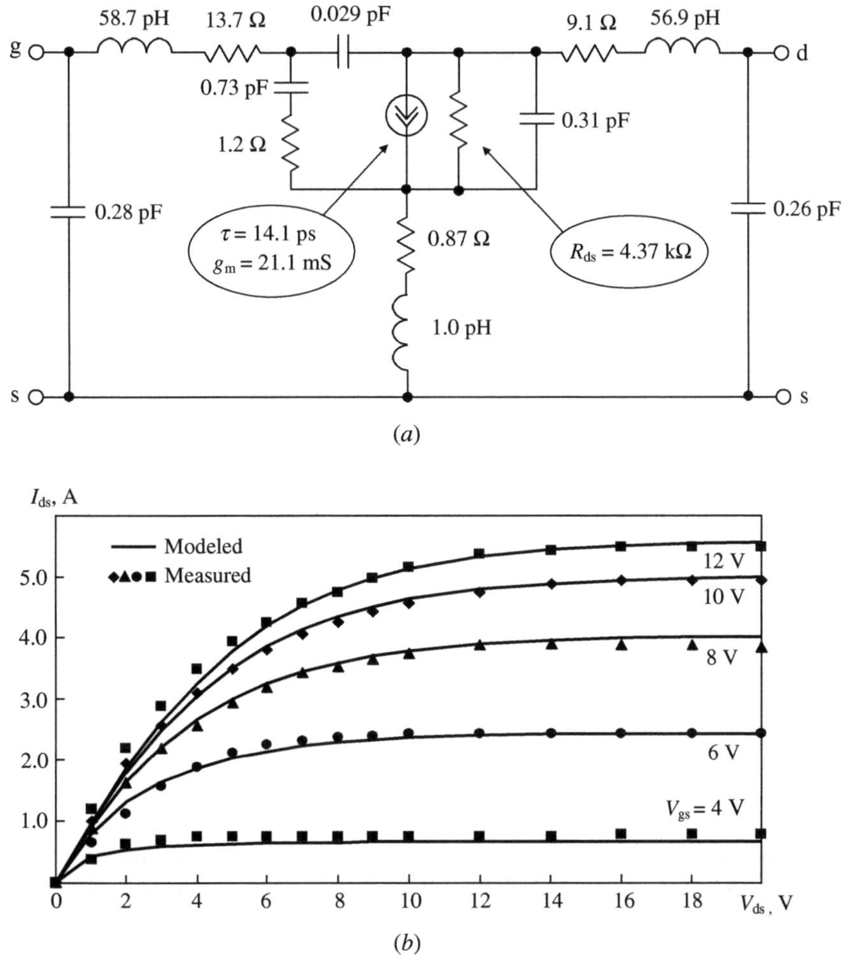

Figure 7.19 Small-signal LDMOSFET equivalent circuit and output *I-V* curves.

to choose the peak drain current, which allows us to maximize the drain efficiency by minimizing the saturation voltage. For example, choosing a peak current of 3.5 A results in a dc current of approximately $3.5/\pi \approx 1.1$ A, according to Eqs. (7.20) and (7.22), that leads to a saturation voltage of about 4 V only. As a result, the maximum drain efficiency of about 80% providing a delivery of the output power of more than 20 W into the load can be achieved using a supply voltage of 24 V.

The circuit schematic of the simulated 500-MHz single-stage lumped LDMOSFET power amplifier is shown in Fig. 7.20, where its load network corresponds to that shown in Fig. 7.15(*b*) and their parameters are calculated from Eq. (7.52). In this particular case, the

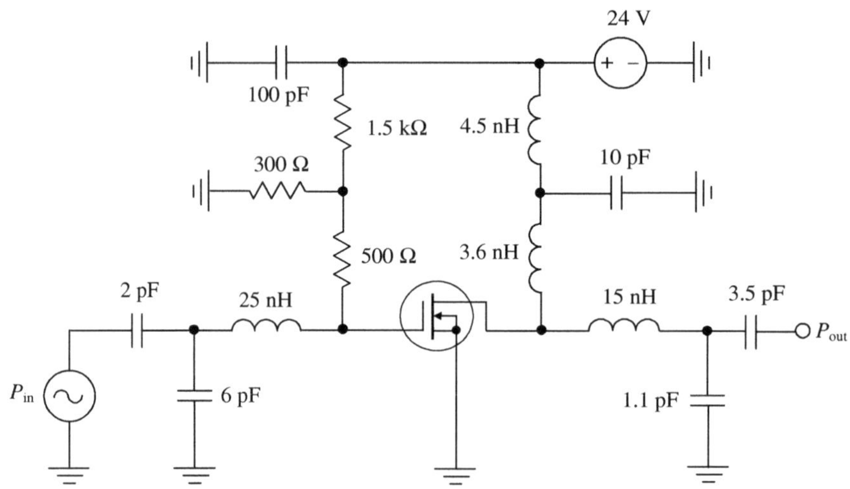

FIGURE 7.20 Simulated lumped LDMOSFET Class-F power amplifier.

total gate width of a high-voltage LDMOSFET device is 7 × 1.44 mm to achieve 8 W of output power. The drain efficiency and power gain of the power amplifier versus input power P_{in} for the case of ideal inductors are given in Fig. 7.21(*a*). The drain efficiency over 75% (curve 2) is obtained due to a short-circuit condition at the second harmonic and open-circuit condition at the third harmonic. Generally, it is important to provide high-impedance conditions at higher-order harmonics that can be readily done by using an output matching circuit with the series inductor as a first element. This shortens the switching time from pinch-off region to voltage-saturation region by better approximating the idealized drain voltage square waveform, as shown in Fig. 7.21(*c*).

As follows from Eq. (7.31) for a symmetrical voltage waveform, the initial phases for the fundamental-frequency and higher-order harmonics should be equal, which is easy to realize by short- and open-circuit conditions. However, according to Eq. (7.32) for a half-sinusoidal current waveform, the phases for any higher-order harmonic component should differ from the phase for the fundamental frequency by 90°. This condition is easily realized in a Class-B load network, where the fundamental component of the drain voltage is in phase with the fundamental component of the drain current, but, for all higher-order current harmonics, the impedance of the resonant circuit will be capacitive because the drain current harmonics mostly flow through the shunt capacitor. Therefore, the accurate harmonic phasing is very important to improve effectiveness of a Class-F load network. The amplifier drain efficiency and power gain will be significantly reduced if the values of the quality factor of the load-network

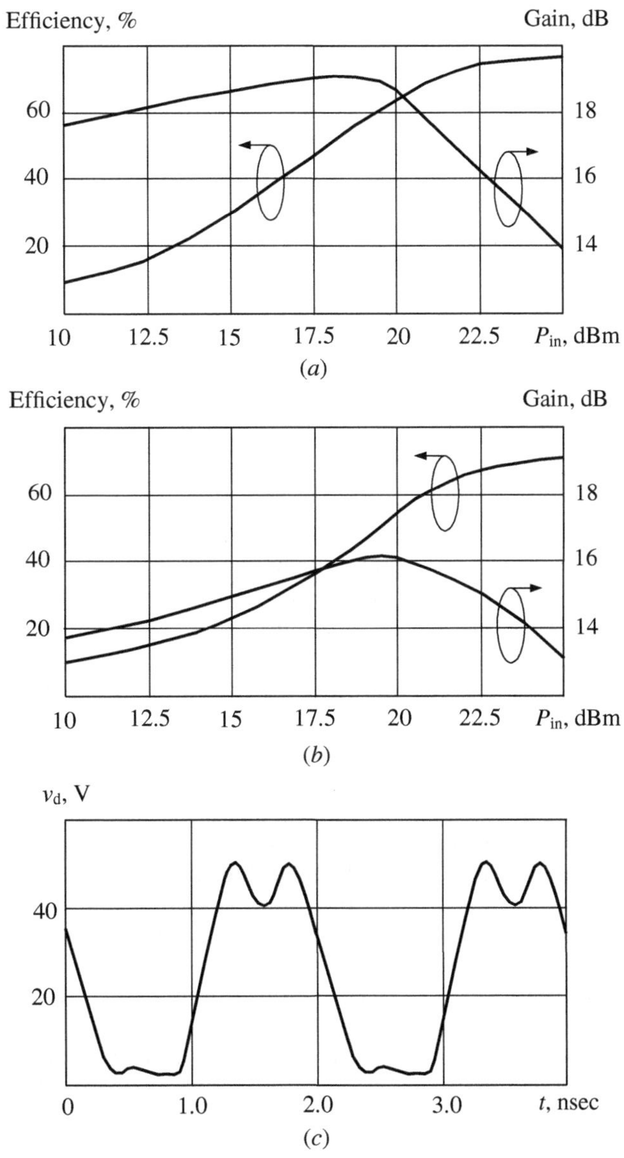

FIGURE 7.21 Drain efficiency, power gain, and voltage waveform.

inductors are sufficiently small. For example, the maximum value of the drain efficiency can reach only 71% when an inductor quality factor at the fundamental frequency is $Q_{ind} = 30$, as shown in Fig. 7.21(*b*).

Therefore, it is preferred at high power level to use the load networks that use microstrip lines. Figure 7.22 shows the equivalent circuit of a simulated 500-MHz single-stage microstrip LDMOSFET

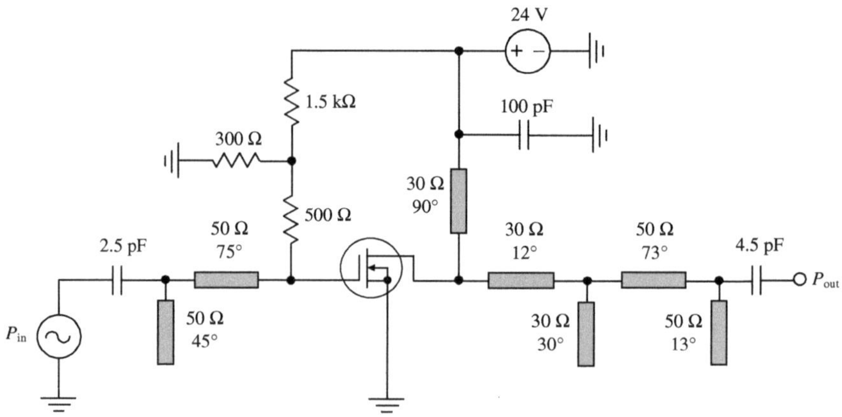

FIGURE 7.22 Simulated microstrip LDMOSFET Class-F power amplifier.

power amplifier using an active device with the same geometry. The input and output matching circuits represent a *T*-type matching circuit each, consisting of a series microstrip line, a parallel open-circuit stub, and a series capacitor. To provide even-harmonic short-circuit termination and third-harmonic peaking for a Class-F mode, an RF grounded quarter-wavelength microstrip line and a combination of the series short-length microstrip line and open-circuit stub with electrical length of 30° at the fundamental frequency are used. Such an output circuit configuration approximates the square drain voltage waveform with a good accuracy, as shown in Fig. 7.23(*a*), and provides the drain efficiency over 75% with a maximum output power of 8 W, as shown in Fig. 7.23(*b*). The resulting smaller value of the drain efficiency compared to the theoretically achievable one can be explained by the non-optimized impedances at higher-order harmonics since, unlike a lumped inductor, the transmission line exhibits an equidistant impedance performance in the frequency domain with consecutive poles and zeros at the characteristic frequencies. This means that using a simple *T*-type transmission-line transformer does not provide high impedance conditions at all higher-order harmonics simultaneously.

Figure 7.24 shows the circuit schematic of a 2-GHz microstrip GaN HEMT power operating in a Class-F mode [16]. The GaN HEMT device on a SiC substrate used in this power amplifier was provided by Cree having a 3.6-mm gate periphery and maximum operating frequency of about 40 GHz. Both input and output matching networks terminate the second, third, and fourth harmonics and some of the higher-order even harmonics using the quarterwave transmission lines. The device output impedance at the fundamental of 70 Ω was chosen for the design as it was a good tradeoff of efficiency and output power. In this case, the load network provides the impedance

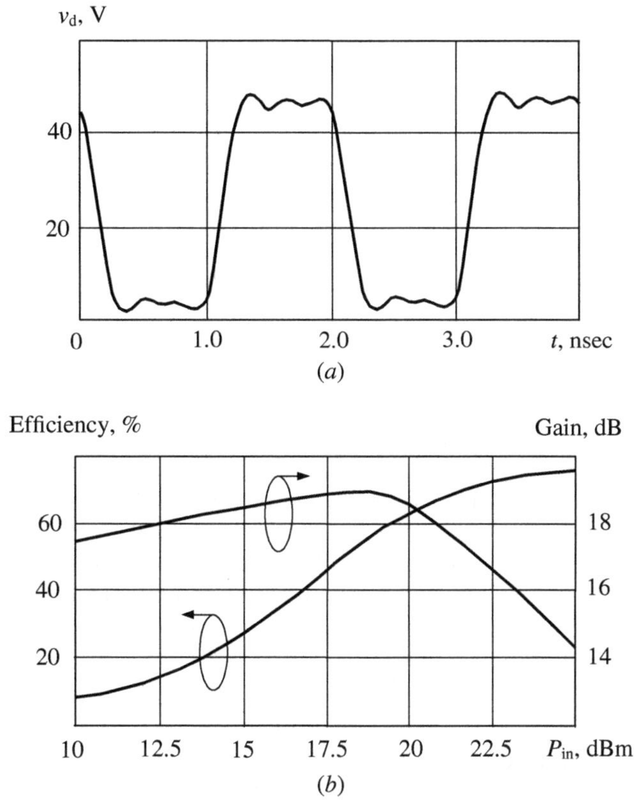

FIGURE 7.23 Drain voltage waveform, efficiency, and power gain.

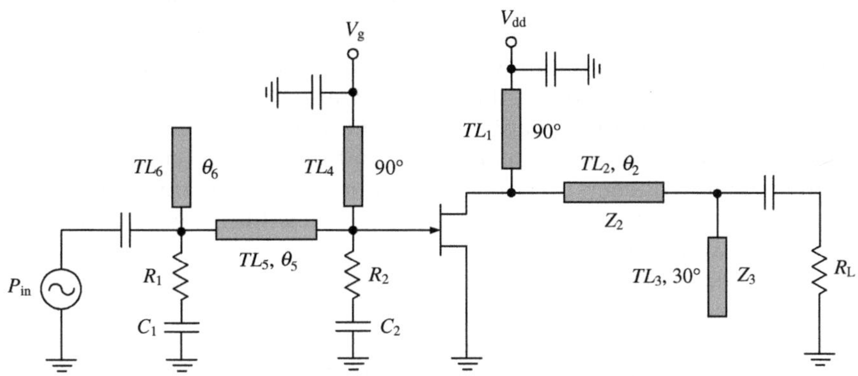

FIGURE 7.24 Circuit schematic of transmission-line Class-F GaN HEMT power amplifier.

matching at the fundamental and the corresponding Class-F harmonic control at the second, third, and fourth harmonics simultaneously. The fundamental matching was provided by choosing the optimum value of the characteristic impedances Z_2 and Z_3. Tuning the output matching network including the device output capacitance resulted in a very high third-harmonic impedance of about 400 Ω, whereas the impedances at the second and fourth harmonics were about 0.5 and 0.7 Ω, respectively. Note that high impedance at the fundamental with corresponding high supply voltage was chosen to minimize the effect of the parasitic bondwire and package inductors in order to provide the near short-circuit Class-F conditions at the second and higher-order even harmonics, whose effect becomes significant at higher operating frequencies. An input matching network was designed to provide a second-harmonic short by using a quarterwave transmission line close to the gate and conjugate matching at the fundamental. Two shunt *RC* networks at the input were added to provide the stability of operation. As a result, the power amplifier achieved the maximum drain efficiency of 87% and *PAE* of 83% at an output power of 17.8 W and a drain supply voltage of 42.5 V, with a maximum power gain of 15.8 dB and its compressed value of 13.4 dB at peak *PAE*. The *PAE* maintained its high value above 80% for drain supply voltages greater than 32.5 V.

7.3 Inverse Class F

The effect of the inclusion of the parallel resonant circuit tuned to the second harmonic and located in series at the anode, as shown in Fig. 7.25(*a*), was first described and analyzed in the early 1940s [5, 17]. It was shown that the symmetrical anode current waveform and level of its depression can be provided with the opposite phase conditions between the fundamental-frequency and second-harmonic components and an optimum value of the ratio between their voltage amplitudes. It was noted that high operation efficiency can be achieved even when impedance of the tank circuit to second harmonic is equal or slightly greater than that of the tank circuit to fundamental frequency. In practical vacuum-tube power amplifiers intended for operation at very high frequencies, the peak output power and anode efficiency can therefore be increased by 1.15 to 1.2 times [18]. In addition, it was later suggested to use an additional resonator, tuned to the fourth harmonic and connected in series with the second-harmonic resonator, as shown in Fig. 7.25(*b*), to maximize the anode efficiency of the vacuum-tube amplifier with approximate square voltage-driving waveform [19].

The simple solution to realize 180° out-of-phase conditions between the voltage fundamental-frequency and second-harmonic components at the device output is to use a second-harmonic tank resonator connected

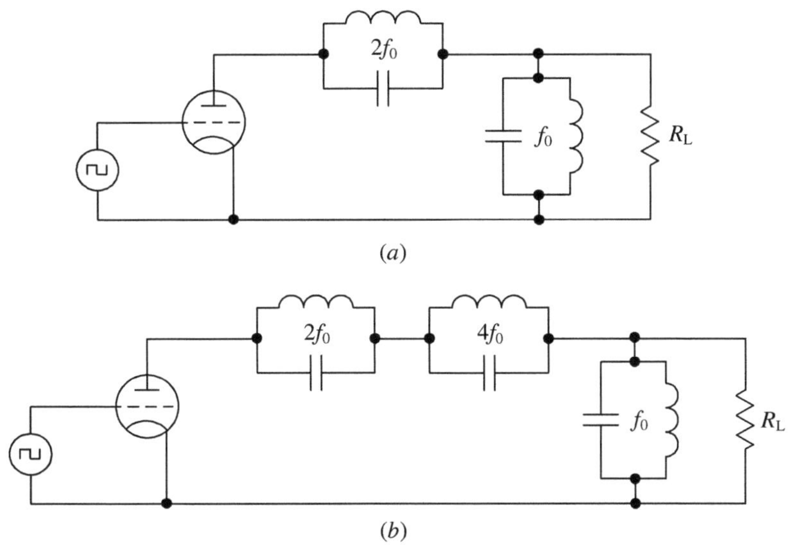

FIGURE 7.25 Biharmonic and polyharmonic power amplifiers.

in series to the device input [17]. Such an approach makes it possible to flatten the anode voltage waveform in active region avoiding the device saturation mode. In this case, the driver stage is loaded by the nonlinear diode-type input grid impedance of the final-stage device providing a flattened grid voltage waveform, which includes the fundamental-frequency and second-harmonic components. The presence of the strong second-harmonic component results in a second-harmonic voltage drop across the resonator. The loaded quality factor of the second-harmonic resonator must be high enough to neglect the voltage drop at the fundamental frequency. As a result, the second-harmonic resonator has no effect on the voltage fundamental-frequency component; however, it provides a phase shift of 180° for the second-harmonic component, as increasing in a voltage drop across the resonator results in decreasing in the voltage drop across the grid-cathode (base-emitter or gate-source) terminals.

Figure 7.26 shows that the shapes of the voltage and current waveforms can be significantly transformed with increased voltage peak factor and current flattening by adding one additional harmonic component with a proper phase. For example, the combination of the fundamental-frequency and third harmonic components with 180° out-of-phase shift at the center of symmetry results in a flattened current waveform with depression in its center, as shown in Fig. 7.26(*a*), which can be minimized by using the proper ratio between the amplitudes of the fundamental and third harmonics. Similarly, the combination of the fundamental and second harmonics, which are in phase at the center of symmetry, sharpens the voltage waveform

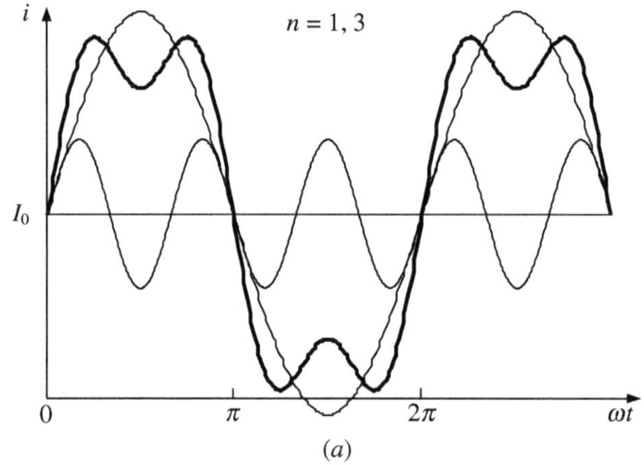

(a)

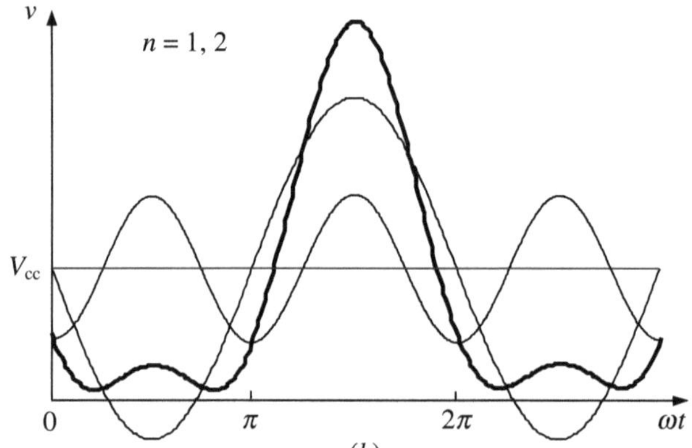

(b)

Figure 7.26 Fourier current and voltage waveforms with third and second harmonics.

corresponding to minimum values of the voltage waveform, as shown in Fig. 7.26(*b*). The optimum ratio between the amplitudes of the fundamental and second current harmonics can maximize the current waveform in one-half of the period and minimize the current waveform during the other half of the period determined by the device saturation resistance in a practical circuit. This means that the power loss due to the active device can be minimized because the results of the integration over the period when minimum current corresponds to maximum voltage will give a much smaller value compared with the power delivered to the load.

7.3.1 Idealized Inverse Class-F Mode

Generally, an infinite number of even-harmonic tank resonators can maintain a square current waveform with a half-sinusoidal voltage

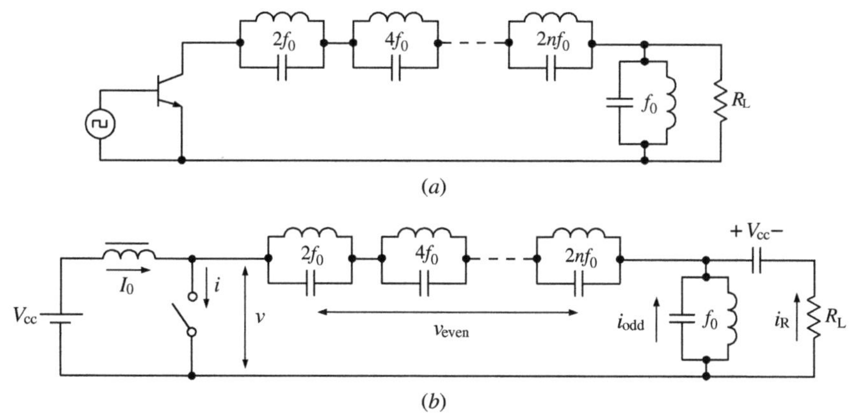

FIGURE 7.27 Basic circuits of inverse Class-F power amplifier with parallel resonant circuits.

waveform at the collector. Figure 7.27(*a*) shows the basic schematic of an inverse Class-F power amplifier with a multiple-resonator output filter to control the harmonic content of its collector (anode or drain) voltage and current waveforms, thereby shaping them to reduce dissipation and to increase efficiency.

The term "inverse" means that collector voltage and current waveforms are interchanged compared to a conventional case under the same idealized assumptions. Consequently, for a purely sinusoidal current flowing into the load, as shown in Fig. 7.28(*a*), the ideal collector current waveform is composed by the fundamental component and odd harmonics approximating a square waveform, as shown in Fig. 7.28(*b*). At the same time, the collector voltage waveform is composed by the fundamental component and even harmonics approximating a half-sinusoidal waveform, as shown in Fig. 7.28(*c*). As a result, the shapes of the collector current and voltage waveforms provide a condition when the current and voltage do not overlap simultaneously, similarly to a conventional Class-F mode. Such a condition, with symmetrical collector voltage and current waveforms, corresponds to an idealized inverse Class-F operation mode with 100% collector efficiency.

Similar analysis of the distribution of voltages and currents in the inverse Class-F load network as for a conventional Class-F mode results in equations for the collector current and voltage waveforms as

$$i(\omega t) = 2I_0 - i(\omega t + \pi) \tag{7.56}$$

where I_0 is the dc current, and

$$v(\omega t) = V_R(\sin\omega t + |\sin\omega t|) \tag{7.57}$$

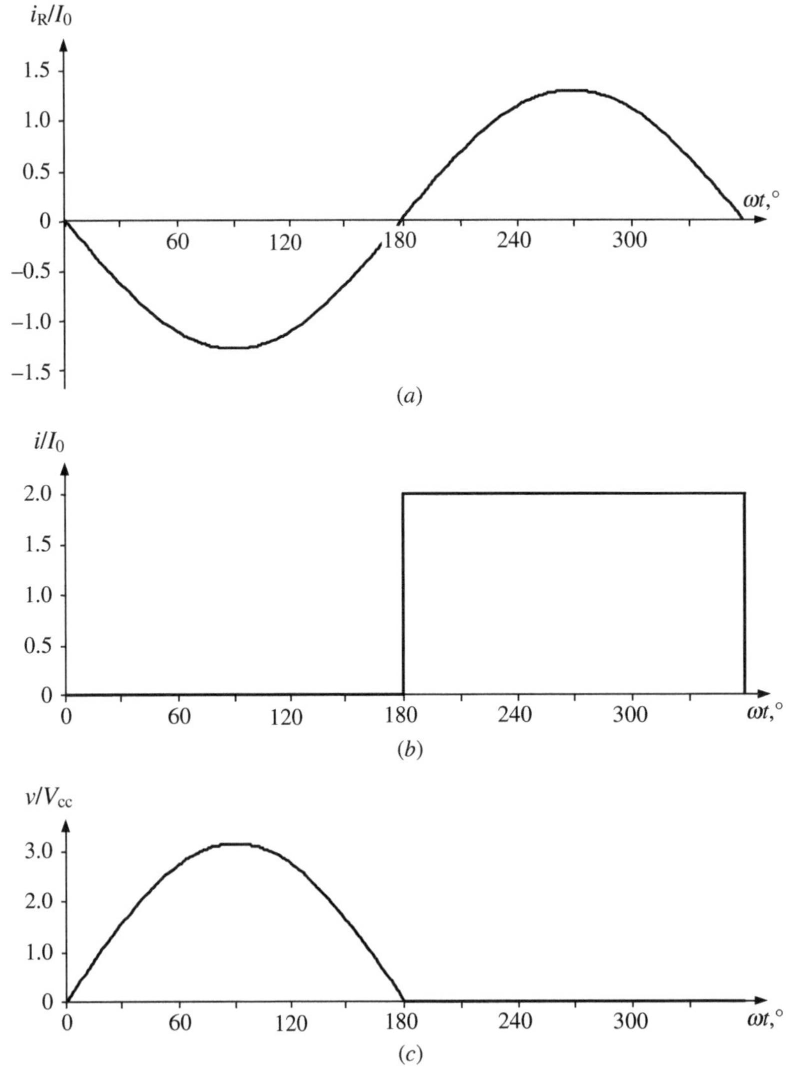

FIGURE 7.28 Ideal waveforms of inverse Class-F power amplifier.

where V_R is fundamental-frequency amplitude at the load. From Eq. (7.56), it follows that maximum value of the collector current cannot exceed that of $2I_0$, and the time duration with maximum amplitude defined as $i = 2I_0$ coincides with that with minimum amplitude defined as $i = 0$. Because the collector current is zero when the switch is turned on, the only possible waveform for the collector current is a square wave composing of only dc, fundamental-frequency, and odd-harmonic components.

By using a Fourier analysis of the current and voltage waveforms, the following equations for the dc voltage, fundamental voltage and current components in the collector voltage and current waveforms can be obtained:

The fundamental current component can be calculated using Eq. (7.56) as

$$I_1 = I_R = \frac{1}{\pi}\int_{\pi}^{2\pi} 2I_0 \sin(\omega t + \pi)\ d\omega t = \frac{4I_0}{\pi} \tag{7.58}$$

The dc voltage V_{cc} can be calculated from Eq. (7.57) as

$$V_{cc} = \frac{1}{2\pi}\int_0^{\pi} 2V_R \sin\omega t\ d\omega t = \frac{2V_R}{\pi} \tag{7.59}$$

The fundamental voltage component can be calculated from Eq. (7.57) as

$$V_1 = \frac{1}{\pi}\int_0^{\pi} 2V_R \sin^2\omega t\ d\omega t = V_R \tag{7.60}$$

Then, the relationship between the dc power P_0 and the output power at the fundamental frequency P_1 can be given as

$$P_1 = \frac{V_1 I_1}{2} = \frac{1}{2}\frac{\pi}{2}\frac{V_{cc}}{}\frac{4I_0}{\pi} = P_0 \tag{7.61}$$

resulting in a theoretical collector efficiency of 100%.

The impedance conditions seen by the device collector for an idealized inverse Class-F mode must be equal to

$$Z_1 = R_1 = \frac{\pi^2}{8}\frac{V_{cc}}{I_0} \tag{7.62}$$

$$Z_{2n+1} = 0 \quad \text{for odd harmonics} \tag{7.63}$$

$$Z_{2n} = \infty \quad \text{for even harmonics} \tag{7.64}$$

7.3.2 Inverse Class F with Quarterwave Transmission Line

An idealized inverse Class-F operation mode can also be represented by using a sequence of the series resonant circuits tuned to the fundamental and odd harmonics, as shown in Fig. 7.29(*a*). In this case, it is assumed that each resonant circuit has zero impedance at the corresponding fundamental frequency f_0 and odd-harmonic components $(2n + 1)f_0$ and infinite impedance at even harmonics $2nf_0$ realizing the idealized inverse Class-F square current and half-sinusoidal voltage waveforms at the device output terminal. As a result, the active device, which is driven to operate as a switch, sees the load resistance R_L at the fundamental frequency, while the odd harmonics are shorted by the series resonant circuits.

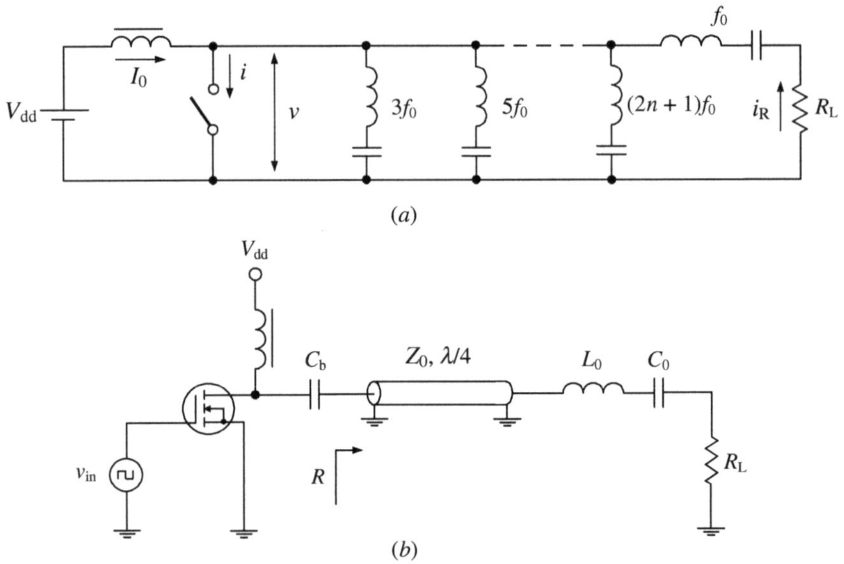

FIGURE 7.29 Inverse Class-F power amplifier with series quarterwave transmission line.

An infinite set of the series resonant circuits tuned to the odd harmonics can be effectively replaced by a quarterwave transmission line with the same operating capability. Such a circuit representation of an inverse Class-F power amplifier with a series quarterwave transmission line loaded by the series resonant circuit tuned to the fundamental frequency is shown in Fig. 7.29(*b*) [7, 20]. The series-tuned output circuit presents a load resistance at the frequency of operation to the transmission line. At the same time, the quarterwave transmission line transforms the load impedance according to

$$R = \frac{Z_0^2}{R_L} \tag{7.65}$$

where Z_0 is the characteristic impedance of a transmission line. For even harmonics, the open circuit on the load side of the transmission line is repeated, thus producing an open circuit at the drain. However, the quarterwave transmission line converts the open circuit at the load to a short circuit at the drain for odd harmonics with resistive load at the fundamental frequency.

Consequently, for a purely sinusoidal current flowing into the load due to infinite loaded quality factor of the series fundamentally tuned circuit, the ideal drain current and voltage waveforms can be represented by the corresponding normalized square and half-sinusoidal waveforms shown in Fig. 7.28(*b*) and 7.28(*c*), respectively. Here, a sum of odd harmonics approximates a square current waveform, and a sum of the fundamental and even harmonics approximates a half-sinusoidal drain voltage waveform. As a result, the shapes of the drain current and voltage waveforms provide a condition when the current and voltage do not overlap simultaneously. The quarterwave transmission line causes the output voltage across the load resistor R_L to be phase-shifted by 90° relative to the fundamental-frequency components of the drain voltage and current.

7.3.3 Load Networks with Lumped and Distributed Parameters

Theoretical results show that the proper control of the second harmonic can significantly increase the collector efficiency of the power amplifier by flattening of the output current waveform and minimizing the product of integration of the voltage and current waveforms. Practical realization of a multielement high-order *LC* resonant circuit can cause a serious implementation problem, especially at higher frequencies and in monolithic integrated circuits, when only three-harmonic components can be effectively controlled. Therefore, it is sufficient to be confined to the three- or four-element resonant circuit composing the load network of the power amplifier. In this case, the operation with the second-harmonic open circuit and third-harmonic short circuit is a promising concept for low-voltage power amplifiers [21].

In addition, it is necessary to take into account that, in practice, both extrinsic and intrinsic transistor parasitic elements such as the output shunt capacitance or serious inductance have a substantial effect on the efficiency. The output capacitance C_{out} can represent the collector capacitance C_c in the case of the bipolar transistor or the sum of drain-source capacitance and gate-drain capacitance, $C_{ds} + C_{gd}$, in the case of the FET device. The output inductance L_{out} is generally composed of the bondwire and lead inductances for a packaged transistor, whose effect becomes significant at higher frequencies.

Figure 7.30 shows the equivalent circuit of the second-harmonic impedance-peaking load network, where the series circuit consisting of an inductor L_1 and a capacitor C_1 creates a resonance at the second harmonic. Because the device output inductance L_{out} and capacitance

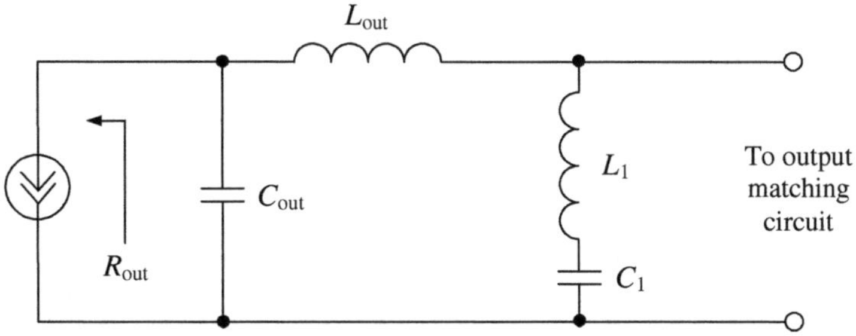

FIGURE 7.30 Second-harmonic impedance-peaking circuit.

C_{out} are tuned to create an open-circuited condition at the second harmonic, the device collector sees resultant high impedance at the second harmonic. To achieve the second-harmonic high impedance, an external inductance may be added to interconnect the device output inductance L_{out} directly at the output terminal (collector or drain) if its value is not sufficient. As a result, the values of the load-network parameters are defined as

$$L_{out} = \frac{1}{4\omega_0^2 C_{out}} \qquad L_1 = \frac{1}{4\omega_0^2 C_1} \tag{7.66}$$

As a first approximation for comparison between different operation classes, the output device resistance R_{out} at the fundamental frequency required to realize an inverse Class-F operation mode with second-harmonic peaking can be estimated as an equivalent resistance $R_{out} = R_1^{(invF)}$ determined at the fundamental frequency for an ideal inverse Class-F mode. For the same supply voltage V_{cc} and output power P_1 at the fundamental, assuming zero saturation voltage and using Eqs. (7.28) and (7.62) yield

$$R_1^{(invF)} = \frac{\pi^2}{8}\frac{V_{cc}}{I_0} = \left(\frac{\pi^2}{8}\right)^2 R_1^{(F)} = \left(\frac{\pi}{2}\right)^2 R_1^{(B)} \tag{7.67}$$

where $R_1^{(F)}$ is the output resistance at the fundamental frequency in a conventional Class-F mode and $R_1^{(B)}$ is the output resistance at the fundamental frequency in an ideal Class-B mode.

The ideal inverse Class-F power amplifier cannot provide all the voltage required by the third- and higher-order odd harmonic short-circuit termination by the use of a single parallel transmission line, as can be easily realized by a quarterwave transmission line for even harmonics in the conventional Class-F power amplifier. In this case, with a sufficiently simple circuit schematic convenient for practical realization, applying the current second-harmonic peaking and voltage third-harmonic shorting can result in a maximum drain efficiency of more than 80% [22]. The output impedance-peaking load network of such a microstrip power amplifier is shown in Fig. 7.31, and its circuit structure is similar to that used to provide a conventional Class-F operation mode.

As it follows from Eq. (7.67), the equivalent output resistance for an ideal inverse Class-F mode is higher by more than 2.4 times compared to a conventional Class-B mode. Therefore, using an inverse Class-F mode simplifies the corresponding load-network design by minimizing the impedance transformation ratio. This is very important for high output power level with a sufficiently small load impedance. However, maximum amplitude of the output voltage waveform can exceed the supply voltage by about three times. In this case, it is required to use the device with high breakdown voltage or to reduce the supply voltage. The latter, however, is not desirable because it may result in lower power gain and efficiency.

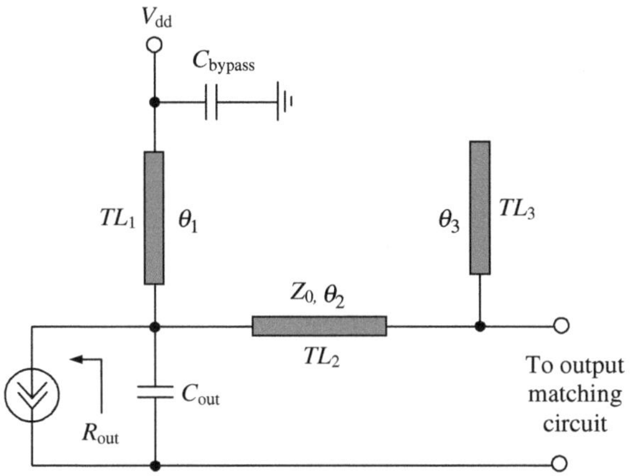

FIGURE 7.31 Transmission-line impedance-peaking circuit.

For such an inverse Class-F microstrip power amplifier, it is necessary to provide the following electrical lengths for the transmission lines at the fundamental frequency:

$$\theta_1 = \frac{\pi}{3} \qquad \theta_2 = \frac{1}{2}\tan^{-1}\left[\left(2Z_0\,\omega_0 C_{out} + \frac{1}{\sqrt{3}}\right)^{-1}\right] \qquad \theta_3 = \frac{\pi}{4} \tag{7.68}$$

where Z_0 is the characteristic impedance of the microstrip lines. The transmission line TL_1 with electrical length $\theta_1 = 60°$ at the fundamental frequency provides a short-circuited condition for the third harmonic and introduces a capacitive reactance at the second harmonic. The open-circuit stub TL_3 with electrical length $\theta_3 = 45°$ creates a short-circuited condition at the end of the transmission line TL_2 at the second harmonic. Thus, the transmission line TL_2 having an inductive reactance is tuned to the parallel resonance condition at the second harmonic with the device output capacitance C_{out} and short-circuited transmission line TL_1.

As an example, Fig. 7.32 shows the frequency-response characteristic of the microstrip impedance-peaking load network, assuming an alumina substrate for microstrip lines, with the device output resistance $R_{out} = 50\ \Omega$ and output capacitance $C_{out} = 2.2$ pF and the microstrip-line characteristic impedance $Z_0 = 50\ \Omega$ and its electrical length $\theta_2 =$ 19°. From Fig. 7.32, it follows that, for the second-harmonic peaking and third-harmonic short circuit termination, an additional output matching at the fundamental frequency $f_0 = 500$ MHz is required, taking into account the reactance introduced by the impedance-peaking load network at the fundamental frequency.

7.3.4 Design Examples of Inverse Class-F Power Amplifiers

The effectiveness of the transmission-line load-network design technique for inverse Class-F application can be demonstrated by the

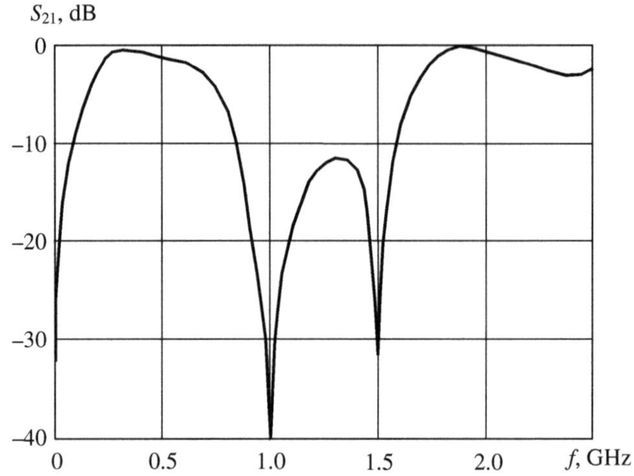

FIGURE 7.32 Frequency response of microstrip impedance-peaking circuit.

example of a 28-V high-power LDMOSFET amplifier with the device of the same geometry as for a conventional Class-F mode, the small-signal equivalent circuit and output current-voltage characteristics of which are shown in Fig. 7.19. The circuit schematic of the simulated 500-MHz single-stage microstrip inverse Class-F power amplifier is shown in Fig. 7.33. In this case, both input and output matching circuits represent the T-type low-pass matching circuits with a series microstrip line, a shunt open-circuit stub having a capacitive reactance, and a series capacitor. To provide the third-harmonic short-circuit termination and second-harmonic peaking corresponding to an inverse Class-F operation mode, the short-circuited microstrip line

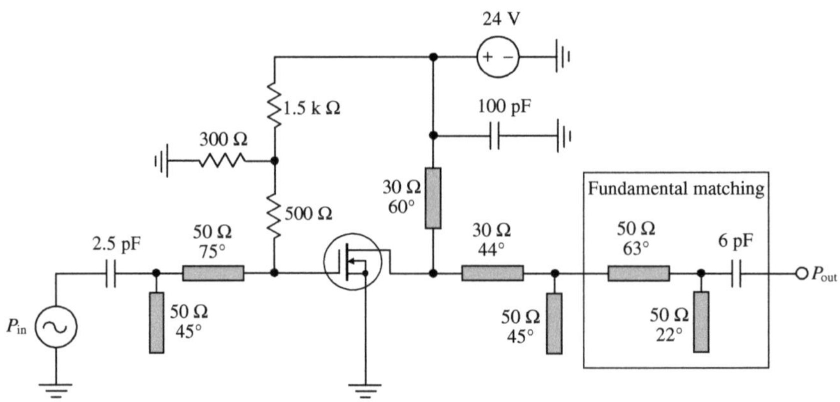

FIGURE 7.33 Simulated 500-MHz single-stage microstrip power amplifier with T-transformer.

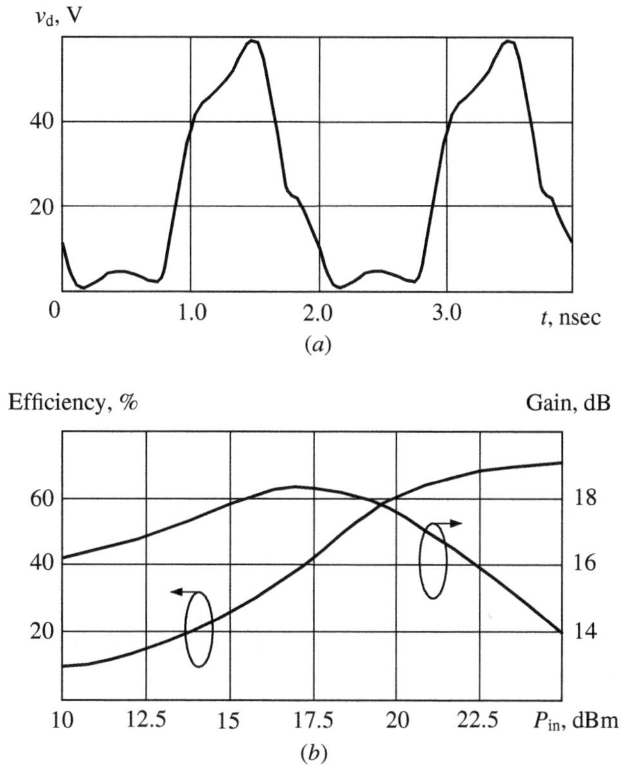

FIGURE 7.34 Drain voltage waveform, efficiency, and power gain.

with electrical length of 60° at the fundamental frequency and combination of a series microstrip line and an open-circuit stub with electrical length of 45° for the corresponding second-harmonic shorting are used.

Figure 7.34(*a*) shows the drain voltage waveform, which is only slightly different from the half-sinusoidal shape due to the effect of the fourth- and higher-order harmonics that are not properly controlled. Nevertheless, a drain efficiency of up to 71% with a maximum output power of 8 W was simulated, as shown in Fig. 7.34(*b*). In this case, the peak value of the drain voltage is more than two times greater than the drain supply voltage of 24 V, resulting in a peak factor approximately equal to $58/24 = 2.4$.

To minimize the number of the circuit elements, the output fundamental matching circuit of such a transmission-line power amplifier can be combined with the harmonic-controlled network. Simulation results indicate that the load network designed to provide the second- and third-harmonic control can perform a function of the matching

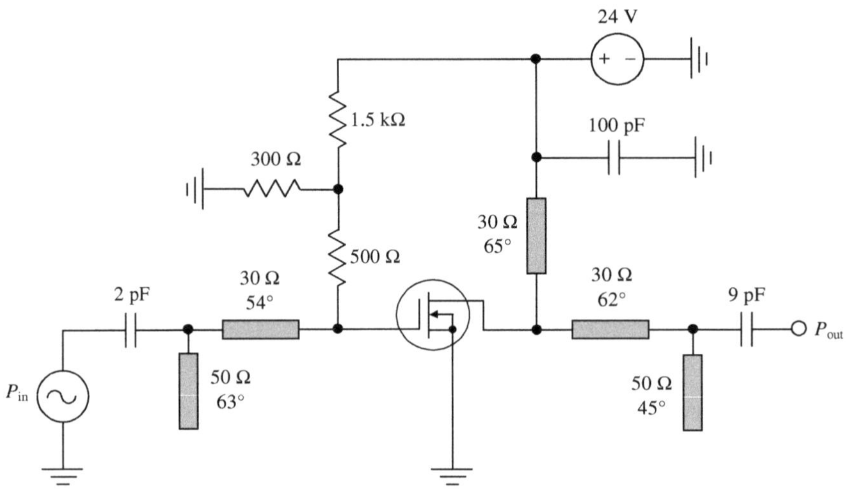

FIGURE 7.35 Schematics of simulated 500-MHz microstrip power amplifier.

circuit as well, together with the series capacitor required for dc blocking. The simplified and slightly modified circuit schematic of the simulated 500-MHz single-stage microstrip inverse Class-F high-power amplifier with a total LDMOSFET channel width of 28 × 1.44 mm is shown in Fig. 7.35.

As a result, the drain efficiency up to 78% for an output power of about 25 W with a power gain of 14 dB can be achieved, as shown in Fig. 7.36(*a*). An analysis of the drain voltage and current waveforms plotted in Fig. 7.36(*b*) indicates that the operation mode obtained is close to an inverse Class-F mode, where the drain current waveform approximates a square wave, while the drain voltage waveform is close to a half-sinusoidal waveform. It should be noted that the negative current values are due to the current flowing through the intrinsic drain-source capacitance when the device multiharmonic voltage-controlled current source is pinched off. Besides, there is a small phase shift between the voltage and current waveforms due to uncompensated phases at the harmonics using such a simple load network, which provides both matching at the fundamental frequency and second- and third-harmonic control. In this case, the maximum drain voltage amplitude does not even reach a value of 60 V.

In a hybrid power amplifier where the packaged device is used, the presence of a transistor output series bondwire and lead inductance L_{out} creates some problems in providing an acceptable second- or third-harmonic open- or short-circuit termination. In this case, it is convenient to use a series transmission line as a first element of the load network connected to the device output, as shown in Fig. 7.37(*a*), where the transmission line TL_1 is placed between the device drain

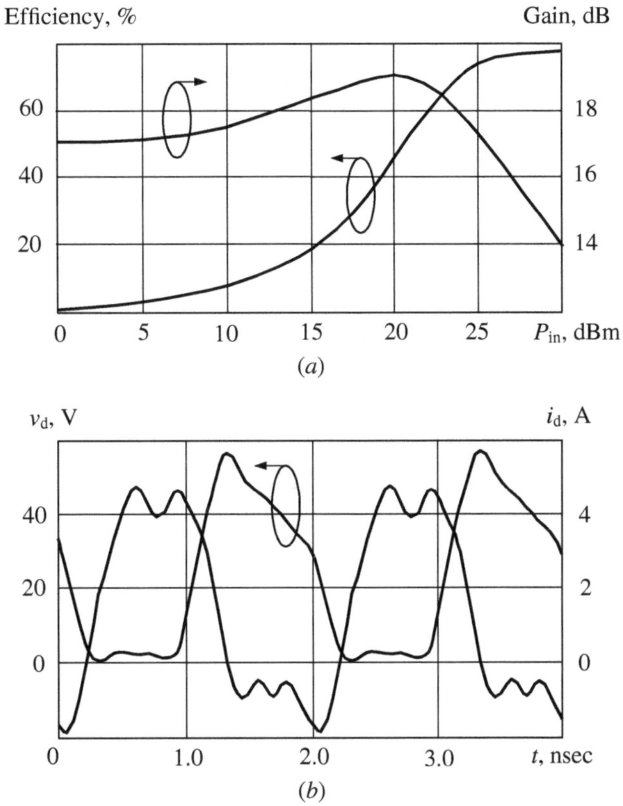

FIGURE 7.36 Drain efficiency, power gain, voltage and current waveforms.

and shunt short-circuited quarterwave transmission line TL_3. However, if the length of combined series transmission line $TL_1 + TL_2$ becomes very long in a Class-F mode with a short circuit at the second harmonic and an open circuit at the third harmonic and additional fundamental-frequency matching circuit is required, then such a load network in an inverse Class-F mode is compact, convenient for harmonic tuning, and very practical.

Figure 7.37(*b*) shows the equivalent circuit of a transmission-line inverse Class-F load network, where the complex-conjugate load matching is provided at the fundamental frequency and both high reactance at the second harmonic and low reactance at the third harmonic are created at the device output by using two series transmission lines TL_1 and TL_2, the electrical lengths of which depend on the values of the device output shunt capacitance C_{out} and series inductance L_{out}, a quarterwave short-circuit stub TL_3, and an open-circuit stub TL_4 with electrical length of 30° [10, 23]. The output shunt capacitance C_{out} can represent both intrinsic bias-dependent drain-source

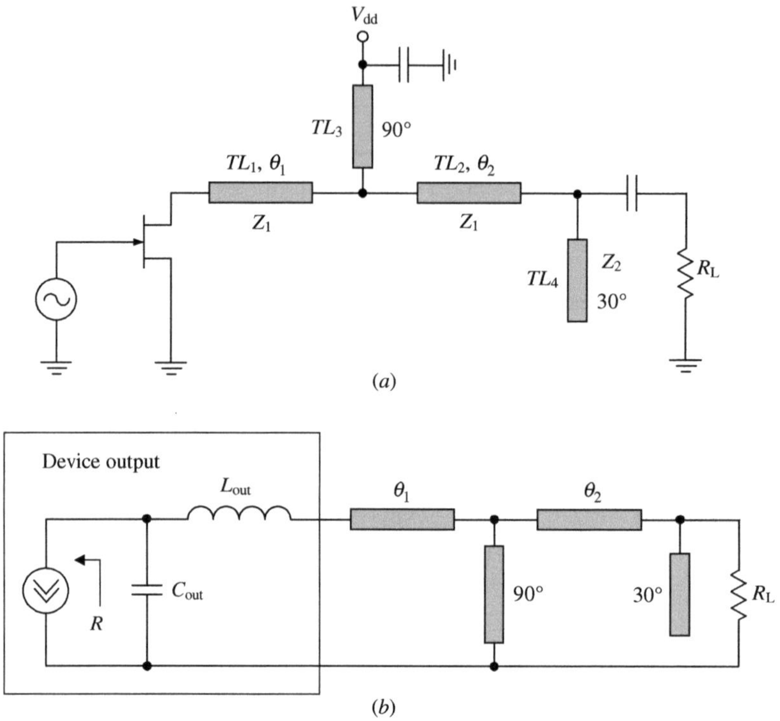

FIGURE 7.37 Transmission-line inverse Class-F power amplifier and its equivalent circuit.

capacitance C_{ds} and extrinsic bias-independent drain pad-contact capacitance C_{dp} of the nonlinear large-signal equivalent circuit for GaN HEMT device, whereas the series output inductance L_{out} is modeled by a combined effect of the metallization, bond wire, and package inductances [24].

The harmonic conditions for an inverse Class-F load network seen by the device multiharmonic current source derived from Eqs. (7.62) to (7.64) for the first three-harmonic components including fundamental are

$$\operatorname{Re} Z_{net}(\omega_0) = R \tag{7.69}$$

$$\operatorname{Im} Z_{net}(2\omega_0) = \infty \tag{7.70}$$

$$\operatorname{Im} Z_{net}(\omega_0) = \operatorname{Im} Z_{net}(3\omega_0) = 0 \tag{7.71}$$

where the load resistance (or equivalent output resistance) R seen by the device output at the fundamental frequency is defined in an ideal inverse Class-F mode by Eq. (7.67).

Figure 7.38(*a*) shows the transmission-line load network seen by the device multiharmonic current source at the fundamental frequency,

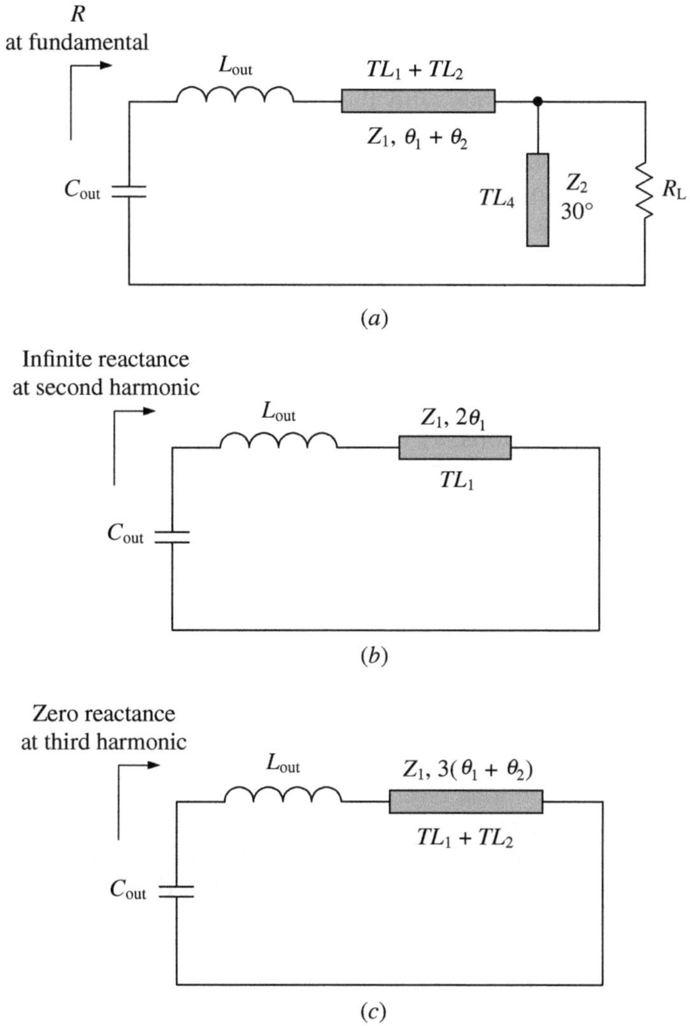

FIGURE 7.38 Load networks seen by the device output at corresponding harmonics.

where the combined series transmission line $TL_1 + TL_2$ (together with an open-circuit capacitive stub TL_4 with electrical length of 30°) provides an impedance matching between the optimum equivalent output device resistance R and the standard load resistance R_L by proper choice of the transmission-line characteristic impedances Z_1 and Z_2, where C_{out} and L_{out} are the elements of the matching circuit. For simplicity of calculation, the characteristic impedances of the transmission lines TL_1 and TL_2 are set to be equal to Z_1.

The load network seen by the device current source at the second harmonic (taking into account the short-circuit effect of the grounded quarterwave transmission line TL_3) is shown in Fig. 7.38(*b*), where the transmission line TL_1 provides an open-circuit condition for the second harmonic at the device output by forming a second-harmonic tank together with the shunt capacitor C_{out} and series inductance L_{out}. Similar load network at the third harmonic is shown in Fig. 7.38(*c*), where the open-circuit effect of the grounded quarterwave transmission line TL_3 and short-circuit effect of the open-circuit stub TL_4 at the third harmonic are used. In this case, the combined transmission line $TL_1 + TL_2$, which is short-circuited at its right-hand side and connected in series with an inductance $L_{out,}$ provides a short-circuit condition for the third harmonic at the device output. Depending on the actual physical length of the device package lead, the on-board adjustment of the transmission lines TL_1 and TL_2 can easily provide the required open-circuit and short-circuit conditions, as well as an impedance matching at the fundamental frequency, because of their series connection to the device output.

By using Eqs. (7.70) and (7.71), the electrical lengths of the transmission lines TL_1 and TL_2, assuming the same characteristic impedance Z_1 for both series transmission-line sections, can be defined from

$$2\omega_0 C_{out} - \frac{1}{2\omega_0 L_{out} + Z_1 \tan 2\theta_1} = 0 \tag{7.72}$$

$$3\omega_0 L_{out} + Z_1 \tan 3(\theta_1 + \theta_2) = 0 \tag{7.73}$$

with the maximum total electrical length $\theta_1 + \theta_2 = \pi/3$ or 60° at the fundamental frequency or 180° at the third harmonic when $L_{out} = 0$.

As a result, the electrical lengths of the transmission lines TL_1 and TL_2 as analytical functions of the device output series inductance L_{out} and shunt capacitance C_{out} are obtained as

$$\theta_1 = \frac{1}{2}\tan^{-1}\frac{1-(2\omega_0)^2 L_{out} C_{out}}{2Z_1\,\omega_0 C_{out}} \tag{7.74}$$

$$\theta_2 = \frac{\pi}{3} - \frac{1}{3}\tan^{-1}\frac{3\omega_0 L_{out}}{Z_1} - \theta_1 \tag{7.75}$$

where the transmission-line characteristic impedance Z_1 can be set in advance.

In order to omit an additional matching section at the fundamental frequency, the inverse Class-F load network can also be used to match the equivalent device fundamental-frequency impedance R with the standard load impedance R_L (usually equal to 50 Ω). In this case, it is necessary to properly optimize both characteristic impedances Z_1 and Z_2. Figure 7.39 shows the equivalent representation of an inverse Class-F load network (including the device output parameters L_{out} and C_{out}) by a lumped low-pass π-type matching circuit where $C = \tan 30°/\omega_0 Z_2$ and $L \approx (Z_1/\omega_0)$

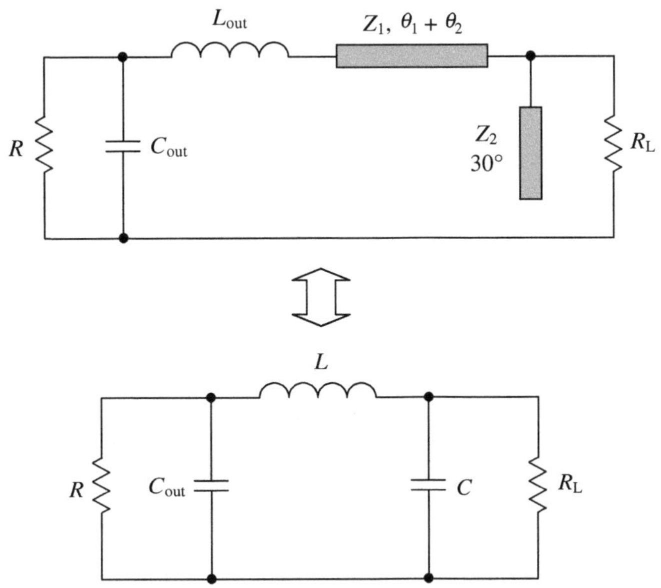

FIGURE 7.39 Equivalent representations of load network at fundamental frequency.

$\sin(\theta_1 + \theta_2) + L_{out}$ due to the sufficiently short length of the combined transmission line $TL_1 + TL_2$, typically much less than 60° at the fundamental frequency depending on the device output parameters. As a result,

$$Z_1 \cong \frac{\omega_0 (L - L_{out})}{\sin(\theta_1 + \theta_2)} \tag{7.76}$$

$$Z_2 = \frac{1}{\omega_0 C\sqrt{3}} \tag{7.77}$$

Figure 7.40(*a*) shows the test board of a transmission-line inverse Class-F power amplifier based on a 28-V 10-W Cree GaN HEMT power transistor CGH40010P and transmission-line load network with the second- and third-harmonic control, as shown in Fig. 7.37(*a*). The input matching circuit provides the fundamental-frequency complex-conjugate matching with the standard 50-Ω source. The parameters of the series transmission line in the load network were optimized for implementation convenience. In this case, the device input and output package leads as external elements were properly modeled to take into account the effect of their inductances, and their models were then added to the simulation setup. The simulation results of a transmission-line inverse Class-F GaN HEMT power amplifier shown in Fig. 7.40(*b*) are based on a nonlinear device model supplied by Cree and technical parameters for a 30-mil RO4350 substrate.

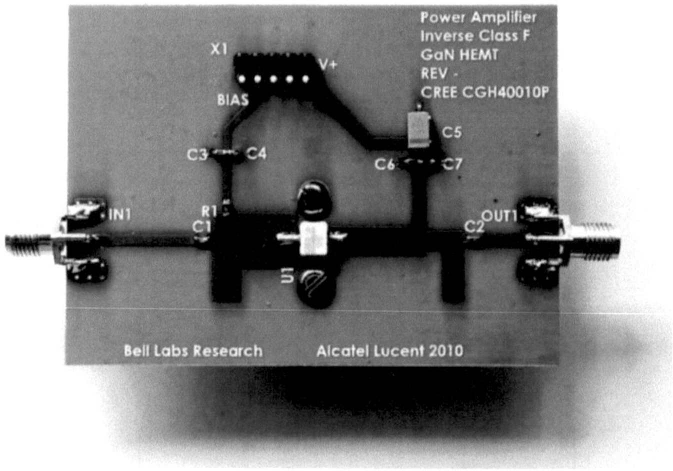

(*a*)

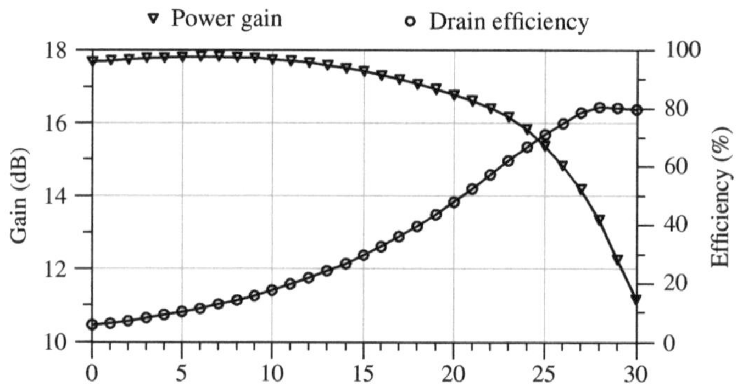

(*b*)

FIGURE 7.40 Transmission-line 10-W inverse Class-F GaN HEMT power amplifier.

The maximum output power of 41.3 dBm, power gain of 13.3 dB (linear gain of about 18 dB), drain efficiency of 80.3%, and *PAE* of 76.5% are achieved at an operating frequency of 2.14 GHz with a supply voltage of 28 V and a quiescent current of 40 mA. The experimental results of the test board shown in Fig. 7.40(*a*) were very close to the simulated results with a maximum output power of 41.0 dBm, a drain efficiency of 76.0%, a *PAE* of 72.2%, and a power gain of 13.0 dB at an operating frequency of 2.14 GHz (gate bias voltage $V_g = -2.8$ V and drain supply voltage $V_{dd} = 28$ V), achieved without any tuning of the input matching circuit and load network [10].

7.4 Class E with Shunt Capacitance

As it was shown many decades ago, using resonant circuits in the load network tuned to the odd and/or even harmonics of the fundamental frequency can generate biharmonic or polyharmonic operation modes of vacuum-tube power amplifiers, which is very effective to increase their operating efficiency. This implies ideally the in-phase or out-of-phase harmonic conditions when the symmetrical flattened voltage or current waveforms can be created. However, as it turned out later, this is not the only way to improve the power amplifier efficiency. Figure 7.41 shows the circuit schematic of a vacuum-tube power amplifier with a parallel-tuned *LC* circuit inserted between the anode and the output matching circuit, which has a resonant frequency equal to about 1.5 times the carrier frequency of the signal to be amplified [25]. In other words, if the carrier signal is transmitting at a fundamental frequency f_0, the parallel resonant circuit will have a resonant frequency of about $1.5f_0$ followed by a filter or output matching circuit to suppress the harmonics of the fundamental frequency and to maximize the output power at the fundamental frequency delivered to the standard load. As a result, an efficiency of 89% was achieved for a 3.2-MHz vacuum-tube high-power amplifier.

Although it was assumed that such a parallel resonant circuit introduces considerable impedance to its own second harmonic, which is the third harmonic $3f_0$ of the carrier frequency and can result in a flattened anode voltage waveform, another interesting and nontrivial conclusion can be derived from this circuit topology. In this case, provided the output π-type matching circuit has purely resistive impedance at the fundamental frequency and capacitive reactances at the harmonic components, the anode of the device sees inductive impedance at the fundamental frequency and capacitive reactances at the second- and higher-order harmonic components. This means that the

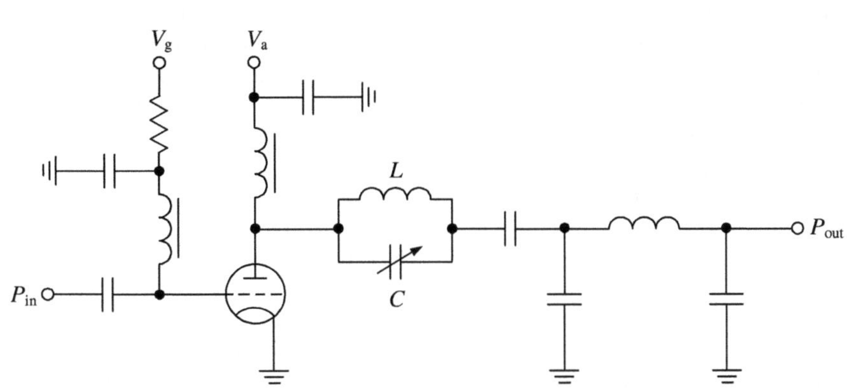

FIGURE 7.41 Class-C power amplifier with detuned resonant circuit.

voltage and current waveforms are not symmetrical anymore, thus representing an alternative approach of the efficiency improvement. Such an effect of increasing efficiency when the output resonant circuit of the vacuum-tube Class-C power amplifier is detuned relatively to the carrier frequency was first described in the early 1960s [26]. In this case, the anode efficiencies of about 92 to 93% were achieved for the phase angles of the output load network in limits of 30° to 40°, resulting in the proper inductive impedance at the fundamental frequency and capacitive reactances at the harmonic components seen by the anode of the active device.

A few years later, it was discovered that very high efficiencies could be obtained with a series resonant *LC* circuit connected to a transistor [27]. The reasons for this high efficiency are that the transistor operates in a pure switching, and the voltage across the transistor and the current flowing through it can both be made equal to zero during the switching transient interval mode when a proper choice of the transistor and circuit parameters are provided. To satisfy such a high-efficiency condition, the transistor current and voltage should be near zero at the time just prior to the conduction interval when the transistor goes into the saturation mode. The series-tuned circuit must appear inductive at the operating frequency. In this case, a loaded quality factor of the series-tuned circuit of about 10 will give a good sinusoidal shape to the load current. As a result, a 20-W, 500-kHz bipolar power amplifier was built having a collector efficiency of 94% with a conduction angle of 180°. The exact theoretical analysis of the single-ended switchmode power amplifier with a shunt capacitance and a series *LC* circuit was then given by Kozyrev [28].

7.4.1 Optimum Load-Network Parameters

However, the single-ended switchmode power amplifier with a shunt capacitance was introduced by Sokals in 1975 as a Class-E power amplifier, and it has found widespread application because of its design simplicity and high operation efficiency [29, 30]. This type of high-efficiency power amplifiers was then widely used in different frequency ranges and with different output power levels ranging from several kilowatts at low RF frequencies up to about 1 W at microwaves [31]. The characteristics of a Class-E power amplifier can be determined by finding its steady-state collector voltage and current waveforms. The basic circuit of a Class-E power amplifier with shunt capacitance is shown in Fig. 7.42(*a*), where the load network consists of a capacitor *C* shunting the transistor, a series inductor *L*, a series fundamentally tuned L_0C_0 circuit, and a load resistor *R*. In a common case, a shunt capacitance *C* can represent the intrinsic device output capacitance and external circuit capacitance added by the load network. The collector of the transistor is connected to the supply voltage

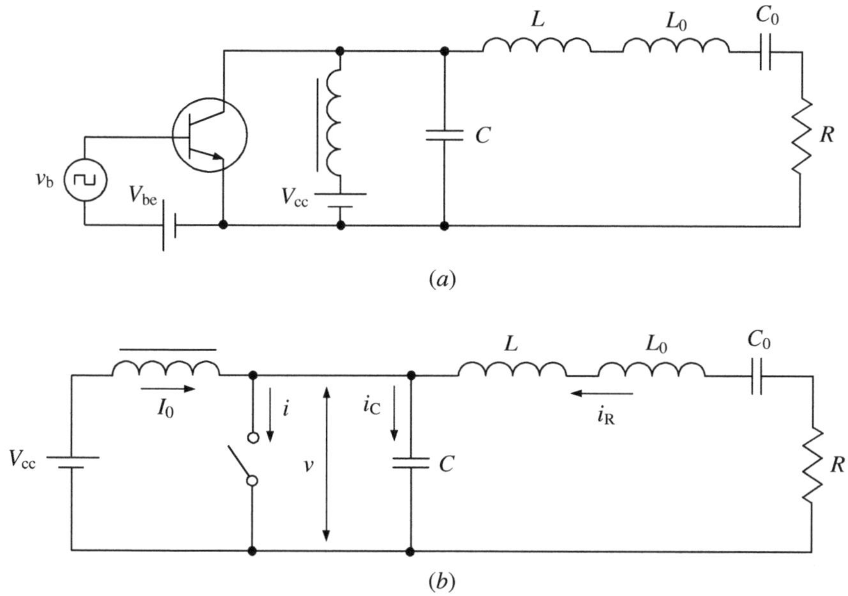

Figure 7.42 Basic circuits of Class-E power amplifier with shunt capacitance.

by an RF choke with high reactance at the fundamental frequency. The transistor is considered an ideal switch that is driven in such a way as to provide the instant device switching between its on-state and off-state operation conditions. As a result, the collector voltage waveform is determined by the switch when it is turned on and by the transient response of the load network when the switch is turned off.

To simplify the analysis of a Class-E power amplifier, whose simplified equivalent circuit is shown in Fig. 7.42(*b*), the following assumptions are introduced:

- The transistor has zero saturation voltage, zero saturation resistance, infinite off-resistance, and its switching action is instantaneous and lossless.
- The total shunt capacitance is independent of the collector and is assumed linear.
- The RF choke allows only a constant dc current and has no resistance.
- The loaded quality factor $Q_L = \omega L_0/R = 1/\omega C_0 R$ of the series resonant L_0C_0 circuit tuned to the fundamental frequency is high enough for the output current to be sinusoidal at the switching frequency.
- There are no losses in the circuit except only in the load R.
- For simplicity, a 50% duty ratio is used.

For a lossless operation mode, it is necessary to provide the following optimum conditions for voltage across the switch (just prior to the start of switch on) at the moment $\omega t = 2\pi$, when transistor is saturated:

$$v(\omega t)\Big|_{\omega t=2\pi} = 0 \tag{7.78}$$

$$\left.\frac{dv(\omega t)}{d\omega t}\right|_{\omega t=2\pi} = 0 \tag{7.79}$$

where $v(\omega t)$ is the voltage across the switch.

A detailed theoretical analysis of a Class E power amplifier with shunt capacitance, for any duty ratio, is given in [32], where the load current is assumed to be sinusoidal,

$$i_{\rm R}(\omega t) = I_{\rm R}\sin(\omega t + \varphi) \tag{7.80}$$

where φ is the initial phase shift.

When the switch is turned on for $0 \le \omega t < \pi$, the current through the capacitance

$$i_{\rm C}(\omega t) = \omega C\frac{dv(\omega t)}{d\omega t} = 0 \tag{7.81}$$

and, consequently,

$$i(\omega t) = I_0 + I_{\rm R}\sin(\omega t + \varphi) \tag{7.82}$$

under the initial on-state condition $i(0) = 0$. Hence, the dc current can be defined as

$$I_0 = -I_{\rm R}\sin\varphi \tag{7.83}$$

and the current through the switch can be rewritten by

$$i(\omega t) = I_{\rm R}\left[\sin(\omega t + \varphi) - \sin\varphi\right] \tag{7.84}$$

When the switch is turned off for $\pi \le \omega t < 2\pi$, the current through the switch $i(\omega t) = 0$, and the current flowing through the capacitor C can be written as

$$i_{\rm C}(\omega t) = I_0 + I_{\rm R}\sin(\omega t + \varphi) \tag{7.85}$$

producing the voltage across the switch by the charging of this capacitor according to

$$\begin{aligned} v(\omega t) &= \frac{1}{\omega C}\int_{\pi}^{\omega t} i_{\rm C}(\omega t)\, d\omega t \\ &= -\frac{I_{\rm R}}{\omega C}\left[\cos(\omega t + \varphi) + \cos\varphi + (\omega t - \pi)\sin\varphi\right] \end{aligned} \tag{7.86}$$

Applying the first optimum condition given by Eq. (7.78) enables the phase angle φ to be determined as

$$\varphi = \tan^{-1}\left(-\frac{2}{\pi}\right) = -32.482° \tag{7.87}$$

Consideration of trigonometric relationships shows that

$$\sin\phi = \frac{-2}{\sqrt{\pi^2+4}} \qquad \cos\phi = \frac{\pi}{\sqrt{\pi^2+4}} \tag{7.88}$$

By using Fourier series expansion and Eqs. (7.83) and (7.88), the expression to determine the supply voltage V_{cc} can be written as

$$V_{cc} = \frac{1}{2\pi}\int_0^{2\pi} v(\omega t)\, d\omega t = \frac{I_0}{\pi\omega C} \tag{7.89}$$

As a result, the normalized steady-state collector voltage waveform for $\pi \le \omega t < 2\pi$ and current waveform for $0 \le \omega t < \pi$ are

$$\frac{v(\omega t)}{V_{cc}} = \pi\left(\omega t - \frac{3\pi}{2} - \frac{\pi}{2}\cos\omega t - \sin\omega t\right) \tag{7.90}$$

$$\frac{i(\omega t)}{I_0} = \frac{\pi}{2}\sin\omega t - \cos\omega t + 1 \tag{7.91}$$

Figure 7.43 shows the normalized (*a*) load current, (*b*) collector voltage waveform, and (*c*) collector current waveforms for an idealized optimum (or nominal) Class-E mode with shunt capacitance. From collector voltage and current waveforms it follows that, when the transistor is turned on, there is no voltage across the switch, and the current $i(\omega t)$, consisting of the load sinusoidal current and dc current, flows through the device. However, when the transistor is turned off, this current flows through the shunt capacitor C. The jump in the collector current waveform at the instant of switching off is necessary to obtain nonzero output power at the fundamental frequency delivered to the load, which can be defined as an integration of the product of the collector voltage and current derivatives over the entire period [33].

As a result, there is no nonzero voltage and current simultaneously, which means a lack of the power losses and gives an idealized collector efficiency of 100%. This implies that the dc power and fundamental-frequency output power delivered to the load are equal,

$$I_0 V_{cc} = \frac{I_R^2}{2} R \tag{7.92}$$

Consequently, the value of the dc supply current I_0 using Eqs. (7.83) and (7.88) can be determined by

$$I_0 = \frac{V_{cc}}{R}\,\frac{8}{\pi^2+4} = 0.577\frac{V_{cc}}{R} \tag{7.93}$$

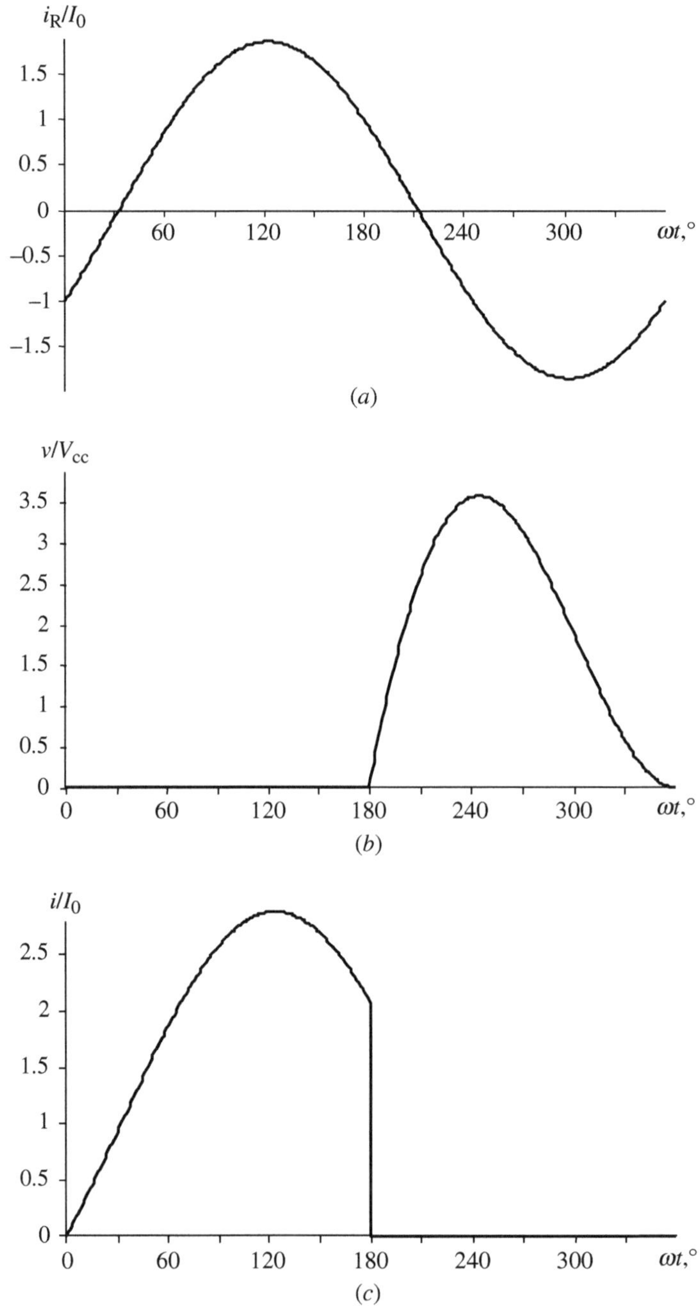

FIGURE 7.43 Normalized (*a*) load current and collector (*b*) voltage and (*c*) current waveforms for idealized optimum Class E with shunt capacitance.

Then, the amplitude of the output voltage $V_R = I_R R$ can be obtained from

$$V_R = \frac{4\,V_{cc}}{\sqrt{\pi^2+4}} = 1.074\,V_{cc} \tag{7.94}$$

The peak collector voltage V_{max} and current I_{max} can be determined by differentiating the appropriate waveforms given by Eqs. (7.90) and (7.91), respectively, and setting the results equal to zero, which gives

$$V_{max} = -2\pi\varphi\,V_{cc} = 3.562\,V_{cc} \tag{7.95}$$

$$I_{max} = \left(\frac{\sqrt{\pi^2+4}}{2}+1\right) I_0 = 2.8621\,I_0 \tag{7.96}$$

The fundamental-frequency voltage $v_1(\omega t)$ across the switch consists of the two quadrature components shown in Fig. 7.44(*a*), whose amplitudes can be found using Fourier formulas and Eq. (7.90) as

$$V_R = -\frac{1}{\pi}\int_0^{2\pi} v(\omega t)\,\sin(\omega t+\varphi)\,d\omega t = \frac{I_R}{\pi\omega C}\left(\frac{\pi}{2}\sin 2\varphi + 2\cos 2\varphi\right) \tag{7.97}$$

$$V_L = -\frac{1}{\pi}\int_0^{2\pi} v(\omega t)\,\cos(\omega t+\varphi)\,d\omega t = -\frac{I_R}{\pi\omega C}\left(\frac{\pi}{2} + \pi\sin^2\varphi + 2\sin 2\varphi\right) \tag{7.98}$$

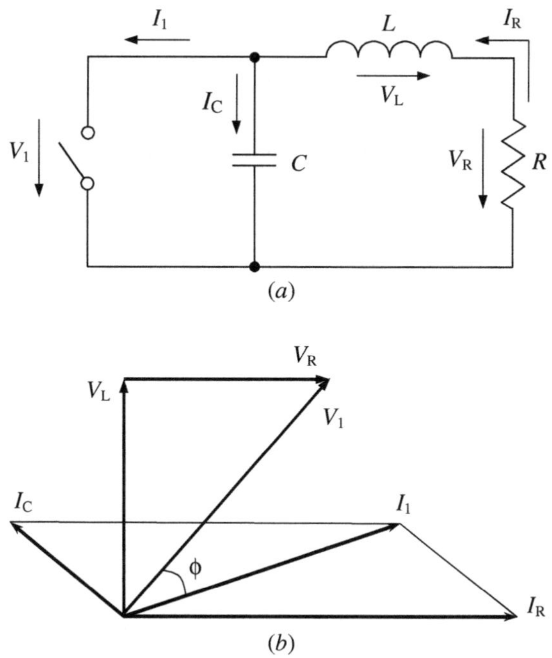

FIGURE 7.44 Equivalent Class-E load network at fundamental frequency.

As a result, the optimum series inductance L and shunt capacitance C can be obtained by

$$\frac{\omega L}{R} = \frac{V_L}{V_R} = 1.1525 \tag{7.99}$$

$$\omega CR = \frac{\omega C}{I_R} V_R = 0.1836 \tag{7.100}$$

The optimum load resistance R for the supply voltage V_{cc} and fundamental-frequency output power P_{out} delivered to the load using Eqs. (7.92) and (7.94) can be obtained by

$$R = \frac{8}{\pi^2 + 4} \frac{V_{cc}^2}{P_{out}} = 0.5768 \frac{V_{cc}^2}{P_{out}} \tag{7.101}$$

Finally, the phase angle of the load network seen by the switch at the fundamental frequency shown in Fig. 7.44(*b*) and required for an idealized optimum (or nominal) Class-E mode with shunt capacitance can be determined through the load network parameters using Eqs. (7.99) and (7.100) as

$$\phi = \tan^{-1}\left(\frac{\omega L}{R}\right) - \tan^{-1}\left(\frac{\omega CR}{1 - \frac{\omega L}{R}\omega CR}\right) = 35.945° \tag{7.102}$$

When realizing a Class-E operation mode, it is very important to know up to which maximum frequency such an idealized efficient operation mode can be extended. In this case, it is important to establish a relationship between a maximum operation frequency f_{max}, a shunt capacitance C, and a supply voltage V_{cc} by using Eqs. (7.89) and Eq. (7.90) when

$$f_{max} = \frac{1}{\pi^2} \frac{1}{\sqrt{\pi^2 + 4} + 2} \frac{I_{max}}{C_{out} V_{cc}} = \frac{I_{max}}{56.5\, C_{out} V_{cc}} \tag{7.103}$$

where $C = C_{out}$ is the device output capacitance limiting the maximum operation frequency of an idealized optimum Class-E power amplifier with shunt capacitance [34].

The high-Q_L assumption for the series resonant L_0C_0 circuit can lead to considerable errors if its value is substantially small in real circuits [35]. For example, for a 50% duty ratio, the values of the load-network parameters for the loaded quality factor Q_L less than unity can differ by several tens of percentages. At the same time, for $Q_L \geq 7$, the errors are found to be less than 10% and they become less than 5% for $Q_L \geq 10$. To match the required optimum Class-E load-network resistance R with a standard load impedance R_L, usually equal to 50 Ω, the series resonant L_0C_0 circuit should be followed (or fully replaced) by the matching circuit, in which the first element represents a series inductor to provide high impedance at the second and higher-order harmonics [28].

7.4.2 Effect of Saturation Resistance, Finite Switching Time, and Nonlinear Shunt Capacitance

In practical power amplifier design, especially when a value of the supply voltage is sufficiently small, it is very important to predict the overall degradation of power amplifier efficiency due to the finite value of the transistor saturation resistance. Figure 7.45(*a*) shows the simplified equivalent circuit of a Class-E power amplifier with shunt capacitance, including the saturation resistance (on-resistance) r_{sat} connected in series to the ideal switch. To obtain a quantitative estimate of the power losses due to the contribution of r_{sat}, the saturated output power P_{sat} can be obtained with a simple approximation when the

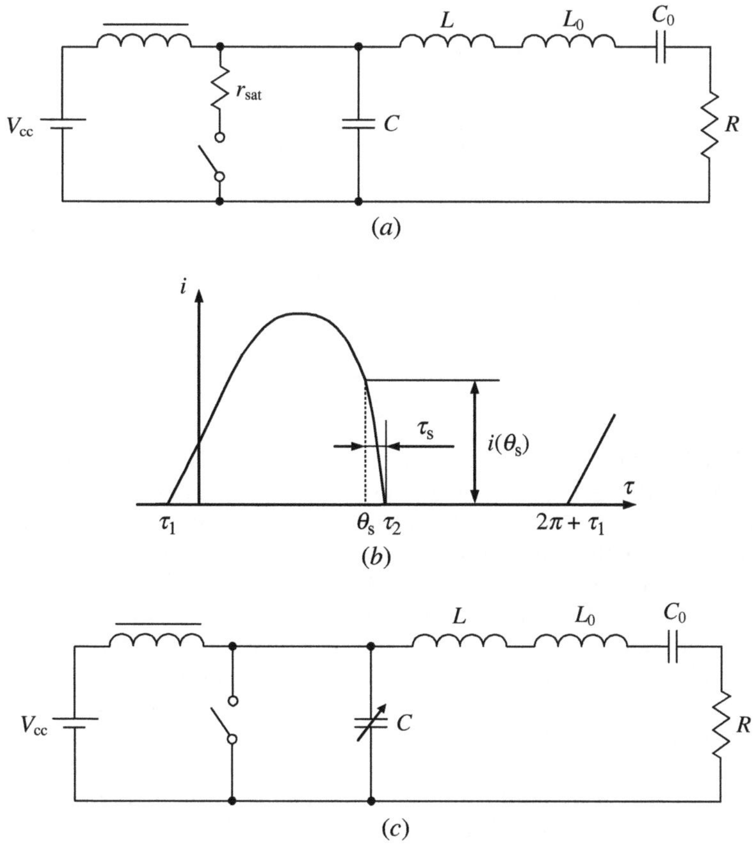

FIGURE 7.45 Equivalent Class-E load networks (*a*) with saturation resistance and (*c*) nonlinear capacitance and (*b*) current waveform with finite time delay.

current $i(\omega t)$ flowing through the saturation resistance r_{sat} is determined in an ideal case by Eq. (7.91).

An analytical expression to calculate the power losses due to the saturation resistance r_{sat}, whose value is assumed constant, can be represented in the normalized form as

$$\frac{P_{sat}}{P_0} = \frac{r_{sat}}{2\pi\, I_0 V_{cc}} \int_0^{\pi} i^2(\omega t)\; d\omega t \tag{7.104}$$

where $P_0 = I_0 V_{cc}$ is the dc power.

By taking into account that

$$\int_0^{\pi} \left(\frac{\pi}{2} \sin \omega t - \cos \omega t + 1\right)^2 d\omega t = \frac{\pi}{8}(\pi^2 + 28)\, I_0^2 \tag{7.105}$$

Eq. (7.104) can be rewritten using Eq. (7.93) as

$$\frac{P_{sat}}{P_0} = \frac{r_{sat}}{2\pi} \frac{I_0}{V_{cc}} \frac{\pi}{8}(\pi^2 + 28) = \frac{r_{sat}}{2R} \frac{\pi^2 + 28}{\pi^2 + 4} = 1.365 \frac{r_{sat}}{R} \tag{7.106}$$

The collector efficiency η can be calculated from

$$\eta = \frac{P_{out}}{P_0} = \frac{P_0 - P_{sat}}{P_0} = 1 - \frac{P_{sat}}{P_0} \tag{7.107}$$

Consequently, the presence of the saturation resistance results in the finite value of the saturation voltage V_{sat}, which can be defined from

$$\frac{V_{sat}}{V_{cc}} = 1 - \frac{1}{1 + 1.365 \dfrac{r_{sat}}{R}} \tag{7.108}$$

where V_{sat} is normalized to the dc supply voltage V_{cc} [36].

More detailed theoretical analysis of the time-dependent behavior of the collector voltage and current waveforms shows that, for the finite value of the saturation resistance r_{sat}, the optimum conditions for idealized operation mode given by Eqs. (7.78) and (7.79) do not correspond anymore to the minimum dissipated power losses, and there are optimum nonzero values of the collector voltage and its derivative at switching time instant corresponding to minimum overall power losses [37]. For example, even for small losses with the normalized loss parameter $\omega C r_{sat} = 0.1$ for a duty ratio of 50%, the optimum series inductance L is almost two times greater, whereas the optimum shunt capacitance C is of about 20% greater than those obtained under nominal conditions given by Eqs. (7.78) and (7.79). However, for collector efficiencies of 90% and greater, both the optimum inductance and optimum capacitance differ by less than 20% from their optimum values for $r_{sat} = 0$ [38]. Thus, generally the switching conditions given by

Eqs. (7.78) and (7.79) can be considered optimum only for an idealized case of a Class-E load network with zero saturation resistance providing the switchmode transistor operation when it operates in pinch-off and saturation regions only. However, they can be considered as a sufficiently accurate initial guess for further design and optimization in a real Class-E power amplifier design.

For an ideal transistor without any memory effects due to intrinsic phase delays, the switching time is equal to zero when the rectangular input drive results in a rectangular output response. Such an ideal case assumes zero device feedback capacitance and zero device input resistance. However, at higher frequencies, it is very difficult to realize the driving signal close to the rectangular form, as it leads to the significant circuit complexity. Fortunately, to realize high-efficiency operation conditions, it is sufficient to drive the power amplifier simply with a sinusoidal signal. The finite-time transition from the saturation mode to the pinch-off mode through the device active mode takes place due to the device inertia when the base (or channel) charge changes to zero with some finite delay time τ_s, as shown in Fig. 7.45(*b*). To minimize the switching time interval, it is sufficient to slightly overdrive the transistor with signal amplitude by 20 to 30% higher than is required for a conventional Class-B power amplifier. As an alternative, the second-harmonic component (approximation of a half-sinusoidal waveform) or third-harmonic component (approximation of a rectangular waveform with close to trapezoidal waveform) with proper phasing can be added to the input driving signal. In both cases, the overall driving waveform will be steeper compared with simply sinusoidal driving signal, thus resulting in a faster switching operation time.

The power dissipated during this on-to-off transition can be calculated assuming zero on-resistance as

$$P_s = \frac{1}{2\pi} \int_{\theta_s}^{\pi} i(\omega t)\, v(\omega t) d\omega t \tag{7.109}$$

where the collector voltage during the transition time $\tau_s = \pi - \theta_s$ is defined as

$$v(\theta_s) = \frac{1}{\omega C} \int_{\theta_s}^{\pi} i_C(\omega t)\; d\omega t \tag{7.110}$$

The short duration of the switching time and the proper behavior of the resulting collector (or drain) waveform allows us to make an additional assumption of a linearly decreasing collector current during fall time $\tau_s = \pi - \theta_s$, starting at $i(\theta_s)$ at time θ_s and decaying to zero at time π, which can be written as

$$i(\omega t) = i(\theta_s)\left(1 - \frac{\omega t - \theta_s}{\tau_s}\right) \tag{7.111}$$

where $i(\theta_s)$ corresponds to the collector current as shown in Fig. 7.45(*b*) [36, 38]. In this case, the capacitor-charging current $i_C(\omega t) = i(\theta_s) - i(\omega t)$, being zero during saturation mode, varies linearly between zero and $i(\theta_s)$ during the on-to-off transition according to

$$i_C(\omega t) = i(\theta_s)\,\frac{\omega t - \theta_s}{\tau_s} \tag{7.112}$$

The collector voltage produces a parabolic voltage waveform during the switching interval according to Eq. (7.110) given by

$$v(\omega t) = \frac{i(\theta_s)}{2\omega C\tau_s}(\omega t - \theta_s)^2 \tag{7.113}$$

As a result, the power dissipated during transition according to Eq. (7.109) is then

$$P_s = \frac{i^2(\theta_s)\,\tau_s^2}{48\pi\omega C} \tag{7.114}$$

For an optimum Class-E power amplifier by assuming in view of a short transition time that $i(\theta_s) = i(\pi)$, from Eq. (7.91) it follows that $i(\pi) = 2I_0$, hence

$$P_s = \frac{I_0^2\tau_s^2}{12\pi\omega C} \tag{7.115}$$

By taking into account Eq. (7.89), the switching loss power P_s normalized to the dc power P_0 can be obtained by

$$\frac{P_s}{P_0} = \frac{I_0\tau_s^2}{12\pi\omega C V_{cc}} = \frac{\tau_s^2}{12} \tag{7.116}$$

As a result, the collector efficiency η can be estimated as

$$\eta = 1 - \frac{P_s}{P_0} = 1 - \frac{\tau_s^2}{12} \tag{7.117}$$

As follows from Eq. (7.116), the power losses due to the nonzero switching time are sufficiently small and, for example, for $\tau_s = 0.35$ or 20°, they are only about 1%, whereas they are approximately equal to 10% for $\tau_s = 60°$. A more exact analysis assuming a linear variation of the collector current during on-to-off transition produces similar results when efficiency degrades to 97.72% for $\tau_s = 30°$ and to 90.76% for $\tau_s = 60°$ [39]. Considering an exponential collector current decay rather than linear during the fall time shows the similar result for $\tau_s =$ 30° when $\eta = 96.8\%$, but the collector efficiency degrades more significantly at longer fall times when, for example, $\eta = 86.6\%$ for $\tau_s = 60°$ [40].

In a common case, the intrinsic output device capacitance is nonlinear, as shown in Fig. 7.45(*c*). If its contribution in overall shunt capacitance is sufficiently large, it is necessary to take into account the nonlinear nature of this capacitance when specifying the breakdown voltage. For example, the collector voltage waveform will rise in the case of the output capacitance described by abrupt diode junction in comparison with the linear capacitance, and its maximum voltage can be greater by about 20% for a 50% duty ratio [28, 41]. However, stronger nonlinearity of the shunt capacitance causes the peak voltages to be higher [42]. At the same time, the deviations of the optimum load-network parameters are insignificant, less than 5% in a wide range of supply voltages. Because the nonlinear capacitance is largest at zero voltage, the collector waveform will rise more slowly than in the linear case. As the collector voltage increases, the capacitance will decrease, and hence the voltage should begin to rise faster than in the linear case. If the shunt capacitance consists of both nonlinear and linear capacitances, the collector voltage waveform is intermediate and located between the two extreme cases of entirely nonlinear or entirely linear capacitance [43].

7.4.3 Optimum, Nominal, and Off-Nominal Class-E Operation

More detailed theoretical analysis of the time-dependent behavior of the collector voltage and current waveforms shows that, for a nonzero value of the saturation resistance r_{sat}, the optimum conditions for an idealized operation mode given by Eqs. (7.78) and (7.79) no longer correspond to the minimum dissipated power losses, and there are optimum nonzero values of the collector voltage and its derivative at switching time instant corresponding to minimum overall power losses [37]. For example, even for small losses with the normalized loss parameter $\omega C r_{sat} = 0.1$ for a duty ratio of 50%, the optimum series inductance L is almost two times greater, and the optimum shunt capacitance C is about 20% greater than those obtained under idealized optimum conditions given by Eqs. (7.78) and (7.79). However, for collector efficiencies of 90% and greater, both optimum inductance and optimum capacitance differ by less than 20% from their optimum values for $r_{sat} = 0$ [38]. It should be noted that if the first zero-voltage condition is satisfied, the power losses close to minimum can be achieved with the second zero-voltage–derivative condition, since the positive voltage derivative results in a positive current jump, which requires greater driving amplitude, whereas the negative voltage derivative with a negative current jump demonstrates reduction in power loss by only 3% [44]. At the same time, operation of the Class-E power amplifier at zero slope and nonzero voltage will result in lower voltage peak factor and higher output power capability with lower efficiency, which reduces with higher voltage at the moment when the switch is turned off [45]. Thus, generally the switching conditions given by Eqs. (7.78) and (7.79) can be considered optimum only for an idealized case of the Class-E load network with zero saturation resistance providing the switchmode transistor operation when it is operated in only pinch-off and saturation regions. However, they can be considered as a sufficiently accurate initial assumption for further design and optimization of the practical high-efficiency Class-E power amplifiers.

The term "optimum" for the idealized Class-E conditions given by Eqs. (7.78) and (7.79) means that the voltage across the switch should be equal to zero and there are no current jumps across the capacitor C (capacitor must be discharged) at the moments when the switch is turned on with further instant transitions from off-state to on-state modes, thus resulting in a maximum achievable collector efficiency. Otherwise, if current starting to flow through the switch at this moment is not equal to zero, the device cannot be considered an ideal switch because of the appearance of the active operation mode of the device between its pinch-off and saturation modes. In this case, the collector current and voltage waveforms overlap each other reducing the collector efficiency because of the power dissipation in the device. Therefore, it is more proper to call the *optimum conditions* given by Eqs. (7.78) and (7.79) for a lossless operation mode with ideal switch as the *nominal (or idealized optimum) conditions.*

Optimum electrical operation is defined as the operating condition that gives the highest possible drain or collector efficiency at a specified output power and peak switch voltage for a given set of parasitic parameters such as the transistor saturation resistance, turn-off and turn-on switching transition times, and finite quality factors of the load-network inductors and capacitors. In this case, each set of components results in a specific "optimum" off-nominal design. The smaller the parasitic series resistances and the larger the parasitic parallel resistances, the closer is the optimum switch-voltage waveform to the nominal waveform. The designers can tradeoff among the power dissipations, so as to minimize the total power dissipation at the specified output power and peak switch voltage.

In a nominal (or idealized optimum) Class-E operation, the load network discharges the device output capacitance prior to turn-on of the device, producing ideally 100% efficiency. Below the maximum frequency, at which the shunt susceptance required for optimum operation is provided by only the device output capacitance, it is generally possible to adjust the series load reactance to achieve almost 100% efficiency. For example, optimum operation can be achieved by adding external shunt capacitance to the device output. However, above the maximum frequency, it is impossible to achieve an ideal 100% efficiency by varying the series load network parameters. As frequency increases, the collector voltage waveform approaches the ramp produced by dc charging of the shunt capacitor. Consequently, the maximum achievable collector (or drain) efficiency decreases as the frequency of operation is increased above the maximum frequency. The maximum possible efficiency η_{max} and the normalized circuit parameter $\omega L/R$ required to produce it, along with the normalized

f/f_{max}	wL/R	V_{max}/V_{cc}	η_{max}, %
1.000	1.152	3.562	100
1.259	1.330	3.198	99.59
1.585	1.053	2.981	96.96
1.995	0.852	2.789	92.16
2.512	0.691	2.632	85.62
3.162	0.561	2.519	77.87

Table 7.2 Suboptimum Operation above Maximum Frequency

peak voltage V_{max}/V_{cc}, for a fixed supply voltage V_{cc}, are shown in Table 7.2 for certain $f/f_{max} \geq 1$ [46].

From Table 7.2, it follows that, above the maximum frequency, efficiency can be maximized by proper selection of the series inductance and load resistance when it looks reasonable. For example, at operating frequency $f = 2.512f_{max}$, the collector efficiency of an ideal Class-E mode remains still high, being even higher than that for a Class-F mode with control of three collector voltage and three collector current harmonic components. The collector efficiency of an ideal Class-E power amplifier drops at $f = 3.162f_{max}$ to 77.87%, which approximately corresponds to the maximum collector efficiency of an ideal Class-B power amplifier of 78.5%.

For off-nominal Class-E operation when only zero-voltage switching condition is satisfied, the designer can use a higher shunt capacitance than for nominal operation at the same switching frequency and load resistance [47]. In this case, both peak voltage and current values are higher for off-nominal operation which occurs for $-0.5\pi < \varphi < 0$ for phase angles $\varphi > \varphi_{opt}$, where φ_{opt} is the optimum phase angle determined by Eq. (7.87) for a nominal Class-E mode at 50% duty ratio. Generally, the peak switch voltage and current values vary with the turn-on switch voltage slope and the duty ratio [48].

7.4.4 Load Network with Transmission Lines

The transmission lines are often preferred over lumped inductors at microwave frequencies because of the convenience of their practical implementation, more predictable performance, less insertion loss, and less effect of the parasitic elements. For example, the matching circuit can be composed with any types of the transmission lines, including open- or short-circuit stubs, to provide the required matching and harmonic suppression conditions. In this case, to approximate the idealized Class-E operation mode of the microwave power amplifier, it is necessary to design the transmission-line load network satisfying the required idealized optimum impedances at the fundamental-frequency and harmonic components. The device output capacitance can fully represent the required shunt capacitance, whose nominal value is defined by Eq. (7.100). Consequently, the main challenge is to satisfy the idealized optimum requirements for the fundamental-frequency impedance $Z_L(\omega_0)$ shown in Fig. 7.46(*a*) and harmonic-component impedances $Z_L(n\omega_0)$ shown in Figs. 7.46(*b*), which can be written using Eq. (7.99) as

$$Z_L(\omega_0) = R + j\omega L = R\left(1 + j\frac{\omega L}{R}\right) = R(1 + j\tan 49.052°) \qquad (7.118)$$

$$Z_L(n\omega_0) = \infty \qquad (7.119)$$

where ω_0 is the radian fundamental frequency and $n = 2, 3, \ldots, \infty$ is the harmonic number.

Generally, it is practically impossible to realize these conditions for an infinite number of harmonic components by using only transmission lines. However, as it turned out from the Fourier-series analysis, a good approximation to Class-E mode may be obtained with the dc, fundamental-frequency, and second-harmonic components of the voltage waveform across the switch [34, 49]. Figure 7.46(*c*) shows the collector (drain) voltage waveform containing these two harmonic components (dashed curve) plotted along with an ideal

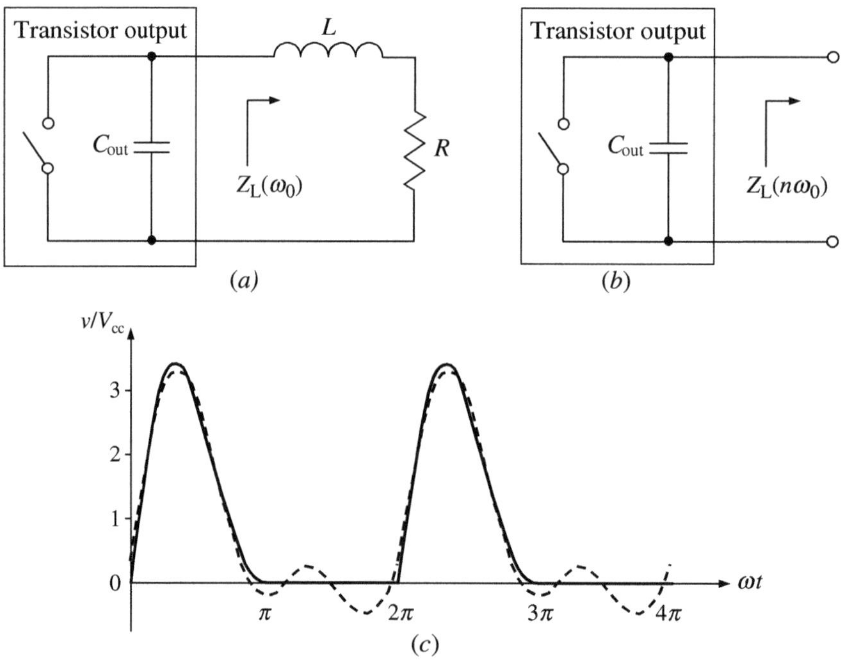

FIGURE 7.46 Optimum load impedance and two-harmonic Class-E voltage waveform.

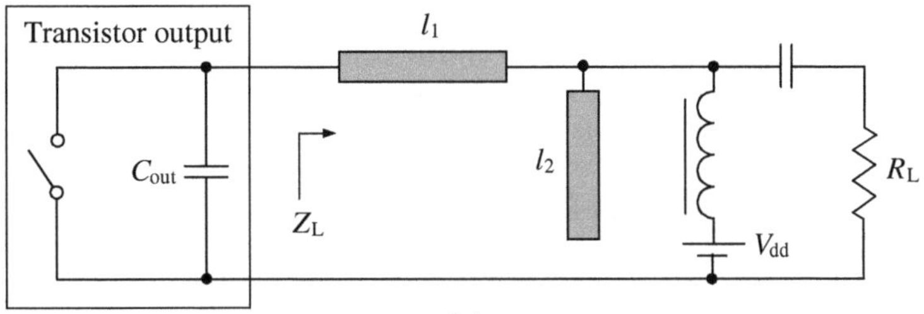

(*a*)

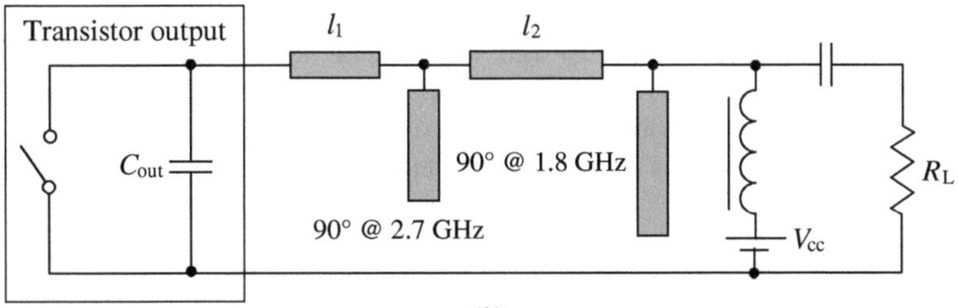

(*b*)

FIGURE 7.47 Equivalent circuits of Class-E power amplifiers with transmission lines.

voltage waveform (solid curve). In practical implementation, the two-harmonic Class-E load network designed for microwave applications will include the series microstrip line l_1 and open-circuit stub l_2, as shown in Fig. 7.47(*a*). The electrical lengths of microstrip lines l_1 and l_2 are chosen to be of about 45° at the fundamental frequency to provide an open-circuit condition seen from the device output at the second harmonic, according to Eq. (7.119). Their characteristic impedances are calculated to satisfy the required inductive impedance condition at the fundamental frequency given by Eq. (7.118). In the case of a packaged active device, its output lead inductance can be accounted for by shortening the length of l_1.

In some cases, a value of the device output capacitance exceeds the required nominal value for a Class-E mode with shunt capacitance. In this situation, it is possible to approximate Class-E mode with high efficiency by setting a properly optimized load at the fundamental frequency and strong reactive load at the second- and third-harmonic components [50]. Such a harmonic-control network consists of open-circuit quarterwave stubs at the second- and third-harmonic components separately, as shown in Fig. 7.47(*b*), where the third-harmonic quarterwave stub is located before the second-harmonic quarterwave stub. As a result, very high collector efficiency can be achieved even with values of the device output capacitance higher than conventionally required at the expense of lower output power, keeping the load at the second and third harmonics strictly inductive (inverse mode). Maximum collector efficiency over 90% with an output power of 1.5 W for the test power amplifier using a commercial bipolar transistor MRF557 was measured at 900 MHz.

7.4.5 Practical Class-E Power Amplifiers

A high level of output power with very high operational efficiency can be easily achieved in a Class-E mode by using high-voltage power MOSFET devices at sufficiently low frequencies. Figure 7.48(*a*) shows the circuit schematic of a 13.56-MHz, 400-W Class-E power amplifier providing a drain efficiency of 82% with an input sinusoidal drive of 12 W at a supply voltage of 120 V [51]. The series inductor L_s and capacitor C_s form the resonant network that produces the rising and falling voltage waveform required for a Class-E operation. The series tank circuit at the load composed of a capacitor C_1 and an inductor L_1

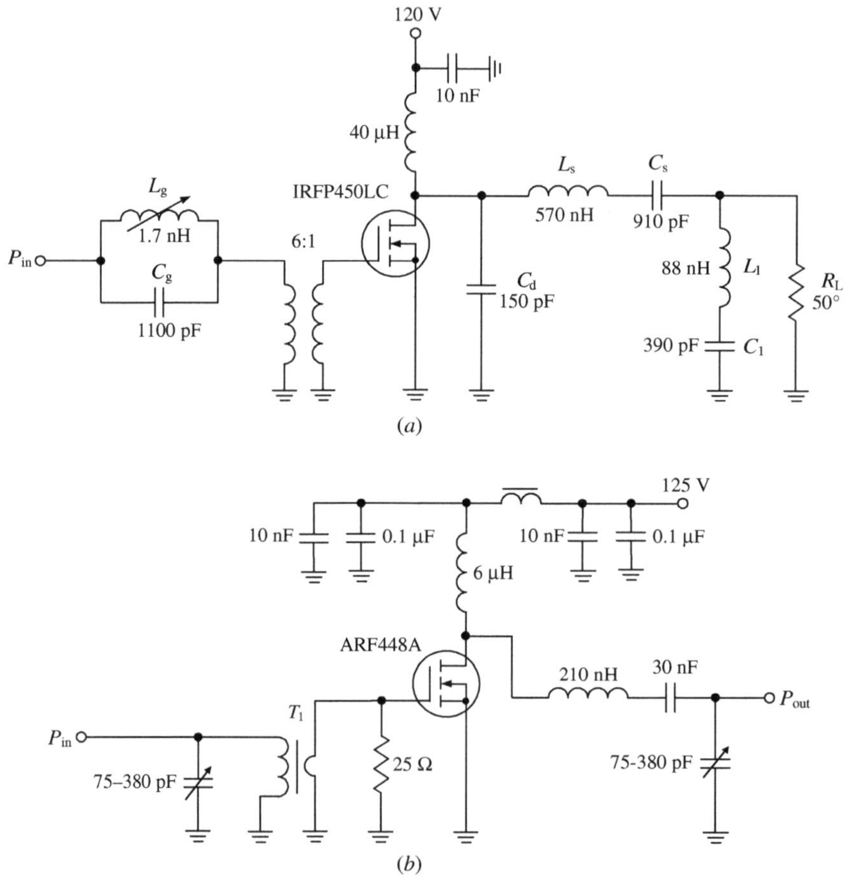

Figure 7.48 High-power high-frequency Class-E MOSFET power amplifiers.

is a trap for the second-harmonic component, which contributes to the overall level of harmonic suppression of more than 40 dB below the carrier. Because at the fundamental frequency this second-harmonic resonant circuit represents a capacitive reactance, it transforms together with a part of the series inductance L_s the standard load of 50 Ω to around 13 Ω required for a nominal Class-E mode. The impedance of the gate is small with a real part of about 3 Ω and an inductive reactance of about 4 Ω. The input transformer with a voltage transformation ratio of 6:1 is used to step up from the input gate impedance to the driving source impedance of 50 Ω. This transformer also sets the dc gate bias to 0 V and ensures the transistor is turned off when it is not driven, as this is far below the threshold voltage of 4 V. The capacitor C_g with the variable inductor L_g is used to compensate for the input inductive reactance of the transistor providing the input *VSWR* of 1.6:1.

Figure 7.48(*b*) shows the circuit schematic of a 27.12-MHz, 500-W Class-E MOSFET power amplifier with a drain efficiency of 83% at a supply voltage of 125 V [52]. The input ferrite transformer provides the 2:1 transformation voltage ratio to match the gate impedance, which is represented by the parallel equivalent circuit with a capacitance of 2200 pF and a resistance of 210 Ω. Use of the external parallel resistor of 25 Ω simplifies the matching procedure and improves the amplifier stability conditions. The transformer secondary winding provides an inductance of 19 nH, which is required to compensate for the device input capacitance at the fundamental. High-quality passive components are necessary to use in the low-pass *L*-type output network, where the quality factor of the bare copper wire inductor was equal to 375. The series blocking capacitor consists of three parallel disc ceramic capacitors. To realize a Class-E operation with shunt capacitance, it is sufficient to be limited to only the output device capacitance with a value of 125 pF. This is just slightly larger than that required to obtain the idealized optimum drain voltage and current Class-E waveforms.

Silicon LDMOSFET devices made it possible to achieve a high output power level in a Class-E operation with a sufficiently high efficiency at higher frequencies. Figure 7.49(*a*) shows the simplified circuit schematic of a high-power VHF LDMOSFET Class-E power amplifier achieving a drain efficiency of 70% for maximum output power of 54 W at an operating frequency of 144 MHz with an input drive of 5 W [53]. The drain efficiency can be increased to 88% if the output power level is reduced to 14 W by an appropriate increase in the series inductance in the load network. The input device impedance is sufficiently low; therefore, a ferrite transformer and a series inductor are used at the input. At the output, the standard load impedance of 50 Ω is transformed to the load resistance of 1.5 Ω required for an idealized optimum Class-E mode by a lumped *L*-transformer with a series inductor, whose value is included with the inductance of 24 nH, and a shunt 100-pF capacitor. The required

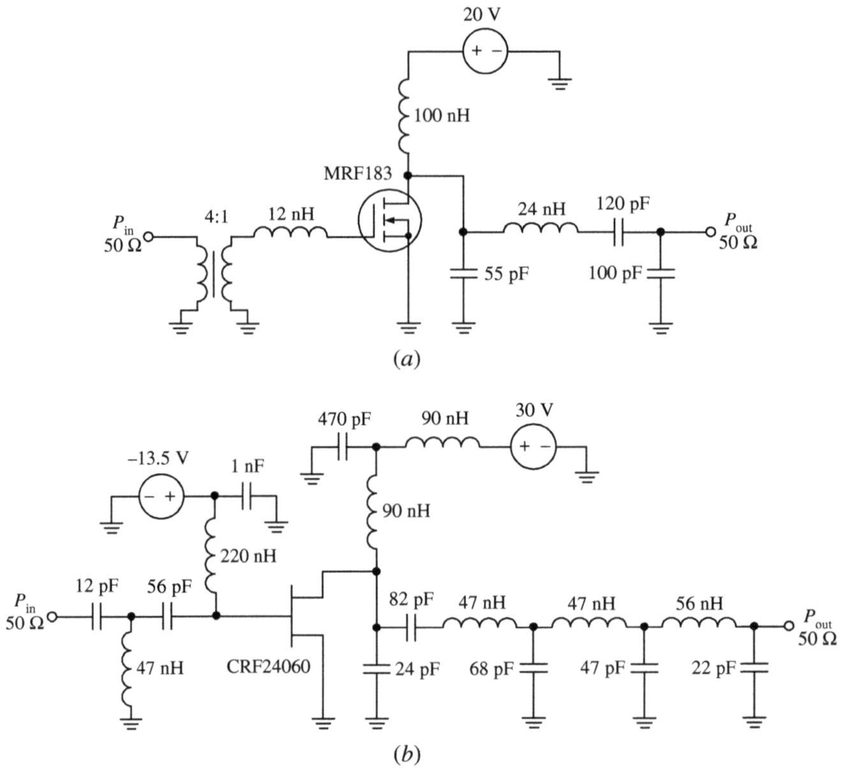

FIGURE 7.49 High-power VHF Class-E power amplifiers.

value of a shunt switching capacitance is provided by the values of the intrinsic device output capacitance of 38 pF and external capacitance of 55 pF. The loaded quality factor of the resonant circuit was chosen at a sufficiently low value of 5 that allows some frequency bandwidth operation to be provided and sensitivity of amplifier performance to the resonant-circuit parameters to be reduced. To decrease the loss in the load network, the inductor was fabricated by using a 5-mm-wide copper ribbon that provides the inductor quality factor of 150 to 250, depending on the distance to the ground plane. By inserting a spacer between the ribbon and the ground plane, the inductance can be tuned at least a factor of 2.

Figure 7.49(*b*) shows the simplified circuit schematic of a silicon carbide (SiC) MESFET Class-E power amplifier that provides a maximum drain efficiency of 86.8% at an output power of 20.5 W at 145 MHz reached at a drain voltage of 30 V, with an input drive power level of 27 dBm [54]. The nominal Class-E impedance of approximately 18 Ω was matched to a 50-Ω load with a low-pass three-section *L*-type matching network to suppress harmonics by at least 60 dB below the carrier. The input of the active device was matched to 50-Ω source by means of a high-pass filter network to prevent the attenuation of the high-frequency harmonic components of the drive signal. Because this power amplifier was designed to provide linear amplification by restoring the input signal envelope with drain amplitude modulation, the drain bias network was built with a low-pass filter that allows drain modulating frequencies of up to a few megahertz to pass through it with minimum attenuation, while at the same time achieving acceptable isolation at the carrier frequency and its harmonics.

The transmission-line Class-E power amplifier topology is shown in Fig. 7.50(*a*), where the electrical lengths of microstrip lines l_3 and l_4 in the load network must be close to 45° so that an approximate open circuit at the second harmonic will be presented to the switch shunt capacitor, which is an equivalent output device capacitance. Microstrip lines with a characteristic impedance of 50 Ω each were fabricated using a substrate with thickness of 2.54 mm and effective dielectric permittivity $\varepsilon_r = 10.5$. For a MESFET device having a drain-source

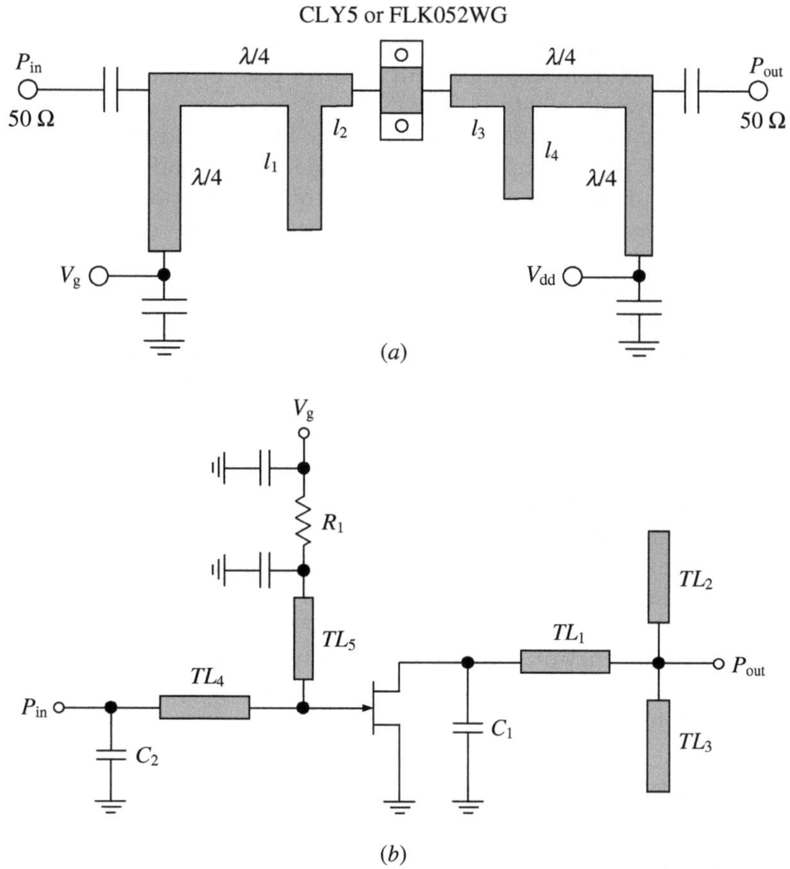

FIGURE 7.50 Circuit topologies of transmission-line Class-E power amplifiers.

capacitance C_{ds} = 2.4 pF, a *PAE* of 80% was achieved at 0.5 GHz with an output power of 0.55 W [49]. In this case, the electrical lengths of microstrip lines are l_1 = 73°, l_2 = 79°, l_3 = 58°, and l_4 = 46°. The power-added efficiency remains above 75% over a 10% bandwidth and above 50% over a 26% bandwidth. Moreover, a *PAE* of 73% can be realized at 1.0 GHz with an output power of 0.94 W. By using a MESFET device with a sufficiently higher transition frequency f_T in such a microstrip Class-E power amplifier, an output power of 0.61 W, a 1-dB compressed power gain of 7.6 dB, a drain efficiency of 81%, and a *PAE* of 72% were achieved at 5 GHz [34]. The power amplifier was fabricated on a substrate with a thickness of 0.508 mm and ε_r= 2.2. As a result, the lengths of 50-Ω microstrip lines (1.6 mm wide) are 9 mm for l_1, 1.8 mm for l_2, 5.3 mm for l_3, and 6.2 mm for l_4, respectively. The power-added efficiency is greater than 70% over a 5% bandwidth and greater than 60% over a 10% bandwidth. A similar design approach can be used to design a monolithic Class-E power amplifier in *X*-band. As a result, the measured performance of a monolithic power amplifier that uses a pHEMT device with a geometry of 0.3 μm × 600 μm showed a peak *PAE* of 63% at 10.6 GHz and a constant output power of greater than 24 dBm together with a power gain of 10 dB over a frequency bandwidth of 9 to 11 GHz [55].

Figure 7.50(*b*) shows the circuit schematic of a *K*-band transmission-line Class-E power amplifier using a single-section load network, which is well suited for monolithic implementation at upper microwave frequencies [56]. The electrical parameters of the capacitive stubs TL_2 and TL_3 were designed to provide low impedances at the second and third harmonics by making the electrical length of the stubs exactly one quarter-wavelength at a particular harmonic. At the same time, the characteristic impedances of the stubs are chosen to provide the desired capacitive reactance for load impedance transformation at the fundamental frequency. The electrical parameters of the series transmission line TL_1 are determined by the requirements to provide the required inductive impedance with a load angle of 49.05° at the fundamental and to transform the low impedance of the stub inputs toward higher reactances at the selected harmonics. As a result, by using a GaAs pHEMT technology and coplanar waveguides for transmission-line implementation, an output power of 20 dBm, a drain efficiency of 59%, and a power gain of 7.5 dB were achieved at an operating frequency of 24 GHz with a supply voltage of 2.4 V when both the second and third harmonics are suppressed by more than 30 and 35 dB, respectively.

7.5 Class E with Finite DC-Feed Inductance

In practice, it is impossible to realize RF choke with infinite impedance at the fundamental frequency and other harmonic components. Moreover, using a finite dc-feed inductance has an advantage of minimizing size, cost, and complexity of the overall circuit. The detailed approach to analyzing the effect of a finite dc-feed inductance on the idealized Class-E mode with shunt capacitance and series filter was first described in Ref. 57. An analysis was based on the Laplace-transform technique to solve a second-order differential equation describing the behavior of a Class E load network with finite dc-feed inductance. Later this approach was extended to the load network with finite Q_L-factor of the series filter and finite device saturation resistance [58, 59]. However, because the results of excessive analytical and numerical calculations are given only for a few particular cases, it is difficult to figure out the basic behavior of the load-network elements and derive simple equations for their parameters. Generally, based on the composing of the circuit equations in the form of a system of the first-order differential equations for currents and voltages and setting the design specifications, the optimum Class-E load network parameters can be numerically calculated taking into account the finite dc-feed inductance, drain current fall time, finite Q_L-factor, nonzero device saturation resistance, and nonlinear operation of any passive element simultaneously [60]. Also, it was analytically shown for a 50% duty ratio based on the Class-E optimum conditions that the series excessive reactance can be either inductive or capacitive depending on the values of the dc-feed inductance and shunt capacitance [61, 62]. Based on a certain number of cases, a Lagrange polynomial interpolation was used to obtain explicit and directly usable design equations for an idealized Class-E mode with small dc-feed inductance and series inductive reactance [63].

7.5.1 General Analysis and Optimum Circuit Parameters

The generalized second-order load network of a switchmode Class E power amplifier with finite dc-feed inductance is shown in Fig. 7.51(*a*) [64 to 66]. The load network consists of a shunt capacitor C, a parallel inductor L, a series inductor L_b, a series reactance X, a series resonant L_0C_0 circuit tuned to the fundamental frequency, and a load resistance R. In a common case, a shunt capacitance C can represent the intrinsic device output capacitance and external circuit capacitance added by the load network; a series inductor L_b can be considered a a bondwire and lead inductance; a parallel inductance L represents the finite dc-feed inductance; and a series reactance X can be positive (inductance), negative (capacitance), or zero depending on the particular Class-E mode. The active device is considered an ideal switch that is driven to provide the device instant switching between its on-state and off-state operation modes. To simplify an analysis of the general-circuit Class-E power amplifier, whose simplified equivalent circuit is shown in Fig. 7.51(*b*), it makes sense to introduce the preliminary assumptions similar to those for the Class-E power amplifier with shunt capacitance,

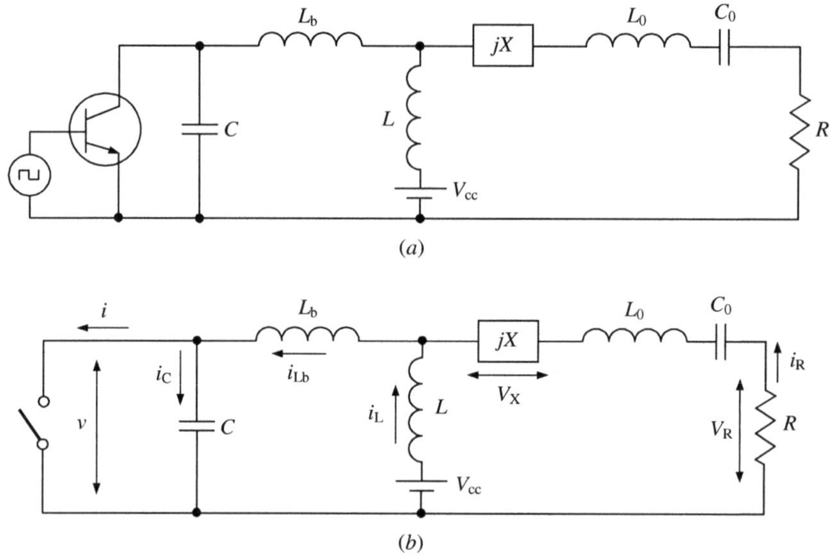

FIGURE 7.51 Equivalent circuits of the Class-E power amplifiers with generalized load network.

assuming that the losses in the reactive circuit elements are negligible, the duty ratio is 50%, the loaded quality factor of the series L_0C_0 circuit is sufficiently high, and also $L_b = 0$. For a lossless operation mode, it is necessary to provide the optimum zero-voltage and zero-voltage-derivative conditions for voltage $v(\omega t)$ across the switch just prior to the start of switch-on, when transistor is saturated, given by Eqs. (7.78) and (7.79).

The output current flowing through the load is written as sinusoidal by

$$i_R(\omega t) = I_R \sin(\omega t + \varphi) \tag{7.120}$$

where I_R is the load current amplitude and φ is the initial phase shift.

When the switch is turned on for $0 \le \omega t < \pi$, the voltage on the switch is $v(\omega t) = V_{cc} - v_L(\omega t) = 0$, the current flowing through the capacitance is $i_C(\omega t) = \omega C(dv_L/d\omega t) = 0$, and

$$i\,(\omega t) = i_L\,(\omega t) + i_R\,(\omega t) = \frac{1}{\omega L}\int_0^{\omega t} V_{cc}d\omega t + i_L\,(0) + I_R \sin(\omega t + \varphi)$$

$$= \frac{V_{cc}}{\omega L}\omega t + I_R\left[\sin(\omega t + \varphi) - \sin\varphi\right] \tag{7.121}$$

where the initial value for the current $i_L(\omega t)$ flowing through the dc-feed inductance L at $\omega t = 0$ can be found using Eq. (7.120) for $i(0) = 0$ as $i_L(0) = -I_R\sin\varphi$.

When the switch is turned off for $\pi \le \omega t < 2\pi$, the switch current $i(\omega t) = 0$, and the current $i_C(\omega t) = i_L(\omega t) + i_R(\omega t)$ flowing through the capacitance C can be rewritten as

$$\omega C \frac{dv\,(\omega t)}{d\omega t} = \frac{1}{\omega L} \int_{\pi}^{\omega t} \left[V_{cc} - v(\omega t) \right] d\omega t + i_L(\pi) + I_R \sin(\omega t + \varphi) \quad (7.122)$$

under the initial off-state conditions $v(\pi) = 0$ and

$$i_L(\pi) = i(\pi) - i_R(\pi) = \frac{V_{cc}\pi}{\omega L} - \omega L I_R \sin\varphi$$

Equation (7.122) can be represented in the form of the linear nonhomogeneous second-order differential equation

$$\omega^2 LC \frac{d^2 v\,(\omega t)}{d(\omega t)^2} + v(\omega t) - V_{cc} - \omega L I_R \cos(\omega t + \varphi) = 0 \quad (7.123)$$

the general solution of which can be obtained in the normalized form

$$\frac{v\,(\omega t)}{V_{cc}} = C_1 \cos(q\omega t) + C_2 \sin(q\omega t) + 1 - \frac{q^2 p}{1-q^2} \cos\,(\omega t + \varphi) \quad (7.124)$$

where

$$q = \frac{1}{\omega\sqrt{LC}} \quad (7.125)$$

$$p = \frac{\omega L I_R}{V_{cc}} \quad (7.126)$$

and the coefficients C_1 and C_2 are determined from the initial off-state conditions by

$$C_1 = -(\cos q\pi + q\pi \sin q\pi)$$

$$- \frac{qp}{1-q^2} \left[q \cos\varphi \; \cos q\pi - (1-2q^2)\sin\varphi \; \sin q\pi\right] \quad (7.127)$$

$$C_2 = (q\pi \cos q\pi - \sin q\pi)$$

$$- \frac{qp}{1-q^2} \left[q \cos\varphi \; \sin q\pi + (1-2q^2) \sin\varphi \; \cos q\pi\right] \quad (7.128)$$

The dc supply current I_0 can be found using Fourier formula and Eq. (7.121) by

$$I_0 = \frac{1}{2\pi} \int_0^{2\pi} i\,(\omega t)\; d\omega t = \frac{I_R}{2\pi} \left(\frac{\pi^2}{2p} + 2\cos\varphi - \pi \sin\varphi \right) \quad (7.129)$$

In an idealized Class-E operation mode, there is no nonzero voltage and current simultaneously that means a lack of power losses and gives an idealized collector efficiency of 100%. This implies that the dc power P_0 and fundamental output power P_{out} are equal,

$$I_0 \, V_{cc} = \frac{V_R^2}{2R} \tag{7.130}$$

where $V_R = I_R R$ is the fundamental voltage amplitude across the load resistance R.

As a result, by using Eqs. (7.129) and (7.130) and taking into account that $R = V_R^2/2P_{out}$, the optimum load resistance R for the specified values of a supply voltage V_{cc} and a fundamental output power P_{out} can be obtained by

$$R = \frac{1}{2}\left(\frac{V_R}{V_{cc}}\right)^2 \frac{V_{cc}^2}{P_{out}} \tag{7.131}$$

where

$$\frac{V_R}{V_{cc}} = \frac{1}{\pi}\left(\frac{\pi^2}{2p} + 2\cos\varphi - \pi\sin\varphi\right) \tag{7.132}$$

The normalized load-network inductance L and capacitance C can be appropriately defined using Eqs. (7.125), (7.126), and (7.129) by

$$\frac{\omega L}{R} = p/\left(\frac{\pi}{2p} + \frac{2}{\pi}\cos\varphi - \sin\varphi\right) \tag{7.133}$$

$$\omega CR = 1/\left(q^2 \frac{\omega L}{R}\right) \tag{7.134}$$

The series reactance X, which may have an inductive, capacitive, or zero reactance in special particular cases depending on the load-network parameters, can be generally calculated using two quadrature fundamental-frequency voltage Fourier components

$$V_R = -\frac{1}{\pi}\int_0^{2\pi} v(\omega t)\sin(\omega t + \varphi)\, d\omega t \tag{7.135}$$

$$V_X = -\frac{1}{\pi}\int_0^{2\pi} v(\omega t)\cos(\omega t + \varphi)\, d\omega t \tag{7.136}$$

The fundamental-frequency current flowing through the switch consists of two quadrature components, whose amplitudes can be found using Fourier formulas and Eq. (7.121) by

$$I_R = \frac{1}{\pi}\int_0^{2\pi} i(\omega t)\sin(\omega t + \varphi)\, d\omega t$$

$$= \frac{I_R}{\pi}\left[\frac{\pi\cos\varphi - 2\sin\varphi}{p} + \frac{\pi}{2} - \sin 2\varphi\right] \tag{7.137}$$

$$I_X = -\frac{1}{\pi}\int_0^{2\pi} i(\omega t)\cos(\omega t+\varphi)\, d\omega t$$
$$= \frac{I_R}{\pi}\left[\frac{\pi\sin\varphi + 2\cos\varphi}{p} - 2\sin^2\varphi\right] \tag{7.138}$$

Generally, Eq. (7.124) for normalized collector voltage contains three unknown parameters q, p, and φ, which must be analytically or numerically determined. In a common case, the parameter q can be considered a variable, and the other two parameters p and φ are calculated from a system of the two equations resulting from applying two optimum zero-voltage and zero-voltage–derivative conditions given by Eq. (7.78) and (7.79) to Eq. (7.124). Figure 7.52 shows the dependences of the optimum parameters p and φ versus q for a Class E with finite dc-feed inductance.

Based on the calculated optimum parameters p and φ as functions of q, the optimum load-network parameters of the Class-E load network with finite dc-feed inductance can be determined using Eqs. (7.131) to (7.134). The series reactance X can be calculated through the ratio of two quadrature fundamental-frequency voltage Fourier components given in Eqs. (7.135) and (7.136) as

$$\frac{X}{R} = \frac{V_X}{V_R} \tag{7.139}$$

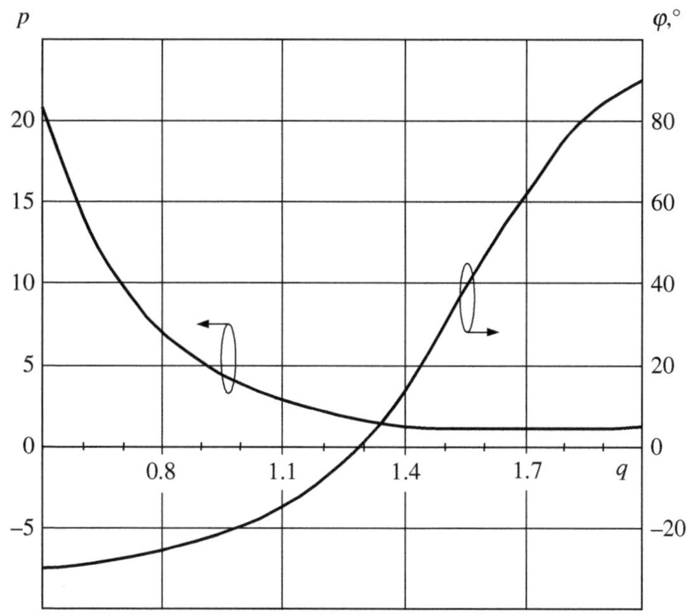

FIGURE 7.52 Optimum Class-E parameters p and φ versus q.

The dependences of the normalized optimum dc-feed inductance $\omega L/R$ and series reactance X/R are shown in Fig. 7.53(*a*), whereas the dependences of the normalized optimum shunt capacitance ωCR and load resistance RP_{out}/V_{cc}^2 are plotted in Fig. 7.53(*b*). Here, the value of

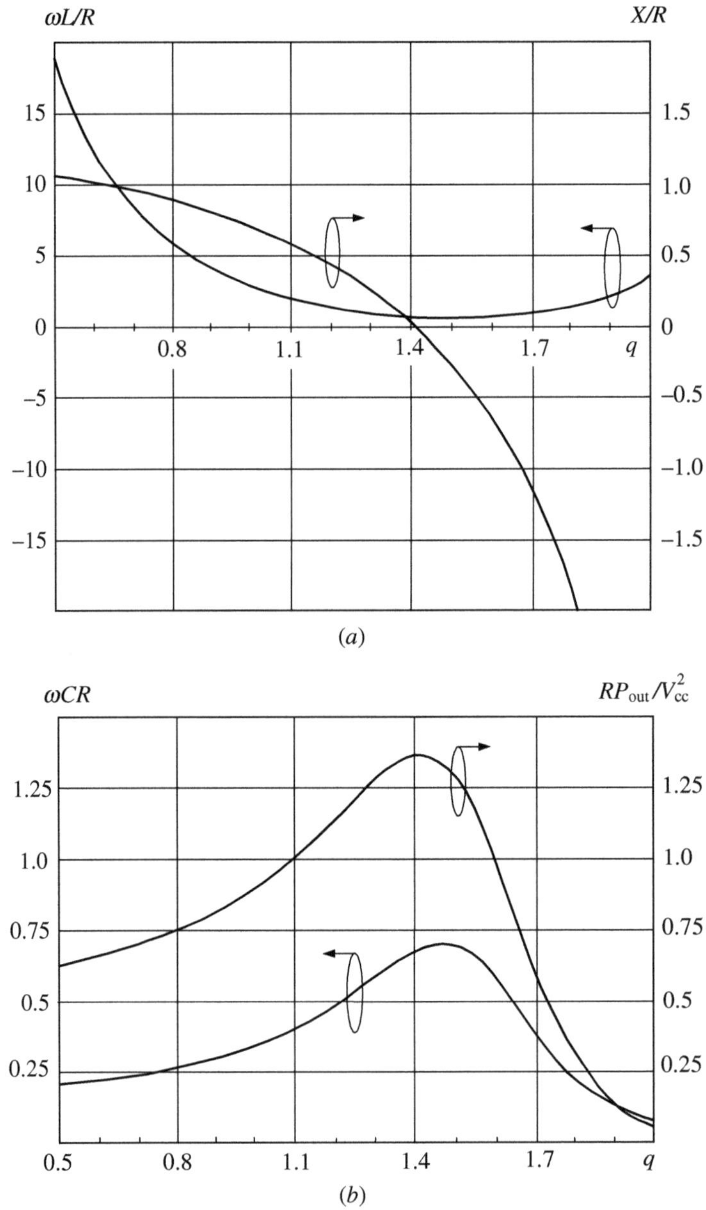

FIGURE 7.53 Normalized optimum Class-E load network parameters.

the series reactance X changes its sign from positive to negative, which means that the inductive reactance is followed by the capacitive reactance. As a result, there is a special case of the load network with a parallel circuit and a load resistor only when $X = 0$ at $q = 1.412$. In this case, the maximum value of the optimum load resistance R can be provided for the same supply voltage and output power, thus simplifying the matching with the standard load of 50 Ω. In addition, the values of a dc-feed inductance L become sufficiently small, making Class-E mode with a parallel circuit very attractive for monolithic applications. The maximum operation frequency f_{max} is realized at $q = 1.468$, where the normalized optimum shunt capacitance ωCR reaches its maximum.

The graphical solutions for the optimum load-network parameters can be replaced by the analytical design equations represented in terms of the simple second- and third-order polynomial functions given by Tables 7.3 and 7.4 for different ranges of the parameter q [66]. The maximum difference between the polynomial approximations and exact numerical solutions given in the graphic form is about 2%.

Parameter	Design Equation
$\frac{\omega L}{R}$	$44.93q^2 - 94.32q + 52.46$
ωCR	$0.426q^2 - 0.379q + 0.3$
$\frac{X}{R}$	$-0.73q^2 + 0.411q + 1.03$
$\frac{P_{out}R}{V_{cc}^2}$	$0.74q^2 - 0.6q + 0.76$

Table 7.3 Load Network Parameters for $0.6 < q < 1.0$

Parameter	Design Equation
$\frac{\omega L}{R}$	$8.08q^2 - 24.53q + 19.23$
ωCR	$-6.97q^3 + 25.93q^2 - 31.071q + 12.48$
$\frac{X}{R}$	$-2.9q^3 + 8.8q^2 + 10.2q + 5.02$
$\frac{P_{out}R}{V_{cc}^2}$	$11.9q^3 + 42.753q^2 - 49.63q + 19.7$

Table 7.4 Load Network Parameters for $1.0 < q < 1.65$

7.5.2 Parallel-Circuit Class E

The theoretical analysis of a switchmode parallel-circuit Class-E power amplifier with a series filter, whose basic circuit schematic is shown in Fig. 7.54(*a*), was first published by Kozyrev with calculation of the voltage and current waveforms and some graphical results [28, 67]. The load network consists of a finite dc-feed inductor *L*, a shunt capacitor *C*, a series L_0C_0 resonant circuit tuned to the fundamental frequency, and a load resistor *R*. In this case, the switch sees a parallel connection of the load resistor *R* and parallel *LC* circuit at the fundamental frequency, as shown in Fig. 7.54(*b*), where also the real and imaginary collector fundamental-frequency current components I_X and I_R and the real collector fundamental-frequency voltage component V_R are indicated.

In the case of a parallel-circuit Class-E load network without series phase-shifting reactance, because the parameter *q* is unknown a priori, generally it is necessary to solve a system of three equations to define the three unknown parameters *q*, *p*, and *φ*. The first two equations are the result of applying the two optimum zero-voltage and zero-voltage-derivative conditions given by Eq. (7.78) and (7.79) to Eq. (7.124). Because the fundamental-frequency collector voltage is fully applied to the load, this means that its reactive part must have zero value, resulting in an additional equation

$$V_X = -\frac{1}{\pi}\int_0^{2\pi} v(\omega t)\cos(\omega t + \phi)\, d\omega t = 0 \tag{7.140}$$

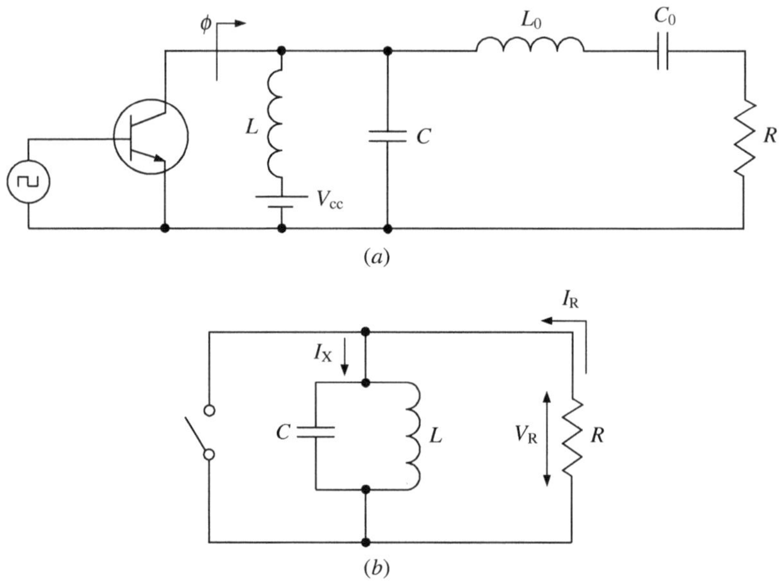

FIGURE 7.54 Equivalent circuits of parallel-circuit Class-E power amplifier.

Solving the system of three equations with three unknown parameters numerically gives the following values [68, 69]:

$$q = 1.412 \tag{7.141}$$

$$p = 1.210 \tag{7.142}$$

$$\phi = 15.155° \tag{7.143}$$

Figure 7.55 shows the normalized (*a*) load current and collector (*b*) voltage, and (*c*) current waveforms for an idealized optimum parallel-circuit Class-E operation mode. From collector voltage and current waveforms, it follows that, similar to other Class-E subclasses, there is no nonzero voltage and current simultaneously. When this happens, no power loss occurs and an idealized collector efficiency of 100% is achieved.

By using Eqs. (7.131) through (7.134), the idealized optimum (or nominal) load resistance R, parallel inductance L, and parallel capacitance C can be appropriately obtained by

$$R = 1.365 \frac{V_{cc}^2}{P_{out}} \tag{7.144}$$

$$L = 0.732 \frac{R}{\omega} \tag{7.145}$$

$$C = \frac{0.685}{\omega R} \tag{7.146}$$

The dc supply current I_0 can be calculated from Eq. (7.129) as

$$I_0 = 0.826\, I_R \tag{7.147}$$

The phase angle ϕ seen from the device collector at the fundamental frequency can be represented either through the two quadrature fundamental-frequency current Fourier components I_X and I_R or as a function of the load-network elements as

$$\phi = \tan^{-1}\left(\frac{R}{\omega L} - \omega RC\right) = 34.244° \tag{7.148}$$

If the calculated value of the optimum Class-E resistance R is too small or differs significantly from the standard load impedance (usually equal to 50 Ω), it is necessary to use an additional matching circuit to deliver maximum output power to the load. It should be noted that, among a family of the Class-E load networks, a parallel-circuit Class-E load network offers the largest value of R, thus simplifying the final matching design procedure. In this case, the first series element of such matching circuits should be the inductor to provide high impedance conditions for harmonics, as shown in Fig. 7.56.

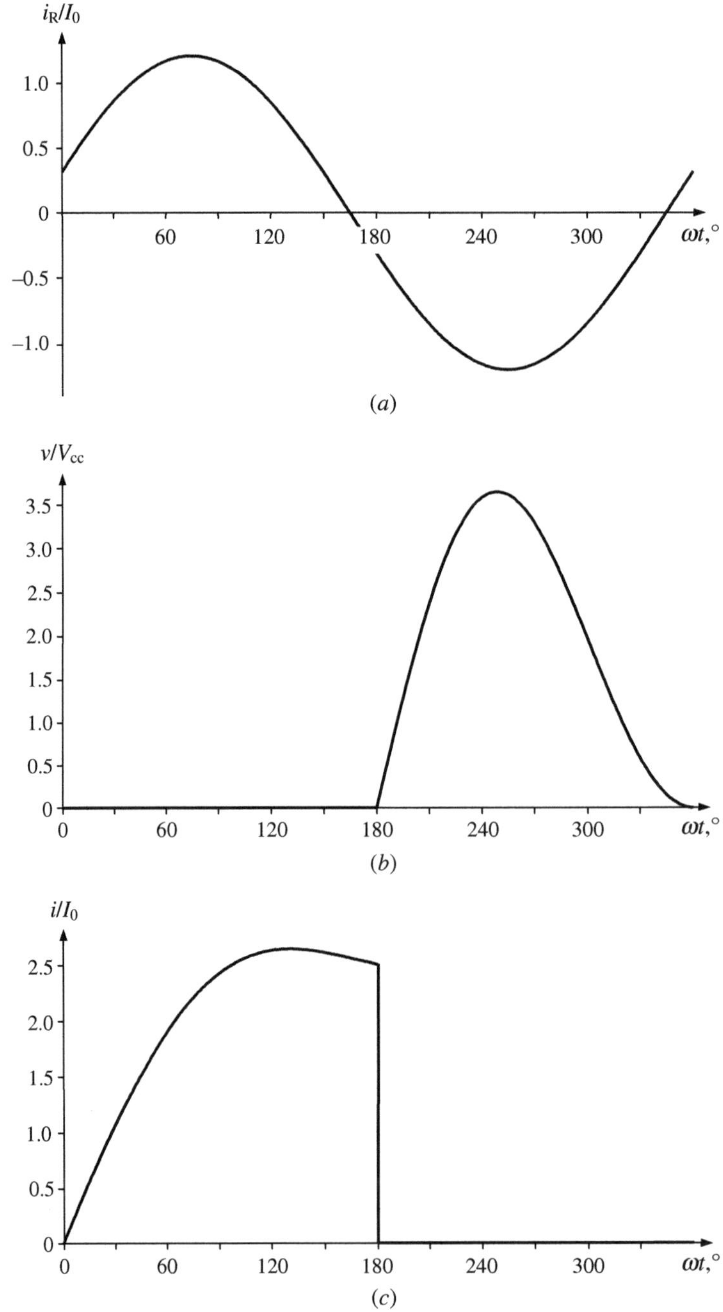

FIGURE 7.55 Normalized (*a*) load current and collector (*b*) voltage and (*c*) current waveforms for idealized optimum parallel-circuit Class E.

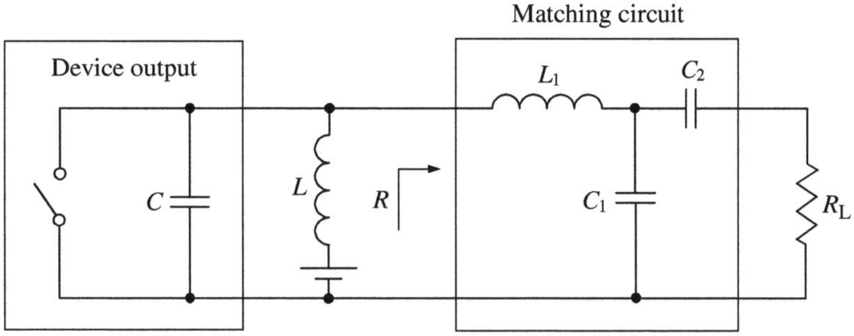

Figure 7.56 Parallel-circuit Class-E power amplifier with lumped matching circuit.

The peak collector current I_{max} and peak collector voltage V_{max} can be determined from Eqs. (7.121), (7.124), and (7.147) as

$$I_{max} = 2.647\, I_0 \tag{7.149}$$

$$V_{max} = 3.647\, V_{cc} \tag{7.150}$$

The maximum frequency f_{max} can be calculated using Eq. (7.144) and (7.146) when $C = C_{out}$, where C_{out} is the device output capacitance, as

$$f_{max} = 0.0798\, \frac{P_{out}}{C_{out} V_{cc}^2} \tag{7.151}$$

which is 1.4 times higher than maximum operation frequency for an optimum Class-E power amplifier with shunt capacitance [70].

7.5.3 Load Networks with Transmission Lines

At microwave frequencies, the parallel inductance L can be replaced by a short-length short-circuited transmission line TL, as shown in Fig. 7.57(*a*), according to

$$Z_0 \tan\theta = \omega L \tag{7.152}$$

where Z_0 and θ are the characteristic impedance and electrical length of such a transmission line, respectively [71]. By using Eq. (7.145), defining the idealized optimum (or nominal) parallel inductance L for a parallel-circuit Class-E mode, Eq. (7.152) can be rewritten as

$$\tan\theta = 0.732 \frac{R}{Z_0} \tag{7.153}$$

Generally, the load-network circuit can be composed with any types of transmission lines, including open-or short-circuit stubs to provide the required matching and harmonic-suppression conditions. In some cases, for example, for compact small-size power-amplifier modules designed for handset wireless transmitters, the series microstrip

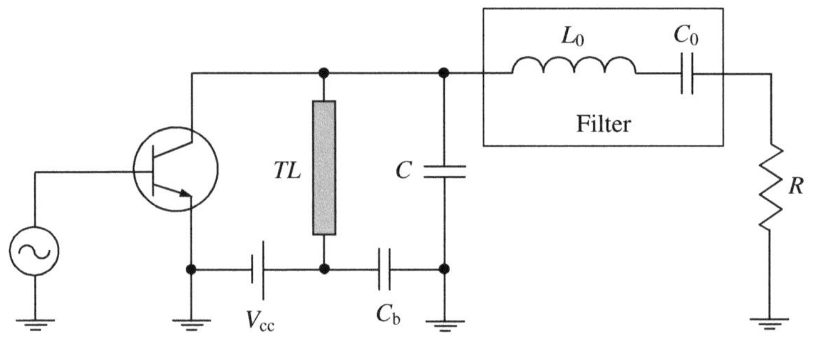

(a)

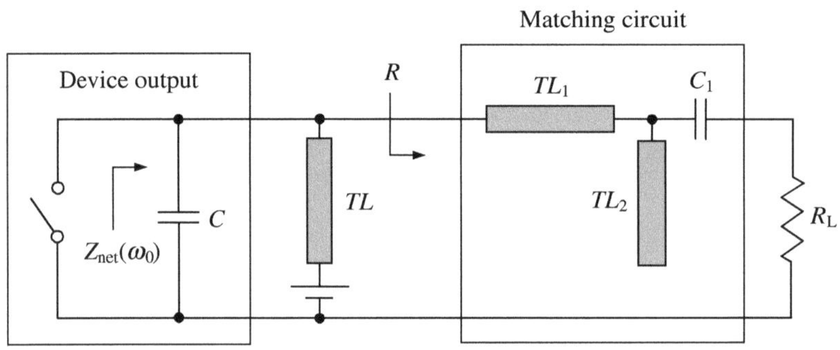

(b)

FIGURE 7.57 Equivalent circuits of transmission-line parallel-circuit Class-E power amplifier.

lines and shunt chip capacitors are usually used in the external output matching circuits. However, to maintain the optimum-switching conditions at the fundamental frequency, such an output matching circuit should contain the series transmission line as the first element, as shown in Fig. 7.57(*b*).

Figure 7.58(*a*) shows an example of the transmission-line Class-E load network of a two-stage 1.75-GHz GaAs HBT power amplifier with an output power of 33 dBm, which was designed for a cellular handset transmitter, and includes the series microstrip line with two shunt chip capacitors [68]. However, because of the fixed certain electrical lengths of the transmission lines, it is impossible to realize simultaneously the required inductive impedance at the fundamental frequency with the purely capacitive reactances at higher-order harmonics. For example, at the second harmonic, the real part of the load network impedance $Z_{net}(2\omega_0)$ is sufficiently high, as shown in Fig. 7.58(*b*). Nevertheless, even such an approximation provides a good proximity to the parallel-circuit Class-E operation mode, resulting in a high operating efficiency of the power amplifier. In this case, there is no need to use an additional RF choke for dc supply current,

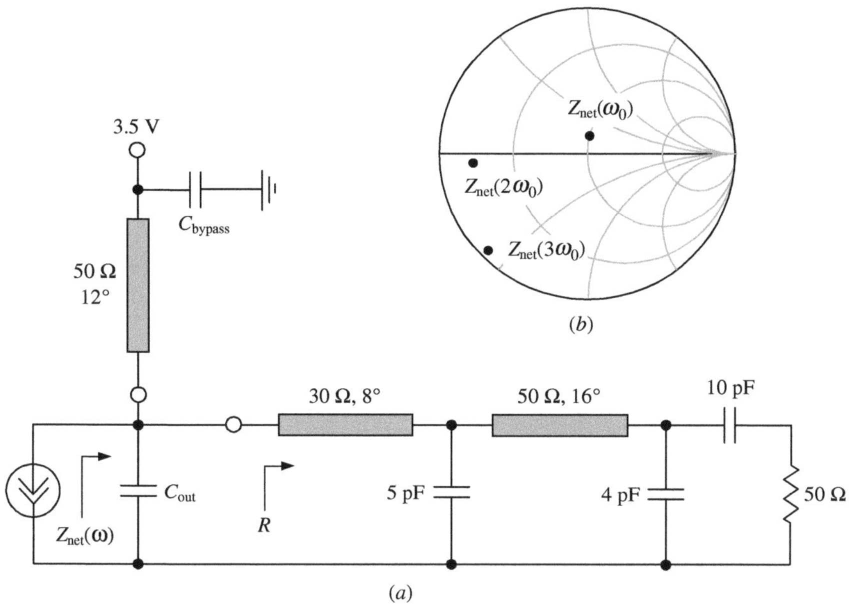

FIGURE 7.58 Transmission-line load network of parallel-circuit Class-E power amplifier for handset application.

because its function can be performed by the same short-length parallel microstrip line required to provide an optimum inductive impedance at the fundamental frequency.

The circuit schematic of a two-stage InGaP/GaAs HBT power amplifier intended to operate in the WCDMA handset transmitters is shown in Fig. 7.59(*a*) [72]. The MMIC part of this power amplifier contains the transistors with emitter areas of the first and second stage as large as 540 μm^2 and 3600 μm^2, input matching circuit, interstage matching circuit, and bias circuits on a die with dimensions of less than 1 mm^2. Without any tuning of the output matching circuit, a saturated output power greater than 30 dBm and a *PAE* greater than 50% were obtained. Using high-*Q* capacitors in output matching circuit can improve the power-added efficiency by about 8%. At the same time, this power amplifier without any additional tuning can provide the high-linearity performance for WCDMA band (1920–1980 MHz) at a 3.5-dB backoff output power of 27 dBm with a power gain of 22.6 dB and a sufficiently high efficiency. The measured *PAE* reached value of 38.3% at center bandwidth frequency of 1.95 GHz with an adjacent channel leakage power ratio (*ACLR*) of –37 dBc at a 5-MHz offset.

Figure 7.59(*b*) shows the circuit schematic of a 1-GHz 12-V LDMOSFET parallel-circuit Class-E power amplifier with a drain

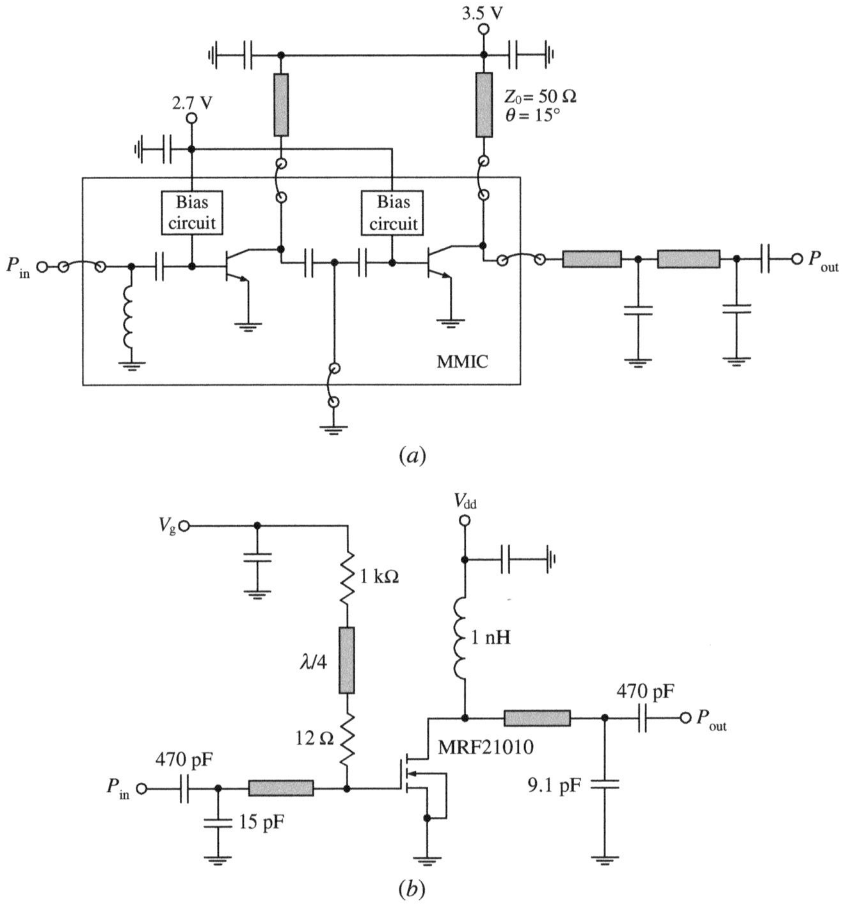

FIGURE 7.59 Schematics of Class-E power amplifiers with transmission-line matching.

efficiency of 70.4% and an output power of more than 38 dBm [73]. In this case, the series *LC* resonant circuit is replaced by a low-pass *L*-type output matching circuit with a series transmission line to match the low optimum Class-E resistance to a 50-Ω load, having almost zero series excessive reactance *X*. The quarterwave transmission line in the gate bias circuit provides RF isolation from the dc-voltage supply, and the 12-Ω gate resistor is required for stability reason.

References

1. D. M. Snider, "A Theoretical Analysis and Experimental Confirmation of the Optimally Loaded and Overdriven RF Power Amplifier," *IEEE Trans. Electron Devices*, vol. ED-14, pp. 851–857, Dec. 1967.
2. D. C. Prince, "Vacuum Tubes as Power Oscillators, Part III," *Proc. IRE*, vol. 11, pp. 527–550, Sep. 1923.

3. J. Zenneck and H. Rukop, *Lehrbuch der Drahtlosen Telegraphie*, Stuttgart: Ferdinand Enke, 1925.
4. I. N. Fomichev, "A New Method to Increase Efficiency of the Radio Broadcasting Station (in Russian)," *Elektrosvyaz*, pp. 58–66, Jun. 1938.
5. R. I. Sarbacher, "Power-Tube Performance in Class C Amplifiers and Frequency Multipliers as Influenced by Harmonic Voltage," *Proc. IRE*, vol. 31, pp. 607–625, Nov. 1943.
6. H. J. Round, "Wireless Telegraph and Telephone Transmission," U.S. Patent 1,564,627, Dec. 1925.
7. F. H. Raab, "An Introduction to Class-F Power Amplifiers," *RF Design*, vol. 19, pp. 79–84, May 1996, p. 14, Jul. 1996.
8. F. H. Raab, "Class-F Power Amplifiers with Maximally Flat Waveforms," *IEEE Trans. Microwave Theory Tech.*, vol. MTT-45, pp. 2007–2012, Nov. 1997.
9. F. H. Raab, "FET Power Amplifier Boosts Transmitter Efficiency," *Electronics*, vol. 49, pp. 122–126, Jun. 1976.
10. A. Grebennikov, "Load Network Design Technique for Class F and Inverse Class F Power Amplifiers," *High Frequency Electronics*, vol. 10, pp. 58–76, May 2011.
11. V. A. Borisov and V. V. Voronovich, "Analysis of Switching Transistor Amplifier with Parallel Forming Transmission Line (in Russian)," *Radiotekhnika i Elektronika*, vol. 31, pp. 1590–1597, Aug. 1986.
12. A. V. Grebennikov, "Circuit Design Technique for High Efficiency Class F Amplifiers," *2000 IEEE MTT-S Int. Microwave Symp. Dig.*, pp. 771–774.
13. A. V. Grebennikov, "Effective Circuit Design Techniques to increase MOSFET Power Amplifier Efficiency," *Microwave J.*, vol. 43, pp. 64–72, Jul. 2000.
14. C. Trask, "Class-F Amplifier Loading Networks: A Unified Design Approach," *1999 IEEE MTT-S Int. Microwave Symp. Dig.*, pp. 351–354.
15. A. V. Grebennikov and F. Lin, "An Efficient CAD-Oriented Large-Signal MOSFET Model," *IEEE Trans. Microwave Theory Tech.*, vol. MTT-48, pp. 1732–1742, Oct. 2000.
16. D. Schmelzer and S. I. Long, "An Efficient CAD-Oriented Large-Signal MOSFET Model," *IEEE J. Solid-State Circuits*, vol. SC-42, pp. 21302–2136, Oct. 2007.
17. A. I. Kolesnikov, "A New Method to Improve Efficiency and to Increase Power of the Transmitter (in Russian)," *Master Svyazi*, pp. 27–41, Jun. 1940.
18. E. S. Glazman, L. B. Kalinin, and Y. I. Mikhailov, "Improving VHF Transmitter Efficiency by Using the Biharmonic Mode," *Telecommunications and Radio Engineering*, Part 1, vol. 30, pp. 46–51, Jul. 1975.
19. V. J. Tyler, "A New High-Efficiency High-Power Amplifier," *Marconi Review*, vol. 21, pp. 96–109, Fall 1958.
20. M. K. Kazimierczuk, "A New Concept of Class F Tuned Power Amplifier," *Proc.* 27th *Midwest Circuits and Systems Symp.*, pp. 425–428, 1984.
21. P. Heymann, R. Doerner, and M. Rudolph, "Harmonic Tuning of Power Transistors by Active Load-Pull Measurement," *Microwave J.*, vol. 43, pp. 22–37, Jun. 2000.
22. C. J. Wei, P. DiCarlo, Y.A. Tkachenko, R. McMorrow, and D. Bartle, "Analysis and Experimental Waveform Study on Inverse Class-F Mode of Microwave Power FETs," *2000 IEEE MTT-S Int. Microwave Symp. Dig.*, pp. 525–528.
23. A. Grebennikov, "High-Efficiency Transmission-Line Inverse Class F Power Amplifiers for 2-GHz WCDMA Systems," *Int. J. RF and Microwave Computer-Aided Eng.*, vol. 21, pp. 446–456, Jul. 2011.
24. A. Jarndal, P. Aflaki, R. Negra, A. K. Kouki, and F. Ghannouchi, "Large-Signal Modeling Methodology for GaN HEMTs for RF Switching-Mode Power Amplifiers Design," *Int. J. RF and Microwave Computer-Aided Eng.*, vol. 21, pp. 45–51, Jan. 2011.
25. J. W. Wood, "High Efficiency Class C Amplifier," U.S. Patent 3,430,157, Feb. 1969.
26. E. P. Khmelnitsky, *Operation of Vacuum-Tube Generator on Detuned Resonant Circuit* (in Russian), Moskva: Svyazizdat, 1962.
27. G. D. Ewing, *High-Efficiency Radio-Frequency Power Amplifiers*, Ph.D. Dissertation, Oregon State University, Jun. 1964.

28. V. B. Kozyrev, "Single-Ended Tuned Switching Power Amplifier with Filtering Resonant Circuit (in Russian)," *Poluprovodnikovye Pribory v Tekhnike Svyazi*, vol. 6, pp. 152–166, 1971.
29. N. O. Sokal and A. D. Sokal, "Class E—A New Class of High-Efficiency Tuned Single-Ended Switching Power Amplifiers," *IEEE J. Solid-State Circuits*, vol. SC-10, pp. 168–176, Jun. 1975.
30. N. O. Sokal and A. D. Sokal, "High-Efficiency Tuned Switching Power Amplifier," U.S. Patent 3,919,656, Nov. 1975.
31. N. O. Sokal, "Class E High-Efficiency Power Amplifiers, from HF to Microwave," *1998 IEEE MTT-S Int. Microwave Sym. Dig.*, vol. 2, pp. 1109–1112.
32. F. H. Raab, "Idealized Operation of the Class E Tuned Power Amplifier," *IEEE Trans. Circuits and Systems*, vol. CAS-24, pp. 725–735, Dec. 1977.
33. B. Molnar, "Basic Limitations on Waveforms Achievable in Single-Ended Switching-Mode Tuned (Class E) Power Amplifiers," *IEEE J. Solid-State Circuits*, vol. SC-19, pp. 144–146, Feb. 1984.
34. T. B. Mader, E. W. Bryerton, M. Marcovic, M. Forman, and Z. Popovic, "Switched-Mode High-Efficiency Microwave Power Amplifiers in a Free-Space Power-Combiner Array," *IEEE Trans. Microwave Theory Tech.*, vol. MTT-46, pp. 1391–1398, Oct. 1998.
35. M. Kazimierczuk and K. Puczko, "Exact Analysis of Class E Tuned Power Amplifier at Any *Q* and Switch Duty Cycle," *IEEE Trans. Circuits and Systems*, vol. CAS-34, pp. 149–158, Feb. 1987.
36. F. H. Raab and N. O. Sokal, "Transistor Power Losses in the Class E Tuned Power Amplifier," *IEEE J. Solid-State Circuits*, vol. SC-13, pp. 912–914, Dec. 1978.
37. A. N. Bruevich, "About Optimum Parameters of Switching-Mode Tuned Power Amplifier with Filtering Resonant Circuit (in Russian)," *Poluprovodnikovaya Elektronika v Tekhnike Svyazi*, vol. 18, pp. 43–48, 1977.
38. I. A. Popov (Ed.), *Transistor Generators of Harmonic Oscillations in Switching Mode* (in Russian), Moskva: Radio i Svyaz, 1985.
39. M. Kazimierczuk, "Effect of the Collector Current Fall Time on the Class E Tuned Power Amplifier," *IEEE J. Solid-State Circuits*, vol. SC-18, pp. 181–193, Apr. 1983.
40. J. A. Blanchard and J. S. Yuan, "Effect of Collector Current Exponential Decay on Power Efficiency for Class E Tuned Power Amplifier," *IEEE Trans. Circuits and Systems-I: Fundamental Theory Appl.*, vol. CAS-I-41, pp. 69–72, Jan. 1994.
41. M. J. Chudobiak, "The Use of Parasitic Nonlinear Capacitors in Class E Amplifiers," *IEEE Trans. Circuits and Systems-I: Fundamental Theories Appl.*, vol. CAS-I-41, pp. 941–944, Dec. 1994.
42. P. Alinikula, K. Choi, and S. I. Long, "Design of Class E Power Amplifier with Nonlinear Parasitic Output Capacitance," *IEEE Trans. Circuits and Systems-II: Analog and Digital Signal Process.*, vol. CAS-II-46, pp. 114–119, Feb. 1999.
43. T. Suetsugu and M. K. Kazimierczuk, "Analysis and Design of Class E Amplifier with Shunt Capacitance Composed of Nonlinear and Linear Capacitances," *IEEE Trans. Circuits and Systems-I: Regular Papers*, vol. CAS-I–51, pp. 1261–1268, Jul. 2004.
44. V. B. Kozyrev and V. V. Shkvarin, "Optimum Operation Mode of Single-Ended Tuned Switching Power Amplifier with Forming Resonant Circuit (in Russian)," *Radiotekhnika*, vol. 37, pp. 90–93, Oct. 1982.
45. M. Acar, A. J. Annema, and B. Nauta, "Variable-Voltage Class-E Power Amplifiers," *2007 IEEE MTT-S Int. Microwave Sym. Dig.*, pp. 1095–1098.
46. F. H. Raab, "Suboptimum Operation of Class-E Power Amplifiers," *Proc. RF Technology Expo '89*, Santa Clara, CA, pp. 85–98, Feb. 1989.
47. F. H. Raab, "Effects of Circuit Variations on the Class E Tuned Power Amplifier," *IEEE J. Solid-State Circuits*, vol. SC-13, pp. 239–246, Apr. 1978.
48. T. Suetsugu and M. K. Kazimierczuk, "Design Procedure of Class-E Amplifier for Off-Nominal Operation at 50% Duty Cycle," *IEEE Trans. Circuits and Systems-I: Regular Papers*, vol. CAS-I-53, pp. 1468–1476, Jul. 2006.
49. T. B. Mader and Z. B. Popovic, "The Transmission-Line High-Efficiency Class-E Amplifier," *IEEE Microwave and Guided Wave Lett.*, vol. 5, pp. 290–292, Sep. 1995.

50. F. J. Ortega-Gonzalez, J. L. Jimenez-Martin, A. Asensio-Lopez, and G. Torregrosa-Penalva, "High-Efficiency Load-Pull Harmonic Controlled Class-E Power Amplifier," *IEEE Microwave and Guided Wave Lett.*, vol. 8, pp. 348–350, Oct. 1998.
51. J. F. Davis and D. B. Rutledge, "A Low-Cost Class-E Power Amplifier with Sine-Wave Drive," *1998 IEEE MTT-S Int. Microwave Symp. Dig.*, vol. 2, pp. 1113–1116.
52. R. Frey, "500 W, Class E 27.12 MHz Amplifier Using a Single Plastic MOSFET," *1999 IEEE MTT-S Int. Microwave Symp. Dig.*, pp. 359–362.
53. H. Zirath and D. Rutledge, "An LDMOS VHF Class E Power Amplifier Using a High Q Novel Variable Inductor," *1999 IEEE MTT-S Int. Microwave Symp. Dig.*, pp. 367–370.
54. M. Franco and A. Katz, "Class-E Silicon Carbide VHF Power Amplifier," *2007 IEEE MTT-S Int. Microwave Symp. Dig.*, pp. 19–22.
55. R. Tayrani, "A Monolithic X-Band Class-E Power Amplifier for Space Based Radar Systems," *RF Design*, vol. 26, pp. D14–D19, Nov. 2003.
56. R. Negra, F. M. Ghannouchi, and W. Baechtold, "Study and Design Optimization of Multiharmonic Transmission-Line Load Networks for Class-E and Class-F *K*-band MMIC Power Amplifiers," *IEEE Trans. Microwave Theory Tech.*, vol. MTT-55, pp. 1390–1397, Jun. 2007.
57. R. E. Zulinski and J. W. Steadman, "Class E Power Amplifiers and Frequency Multipliers with Finite DC-Feed Inductance," *IEEE Trans. Circuits and Systems*, vol. CAS-34, pp. 1074–1087, Sept. 1987.
58. C. P. Avratoglou, N. C. Voulgaris, and F. I. Ioannidou, "Analysis and Design of a Generalized Class E Tuned Power Amplifier," *IEEE Trans. Circuits and Systems*, vol. CAS-36, pp. 1068–1079, Aug. 1989.
59. G. H. Smith and R. E. Zulinski, "An Exact Analysis of Class E Power Amplifiers with Finite DC-Feed Inductance at Any Output *Q*," *IEEE Trans. Circuits and Systems*, vol. CAS-37, pp. 530–534, Apr. 1990.
60. H. Sekiya, I. Sasase, and S. Mori, "Computation of Design Values for Class E Amplifiers without Using Waveform Equations," *IEEE Trans. Circuits and Systems-I: Fundamental Theory Appl.*, vol. CAS-I-49, pp. 966–978, Jul. 2002.
61. C.-H. Li and Y.-O. Yam, "Maximum Frequency and Optimum Performance of Class E Power Amplifiers," *IEE Proc. Circuits Devices Syst.*, vol. 141, pp. 174–184, Jun. 1994.
62. C. K. Ho, H. Wong, and S. W. Ma, "Approximation of Non-Zero Transistor ON Resistance in Class-E Amplifiers," *Proc. 5th IEEE Int. Caracas Conf. Devices Circuits Syst.*, pp. 90–93, 2004.
63. D. Milosevic, J. van der Tang, and A. van Roermund, "Explicit Design Equations for Class-E Power Amplifiers with Small DC-Feed Inductance," *Proc. 2005 Europ. Conf. Circuit Theory and Design*, vol. 3, pp. 101–104.
64. A. Grebennikov, "Load Network Design Techniques for Class E RF and Microwave Amplifiers," *High Frequency Electronics*, vol. 3, pp. 18–32, Jul. 2004.
65. A. Grebennikov and N. O. Sokal, *Switchmode RF Power Amplifiers*, New York: Newnes, 2007.
66. M. Acar, A. J. Annema, and B. Nauta, "Analytical Design Equations for Class-E Power Amplifiers," *IEEE Trans. Circuits and Systems-I: Regular Papers*, vol. CAS-I-54, pp. 2706–2717, Dec. 2007.
67. A. Grebennikov, "Class E High-Efficiency Power Amplifiers: Historical Aspect and Future Prospect," *Applied Microwave & Wireless*, vol. 14, pp. 64–71, Jul. 2002, pp. 64–72, Aug. 2002.
68. A. V. Grebennikov and H. Jaeger, "Class E with Parallel Circuit—A New Challenge for High-Efficiency RF and Microwave Power Amplifiers," *2002 IEEE MTT-S Int. Microwave Symp. Dig.*, vol. 3, pp. 1627–1630.
69. A. Grebennikov, "Switched-Mode RF and Microwave Parallel-Circuit Class E Power Amplifiers," *Int. J. RF and Microwave Computer-Aided Eng.*, vol. 14, pp. 21–35, Jan. 2004.
70. M. K. Kazimierczuk and W. A. Tabisz, "Class C-E High-Efficiency Tuned Power Amplifier," *IEEE Trans. Circuits and Systems*, vol. CAS-36, pp. 421–428, Mar. 1989.

71. A. V. Grebennikov and H. Jaeger, "High Efficiency Transmission Line Tuned Power Amplifier," U.S. Patent 6,552,610, Apr. 2003.
72. H. Jaeger, A. V. Grebennikov, E. P. Heaney, and R. Weigel, "Broadband High-Efficiency Monolithic InGaP/GaAs HBT Power Amplifiers for 3G Handset Applications," *2002 IEEE MTT-S Int. Microwave Symp. Dig.*, vol. 2, pp. 1035–1038.
73. Y. Xu, X. Zhu, and C. You, "Analysis and Design of Class-E Power Amplifier Using Equivalent LDMOS Model with Drift Region Effect," *Microwave and Optical Technology Lett.*, vol. 52, pp. 1836–1842, Aug. 2010.

CHAPTER 8

Broadband Power Amplifiers

In many telecommunication, radar, or testing systems, the transmitters operate in a very wide frequency range, for example, 1.5 to 30 MHz in high-frequency (HF) transceivers, 225 to 400 MHz in military frequency-agility systems, 470 to 860 MHz in ultra–high-frequency (UHF) TV transmitters, or 2 to 8 GHz and 6 to 18 GHz in microwave applications. The power amplifier design based on a broadband concept provides some advantages when there is no need to tune resonant circuits, and it is possible to realize fast frequency agility or to transmit a wide multimode signal spectrum. However, there are many factors that restrict the frequency bandwidth depending on the active device parameters. For example, it is quite easy to provide multioctave amplification from very low frequencies up to UHF band using the power MOSFET devices when lossy gain compensation is easily provided. This can be possible due to the some margin in a power gain at lower frequencies for these devices, because its value decreases with frequency by approximately 6 dB per octave. Besides, lossy gain-compensating networks can provide lower input reflection coefficients, smaller gain ripple, more predictable amplifier design, and can contribute to the amplifier stability factors that are superior to those of lossless matching networks. At higher frequencies when the device input impedance is significantly smaller and influence of its internal feedback and parasitic parameters is substantially higher, it is necessary to use multisection matching networks with lumped and distributed elements.

Generally, the matching design procedure is based on the methods of circuit analysis, optimization, and synthesis. According to the first one, the circuit parameters are calculated at one frequency chosen in advance (usually the center or high bandwidth frequency), and then the power amplifier performance is analyzed across the entire frequency bandwidth. In order to synthesize the broadband matching/compensation network, it is necessary to choose the maximum attenuation level or reflection coefficient magnitude inside the operating frequency bandwidth and then to obtain the parameters of matching networks by using special tables and formulas to convert the lumped element into distribute ones. For push-pull power amplifiers, it is very convenient to use both lumped and distributed parameters when the lumped capacitors are connected in parallel to the microstrip lines due to the effect of virtual ground.

8.1 Bode-Fano Criterion

Generally, the design for a broadband matching circuit should solve a problem with contradictory requirements when wider matching bandwidth is required with minimum reflection coefficient, or how to minimize the number of the matching network sections for a given wideband specification. The necessary requirements are determined by the Bode-Fano criterion, which gives (for certain canonical types of load impedances) a theoretical limit on the maximum reflection coefficient magnitude that can be obtained with an arbitrary matching network [1, 2].

For the lossless matching networks with a parallel *RC* load shown in Fig. 8.1(*a*) and with a series *LR* load shown in Fig. 8.1(*b*), the Bode-Fano criterion states that

$$\int_0^{\infty} \ln \frac{1}{|\Gamma(\omega)|} d\omega \le \frac{\pi}{\tau} \tag{8.1}$$

where $\Gamma(\omega)$ is the input reflection coefficient seen looking into the arbitrary lossless matching network and $\tau = RC = L/R$.

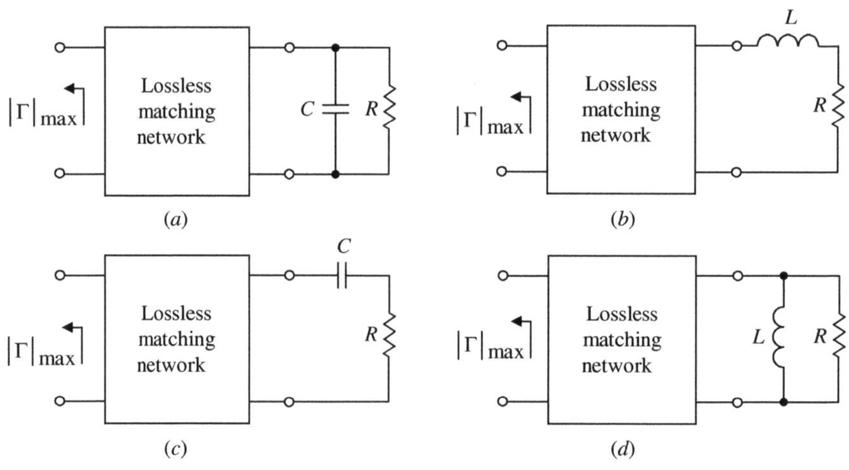

Figure 8.1 Loaded lossless matching circuits.

For the lossless matching networks with a series *RC* load shown in Fig. 8.1(*c*) and with a parallel *LR* load shown in Fig. 8.1(*d*), the Bode-Fano integral is written as

$$\int_0^\infty \omega^{-2} \ln\frac{1}{|\Gamma(\omega)|} d\omega \le \pi\tau \tag{8.2}$$

The mathematical relationships expressed by Eqs. (8.1) and (8.2) reflect the flat responses of an ideal filter over the required frequency bandwidth, as shown in Fig. 8.2 for two different cases. For the same load, both plots illustrate the important tradeoff: the wider the matching network bandwidth, the worse the reflection coefficient magnitude. From Eq. (8.1) it follows that, when $|\Gamma(\omega)|$ is constant and equal to $|\Gamma|_{max}$ over a frequency band of width $\Delta\omega$ and $|\Gamma(\omega)| = 1$ otherwise,

$$\int_0^\infty \ln\frac{1}{|\Gamma(\omega)|} d\omega = \int_{\omega_1}^{\omega_2} \ln\frac{1}{|\Gamma(\omega)|} d\omega = \Delta\omega \ \ln\frac{1}{|\Gamma|_{max}} \le \frac{\pi}{\tau} \tag{8.3}$$

As a result,

$$|\Gamma|_{max} = \exp\left(\frac{-\pi}{\Delta\omega\tau}\right) \tag{8.4}$$

where $\Delta\omega = \omega_2 - \omega_1$.

Similarly, for the lossless network with a series *RC* load and with a parallel *LR* load,

$$|\Gamma|_{max} = \exp\left(\frac{-\pi\omega_0^2\tau}{\Delta\omega}\right) \tag{8.5}$$

where $\omega_0 = \sqrt{\omega_1\omega_2}$ is the center bandwidth frequency. It should be noted that the theoretical bandwidth limits can be realized only with an infinite number of matching network sections. The frequency bandwidth with a maximum reflection coefficient magnitude is determined

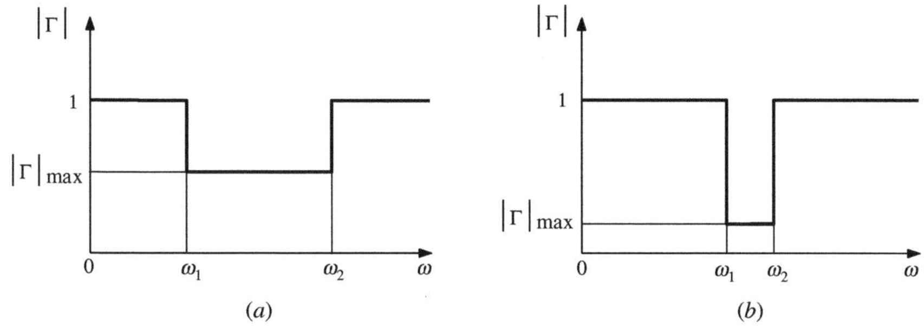

Figure 8.2 Ideal filter flat responses.

by a loaded quality factor $Q_L = \omega_0 \tau$ for the series *RL* or parallel *RC* circuit and by $Q_L = 1/(\omega_0 \tau)$ for the parallel *RL* or series *RC* circuit, respectively. The Chebyshev matching transformer with a finite number of sections can be considered as a close approximation to the ideal passband network when the ripple of the Chebyshev response is made equal to $|\Gamma|_{max}$. By combining matching theory with the closed formulas for the element values of a Chebyshev low-pass filter, explicit formulas for optimum matching networks can be obtained in certain simple but common cases [3]. For example, analytic closed-form solutions for the design of optimum matching networks up through order $n = 4$ can be derived [4].

Generally, Eqs. (8.4) and (8.5) can be rewritten in a simplified form

$$|\Gamma|_{max} = \exp\left(-\pi \frac{Q_0}{Q_L}\right) \tag{8.6}$$

where $Q_0 = \omega_0/\Delta\omega$.

8.2 Matching Networks with Lumped Elements

To correctly design the broadband matching circuits for the transistor power amplifiers, it is necessary to transform and match the device complex impedances with the source and load impedances, which are usually resistive and equal to 50 Ω. For high-power or low-supply voltage cases, the device impedances may be small enough, and it needs to include an ideal transformer (*IT*) together with a matching circuit, as shown in Fig. 8.3. In this case, such an ideal transformer provides only a required transformation between the source resistance R_S and the input impedance of the matching circuit, and does not have any effect on the circuit frequency characteristics.

To implement such an ideal transformer to the impedance-transforming circuit, it is useful to operate with the Norton transform. As a result, an ideal transformer with two capacitors C_1 and C_2, which is shown in Fig. 8.4(*a*), can be equivalently replaced by three capacitors

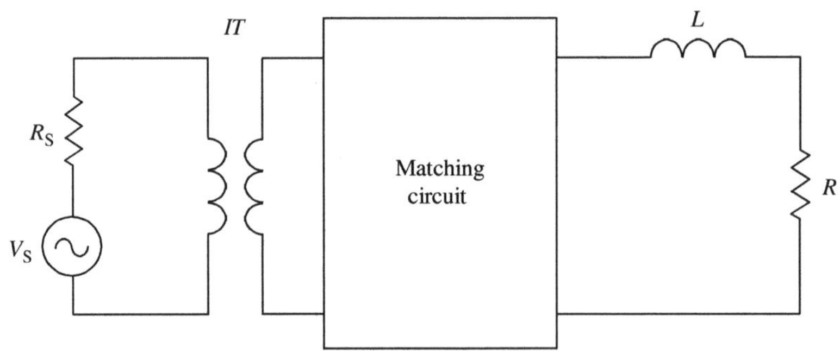

FIGURE 8.3 Matching circuit with ideal transformer.

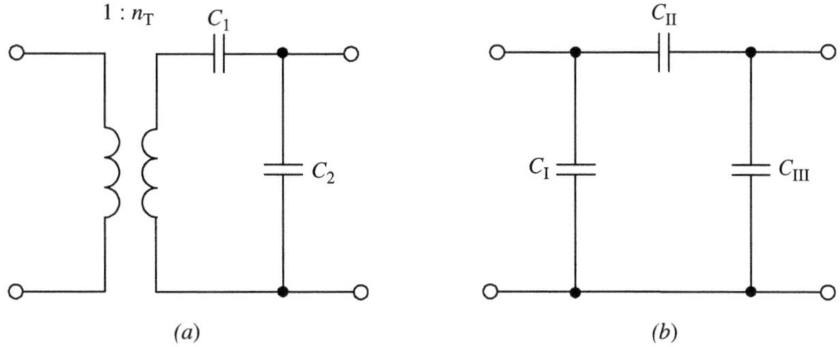

Figure 8.4 Capacitive impedance-transforming circuits.

C_I, C_{II}, and C_{III} connected in the form of a π-transformer, as shown in Fig. 8.4(*b*). Their values are determined by

$$C_I = n_T(n_T - 1)C_1 \tag{8.7}$$

$$C_{II} = n_T C_1 \tag{8.8}$$

$$C_{III} = C_2 - (n_T - 1)C_1 \tag{8.9}$$

where n_T is the transformation coefficient. In this case, all of the parameters of these two-port networks are assumed identical at any frequency. However, such a replacement is possible only if the capacitance C_{III} obtained by Eq. (8.9) is positive and, consequently, physically realizable.

Similarly, an ideal transformer with two inductors L_1 and L_2, as shown in Fig. 8.5(*a*), can be replaced by three inductors L_I, L_{II}, and L_{III} connected in the form of a *T*-transformer, as shown in Fig. 8.5(*b*), with values determined by

$$L_I = n_T(n_T - 1)L_2 \tag{8.10}$$

$$L_{II} = n_T L_2 \tag{8.11}$$

$$L_{III} = L_1 - (n_T - 1)L_2 \tag{8.12}$$

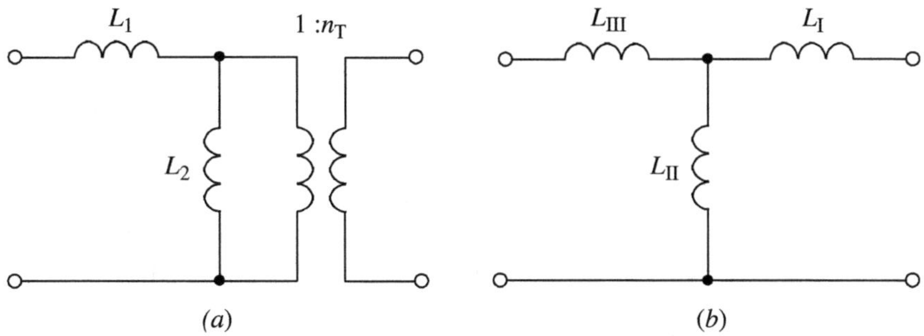

Figure 8.5 Inductive impedance-transforming circuits.

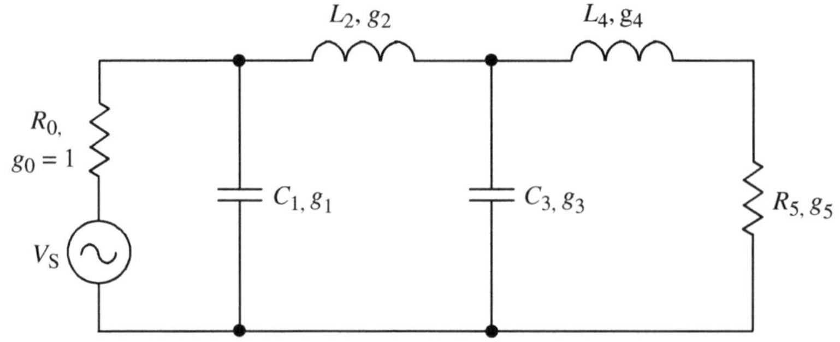

FIGURE 8.6 Two-section impedance-transforming circuit.

Again, this replacement is possible only if the inductance L_{III} defined by Eq. (8.12) is positive and, consequently, physically realizable.

The broadband impedance-transforming circuits generally represent the transforming bandpass filters when the in-band matching requirements with specified ripple must be satisfied. In this case, the out-of-band mismatching can be very significant. One of the design methods of such matching circuits is based on the theory of transforming the low-pass filters of a ladder configuration of series inductors alternating with shunt capacitors, whose two-section equivalent representation is shown in Fig. 8.6. For a large ratio of R_0/R_5, mismatching at zero frequency is sufficiently high, and such a matching circuit can be treated as a bandpass impedance-transforming filter.

Table 8.1 gives the maximum passband ripples and coefficients g_1 and g_2 required to calculate the parameters of a two-section low-pass Chebyshev filter for different transformation ratios $r = R_0/R_5$ and frequency bandwidths $w = 2(f_2 - f_1)/(f_2 + f_1)$, where f_2 and f_1 are the high- and low-bandwidth frequencies, respectively [5]. The coefficients g_3 and g_4 are calculated as $g_3 = rg_2$ and $g_4 = g_1/r$, respectively, and the circuit elements can be obtained by

$$C_1 = \frac{g_1}{\omega_0 R_0} \qquad C_3 = \frac{g_3}{\omega_0 R_0} \tag{8.13}$$

$$L_2 = \frac{g_2 R_0}{\omega_0} \qquad L_4 = \frac{g_4 R_0}{\omega_0} \tag{8.14}$$

where $\omega_0 = \sqrt{\omega_1 \omega_2}$ is the center bandwidth frequency.

As an example, consider the design of a broadband input matching circuit in the form of a two-section low-pass transforming filter shown in Fig. 8.6, with a center bandwidth frequency $f_0 = 3$ GHz, to match the source impedance $R_S = R_0 = 50\ \Omega$ with the device input impedance $Z_{in} = R_{in} + j\omega_0 L_{in}$, where $R_{in} = R_5 = 2\ \Omega$, $L_{in} = L_4 = 0.223$ nH,

r	w	ripple, dB	g_1	g_2
	0.1	0.000087	1.26113	0.709217
5	0.2	0.001389	1.27034	0.704050
	0.3	0.007023	1.28561	0.695548
	0.4	0.022109	1.30687	0.638859
	0.1	0.000220	1.60350	0.591627
10	0.2	0.003516	1.62135	0.585091
	0.3	0.017754	1.65115	0.574412
	0.4	0.055746	1.69304	0.559894
	0.1	0.000625	2.11734	0.462747
25	0.2	0.009993	2.15623	0.454380
	0.3	0.050312	2.22189	0.440863
	0.4	0.156725	2.31517	0.422868
	0.1	0.001303	2.57580	0.384325
50	0.2	0.020801	2.64380	0.374422
	0.3	0.104210	2.75961	0.358638
	0.4	0.320490	2.92539	0.338129

TABLE 8.1 Two-Section Low-Pass Chebyshev Filter Parameters

and $\omega_0 = 2\pi f_0$. The value of the series input device inductance $L_{\text{in}} = L_4$ is chosen to satisfy the requirements of Table 8.1 for $r = 25$ and $w = 0.4$ with maximum ripple of 0.156725 when $g_1 = 2.31517$ and $g_2 = 0.422868$. From Eq. (8.14), it follows that

$$L_4 = \frac{g_4 R_0}{\omega_0} = \frac{g_1 R_0}{\omega_0 r} = 0.223 \text{ nH}$$

As a result, the circuit parameters shown in Fig. 8.7(*a*) are calculated from Eqs. (8.13) and (8.14), thus resulting in the corresponding circuit frequency response shown in Fig. 8.7(*b*) with the required passband from 2.6 to 3.4 GHz. The particular value of the inductance L_{in} is chosen for the design convenience. If this value differs from the required value, it means that it is necessary to change the maximum frequency bandwidth, the power ripple, or the number of ladder sections.

Another approach is based on the transformation from the low-pass impedance-transforming prototype filters, the simple *L*-, *T*-, and π-type equivalent circuits of which are shown in Fig. 8.8, to the bandpass impedance-transforming filters. Table 8.2 gives the parameters of the low-pass impedance-transforming Chebyshev filters-prototypes for different maximum in-band ripples and number of

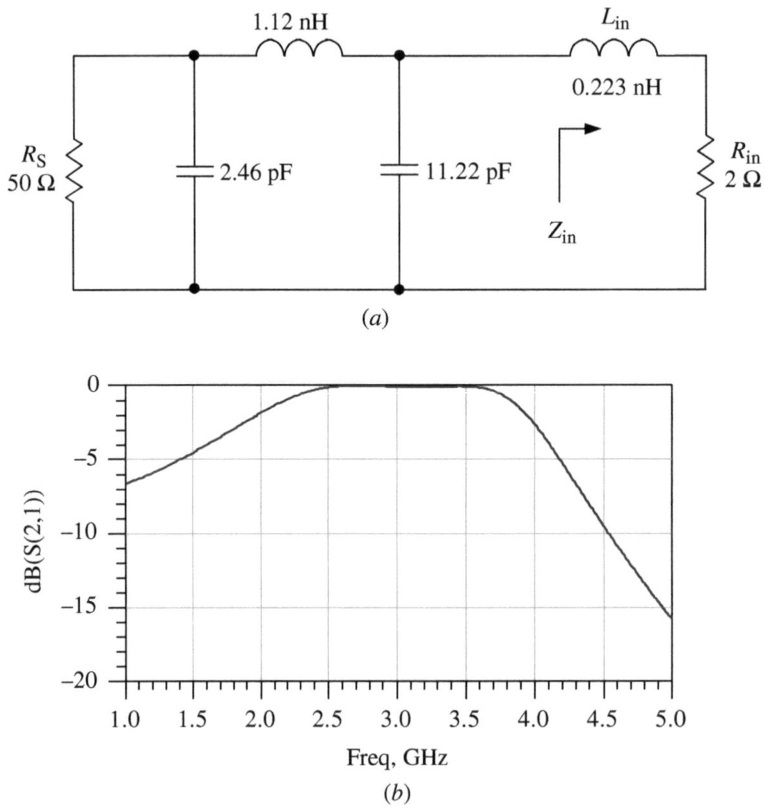

Figure 8.7 Two-section broadband low-pass matching circuit and its frequency response.

elements n [6]. This transformation can be obtained using the frequency substitution as

$$\omega \rightarrow \frac{\omega_0}{\Delta\omega}\left(\frac{\omega}{\omega_0} - \frac{\omega_0}{\omega}\right) \tag{8.15}$$

where $\omega_0 = \sqrt{\omega_1\omega_2}$ is the center bandwidth frequency, $\Delta\omega = \omega_2 - \omega_1$ is the passband, ω_1 and ω_2 are the low and high edges of the passband, respectively.

As a result, a series inductor L_k is transformed into a series LC circuit according to

$$\omega L_k = \frac{\omega_0}{\Delta\omega}\left(\frac{\omega}{\omega_0} - \frac{\omega_0}{\omega}\right)L_k = \omega L_k' - \frac{1}{\omega C_k'} \tag{8.16}$$

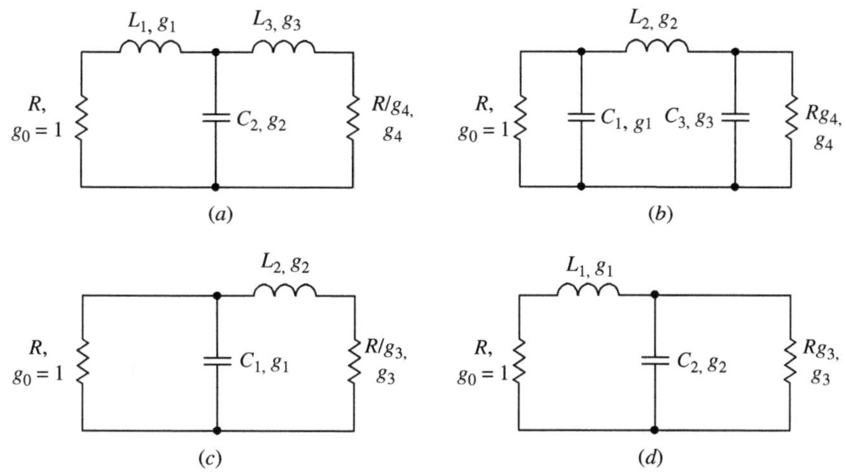

FIGURE 8.8 Lumped *L*-, p- and *T*-type impedance-transforming circuits.

where

$$L'_{\mathrm{k}} = \frac{L_{\mathrm{k}}}{\Delta\omega} \qquad C'_{\mathrm{k}} = \frac{\Delta\omega}{\omega_0^2 L_{\mathrm{k}}} \tag{8.17}$$

Similarly, a shunt capacitor C_{k} is transformed into a shunt *LC*-circuit as

$$\omega C_{\mathrm{k}} = \frac{\omega_0}{\Delta\omega}\left(\frac{\omega}{\omega_0} - \frac{\omega_0}{\omega}\right)C_{\mathrm{k}} = \omega C'_{\mathrm{k}} - \frac{1}{\omega L'_{\mathrm{k}}} \tag{8.18}$$

Ripple, dB	*n*	g_1	g_2	g_3	g_4
	1	0.0960	1.0000		
0.01	2	0.4488	0.4077	1.1007	
	3	0.6291	0.9702	0.6291	1.0000
	1	0.3052	1.0000		
0.1	2	0.8430	0.6220	1.3554	
	3	1.0315	1.1474	1.0315	1.0000
	1	0.4342	1.0000		
0.2	2	1.0378	0.6745	1.5386	
	3	1.2275	1.1525	1.2275	1.0000
	1	0.6986	1.0000		
0.5	2	1.4029	0.7071	1.9841	
	3	1.5963	1.0967	1.5963	1.0000

TABLE 8.2 Parameters of Low-Pass Chebyshev Filters-Prototypes

where

$$C'_k = \frac{C_k}{\Delta\omega} \qquad L'_k = \frac{\Delta\omega}{\omega_0^2 C_k} \tag{8.19}$$

The low-pass impedance-transforming prototype filter will be transformed to the bandpass impedance-transforming filter when all its series elements are replaced by the series resonant circuits and all its parallel elements are replaced by the parallel resonant circuits, where each of them are tuned to the center bandwidth frequency ω_0. The bandpass filter elements can be calculated from

$$\Delta\omega C_k = \frac{g_k}{R} \tag{8.20}$$

$$\Delta\omega L_k = g_k R \tag{8.21}$$

where k is an element serial number for the low-pass prototype filter and g_k are the appropriate coefficients given by Table 8.2.

Consider the design of the low-pass prototype filter for a given maximum ripple level in a frequency range up to $\omega_{2(2)}$ for two-element filter and up to $\omega_{2(3)}$ for three elements, as shown in Fig. 8.9(*a*). Then, for a selected arbitrary frequency ω_0, a series capacitance is added to each inductance and a parallel inductance is added to each capacitance on the assumption that all these resonant circuits are tuned to the selected frequency ω_0. As a result, a new bandpass filter will be realized with the same ripple, as shown in Fig. 8.9(*b*) for $n = 2$ and $n = 3$ with the passbands $\Delta\omega_{(2)}$ and $\Delta\omega_{(3)}$, respectively. Their elements are calculated according to Eqs. (8.20) and (8.21).

The maximum ripple level shown in Fig. 8.9 determines the insertion loss IL (or power loss ratio P_{LR}) through the magnitude of the reflection coefficient Γ as

$$IL = 10\ \log_{10} P_{LR} = -10\ \log_{10}(1 - |\Gamma(\omega)|^2) \tag{8.22}$$

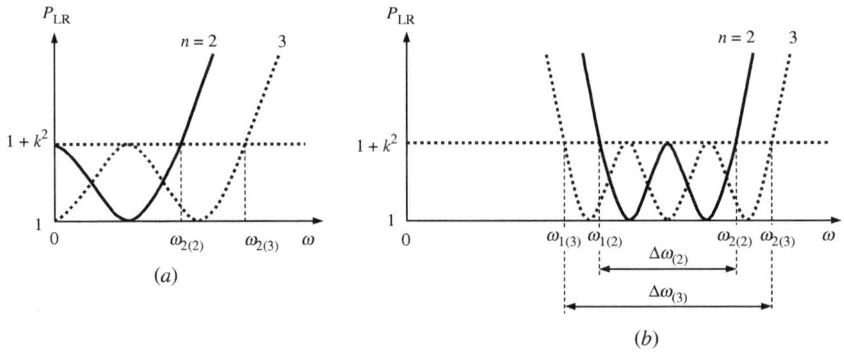

FIGURE 8.9 Maximum ripple level versus frequency bandwidth.

For an n-order Chebyshev low-pass filter, P_{LR} can be obtained by

$$P_{LR} = 1 + k^2 T_n^2\left(\frac{\omega}{\omega_c}\right) \quad (8.23)$$

where ω_c is the cutoff frequency. The passband response has equal ripples of amplitude $1 + k^2$, and the n-order Chebyshev polynomials are

$$T_1(x) = x \quad (8.24)$$

$$T_2(x) = 2x^2 - 1 \quad (8.25)$$

$$T_3(x) = 4x^3 - 3x \quad (8.26)$$

$$T_4(x) = 8x^4 - 8x^2 + 1 \quad (8.27)$$

Higher-order polynomials can be found using recurrence form of

$$T_n(x) = 2xT_{n-1}(x) - T_{n-2}(x) \quad (8.28)$$

where $x = \omega/\omega_c$.

Generally, the low-pass prototype filters obtained on their basis bandpass filters do not perform an impedance transformation. The input and output resistances are either equal for the symmetric T- or π-type filters shown in Fig. 8.8(a, b) where $g_4 = 1$ or their ratio is too small for L-type filters, as those shown in Fig. 8.8(c, d), where $g_3 < 2$. Therefore, in this case, it is necessary to use an ideal transformer concept. This approach is based on using the existing data tables, from which the parameters of such impedance-transforming networks can be easily calculated for a given quality factor of the device input or output circuit. However, they can also be very easily verified or optimized by using a computer-aided design (CAD) optimization procedure incorporating in any comprehensive circuit simulator.

Consider the design example of the broadband interstage impedance-transforming filter with the center bandwidth frequency of 1 GHz to match the output driver-stage circuit with the input final-stage circuit of the power amplifier, as shown in Fig. 8.10(a) [7]. In this case, it is convenient initially to convert the parallel connection of the device output resistance R_{out} and capacitance C_{out} into the corresponding series connection at the center bandwidth frequency ω_0 according to

$$R'_{out} = \frac{R_{out}}{1 + (\omega_0 R_{out} C_{out})^2} \quad (8.29)$$

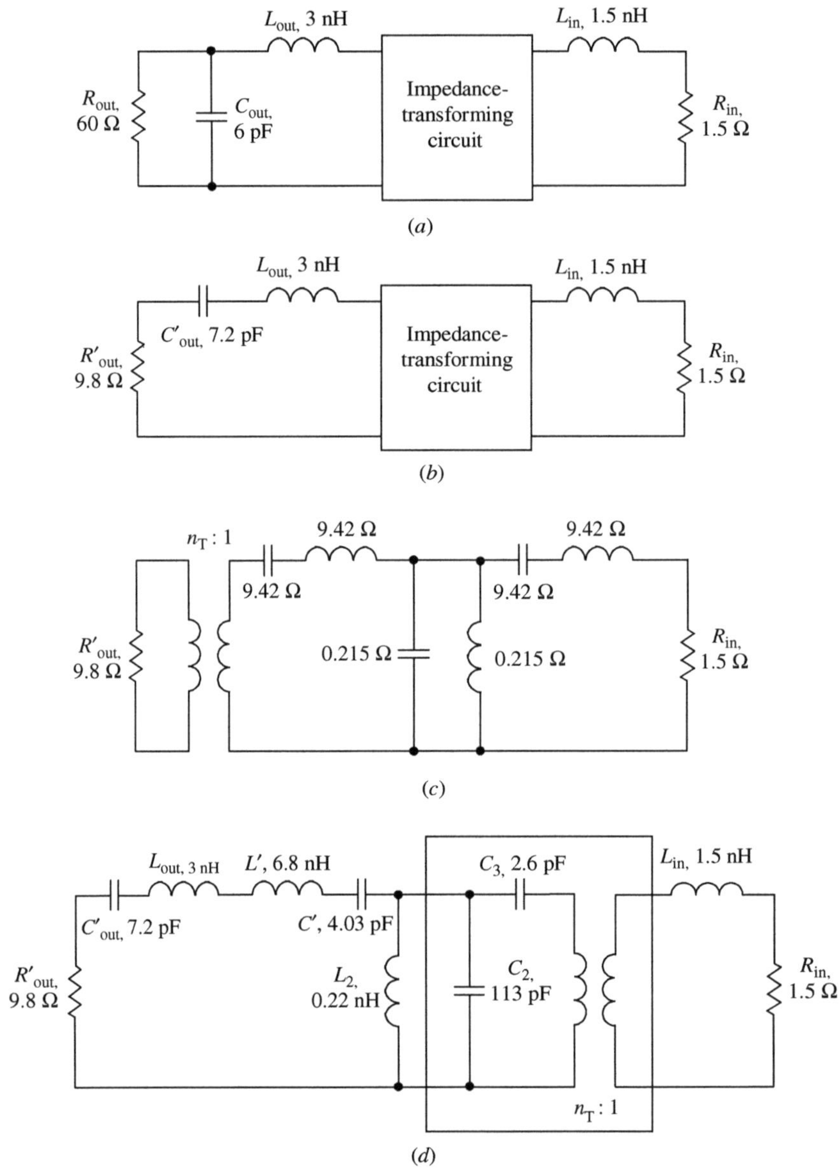

FIGURE 8.10 Impedance-transformer design procedure using low pass filter-prototype.

$$C'_{\text{out}} = \frac{1+(\omega_0 R_{\text{out}} C_{\text{out}})^2}{(\omega_0 R_{\text{out}})^2 C_{\text{out}}} \tag{8.30}$$

as shown in Fig. 8.10(*b*).

For the three-element low-pass impedance-transforming prototype filter shown in Fig. 8.8(*a*) with a maximum in-band ripple of

0.1 dB, we can obtain $g_1 = g_3 = 1.0315$, $g_2 = 1.1474$, and $g_4 = 1$ for $n = 3$ from Table 8.2. According to Eq. (8.21), the relative frequency bandwidth in this case is defined as

$$\frac{\Delta\omega}{\omega_0} = \frac{g_1 R_{\text{in}}}{\omega_0 L_{\text{in}}} = 16.5\%$$

based on a value of which the shunt capacitance C_2 can be calculated using Eq. (8.20), thus resulting in a capacitive reactance equal to 0.215 Ω. The inductive reactance corresponding to a series inductance L_{in} is equal to 9.42 Ω.

To convert the low-pass filter to its bandpass prototype, it is necessary to connect the capacitor in series to the input inductor and the inductor in parallel to the shunt capacitor and calculate with the same reactances to resonate at the center bandwidth frequency ω_0, as shown in Fig. 8.10(*c*), where an ideal transformer *IT* with the transformation coefficient $n_{\text{T}} = \sqrt{9.8/1.5} = 2.556$ is included. Here, the reactances for each series element are equal to 9.42 Ω, whereas those for each parallel element are equal to 0.215 Ω, respectively. Then, moving the corresponding elements with transformed parameters (each inductive and capacitive reactance is multiplied by n_{T}^2) to the left-hand side of *IT* in order to apply a Norton transform gives the circuit shown in Fig. 8.10(*d*), where the required series elements with reactances of $9.42n_{\text{T}}^2$ Ω are realized by the inductance L_{out}, converted device output capacitance C'_{out}, and additional elements L' and C'. Finally, by using a Norton transform shown in Fig. 8.4 with the ideal transformer *IT* and two capacitors, the resulting impedance matching bandpass filter is obtained, as shown in Fig. 8.11(*a*).

The frequency response of the filter with minimum in-band ripple and significant out-of-band suppression is shown in Fig. 8.11(*b*). In the case of serious difficulties with practical implementation of a very small inductance of 0.22 nH or a very large capacitance of 109 pF, it is possible to design a multisection low-pass impedance-transforming circuit.

Figure 8.12(*a*) shows the circuit schematic of a microwave broadband amplifier using a 1-μm GaAs FET (field effect transistor) packaged transistor, where the input multisection matching circuit is designed to provide the required gain taper and both input and output matching circuits are optimized to provide broadband impedance transformation [8]. As a result, a nominal power gain of 8 dB with a maximum deviation of ±0.07% in a frequency range of 7 to 14 GHz was achieved. In the first monolithic broadband GaAs FET amplifier, the input and output matching circuits were based on lumped elements fabricated together with the FET device on a semi-insulating high-resistivity gallium-arsenide substrate with a total size of 1.8 × 1.2 mm², providing a power gain of 4.5 ± 0.9 dB with an output power of 11 dBm at 1-dB gain compression from 7.0 to 11.7 GHz [9].

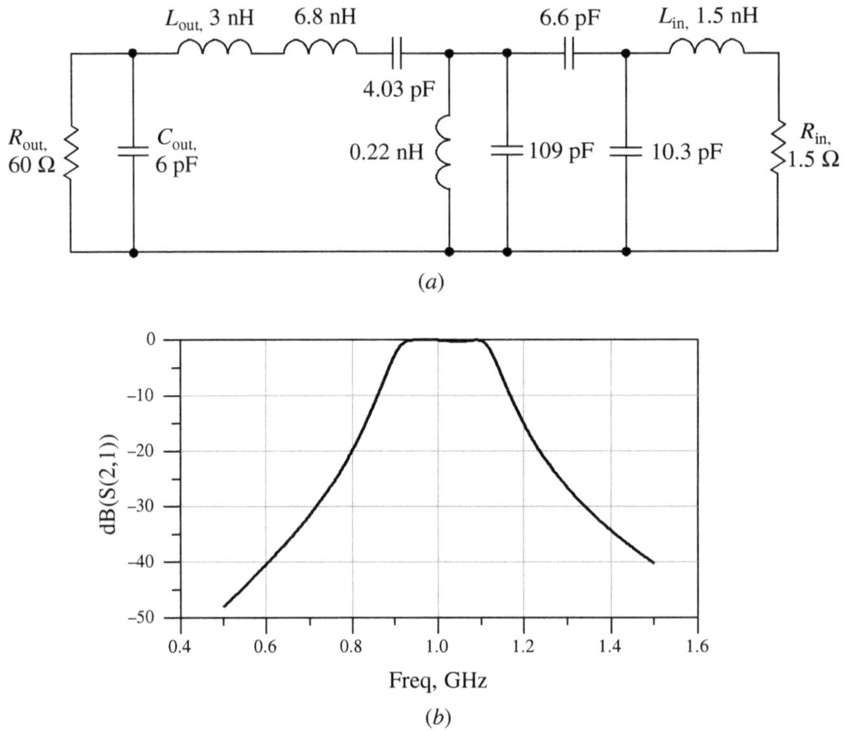

FIGURE 8.11 Impedance-transforming bandpass filter and its frequency response.

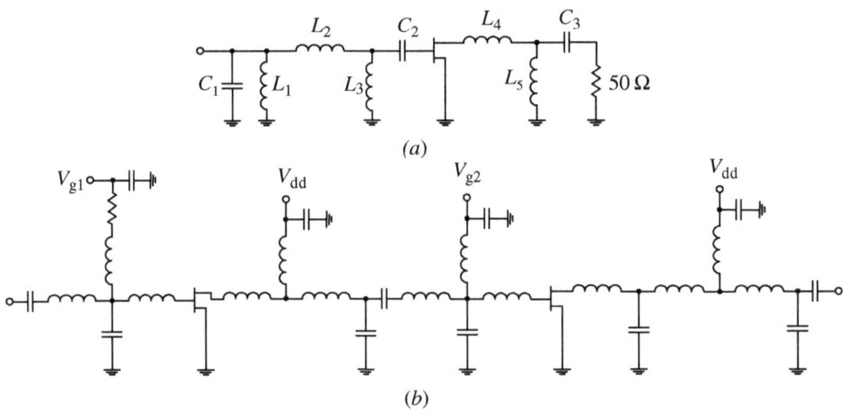

FIGURE 8.12 Schematics of broadband lumped-element microwave FET amplifiers.

The circuit diagram of a two-stage pHEMT MMIC power amplifier for *Ku*-band applications is shown in Fig. 8.12(*b*), where the lumped components were used in the input, interstage, and output matching circuits to minimize the overall chip size [10]. Here, the topology of each matching network represents a double-resonant circuit to form a broadband impedance transformer, which includes a shunt inductor in series with a bypass capacitor to provide a dc path, a series blocking capacitor, and a low-pass *L*-section transformer. In this case, for a 8.4-mm driver-stage pHEMT and a 16.8-mm power-stage pHEMT, a saturated output power of 38.1 dBm (6.5 W), a small-signal gain of 10.5 dB, and a peak *PAE* of 24.6% from 13.6 to 14.2 GHz were achieved with a chip size of MMIC as small as 3.64×2.35 mm^2. Based on the *T*-shape combining transformers with three individual inductors implemented in a 0.15-μm pHEMT technology, a broadband MMIC power amplifier combining two pHEMT devices with an overall 400-μm gate-width size achieved a saturated output power of 22 to 23.5 dBm and a power gain over 10 dB from 17 to 35 GHz [11]. In a 90-nm standard CMOS process, a canonical doubly terminated third-order bandpass network was converted to the output matching topology, which provides both impedance transformation and differential-to-single-ended power combining [12]. The power amplifier achieved a 3-dB bandwidth from 5.2 to 13 GHz with a 25.2-dBm peak saturated output power and a peak *PAE* of 21.6%.

8.3 Matching Networks with Mixed Lumped and Distributed Elements

The matching circuits, which incorporate mixed lumped and transmission-line elements, are widely used both in hybrid and monolithic design techniques. Such matching circuits are very convenient when designing the push-pull power amplifiers with effect of virtual grounding, where the shunt capacitors are connected between two series microstrip lines. According to the quasi-linear transformation technique, the basic four-step design procedure consists of an appropriate choice of the lumped prototype schematic, resulting in near-maximum gain across the required frequency bandwidth, its decomposition into subsections, their replacement by almost equivalent distributed circuits, and then the application of an optimization technique to minimize power variation over the operation frequency bandwidth [13].

A periodic lumped *LC* structure in the form of a low-pass ladder π-network is used as a basis for the lumped matching prototype. Then, the lumped prototype should be split up into individual π-type sections with equal capacitances by consecutive step-by-step process and replaced by their equivalent distributed network counterparts. Finally, the complete mixed matching structure is optimized to improve the overall performance by employing standard nonlinear optimization routine on the element values. Note that generally the lumped prototype structure can be decomposed into different subnetworks including also L-type matching sections and individual capacitors or inductors.

For a single-frequency equivalence between lumped and distributed elements, the low-pass lumped π-type ladder section can be made equivalent to a symmetrically loaded transmission line at the certain frequency, as shown in Fig. 8.13(*a*). The transmission *ABCD*-matrices of these lumped and distributed ladder sections can be written, respectively, as

$$[ABCD]_{\mathrm{L}} = \begin{bmatrix} 1-\omega_0^2 LC & j\omega_0 L \\ j\omega_0 C(2-\omega_0^2 LC) & 1-\omega_0^2 LC \end{bmatrix} \tag{8.31}$$

$$[ABCD]_{\mathrm{T}} = \begin{bmatrix} \cos\theta_0 - \omega_0 C_{\mathrm{T}} Z_0 \sin\theta_0 & jZ_0 \sin\theta_0 \\ \dfrac{j}{Z_0}(2\omega_0 C_{\mathrm{T}} Z_0 \cos\theta_0 + \sin\theta_0 - \omega_0^2 C_{\mathrm{T}}^2 Z_0^2 \sin\theta_0) & \cos\theta_0 - \omega_0 C_{\mathrm{T}} Z_0 \sin\theta_0 \end{bmatrix} \tag{8.32}$$

where θ_0 is the electrical length of a transmission line at the center bandwidth frequency ω_0.

Consequently, because these two circuits are equivalent, equal matrix elements $A_{\mathrm{L}} = A_{\mathrm{T}}$ and $B_{\mathrm{L}} = B_{\mathrm{T}}$ can be rewritten as

$$1-\omega_0^2 LC = \cos\theta_0 - \omega_0 C_{\mathrm{T}} Z_0 \sin\theta_0 \tag{8.33}$$

$$j\omega_0 L = jZ_0 \sin\theta_0 \tag{8.34}$$

After Eqs. (8.33) and (8.34) are solved, the characteristic impedance Z_0 and shunt capacitance C_{T} can be explicitly calculated by

$$Z_0 = \frac{\omega_0 L}{\sin\theta_0} \tag{8.35}$$

$$C_{\mathrm{T}} = \frac{\cos\theta_0 + \omega_0^2 LC - 1}{\omega_0^2 L} \tag{8.36}$$

To provide the design method using a single-frequency equivalent technique, the following consecutive design steps can be performed:

Designate the lumped π-type C_1-L_1-C_2 section to be replaced.

From a chosen π-type C_1-L_1-C_2 section, form the symmetrical C-L-C ladder section with equal capacitances C, as shown in Fig. 8.13(*b*).

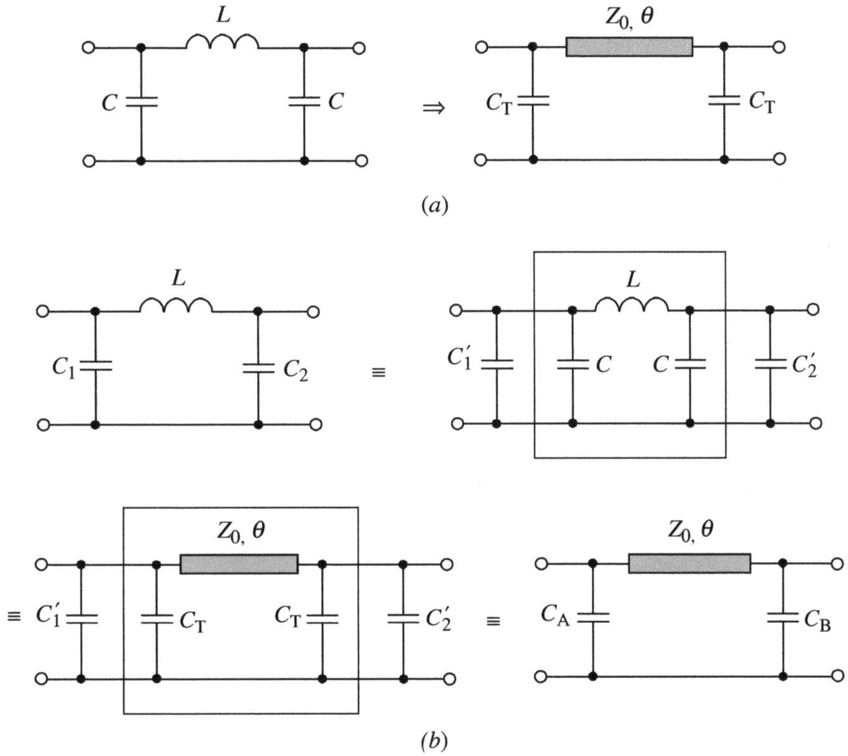

FIGURE 8.13 Transforming design procedure for lumped and distributed matching circuits.

The choice of capacitances is arbitrary, but their values cannot exceed the minimum of (C_1, C_2).

Calculate the parameters of the symmetrical C_T-TL-C_T section using the parameters of the lumped equivalent π-section by setting the electrical length θ_0 of the transmission line according to Eqs. (8.35) and (8.36). Here, it is assumed that the minimum of the capacitances C_1 and C_2 should be C_T or greater so that C_T can be readily embedded in the new C_T-TL-C_T section.

Finally, replace the π-type C_1-L_1-C_2 ladder section by the equivalent symmetrical C_T-TL-C_T section and combine adjacent shunt capacitances, as shown in Fig. 8.13(*b*), where the loaded shunt capacitances C_A and C_B are given as $C_A = C_1' + C_T$ and $C_B = C_2' + C_T$.

Figure 8.14(*a*) shows the circuit schematic of a simulated broadband 28-V LDMOSFET power amplifier. To provide an output power of about 15 W with a power gain of greater than 10 dB in a frequency range of 225 to 400 MHz, an LDMOSFET device with a gate geometry

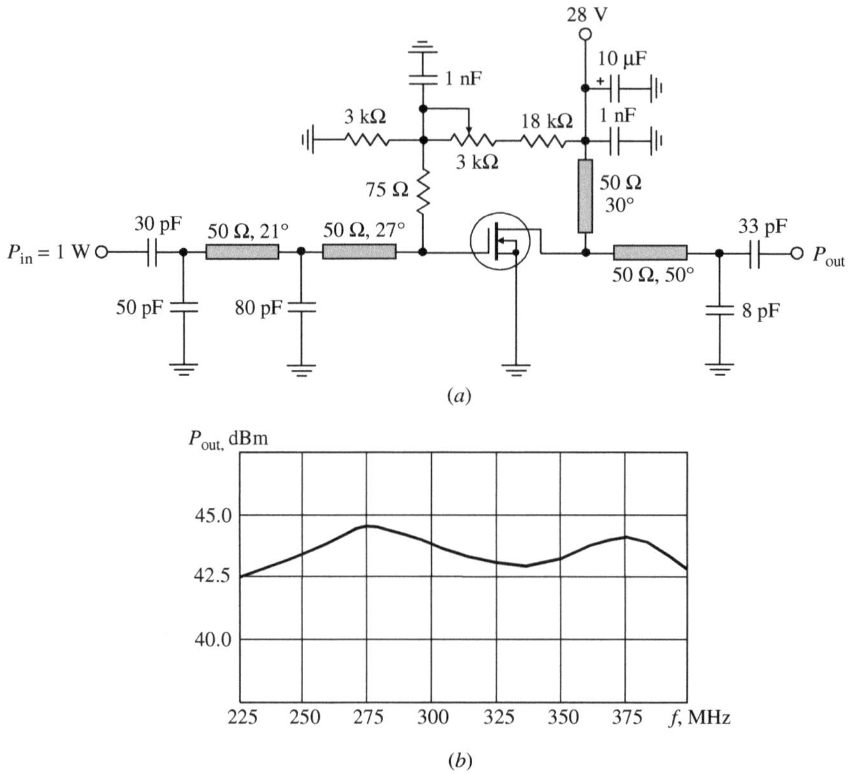

Figure 8.14 Circuit schematic and performance of broadband LDMOSFET power amplifier.

of $1.25\ \mu m \times 40$ mm was chosen. In this case, the matching design technique is based on using multisection low-pass networks, with two π-type sections for the input matching circuit and one π-type section for the output matching circuit. The sections adjacent to the device input and output terminals incorporate the corresponding internal input gate-source and output drain-source device capacitances. Because a ratio between the device equivalent output resistance at the fundamental for several tens of watts of output power and the load resistance of 50 Ω is not significant, it is sufficient to be limited to only one matching section for the output matching network.

Once a matching network structure is chosen, based on the requirements for the electrical performance and frequency bandwidth, the simplest and fastest way is to apply an optimization procedure using CAD simulators to satisfy certain criteria. For such a broadband power amplifier, these criteria can be the minimum output power ripple and input return loss with maximum power gain and efficiency. To minimize the overall dimensions of the power amplifier board, the shunt microstrip line in the drain circuit can be treated as an element of the output matching circuit and its electrical length can be considered as a variable to be optimized. Applying a nonlinear broadband CAD optimization technique implemented in any high-level circuit simulator and setting the ranges of electrical length of the transmission lines between 0 and 90° and parallel capacitances from 0 to 100 pF, we can potentially obtain the parameters of the input and output matching circuits. The characteristic impedances of all transmission lines can be set to 50 Ω for simplicity and convenience of the circuit implementation. However, to speed up this procedure, it is best to optimize circuit parameters separately for input and output matching circuits with the device equivalent input and output impedances: a series *RC* circuit for the device input and a parallel *RC* circuit for the device output. It is sufficient to use a fast linear optimization process, which will take only a few minutes to complete the matching circuit design. Then, the resulting optimized values are incorporated into the overall power amplifier circuit for each element and final optimization is performed using a large-signal active device model. In this case, the optimization process is finalized by choosing the nominal level of input power with optimizing elements in much narrower ranges of their values of about 10 to 20% for most critical elements. Figure 8.14(*b*) shows the simulated broadband power amplifier performance, with an output power of 43.5 ± 1.0 dBm and a power gain of 13.5±1.0 dB in a frequency bandwidth of 225 to 400 MHz.

8.4 Matching Networks with Transmission Lines

The lumped or mixed matching networks generally work well at sufficiently low frequencies (up to one or several gigahertz). However, the lumped elements such as inductors and capacitors are difficult to implement at microwave frequencies where they can be treated as distributed elements. In addition, the quality factors for inductors are sufficiently small that they contribute to additional losses.

Generally, the design of a practical distributed filter circuit is based on some approximate equivalence between lumped and distributed elements, which can be established by applying Richards's transformation [14]. This implies that the distributed circuits composed of equal-length open- and short-circuited transmission lines can be treated as lumped elements under the transformation

$$s = j\tan\frac{\pi\omega}{2\omega_0} \tag{8.37}$$

where $s = j\omega/\omega_c$ is the conventional normalized complex frequency variable and ω_0 is the radian frequency for which the transmission lines are a quarter wavelength [15].

As a result, the one-port impedance of a short-circuited transmission line corresponds to the reactive impedance of a lumped inductor Z_L as

$$Z_L = sL = j\omega L = jL\tan\frac{\pi\omega}{2\omega_0} \tag{8.38}$$

Similarly, the one-port admittance of an open-circuited transmission line corresponds to the reactive admittance of a lumped capacitor Y_C as

$$Y_C = sC = j\omega C = jC\tan\frac{\pi\omega}{2\omega_0} \tag{8.39}$$

The results given by Eqs. (8.38) and (8.39) show that an inductor can be replaced with a short-circuited stub of the electrical length $\theta = \pi\omega/(2\omega_0)$ and characteristic impedance $Z_0 = L$, whereas a capacitor can be replaced with an open-circuited stub of the electrical length $\theta = \pi\omega/(2\omega_0)$ and characteristic impedance $Z_0 = 1/C$ when a unity-filter characteristic impedance is assumed.

From Eq. (8.37), it follows that, for a low-pass filter prototype, the cutoff occurs when $\omega = \omega_c$, resulting in

$$\tan\frac{\pi\omega_c}{2\omega_0} = 1 \tag{8.40}$$

which gives a stub length $\theta = 45°$ (or $\pi/4$) with $\omega_c = \omega_0/2$. Hence, the inductors and capacitors of a lumped-element filter can be replaced with the short- and open-circuited stubs, as shown in Fig. 8.15. Because the lengths of all stubs are the same and equal to $\lambda/8$ at the cutoff frequency ω_c, these lines are called the *commensurate lines.* At the frequency $\omega = \omega_0$, the transmission lines will be a quarter-wavelength long, resulting in an attenuation pole. However, at any frequency away from ω_c, the impedance of each stub will no longer match the original lumped-element impedances, and the filter response will differ from the desired filter prototype response. Note that the response will be periodic in frequency, repeating every $4\omega_c$.

Because the transmission line generally represents a four-port network, it is very convenient to use a matrix technique for a filter design. In the case of cascade of several networks, the rule is that the overall matrix of the new network is simply the matrix product of the matrices for the individual networks taken in the order of connection [16]. In terms of Richards's variable, an *ABCD* matrix for a transmission line with the characteristic impedance Z_0 can be written as

$$\begin{bmatrix} A & B \\ C & D \end{bmatrix} = \frac{1}{\sqrt{1-s^2}}\begin{bmatrix} 1 & sZ_0 \\ \dfrac{s}{Z_0} & 1 \end{bmatrix} \tag{8.41}$$

representing a unit element which has a half-order transmission zero at $s = \pm 1$. The matrix of the unit element is the same as that of a transmission line of the electrical length θ and characteristic impedance Z_0. Unit elements are usually introduced to separate the circuit elements

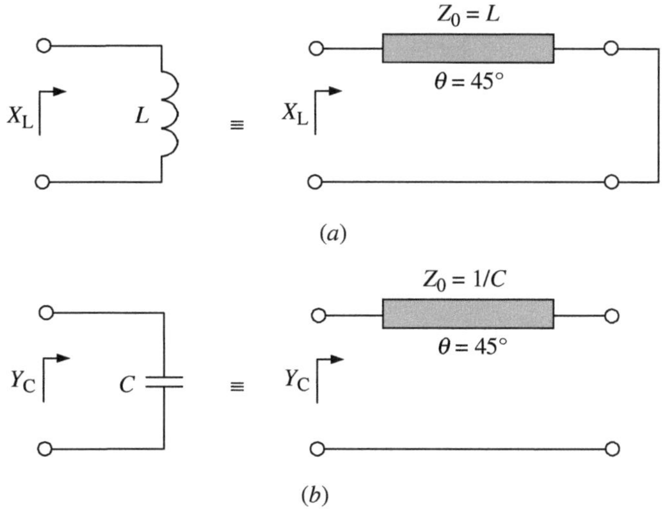

FIGURE 8.15 Equivalence between lumped elements and transmission lines.

in transmission-line filters, which are otherwise located at the same physical point.

The application of Richards's transformation provides a sequence of the short- and open-circuited stubs, which are then converted to a more practical circuit implementation. This can be done based on a series of equivalent circuits known as *Kuroda identities*, which allows these stubs to be physically separated, transforming the series stub into the shunt and changing impractical characteristic impedances into more realizable impedances [17]. The Kuroda identities use the unit elements, and these unit elements are thus commensurate with the stubs used to implement inductors and the capacitors of the prototype design. Connecting the unit element with the characteristic impedance Z_0 to the same load impedance Z_0 does not change the input impedance. The four Kuroda identities are illustrated in Fig. 8.16, where the combinations of unit elements with the characteristic impedance Z_0 and electrical length $\theta = 45°$, the reactive elements, and the relationships between them are given.

To prove the equivalence, consider two circuits of identity at the first row in Fig. 8.16 when *ABCD* matrix for the entire left-hand circuit can be written as

$$\begin{bmatrix} A & B \\ C & D \end{bmatrix}_{\mathrm{L}} = \frac{1}{\sqrt{1-s^2}} \begin{bmatrix} 1 & 0 \\ sC & 1 \end{bmatrix} \begin{bmatrix} 1 & sZ_1 \\ \dfrac{s}{Z_1} & 1 \end{bmatrix} = \frac{1}{\sqrt{1-s^2}} \begin{bmatrix} 1 & sZ_1 \\ s\left(C + \dfrac{1}{Z_1}\right) & 1 + s^2 Z_1 C \end{bmatrix} \tag{8.42}$$

where Z_1 is the characteristic impedance of the left-hand unit element.

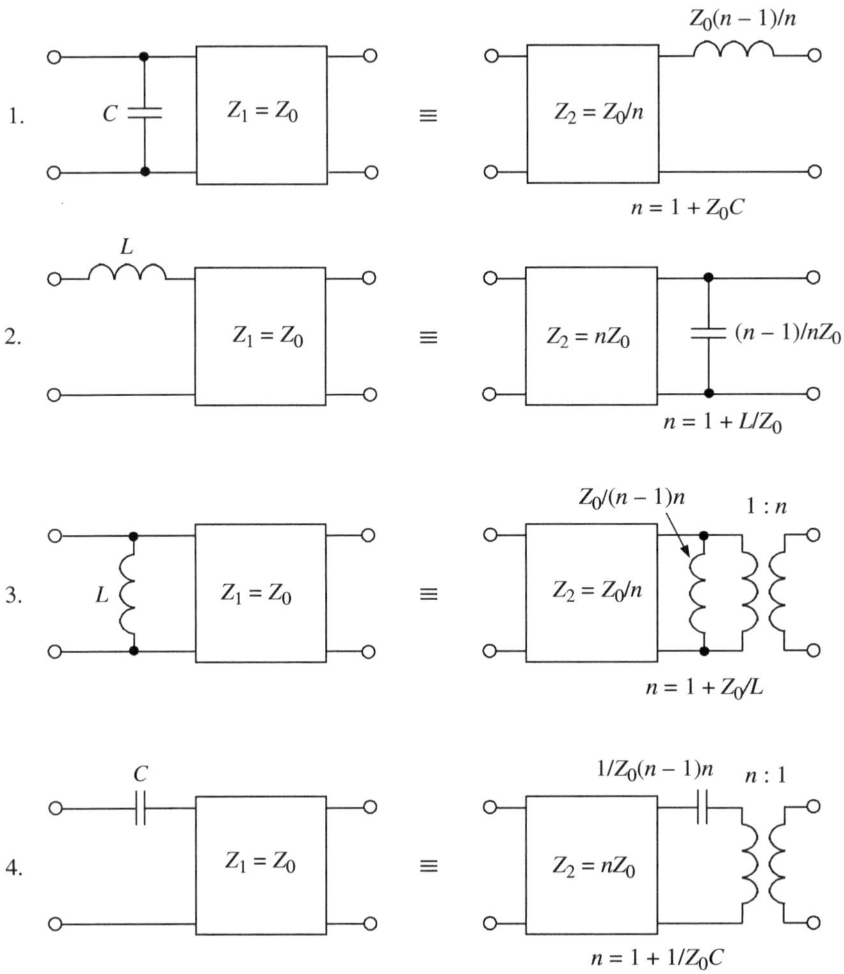

FIGURE 8.16 Four Kuroda identities.

Similarly, for the right-hand circuit,

$$\begin{bmatrix} A & B \\ C & D \end{bmatrix}_R = \frac{1}{\sqrt{1-s^2}} \begin{bmatrix} 1 & sZ_2 \\ \frac{s}{Z_2} & 1 \end{bmatrix} \begin{bmatrix} 1 & sL \\ 0 & 1 \end{bmatrix}$$

$$= \frac{1}{\sqrt{1-s^2}} \begin{bmatrix} 1 & s(Z_2+L) \\ \frac{s}{Z_2} & 1+\frac{s^2L}{Z_2} \end{bmatrix} \tag{8.43}$$

where Z_2 is the characteristic impedance of the right-hand unit element.

The results in Eqs. (8.42) and (8.43) are identical if

$$Z_1 = Z_2 + L \qquad \frac{1}{Z_1} + C = \frac{1}{Z_2} \qquad \frac{L}{Z_2} = Z_1 C$$

or

$$Z_2 = \frac{Z_1}{n} \qquad L = \frac{n-1}{n} Z_1 \tag{8.44}$$

where $n = 1 + Z_1C$.

As an example, consider the design of the broadband input transmission-line matching circuit based on a lumped two-section low-pass transforming filter shown in Fig. 8.6, with a center bandwidth frequency $f_0 = 3$ GHz to match a 50-Ω source impedance with the device input impedance $Z_{in} = R_{in} + j\omega_0 L_{in}$, where $R_{in} = 2\ \Omega$ and $L_{in} = 0.223$ nH. The value of the series input device inductance is chosen to satisfy Table 8.1 when, for $n = 4$, $r = 25$, $w = 0.4$, maximum ripple of 0.156725, and $g_1 = 2.31517$, from Eq. (8.14) it follows that

$$L_4 = \frac{g_4 R_0}{\omega_0} = \frac{g_1 R_0}{\omega_0 r} = 0.223 \text{ nH}$$

From Table 8.1, we obtain $g_2 = 0.422868$, which gives from Eqs.(8.13) and (8.14) the circuit parameters shown in Fig. 8.17. The inductance value is chosen for the design convenience. If this value differs from the required value, it means that it is necessary to change the maximum frequency bandwidth, the power ripple, or the number of ladder sections.

Figure 8.18 shows the design transformation of a lumped low-pass transforming filter to a microstrip one using the Kuroda identities. The first step, which is shown in Fig. 8.18(*a*), is to add a 50-Ω unit element at the end of the circuit and convert a shunt capacitor to a series inductor using the second Kuroda identity, as shown in Fig. 8.18(*b*). Then, adding another unit element and applying the first

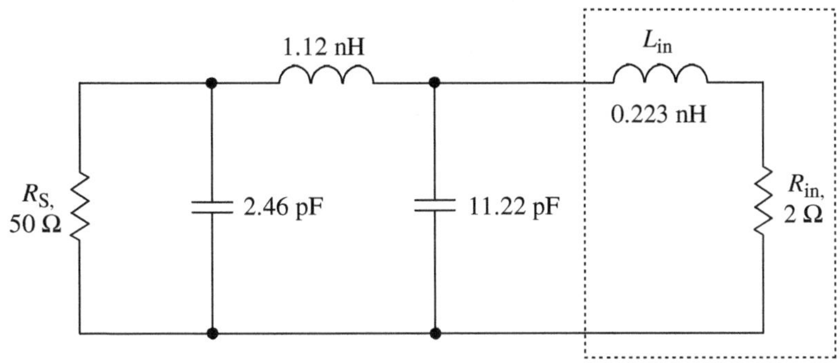

FIGURE 8.17 Two-section broadband matching circuit.

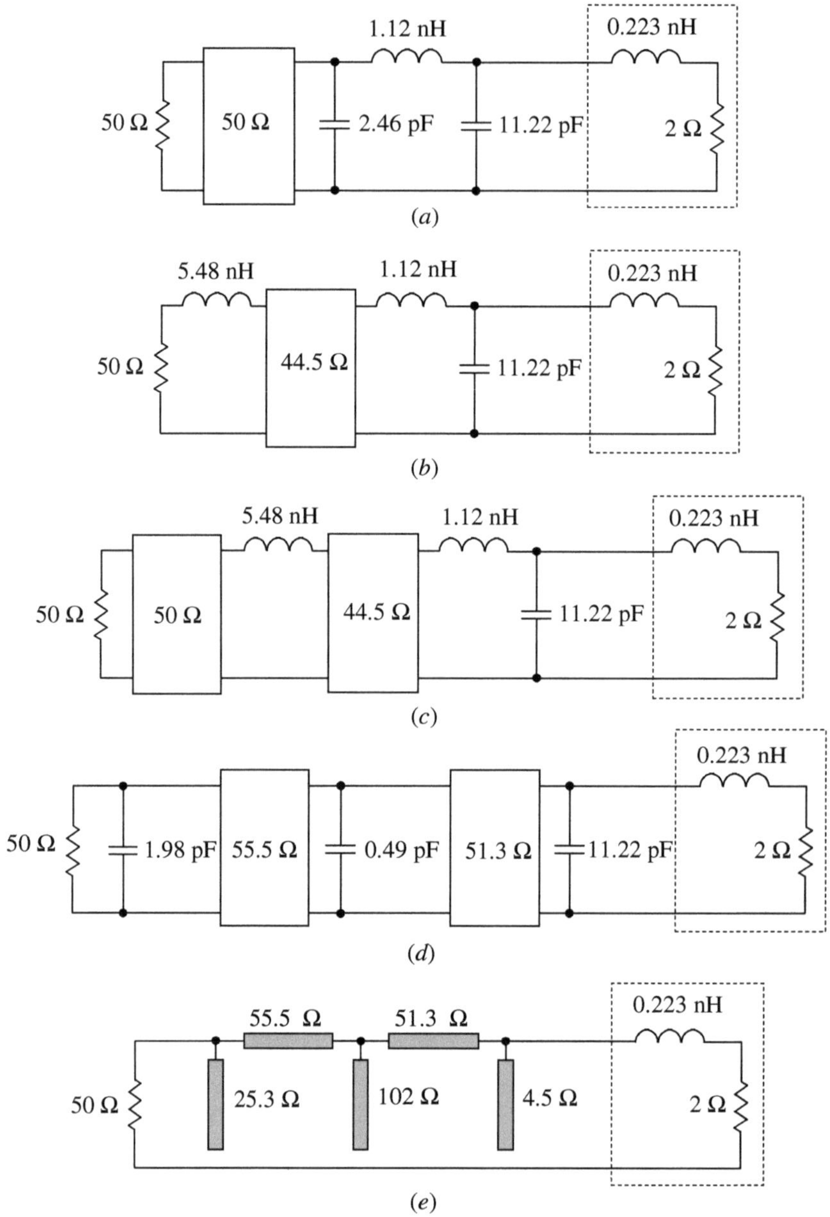

FIGURE 8.18 Design transformation from lumped low-pass to microstrip transforming filter.

Kuroda identity, as shown in Fig. 8.18(*c*), result in the circuit with two unit elements and three shunt capacitors shown in Fig. 8.18(*d*). To keep the same physical dimensions during the calculation of the circuit parameters, the inductance should be taken in nanohenri, and the capacitance is measured in nanofarads if the operating frequency is measured in gigahertz. Finally, Richards's transformation is used to convert the shunt capacitors to the corresponding transmission-line stubs. According to Eq. (8.39), the normalized characteristic impedance of a shunt stub is $1/C$, which is necessary to multiply by 50 Ω.

Figure 8.18(*e*) shows the microstrip layout of the final low-pass transforming circuit, where the lengths of the shunt stubs are $\lambda/8$ at the cutoff frequency f_c, as well as the lengths of each unit element representing the series stubs. If the normalized frequency bandwidth and center bandwidth frequency are chosen to be $w = 0.4$ and $f_0 =$ 3 GHz, respectively, the cutoff frequency becomes equal to

$$f_c = f_0\left(1+\frac{w}{2}\right) = 3.6\text{ GHz}$$

In practical design of a microwave bipolar or GaAs FET amplifier, it is necessary to take into account that the intrinsic device generally exhibits a small-signal gain rolloff with increasing frequency at approximately 6 dB per octave [18, 19]. Therefore, to maintain a constant gain across the design frequency band, the matching network must be designed for maximum gain at the highest frequency of interest [20]. In this case, reflective mismatching conditions are provided to compensate for the increase in the intrinsic gain of an FET when the frequency is decreased. As a result, it is necessary to selectively mismatch the input of the transistor by employing the gain-tapered input matching circuit so that the overall gain of the amplifier will be flat [21]. Alternatively, the gain tapering could be done in the output network with input flat matching conditions. In a simplified practical implementation when two impedance-transforming *L*-sections with series microstrip lines and shunt microstrip stubs in the input matching circuit and a single impedance-transforming *T*-section with two series microstrip lines and one shunt microstrip stub in the output matching network are used, a flat gain of about 6 dB was achieved across the octave band of 4 to 8 GHz for a single-cell GaAs FET amplifier with the device transconductance $g_m = 55$ mS [22].

Figure 8.19 shows the matching circuit design steps and the circuit schematic of a broadband microwave GaN HEMT power amplifier [23]. In this case, the first step to design the octave-band power amplifier intended to operate across the frequency bandwidth of 2 to 4 GHz was to find the optimum source and load impedances that maximize the performance of the device in terms efficiency in the required bandwidth. In view of a GaN HEMT Cree CGH60015DE device, because the optimum impedances were relatively close to each other across the band and acceptable level of degradation in *PAE*

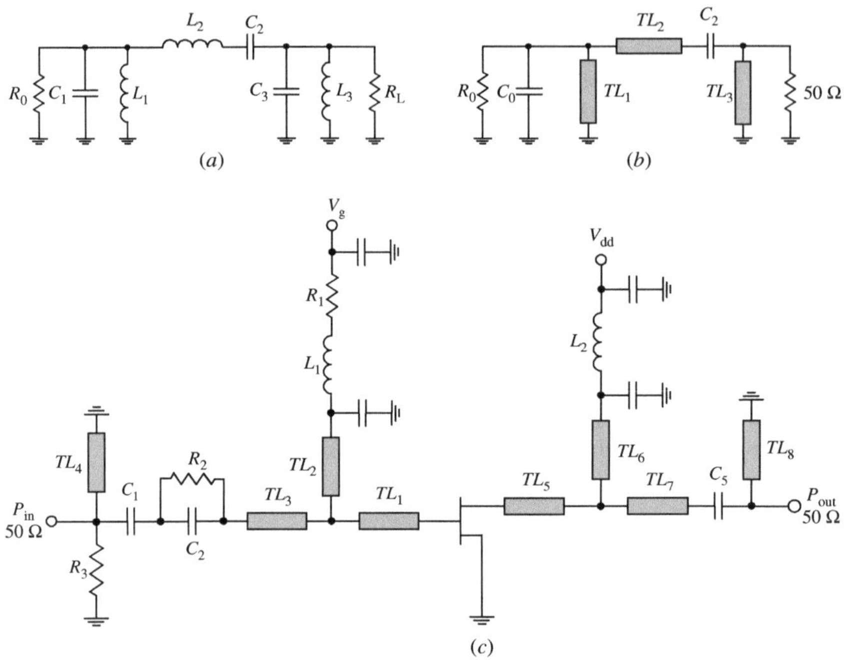

FIGURE 8.19 Schematics of broadband microwave GaN HEMT power amplifier.

was estimated to be less than 8%, the task was simplified to provide the optimum impedances seen by the device input and output at the center bandwidth frequency across the entire bandwidth. The bandpass matching network shown in Fig. 8.19(*a*) was derived from a lowpass prototype matching circuit, assuming that the transistor output can be approximated by an ideal current source with a parallel *RC* network, where R_0 is the source resistance corresponding to device equivalent output resistance at the fundamental (or load-line resistance) and the capacitance C_0 is the total drain-source capacitance. In order to scale the obtained terminating resistor R_L upward to 50 Ω, a Norton transformation of an ideal transformer ($n_T = 1.173$) with two series-shunt capacitors to an arrangement of three capacitors, as shown in Fig. 8.4, was used. Then, based on the transforms between lumped and distributed elements, the two resonant parallel *LC* circuits were approximated by the corresponding grounded shunt quarterwave transmission lines TL_1 and TL_3 with the characteristic impedance of each line equal to the reactance of the inductor or capacitor multiplied by $\pi/4$, whereas the lumped π-network with a series inductor and two shunt capacitors was approximated by the series transmission line TL_2, as shown in Fig. 8.19(*b*). Similar approach can be applied to the design of the input matching circuit, which also includes lossy elements for better input return loss and stability.

The entire circuit schematic of the designed broadband GaN HEMT power amplifier is shown in Fig. 8.19(*c*), where the two series tapered transmission lines TL_1 and TL_5 are added at the input and output of the device. As a result, an output power of 41 ± 1 dBm with a power gain of 10 ± 1 dB and a drain efficiency of 52 to 72% was achieved across the frequency bandwidth of 1.9 to 4.3 GHz.

An alternative impedance matching technique is based on the multisection matching transformers consisting of the stepped transmission-line sections with different characteristic impedances and electrical lengths [24]. These transformers, in contrast to the continuously tapered transmission-line transformers, are significantly shorter and provide broader performance. Figure 8.20(*a*) shows the schematic structure of a stepped transmission-line transformer, which consists of a cascaded connection of n uniform sections of equal quarterwave lengths $l = \lambda_0/4$, where λ_0 is the wavelength corresponding to the center bandwidth frequency. Such a stepped transmission-line transformer represents an antimetric structure, for which the ratio between the characteristic impedances of its transmission-line sections can be written in the general form as

$$Z_i Z_{n+1-i} = Z_S Z_L \tag{8.45}$$

where $i = 1, 2, \ldots, n$ and n is the number of sections, Z_S is the source impedance and Z_L is the load impedance [25].

In Fig. 8.21, as a practical example, the minimum possible *VSWR* is plotted for the five-step transmission-line impedance transformer with a total characteristic-impedance variation of 8:1, which was designed for maximum *VSWR* of 1.021 in an octave frequency bandwidth and where each section is of a quarterwave electrical length [26].

The main drawback to the stepped quarterwave transformers is their significant total length of $L = n\lambda_0/4$. However, it is possible to reduce the overall transformer length by applying other profiles of its structure. The stepped transformers using n cascaded uniform transmission-line sections of various lengths with alternating impedances are shorter by 1.5 to 2 times. In this case, the number of sections n is always an even number and the section impedances can be equal to the source and load impedances to be matched, as shown in Fig. 8.20(*b*). For example, the input and output matching circuits of a microwave GaAs MESFET power amplifier, which was designed to operate in a frequency bandwidth of 4 to 8 GHz, were composed of the stepped microstrip lines where all the high-impedance sections were made 50 Ω and all low-impedance sections were made 10 Ω [22].

To define the unknown section lengths, the optimization approach to achieve the global minimum of the objective function $|\Gamma(\theta, A)|$ can be used, which is written as

$$\min_{A} \max_{\theta \in [\theta_1, \theta_2]} \left|\Gamma(\theta, A)\right| \tag{8.46}$$

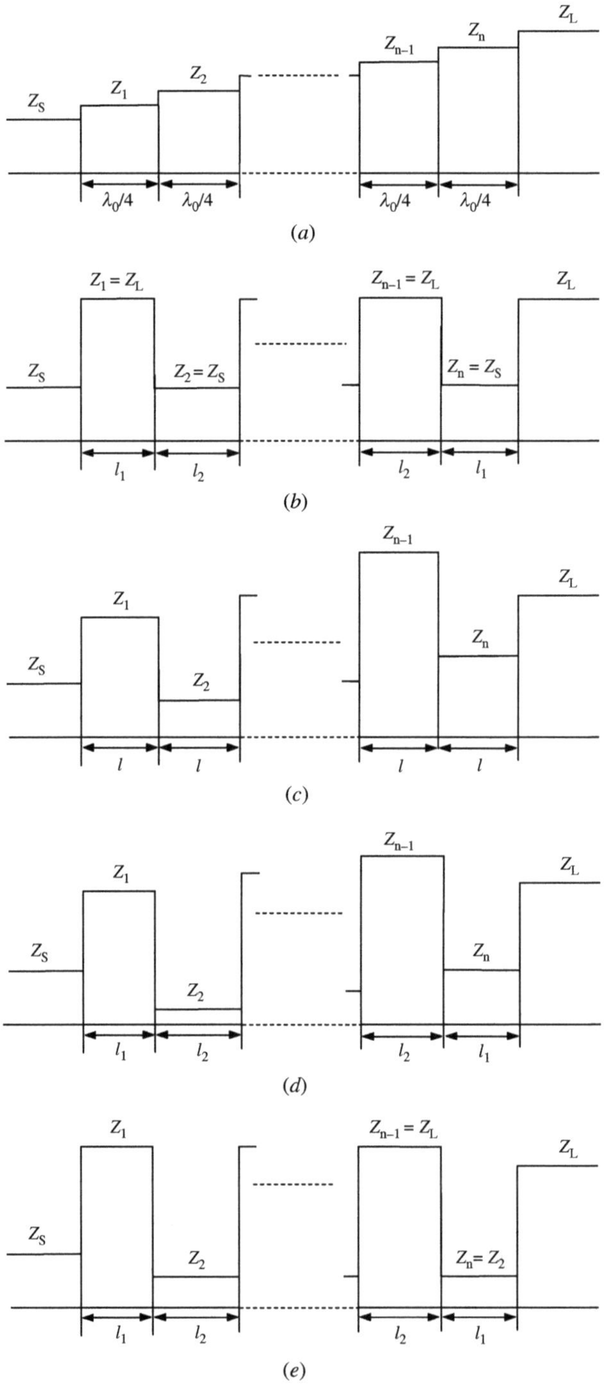

Figure 8.20 Schematic structures of different stepped transmission-line transformers.

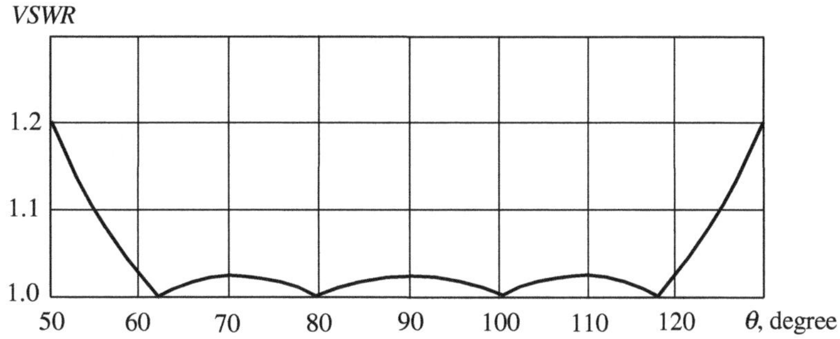

FIGURE 8.21 Theoretical frequency bandwidth of five-step transformer.

where θ_1 and θ_2 are the electrical lengths at the low- and high-frequency bandwidth edges, respectively, and the vector $A = (A_1, A_2, \ldots, A_n)$ consists of the normalized section lengths $L_i = l_i/\lambda_0$ as components [27]. By solving Eq. (8.46) numerically, the optimum Chebyshev characteristics can be provided by the stepped transmission-line structure with

$$l_i = l_{n+1-i} \tag{8.47}$$

where $i = 1, 2, \ldots, n/2$.

The total length of such a stepped transmission-line transformer can be further reduced by using the structure representing the cascade connection of n transmission-line sections of the same length $l < \lambda_0/4$ with

$$\begin{aligned} Z_1 < Z_3 < \cdots < Z_{n-1} \\ Z_2 < Z_4 < \cdots < Z_n \end{aligned} \tag{8.48}$$

where n is an even number and $Z_1 > Z_n$ when $Z_S < Z_L$, as shown in Fig. 8.20(*c*) [28]. An example of the stepped transmission-line transformer to match the source impedance of 25 Ω with the load impedance of 50 Ω is shown in Fig. 8.22(*a*), where the electrical length of each section is equal to $\lambda_0/12$. In this case, the total transformer length is shorter by three times compared to the basic structure with the quarterwave sections, and an octave passband from 2 to 4 GHz for the lossless ideal transmission-line sections is provided with an input return loss better than 25 dB, as shown in Fig. 8.22(*b*). However, it requires the use of a high impedance ratio for its sections reaching 30 to 50 when the source and load impedances differ significantly.

To reduce a high impedance ratio of the stepped transformers with a short total length, their generalized structure representing cascaded even n sections of different lengths l_i and impedances Z_i, as shown in Fig. 8.20(*d*), can be used. The optimum Chebyshev characteristics for

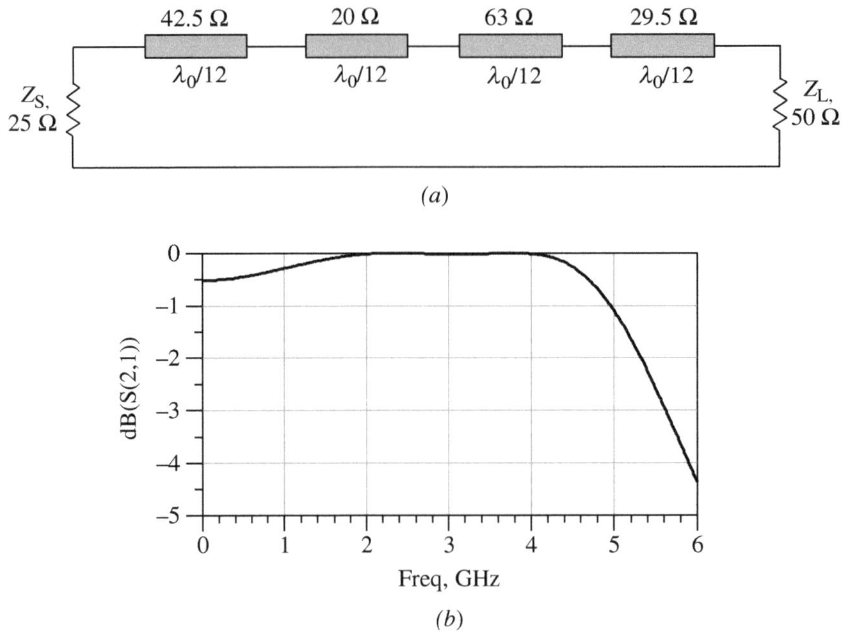

FIGURE 8.22 Stepped transmission-line transformer with equal-length sections.

this structure can be provided with the ratios between the lengths and characteristic impedances of its sections according to

$$l_i = l_{n+1-i}$$
$$Z_i Z_{n+1-i} = Z_S Z_L \tag{8.49}$$

where $i = 1, 2, \dots, n/2$, and

$$Z_{n-1} > Z_{n-3} > \cdots > Z_1 > Z_n > Z_{n-2} > \cdots > Z_2 \tag{8.50}$$

where the impedances of both even and odd sections decrease in the direction from higher impedance Z_L to lower impedance Z_S and the impedance of any odd section is always larger than that of any even section [27]. The lengths of even sections decrease in the direction from the transmission line of a smaller impedance, whereas those of odd sections increase in the same direction.

Another structure of the stepped transmission-line transformer with the reduced total electrical length is shown in Fig. 8.20(*e*), for which Eq. (8.47) can be applied and for which the same characteristic impedances for odd and even sections differ from the source and load impedances according to

$$Z_1 = Z_3 = \cdots = Z_{n-1}$$
$$Z_2 = Z_4 = \cdots = Z_n \tag{8.51}$$

$\lvert\Gamma\rvert_{max}$	Equal-Length Structure of Fig. 8.20(c)					
	$L_{1,2,3,4}$	Z_1, Ω	Z_2, Ω	Z_3, Ω	Z_4, Ω	L
0.065	0.0833	42.38	19.80	63.13	29.49	0.3330
0.071	0.0625	55.75	13.58	92.03	22.42	0.2500
0.074	0.0418	82.58	8.45	148.01	12.14	0.1670
0.076	0.0313	109.64	6.17	202.48	11.40	0.1 250

	Generalized Structure of Fig. 8.20(d)						
	$L_{1,2}$	$L_{3,4}$	Z_1, Ω	Z_2, Ω	Z_3, Ω	Z_4, Ω	L
0.068	0.0625	0.0833	51.75	18.20	68.73	24.15	0.2916
0.070	0.0525	0.0725	62.00	15.28	81.85	20.15	0.2500
0.075	0.0320	0.0510	103.00	10.13	123.39	12.14	0.1660
0.076	0.0205	0.0420	152.90	7.78	160.95	8.18	0.1250

	New Structure of Fig. 8.20(e)				
	$L_{1,2}$	$L_{3,4}$	Z_1, Ω	Z_2, Ω	L
0.064	0.0479	0.1171	52.38	23.86	0.3330
0.070	0.0405	0.0841	72.91	17.14	0.2500
0.074	0.0282	0.0553	114.55	10.91	0.1670
0.075	0.0213	0.0412	155.67	8.03	0.1250

TABLE 8.3 Optimum Parameters for Different Four-Section Transformers

where $Z_1Z_2 = Z_SZ_L$, $Z_n < Z_S$, and $Z_{n-1} > Z_L$ [27]. In particular situations of high impedance-matching ratio at microwave frequencies when it is necessary to match the standard source load impedance of 50 Ω with the device input and output impedance of 1 Ω and smaller, both the length and width of the microstrip-line sections can be optimized.

Table 8.3 gives the optimum parameters for different four-section transformers ($n = 4$) designed to match the transmission lines with impedances $Z_S = 25$ Ω and $Z_L = 50$ Ω in an octave frequency range, where the section lengths L_i and total length L are normalized to λ_0 [27].

8.5 Power Amplifiers with Lossy Compensation Networks

Dissipative or lossy gain compensation matching circuits can achieve the important tradeoff between gain, reflection coefficient, and operating frequency bandwidth. Moreover, the resistive nature of such a simple matching circuit may also improve amplifier stability and reduce its size and cost. For the first time, use of the attenuation equalizing circuits was suggested to maintain a high-quality transmission in long telephone circuits by H. W. Bode in the mid-1930s [29]. For example, such an attenuation equalizer can represent a four-terminal

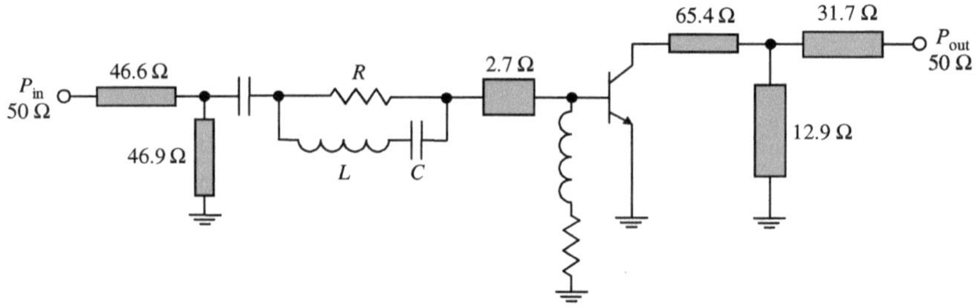

FIGURE 8.23 Schematic of octave-band microstrip lossy match bipolar power amplifier.

frequency-selective network together with connected source and load impedances having the constant-resistance image impedances at its input and output terminals.

Because it was impossible to provide broadband input matching of the bipolar power transistors with low-value frequency-varying input impedances, initially a circuit arrangement composed of a resistor in addition to pure reactances was implemented to keep the input reflection coefficient at low values across large frequency bandwidths [30]. Such an impedance network was so dimensioned that the amplification slope of approximately 6 dB is compensated as much as possible within the bandwidth of one octave. Figure 8.23 shows the circuit schematic of an octave-band single-stage microstrip bipolar amplifier covering the frequency range from 500 to 1000 MHz with a maximum output power of about 3 W, a power gain of around 7 dB, and an input *VSWR* less than 2.5, where a lossy compensation circuit consists of a series resistor *R* shunted by the series resonant circuit composed of a capacitor *C* and an inductor *L*, whose parameters were properly optimized.

8.5.1 Lossy Match Design Techniques

In many practical cases, to provide broadband matching with minimum gain flatness and input reflection coefficient, it is sufficient to use the resistive shunt element at the transistor input. An additional matching improvement with reference to upper frequencies can be achieved by using inductive reactive elements in series to the resistor. The resistive nature of this type of network may also improve amplifier stability and distortion. To provide a broadband performance for microwave GaAs MESFET power amplifiers, a resistively loaded shunt network, where the resistor is connected in series with a short-circuited quarterwave microstrip line to decrease the loaded quality factor without greatly reducing the maximum available gain, was used in the load network to provide a flat gain over 8 to 12 GHz, or in the input matching circuit to cover a frequency bandwidth of 2 to 6.2 GHz [31, 32]. For ultra-broadband high-gain multistage amplifiers, using a simple lossy compensation shunt circuit with a resistor in series with an inductor placed at the input and output of each transistor in parallel with the second-order *LC* circuits allows the gain of 12 ± 1.5 dB with a *VSWR* of less than 2.5 from 150 MHz to 16 GHz to be achieved for a three-stage GaAs MESFET amplifier [33]. A 14-dB gain was obtained over the 3-dB bandwidth from 700 kHz to 6 GHz for a two-stage microstrip GaAs MESFET power amplifier, where a flat gain performance was achieved by using a shunt lossy gain-compensation circuit with a resistor in series with a short-circuited microstrip line placed at the input and output of the first-stage transistor in parallel to the input and interstage *LC* matching circuits [34].

A bandstop/bandpass diplexing *RLC* network is more useful than a simple lossy *RL* gain-compensation circuit because it provides an exact match at one frequency and an arbitrary amount of attenuation at any other frequency. Diplexing networks can be used in either input or output networks of the amplifier depending on noise figure, power output, and other amplifier constraints. Figure 8.24(*a*) shows the resonant diplexer *LC* network for lossy gain compensation, where the series L_sC_s and parallel L_pC_p resonant circuits are tuned to high bandwidth frequency and $R_L = R_0$ [35]. Here, the series capacitance C_s and shunt inductance L_p are obtained as $C_s = BW/\omega_h R_L$ and $L_p = BW(R_L/\omega_h)$, where BW is the normalized frequency bandwidth and $\omega_h = 2\pi f_h$ is the high-bandwidth radian frequency. The distributed form of a lossy gain-compensation network with additional input low-pass matching section is shown in Fig. 8.24(*b*), where $Z_p = 4\omega_h L_p/\pi$, $Z_s = \omega_h L_s/\tan\theta_s$, and θ_s is the electrical length of the series transmission line.

Figure 8.25(*a*) shows the basic block of a microwave lossy match GaAs MESFET amplifier, where an input matching circuit and an open-circuit shunt stub cascaded with a series transmission line at the device drain terminal are included to provide the amplifier desired frequency response [36]. For frequencies up to 1 GHz, the reactive elements of the transistor equivalent model have relatively little influence on the gain magnitude and reflection coefficients. As a result, the transistor described by *S*-parameters can be represented by its low-frequency model and the amplifier circuit can be significantly reduced to a simple network, where $S_{12} = 0$ and both S_{11} and S_{22} have negligible imaginary components. Then, the amplifier gain can be derived as

$$\text{Gain} = |S_{21}|^2 = \left[\frac{g_m Z_0}{2}(1+S_{11})(1+S_{22})\right]^2 \tag{8.52}$$

which clearly expresses the tradeoffs between the gain and the reflection coefficients, where g_m is the device transconductance and Z_0 is the characteristic impedance [36, 37]. The schematic of a multistage lossy match amplifier can be divided into three basic circuit functions: input matching, amplification, and interstage matching. Figure 8.25(*b*) shows

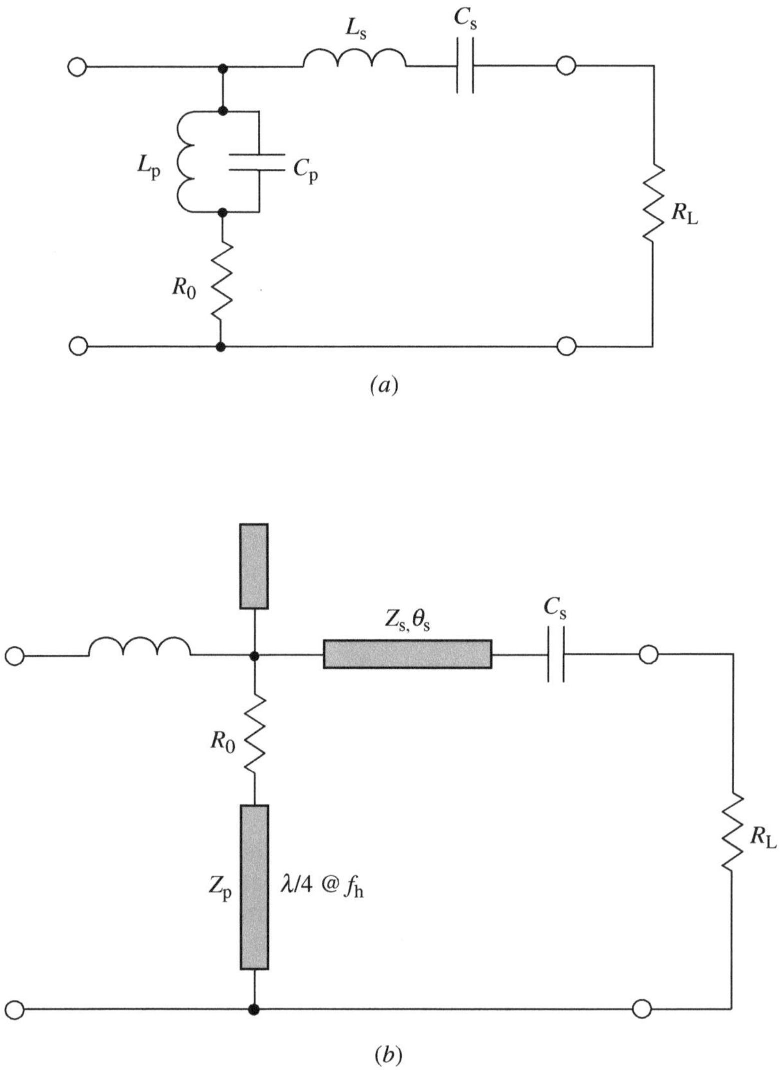

Figure 8.24 Circuit schematics of lossy gain-compensation circuits.

the lossy match two-stage GaAs MESFET amplifier with optimum values of the gate and drain shunt resistances to achieve flat gain performance over the frequency bandwidth of 2 to 8 GHz.

For a broadband lossy match silicon MOSFET high-power amplifier, it is sufficient to use a simple gain-compensation network with a resistor connected in series with a lumped inductor when operating frequencies are low enough compared to the device transition frequency f_T [38]. In this case, it is very important to optimize the elements of a lossy matching circuit to achieve minimum gain flatness

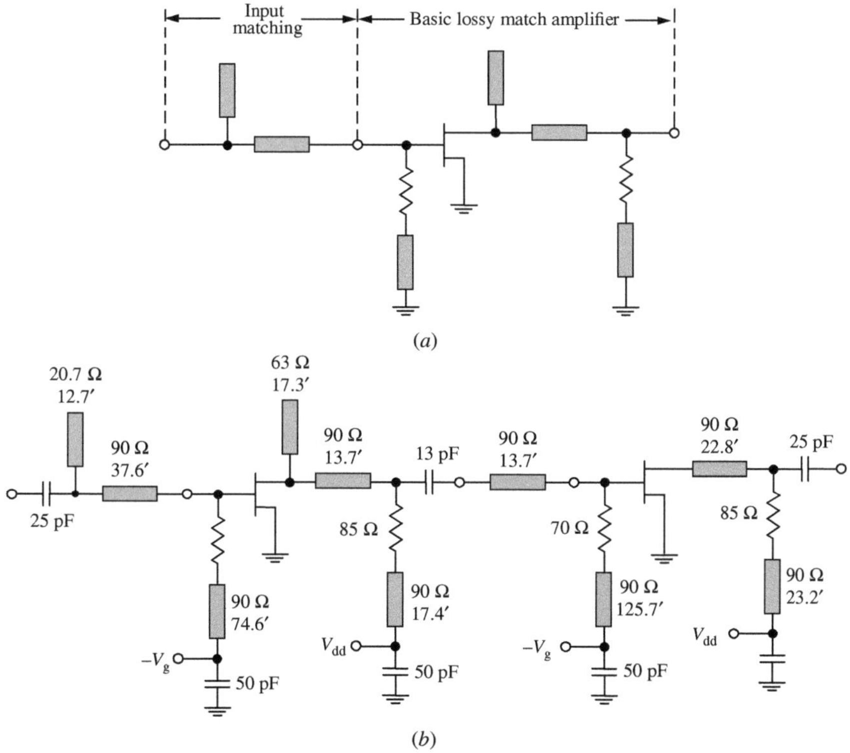

FIGURE 8.25 Circuit topologies of microstrip lossy match MESFET amplifiers.

over maximum frequency bandwidth. Let us consider the small-signal silicon MOSFET equivalent circuit, which is shown in Fig. 8.26. When the load resistor R_L is connected between the drain and source terminals, an analytical expression for the input device impedance Z_{in} can be obtained as

$$Z_{in} = \frac{R_g + \left(R_{gs} + \dfrac{1}{j\omega C_{gs}}\right)}{\left[1 + \dfrac{C_{gd}}{C_{gs}} \dfrac{(1 + j\omega\tau_g)(1 + j\omega C_{ds} R_{L0}) + g_m R_{L0}}{1 + j\omega R_{L0}(C_{ds} + C_{gd})}\right]} \tag{8.53}$$

where $R_{L0} = (R_L + R_d)/[1 + (R_L + R_d)/R_{ds}]$ and $\tau_g = R_{gs}C_{gs}$

The modified circuit shown in Fig. 8.27(*a*) describes adequately the frequency behavior of such input impedance of Fig. 8.26. In Eq. (8.53), the series source resistance R_s and transit time τ are not taken into account due to their sufficiently small values for high-power MOSFETs in a frequency range of $f \le 0.3f_T$, where $f_T = g_m/2\pi C_{gs}$. When $\omega\tau_g \le 0.3$ and the device output capacitive impedance is inductively compensated, the input

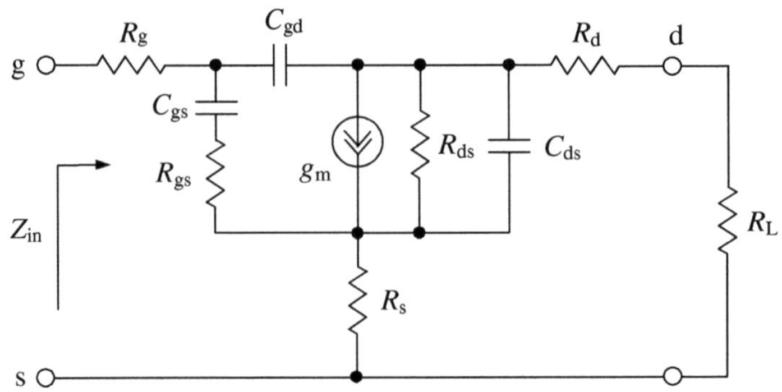

FIGURE 8.26 Small-signal silicon MOSFET equivalent circuit.

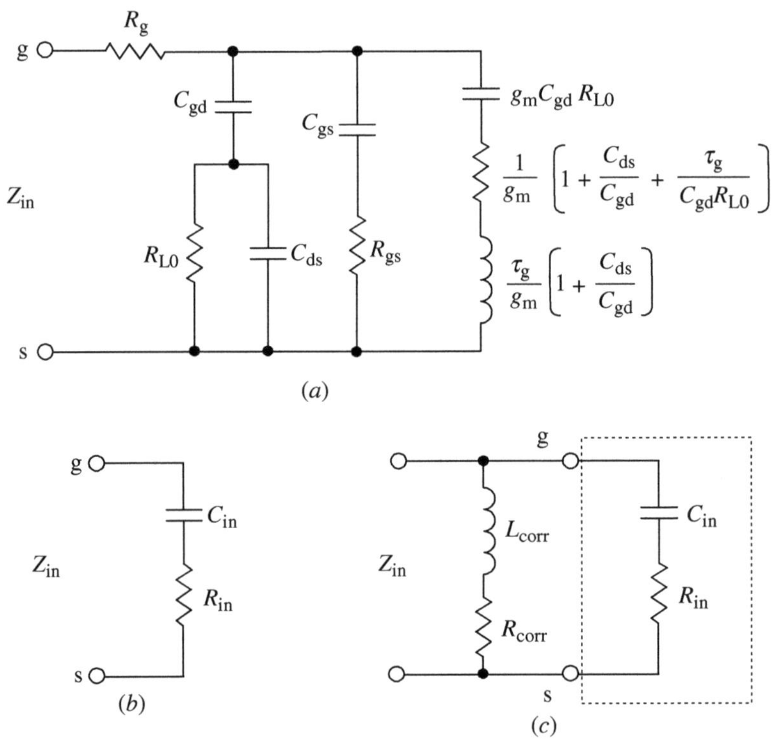

FIGURE 8.27 Equivalent circuits characterizing device input impedance.

equivalent circuit simplifies significantly and can represent a capacitor and a resistor connected in series, as shown in Fig. 8.27(*b*), where

$$R_{\text{in}} \cong R_{\text{g}} + R_{\text{gs}} \tag{8.54}$$

$$C_{\text{in}} \cong C_{\text{gs}} + C_{\text{gd}}\left[1 + g_{\text{m}}\frac{R_{\text{L}} + R_{\text{d}}}{1 + R_{\text{L}}/(R_{\text{ds}} + R_{\text{d}})}\right] \tag{8.55}$$

To provide a constant real part of the input impedance Z_{in} in a frequency range up to $0.1f_{\text{T}}$, it is enough to use a simple lossy compensation circuit consisting of an inductor L_{corr} and a resistor R_{corr} connected in series, as shown in Fig. 8.27(*c*).

The total input impedance of both lossy match gain-compensation circuit and device input circuit is written as

$$Z_{\text{in}} = \frac{R_{\text{corr}} - \omega^2 C_{\text{in}} R_{\text{in}} L_{\text{corr}} + j\omega(L_{\text{corr}} + C_{\text{in}} R_{\text{in}} R_{\text{corr}})}{1 - \omega^2 L_{\text{corr}} C_{\text{in}} + j\omega C_{\text{in}}(R_{\text{corr}} + R_{\text{in}})} \tag{8.56}$$

whose real and imaginary parts can be expressed through the circuit parameters by

$$\operatorname{Re} Z_{\text{in}} = \frac{(1-\omega^2 L_{\text{corr}} C_{\text{in}})(R_{\text{corr}} - \omega^2 C_{\text{in}} R_{\text{in}} L_{\text{corr}}) + \omega^2 C_{\text{in}}(R_{\text{corr}} + R_{\text{in}})(L_{\text{corr}} + C_{\text{in}} R_{\text{in}} R_{\text{corr}})}{(1-\omega^2 L_{\text{corr}} C_{\text{in}})^2 + (\omega C_{\text{in}})^2 (R_{\text{corr}} + R_{\text{in}})^2} \tag{8.57}$$

$$\operatorname{Im} Z_{\text{in}} = \omega\,\frac{L_{\text{corr}}(1-\omega^2 L_{\text{corr}} C_{\text{in}}) - C_{\text{in}}(R_{\text{corr}}^2 - \omega^2 L_{\text{corr}} C_{\text{in}} R_{\text{in}}^2)}{(1-\omega^2 L_{\text{corr}} C_{\text{in}})^2 + (\omega C_{\text{in}})^2 (R_{\text{corr}} + R_{\text{in}})^2} \tag{8.58}$$

Under the condition $R = R_{\text{corr}} = R_{\text{in}}$, the equations for $\text{Re}Z_{\text{in}}$ and $\text{Im}Z_{\text{in}}$ can be reduced to

$$\operatorname{Re} Z_{\text{in}} = R\,\frac{(1-\omega^2 L_{\text{corr}} C_{\text{in}})^2 + 2\omega^2 C_{\text{in}}(L_{\text{corr}} + C_{\text{in}} R^2)}{(1-\omega^2 L_{\text{corr}} C_{\text{in}})^2 + (2\omega C_{\text{in}} R)^2} \tag{8.59}$$

$$\operatorname{Im} Z_{\text{in}} = \omega(1-\omega^2 L_{\text{corr}} C_{\text{in}})\frac{L_{\text{corr}} - C_{\text{in}} R^2}{(1-\omega^2 L_{\text{corr}} C_{\text{in}})^2 + (2\omega C_{\text{in}} R)^2} \tag{8.60}$$

From Eq. (8.60), it follows that the reactive part of the input impedance Z_{in} becomes zero, that is, $\text{Im}Z_{\text{in}} = X_{\text{in}} = 0$, when

$$L_{\text{corr}} = C_{\text{in}} R^2 \tag{8.61}$$

which leads to a pure active input impedance Z_{in} obtained as

$$Z_{\text{in}} = R = R_{\text{in}} \tag{8.62}$$

At microwaves, the short-circuited transmission line can be included instead of an inductor L_{corr} with the same input inductive reactance.

In terms of amplifier circuit parameters, the low-frequency gain in decibels can be calculated as

$$\text{Gain} = 20\log_{10}\left[\frac{g_{\mathrm{m}}\sqrt{R_{\mathrm{corr}}R_{\mathrm{L}}}}{1+(R_{\mathrm{L}}+R_{\mathrm{d}})/R_{\mathrm{ds}}}\right] \tag{8.63}$$

However, when the frequency increases, the voltage amplitude applied to the input capacitance C_{in} decreases. This leads to the appropriate decrease in the operating power gain G_{P} at higher bandwidth frequencies. Because of the small values of R_{in} for high-power MOSFETs, the value of G_{P} may not be high enough. Therefore, it is necessary to provide an additional impedance matching with lossless matching circuits in order to match with the source impedance of 50 Ω or high output impedance of the active device of the previous power-amplifier stage.

Figure 8.28(*a*) shows the circuit schematic of a broadband LDMOSFET power amplifier with device geometry of 1.25 μm × 40 mm. The optimized input three-element lossy matching circuit allows a very broadband operation to be provided with minimum power gain flatness, and a 1:2 output transformer contributes to increase in the output power level. The capacitor of 20 pF connected in parallel with the resistor of 27 Ω provides an additional increase of power gain at higher bandwidth frequencies. The simulation results are shown in Fig. 8.28(*b*), where an output power of 22 to 25 W with a power gain of 13.7 ± 0.3 dB in a frequency range of 5 to 300 MHz can be achieved (curve 1). In this case, the input return loss is greater than 8 dB up to 225 MHz (curve 2). However, when a 50-Ω load is directly connected to the device drain terminal through the blocking capacitor, this results in output power levels in the range of 6 to 7 W.

8.5.2 Practical Examples

For solid-state SSB and AM communication transmitters, it is required to provide a linear amplification across the entire frequency range of 2 to 30 MHz, which firstly was covered by using bipolar technology based on a push-pull amplifier implementation with broadband toroidal transmission-line impedance transformers and combiners and interstage *RLC* gain-compensation networks. In this case, the driver stages are operated in a Class-A mode for increased power gain, whereas the final stages are biased in a Class AB with optimized quiescent currents for better linearity. As a result, the overall four-stage bipolar power amplifier achieved a *PAE* of greater than 31% for two-tone 60-W PEP signal over the entire frequency range of 2 to 30 MHz, with IM_3 equal to −30 dBc or better at output powers of 5 to 60 W [39].

Figure 8.29 shows the circuit schematic of a bipolar broadband high-power amplifier designed for broadcasting VHF FM transmitters in a frequency range of 66 to 108 MHz [40]. When using the

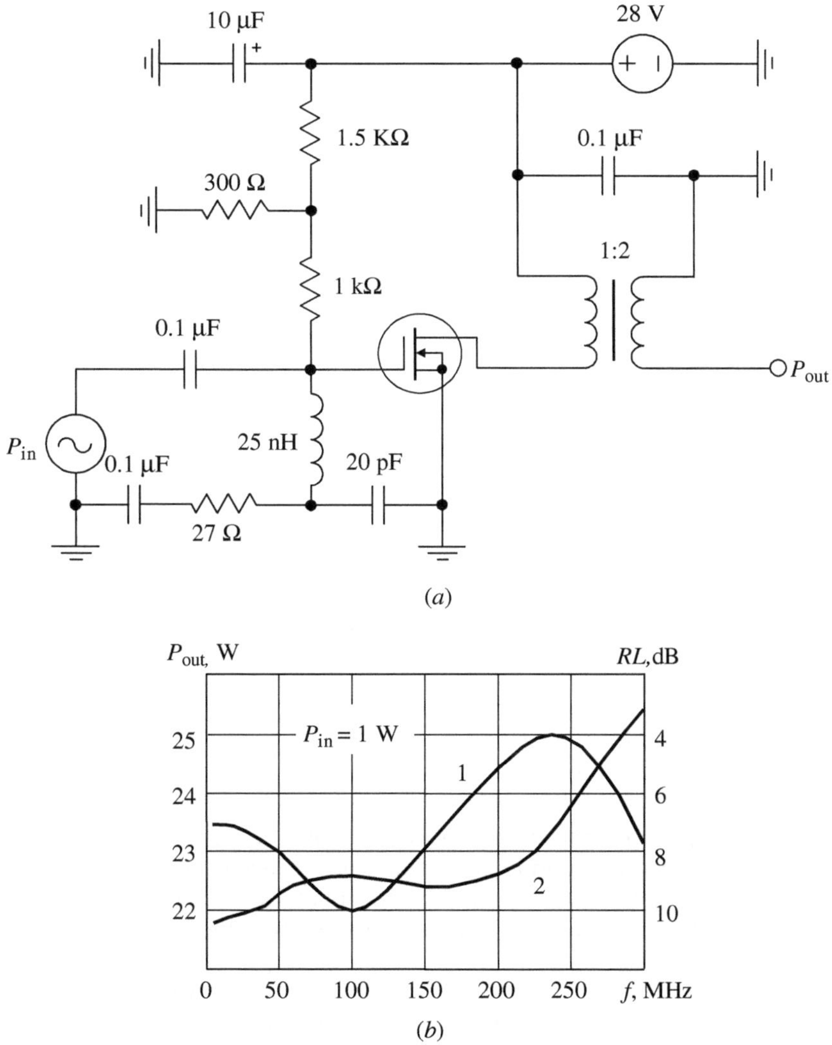

FIGURE 8.28 Schematic and performance of broadband LDMOSFET power amplifier.

200-W balanced VHF-UHF bipolar transistors, such as NEC 2SC3812, biased in a Class-C mode, an output power of 350 W with a power gain of 11 ± 1 dB and a collector efficiency of about 60% can be provided across the entire frequency bandwidth by combining two transistors. An appropriate negative biasing in a Class-C mode is achieved by using a series resistor of 5.1 Ω together with a series inductor of about 15 nH in each bias circuit, which also serves as a lossy match gain-compensation circuit to provide minimum gain and power variations. The asymmetric 1:2 input TL_1 and output TL_8 transformers

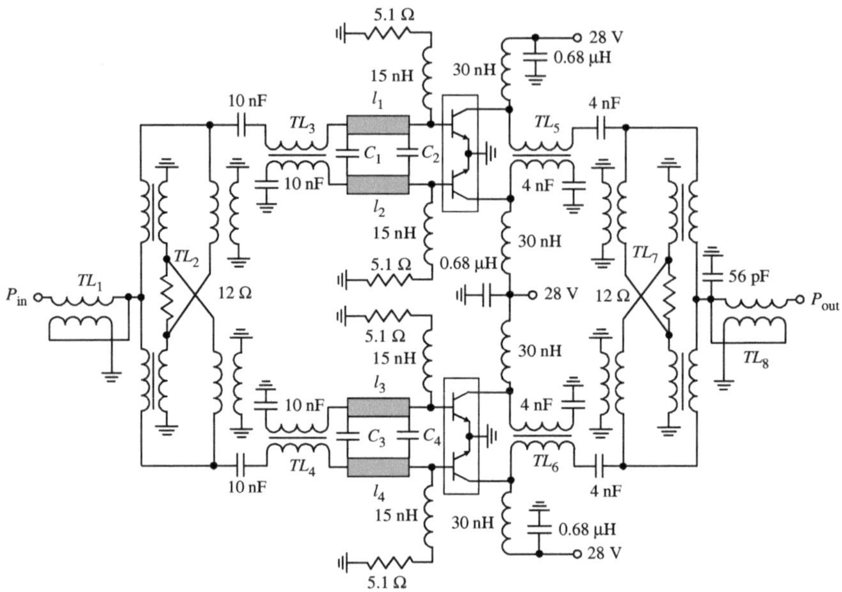

FIGURE 8.29 Bipolar broadband high-power amplifier for VHF FM transmitters.

with the coaxial-cable characteristic impedances of 25 Ω are used to convert 12.5 Ω to standard source and load 50-Ω impedance, respectively. The unbalanced-to-balanced stripline transformers TL_3 to TL_6 with the stripline characteristic impedances of 6 Ω are necessary to provide the 3-Ω source and load impedances for each part of the balanced bipolar transistors. Because of the small value of the device single-ended input impedance of about 1 Ω with inductive component, the additional input two-section *L*-type impedance-matching circuits are used. Here, the series microstrip lines l_1 to l_4 are the inductive elements for the first section and the device lead inductances are the inductive elements for the second section. Power dividing at the input as well as power combining at the output of the high-power amplifier is realized by hybrid power splitters/combiners TL_2 and TL_7, each with 12-Ω ballast resistors and the stripline characteristic impedances of 12.5 Ω. Such a hybrid power splitter/combiner provides an excellent device-to-device and device-to-load isolations, and contributes to amplifier operation stability.

The circuit schematic of the input, interstage, and output networks intended to be implemented in microwave broadband power amplifiers are shown in Fig. 8.30 [41]. A constant-resistance input network shown in Fig. 8.30(*a*) provides the input device impedance to be pure resistive and equal to $Z_{in} = R_{in}$ when $L_1 = C_{gs}R_{in}^2$, $C_1 = L_g/R_{in}^2$, and $R_1 = R_{in}$, thus making wideband transformation of the input resistance to the

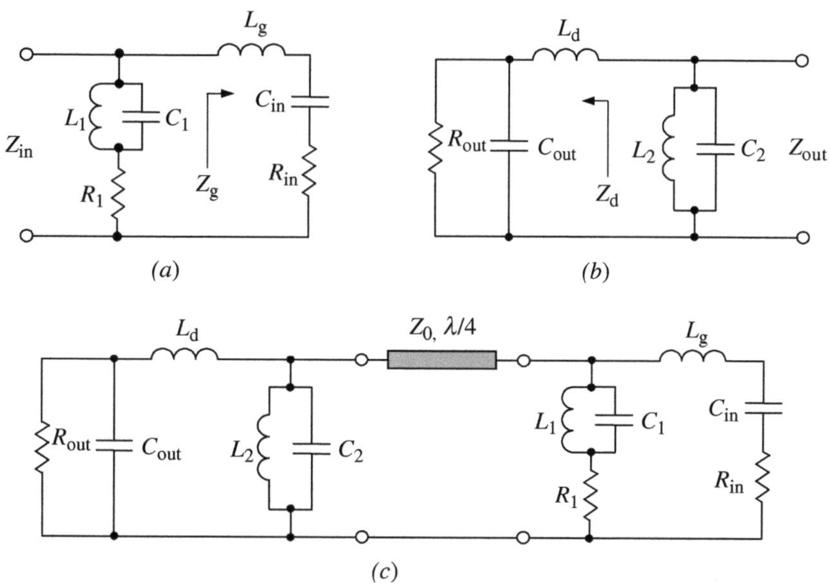

FIGURE 8.30 Schematics of input, output, and interstage broadband matching circuits.

source resistance much easier. In the output network shown in Fig. 8.30(*b*), a value of the drain inductance L_d is properly chosen to compensate for the capacitive device output reactance at the center bandwidth frequency. Then, a resonant frequency of the parallel L_2C_2 circuit is set to be equal to the same center bandwidth frequency. In this case, for lower frequencies where the device output impedance Z_d is capacitive, reactance of the parallel resonant circuit is inductive. On the other hand, for higher frequencies where the impedance Z_d is inductive, reactance of the parallel resonant circuit is capacitive. As a result, the wideband reactance compensation is realized when reactive part of the overall output impedance becomes very small over wide frequency bandwidth. For microwave applications, such a parallel resonant circuit is fabricated by using a quarterwave short-circuit stub. The interstage network, whose circuit schematic is shown in Fig. 8.30(*c*), comprises the input and output networks described previously and a quarterwave microstrip transformer with the characteristic impedance of $Z_0 = \sqrt{R_{in}L_d/R_{out}C_{out}}$.

Figure 8.31 shows the circuit schematic diagram of a two-stage lossy match MESFET power amplifier [41]. By using a 1.05-mm device in the driver stage and two 1.35-mm devices in the final stage, a saturated output power of 27.7 ± 2.7 dBm, a linear power gain of 8.3 ± 2.8 dB, and a drain efficiency of 15.3 ± 8.3% were measured in a frequency range of 4 to 25 GHz. The input and interstage constant-resistance networks are represented by the series connection of a

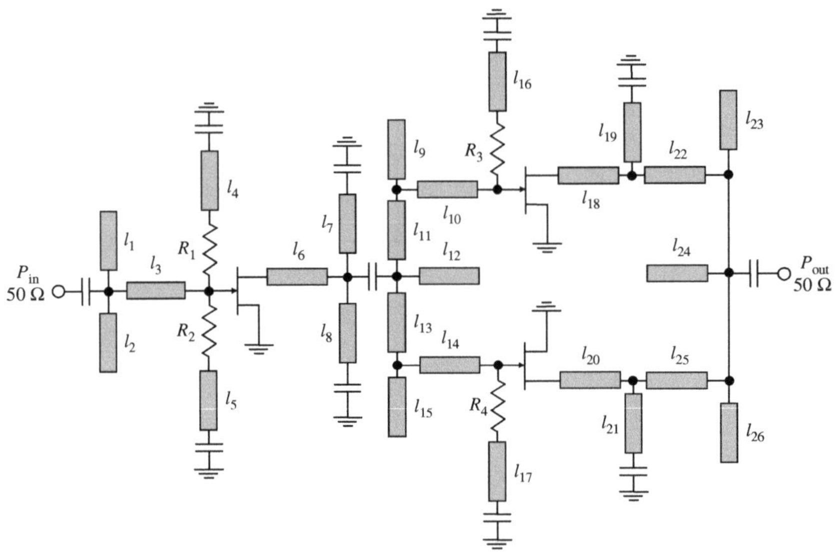

FIGURE 8.31 Microstrip two-stage lossy match MESFET power amplifier.

resistor and a high-impedance microstrip line each. Two such networks connected in parallel provide pure resistive input impedance, where l_4 and l_5 are the series microstrip lines, and R_1 and R_2 are the series resistors. The short-circuited microstrip lines (l_7 and l_8 in the interstage network, l_{19} and l_{21} in the output network) with quarterwave electrical lengths at the center bandwidth frequency serve as the parallel resonant circuits connected at the device output terminals. The microstrip lines l_{10} and l_{14} in the interstage network represent the quarterwave impedance transformers, which provide matching between the output impedance of the driver-stage device and the input impedance of the second-stage devices connected in parallel. The input and output matching circuits are realized in the form of *T*-transformers, where the series microstrip lines and parallel open-circuit microstrip stubs replace the series inductors and shunt capacitors, respectively. To further increase an output power, the number of amplifying stages with lossy input and interstage matching circuits connected in parallel can be increased. As a result, by optimizing the output matching and combining circuits, for a three-stage MMIC 0.25-μm pHEMT power amplifier with a distributed amplifier used as a driver stage and four 1200-μm transistors in the output stage, an output power of 2.4 ± 1.1 W with a small-signal gain of 24 ± 3.5 dB over the frequency range of 6 to 18 GHz was measured [42].

Figure 8.32(*a*) shows the circuit schematic of a broadband GaN HEMT microwave power amplifier implemented in the form of a flip-chip integrated circuit with the device geometry of 0.7 μm × 1 mm,

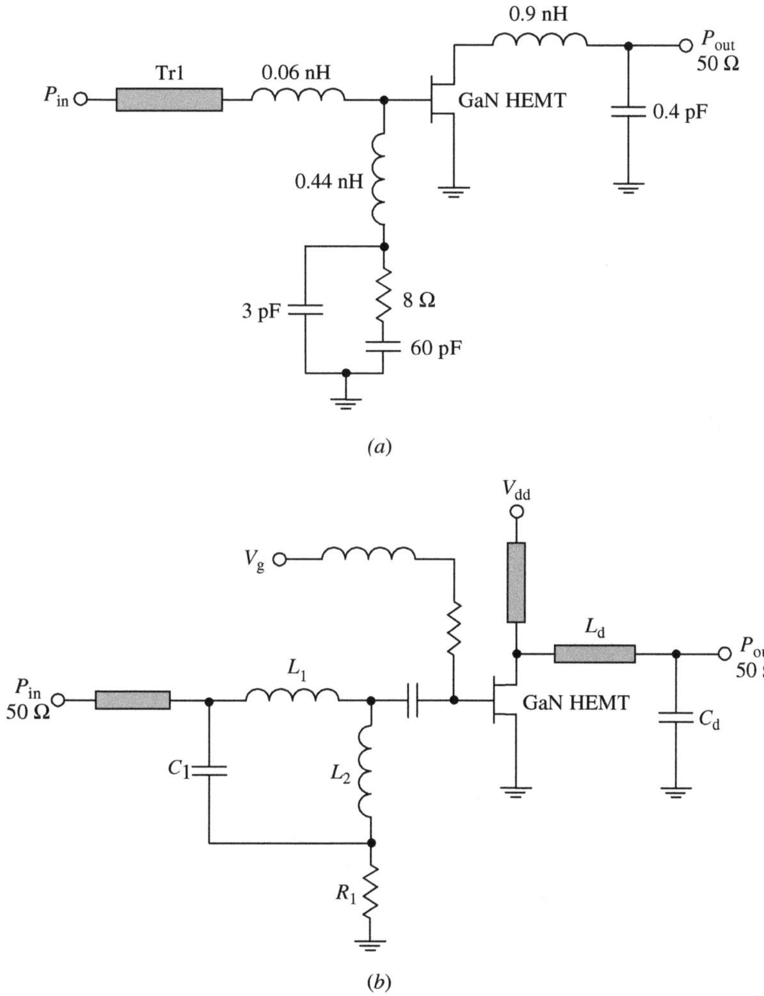

FIGURE 8.32 Schematics of microwave broadband GaN HEMT power amplifiers.

transition frequency f_T = 18 GHz, and maximum frequency f_{max} = 35 GHz [43]. The optimized input three-element lossy *LCR* matching circuit provides a power gain up to 11.5 dB and a low input reflection less than −10 dB over frequency range of 3 to 9 GHz. As the impedance at the input of a lossy match gain-compensation circuit is only of about 10 Ω, this necessitates an additional 50- to 10-Ω broadband impedance transformation (Tr1), which was realized using a few sections of quarterwave coplanar transmission lines with decreasing characteristic impedances. The output network incorporates a low-pass *LC* circuit to compensate for the output device capacitance such that the intrinsic device sees approximately a real load within entire frequency bandwidth. Because the optimum load for this 1-mm device with a supply voltage of 20 V is of about 50 Ω, hence no output impedance transformation is needed. The output power was measured of about 1.6 W with a *PAE* of 14 to 24% across the frequency bandwidth of 4 to 8 GHz. By combining of four such GaN HEMT power amplifiers connected in parallel, the highest output power of 8 W with a *PAE* of about 20% was obtained at 9.5 GHz and the lowest output power of 4.5 W was measured at 4.5 GHz, with a small-signal gain of 7 dB across the frequency bandwidth of 3 to 10 GHz [44].

To provide multi-decade bandwidth with very good input return loss, a compact bridged-*T* all-pass input *RLC* matching network can be used, as shown in Fig. 8.32(*b*), where the resistor R_1 was chosen to be of 50 Ω [45]. In this case, a GaN HEMT periphery of 2.2 mm was chosen to obtain an output power in the range of 10 Ω. A simple two-element matching circuit consisting of a series microstrip line and a shunt capacitor was used at the output to provide optimum load impedance at the upper band edge. The power amplifier was packaged in a ceramic SO8 package, including GaN on SiC device operating at 28 V and GaAs integrated passive matching circuitry. As a result, an output power of 8 W and a power gain of 12 dB were measured over frequency bandwidth from 50 MHz to 2 GHz with a drain efficiency of 36.7 to 65.4%.

8.6 Broadband Class-E Power Amplifiers

The high-efficiency broadband operation of a switchmode Class-E power amplifier using reactance compensation technique can be realized, if a simple network consisting of a series resonant *LC* circuit tuned to the fundamental frequency and a parallel inductor provides a constant load phase angle of 50° in a frequency range of about 50% [46]. From theoretical considerations it was found yet in the mid-1960s that the bandwidth response of a parametric amplifier can be improved using multiple-resonant bandpass filters for the signal and idling circuits rather than simple resonant circuits [47, 48]. At the same time, it was analytical calculated that the added resonant circuits should have an appropriate Q_L-factor to optimally reduce the rate of change of reactance of both the signal and idling circuits [49]. Adding additional resonators can increase the potential amplifier bandwidth even further, but the amount of improvement per additional resonator will decrease rapidly as the number of resonators is increased.

8.6.1 Reactance Compensation Technique

To describe reactance compensation circuit technique, let us consider the simplified equivalent load networks, one with a shunt resonant L_pC_p circuit followed by a series resonant L_sC_s circuit shown in Fig. 8.33(*a*) and the other with a series resonant L_sC_s circuit followed

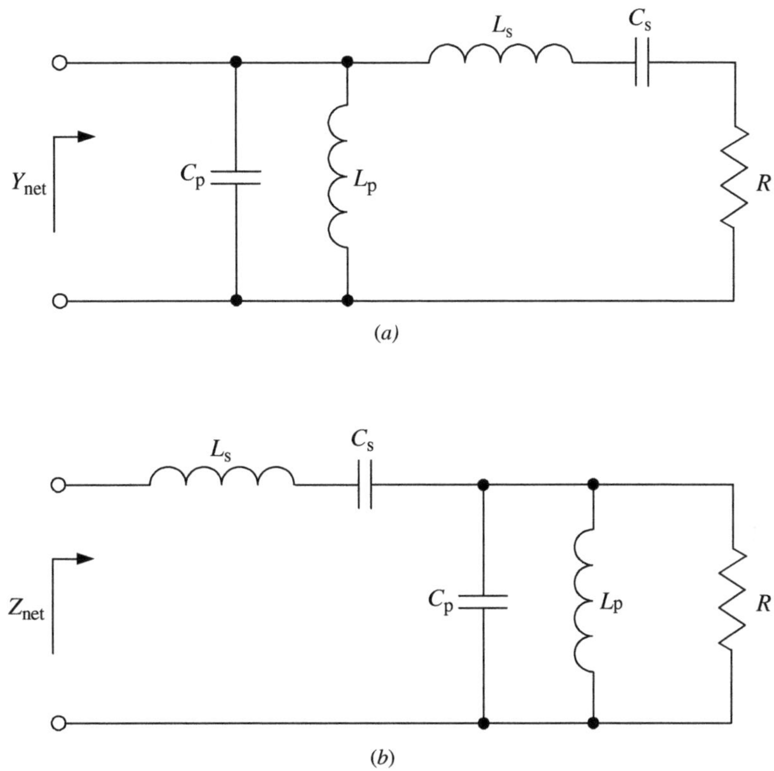

FIGURE 8.33 Single-reactance compensation circuits.

by a shunt resonant L_pC_p circuit shown in Fig. 8.33(*b*). In this case, all resonant circuits are tuned to the fundamental frequency and R is the load resistance. The reactances of the series and shunt resonant circuits vary with frequency, increasing in the case of a series resonant circuit and reducing in the case of a loaded parallel resonant circuit near the resonant frequency. As a result, near the resonant frequency of the series circuit with positive slope of its reactance, the slope of a shunt circuit reactance is negative that reduces the overall reactance slope of the load network. By correctly choosing the components in the shunt circuit, the rate of change of reactance with frequency can be made exactly opposite to that of the series circuit, thus producing a zero total variation over a wide frequency bandwidth.

Consider the load-network admittance Y_{net} corresponding to a single-reactance compensation circuit shown in Fig. 8.33(*a*), which can be written as

$$Y_{net}(\omega) = \left(j\omega C_p + \frac{1}{j\omega L_p} + \frac{1}{R + j\omega' L_s} \right) \tag{8.64}$$

where

$$\omega' = \omega\left(1 - \frac{\omega_0^2}{\omega^2}\right) \tag{8.65}$$

and $\omega_0 = 1/\sqrt{L_s C_s} = 1/\sqrt{L_p C_p}$ is the radian resonant frequency.

At the resonant frequency when $\omega' = 0$, the load-network admittance $Y_{net}(\omega)$ reduces to

$$Y_{net}(\omega) = \left(j\omega C_p + \frac{1}{j\omega L_p} + G\right) \tag{8.66}$$

where $G = 1/R$ is the load conductance.

The frequency bandwidth with zero susceptance will be maximized if, at a resonant radian frequency ω_0,

$$\left.\frac{dB_{net}(\omega)}{d\omega}\right|_{\omega=\omega_0} = 0 \tag{8.67}$$

where

$$B_{net}(\omega) = \operatorname{Im} Y_{in}(\omega) = \omega C_p - \frac{1}{\omega L_p} - \frac{\omega' L_s}{R^2 + (\omega' L_s)^2} \quad (8.68$$

is the load-network susceptance.

As a result, an additional equation can be written as

$$C_p + \frac{1}{\omega_0^2 L_p} - \frac{2L_s}{R^2} = 0 \tag{8.69}$$

based on which the values of the series components L_s and C_s can respectively be obtained through the values of the shunt components L_p and C_p by

$$L_s = C_p R^2 \tag{8.70}$$

$$C_s = \frac{L_p}{R^2} \tag{8.71}$$

Similarly, it may be shown that, for the load network with a series resonant $L_s C_s$ circuit followed by a shunt resonant $L_p C_p$ circuit shown in Fig. 8.33(*b*), the maximum bandwidth with zero reactance can be achieved if

$$\left.\frac{dX_{net}(\omega)}{d\omega}\right|_{\omega=\omega_0} = 0 \tag{8.72}$$

where

$$X_{net}(\omega) = \operatorname{Im} Z_{net}(\omega) = \omega L_s - \frac{1}{\omega C_s} - \frac{\omega' C_p}{G^2 + (\omega' C_p)^2} \tag{8.73}$$

is the load-network reactance, resulting in Eqs. (8.70) and (8.71). From Eq. (8.70), it follows that the loaded quality factor of the shunt circuit $Q_L = \omega C_p R$ is equal to the loaded quality factor of the series compensating circuit $Q_L = \omega L_s / R$.

Figure 8.34(*a*) shows the example of a susceptance compensation load network, whose conductance $\mathrm{Re}Y_{\mathrm{net}}$ is almost constant across the frequency range of 40% (from 4 to 6 MHz), as shown in Fig. 8.34(*b*). The susceptance $\mathrm{Im}Y_{\mathrm{net}}$ of a shunt circuit varies with frequency, as shown in Fig. 8.34(*c*) by curve 1, with the gradient at ω_0 being equal to $2C_p$. The addition of a series circuit with the same resonant frequency of 5 MHz between the shunt circuit and the load of the shunt circuit gives an additional susceptance term with a negative slope, as shown in Fig. 8.34(*c*) by curve 2. Proper selection of the components of the series circuit enables the magnitude of the two slopes to be made identical, so that the total susceptance slope around resonance is zero in an octave frequency range from 3.5 to 7 MHz, as shown in Fig. 8.34(*c*) by curve 3.

The load network that provides reactance compensation is shown in Fig. 8.35(*a*), where the shunt resonant circuit is connected between the series resonant circuit and the load. In this case, the resistance and reactance curves, whose frequency behavior is similar to that for the conductance and susceptance curves characterizing the behavior of a susceptance compensation load network, are shown in Fig. 8.35(*b*) and 8.35(*c*), respectively. Here, the reactance of a series resonant circuit with a positive slope is shown by curve 1, the reactance of a shunt resonant circuit with a negative slope is shown by curve 2, and the total reactance slope shown by curve 3 is zero from 3.5 to 7 MHz.

Wider frequency bandwidth can be achieved using a double-susceptance compensation circuit shown in Fig. 8.36(*a*), where L_sC_s and L_1C_1 are the series and parallel compensating circuits, respectively. In this case, a system of two additional equations to maximize the frequency bandwidth can be used, where the first and the third derivatives are set to zero according to

$$\left.\frac{dB_{\mathrm{net}}(\omega)}{d\omega}\right|_{\omega=\omega_0} = \left.\frac{d^3B_{\mathrm{net}}(\omega)}{d\omega^3}\right|_{\omega=\omega_0} = 0 \tag{8.74}$$

as the second derivative cannot provide an appropriate analytical expression.

To determine the load-network parameters for a double-susceptance compensation circuit with the load-network susceptance

$$B_{\mathrm{net}}(\omega) = \omega C_p - \frac{1}{\omega L_p} + \omega' \frac{C_1R^2[\,1-(\omega')^2L_sC_1] - L_s}{R^2[\,1-(\omega')^2L_sC_1]^2 + (\omega' L_s)^2} \tag{8.75}$$

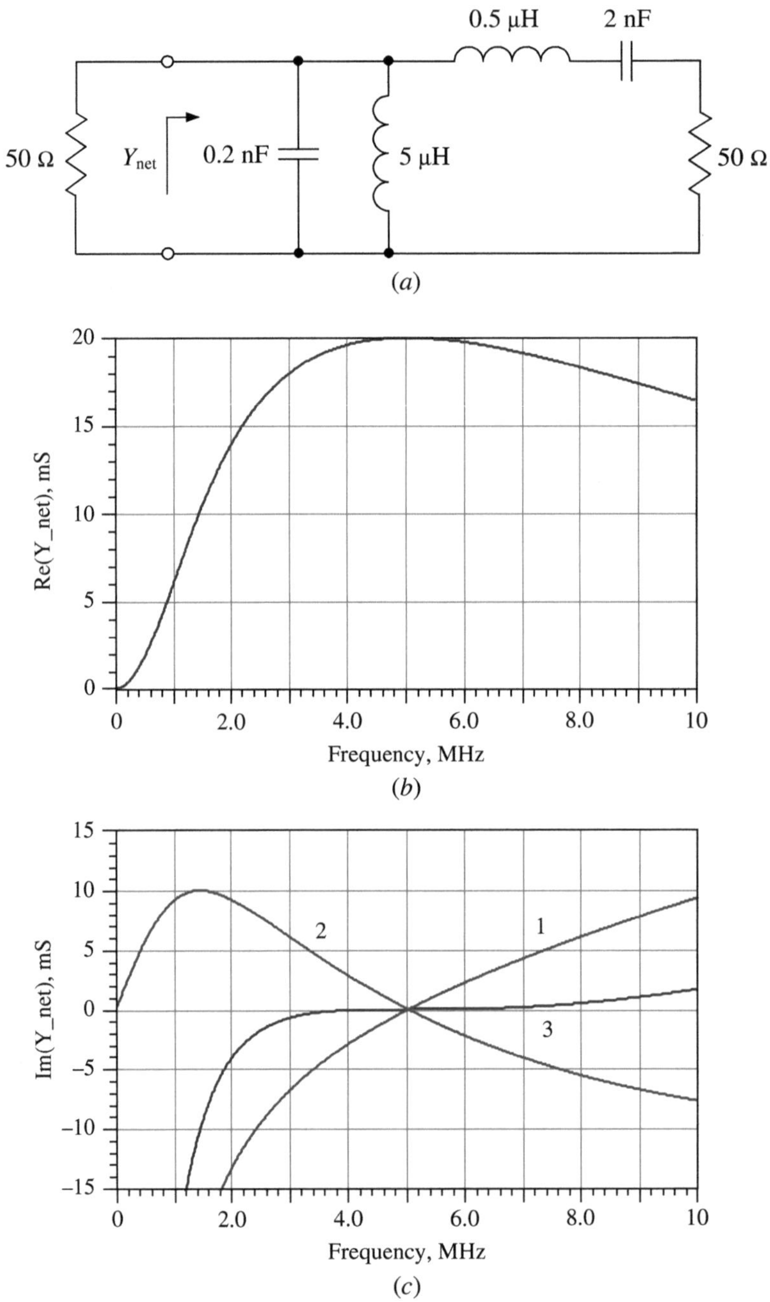

FIGURE 8.34 Single-susceptance compensation circuit and admittances.

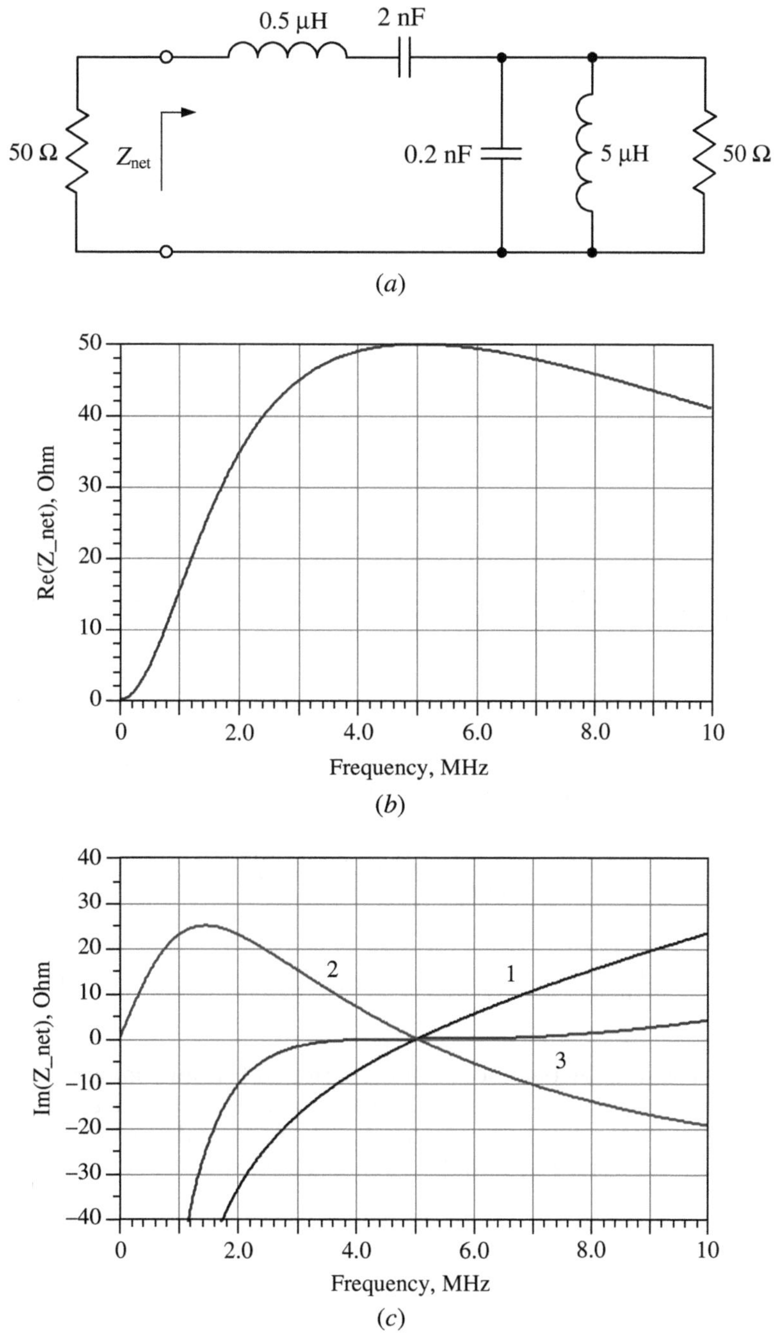

Figure 8.35 Single-reactance compensation circuit and impedances.

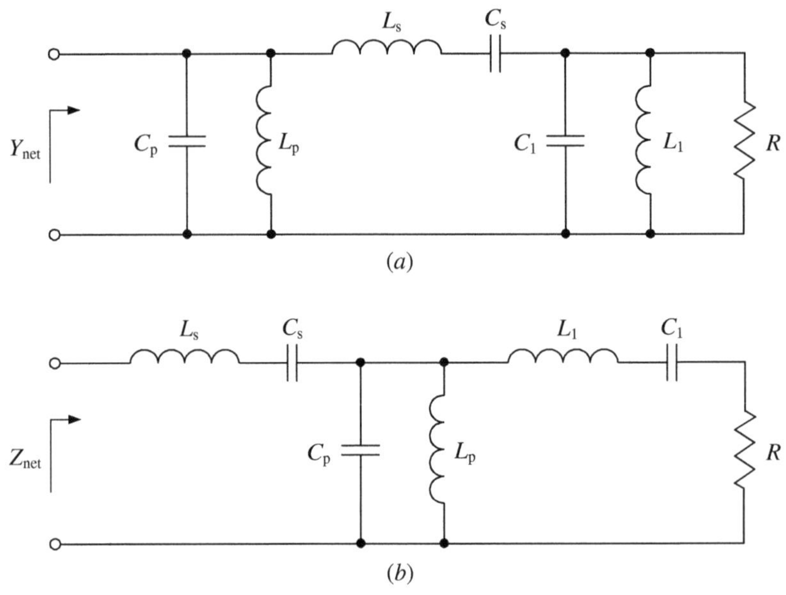

FIGURE 8.36 Double-reactance compensation circuits.

where $B_{net} = \text{Im}Y_{net}$, it is necessary to solve simultaneously the two following equations at the resonant frequency ω_0:

$$C_{p1} + \frac{1}{\omega_0^2 L_p} - 2\,\frac{C_1 R^2 - L_s}{R^2} = 0 \tag{8.76}$$

$$\frac{1}{\omega_0^2 L_p} + \frac{C_1 R^2 - L_s}{R^2} - 8\omega_0^2 L_s \left[C_1^2 + \frac{(C_1 R^2 - L_s)(L_s - 2C_1 R^2)}{R^4}\right] = 0 \tag{8.77}$$

As a result, the parameters of the series and shunt compensating resonant circuits with the corresponding loaded quality factors $Q_s = \omega_0 L_s / R$ and $Q_1 = \omega_0 C_1 R$, which are close to unity and greater, can be calculated as a starting point for circuit optimization from

$$L_s = \frac{R}{\omega_0}\,\frac{2}{\sqrt{5}-1} \qquad C_s = \frac{1}{\omega_0^2 L_s} \tag{8.78}$$

$$C_1 = \frac{L_s}{R^2}\,\frac{3-\sqrt{5}}{2} \qquad L_1 = \frac{1}{\omega_0^2 C_1} \tag{8.79}$$

Similarly, the elements for the double-reactance compensation load network shown in Fig. 8.36(*b*) can be calculated from

$$L_1 = L_s \frac{\sqrt{5}-1}{2} \qquad C_1 = C_s \frac{2}{\sqrt{5}-1} \tag{8.80}$$

$$L_p = C_s \frac{2R^2}{\sqrt{5}+1} \qquad C_p = L_s \frac{\sqrt{5}+1}{2R^2} \tag{8.81}$$

where an inductance L_s and a capacitance C_s are known in advance [50]. An example of the load network that provides double-reactance compensation is shown in Fig. 8.37(*a*), whose resistance $\mathrm{Re}Z_{\mathrm{net}}$ shown in Fig. 8.37(*b*) by curve 1 provides less deviation from 50 Ω in a slightly wider frequency bandwidth compared to the single-resonance load network (curve 2) with $L_s = 0.5$ μH, $C_s = 2$ nH, $L_p = 5$ μH, and $C_p = 0.2$ nF. The reactance $\mathrm{Im}Z_{\mathrm{net}}$ of a double-reactance compensation circuit shown in Fig. 8.37(*c*) by curve 1 is close to zero near resonance across the frequency range from 3 to 8 MHz, which is wider enough than that for a single-resonance compensation circuit (curve 2).

The reactance compensation circuit technique can also be used for bandwidth improvement of microwave transistor amplifiers because the input and output transistor impedances generally can be represented by series or shunt *RLC* circuits. For compensating the reactive part and transforming the real part of the equivalent output transistor impedance to the conventional load impedance at the fundamental frequency, the quarter- and half-wavelength transmission lines can be used. For the first time, a quarter-wavelength transmission-line transformer was used for active reactance compensation when, by connecting two identical active devices together with a quarter-wavelength transformer, the inverted impedance of one device compensates the impedance of the other one by reducing the total circuit reactance [51].

Let us consider the characteristics of the transmission line as an element of a susceptance compensation circuit shown in Fig. 8.38. For a parallel equivalent circuit, which represents the device output, the load-network input susceptance $B_{\mathrm{net}} = \mathrm{Im}Y_{\mathrm{net}}$ can be defined as

$$B_{\mathrm{net}}(\omega) = \omega L_p C_p \left(1 - \frac{\omega_0^2}{\omega^2}\right) + \frac{\tan\theta}{Z_0} \frac{R_L^2 - Z_0^2}{R_L^2 + Z_0^2 \tan^2\theta} \tag{8.82}$$

where

$$\theta = \frac{\pi}{2} \frac{f}{f_0} k \tag{8.83}$$

is the transmission-line electrical length, Z_0 is the transmission-line characteristic impedance, $f_0 = \omega_0 / 2\pi$ is the transmission-line resonant frequency, $k = 1, 2, \ldots, \infty$.

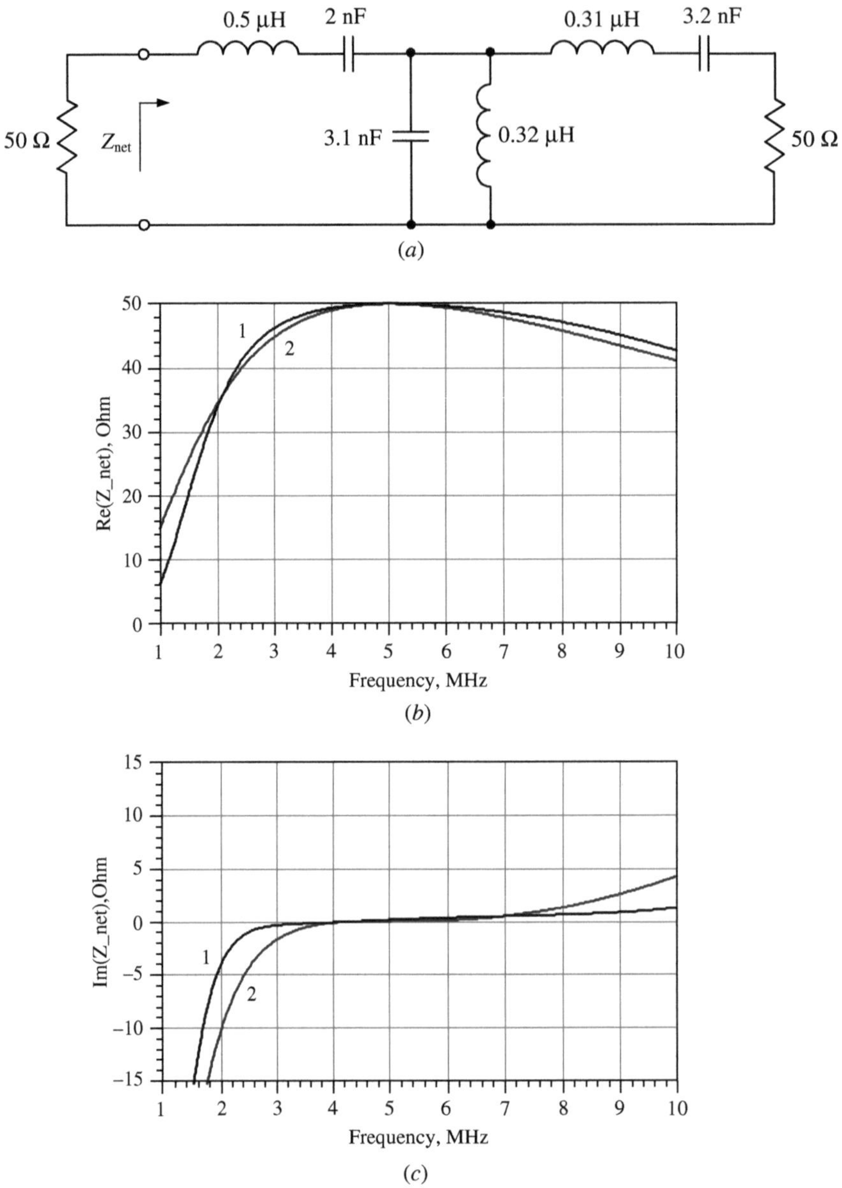

FIGURE 8.37 Double-reactance compensation circuit and impedances.

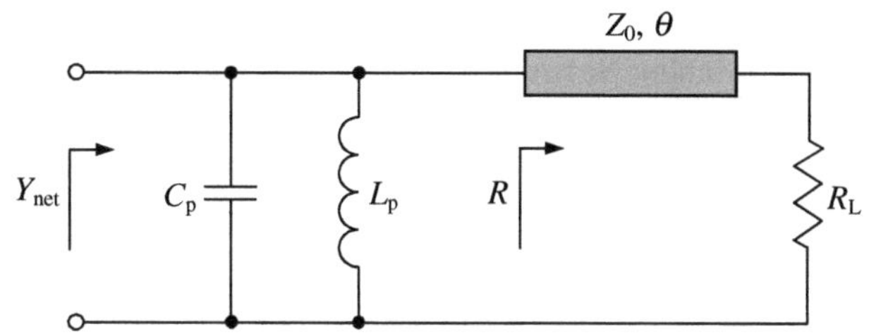

Figure 8.38 Transmission-line susceptance compensation circuit.

Applying the zero susceptance-derivative condition given by Eq. (8.67) allows us to obtain the susceptance-compensation circuit parameters for different electrical lengths of a transmission line in accordance with

$$2C_p + \frac{\pi}{2Z_0\omega_0}\,\frac{R_L^2 - Z_0^2}{\cos^2\theta}\,\frac{R_L^2 - Z_0^2\tan^2\theta}{(R_L^2 + Z_0^2\tan^2\theta)^2} = 0 \tag{8.84}$$

For a quarter-wavelength transmission line when $k = 1$ and $\theta = \pi/2$, the susceptance compensation will be performed under the condition $Z_0 < R_L$ with the characteristic impedance Z_0 defined from a quadratic equation

$$Z_0^2 + 4\,\frac{QR_L}{\pi}Z_0 - R_L^2 = 0 \tag{8.85}$$

where $Q = \omega_0 C_p R$ and $R = Z_0^2/R_L$.

As a result, the required value of the characteristic impedance Z_0 is obtained by

$$Z_0 = R_L\left(-\frac{2Q}{\pi} + \sqrt{\left(\frac{2Q}{\pi}\right)^2 + 1}\right) \tag{8.86}$$

or

$$Z_0 = R\Big/\left(-\frac{2Q}{\pi} + \sqrt{\left(\frac{2Q}{\pi}\right)^2 + 1}\right) \tag{8.87}$$

By using the quarter- and half-wavelength transformers, the reactance- or susceptance-compensation load network generally can be realized differently for shunt and series equivalent output transistor circuits, as shown in Table 8.4 along with respective design equations [52, 53]. The two most important device parameters in the equations are the loaded quality factor Q and the real part R of the equivalent device output impedance. Depending on the values of the transmission-line characteristic impedances Z_1 and Z_2, each circuit provides either positive or negative parallel-resonant slope-reactance compensation.

Output Circuit Type	Matching Network	Design Equation
$Q = \omega_0 RC$ 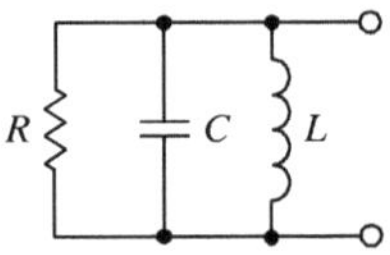	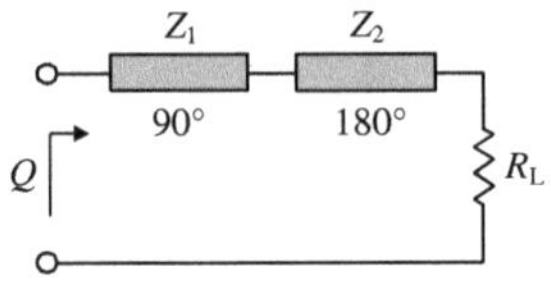	$Z_1 = \sqrt{RR_L}$ $T = \left(-\frac{2Q}{\pi} - \frac{1}{2}\right)\left(\frac{R_L}{Z_1} - \frac{Z_1}{R_L}\right)$ $Z_2 = -\frac{TR_L}{2} + \sqrt{\left(\frac{TR_L}{2}\right)^2 + R_L^2}$
	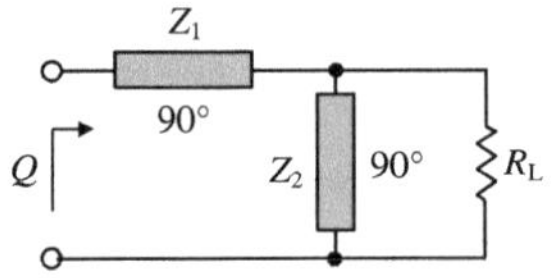	$Z_1 = \sqrt{RR_L}$ $Z_2 = \frac{\pi Z_1 R_L^2}{\pi Z_1^2 - \pi R_L^2 - 4QZ_1R_L}$
$Q = 1/\omega_0 RC$ 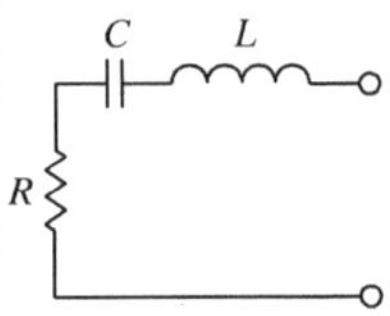	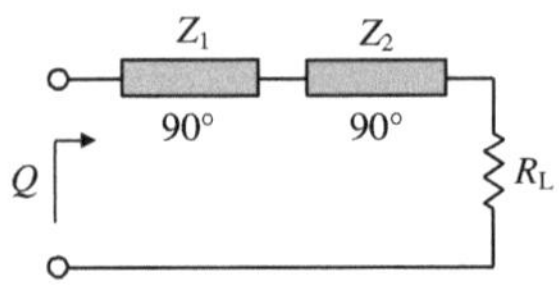	$Z_1 = AZ_2$ $Z_2 = BR_L$ $A = \sqrt{R/R_L}$ $B = \frac{A}{1+A}\left[-\frac{2Q}{\pi} + \sqrt{\left(\frac{2Q}{\pi}\right)^2 + \frac{(1+A)^2}{A}}\right]$
	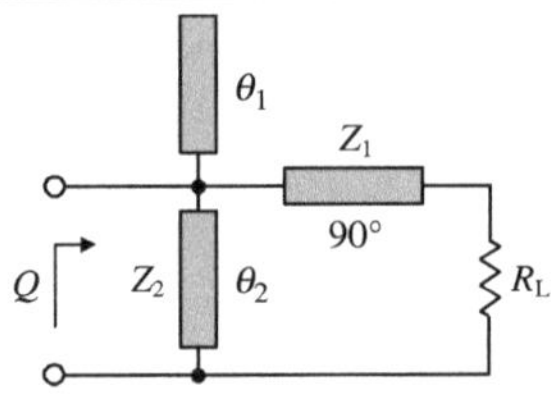	$Z_2 = \sqrt{RR_L}$ $Z_1 = \left(\frac{TZ_2}{R_L}\right)^2 \frac{1}{\left(\frac{R_L}{Z_2} - \frac{Z_2}{R_L}\right) + \frac{4Q}{\pi}}$ $T = \frac{2}{\pi \cos\theta_2}(\theta_1 + \theta_2)$ $\theta_1 + \theta_2 = \frac{n\pi}{2} \rightarrow n = 1 \text{ or } 3$

Table 8.4 Transmission-Line Reactance Compensation Circuits and Design Equations

Figure 8.39(*a*) shows the example of a single-susceptance compensation load network with a series quarterwave transmission line having a characteristic impedance of 61.2 Ω to match a 50-Ω real part of the device equivalent output admittance to a 75-Ω load and an electrical length of 90° at 50 MHz. The combination of the resistances of a shunt *LC* circuit (curve 1) and a series quarterwave transmission line (curve 2) provides minimum variations of the total resistance $\text{Re}Z_{net}$ shown in Fig. 8.39(*b*) by curve 3 around 50 Ω in a very wide frequency range. The susceptance $\text{Im}Y_{net}$ of a shunt circuit having a resonant frequency of 50 MHz varies with frequency with a positive slope, as shown in Fig. 8.39(*c*) by curve 1. The addition of a series quarter-wavelength transmission-line transformer between the shunt circuit and the load results in a negative slope providing by an additional susceptance, as

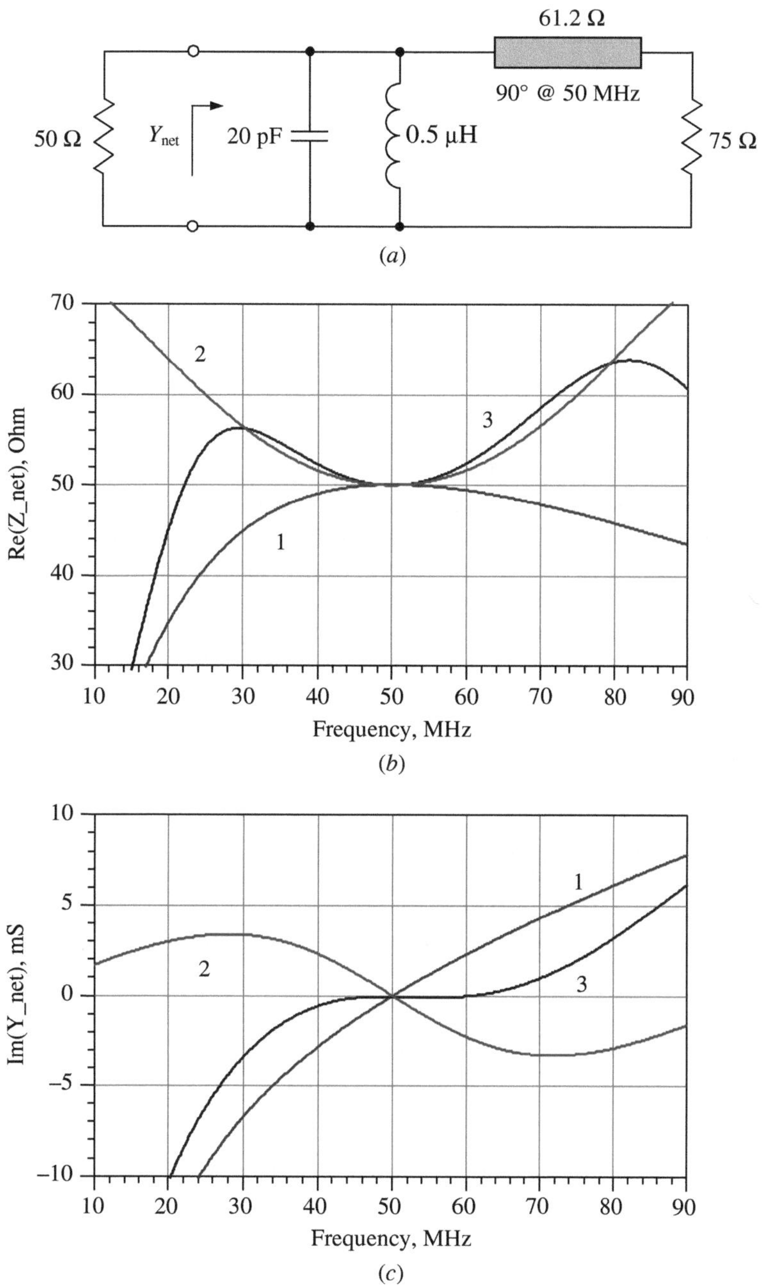

FIGURE 8.39 Susceptance compensation circuit with quarter-wavelength transmission line.

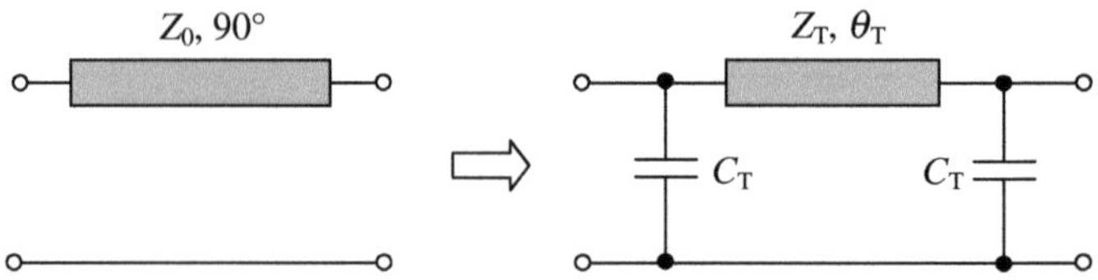

FIGURE 8.40 Transmission-line single-frequency equivalence technique.

shown in Fig. 8.39(*c*) by curve 2. Selection of the proper characteristic impedance of the series quarterwave transmission line and the load resistance enables the magnitude of two slopes to be made identical, so that the total susceptance slope around resonance is zero in a frequency range from 45 to 65 MHz, as shown in Fig. 8.39(*c*) by curve 3.

From Eq. (8.110), it follows that the maximum value of the characteristic impedance Z_0 is limited by the load resistance R_L, and its value in some cases, especially for high value of Q, can be substantially smaller than 50 Ω, which causes a problem in the practical implementation of a transmission line. In this case, it is best to apply a single-frequency equivalence technique when a quarterwave transmission line can be replaced by a symmetrical π-type low-pass transmission-line section with two equal shunt capacitances at a frequency ω_0, as shown in Fig. 8.40.

The transmission A-matrix (or $ABCD$-matrix) for a quarterwave transmission line can be written as

$$A_{90^\circ} = \begin{bmatrix} \cos 90^\circ & jZ_0 \sin 90^\circ \\ j\dfrac{\sin 90^\circ}{Z_0} & \cos 90^\circ \end{bmatrix} = \begin{bmatrix} 0 & jZ_0 \\ j\dfrac{1}{Z_0} & 0 \end{bmatrix} \tag{8.88}$$

whereas, for a π-type low-pass transmission-line section, we can write

$$A_\pi = \begin{bmatrix} 1 & 0 \\ j\omega C_T & 1 \end{bmatrix} \begin{bmatrix} \cos\theta_T & jZ_T \sin\theta_T \\ j\dfrac{\sin\theta_T}{Z_T} & \cos\theta_T \end{bmatrix} \begin{bmatrix} 1 & 0 \\ j\omega C_T & 1 \end{bmatrix}$$

$$= \begin{bmatrix} \cos\theta_T - \omega C_T Z_T \sin\theta_T & jZ_T \sin\theta_T \\ \dfrac{j}{Z_T}\left(2Z_T\omega C_T \cos\theta_T + \sin\theta_T - Z_T^2\omega^2 C_T^2 \sin\theta_T\right) & \cos\theta_T - \omega C_T Z_T \sin\theta_T \end{bmatrix} \tag{8.89}$$

Hence, equating A and B elements from each matrix yields

$$Z_T = \frac{Z_0}{\sin\theta_T} \tag{8.90}$$

$$C_T = \frac{\cos\theta_T}{\omega Z_0} \tag{8.91}$$

As a result, the electrical length of the transmission line can be reduced significantly with the increase in its characteristic impedance. Also, such a transformation is very important when the value of the device output capacitance exceeds the required optimum value for the optimum Class-E operation. In this case, the excess capacitance can be used as a part or entire shunt capacitance in the π-type low-pass section, and the optimum switching Class-E conditions will be completely satisfied at the fundamental frequency.

8.6.2 Broadband Class E with Shunt Capacitance

In the basic circuit of a Class-E power amplifier with shunt capacitance shown in Fig. 8.41(*a*), the harmonic impedance of the series fundamentally tuned L_0C_0 circuit is assumed to be high due to its high loaded quality factor. The value of the shunt capacitor C must also be correct to produce the correct voltage when the switch is turned off to satisfy the steady-state switching conditions. In this case, the load phase angle of the series-tuned circuit composed of the total inductor

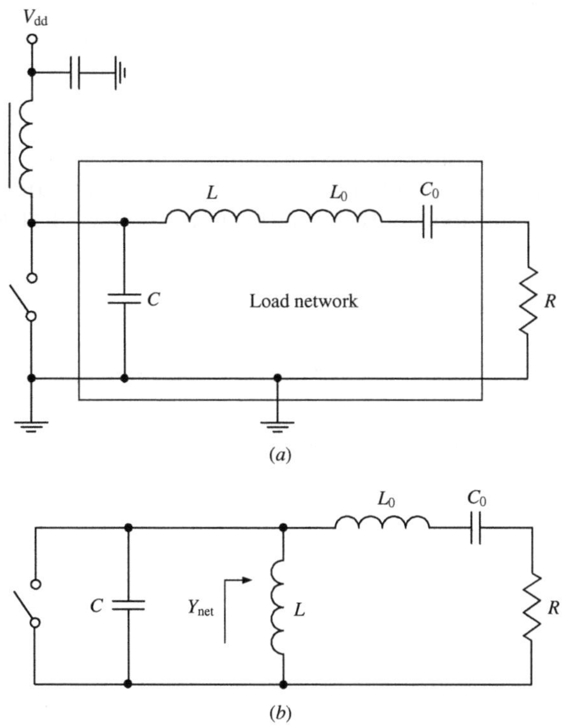

FIGURE 8.41 Load networks of Class-E with shunt capacitance.

$(L + L_0)$ and capacitor C_0, which determines the optimum angle for producing the correct voltage waveform, can be obtained according to Eq. (7.99) given in Chap. 7 at the resonant radian frequency $\omega_0 = 1/\sqrt{L_0 C_0}$ as

$$\theta = \tan^{-1}\left(\frac{\omega_0 L}{R}\right) = \tan^{-1} 1.1525 = 49.052^\circ \tag{8.92}$$

If the load network is designed without incorporating the shunt capacitance, a simple broadband network with an optimum load angle $\theta = 49.052^\circ$ given in Eq. (8.92) can be designed. Then, this phase angle reduces to the required angle $\phi = 35.945^\circ$ given by Eq. (7.102) in Chap. 7 when a shunt capacitance is added. The circuit schematic of a simple load network capable of presenting a constant load angle over a very large bandwidth is shown in Fig. 8.41(*b*) [46]. The load network consists of a low-Q series L_0C_0 circuit connected in parallel with an inductance L that allows a constant susceptance to be maintained over a wide bandwidth. The frequency behavior of the conductance $\mathrm{Re}Y_{net}$ and susceptance $\mathrm{Im}Y_{net}$ of this load network with parameters L = 42 nH, L_0 = 30 nH, and C_0 = 35 pF are shown in Figs. 8.42(*a*) and 8.42(*b*), respectively, where combination of the susceptance of the series resonant circuit with negative slope (curve 1) and the susceptance of the shunt inductance with positive slope (curve 2) provides a constant total susceptance over a very wide frequency range (curve 3).

In order to maintain the load angle constant in a wide frequency range, the slope of the susceptance provided by the inductance L should be cancelled by the slope provided by the resonant L_0C_0 circuit. The load-network admittance of Fig. 8.41(*b*) can be written as

$$Y_{net} = -\frac{j}{\omega L} + \frac{1}{R + j\left(\omega L_0 - \dfrac{1}{\omega C_0}\right)} \tag{8.93}$$

which reduces at the resonant frequency to

$$Y_{net} = \frac{1}{R} - \frac{j}{\omega L} \tag{8.94}$$

For slope cancellation, it is necessary to apply a zero-derivative condition of Eq. (8.67) to Eq. (8.93). As a result,

$$\frac{1}{\omega_0^2 L} = \frac{2}{\omega_0^2 C_0 R^2} \tag{8.95}$$

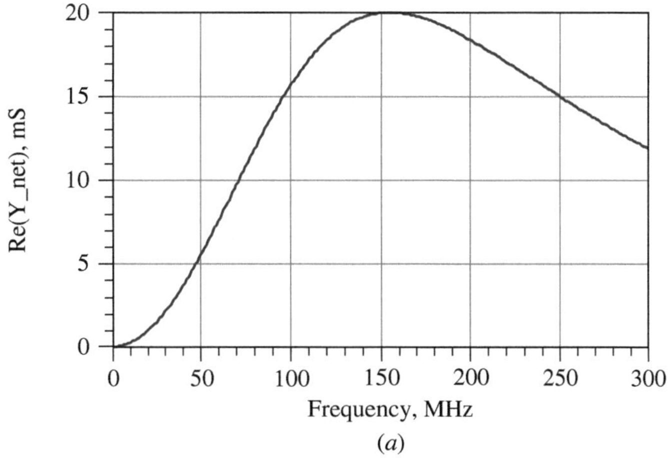

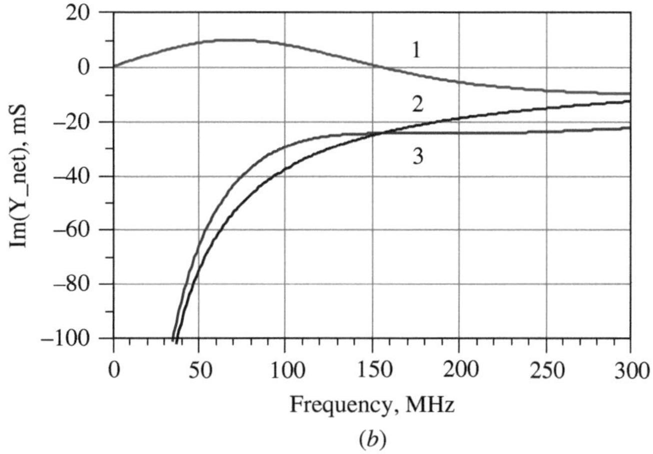

FIGURE 8.42 Conductance and susceptance of broadband Class-E circuit.

Thus, the design equations to calculate the parameters of a broadband Class-E load network providing maximum flatness can be calculated from

$$L = \frac{R}{\omega_0 \tan\theta} \tag{8.96}$$

$$C_0 = \frac{2L}{R^2} \tag{8.97}$$

$$L_0 = \frac{1}{\omega_0^2 C_0} \tag{8.98}$$

To reduce the output power at the harmonics, such a simple load network can be combined with a broadband matching network and a bandpass filter. As an example, a complete circuit based on a low-pass L-type matching section and a third-order Chebyshev bandpass filter, as shown in Fig. 8.43(*a*), was designed to deliver 12 W into a 50-Ω load across the frequency bandwidth from 130 to 180 MHz using a 12-V power supply [46]. From Fig. 8.43(*b*), it follows that this load network presents a constant magnitude of input impedance of 12 Ω (curve 1) and a load phase angle of around 36° (curve 2) over the required wide frequency range. As a result, the broadband MOSFET Class-E power amplifier was capable to provide a fairly constant efficiency at approximately 60% with suppression of the second, third, and fourth harmonics better than 45 dB below fundamental. The drain efficiency

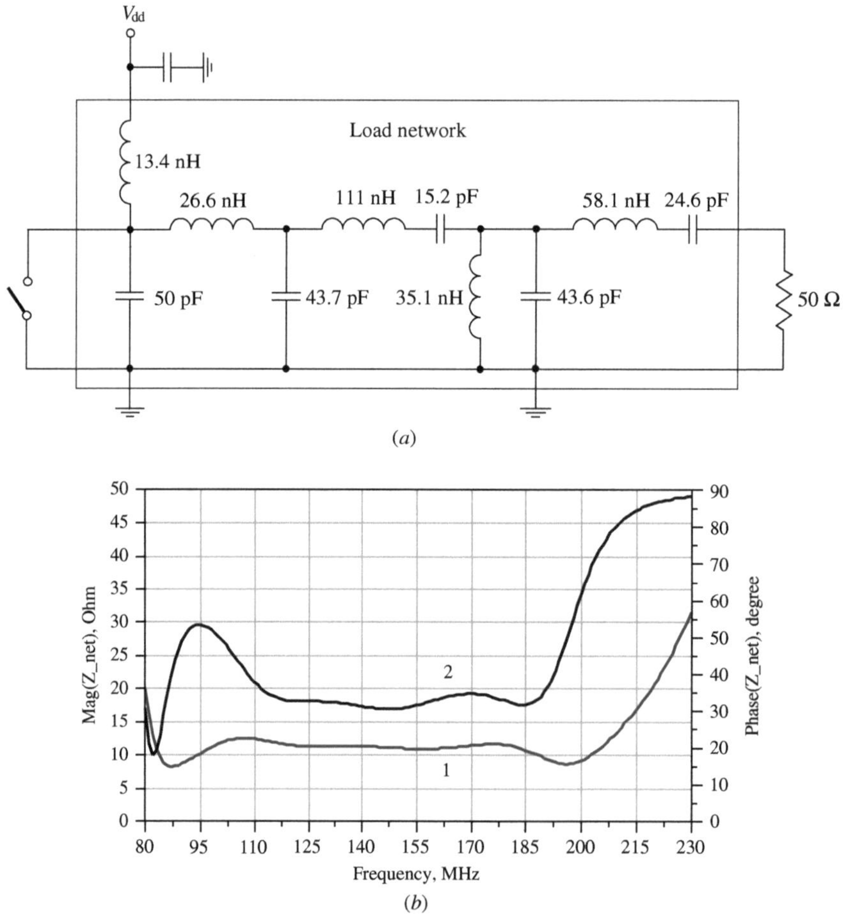

FIGURE 8.43 Broadband Class-E load network with bandpass filter and impedance.

of a GaN HEMT power amplifier with a Butterworth bandpass filter in the load network can be increased to greater than 80% in a frequency bandwidth from 600 to 800 MHz with an output power greater than 45 dBm [54, 55]. To provide the frequency bandwidth of 30% around the center bandwidth frequency of 1 GHz, the load network can be composed of a series transmission line and a shunt open-circuit stub [56].

Figure 8.44(*a*) shows the example of a reactance-compensation load network for Class-E power amplifier with shunt capacitance including a series transmission line and a parallel resonant circuit. In this case, the reactance of a Class-E load network with shunt capacitance and series inductance varies similar to that of the series resonant circuit with positive slope, whereas the required negative slope is provided by the parallel resonant circuit. Selection of the proper characteristic impedance and electrical length of the series transmission line enables the magnitude of two slopes to be made identical, so as to achieve a constant total real part and phase of the load network impedance Z_{net} over a wide frequency range. The simulation results at the fundamental frequency show that the resistance ReZ_{net} varies

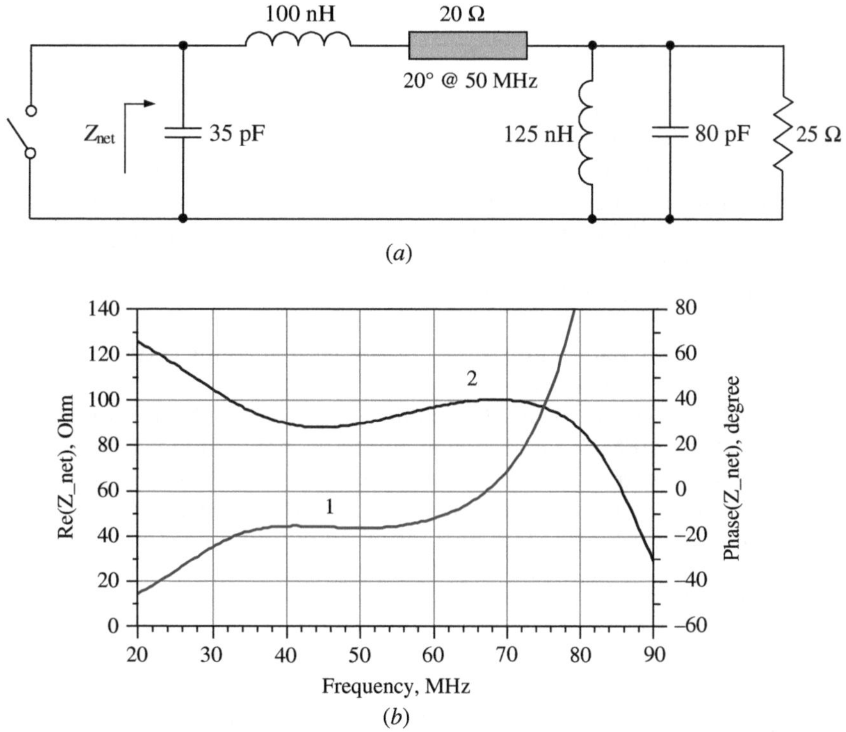

Figure 8.44 Class-E reactance compensation circuit with lumped elements and transmission line.

from 35 Ω at 30 MHz to 68 Ω at 70 MHz, as shown in Fig. 8.44(*b*) by curve 1, whereas the load-network phase varies between 27 and 40° in more than octave bandwidth from 33 to 80 MHz (curve 2), which corresponds to a constant inductive impedance at the fundamental across this frequency range. It is also very important for high-efficiency Class-E approximation that the reactances at the second- and higher-order harmonics will be capacitive, although they are not properly optimized over the entire frequency bandwidth.

Figure 8.45(*a*) shows the idealized simulation setup of a 10-W 28-V broadband Class-E power-amplifier circuit designed to operate over a frequency bandwidth from 1.7 to 2.7 GHz and based on a GaN HEMT CGH40010 device, where both the input matching circuit and load network are composed of ideal transmission lines. To provide an input broadband matching, it is possible to use a multisection matching transformer consisting of the stepped transmission-line sections with different characteristic impedances and electrical lengths. Such an input matching structure is convenient in practical implementation because there is no need to use any tuning capacitors. The nominal Class-E load resistance can be calculated for $P_{out} = 15$ W, $V_{dd} = 28$ V, and $V_{sat} = 2.5$ V according to Eq. (7.101) given in Chap. 7 as

$$R = 0.5768\,\frac{(V_{dd} - V_{sat})^2}{P_{out}} = 25\ \Omega \tag{8.99}$$

where P_{out} is the output power at the fundamental frequency, V_{dd} is the drain supply voltage, and V_{sat} is the saturation voltage defined from the device output current-voltage characteristics. In this case, the parallel resonant circuit in the broadband Class-E load network connected in parallel to a 25-Ω load is represented by the open- and short-circuit stubs, each having a characteristic impedance of 50 Ω and electrical length of 45° at 2 GHz. Simulation results show that drain efficiencies of 75% or greater can be achieved over whole required frequency bandwidth with a power gain of about 11 dB and an output power greater than 42 dBm.

Figure 8.45(*b*) shows the implementation of an idealized circuit of a broadband GaN HEMT Class-E power amplifier shown in Fig. 8.45(*a*) into a RO4360 substrate, where an additional series transmission line with low characteristic impedance is used to match an idealized 25-Ω load with a standard 50-Ω load. As a result, an output power around 42 dBm with a power gain of more than 10 dB was simulated for an input power of 31 dBm, as shown in Fig. 8.46(*a*). In this case, the drain efficiency over 72% was achieved across the required frequency range from 1.7 to 2.7 GHz, as shown in Fig. 8.46(*b*). Previously, a *PAE* above 60% was achieved between 1.87 and 2.11 GHz with an output power varying from 20 to 23 dBm for a medium-power broadband pHEMT Class-E power amplifier using a transmission-line parallel resonant circuit with short- and open-circuit stubs [57]. For a harmonically tuned

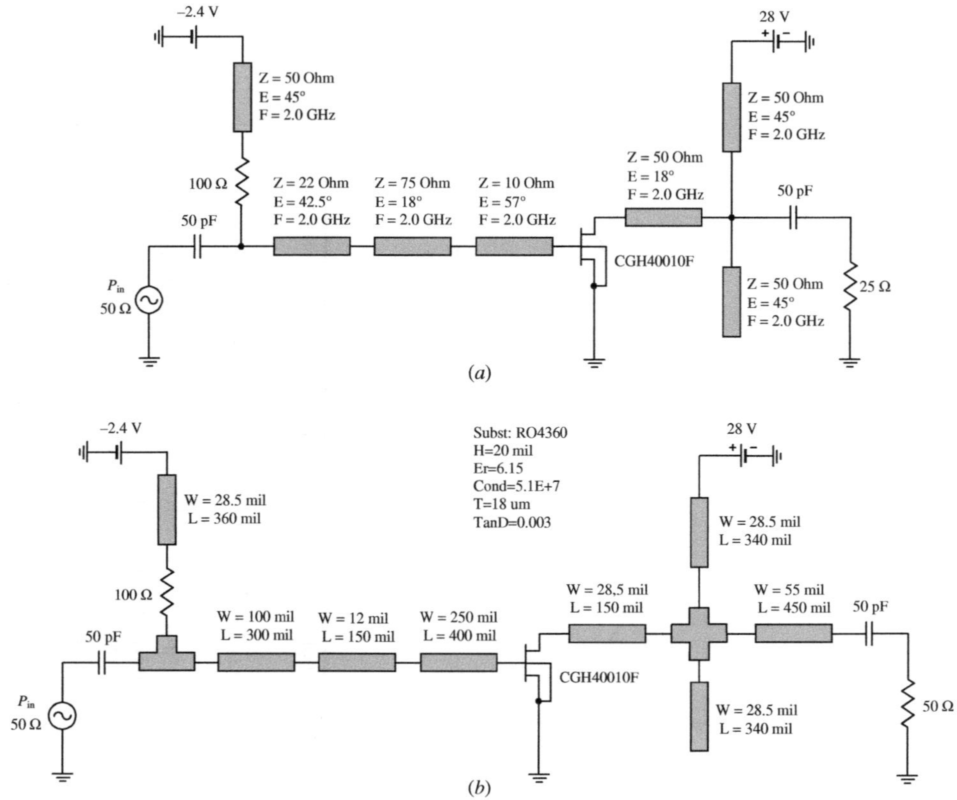

FIGURE 8.45 Circuit schematics of broadband GaN HEMT Class-E power amplifier.

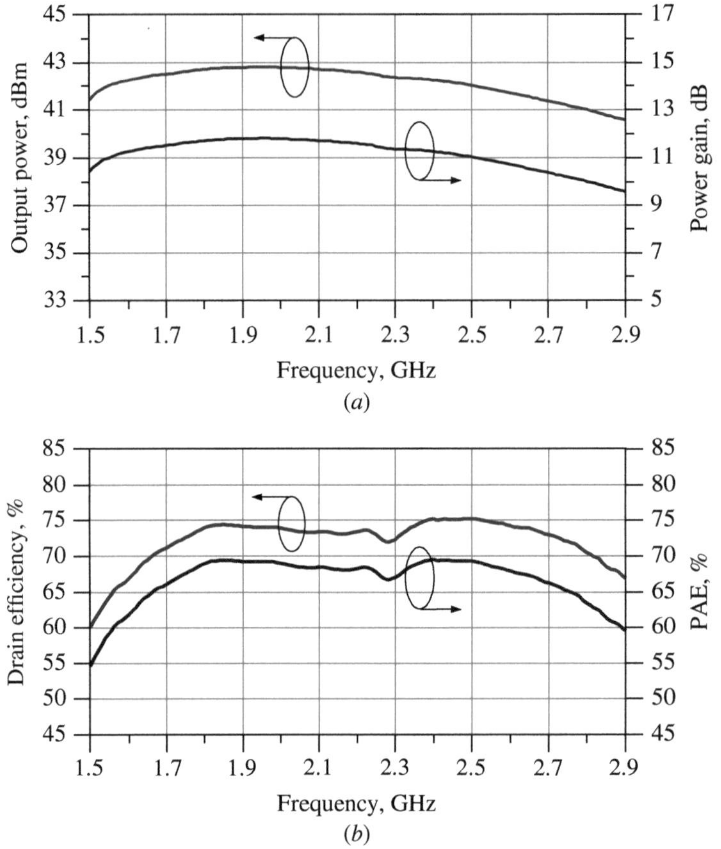

Figure 8.46 Output power, power gain, and efficiency versus frequency.

GaN HEMT broadband power amplifier using open-circuit stubs in the load network incorporating a three-section bandpass filter, an output power of around 100 W with a drain efficiency of greater than 65% in a frequency bandwidth from 1.55 to 2.25 GHz was achieved [58].

Ideally, the requirements to the output matching network for a broadband Class-E power amplifier should include not only achieving the inductive fundamental-frequency impedance across the desired bandwidth, but also necessarily providing high reactance at harmonics. In this case, to provide an impedance matching with high transformation ratio and satisfy Class-E requirements over octave bandwidth with minimum in-band ripple, at least three stages for the low-pass ladder-type matching network are needed. Here, the series inductor as a first matching element can provide high-impedance condition at harmonics and the device output capacitance should provide the required capacitive harmonic reactances. By using a three-stage six-order low-pass filter-matching load network in a GaN HEMT

Class-E power amplifier where the series inductors are replaced by the short-length high-impedance transmission-line sections and the shunt capacitors are replaced by the open-circuit low-impedance stubs, a drain efficiency of 63 to 89% with an output power of 10 to 20 W and a power gain of 10 to 13 dB was measured in a frequency bandwidth from 0.9 to 2.2 GHz at a supply voltage of 26 V [59].

8.6.3 Broadband Parallel-Circuit Class E

The susceptance compensation technique can be directly applied to the switchmode parallel-circuit Class-E power amplifier because its load-network configuration has an exactly the same structure with shunt and series resonant circuits, as shown in Fig. 8.47(*a*) [60, 61]. In this case, the nominal load resistance R and phase angle ϕ of the parallel-circuit Class-E load network can be obtained from Eqs. (7.144) and (7.148) in Chap. 7, respectively. The parallel inductance L and shunt capacitance C required for an idealized optimum parallel-circuit Class-E operation are calculated as functions of the load resistance R at the operating frequency from Eqs. (7.145) and (7.146) in Chap. 7, respectively. The parameters of the series resonant L_0C_0 circuit must be chosen to provide a constant phase angle of the load network over a required wide frequency bandwidth.

As a result, by substituting Eqs. (7.145) and (7.146) into Eq. (8.69), the series capacitance C_0 and inductance L_0 can be calculated at the center bandwidth frequency ω_0 by

$$L_0 = 1.026\frac{R}{\omega_0} \tag{8.100}$$

$$C_0 = \frac{1}{\omega_0^2 L_0} \tag{8.101}$$

Wider frequency bandwidth with high-efficiency performance can be achieved using a double-susceptance compensation circuit shown in Fig. 8.47(*b*), where L_0C_0 and L_1C_1 are the series and parallel resonant circuits, respectively [62]. In this case, similarly to the broadband design in a Class-E mode with shunt capacitance using a double-susceptance compensation, the parameters of the series and shunt resonant circuits for the broadband design in a parallel-circuit Class-E mode with the corresponding loaded quality factors $Q_0 = \omega_0 L_0/R$ and $Q_1 = \omega_0 C_1 R$, which are close to unity and greater, can approximately be calculated from Eqs. (8.102) and (8.103), where the load angle $\theta = \tan^{-1}(R/\omega_0 L)$ using Eq. (7.145) from Chap. 7 is taken into account. Such a load network can be considered as a broadband matching-forming circuit that provides simultaneously the Class-E switching conditions and matching with a standard 50-Ω load over wide frequency bandwidth [63].

The circuit simulations for these two types of susceptance compensation load networks were performed at a center bandwidth frequency

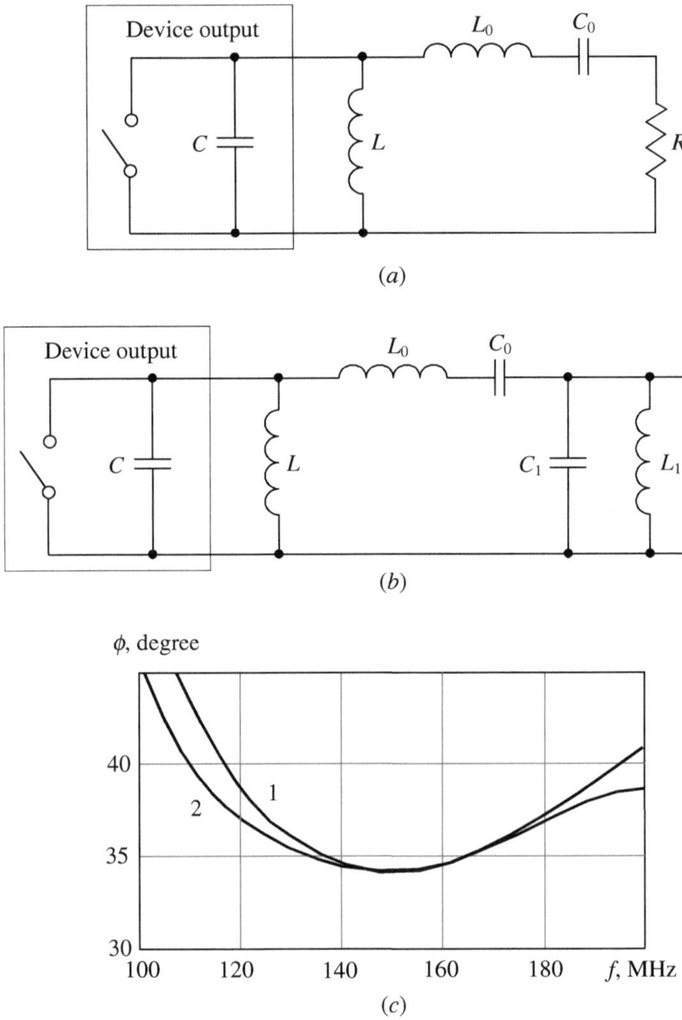

FIGURE 8.47 Single- and double-susceptance compensation circuits.

f_0 = 150 MHz for a standard load resistance R = 50 Ω. Figure 8.47(*c*) shows the frequency dependences of the load-network phase angle ϕ for the single- (curve 1) and double-susceptance (curve 2) compensation circuits, demonstrating their very broadband operation capability. Using just a single-susceptance load network yields a significant widening of the operating frequency bandwidth with a minimum deviation of the magnitude and phase of the load-network impedance. A double-susceptance compensation load network obtains a maximum deviation from the optimum value of about 34° by only 3° in a frequency range from 120 to 180 MHz.

To achieve the high-efficiency broadband operation mode with a high-power gain in VHF frequency band, it is best to design the power amplifier based on silicon LDMOSFET devices. It is easy to provide a very broadband input matching using lossy-matching circuit, especially at operating frequencies about 10 times lower than the device transition frequency f_T. Figure 8.48 shows the circuit schematic of an LDMOSFET power amplifier designed for operation in a 2:1 frequency bandwidth from 100 to 200 MHz using a double-susceptance compensation load network with broadband matching properties at the fundamental frequency. The input lossy-matching circuit includes a simple *L*-transformer connected in parallel with a series circuit consisting of an inductor of 20 nH and a resistor of 50 Ω. This provides a minimum input return loss at 200 MHz of about 15 dB and an input *VSWR* less than 1.4 over the entire frequency bandwidth from 100 to 200 MHz.

From Fig. 8.49(*a*), it follows that for such an octave-band VHF Class-E power amplifier with an input power of 1 W using a 1.25-μm LDMOSFET device with a total gate width of 28 × 1.44 mm, a power gain of 10 dB with deviation of only ±0.5 dB (curve 2) can be achieved with a drain efficiency of about 70% or higher (curve 1). An analysis of the simulated drain voltage and current waveforms at the center bandwidth frequency of 150 MHz shown in Fig. 8.49(*b*) demonstrates that the broadband operating mode is very close to a nominal parallel-circuit Class-E operation mode, although the impedance conditions at higher harmonics are not controlled properly. As seen from the plots when the transistor is turned on, high values of drain current (up to 1.3 A) are

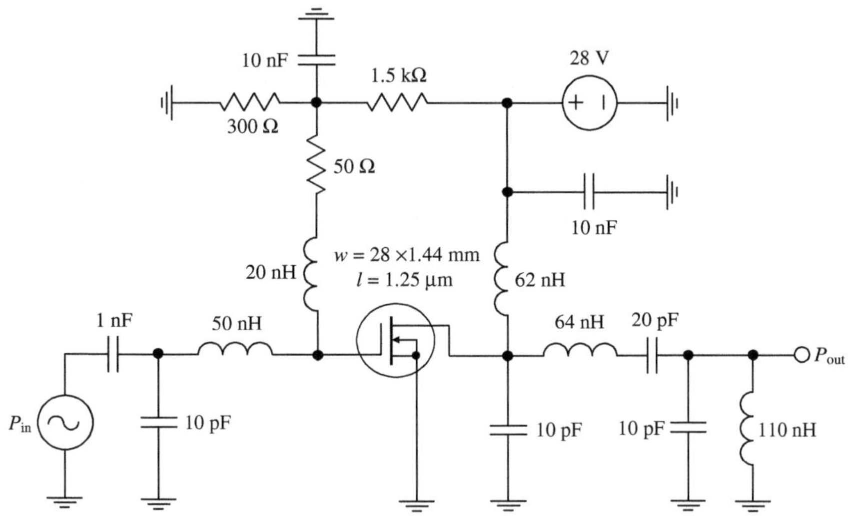

FIGURE 8.48 Simulated broadband Class-E LDMOSFET power amplifier.

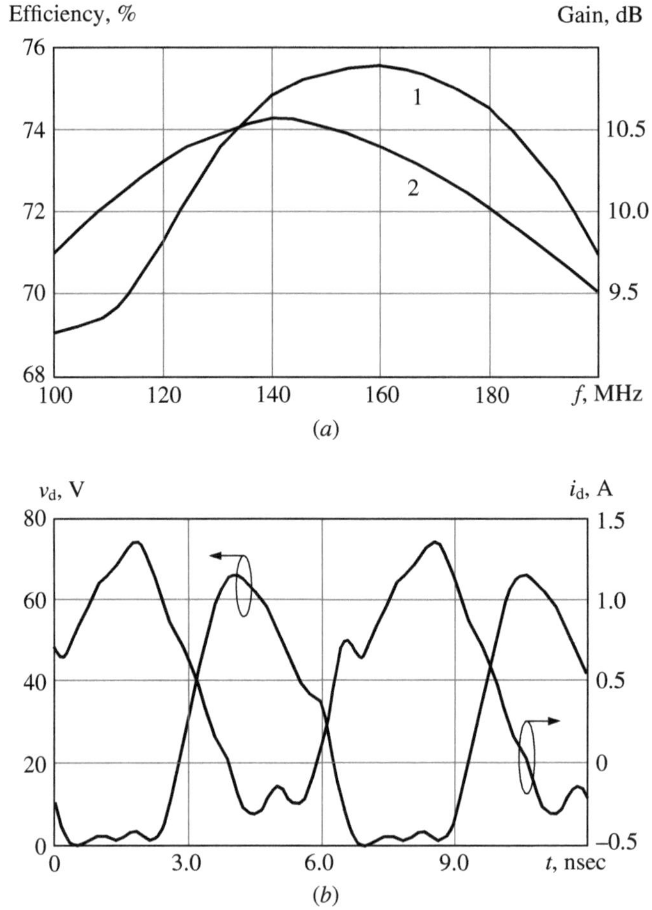

FIGURE 8.49 Broadband performance of Class-E LDMOSFET power amplifier.

achieved with small saturation voltages of 0 to 4 V. On the other hand, when the transistor is turned off, the drain current continues to flow, but now through the device gate-drain capacitance C_{gd} and drain-source capacitance C_{ds} but not through the active channel. A drain efficiency of 74% with an output power of 8 W across the frequency range from 136 to 174 MHz with a power flatness of 0.7 dB was measured for a parallel-circuit Class-E LDMOSFET power amplifier with a low supply voltage of 7.2 V [64]. A power-added efficiency can be increased to 80% or greater in a frequency range of 140 to 180 MHz with an output power of 34.4 ± 1.5 dBm using a GaN HEMT device [65].

Similarly, the transmission-line susceptance compensation technique can also be applied to a parallel-circuit Class-E power amplifier where the series transmission line of a quarter wavelength at the center

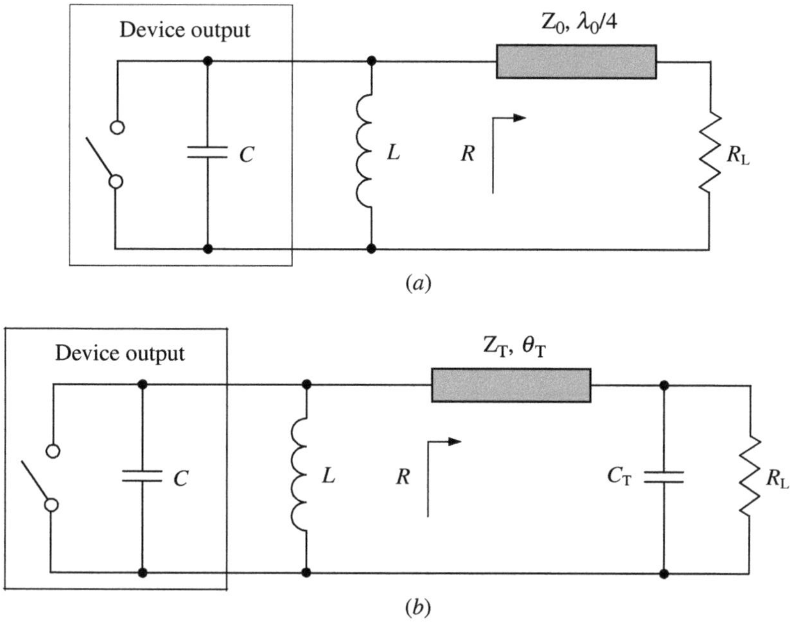

FIGURE 8.50 Transmission-line susceptance compensation circuit.

bandwidth frequency can be used instead of a series L_0C_0 resonant circuit, as shown in Fig. 8.50(*a*). In some practical cases, the series quarterwave line can be replaced by an equivalent low-pass π-type circuit consisting of a series transmission line with higher characteristic impedance and electrical length much less than 90° and two shunt capacitors when the capacitance adjacent to the device output can be counted within the total shunt capacitance required for a nominal parallel-circuit Class-E mode. If it is necessary to additionally provide an output matching between the nominal Class-E resistance R and standard load $R_L = 50\ \Omega$, a series quarterwave line can be replaced by a low-pass L-type matching circuit with a series transmission line and a shunt capacitor, as shown in Fig. 8.50(*b*).

Figure 8.51(*a*) shows the example of a transmission-line broadband Class-E load network, where the parallel inductor is replaced by a short-length short-circuited transmission, line, which can be easily implemented on printed-circuit board to minimize insertion losses. The electrical lengths of the transmission lines are given at the center bandwidth frequency of 300 MHz. In this case, the input load-network resistance varies from 17 Ω at 225 MHz to 47.5 Ω at 400 MHz, as shown in Fig. 8.51(*b*) by curve 1, with much less variation from 18.5 to 27 Ω in a frequency range from 250 to 350 Ω. The phase stays almost constant around 33° in a frequency range from 250 to 350 Ω and varies from 22.5° at 225 MHz to 39.5° at 400 MHz (curve 2).

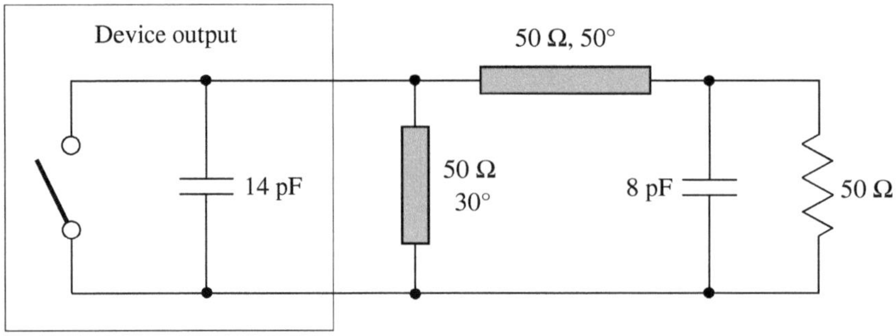

(*b*)

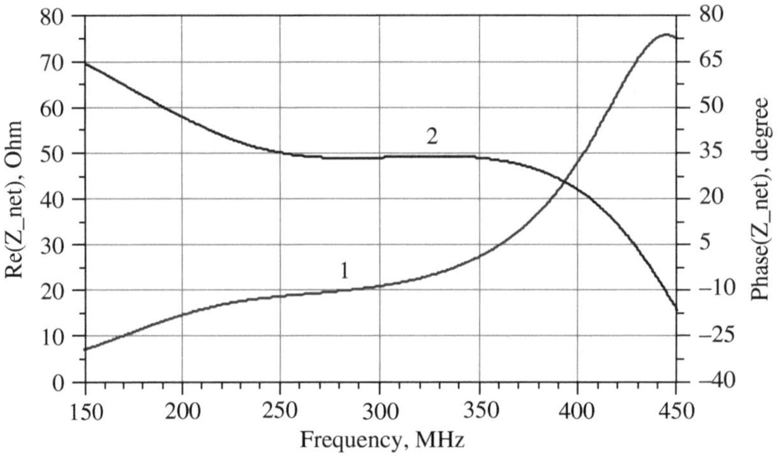

(*b*)

FIGURE 8.51 Transmission-line Class-E load network with susceptance compensation.

Figure 8.52(*a*) shows the circuit schematic of a broadband high-efficiency microstrip LDMOSFET power amplifier with an output power of around 20 W and a power gain of greater than 12 dB in a frequency range from 225 to 400 MHz at a dc-supply voltage of 28 V. Here, to approximate the parallel-circuit Class-E mode in a wide frequency range, the load network was designed to realize a single-susceptance compensation technique using a parallel short-length transmission line in conjunction with a single *L*-type transmission-line transformer, because a ratio between the device equivalent output resistance required for an optimum Class-E operation and the standard load of 50 Ω is not significant. The input matching circuit includes the two low-pass *L*-type matching sections to compensate for the device input capacitance over the entire frequency range. A lossy parallel resistance of 75 Ω is necessary to simplify the

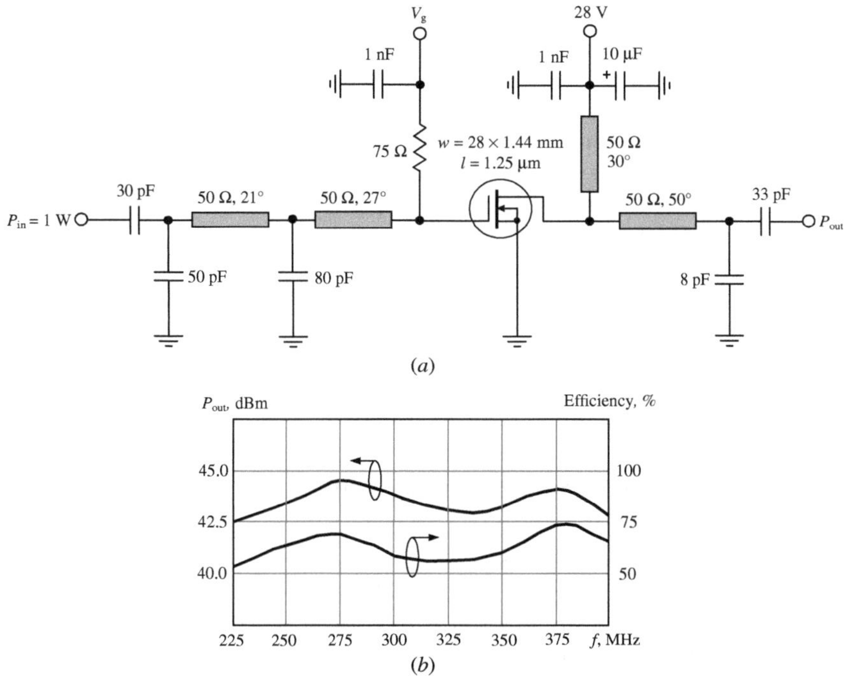

Figure 8.52 Broadband high-efficiency microstrip LDMOSFET power amplifier.

matching procedure and improve the input return loss. As a first step, each matching network structure is calculated at the center-band frequency based on the technical requirements and device equivalent circuit parameters. Then, to optimize the power amplifier performance over the entire frequency band, the simplest and fastest way is to apply an optimization procedure using computer simulators to satisfy certain criteria. For such a broadband power amplifier, the minimum output power ripple and input return loss with maximum power gain and efficiency can be chosen as the criteria. Generally, by applying a nonlinear broadband optimization technique and setting the ranges of electrical length of the transmission lines between 0 and 90° and parallel capacitances from 0 to 100 pF, we can obtain the parameters of the input matching circuit and output load network.

However, to speed up this procedure, it is best to optimize circuit parameters separately for the input and output circuits. In this case, the input matching circuit is loaded by the device equivalent input series *RC* circuit, consisting of its gate resistance and gate-source capacitance. The load network must include at its input the device equivalent output shunt *RC* circuit consisting of an optimum Class-E load resistance required for a specified output power and supply voltage and drain-source capacitance. In this case, it is sufficient to use a fast linear optimization process, which will take only a few minutes to complete the circuit design procedure. Finally, the resulting optimized values are incorporated into the overall power amplifier circuit for each element and final optimization is performed using a nonlinear active device model. The optimization process is finalized by choosing the nominal level of input power with optimizing elements in narrower ranges of their values of about 10 to 20% for most critical elements. For practical convenience, it is advisable to choose the characteristic impedances of all transmission lines of 50 Ω. Figure 8.52(*b*) shows the simulated broadband high-efficiency power amplifier performance achieving an output power of 42.5 to 44.5 dBm, a power gain of 13.5 ± 1 dB, and a drain efficiency of 64 ± 10% in a frequency bandwidth from 225 to 400 MHz.

8.6.4 Monolithic Microwave Broadband Class-E Power Amplifiers

Generally, by providing an open-circuit termination for the second- and third-harmonic components, the collector efficiency of a microwave Class-E power amplifier can be increased by 10% [66]. In this case, the second-harmonic termination has the most impact on the collector efficiency, whereas effect of an open-circuit termination for the fourth harmonic is negligible. Moreover, the variation of the second-harmonic load reflection coefficient by 10% in magnitude from 1 to 0.9 and ±20° in phase angle results in an insignificant efficiency variation within 1% only. Figure 8.53(*a*) shows the circuit schematic of a monolithic broadband Class-E power amplifier with a chip size of 2 mm × 2.2 mm. The load network is a compromise solution between having a low insertion loss and meeting the necessary requirements for the optimum Class-E operation with nonzero voltage and voltage-derivative switching conditions [67]. This is accomplished by using two open-circuit stubs in conjunction with a shunt capacitor, where the first open-circuit stub in combination with the series transmission line presents broadband high-impedance terminations for the second harmonics within 18 to 22 GHz, whereas the combination of the second open-circuit stub and the shunt capacitor presents broadband low impedances at the third harmonics within 27 to 33 GHz and also transforms the optimum load impedances at the fundamental frequencies. The simulated loading conditions presented to the output of the device at the fundamental (inductive impedance), second- (high impedance), and third-harmonic (low impedance) frequencies are shown in Fig. 8.53(*b*), which are proved to be adequate for broadband Class-E power amplifiers. As a result, using an indium phosphide (InP) double HBT (DHBT) technology, a *PAE* of 49 to 65% with an output power of 18 to 22 dBm was achieved over the frequency bandwidth from 9 to 11 GHz [66]. Based on an InP

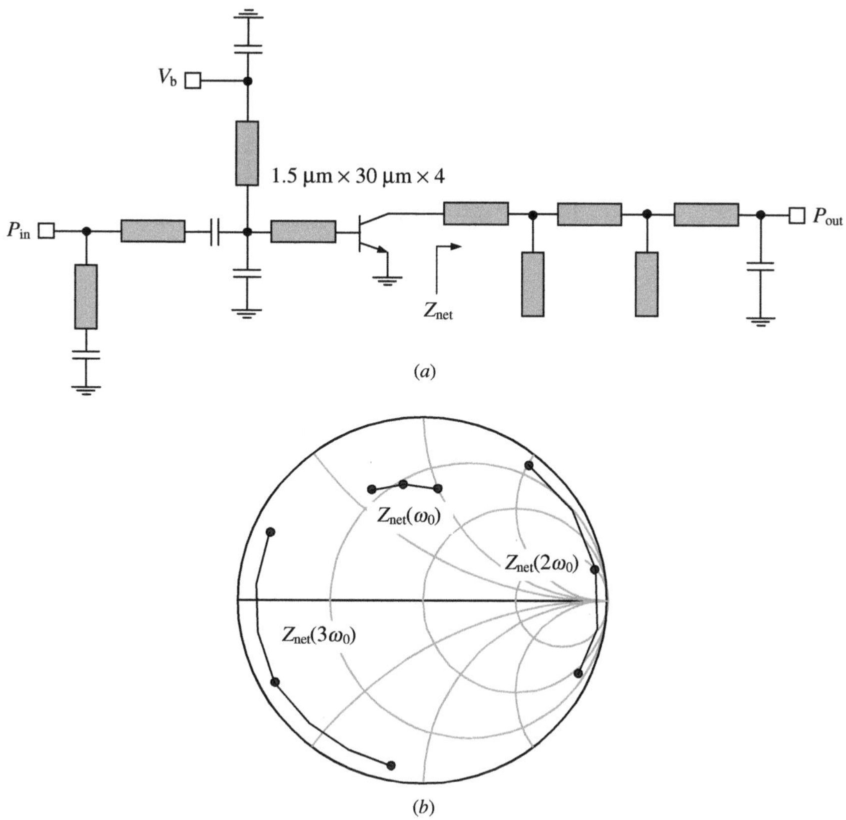

FIGURE 8.53 Broadband *X*-band In DHBT Class-E power amplifier and impedance conditions.

DHBT technology, a single-stage broadband *X*-band Class-E power amplifier can also achieve a *PAE* of 45 to 60% with an output power of 19 to 21.5 dBm and a power gain of 9 to 11.5 dB over a 34% bandwidth, from 8.2 to 11.6 GHz [68].

To increase the overall efficiency of a two-stage power amplifier, it may be assumed that it is worthwhile to optimize both amplifying stages to operate in a Class-E mode. For example, for a hybrid microwave GaAs MESFET Class-E power amplifier using the same devices in both stages, the maximum two-stage power-added efficiency was achieved as high as 52% (including connector loss) with a corresponding power gain of 16 dB and an output power of 20 dBm at a carrier frequency of 10 GHz and a supply voltage of 4.2 V [69]. However, because of Class-E operation mode of the driver stage, the overall power gain is sufficiently small, thus affecting the overall efficiency. Therefore, by using a Class-AB driver stage, similar efficiency can be achieved with substantially higher power gain. As a result, for

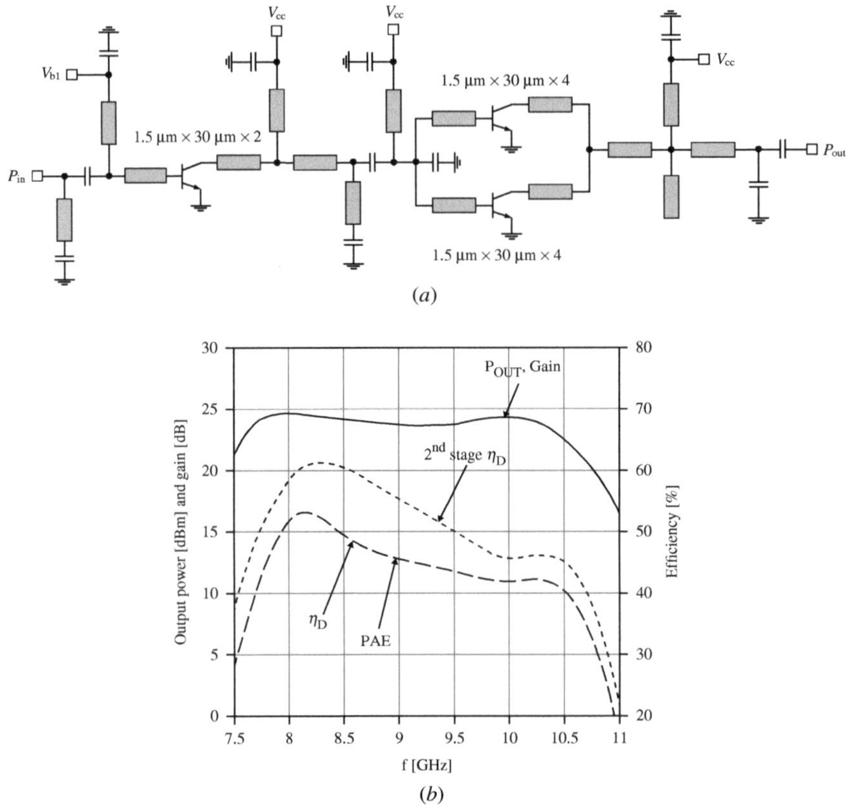

FIGURE 8.54 Broadband two-stage *X*-band In DHBT Class-E power amplifier and its performance.

a monolithic microwave two-stage high-efficiency InP DHBT power amplifier shown in Fig. 8.54(*a*), where the driver stage is operated in a Class-AB mode and the output stage is operated in a Class-E mode, a *PAE* of 52% with an output power of 24.6 dBm and a power gain of 24.6 dB was achieved at a carrier frequency of 8 GHz and a supply voltage of 4 V. The total emitter area of the driver-stage device was chosen to be 90 µm^2, providing a *PAE* of the driver stage above 40% and an adequate power to push the output stage deep into compression, as required for a switchmode Class-E operation. The output stage consists of two active devices with a total emitter area of 360 µm^2 combined in parallel reactively, taking care to provide odd-mode instability suppression resistors between the base and collector of each transistor. The power-added efficiency is maintained greater than 40% over a frequency bandwidth from 7.7 to 10.5 GHz, as shown in Fig. 8.54(*b*) [69].

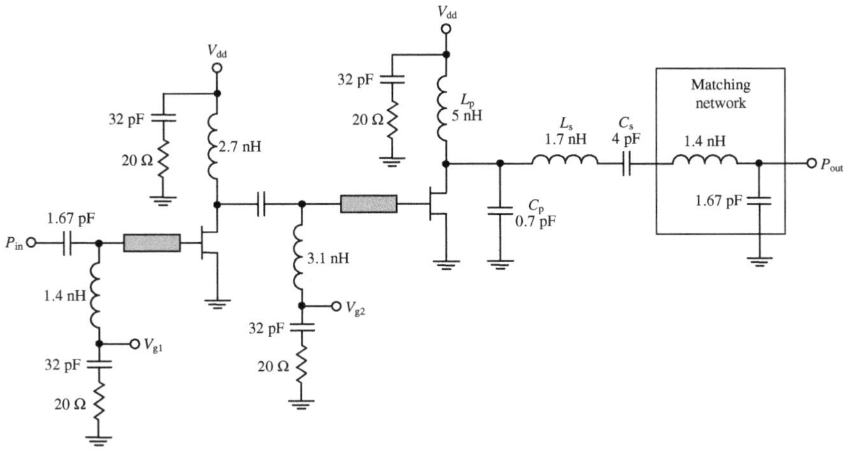

FIGURE 8.55 Circuit schematic of broadband Class-E pHEMT power amplifier.

Figure 8.55 shows the circuit schematic of a two-stage broadband Class-E power amplifier implemented in a 0.5-μm enhancement/depletion pHEMT process with a chip size of 2 × 2 mm², which is intended to operate in a frequency range from 1.5 to 3.8 GHz with a *PAE* better than 62% and an output power of more than 27 dBm at $V_{dd} = 6$ V [70]. In this case, to provide high operation efficiency in a wide frequency range, the Class-E load network with reactance compensation technique followed by the low-pass matching network is used. The driver stage is designed to operate in a Class-AB mode with a small quiescent current for high gain and high efficiency when both input and interstage matching circuits are conjugately matched. For a 0.5-μm pHEMT two-stage broadband Class-E power amplifier with a chip size of 5.25 × 2.8 mm², a *PAE* above 50% with an output power over 36 dBm at a drain supply voltage of 6 V was obtained in a frequency range of 3.0 to 3.75 GHz [71].

8.6.5 Broadband CMOS Class-E Power Amplifiers

The use of cascode topologies is extremely attractive for CMOS power amplifiers, especially at high output powers and dc supply voltages. Optimizing the cascode topology requires setting the bias voltage V_g of the common-gate transistor shown in Fig. 8.56(*a*) to minimize the voltage drop across the oxide of each transistor M_1 and M_2 when these voltage drops become equal allowing the use of approximately twice the supply voltage [72, 73]. However, there is an additional power loss mechanism as a specific property of a cascode configuration in a switching Class-E mode when the common-source device M_1 is turned off, which is associated with charging and discharging processes of the

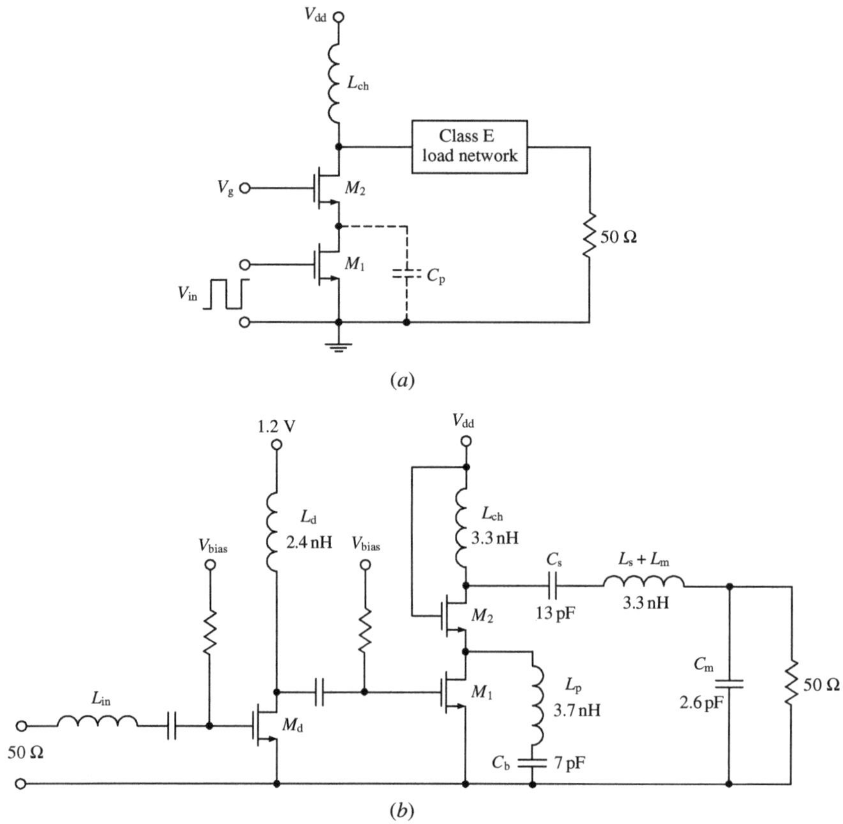

FIGURE 8.56 Broadband cascode Class-E power amplifier with compensating inductor.

shunt parasitic capacitor C_p consisting of the drain-bulk capacitance of the device M_1 and gate-source and source-bulk capacitances of the device M_2. This results in a finite switching time of a common gate device M_2 when it cannot be instantly switched from the saturation mode to the pinch-off mode, and operates in the active region when simultaneously output current and output voltage are positive with the output power dissipation within the device. The parasitic capacitance C_p can be three to four times larger than the drain-bulk capacitance of the device M_2 resulting in a power loss as large as 20% of the output power. A simple and effective way to minimize this power loss contribution is to use a parallel inductor L_p resonating the parasitic capacitor C_p at the operating frequency, as shown in Fig. 8.56(*b*), where C_b is the blocking capacitor. The series resonant circuit required to provide a sinusoidal current flowing to the load is replaced by the series inductor L_m and shunt capacitor C_m forming an *L*-type lumped transformer to match the optimum Class-E load resistance with a standard load of 50 Ω. As a result, the two-stage cascode Class-E power amplifier with a compensating inductor implemented in a 0.13-μm CMOS process achieved a drain efficiency of 71% and a *PAE* of 67% when delivering an output power of 23 dBm at an operating frequency of 1.7 GHz with a supply voltage of 2.5 V. The driving stage with a supply voltage of 1.2 V is biased in a Class C. The value of an inductor L_d is chosen to compensate for the gate-source capacitance of the device M_1. The power-added efficiency higher than 60% was measured over the frequency bandwidth of 1.4 to 2.0 GHz.

To realize a broadband high-efficiency operation of the fully integrated CMOS Class-E power amplifier, a broadband and low-loss 1:4 Ruthroff-type transmission-line transformer based on the broadside-coupled transmission lines can provide an impedance transformation from 12.2 ± 0.1 to 50 Ω [74]. In a six-layer 0.18-μm CMOS process, the thickest top metal 6 is used as the primary winding, the identical thick metal stacked from metal 1 to 4 is used as the secondary winding to improve insertion loss, and both windings are wound in loops keeping the 1:1 turns ratio to reduce the transformer size. Figure 8.57(*a*) shows the circuit schematic of a 0.18-μm CMOS Class-E power amplifier composed of the two nMOS transistors in a cascode configuration and one shunt capacitor in the load network required for optimum Class-E operation. Here, the series *LC* resonant circuit at the fundamental frequency of the Class-E power amplifier is replaced by the 1:4 transmission-line transformer operating as a broadband bandpass filter. To enhance the reliability of the transistors, the thick-oxide transistor M_2 is used for the common-gate stage, and the thin-oxide transistor M_1 is used for the common-source stage. The fully integrated CMOS Class-E power amplifier with a 1:4 transmission-line transformer exhibits a broadband output power level of 24 ± 0.2 dBm from 2.4 to 3.5 GHz at a supply voltage of 3.6 V, with a maximum *PAE* of 33.2% at 2.6 GHz.

Figure 8.57(*b*) shows the circuit schematic of a two-stage broadband Class-E CMOS power amplifier, where the power output stage is formed by a high-voltage, extended-drain, thick-oxide nMOS device implemented in a standard 65-nm CMOS technology [75]. The total gate width of the transistor is 3.84 mm and the channel length is 0.28 μm, realizing an on-resistance of 0.7 Ω, an off-resistance of 10 kΩ, and a drain-source capacitance of approximately 4.14 pF. To drive the output stage as a switch, a square-wave signal is generated by an inverter-based driver implemented using standard thick-oxide MOS devices with a gate length of 0.28 μm. To reduce the peak drain voltage and improve reliable operation, a suboptimum Class-E operation is applied. The broadband load network represents an off-chip two-section *LC* ladder circuit. As a result, a measured output power of 30.5 ± 0.5 dBm, a power gain of 16.5 ± 0.5 dB, a drain efficiency above 67%, and a *PAE* above 52% are achieved across the frequency bandwidth from 550 to 1050 MHz.

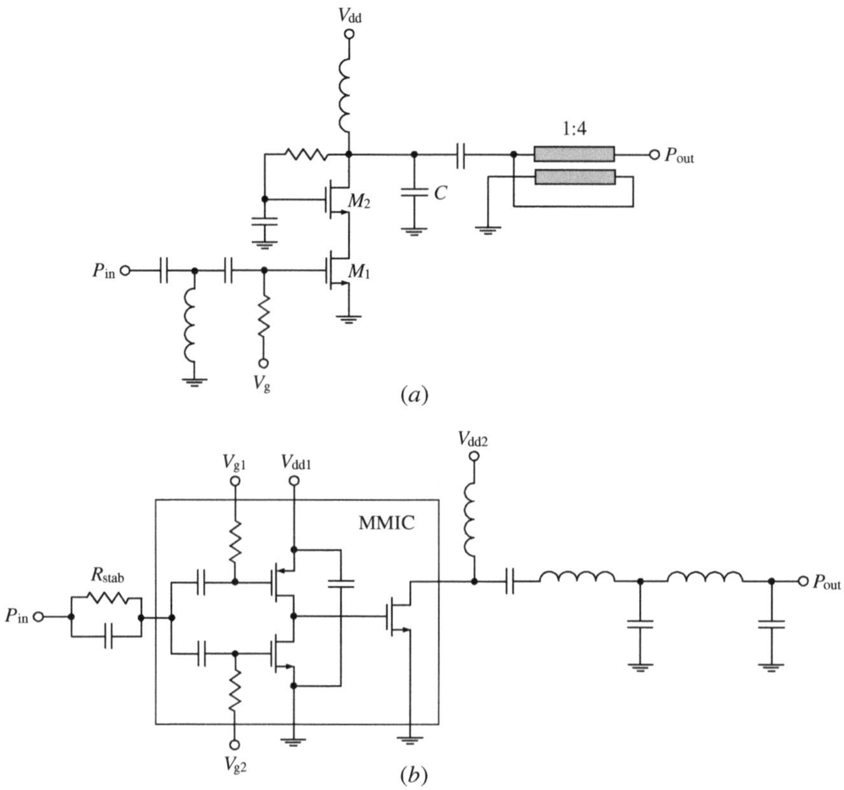

FIGURE 8.57 Circuit schematics of broadband Class-E CMOS power amplifiers.

8.7 Practical Broadband RF and Microwave Power Amplifiers

Multisection matching networks based on the low- and high-pass *L*-transformers for input and output matching circuits can provide a wide frequency bandwidth with minimum power gain ripple and significant harmonic suppression. Such a multisection matching circuit configuration using lumped elements was applied for the design of a 60-W power amplifier operating in the frequency bandwidth of 140 to 180 MHz. The complete circuit schematic of the power amplifier is shown in Fig. 8.58 [76]. To realize such technical requirements, an internally matched bipolar transistor for VHF applications, which provides a 100-W output power level at a supply voltage of 28 V, was used. According to the device data sheet, the input device impedance at the center bandwidth frequency $f_0 = \sqrt{140 \times 180} = 159$ MHz is equal to $Z_{in} = (0.9 + j1.8)\ \Omega$. Therefore, the input matching circuit was designed

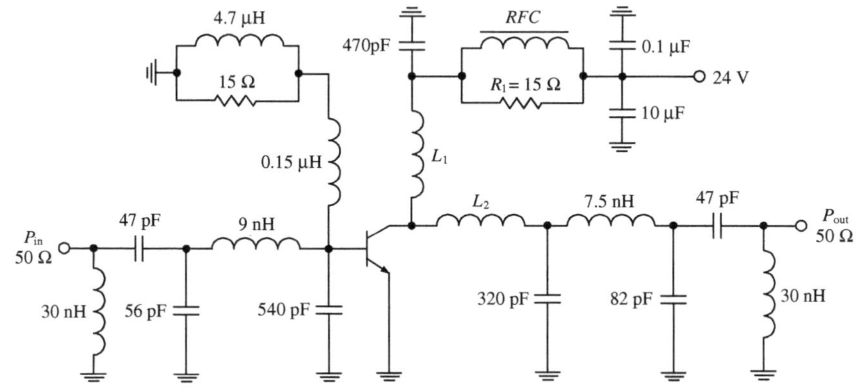

FIGURE 8.58 Circuit schematic of broadband high-power VHF bipolar amplifier.

as a three-section network with two low-pass sections and one high-pass section to minimize the circuit quality factor Q. In this case, the device input lead inductance of $1.8/(2\pi \times 0.159) = 1.8$ nH was considered as a series inductive element of the second low-pass section with a shunt capacitor of 540 pF. This power amplifier is operated in a Class C due to the base bias circuit composed of the two inductors and a 15-Ω resistor, which also provides low-frequency stability.

A similar design philosophy was used to design the output matching circuit when the three-section network maintains a value of the quality factor close to unity or within $Q = 1$ circle on a Smith chart. The output device impedance is practically resistive of 1.65 Ω because the output device capacitive reactance is compensated by the device lead inductance. The series inductance L_2 of the first matching low-pass section adjacent to the collector terminal according to the Smith chart can be realized as a section of a 50-Ω microstrip line with the electrical length of $0.011\lambda_0$, where λ_0 is the wavelength corresponding to the center bandwidth frequency f_0. The physical length of this microstrip line for a 1/16-in Teflon fiberglass with a dielectric permittivity $\varepsilon_r = 2.55$ must be of 0.51 in, whereas its width is equal to 0.4 in. The collector feed is provided through the combination of an inductor L_1, a resistor $R_1 = 15$ Ω, and an RF choke (*RFC*), which behaves as a high-impedance circuit at the operating frequencies but offers a very low resistance at dc. As a result, the designed broadband power amplifier achieved a power gain of at least 8 dB with a gain ripple of less than 3 dB, a collector efficiency of greater than 50%, and an input *VSWR* below 3:1 [76]. As an alternative, the broadband input and output matching circuits can be composed of a single low-pass matching section followed by a 4:1 transmission-line transformer each. In this case, an output power of more than 25 W with the collector efficiency close to 70% was achieved across the frequency range of 118 to 136 MHz for the input power of 2 W using a 12.5-V bipolar device [77].

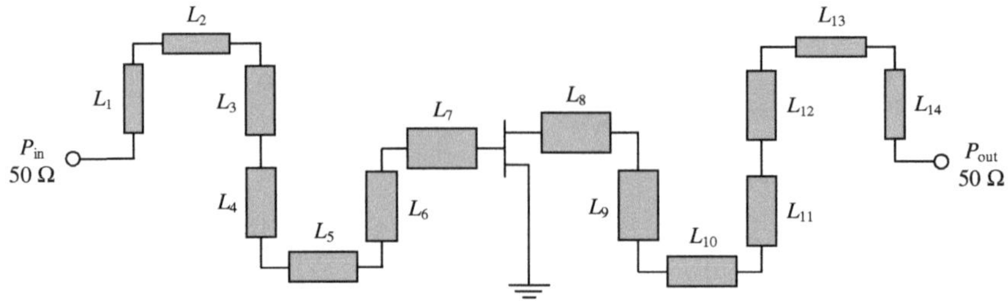

FIGURE 8.59 Microstrip broadband 15-W GaAs MESFET power amplifier.

At microwave frequencies, the amplifier bandwidth performance can also be improved by using an increased number of the transmission-line transformer sections. For example, with the use of a multisection transformer with seven quarterwave transmission lines of different characteristic impedances, a power gain of 9 ± 1 dB and a *PAE* of 37.5 ± 7.5% over 5 to 10 GHz were achieved for a 15-W GaAs MESFET power amplifier [78]. The simplified schematic diagram of this microwave octave-band power amplifier is shown in Fig. 8.59. To achieve minimum output power flatness, the number of sections of the output matching circuit is determined based on load-pull measurements. At the same time, the number of sections of the input matching circuit to compensate for the frequency-dependent power gain is chosen based on the small-signal *S*-parameter measurements. For a 5.25-mm GaAs MESFET device, the values of the input and output impedances at the fundamental derived from its large-signal model were assumed resistive and equal to $Z_{in} = 0.075\ \Omega$ and $Z_{out} = 1.32\ \Omega$, respectively. To achieve minimum gain flatness, the length of each microstrip section initially was chosen as a quarter-wavelength at the highest frequency of 10 GHz. However, because the input and output device impedances are not purely resistive in practical implementation, the final optimized length of each microstrip section was reduced to be a quarter-wavelength at around 15 GHz. The microstrip transformer sections $L_1 \ldots L_6$ and $L_{10} \ldots L_{14}$ were fabricated on alumina substrate with a dielectric permittivity $\varepsilon_r = 9.8$ and a thickness of 0.635 mm for L_1 and L_2, 0.2 mm for $L_3 \ldots L_6$ and $L_{10} \ldots L_{12}$, and 0.38 mm for $L_{13} \ldots L_{14}$. The microstrip section L_7 was realized on a high-dielectric substrate with $\varepsilon_\tau = 38$ and thickness of 0.18 mm, whereas the microstrip sections L_8 and L_9 were fabricated on a high-dielectric substrate with $\varepsilon_r = 89$ and thickness of 0.15 mm. The final power amplifier represents a balanced configuration of the two 5.25-mm GaAs MESFETs with hybrid quadrature couplers.

The circuit schematic of a high-power amplifier intended for applications in TV transmitters based on a balanced bipolar transistor BLV861 is shown in Fig. 8.60 [79]. In a Class-AB operation with a quiescent current of 100 mA, it covers the frequency bandwidth

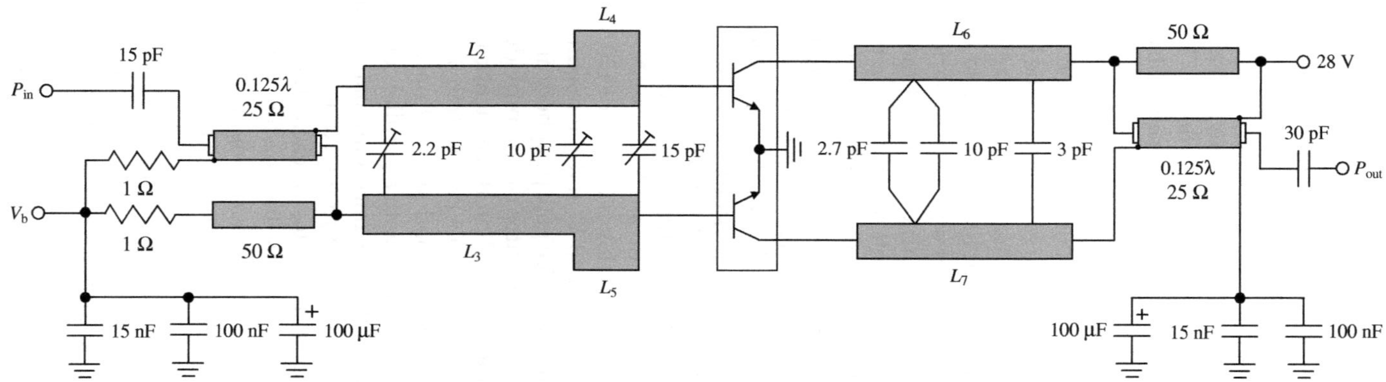

FIGURE 8.60 Bipolar high-power UHF amplifier for TV transmitters.

of 470 to 860 MHz with an output power of 100 W, a power gain of about 9.5 dB with a gain ripple of ± 0.5 dB, and a collector efficiency of 55%. The nominal device input and load impedances at 663 MHz are equal to $Z_{in} = (4.4 + j7.9)\ \Omega$ and $Z_L = (8.8 - j3.65)\ \Omega$, respectively. In this case, the three-section input matching circuit and two-section output matching circuit contain mixed microstrip-lumped elements to transform each terminal impedance level to approximately 25 Ω. The balanced-to-unbalanced transformation to 50 Ω is obtained by the transmission-line baluns, each represented by a 25-Ω semi-rigid coaxial cable with an electrical length of 45° at the midband and a diameter of 1.8 mm, soldered over the whole length on top of the same length microstrip line. For low-frequency stability enhancement, the input balun stubs are connected to the bias point by means of 1-Ω series resistors. The large-value electrolytic capacitors are added at the input and output biasing points to improve the amplifier video response. The power amplifier is fabricated on a laminate substrate with $\varepsilon_r = 2.55$ and a thickness of 0.51 mm (20 mils).

Figure 8.61 shows the circuit schematic of a two-stage reactively matched GaN HEMT MMIC power amplifier, which operates as a driver amplifier of an ultra-wideband high-power transmit modules for multifunctional active electronically scanned antenna radar systems [80]. MMICs based on GaN HEMT technology can provide wider bandwidth, higher output power density, improved reliability at high junction temperature, better thermal properties, higher breakdown voltage, and higher operating efficiency compared to MMICs based on GaAs technology. For a 0.25-μm GaN HEMT technology using SiC substrate, the breakdown voltage of 120 V allows operation with a supply voltage up to 40 V, and the maximum output power density of 5.6 W/mm for device gate periphery and capacitance sheet of 250 pF/mm^2 for MIM capacitor can be provided. In this case, the MMIC driver amplifier is based on three identical GaN HEMT cells, each with 8 × 100-μm gate periphery (one transistor in the first stage and two transistors in the second stage), to achieve the maximum output power of about 36 dBm with a parallel connection of two second-stage amplifying paths. The unconditional stability of the MMIC driver amplifier from 100 MHz to 6 GHz is provided by applying parallel *RC* networks at the gates of each transistor cell. The integrated resistors are also used in the gate bias circuits of each device cell to ensure stability without sacrificing gain or efficiency. The dc-feed paths, which consist of narrow microstrip lines to provide the corresponding inductive reactances and bypass MIM capacitors to provide isolation between the dc and RF paths, are constituent parts of the input, interstage, and output matching circuits, which are realized in the form of low-pass *L*- and *T*-transformers with the series microstrip lines and shunt MIM capacitors. The matching networks provide impedance transformation with low *Q*-factors enabling an increased frequency bandwidth.

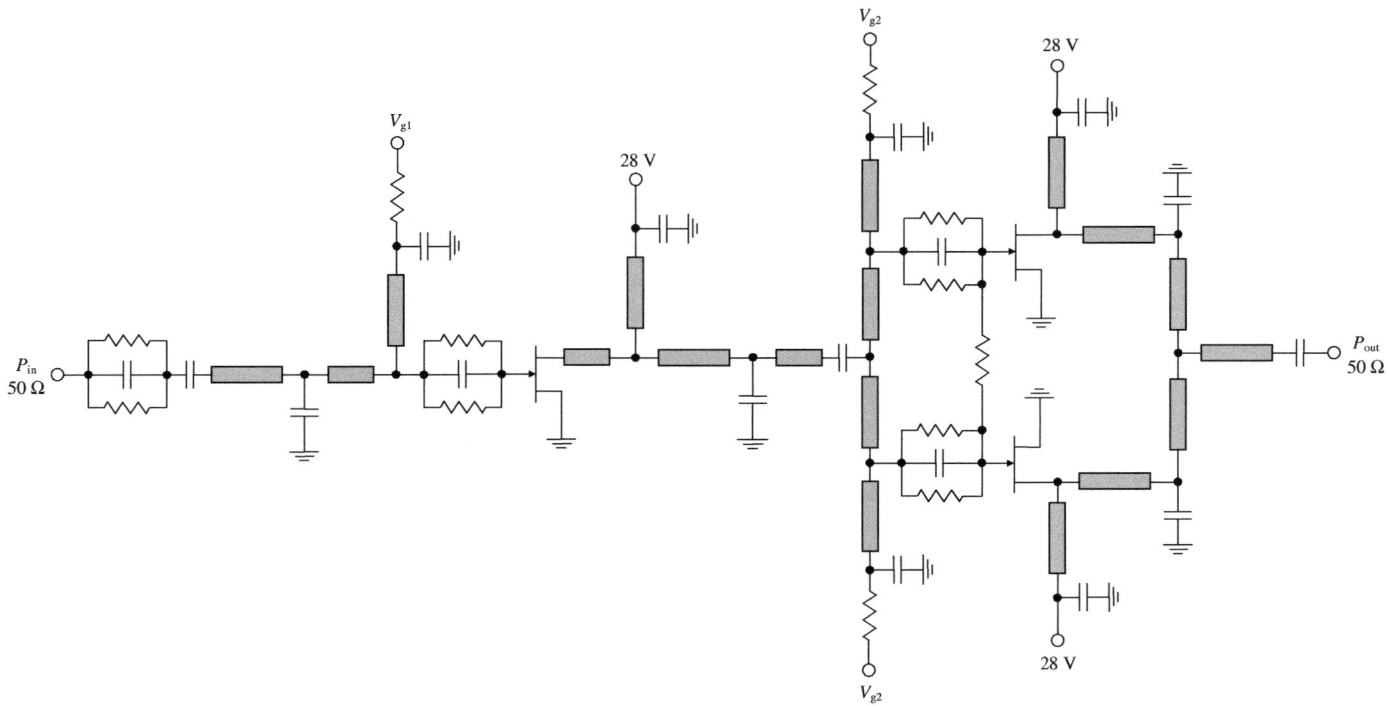

FIGURE 8.61 Circuit schematic of broadband GaN HEMT MMIC power amplifier.

Figure 8.62(*a*) shows a photograph of the reactively matched MMIC driver amplifier with a chip size of 3 × 4 mm². When driven with a 20-dBm input power, the maximum measured output power achieves 4 W with a typical output power of 1.8 W and a worst-case return loss of 7.5 dB in the frequency range from 6 to 18 GHz, as

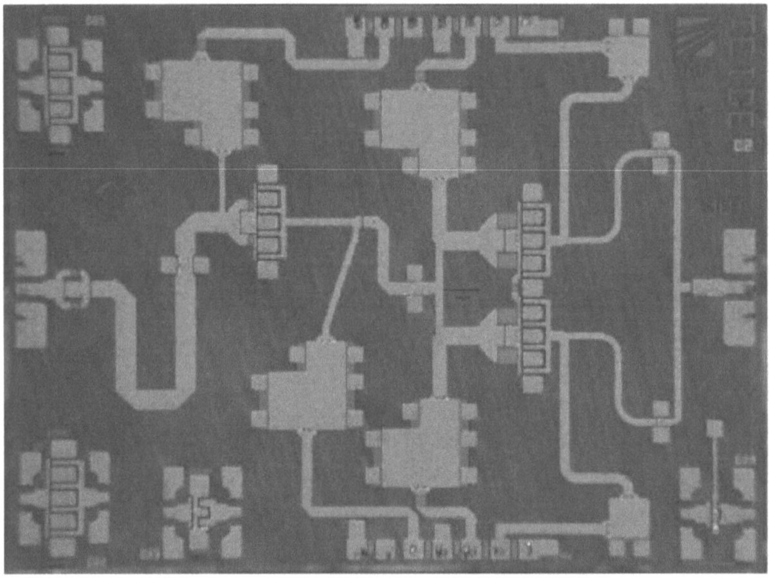

(*a*)

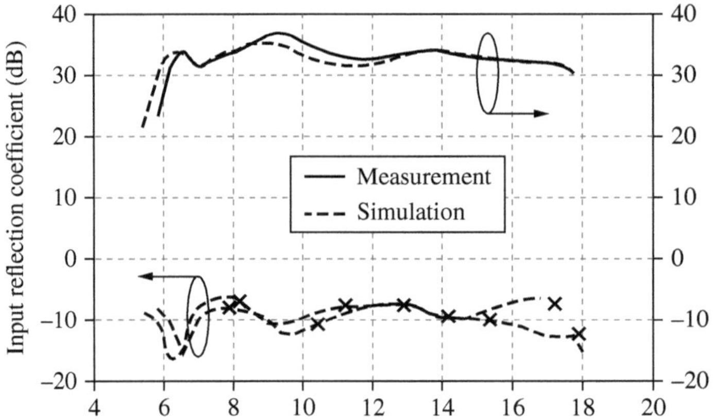

(*b*)

FIGURE 8.62 Circuit schematic of reactively matched two-stage broadband GaN HEMT MMIC power amplifier (*Courtesy of Cassidian*).

shown in Fig. 8.62(*b*). The output power of 10 W was achieved for a three-stage reactively matched GaN HEMT power amplifier, with four similar device cells in the final stage [80]. When driven with a 28-dBm input power, the maximum measured output power achieves 15.6 W with an average output power of 10.6 W and a worst-case return loss of 7.5 dB in the frequency range from 6.4 to 18.4 GHz. Over the complete frequency range from 6 to 18 GHz, average values for *PAE* of 18% are the same for simulations and measurements, with typical value of 20% across the frequency range from 6 to 12 GHz.

References

1. H. W. Bode, *Network Analysis and Feedback Amplifier Design*, New York: Van Nostrand, 1945.
2. R. M. Fano, "Theoretical Limitations on the Broad-Band Matching of Arbitrary Impedances," *J. Franklin Institute*, vol. 249, pp. 57–83, Jan. 1950, pp. 139–154, Feb. 1950.
3. R. Levy, "Explicit Formulas for Chebyshev Impedance-Matching Networks, Filters and Interstages," *Proc. IEE*, vol. 111, pp. 1099–1106, Jun. 1964.
4. D. E. Dawson , "Closed-Form Solutions for the Design of Optimum Matching Networks," *IEEE Trans. Microwave Theory Tech.*, vol. MTT-57, pp. 121–129, Jan. 2009.
5. G. L. Matthaei, "Tables of Chebyshev Impedance-Transforming Networks of Low-Pass Filter Form," *Proc. IEEE*, vol. 52, pp. 939–963, Aug. 1964.
6. G. L. Matthaei, L. Young, and E. M. T. Jones, *Microwave Filters, Impedance-Matching Networks, and Coupling Structures*, New York: Artech House, 1980.
7. O. A. Chelnokov (Ed.), *Radio Transmitter Devices* (in Russian), Moskva: Radio i Svyaz, 1982.
8. W. H. Ku, M. E. Mokari-Bolhassan, W. C. Petersen, A. F. Podell, and B. Kendall, "Microwave Octave-Band GaAs FET Amplifiers," *1975 IEEE MTT-S Int. Microwave Symp. Dig.*, pp. 69–72.
9. R. S. Pengelly and J. A. Turner, "Monolithic Broadband GaAs F.E.T. Amplifiers," *Electronics Lett.*, vol. 12, pp. 251–252, May 1976.
10. C. H. Lin, H. Z. Liu, C. K. Chu, H. K. Huang, C. C. Liu, C. H. Chang, C. L. Wu, et al., "A Compact 6.5-W PHEMT MMIC Power Amplifier for Ku-Band Applications," *IEEE Microwave and Wireless Components Lett.*, vol. 17, pp. 154–156, Feb. 2007.
11. P. C. Huang, Z. M. Tsai, K. Y. Lin, and H. Wang , "A 17-35 GHz Broadband, High Efficiency PHEMT Power Amplifier Using Synthesized Transformer Matching Technique," *IEEE Trans. Microwave Theory Tech.*, vol. MTT-60, pp. 112–119, Jan. 2012.
12. H. Wang, C. Sideris, and A. Hajimiri, "A CMOS Broadband Power Amplifier with a Transformer-Based High-Order Output Matching Network," *IEEE J. Solid-State Circuits*, vol. SC-45, pp. 2709–2722, Dec. 2010.
13. B. S. Yarman and A. Aksen, "An Integrated Design Tool to Construct Lossless Matching Networks with Mixed Lumped and Distributed Elements," *IEEE Trans. Circuits and Systems-I: Fundamental Theory Appl.*, vol. CAS-I-39, pp. 713–723, Sep. 1992.
14. P. I. Richards, "Resistor-Transmission Line Circuits," *Proc. IRE*, vol. 36, pp. 217–220, Feb. 1948.
15. R. Saal and E. Ulbrich, "On the Design of Filters by Synthesis," *IRE Trans Circuit Theory*, vol. CT-5, pp. 284–327, Dec. 1958.
16. P. I. Richards, "Applications of Matrix Algebra to Filter Theory," *Proc. IRE*, vol. 34, pp. 145–150, Mar. 1946.
17. H. Ozaki and J. Ishii, "Synthesis of Transmission-Line Networks and the Design of UHF Filters," *IRE Trans Circuit Theory*, vol. CT-2, pp. 325–336, Dec. 1955.

18. H. F. Cooke, "Microwave Transistors: Theory and Design," *Proc. IEEE*, vol. 59, pp. 1163–1181, Aug. 1971.
19. R. S. Tucker, "Gain-Bandwidth Limitations of Microwave Transistor Amplifiers," *IEEE Trans. Microwave Theory Tech.*, vol. MTT-21, pp. 322–327, May 1973.
20. C. A. Liechti and R. L. Tillman, "Design and Performance of Microwave Amplifiers with GaAs Schottky-Gate Field-Effect Transistors," *IEEE Trans. Microwave Theory Tech.*, vol. MTT-22, pp. 510–517, May 1974.
21. W. H. Ku and W. C. Petersen, "Optimum Gain-Bandwidth Limitations of Transistor Amplifiers as Reactively Constrained Active Two-Port Networks," *IEEE Trans. Circuits and Systems*, vol. CAS-22, pp. 523–533, Jun. 1975.
22. R. E. Neidert and H. A. Willing, "Wide-Band Gallium Arsenide Power MESFET Amplifiers," *IEEE Trans. Microwave Theory Tech.*, vol. MTT-24, pp. 342–350, Jun. 1976.
23. P. Saad, C, Fager, H. Cao, H. Zirath, and K. Andersson, "Design of a Highly Efficient 2-4-GHz Octave Bandwidth GaN-HEMT Power Amplifier," *IEEE Trans. Microwave Theory Tech.*, vol. MTT-58, pp. 1677–1685, Jul. 2010.
24. H. Q. Tserng, V. Sokolov, H. M. Macksey, and W. R. Wisseman, "Microwave Power GaAs FET Amplifiers," *IEEE Trans. Microwave Theory Tech.*, vol. MTT-24, pp. 936–943, Dec. 1976.
25. H. J. Riblet, "General Synthesis of Quarter-Wave Impedance Transformers," *IRE Trans. Microwave Theory Tech.*, vol. MTT-5, pp. 36–43, Apr. 1957.
26. S. B. Cohn, "Optimum Design of Stepped Transmission-Line Transformers," *IRE Trans. Microwave Theory Tech.*, vol. MTT-3, pp. 16–21, Apr. 1955.
27. V. P. Meschanov, I. A. Rasukova, and V. D. Tupikin, "Stepped Transformers on TEM-Transmission Lines," *IEEE Trans. Microwave Theory Tech.*, vol. MTT-44, pp. 793–798, Jun. 1996.
28. G. L. Matthaei, "Short-Step Chebyshev Impedance Transformers," *IEEE Trans. Microwave Theory Tech.*, vol. MTT-14, pp. 372–383, Aug. 1966.
29. H. W. Bode, "Attenuation Equalizer," U.S. Patent 2,096,027, Oct. 1937 (filed Jan. 1936).
30. G. Loeber, H. Overbeck, and W. Schlotterbeck, "Transistorized Microwave Broadband Power Amplifiers Covering the Frequency Range from 500 to 1000 MHz," *Proc. 1st Europ. Microwave Conf.*, pp. 439–442, 1969.
31. C. A. Liechti and R. L. Tillman, "Design and Performance of Microwave Amplifiers with GaAs Schottky-Gate Field-Effect Transistors," *IEEE Trans. Microwave Theory Tech.*, vol. MTT-22, pp. 510–517, May 1974.
32. D. P. Hornbuckle and L. J. Kuhlman, Jr., "Broad-Band Medium-Power Amplification in the 2-12.4-GHz Range with GaAs MESFET's," *IEEE Trans. Microwave Theory Tech.*, vol. MTT-24, pp. 338–342, Jun. 1976.
33. J. Obregon, R. Funck, and S. Barvet, "A 150MHz-16GHz FET Amplifier," *1981 IEEE Int. Solid-State Circuits Conf. Dig.*, pp. 66–67.
34. K. Honjo and Y. Takayama, "GaAs FET Ultrabroad-Band Amplifiers for Gbit/s Data Rate Systems," *IEEE Trans. Microwave Theory Tech.*, vol. MTT-29, pp. 629–633, Jul. 1981.
35. N. Riddle and R. J. Trew, "A Broad-Band Amplifier Output Network Design", *IEEE Microwave Theory Tech.*, vol. MTT-30, pp. 192–196, Feb. 1982.
36. K. B. Niclas, "On Design and Performance of Lossy Match GaAs MESFET Amplifiers," *IEEE Trans. Microwave Theory Tech.*, vol. MTT-30, pp. 1900–1906, Nov. 1982.
37. K. B. Niclas, "Multi-Octave Performance of Single-Ended Microwave Solid-State Amplifiers," *IEEE Trans. Microwave Theory Tech.*, vol. MTT-32, pp. 896–908, Aug. 1984.
38. V. V. Nikiforov, T. T. Kulish, and I. V. Shevnin, "Broadband HF-VHF MOSFET Power Amplifier Design (in Russian)," *Poluprovodnikovaya Elektronika v Tekhnike Svyazi*, vol. 23, pp. 27–36, 1983.
39. O. Pitzalis, Jr., R. E. Horn, and R. J. Baranello, "Broadband 60-W HF Linear Amplifier," *IEEE J. Solid-State Circuits*, vol. SC-6, pp. 93–103, Jun. 1971.
40. A. V. Grebennikov, V. V. Nikiforov, and A. B. Ryzhikov, "The Powerful Transistor Amplifier Modules for VHF FM and TV Broadcasting (in Russian)," *Elektrosvyaz*, pp. 28–31, Mar. 1996.

41. Y. Ito, M. Nii, Y. Kohno, M. Mochizuki, and T. Takagi, "A 4 to 25 GHz 0.5 W Monolithic Lossy Match Amplifier," *1994 IEEE MTT-S Int. Microwave Symp. Dig.*, pp. 257–260.
42. A. R. Barnes, M. T. Moore, and M. B. Allenson, "A 6-18 GHz Broadband High Power MMIC for EW Applications," *1997 IEEE MTT-S Int. Microwave Symp. Dig.*, pp. 1429–1432.
43. Y. F. Wu, R. A. York, S. Keller, B. P. Keller, and U. K. Mishra, "3-9-GHz GaN-Based Microwave Power Amplifiers with L-C-R Broad-Band Matching," *IEEE Microwave and Guided Wave Lett.*, vol. 9, pp. 314–316, Aug. 1999.
44. J. J. Xu, S. Keller, G. Parish, S. Heikman, U. K. Mishra, and R. A. York, "A 3-10-GHz GaN-Based Flip-Chip Integrated Broad-Band Power Amplifier," *IEEE Trans. Microwave Theory Tech.*, vol. MTT-48, pp. 2573–2578, Dec. 2000.
45. K. Krishnamurthy, D. Wang, B. Landberg, and J. Martin, "RLC Matched GaN HEMT Power Amplifier with 2 GHz Bandwidth," *2008 IEEE Compound Semiconductor Integrated Circuits Symp. Dig.*, pp. 1–4.
46. J. K. A. Everard and A. J. King, "Broadband Power Efficient Class E Amplifiers with a Non-Linear CAD Model of the Active MOS Device," *J. IERE*, vol. 57, pp. 52–58, Mar. 1987.
47. G. L. Matthaei, "A Study of the Optimum Design of Wide-band Parametric Amplifiers and Up-Converters," *IRE Trans. Microwave Theory Tech.*, vol. MTT-9, pp. 23–38, Jan. 1961.
48. J. T. DeJaeger, "Maximum Bandwidth Performance of a Nondegenerate Parametric Amplifier with Single-Tuned Idler Circuit," *IEEE Trans. Microwave Theory Tech.*, vol. MTT-12, pp. 459–467, Jul. 1964.
49. B. L. Humphreys, "Characteristics of Broadband Parametric Amplifiers Using Filter Networks," *Proc. IEE*, vol. 111, pp. 264–274, Feb. 1964.
50. R. V. Gelsthorpe and C. S. Aitchison, "Analytical Evaluation of the Components Necessary for Double Reactance Compensation of an Oscillator," *Electronics Lett.*, vol. 12, pp. 485–486, Sep. 1976.
51. A. G. Chapman and C. S. Aitchison, "Circuit Technique for Broadband Impedance Matching of Passive Loads," *IEE J. Microwaves Optics Acoustics*, vol. 3, pp. 43–50, Mar. 1979.
52. E. Camargo and D. Consoni, "Reactance Compensation Matches FET Circuits," *Microwaves*, vol. 24, pp. 93–95, Jun. 1985.
53. R. Soares, *GaAs MESFET Circuit Design*, Boston: Artech House, 1988.
54. A. Al Tanany, A. Sayed, and G. Boeck, "Broadband GaN Switch Mode Class E Power Amplifier for UHF Applications," *2009 IEEE MTT-S Int. Microwave Symp. Dig.*, pp. 761–764.
55. A. Al Tanany, A. Sayed, O. Bengtsson, and G. Boeck, "Time Domain Analysis of Broadband GaN Switch Mode Class-E Power Amplifier," *Proc. 5th German Microwave Conf.*, pp. 254–257, 2010.
56. V. S. Rao Gudimetla and A. Z. Kain, "Design and Validation of the Load Networks for Broadband Class E Amplifiers Using Nonlinear Device Models," *1999 IEEE MTT-S Int. Microwave Symp. Dig.*, pp. 823–826.
57. Y. Qin, S. Gao, P. Butterworth, E. Korolkiewicz, and A. Sambell, "Improved Design Technique of a Broadband Class-E Power Amplifier at 2GHz," *Proc. 35th Europ. Microwave Conf.*, vol. 1, pp. 1–4, 2005.
58. A. Al Tanany, D. Gruner, and G. Boeck, "Harmonically Tuned 100 W Broadband GaN HEMT Power Amplifier with more than 60% PAE," *Proc. 41st Europ. Microwave Conf.*, pp. 159–162, 2011.
59. K. Chen and D. Peroulis, "Design of Highly Efficient Broadband Class-E Power Amplifier Using Synthesized Low-Pass Matching Networks," *IEEE Trans. Microwave Theory Tech.*, vol. MTT-59, pp. 3162–3173, Dec. 2011.
60. H. Jaeger, A. V. Grebennikov, E. P. Heaney, and R. Weigel, "Broadband High-Efficiency monolithic InGaP/GaAs HBT Power Amplifiers for 3G Handset Applications," *2002 IEEE MTT-S Int. Microwave Symp. Dig.*, vol. 2, pp. 1035–1038.
61. H. Jaeger, A. V. Grebennikov, E. P. Heaney, and R. Weigel, "Broadband High-Efficiency Monolithic InGaP/GaAs HBT Power Amplifiers for Wireless Applications," *Int. J. RF and Microwave Computer-Aided Eng.*, vol. 13, pp. 496–519, Jun. 2003.

62. A. Grebennikov, "Simple Design Equations for Broadband Class E Power Amplifiers with Reactance Compensation," *2001 IEEE MTT-S Int. Microwave Symp. Dig.*, vol. 3, pp. 2143–2146.
63. V. I. Degtev and V. B. Kozyrev, "Transistor Single-Ended Switching-Mode Power Amplifier with Forming Circuit (in Russian)," *Poluprovodnikovaya Elektronika v Tekhnike Svyazi*, vol. 26, pp. 178–188, 1986.
64. N. Kumar, C. Prakash, A. Grebennikov, and A. Mediano, "High-Efficiency Broadband Parallel-Circuit Class E Power Amplifier with Reactance-Compensation Technique," *IEEE Trans. Microwave Theory Tech.*, vol. MTT-56, pp. 604–612, Mar. 2008.
65. E. Khansalee, N. Puangngernmak, and S. Chalermwisutkul, "Design of 140-170 MHz Class E Power Amplifier with Parallel Circuit on GaN HEMT," *Proc. Int. Electrical Eng./Electron. Computer Telecom. Inform. Technol. Conf.*, pp. 570–574, 2010.
66. T. K. Quach, P. M. Watson, W. Okamura, E. N. Kaneshiro, A. Gutierrez-Aitken, T. R. Block, J. W. Eldredge, et al., "Ultra-High Efficiency Power Amplifier for Space Radar Applications," *IEEE J. Solid-State Circuits*, vol. SC-37, pp. 1126–1134, Sep. 2002.
67. P. Watson, R. Neidhard, L. Kehias, R. Welch, T. Quach, R. Worley, M. Pacer, et al., "Ultra-High Efficiency Operation Based on an Alternative Class-E Mode," *2000 IEEE GaAs IC Symp. Dig.*, pp. 53–56.
68. P. Watson, T. Quach, H. Axtel, A. Gutierrez-Aitken, E. Kaneshiro, W. Lee, A. Mattamana, A. Oki, P. Orlando, V. Patel, and D. Sawdai, "An Indium Phosphide X-Band Class-E Power MMIC with 40% Bandwidth," *2005 IEEE CSIC Symp. Dig.*, pp. 220–223.
69. S. Pajic, N. Wang, P. M. Watson, T. K. Quach, and Z. Popovic, "X-Band Two-Stage High-Efficiency Switched-Mode Power Amplifiers," *IEEE Trans. Microwave Theory Tech.*, vol. MTT-53, pp. 2899–2907, Sep. 2005.
70. C.-H. Lin and H.-Y. Chang, "A High Efficiency Broadband Class-E Power Amplifier Using a Reactance Compensation Technique," *IEEE Microwave and Wireless Components Lett.*, vol. 20, pp. 507–509, Sep. 2010.
71. M. van Wanum, R. van Dijk, P. de Hek, and F. E. van Vliet, "Broadband S-Band Class E HPA," *Proc. 4th Europ. Microwave Integrated Circuits Conf.*, pp. 29–32, 2009.
72. A. Mazzanti, L. Larcher, R. Brama, and F. Svelto, "A 1.4 GHz–2 GHz Wideband CMOS Class-E Power Amplifier Delivering 23dBm Peak with 67% PAE," *2005 IEEE RFIC Symp. Dig.*, pp. 425–428.
73. A. Mazzanti, L. Larcher, R. Brama, and F. Svelto, "Analysis of Reliability and Power Efficiency in Cascode Class-E PAs," *IEEE J. Solid-State Circuits*, vol. SC-41, pp. 1222–1229, May 2006.
74. H.-Y. Liao, M.-W. Pan, and H.-K. Chiou, "Fully-Integrated CMOS Class-E Power Amplifier Using Broadband and Low-Loss 1:4 Transmission-Line Transformer," *Electronics Lett.*, vol. 46, pp. 1490–1491, Oct. 2010.
75. R. Zhang, M. Acar, M. P. van der Heijden, M. Apostolidou, L. C. N. de Vreede, and M. W. Leenaerts, "A 550-1050MHz +30dBm Class-E Power Amplifier in 65 nm CMOS," *2011 IEEE RFIC Symp. Dig.*, pp. 289–292.
76. A. Tam, "Network Building Blocks Balance Power Amp Parameters," *Microwaves & RF*, vol. 23, pp. 81–87, Jul. 1984.
77. B. Becciolini, "Impedance Matching Networks Applied to RF Power Transistors," *Application Note AN721*, Freescale Semiconductor, 2005.
78. Y. Ito, M. Mochizuki, M. Kohno, H. Masuno, T. Takagi, and Y. Mitsui, "A 5-10 GHz 15-W GaAs MESFET Amplifier with Flat Gain and Power Responses," *IEEE Microwave and Guided Wave Lett.*, vol. 5, pp. 454–456, Dec. 1995.
79. "A Broadband 100 W Push Pull Amplifier for Band IV & V TV Transmitters Based on the BLV861," *Application Note AN98033*, Philips Semiconductors, Mar. 1998.
80. U. Schmid, H. Sledzik, P. Schuh, J. Schroth, M. Oppermann, P. Brueckner, F. van Raay, et al., "Ultra-Wideband GaN MMIC Chip Set and High Power Amplifier Module for Multi-Function Defense AESA Applications," *IEEE Trans. Microwave Theory Tech.*, vol. MTT-61, pp. 3043–3051, Aug. 2013.

CHAPTER 9

Linearization and Efficiency Enhancement Techniques

In modern wireless communication systems, it is very important to realize simultaneously high-efficiency and linear operation of the power amplifiers. This chapter describes a variety of techniques and approaches that can improve the power amplifier performance. To increase efficiency over power backoff range, the outphasing and envelope-tracking power amplifier architectures, as well as the switched-path and variable-load power amplifier configurations are discussed and analyzed. There are several linearization techniques that provide linearization of both entire transmitter system and individual power amplifier. Feedforward linearization techniques is available technology for satellite and cellular base station applications with achieving very high linearity levels. Its practical realization is quite complicated and very sensitive to both the feedback loop imbalance and the parameters of its individual components. Analog predistortion linearization technique is the simplest form of power amplifier linearization and can be used for handset applications, although significant linearity improvement is difficult to realize. The choice of a proper high-efficiency approach or linearity correction scheme depends on performance tradeoffs, as well as manufacturing capabilities. Finally, the circuit schematics and monolithic implementation of the high-efficiency GaAs HBT and CMOS power amplifiers for handset applications are shown and described.

9.1 Feedforward Amplifier Architecture

In the mid-1920s, H. S. Black firstly proposed the method of suppressing even- and odd-order distortion components produced in a nonlinear transmitting system [1]. However, interest in this invention was limited at that time due to success of the competitive feedback approach invented later by him as well, because of the simplicity and effectiveness of the latter. Only almost three decades later W. D. Lewis extended this approach to microwave frequencies by using the waveguide sections for delay lines, branch-line hybrid junctions, and directional couplers [2]. Since then, the interest in the use of a feedforward correction in RF and microwave applications had become significant to satisfy simultaneously strong requirements in high output powers, extremely high degrees of linearity, good time stability, and broad bandwidths. H. Seidel described in detail the application of a feedforward compensated circuit, in which the amplified signal is compared with a time-shifted reference signal [3]. In this case, the error component, which includes both noise and distortion components introduced by the main amplifier, is then amplified by means of a high-quality linear subsidiary amplifier and added to the time-shifted amplified signal in such a phase as to minimize the error in the output signal. To minimize errors due to impedance mismatch in the amplifier circuit, the hybrid-coupler power dividers can be used. At the same time, to minimize noise in the output signal due to the subsidiary amplifier, the portion of input signal coupled to the subsidiary amplifier must be larger than that of coupled to the main amplifier. Most efficient utilization of the power in the amplified signal and the error signal can be realized by using a reactive three-port network to match the main signal path and the error signal path to the output load. As a part of a test to determine its applicability to coaxial repeaters, a feedforward error-controlled system was applied to a coaxial flat-gain amplifier operating in the frequency range of 0.5 to 20 MHz. As a result, a modulation product reduction of greater than 35 dB over a 40:1 bandwidth was achieved [4, 5]. The use of a feedforward architecture could result in up to 20-dB distortion improvement in a feedback amplifier operating over the whole frequency decade bandwidth of 30 to 300 MHz [6]. In a practical 2.2-GHz feedforward amplifier system with a power gain of 30 dB and an output power of 1.25 W, the suppression of the intermodulation distortion products of at least 50 dB from the carrier level was achieved [7].

Figure 9.1 shows the principle of operation and basic structure of the feedforward linearization amplifier, which consists of two cancellation loops and generally includes the main power amplifier, three couplers, two phase shifters, and an auxiliary error amplifier. The operation of the feedforward linearization circuit is based on the subtraction of two equal signals with subsequent cancellation of the error signal in the amplifier output spectrum. Its operation principle can be seen clearly from the two-tone test spectra at various points in the block diagram. The input signal is split by input coupler-splitter into two identical parts, although in a common case the ratio used in the splitting process does not need to be equal, with one part going to the

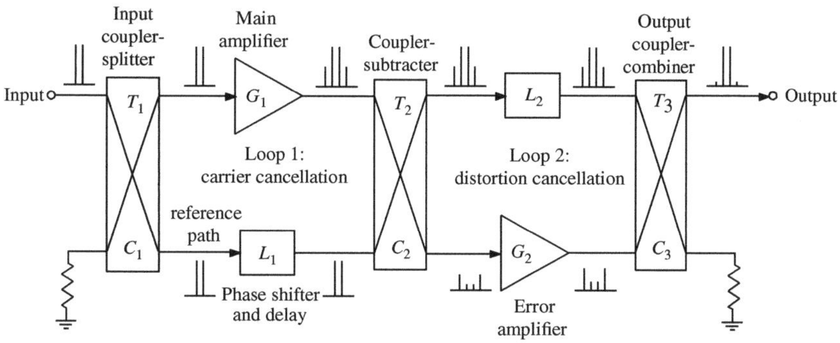

FIGURE 9.1 Basic structure and operation principle of feedforward amplifier.

main power amplifier, while the other part goes to a delay element. The signal in the top path is amplified by the main amplifier, whose inherent nonlinear behavior contributes to the intermodulation and harmonic distortion components that are added to the original signal. This signal is sampled and scaled by the coupler-subtracter before being combined with the delayed distortion-free portion of the input signal. The resulting error signal ideally contains only the distortion components provided by the main amplifier. The error signal is then amplified linearly by a low-power high-linearity error amplifier to the level required to cancel the distortion in the main amplifying path, and is then fed to the output directional coupler-combiner, on the other input of which a time-delayed and out-of-phase main-path signal is forwarded. The resulting signal at the output of the feedforward linearization system represents an error-free signal in an ideal case or essentially an amplified version of the original input signal in practice.

The operation quality of the feedforward amplifier system obviously depends significantly on cancellation accuracy at coupler-subtracter and output coupler-combiner. The level of distortion reduction is determined by the cancellation occurring at output coupler-combiner, and cancellation of the fundamental signals at coupler-subtracter is required to prevent subtraction of the fundamentals at output coupler-combiner and consequent gain loss. At the same time, cancellation of the fundamentals at coupler-subtracter is also important in order to prevent large amplitude error signals from entering error amplifier and possibly causing significant distortion in that amplifier. Generally, in the first carrier-cancellation loop the precision in cancellation is required only to such a degree as to avoid any substantial degradation of linearity in the error amplifier [5]. On the other hand, because the second distortion-cancellation loop controls the entire linearity improvement of the feedforward system, the degree of its balance should always be at the highest level [7].

To analyze the effect of imperfect magnitude and phase equalization in the amplifier and delay line paths at any particular frequency, consider the signal in each upper and lower paths of the first cancellation loop to be cosinusoidal as

$$v_1 = V\cos\omega t \tag{9.1}$$

$$v_2 = (V \pm \Delta V)\cos(\omega t \pm \theta) \tag{9.2}$$

where ΔV is the amplitude imbalance and θ is the phase imbalance.

After subtraction of these signals in a coupler-subtracter, we have in a normalized form

$$\begin{aligned}\frac{\Delta v}{V} &= \cos\omega t - \left(1 \pm \frac{\Delta V}{V}\right)\cos(\omega t \pm \theta)\\ &= (1-\alpha\cos\theta)\cos\omega t \pm \alpha\sin\theta\sin\omega t\end{aligned} \tag{9.3}$$

where $\Delta v = v_1 - v_2$ and

$$\alpha = 1 \pm \frac{\Delta V}{V} \tag{9.4}$$

As a result, for a total imbalance magnitude,

$$\frac{|\Delta v|}{V} = \sqrt{(1-\alpha\cos\theta)^2 + (\alpha\sin\theta)^2} \tag{9.5}$$

Because cancellation achieved by the second loop can be analyzed ideally in a similar way, the cancellation result achieved by the first and second loops independently can be rewritten in the corresponding forms (in decibels) as

$$CANC_1 = 10\log_{10}(1+\alpha_1^2 - 2\alpha_1\cos\theta_1) \tag{9.6}$$

$$CANC_2 = 10\log_{10}(1+\alpha_2^2 - 2\alpha_2\cos\theta_2) \tag{9.7}$$

where α_1 and θ_1 are the amplitude and phase imbalance in the first loop, and α_2 and θ_2 are the amplitude and phase imbalance in the second loop [8].

Figure 9.2 shows the distortion cancellation as a function of amplitude and phase imbalance. From these curves it follows that it is necessary to maintain extremely small amounts of amplitude imbalance in order to obtain a high degree of cancellation accuracy. For example, a 40 dB of cancellation would require a phase imbalance of less than 1° and an amplitude imbalance of less than 0.1 dB. However, the demand for a high degree of linearity improvement will cause the system to become sensitive to circuit parameter variations due to temperature change. To achieve temperature stability in a practical system, the degree of linearity improvement should be kept at a reasonably low level. For example, a 30 dB of cancellation would

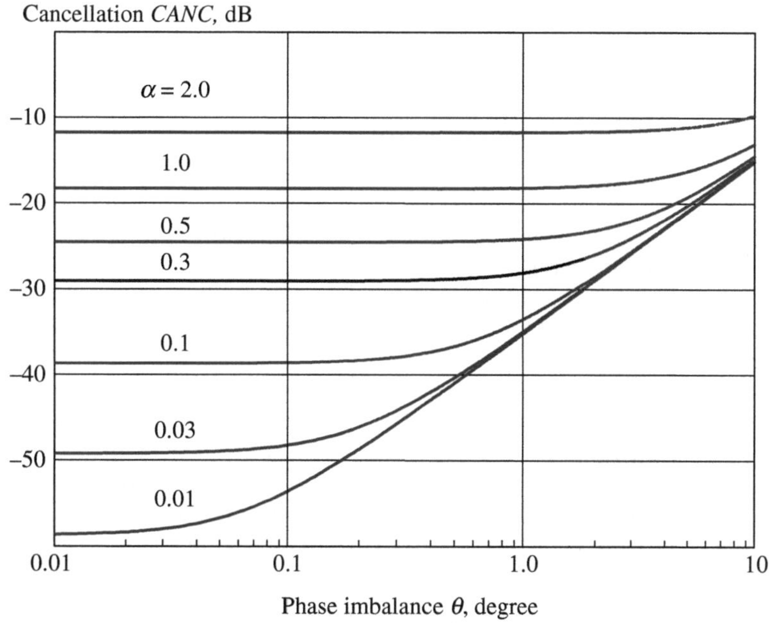

FIGURE 9.2 Cancellation as function of amplitude and phase imbalance.

require only an amplitude imbalance of 0.25 dB and a phase imbalance of 1.8°. To improve the temperature stability characteristic, it is better to implement both main and error amplifiers using the same technology, similar components, and assembly techniques. However, if a higher degree of balance is to be maintained at all times, an automatic adaptive control system must be employed. Besides, an additional transmission-line delay mismatch can be taken into account when using transmission lines in high-frequency feedforward linearization systems. For example, if the difference in wavelength between the transmission lines in upper and lower paths at the center bandwidth frequency f_0 is equal to $0.1f_0$, the phase imbalance should be maintained within approximately 1.0° in order to obtain a 30 dB of cancellation with a 30-MHz bandwidth at 800 MHz for $\alpha = 0.1$ dB [9].

It is of great importance for telecommunication system to minimize its nonlinear distortion level, and the main factor of its linearity is a level of the third-order intermodulation products at the system output. In this case, consider the cancellation provided by both first and second loop through the parameters of the feedforward system [10]. At the output of the coupler-subtracter with suppressed carrier P_{supp}, the cancellation of the first loop is defined as

$$CANC_1 = \frac{P_{supp}}{C_2 P_{main}} \tag{9.8}$$

where C_2 is the coupling coefficient of the second coupler-subtracter and P_{main} is the carrier power level of the main amplifier. On the other hand, the cancellation achieved in the second loop is

$$CANC_2 = \frac{P_{IM3supp}}{P_{IM3main}} \frac{1}{T_2 L_2 T_3} \tag{9.9}$$

where $P_{IM3main}$ is the power level of the third-order intermodulation component, $P_{IM3supp}$ is the power level of the third-order intermodulation component of the main amplifier suppressed at the linearizer output due to the corrective action of the second loop, L_2 is the delay-line loss in the second loop, and T_2 and T_3 are the transmission losses in coupler-subtracter and output coupler-combiner, respectively.

The effective cancellation of the overall feedforward linearization system is defined as the ratio of the power level of all intermodulation components at the feedforward system output over the power level of the intermodulation products for open-loop configuration. As a result, for in-phase addition of the intermodulation components of the main and error amplifiers, the effective cancellation can be expressed (in decibels) by

$$CANC_{eff} = 20\log_{10}\left[\sqrt{CANC_2} + \sqrt{CANC_1^{\,3}\left(\frac{IP_{3main}}{IP_{3error}}\right)^2 \frac{T_2^2 L_2^2}{\alpha_2^3}\left(\frac{T_3}{C_3}\right)^2}\,\right] \tag{9.10}$$

where the amplitude imbalance α_2 is defined as the ratio of the power gains of two paths as

$$\alpha_2 = \frac{T_2 L_2 T_3}{C_2 G_2 C_3} \tag{9.11}$$

where G_2 is the power gain of the error amplifier, C_3 is the coupling coefficient of the output coupler-combiner, and IP_{3main} and IP_{3error} are the third-order intercept points of the main and error amplifiers, respectively. The first term in Eq. (9.10) depends on the balance level achieved in the second loop, whereas the second term defines the possible imbalance created by the first loop and some other feedforward circuit parameters. In particularly, the error amplifier with a sufficiently low power capability having too small value of IP_{3error} or too big coupling coefficient C_3 of the output coupler-combiner and loss $(T_2 L_2 T_3)$ through the main path increases the effect of the amplitude and phase imbalance.

The relationship between the overall feedforward system efficiency η and the efficiencies of the two amplifiers, η_{main} for main amplifier and η_{error} for error amplifier, when the losses $(T_2 L_2 T_3)$ through the main path are considered negligible, can be written as

$$\eta = \frac{\eta_{main}\eta_{error}C_3(1-C_3)}{\eta_{error}C_3 + \eta_{main} f_{main}(1-C_3)} \tag{9.12}$$

where $\log_{10}(f_{\text{main}}) = -(C/I)_{\text{main}}/10$, $(C/I)_{\text{main}}$ is the ratio of the carrier to third-order intermodulation product of the main amplifier [11, 12]. Provided the optimum value of C_3, which maximizes the overall efficiency η when the other system parameters are fixed, the maximum $\eta_{\max}$ can be obtained by

$$\eta_{\max} = \eta_{\text{main}} \Big/ \left(1+\sqrt{\frac{\eta_{\text{main}}}{\eta_{\text{error}}} f_{\text{main}}}\right)^2 \tag{9.13}$$

which shows the efficiency degradation due to the linearization system [13]. For example, for a typical 10-dB coupling ratio of the output coupler-combiner, only 10% of the power from the error amplifier reaches the load, which means that the error amplifier must produce 10 times the power of the distortion in the main amplifier. In this case, it should operate in an inefficient linear mode in order not to disturb the error signal. As a result, the dc power consumed by the error amplifier can represent a significant part of that of the main amplifier. In addition, it needs to take into account the fact that, despite its excellent distortion cancellation property, generally the feedforward amplifier linearization system requires well-equalized circuitry and is characterized by substantially increased complexity and cost.

The efficiency of the conventional feedforward linearization system using a balanced configuration of the main amplifier shown in Fig. 9.3(*a*) can be improved by providing some restructuring of the system. As a result, the modified feedforward system consists of three major loops shown in Fig. 9.3(*b*): carrier cancellation loop, balanced power amplifier loop, and error-injection loop [14]. In this case, the carrier cancellation loop extracts the error signal from the amplified signal at the output of top power amplifier, whereas the error-injection loop provides an injection of the amplitude-adjusted and properly phased distortion into the output of bottom power amplifier. Finally, the amplified signals in balanced paths are combined in output hybrid combiner with corresponding distortion cancellation. Unlike the conventional feedforward system, in its balanced version each power amplifier sees only one coupler, either coupler-subtracter or output coupler-combiner, which means that there is no additional insertion loss due to output coupler-combiner as in the conventional feedforward system. As a result, for a four-carrier WCDMA signal with a *PAR* of 10 dB, an efficiency improvement was 2% at an average output power of 40 dBm with an improvement in *ACLR* (5 MHz offset) of about 18.6 dB by cancellation at the center bandwidth frequency of 2.14 GHz.

Generally, it is a serious problem for the conventional feedforward linearization system to maintain the great accuracy in amplitude and phase balance over time, temperature, supply voltage, or input source and load variations. In practice, some forms of the gain and phase adjustments are essential to achieve an acceptable low

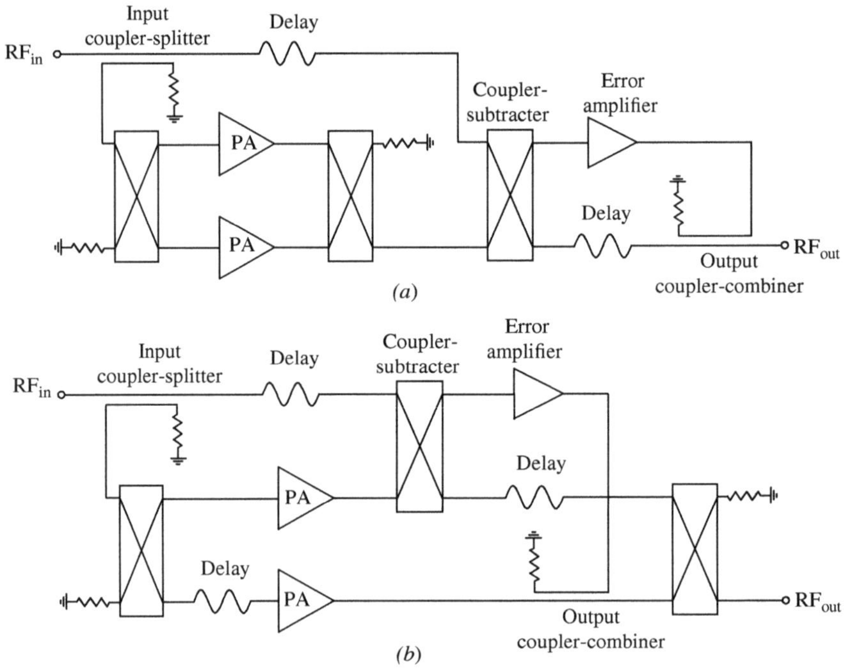

FIGURE 9.3 Balanced feedforward amplifier topologies.

level of intermodulation distortion. Figure 9.4(*a*) shows the block schematic of an analog adaptive feedforward linearization system that includes a feedback network for adaptively adjusting the performance of the overall feedforward system to compensate for uncontrolled variations of its component parameters [15]. The feedback network provides a control of the carrier and distortion cancellation loops by comparing the signals sampled at their inputs, and outputs, and adaptively adjusts the corresponding vector modulators to minimize the amplitude and phase imbalance when it is necessary. Different adaptation algorithms using optimization techniques can be implemented to improve the cancellation results for an analog adaptive feedforward linearization system [16].

Using a digital signal processing (DSP) creates a good opportunity to provide correction of amplitude and phase imbalance in the feedforward linearization system at baseband level, thus making this procedure more predictable and fast and overcoming the problems with mixer dc offset and masking of strong signals by weaker ones than can compromise analog adaptive implementations [17]. To compensate for component frequency response and nonadaptive nature of the delay lines, a hybrid of the conventional feedforward linearizer and a digital signal processor can be used, as shown in Fig. 9.4(*b*),

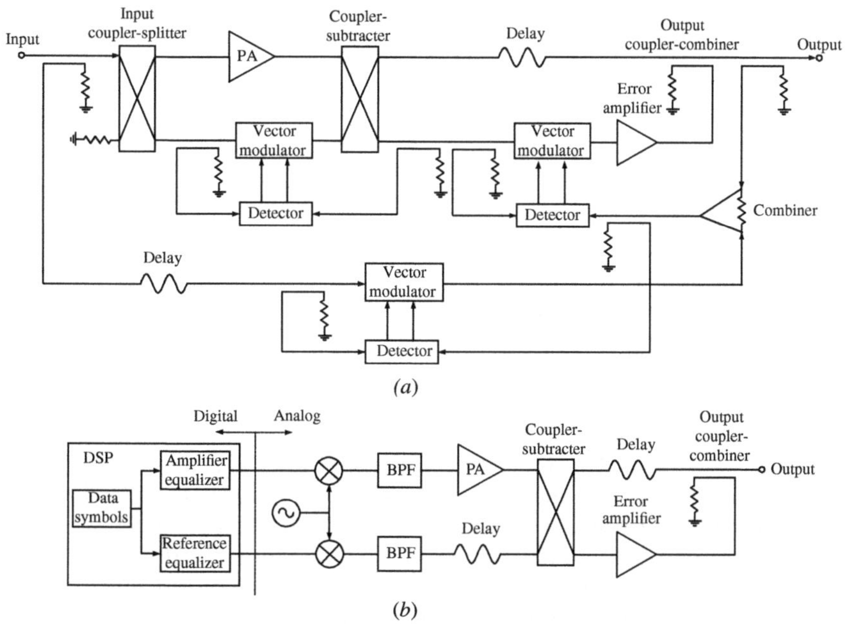

FIGURE 9.4 Adaptive analog and digital feedforward amplifier linearizers.

where both the amplifier input signal and the reference signal are generated by DSP [18]. The reference signal is then used to cancel the linearly amplified component of the distorted amplifier output signals, leaving an error signal containing only the main amplifier distortion. By generating the reference signal in the DSP, rather than using an analog splitter, some of the analog hardware can be moved into a simpler digital implementation, with independent control of the main amplifier and reference signals by using equalizers. In this case, the amplitude and reference equalizers correct the phase shift, time delay, and nonideal response of the analog components to achieve the proper distortion cancellation. By improving the cancellation of the first loop, a more accurate error signal is generated that consists only of the distortion from the main amplifier. Generally, the use of amplifier and reference equalizers in the first loop has an advantage in that the tuning, previously done manually, has now been moved back into the DSP where it can be done adaptively.

9.2 Predistortion Linearization

To achieve simultaneously high-efficiency and low-distortion operation conditions of the power amplifier when the linearity requirements are not extremely high, it is possible to use an analog predistortion linearizer that provides the positive amplitude and negative phase

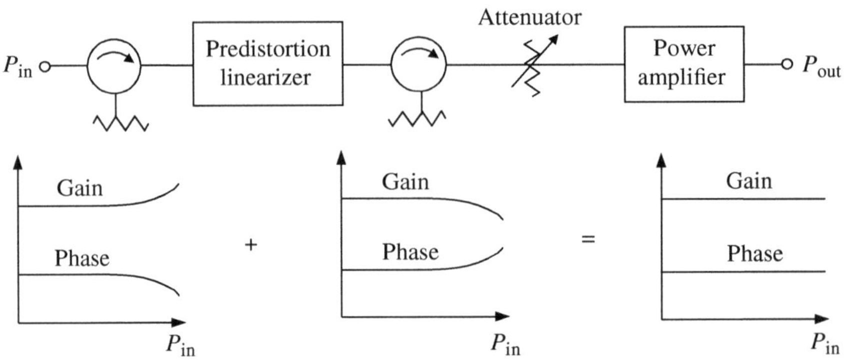

FIGURE 9.5 Block diagram of power amplifier with predistortion linearizer.

deviations for input RF signal to compensate for the active device nonlinearity whose nonlinear behavior (when a power amplifier is operated close to saturation) usually represents the opposite behavior of its amplitude and phase characteristics. Historically, the initial idea to compensate for the third-order intermodulation products arising in a vacuum-tube amplifier was to use the linearization scheme where the nonlinear amplifier having a compressing characteristic is followed by a nonlinear element having an expanding characteristic and producing the third-order distortion of opposite sign to that of the amplifier [19].

The block diagram of the linearized power amplifier system with a predistortion linearizer with indication of the appropriate amplitude and phase dependences at each stage of the system is shown in Fig. 9.5, where also a variable attenuator for adjusting the amplitude level of the input signal is included. At microwaves, a linearized power amplifier very often includes two isolators for stable operation conditions. The conventional predistortion linearizer circuits generally use either diodes or transistors as sources of intermodulation [20, 21].

As an interesting fact, yet at the beginning of the 1920s it was claimed that, by using similar vacuum tubes in both stages of a two-stage power amplifier with similar signals at their inputs, the even harmonics generated by the first amplification are neutralized by the even harmonics generated by the second amplification because they are similar in amplitude and opposite in phase at the output of the second vacuum tube [22]. Indeed, as it turned out regarding modern transistor power amplifiers using GaAs pHEMT devices, it is enough to properly choose a bias point of the driver-stage device in a two-stage amplifier to provide a negative phase deviation to compensate for the positive phase deviation of the final stage [23]. In this case, the quiescent current of the driver-stage device, whose size is three times smaller than that of a final-stage device, is sufficiently small.

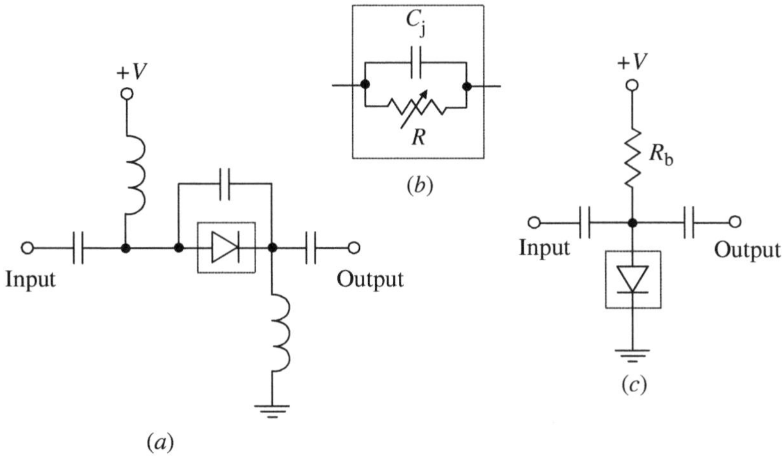

FIGURE 9.6 Simple diode-based predistortion linearizers.

As a result, for a quiescent current equal to 1.25% of the device dc saturated current, an improvement of greater than 5 dB in *ACLR* of an entire high-efficiency two-stage cellular-phone WCDMA power amplifier operating at 1.95 GHz can be achieved at backoff output powers close to the saturation power.

Figure 9.6(*a*) shows the schematic of a simple diode linearizer composed of a series Schottky diode and a parallel capacitor with two RF chokes for dc feed and two blocking capacitors that provides positive amplitude and negative phase deviations when input power increases [24]. The equivalent circuit of the series diode is shown in Fig. 9.6(*b*), where R is the diode equivalent resistance and C_j is the junction capacitance. With the increase in an incident input signal power, the forward diode current increases, which leads to the decrease in the diode resistance R. In this case, the positive amplitude and negative phase deviations can be achieved under low forward-bias conditions when the diode current ranges from 0.1 to 1.0 mA, and, in the latter case, the phase deviation can reach a value of –30°. Applying such a linearizer to a 1.9-GHz MMIC (monolithic microwave integrated circuit) power amplifier with a saturated output power of 22.5 dBm, an improvement of *ACLPR* by 5 dB can be achieved for the QPSK modulated signal for output powers less than 15 dBm.

A similar improvement of *ACLR* can be achieved by using a linearizer based on a parallel Schottky diode with the bias feed resistor R_b, which is shown in Fig. 9.6(*c*) [25]. With the increase in input power, the bias point of a diode changes due to the voltage drop across the resistor R_b caused in turn by the increased diode forward current. As a result, because of the decreased diode resistance R, the linearizer achieves positive gain and negative phase deviations. By applying

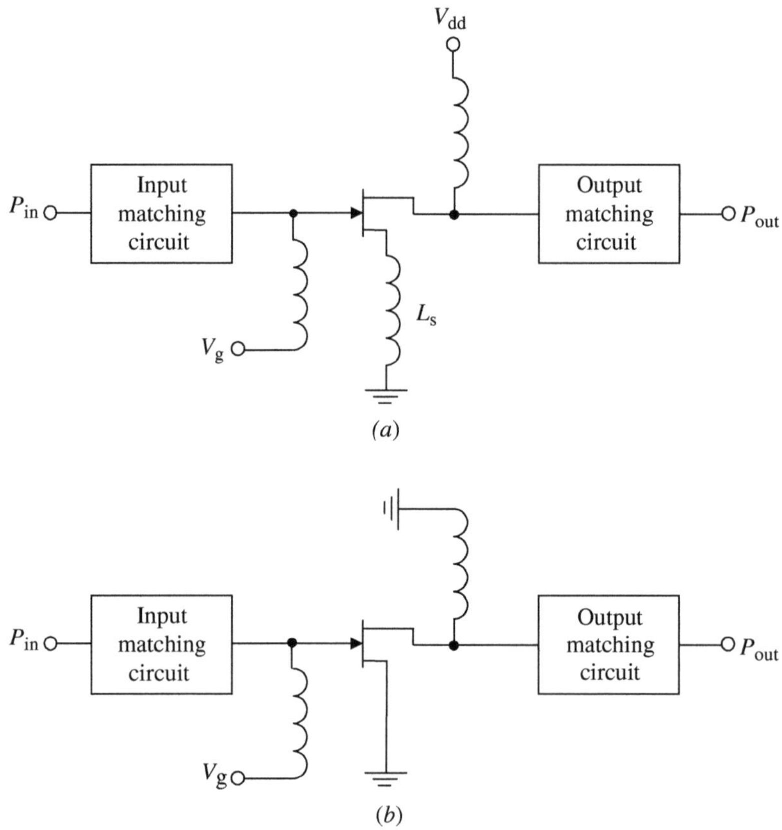

FIGURE 9.7 Transistor-based linearizers.

such a linearizer to a 2.7-GHz power amplifier, a maximum improvement of 5 dB was achieved for low quiescent current conditions at an output power of 34 dBm.

Positive amplitude deviation with negative phase deviation can also be achieved using a series-feedback GaAs MESFET amplifier with a large source inductance L_s, whose block diagram (including matching circuits) is shown in Fig. 9.7(*a*) [26]. The required amplitude and phase deviations are due to nonlinearities of the device transconductance g_m, gate-source capacitance C_{gs}, and differential drain-source resistance R_{ds}. For the device with a gate width of 1.2 mm, a nonlinearity of g_m contributes to the positive amplitude deviation when L_s = 20 nH. At the same time, nonlinearities of both g_m and R_{ds} contribute to the negative phase deviation when $L_s \geq 3$ nH. A nonlinearity of C_{gs} has a negligibly small effect on both amplitude and phase deviations. As a result, for a linearizer with L_s = 16 nH at an operating frequency of 1.9 GHz, the positive amplitude and negative phase deviations were obtained across the input power dynamic range from 5 to 18 dBm, with amplitude deviation of 2.5 dB and phase deviation of 30° at 18-dBm input power. The GaAs MESFET device was biased in a Class-AB mode with a drain-source supply voltage of 2 V providing a quiescent current of 78 mA. By applying this linearizing technique to a 1.9-GHz MMIC power amplifier with 1-dB compressed power of 17 dBm, an improvement of *ACLR* up to 7 dB was achieved for a $\pi/4$-shifted QPSK signal.

As an alternative, it is also possible to achieve positive amplitude and negative phase deviations using a source-grounded MESFET device with zero drain-source supply voltage [27]. The schematic diagram of such a linearizer is shown in Fig. 9.7(*b*). In this case, for the device with a gate width of 240 μm at the saturation power of 20 mW under the gate bias condition of $V_g = -0.4$ V, the 3-dB increased power gain and of about 30° negative phase were achieved due to the varying drain-source resistance. Because of its simplicity, such a linearizer can operate from 2 to 12 GHz with good thermal stability. When it was implemented into a 50-W solid-state power amplifier system at an operating frequency of 7 GHz, the system noise power ratio was improved over 15-dB dynamic range, in particular by 2 dB at the 3-dB output power backoff point.

More advanced configurations of the predistortion linearizer is based on the splitting of the input signal into nonlinear and linear paths using a directional coupler or a hybrid divider with subsequent subtraction of resulting signals in output coupler-subtracter. The block diagram of such a predistortion linearizer, which uses two power amplifiers in a balanced configuration using two 90° hybrids, is shown in Fig. 9.8(*a*) [21]. In this case, the upper power amplifier is

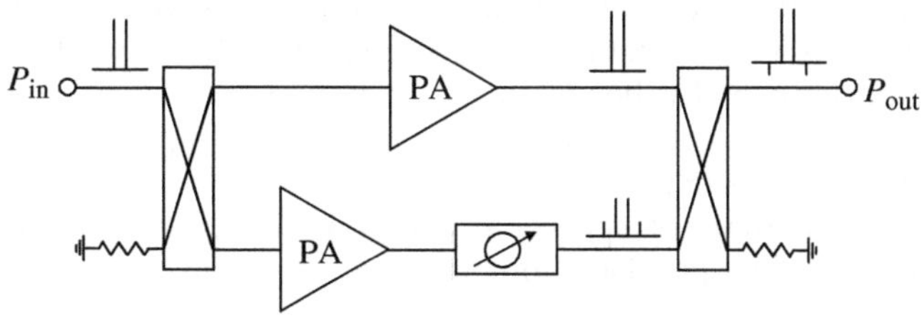

(*a*)

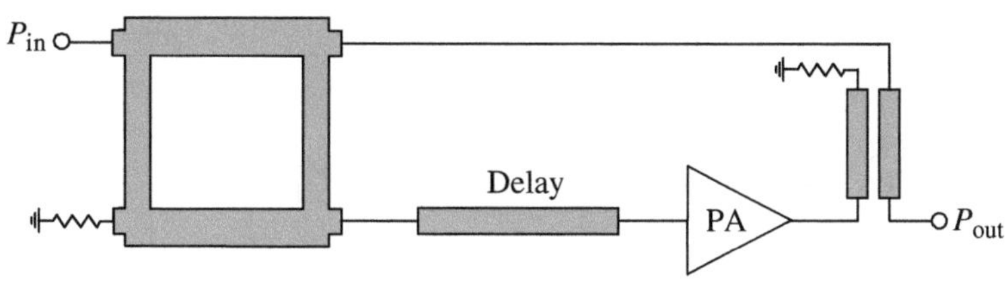

(*b*)

FIGURE 9.8 Block diagrams of power amplifier linearizers with input power splitting.

operated in a linear Class-A mode, whereas the lower power amplifier is biased in a nonlinear Class-AB or Class-B mode to generate the proper intermodulation products by controlling the input power and device bias conditions. The phase shifter in a lower amplifying path is necessary to optimize the level of the fundamental components in the resulting output spectrum. Because both devices present approximately the same input impedances, a low input return loss is provided because most of the reflected power flows into the isolated port. Figure 9.8(*b*) shows the practical microwave microstrip implementation of a two-path predistortion linearizer with an input 90° branch-line hybrid coupler, a nonlinear power amplifier in a lower path, and an output directional coupler [28]. The microstrip transmission line in a lower amplifying path having a required electrical length compensates for the additional phase shift provided by the active device, whereas the required amplitude conditions are realized with the coupling coefficient of output coupler-subtracter to be chosen. As a result, for a *Ku*-band multicarrier 4.5-W power amplifier, the phase deviation of a 12-dBm signal at the linearizer output up to –10° was achieved with a 22-dBm signal at the linearizer input.

Figure 9.9 shows the modified three-path predistortion linearizer structure, where the balanced configuration with a nonlinear power amplifier is adjusted for suppression of the fundamental components with the resulting error signal. Then, the amplitude-adjusted and properly phased error signal is amplified by an error amplifier and added to the linear component in the upper path, which is a delayed portion of the input signal. However, it is very difficult to match the nonlinear characteristics of the predistorter and the main power amplifier because generally they differ both in size and number of stages that can only result in less than 10-dB improvement of *ACLR* at 5-MHz offset from the center bandwidth frequency [29]. Therefore, it is very important for further linearity improvement to use similar devices in the predistorter and in the main power amplifier with a preferred balanced structure.

The block schematic of a power amplifier module, which includes a three-path or mirror predistortion linearizer shown in Fig. 9.9 and a

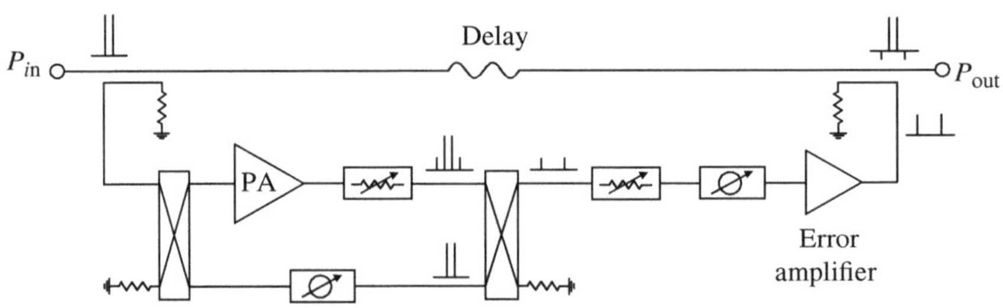

Figure 9.9 Modified three-path predistortion linearizer.

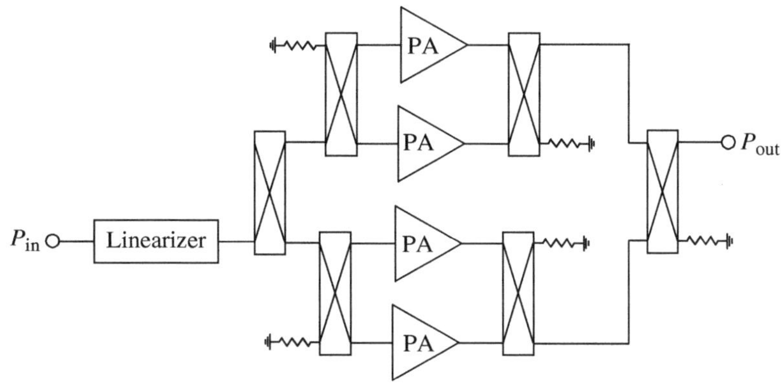

FIGURE 9.10 Power amplifier module with linearizer.

main power amplifier based on four power amplifiers configured into a balanced structure, is shown in Fig. 9.10. In this case, the identical power amplifiers operated in a Class-AB mode and based on the same Freescale MW6S004NT LDMOS devices were used both in the linearizer and power amplifier module. As a result, a significant improvement of *ACLR* for a 2.14-GHz WCDMA signal with a *PAR* of 6.5 dB was achieved, from –42 to –57 dBc for a 32-dBm channel power and from –37 to –49 dBc for a 36-dBm channel power [30]. A hybrid power amplifier module using similar mirror predistortion linearization technique with five identical power amplifiers based on AMCOM MMIC power amplifiers (AM204437WM-BM) was used for WiMAX application achieving a 23-dB improvement of the third-order intermodulation distortion at a two-tone total output power of 34 dBm, which is 7.5 dB backoff from a 1-dB compressed output power of 41.5 dBm, with the two-tone frequency separation by 10 MHz [31].

The concept of a feedforward loop with its high cancellation performance can also be used for a predistortion linearizer implementation. Because the feedforward loop is placed in front of the main amplifier, the linearity and power requirements to the error amplifier are reduced significantly compared to the conventional feedforward system. In this case, the delay-line and coupler losses are not so significant factors affecting the amplifier module performance. Figure 9.11 shows the simplified schematic diagram of a power amplifier module with the feedforward distortion linearization using five identical power amplifiers based on MRF5S21090 LDMOS devices [32]. For a forward-link four-carrier WCDMA signal at 2.35 GHz, the *ACLR* was enhanced by about 7 dB at 5-MHz offset and the total efficiency of 12.7% was achieved at an average output power of 47.8 dBm, backed off by 10.8 dB from the total peak power of 720 W.

Figure 9.12 shows the block schematic of a digital predistortion (DPD) linearizer where the predistortion algorithm is based on an initially measured PA amplitude-to-amplitude modulation (*AM-AM*)

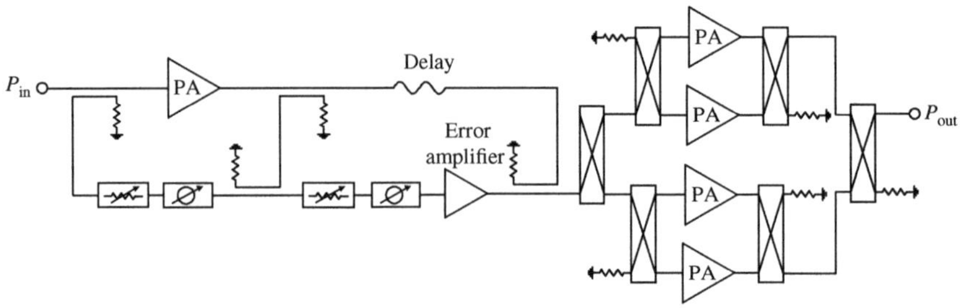

FIGURE 9.11 Power amplifier module with feedforward predistortion linearizer.

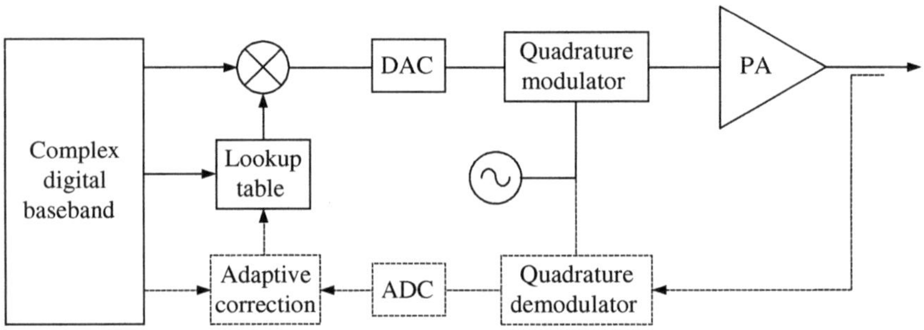

FIGURE 9.12 Digital predistortion (DPD) system.

and amplitude-to-phase modulation (*AM-PM*) responses extracted from the *S*-parameter measurements by a vector network analyzer (VNA) [33]. The amplitude and phase characteristics are interpolated using splines, which are continuous piecewise cubic functions with continuous first and second derivatives. The interpolated amplitude and phase characteristics are then used to compute the appropriate predistortion coefficients representing a lookup table (LUT), which are multiplied with the original IS-95 signal to generate the desired predistorted baseband signal. The results show the limitations of this technique when the LDMOSFET power amplifier operation conditions are close to saturation when better than 6-dB improvement in *ACLR* can only be achieved. Generally, an adaptive correction mechanism is required to maintain the linearized power amplifier performance over varying load, supply voltage, or temperature conditions. This means that the LUT needs to be updated continuously to keep difference between the source signal and the transmitted signal sufficiently small. This can be realized by downconverting the portion of the transmitted signal and comparing it with the source signal. In this case, it is important to provide the optimization of the word lengths required in different parts of the predistortion linearizer to reduce power consumption and increase bandwidth for the required adjacent channel interference level [34]. The feedback complexity can be reduced with special adaptation algorithm when a single mixer and an analog-to-digital converter (ADC) may be used in the feedback path instead of the full quadrature demodulation [35]. In addition, a noniterative adaptation method can be used to eliminate the convergence constrains usual for iterative methods [36].

9.3 Outphasing Power Amplifiers

The outphasing modulation technique was invented in the mid-1930s in order to improve both the efficiency and linearity of AM-broadcast transmitters [37]. Substantially later in the 1970s, its application was extended up to microwave frequencies under the name LINC (*linear amplification using nonlinear components*) [38]. An outphasing transmitter operates as a linear power amplifier system for amplitude-modulated signals having a linear transfer function over a wide range of the input signal levels by combining the outputs of two nonlinear power amplifiers that are driven with signals of constant amplitude but different time-varying phases corresponding to the envelope of the input signal.

A simple outphasing power amplifier system is shown in Fig. 9.13(*a*) [39]. The signal component separator (SCS) generates from the input amplitude-modulated signal two sinewave signals of constant envelopes with different phases $+\phi(t)$ and $-\phi(t)$. These two signals are then separately amplified by the identical nonlinear power amplifiers each and combined to produce the output amplitude-modulated signal. The peak output power is obtained with $\phi = 90°$ when currents from power amplifiers with equal amplitudes $I_L = I_1 = I_2$ are added in phase, similar to a push-pull operation. Zero output power corresponds to the signal with $\phi = 0°$ when equal currents from power amplifiers cancel each other resulting in $I_L = 0$. Intermediate values of phase in the range of $0° < \phi < 90°$ produce intermediate values of the output voltage amplitude. As shown in Fig. 9.13(*b*), the time-varying phase ϕ can be written using the vector sum of the output voltages V_1 and V_2 as

$$\phi = \sin^{-1}\left(\frac{V_L}{V_{LPEP}}\right) \tag{9.14}$$

where $V_L = I_L R_L$ is the output voltage amplitude across the load resistance R_L and V_{LPEP} is the maximum output voltage amplitude at peak envelope power.

The instantaneous collector efficiency of a simple outphasing system with Class-B power amplifiers can be calculated from

$$\eta = \frac{\pi}{4}\frac{V_L}{V_{LPEP}} \tag{9.15}$$

having a maximum value of 78.5% in saturation when $V_L = V_{LPEP}$ with $\phi = 90°$ and zero value when $V_L = 0$ with $\phi = 0°$. Thus, the efficiency of

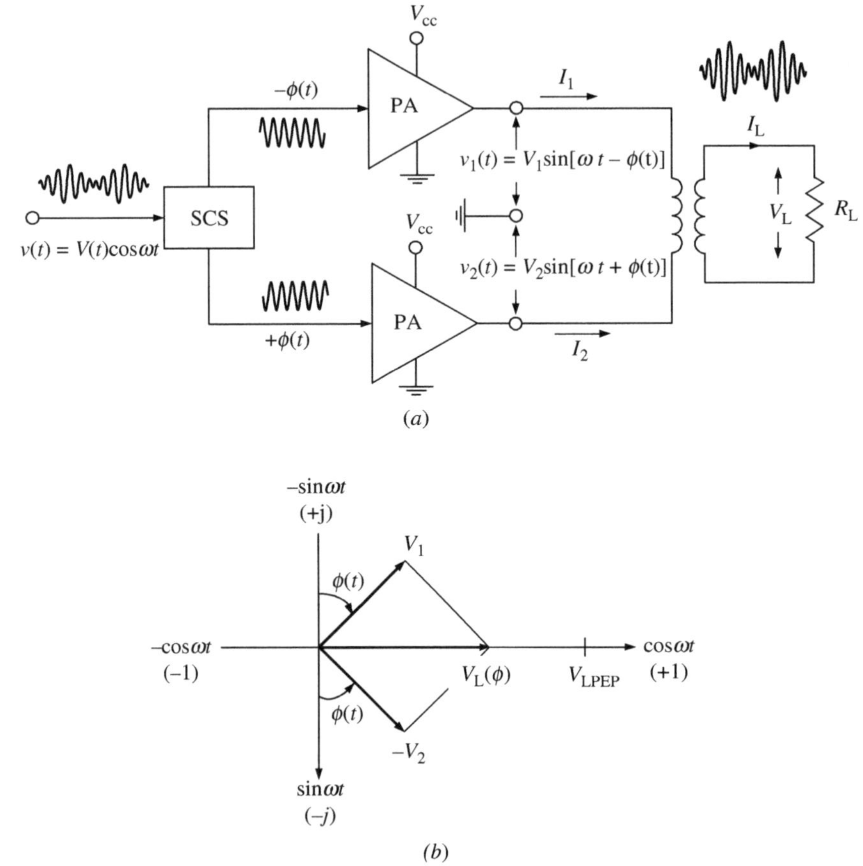

FIGURE 9.13 Simple outphasing power amplifier system.

a simple power amplifier outphasing system is the same as that of an ideal Class-B power amplifier, reducing linearly with the output voltage amplitude. In this case, to perform accurately the required signal component separation, it is necessary to use the DSP technique [40].

The outphasing power amplifier systems used at microwave frequencies use hybrid combiners to isolate the two power amplifiers from each other, allowing them to see resistive loads at all signal levels, as shown in Fig. 9.14(*a*). Typical structures of the hybrid combiners represent a quadrature branch-line coupler or an in-phase Wilkinson combiner, which are fully matched, and lossy combining structures with high isolation between the combined amplifying paths for properly phased and equal signal powers. However, because both power amplifiers deliver full power all of the time, the efficiency of such a hybrid-coupled microwave LINC transmitter varies with the output power, resulting in its significant degradation at lower power levels.

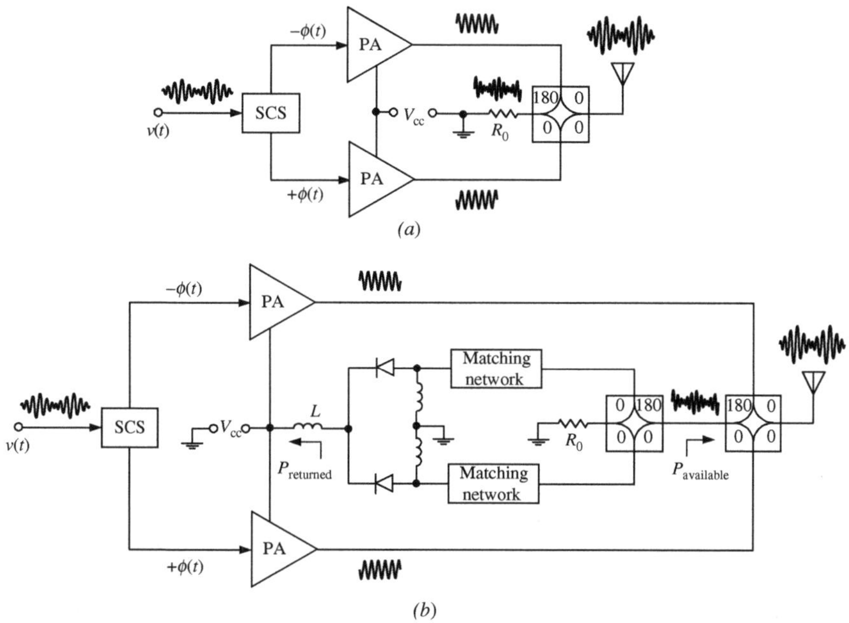

FIGURE 9.14 Outphasing power amplifier system with hybrid combiner.

This is because most of the output power is dissipated in the ballast resistor R_0 of the combining network when the two power amplifiers are operated substantially out of phase.

Figure 9.14(*b*) shows the power recycling schematic where an RF-to-dc converter is implemented with high-speed Schottky diodes and optimized matching networks [41]. To achieve better efficiency, the diodes should be switched fully having minimal series on-resistances. As a result, at the operating frequency of 1.95 GHz, the measured power reuse efficiency, which is defined as a ratio of the returned power $P_{returned}$ to the power available from the hybrid $P_{available}$, was found to be approximately 63% at power levels varied with supply voltage. The amount of the overall system efficiency improvement depends on the modulation scheme and could be compromised for modulation schemes that exhibit very deep variations in envelope power on a regular basis. By eliminating the ballast resistor in a Wilkinson combiner resulting in a simple lossless *T*-type combiner, efficiency can also be increased due to varying load impedance seen by each device [42]. However, there is a tradeoff between efficiency and linearity when a matched combiner provides greater linearity performance compared to the lossless combiner, while the lossless combiner shows equal or better efficiency performance compared to the matched combiner.

In a practical LINC transmitter, there are three main mechanisms that degrade the overall performance: power gain and phase imbalance between two RF amplifying paths and different nonlinear characteristics of both power amplifiers [43]. For example, if the gain imbalance between paths is about 1%, the output rejection may reduce to 45 dB and even further to 28 dB only depending on the relative level of the input modulating signal amplitude. On the other hand, a phase error between amplifying paths as low as 2° diminishes the undesired response rejection to only 33 dB for the best case. At the same time, effect of the imbalance of the power amplifier nonlinearities is less meaningful. Therefore, the practical implementation of the entire LINC outphasing power amplifier system is very difficult because of its inherent strong sensitivity to the amplitude and phase errors caused, first of all, by the difference in electrical lengths between the two power amplifier branches.

Figure 9.15 shows the block diagram of an outphasing LINC architecture incorporating a feedback loop to compensate for phase errors [44]. In this case, the phase difference between the two branches is detected by a multiplier that allows a phase control of one branch by adding or subtracting a certain phase increment. As a result, the output power of 7.5 W with a power amplifier efficiency of 21% (including hybrid and isolator losses) was achieved at 900 MHz. An additional linearity improvement can be achieved by providing an adaptive amplitude adjustment as well, by using a feedback loop from the output to measure output signal in adjacent channels and find optimal gain and phase correction by optimization algorithm [45]. To provide high amplitude and phase accuracy of the LINC system, a DSP-based architecture can be developed where the compensation of the amplitude and phase imbalances can be accomplished using calibration schemes [46]. As an example, for an LINC transmitter

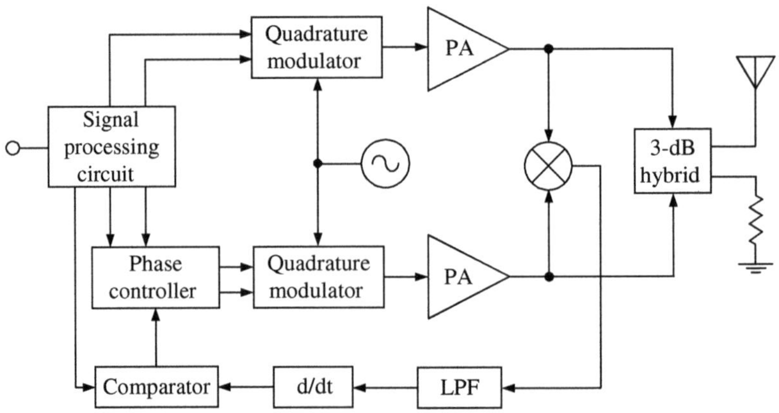

Figure 9.15 LINC transmitter with phase error compensating loop.

amplifying a $\pi/4$-shifted differential QPSK signal, an *ACLR* of –65 dBc can be obtained without predistortion when the amplifier branch phase imbalance range is less than ± 0.6° and the gain imbalance range is less than ± 0.07 dB. With adaptive DPD the same *ACLR* level can be achieved over a phase imbalance range of ± 7° and a gain imbalance range of ± 1 dB [47].

The efficiency of an outphasing LINC system at lower output voltages can be significantly improved by using a lossless Chireix combiner at the output of the outphasing power amplifier system shown in Fig. 9.16(*a*), which includes additional series quarterwave transmission lines and shunt reactances. Phasor analysis of the load network for the time-varying phase ϕ with an impedance-transforming quarterwave transmission line results in

$$Y_3 = \frac{2R_L}{Z_0^2}\frac{V_L}{V_{LPEP}}(\sin\phi + j\cos\phi) \tag{9.16}$$

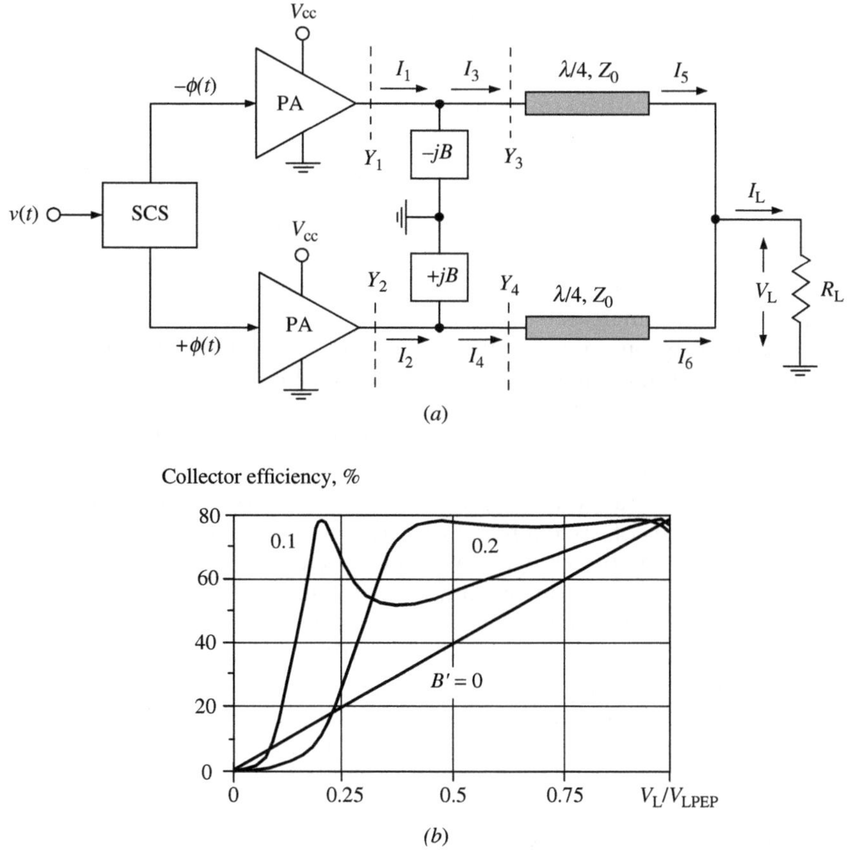

FIGURE 9.16 Chireix outphasing power amplifier system and instantaneous efficiencies.

$$Y_4 = \frac{2R_L}{Z_0^2}\frac{V_L}{V_{LPEP}}(\sin\phi - j\cos\phi) \tag{9.17}$$

where Z_0 is the characteristic impedance of the transmission lines [39]. From Eqs. (9.16) and (9.17), it follows that the admittances Y_3 and Y_4 are purely resistive only for $\phi = 90°$, corresponding to the case of in-phase output currents. However, for most values of phase ϕ, the power amplifiers have highly reactive loads that become completely reactive when $\phi = 0°$ with 180° out-of-phase output currents. The effect of the reactive loads can be partially compensated by adding the corresponding shunt susceptances $-B$ and $+B$, respectively. In this case, the reactive parts of the admittances $Y_1 = Y_3 - jB$ and $Y_2 = Y_4 + jB$ can be zeroed at one specific output voltage amplitude by setting

$$B = \frac{2R_L}{Z_0^2}\frac{V_L}{V_{LPEP}}\sqrt{1-\left(\frac{V_L}{V_{LPEP}}\right)^2} \tag{9.18}$$

which can be obtained by substituting Eq. (9.14) into Eqs. (9.16) and (9.17). As a result, for the case of a purely resistive load, the instantaneous collector efficiency of a Chireix outphasing system with ideal Class-B power amplifiers can reach the maximum value of

$$\eta = \frac{\pi}{4} \tag{9.19}$$

The instantaneous efficiencies of the Chireix outphasing system for different values of the normalized shunt susceptance $B' = BZ_0^2/2R_L$ are shown in Fig. 9.16(*b*), from which it follows that the selection of a proper value of B increases efficiency at a specified medium level of the output voltage amplitude however, it is degraded at low and high amplitudes. Using a value $B' = 0.2$ can provide high efficiency over the upper 6 dB of the output voltage range. The case when $B' = 0$ corresponds to the collector efficiency variations of an ideal Class-B power amplifier. An improvement in the average efficiency calculated over a wide range of output voltages for various amplitude-modulated signals of up to a factor of 2 over that of an ideal Class-B power amplifier can be achieved by properly selecting the shunt susceptances in outphasing power amplifier system. On the whole, to design such an outphasing system, it is necessary to consider simultaneously such factors as a complexity of the SCS circuit and sensitivity of the power amplifiers to the wide range of load impedances. In terms of the Chireix power amplifier parameters, it is very important to minimize the combiner impedance mismatching, especially at microwave frequencies where the Chireix outphasing combiner should consist of two shunt stubs of equal and opposite reactances and two series quarterwave transmission lines [48].

For a Chireix outphasing system with a transmission-line combiner, an average efficiency of 30% with an *ACLR* of –45 dBc was achieved for a single-carrier WCDMA signal in a frequency range of 2.11 to 2.17 GHz, measured at a channel output power of 20 W [49]. Further efficiency improvement can potentially be achieved by using high-efficiency power amplifiers operating in different switching modes or their approximations. For example, a drain efficiency of a 800-MHz outphasing system based on the voltage-mode Class-D amplifiers implemented in a 0.18-μm SiGe BiCMOS process and transmission-line Chireix combiner was improved for a CDMA IS-95 signal with a $PAR = 5.5$ dB from 38.6 to 48% at an output power of 15.4 dBm [50]. Applying a single-carrier WCDMA signal, a drain efficiency of 51% with an output power of 21.6 dBm were measured for a 1.92-GHz outphasing system based on the voltage-mode Class-D amplifiers and integrated lumped Chireix combiner fully implemented in a 0.13-μm CMOS process [51].

Using a push-pull configuration with a rat-race balun for each saturated Class-B power amplifier based on 0.25-μm pHEMT devices in an outphasing system with a Chireix microstrip combiner had contributed to a system efficiency of 42.2% with a channel power of 31.2 dBm (–7 dB backoff) for a single-carrier 2.14-GHz WCDMA signal, which is more than two times improvement over the Wilkinson combiner system [52]. Note that the Chireix combiner when used with ideal sources leads to a linear LINC system. However, for 10-W GaN HEMT power amplifiers operating in a Class-F mode, greater than 4-dB expansion in gain and 14° compression in phase were obtained in a Chireix outphasing system due to load-pulling effect [53]. Because the concept of an outphasing system using an asymmetric Chireix transmission-line combiner with different electrical lengths and additional input phase adjustment where both power amplifiers are based on 15-W LDMOSFET transistors operating in an inverse Class-F mode, a drain efficiency of 48% at 6-dB output power backoff was achieved at an operating frequency of 2.14 GHz [54].

Figure 9.17(*a*) shows the asymmetric architecture for the outphasing of Class-E power amplifiers, where the transmission lines TL_1 and TL_2 have electrical lengths of $\theta + \delta$ and $\theta - \delta$, respectively [55, 56]. In this case, the transmission lines TL_1 and TL_2 rotate the impedance loci on the Smith chart to center them on the line at 65°, which corresponds to the maximum efficiency of the Class-E power amplifiers when $\theta = 147.5°$. As a result of centering the impedance loci, the amplitudes of the power amplifier outputs ideally vary identically with difference in phase between the drive signals. The phases of the output signals are nearly the same over most of the amplitude range of $\Delta\phi$. The value of a differential line length δ can be chosen to alter the instantaneous efficiency characteristics to optimize the average efficiency for a given signal. The circuit schematic of a lumped MOSFET prototype of the Chireix outphasing system with asymmetric combiner operating at 1.82 MHz is shown in Fig. 9.17(*b*). Here, the two identical Class-E power amplifiers using IRF510 MOSFETs and

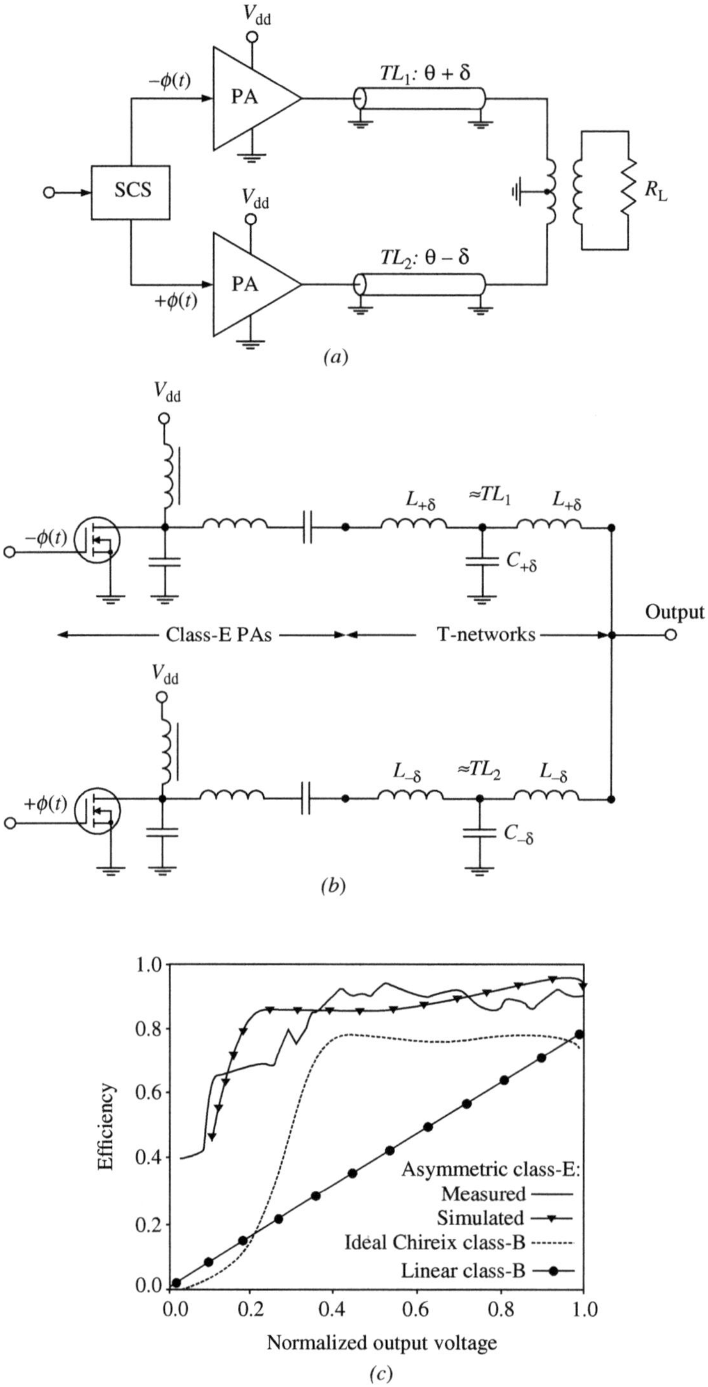

Figure 9.17 Asymmetric Chireix outphasing systems with Class-E power amplifiers.

achieving a drain efficiency of 95% with an output power of 14 W each were implemented using lumped shunt capacitors and series lumped inductors. The asymmetric Chireix combiner represents a lumped low-pass *T*-network replacing the transmission lines TL_1 and TL_2, where the values of the *LC* elements can be determined analytically using a single-frequency equivalence between the lumped and distributed circuits.

The chain matrix $[ABCD]_{TL}$ for a transmission line with electrical lengths of $\theta \pm \delta$ and the chain matrix $[ABCD]_{LC}$ for a low-pass *T*-type lumped circuit, consisting of a shunt capacitor and two series inductors, are written respectively as

$$[ABCD]_{TL} = \begin{bmatrix} \cos(\theta \pm \delta) & jZ_0 \sin(\theta \pm \delta) \\ j\dfrac{\sin(\theta \pm \delta)}{Z_0} & \cos(\theta \pm \delta) \end{bmatrix} \tag{9.20}$$

$$[ABCD]_{LC} = \begin{bmatrix} 1 & j\omega L \\ 0 & 1 \end{bmatrix} \begin{bmatrix} 1 & 0 \\ j\omega C & 1 \end{bmatrix} \begin{bmatrix} 1 & j\omega L \\ 0 & 1 \end{bmatrix}$$

$$= \begin{bmatrix} 1 - \omega^2 LC & j\omega L\,(2 - \omega^2 LC) \\ j\omega C & 1 - \omega^2 LC \end{bmatrix} \tag{9.21}$$

Equating the corresponding elements of both matrices yields the simple equations to determine the corresponding parameters of the lumped *T*-network through the transmission-line parameters as

$$C_{\pm\delta} = \frac{\sin(\theta \pm \delta)}{\omega Z_0} \tag{9.22}$$

$$L_{\pm\delta} = \frac{Z_0}{\omega} \tan\frac{\theta \pm \delta}{2} \tag{9.23}$$

where Z_0 is the transmission-line characteristic impedance.

Figure 9.17(*c*) shows the simulated and measured results of a Class-E outphasing efficiency versus normalized output voltage for $\theta = 147.5°$ and $\delta = 17.4°$, from which it follows that the drain efficiency of 85% or better for amplitude range of 15 dB from 0.8 W to full output of 27.5 W in simulation and for amplitude range of 10 dB in measurement is maintained. In contrast, the drain efficiency for the outphasing with ideal Class-B power amplifiers is no better than 78.5%. The value of the difference δ in a transmission-line length (or its equivalent *LC* *T*-network) does not appear to be critical, and high efficiency is obtained for δ equals to 10°, 18.4°, 22°, and 48°.

The schematic of a CMOS outphasing system with a transformer-based power combiner intended for broadband operation is shown in Fig. 9.18(*a*) [57]. Here, the series resonators of both Class-E power amplifiers are shifted to the secondary side of the combiner, in series with the transformer leakage inductance defined as $L_{leak} = L_p(1 - k^2)/k^2$, where *k* is the transformer coupling coefficient. The leakage inductance

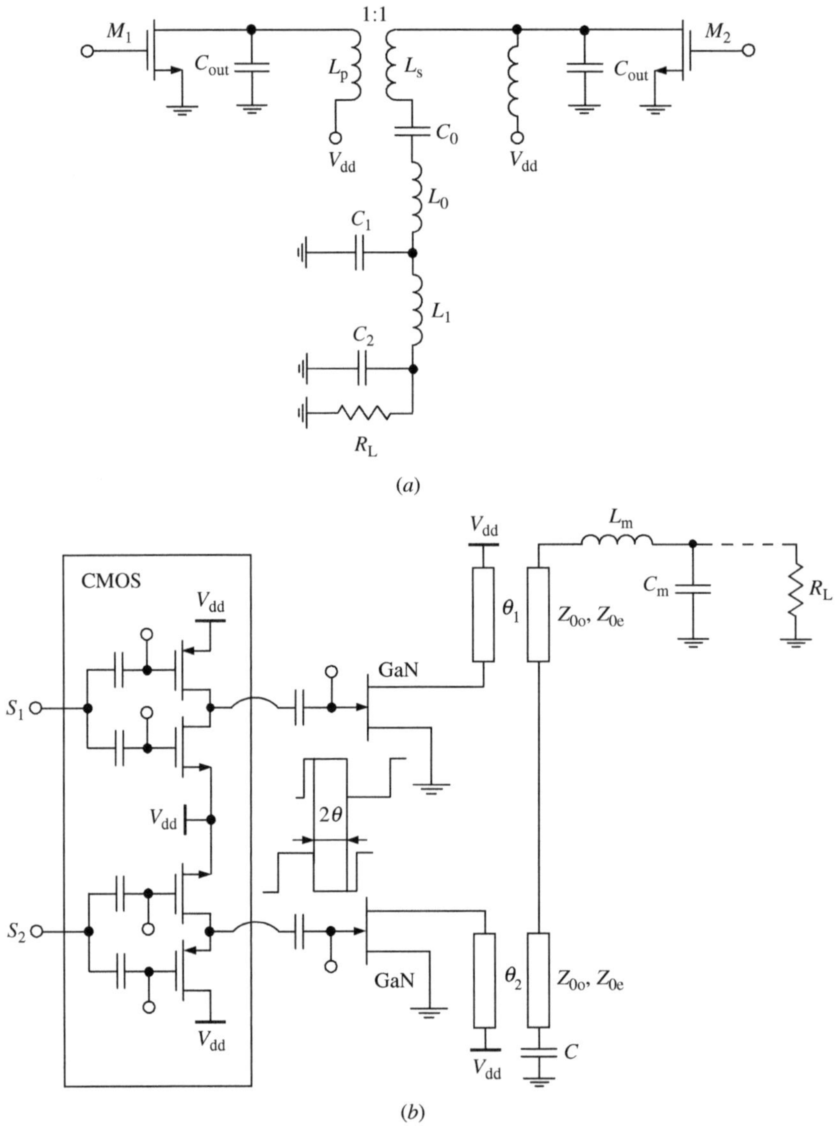

FIGURE 9.18 CMOS Chireix outphasing systems with Class-E power amplifiers.

is merged with the resonator inductor and inductor of the wideband two-section output matching circuit. To achieve a transformation ratio equal to 1, the primary inductance L_p and secondary inductance L_s must relate to each other as $L_p/L_s = 1/k$. The compensation susceptance of the transistor M_2 is created by changing the value of the dc-feed inductance to a larger value. The primary transformer winding serves as a dc-feed inductance for the transistor M_1 and a compensation susceptance. The resonator at the secondary side has a low loaded quality factor to enable a high-bandwidth operation. For a larger bandwidth, the inductance L_p must be small and coupling coefficient k must be close to 1. As a result, such a Chireix outphasing system with a combining circuit implemented on a three-layer PCB and two 65-nm CMOS Class-E power amplifiers achieved a peak power of greater than 30 dBm and a drain efficiency of greater than 46.4% at 6-dB power backoff over a frequency range of 600 to 800 MHz.

Because a lumped-element combining circuit is difficult to implement for high powers at microwave frequencies, coupled lines can be used to combine the outputs like in a Marchand balun. Figure 9.18(*b*) shows the circuit schematic of a two-stage Chireix outphasing system where the push-pull drivers are fabricated in a standard 65-nm CMOS technology and the Class-E power amplifiers are based on a 28-V GaN HEMT technology [58]. The CMOS drivers are ac-coupled to overcome the negative bias voltage required for GaN HEMT devices. The electrical lengths of the coupled lines with odd-mode characteristic impedance $Z_{0o} = 1/Y_{0o}$ and even-mode characteristic impedance $Z_{0e} = 1/Y_{0e}$ are unequal and the terminating capacitance is defined as

$$C = \frac{Y_{0o} + Y_{0e}}{2\omega} \cot\theta_2 \tag{9.24}$$

that tunes out the leakage inductance of the coupled-line transformer. A fourth-order Butterworth matching filter transforms the 50-Ω antenna impedance to the required Class-E load and sets the bandwidth property. As a result, a drain efficiency of 65.1% and a total system efficiency of 51.6% with an output power of 19 W were measured for a single-carrier WCDMA signal with *PAR* of 7.5 dB at 1.95 GHz. Besides, a peak drain efficiency of greater than 60% can be achieved in a frequency range of 1800 to 2050 MHz at 6-dB power backoff.

A package-integrated Chireix outphasing system with two Class-E power amplifiers based on high-power GaN HEMT devices, where a Chireix power combiner uses bondwires with MIM capacitors and a second-order Butterworth bandpass filter is implemented outside the package, reaches a drain efficiency of 53.5% and an *ACLR* better than –49 dBc with a memoryless digital predistortion DPD at 2.3 GHz for a 5-MHz WCDMA signal with a *PAR* of 9.6-dB [59]. In a single-tone frequency mode, this high-power package-integrated outphasing Chireix system provides an output power of 47.9 ± 0.6 dBm and a drain efficiency of greater than 50% across greater than 260 MHz within 2.1 to 2.4 GHz at 6-dB power backoff with a nominal 28-V drain supply.

The basic advantage of a Chireix combining technique is that the combiner is ideally lossless, and that the real components of the effective load admittances seen by the individual power amplifiers vary with outphasing so as to minimize power amplifier losses when output power reduces. However, the reactive portions of the effective load admittances are only zero for at most two outphasing angles, and become large outside of a limited power range. This limits efficiency, due both to loss associated with added reactive currents and to degradation of power amplifier performance with reactive loading. In this case, multistage technique applied to a conventional Chireix outphasing system can overcome the loss and reactive loading problems, providing an ideally lossless power combining with nearly resistive loading of the individual power amplifiers over a very wide output power range when high average efficiencies are achieved even for large peak-to-average power ratios [60].

Figure 9.19(*a*) shows the simplified four-stage outphasing architecture, where the power amplifiers are modeled as ideal voltage sources. The lossless power combiner has four input ports and one output port connecting to the load, and it comprises reactive elements having specified reactances at the operating frequency. The system behavior can be described based on the relationships between the source voltages and input currents of the network of Fig. 9.19(*a*) according to

$$\begin{bmatrix} I_A \\ I_B \\ I_C \\ I_D \end{bmatrix} = \frac{1}{X_1} \begin{bmatrix} \gamma + j(1-\beta) & -\gamma + j\beta & \gamma & -\gamma \\ -\gamma + j\beta & \gamma - j(\beta+1) & -\gamma & \gamma \\ \gamma & -\gamma & \gamma + j(\beta+1) & -\gamma - j\beta \\ -\gamma & \gamma & -\gamma - j\beta & \gamma + j(\beta-1) \end{bmatrix} \begin{bmatrix} V_A \\ V_B \\ V_C \\ V_D \end{bmatrix} \tag{9.25}$$

where $\gamma = R_L/X_1$, $\beta = X_2/X_1$, and voltage amplitudes are equal having different control angles used for outphasing. In this case, $+X_1$ and $+X_2$ represent inductive reactances due to inductances and $-X_1$ and $-X_2$ represent capacitive reactances due to capacitances. The effective admittance at a combiner input port is the complex ratio of current to voltage at the port with all sources active. The effective admittances represent the admittances seen by the sources when they are operating under outphasing control. A key advantage of this outphasing system is that the susceptive portion of admittance loading the power amplifiers is substantially smaller than that with a conventional Chireix combining.

The drain efficiency of a multistage outphasing system using ideal saturated Class-B power amplifiers can be obtained by

$$\eta = \frac{\pi}{4} \frac{\sum_{j=1}^{N} \mathrm{Re}\,(Y_{\mathrm{eff},j})}{\sum_{j=1}^{N} \left| Y_{\mathrm{eff},j} \right|} \tag{9.26}$$

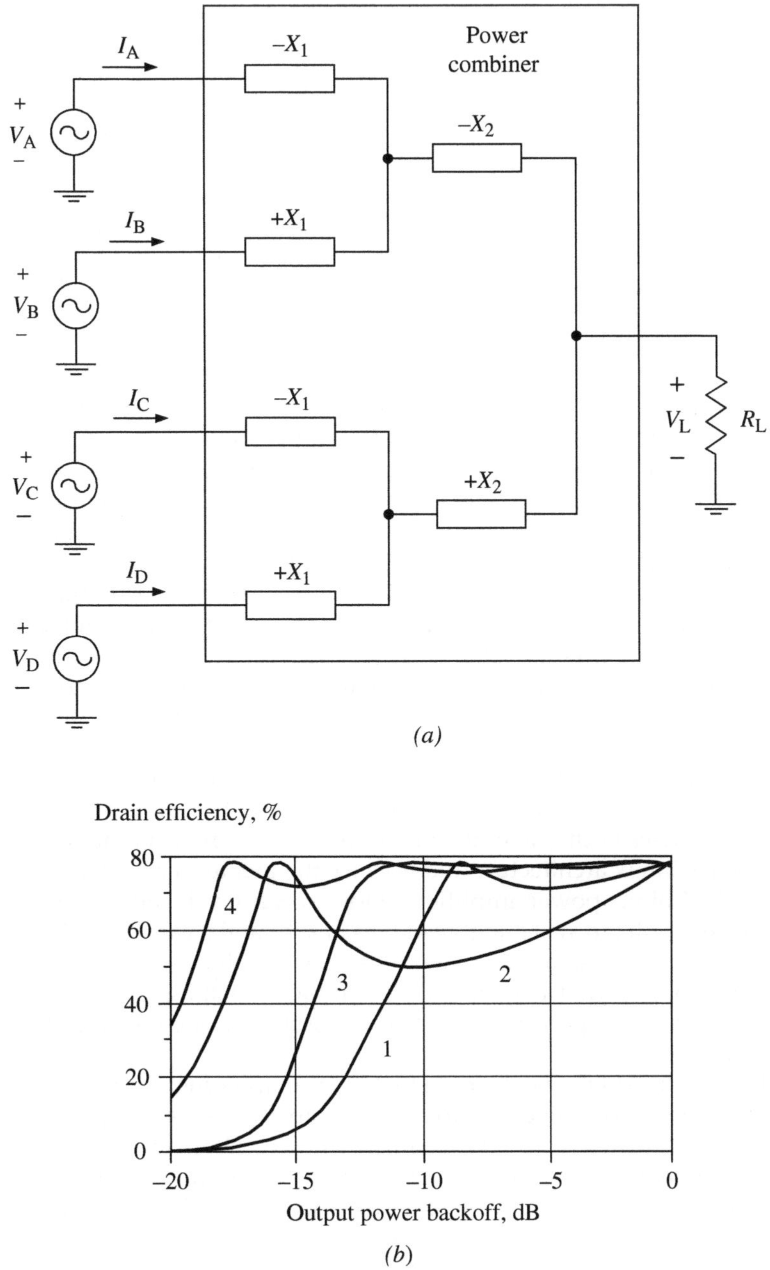

FIGURE 9.19 Four-stage outphasing architecture and instantaneous efficiencies.

where $Y_{eff,j}$ is the effective admittance at a corresponding combiner input port seen by the *j*th source and *N* is the number of ports. For example, for the output probability density function (PDF) representing a WLAN signal with a *PAR* of 9.01 dB, the average drain efficiency remains almost constant for much larger range of backoff output powers for a four-stage outphasing architecture (curves 3 and 4) compared with a conventional Chireix outphasing architecture (curves 1 and 2), as shown in Fig. 9.19(*b*), with average drain efficiencies of 38% and 46.1% corresponding to curves 1 and 2 and average drain efficiencies of 56.9% and 69% corresponding to curves 3 and 4, respectively [60].

Figure 9.20(*a*) shows the implementation of a Chireix outphasing system, where each of the two outphased sources is constructed of a pair of the power amplifiers designed to operate in a Class-E/F_{odd} mode at 27.12 MHz with output powers up to 750 W using ARF521 power MOSFETs. The two pairs of power amplifiers are outphased using a Chireix combiner having $X_c = 13.6\ \Omega$ and supply a load resistance $R_L = 13\ \Omega$. For the four-stage outphasing system shown in Fig. 9.19(*a*), the power combiner is used with a load resistance of 50 Ω and reactance values $X_1 = 35\ \Omega$ and $X_2 = 48.78\ \Omega$. The drain efficiency versus output power for both the four-stage combining system and conventional Chireix system is shown in Fig. 9.20(*b*), where the four-stage outphasing architecture (curve 2) demonstrates superior performance at low power levels compared with a conventional Chireix system (curve 1), with greater than 12% higher efficiency for 300-W output at 10-dB power backoff and greater than 20% higher efficiency at 12.5-dB power backoff. The difference in efficiency arises both due to the higher reactive currents in the conventional Chireix architecture and because the greater reactive loading on the Chireix power amplifiers at low power levels causes them to lose zero-voltage switching, resulting in substantial capacitive discharge loss. At the same time, the four-stage outphasing system is advantageous in that it maintains desirable switching conditions down to far lower power levels than is possible in a conventional Chireix system.

In a practical implementation, a four-stage outphasing system with lumped-element combiner and four microstrip inverse Class-F power amplifiers achieves a peak CW (continuous wave) drain efficiency of 68.9%, with efficiency greater than 55% over a 5.5-dB power range at 2.14 GHz. As a result, an average drain efficiency of 57% at an output power of 42 dBm with an *ACLR* of –36.6 dBc was provided for a 5-MHz WCDMA signal with a *PAR* of 3.47 dB [61]. Here, the inductive branch impedances are synthesized from the series *LC* combinations in order to provide dc blocking, and capacitive branch impedances use both series and parallel capacitor combinations to facilitate tuning and so that all branches have the same physical length.

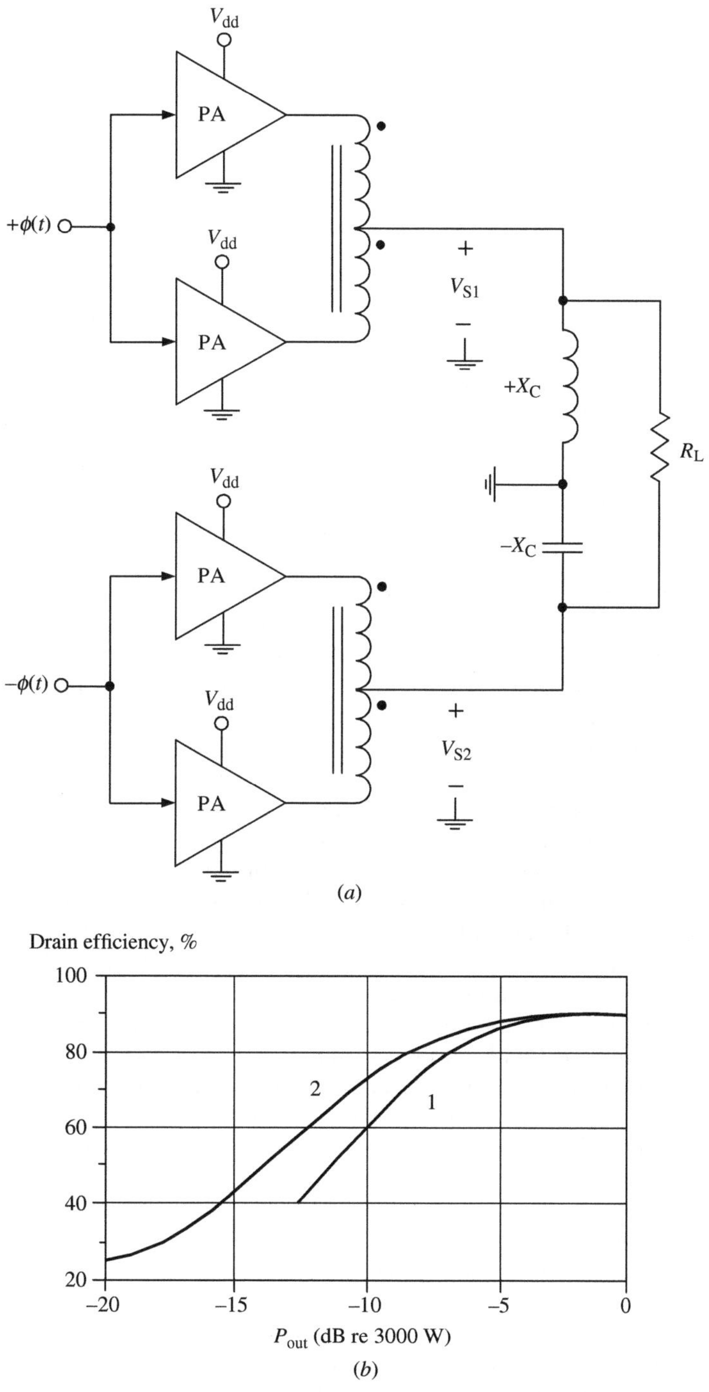

FIGURE 9.20 Chireix outphasing power amplifier system and instantaneous efficiencies.

9.4 Envelope Tracking

For a convenience of further consideration, it is important to represent the collector (drain) efficiency of a power amplifier in an analytical form (excluding losses in the output load network), which can be expressed as

$$\eta = \frac{P_1}{P_0} = \frac{1}{2}\frac{I_1}{I_0}\frac{V_1}{V_{cc}} \tag{9.27}$$

where $P_1 = I_1V$ is the output power at fundamental frequency, $P_0 = I_0V_{cc}$ is the dc supply power, I_1 is the fundamental current amplitude, I_0 is the dc current, V_1 is the fundamental collector voltage amplitude, and V_{cc} is the collector supply voltage. For the operation conditions with the same conduction angle (e.g., the conduction angle in Class B is equal to 180° regardless of the collector current waveform amplitude), the current ratio between the fundamental and dc components I_1/I_0 keeps a constant value. Generally, depending on the conduction angle, the current ratio I_1/I_0 varies from 1 in Class A with a conduction angle of 360° to 2 for ideal limiting case of Class C with a conduction angle of 0°. For a Class-B operation, the current ratio I_1/I_0 is equal to 1.57. Consequently, for a nearly Class-B operation mode when the value of the conduction angle slightly deviates from 180°, the current ratio I_1/I_0 varies within a small range of 10 to 20%. Thus, as follows from Eq. (9.27), the main factor of a collector efficiency improvement at backoff output power levels is the voltage ratio V_1/V_{cc}, which should be kept constant for different output power levels. This can be achieved by reducing the supply voltage V_{cc} using envelope tracking technique or increasing the load resistance.

Because the collector efficiency is proportional to the ratio of the fundamental amplitude to dc supply voltage, it becomes extremely small already at the output powers lower than the peak power by 10 dB, provided the load resistance and dc supply voltage are kept constant. However, usually for CDMA2000, WCDMA, or LTE cellular phone transmitters, the output power can vary in a wide dynamic range of about 80 dB with maximum statistically averaged transmitting power required to deliver signal to the base station of about 15 to 30 dB less than the peak output power. Therefore, the average envelope tracking technique can be very useful to increase power amplifier efficiency in a wide range of output powers of the cellular phone transmitters by varying the dc supply voltage according to the RF signal envelope. When the supply voltage tracks the instantaneous output envelope, it is known as a *wideband envelope* or *envelope-follower tracking* (ET). However, when the supply voltage tracks the long-term average of the output envelope, it is known as an *average envelope* or *power tracking* (APT). The ET is necessary to increase efficiency at fixed output power, whereas APT is required to increase efficiency at backoff powers.

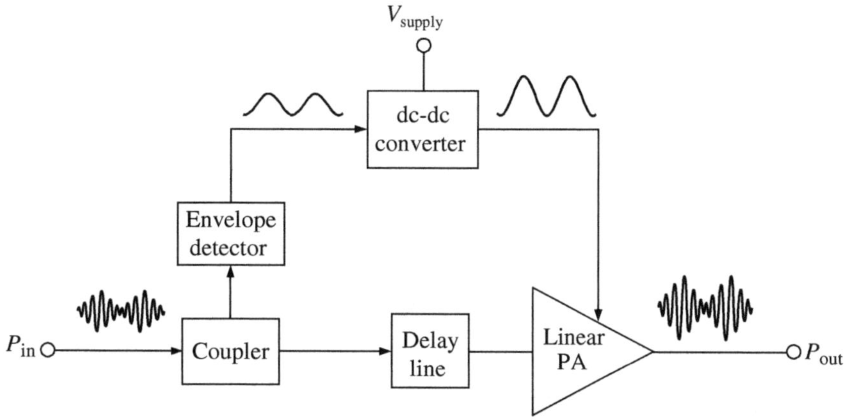

Figure 9.21 Envelope tracking power amplifier architecture with analog control.

Figure 9.21 shows the envelope tracking power amplifier architecture with analog control, where the envelope detector is included at the input to detect the input signal envelope [62, 63]. Here, a dc-dc converter is used to provide the dynamically controlled supply voltage to a linear power amplifier. Though both buck (step-down) or boost (step-up) dc-dc converters can be used, the latter allows operation of the power amplifier from a supply voltage higher than the dc-supply voltage. Such a dc-dc boost converter implemented using AlGaAs/GaAs HBT process with a 10-MHz switching (clock) frequency can provide the efficiency of 74% at maximum dc power [62]. The ripple obtained in the dc-dc converter output results in the spurious in the output signal spectrum of approximately 60 dB lower than the fundamental spectral component. The delay line is necessary to compensate for the phase misalignment between the envelope and RF signal paths. In modern transmitters for wireless applications, the DSP technique is used to provide both the separate envelope signal and predistortion linearization using the feedback loop from the output.

To design wideband high-efficiency envelope amplifier, a linear stage to provide a wideband voltage source and, in parallel, a switching stage to provide an efficient current supply based on the buck dc-dc converter topology can be used [64]. In this case, with the average switching frequency of the nMOS switch of approximately 1.28 MHz, an efficiency of 76.7% for a WCDMA signal (PAR = 6.6 dB) with a peak output voltage of 29.5 V and a root-mean-square (rms) voltage of 12.8 V were achieved. As a result, a PAE of 58% at an average output power of 42 dBm was obtained for a WCDMA ET-based power amplifier for base station applications using a high-voltage GaAs HBT device and a memoryless DPD to achieve an $ACLR$ of –49 dBc [65]. For an envelope amplifier with two switching stages and one linear stage, where the first switching amplifier and the linear amplifier

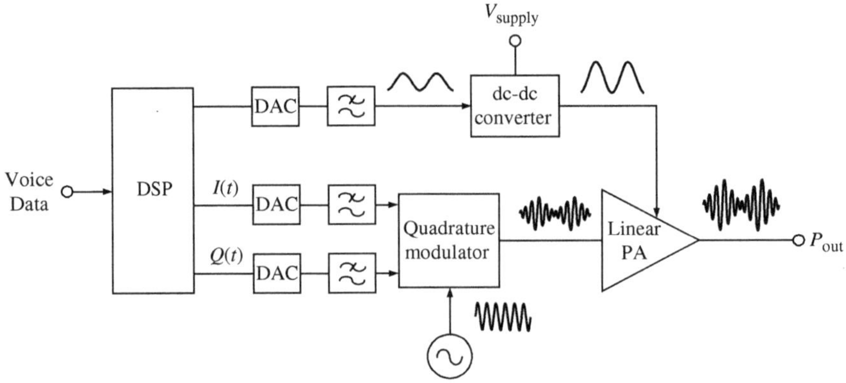

FIGURE 9.22 Envelope-tracking (ET) power amplifier architecture with digital control.

fabricated in a 0.15-μm CMOS process provide wideband and high-efficiency operation, whereas the second switching stage provides a reduced bandwidth dynamically varying envelope signal to the supply terminal of the linear stage to minimize its power loss, a maximum average efficiency is increased to 82% for a 10-MHz LTE signal with a *PAR* of 6 dB at 29.7-dBm output power [66]. With a digital control used to cancel the delay in the output filter of the switching stage, an efficiency of 79% at 29-dBm average output power for a 20-MHz LTE signal with a *PAR* of 6.9 dB was achieved [67].

The linearity of a power amplifier with envelope tracking is usually worse than that of the power amplifier with fixed supply voltage because of the gain and phase variations with supply voltage. In this case, it is possible to use a fixed analog predistortion or digital control to minimize the increased nonlinearity to a considerable extent. Figure 9.22 shows the envelope-tracking power amplifier architecture with a digital control [68]. In addition to providing a proper control voltage for the dc-dc converter according to the signal envelope, the DSP system also computes a predistorted input signal for both the in-phase (*I*) and quadrature (*Q*) channels using amplifier amplitude and phase characteristics. The practical results show that, in this case, the *ACPR* of the IS-95 CDMA input signal can be improved by 8 dB at an output power of 28 dBm. However, the *ACPR* is quite sensitive to the timing relationships between the varying supply voltage and the input signal. For a GaAs MESFET power amplifier with boost converter operated at 950 MHz with maximum output power of 1 W, the power usage efficiency calculated in accordance with a PDF of its output power is 1.64 times higher than the one for just battery operation [63]. However, to meet CDMA IS-95 specifications for *ACPR*, a fast feedback loop regulation scheme and dynamic gate biasing are needed to reduce the intermodulation distortions caused by the gain variation (due to the significant variation of the device transconductance with the supply voltage) and parasitic phase modulation (due to the inherently nonlinear behavior of the intrinsic device capacitances). Efficiency improvement can be achieved for a power amplifier with harmonic tuning instead of a linear power amplifier operating in Class AB using a high-performance GaN HEMT device [69]. In this case, a 2.14-GHz GaN HEMT power amplifier with a drain efficiency of 81% and a peak power of 40 W achieved a 52.5% composite *PAE* with high linearity for a 5-MHz WCDMA signal with a 23-MHz supply modulator bandwidth using the DPD with dynamic deviation reduction demonstrated [70].

It is much easier to provide an APT because power control dynamically changes the supply voltage and current as a function of power at a much slower rate compared to envelope. The Class-S modulator that can provide a long-term dynamically controlled supply voltage is similar in form to a buck dc-dc converter where the width or duty ratio of the pulses is proportional to an input control voltage. The control voltage corresponds to the rms value of the modulated signal. The high switching frequency provides several advantages: reduced value and size of the low-pass filter (LPF), better suppression of the switching frequency, and fast dynamic response. Maximizing the quality factors of the elements of the LPF and minimizing the on-resistances of the switching nMOS and pMOS transistors using larger size of their gate channel widths can result in a Class-S modulator efficiency of approximately 90% at the switching frequency of 5 MHz and slightly less at that of 10 MHz [71]. Figure 9.23 shows the envelope-tracking (ET) power amplifier architecture with a Class-S modulator.

The experimental results for a 2-W ET MESFET power amplifier using a high-efficiency Class-S modulator intended for CDMA cellular handset applications in a frequency range of 824 to 849 MHz are shown in Fig. 9.24 [71]. The phase deviations of the output signal

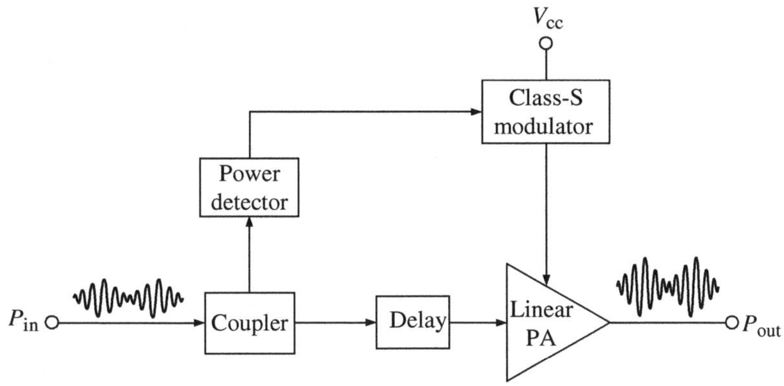

Figure 9.23 Envelope-tracking (ET) power amplifier architecture with Class-S modulator.

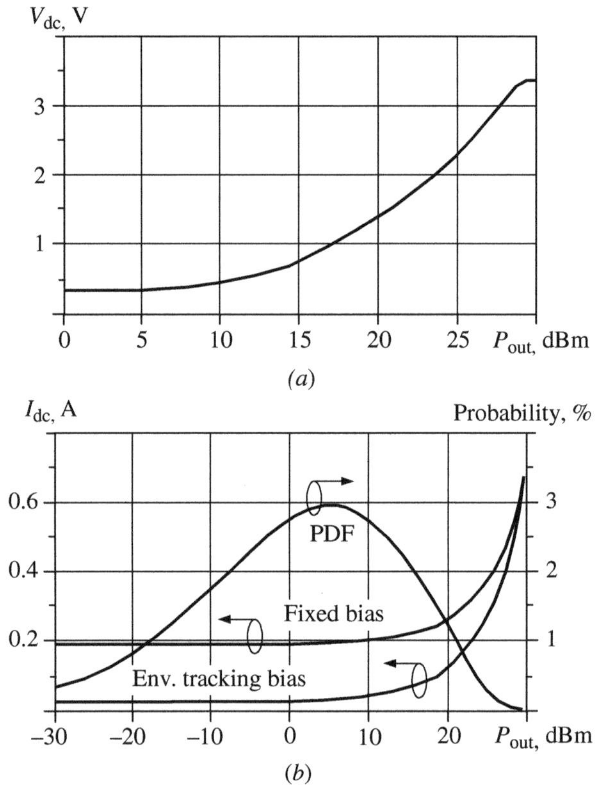

FIGURE 9.24 Envelope-tracking (ET) power amplifier dc supply (*a*) voltage and (*b*) current.

across the dynamic range of the supply voltage were less than 3° up to the 2-dB gain compression point resulting in adjacent and alternate channel power ratios of –46 dBc and –56 dBc, respectively. Each *ACPR* is measured as the ratio of power in a 30-kHz bandwidth offset from the carrier by 885 kHz for adjacent channel and 1.98 MHz for alternate channel. The supply voltage is dynamically stepped down via the Class-S modulator to approximately 0.3 V in a deep backoff with maximum voltage of about 3.3 V due to small voltage drop from the battery dc supply voltage of 3.5 V, as shown in Fig. 9.24(*a*). In this case, more than five times improvement in power usage efficiency, as shown in Fig. 9.24(*b*), was obtained compared to a power amplifier with fixed bias voltage when the significant amount of the quiescent current is still kept over a wide range of output powers.

In handset cellular applications, the dc-dc converters should provide high-efficiency, small-size, and low-cost operation. In this case, a deep submicron SiGe BiCMOS process technology can be used to fabricate the monolithic supply-modulated power amplifier where the power transistor can be implemented using the SiGe HBT process [72].

The size of the passive elements can be reduced to practical values for integration by increasing the switching frequencies to the order of 100 MHz. Switches can be implemented using nMOS and pMOS devices with optimum channel widths to minimize their power losses. The filter capacitor is realized as an MOS capacitor having higher specific capacitance compared with an MIM capacitance. To increase the quality factor of the filter inductance, it is implemented using a thick last metal layer far above the substrate. Such a monolithic 900-MHz power amplifier with a high-speed synchronous buck dc-dc converter can provide substantially higher efficiency compared with the similar power amplifier using constant voltage supply at lower power levels. Because the bandwidth of the ET modulator is related to the bandwidth of the RF signal envelope that can be as high as 20 MHz, it is clear that the design of the modulator and the interface to the power amplifier requires care in design and implementation.

By using a buck-boost dc-dc converter with high efficiency over a wide loading range, the system efficiency can be further improved [73, 74]. The boost converter is only used for the linear envelope stage, while the buck converter is directly connected to the battery with a nominal voltage of 3.4 V. By boosting the supply voltage to the linear stage to a fixed 5 V regardless of the battery voltage variation, it provides a maximum supply voltage of 4.5 V, considering the 0.5-V drop across the modulator. The efficiency degradation due to the boost converter is minimal because a major part of power is generated by the switching stage. In a practical implementation, a 1-W boost-mode hybrid switching supply modulator including the linear stage and both boost and buck converters can occupy a chip size of 2.6×1.7 mm^2, with the maximum efficiency of 76.8% at 3.6 V [75].

Generally, varying the supply voltage in accordance with the signal envelope affects the *AM-AM* and *AM-PM* characteristics of the power amplifier and degrades its linearity. However, the proper voltage control can provide the constant power gain across the input signal dynamic range combined with a corresponding correction of *AM-PM* response, thus minimizing the power amplifier amplitude and phase distortions and extending the useful power amplifier linear dynamic range up to saturation point [76]. In this case, the mapping between the instantaneous RF envelope and the applied supply voltage seriously influences the *AM-AM* and *AM-PM* characteristics, as well as the average output power and efficiency. In an ET system, the contents of a shaping table as a part of a digital baseband in the envelope path, as shown in Fig. 9.25(*a*), determine this mapping [77]. To achieve the so-called "ISOgain" shaping, the mapping between the RF envelope and supply voltage is chosen to achieve a particular constant power amplifier gain, when operating at high supply voltage close to the maximum value of 4.5 V in a compressed region, at medium supply voltages in a transition region, and at low supply voltages in a linear region, as shown in Fig. 9.25(*b*).

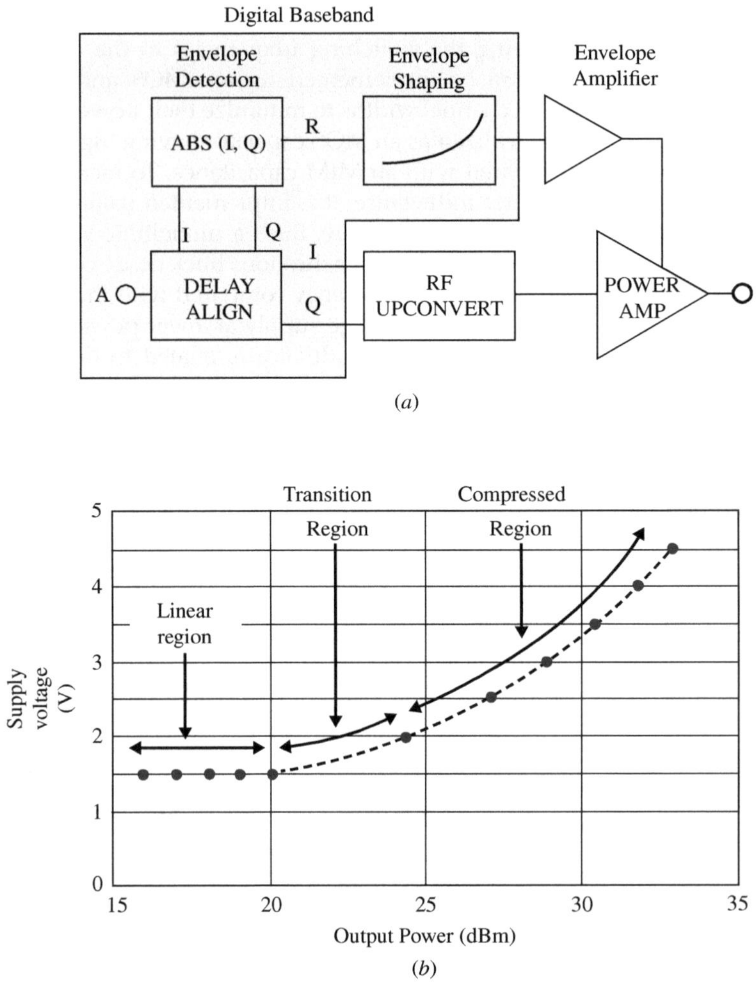

FIGURE 9.25 Envelope-tracking power amplifier and varying supply voltage.

With this mapping, the ET power amplifier system can achieve low *AM-AM* distortion despite operating in compression over much of the envelope cycle, according to Fig. 9.26(*a*). The ET system tradeoff of using the shaping table to linearize the power amplifier is a small loss of efficiency for a substantial improvement in linearity. The choice of shaping function also has a strong influence on the bandwidth requirement of the envelope path. A smooth transition between the linear and compressed regions corresponding to ET optimum efficiency shaping that is shown in Fig. 9.26(*b*) results in a lower bandwidth requirement for the envelope amplifier for a 1 to 2% loss in system efficiency [77]. Note that, unlike with *AM-AM* distortion,

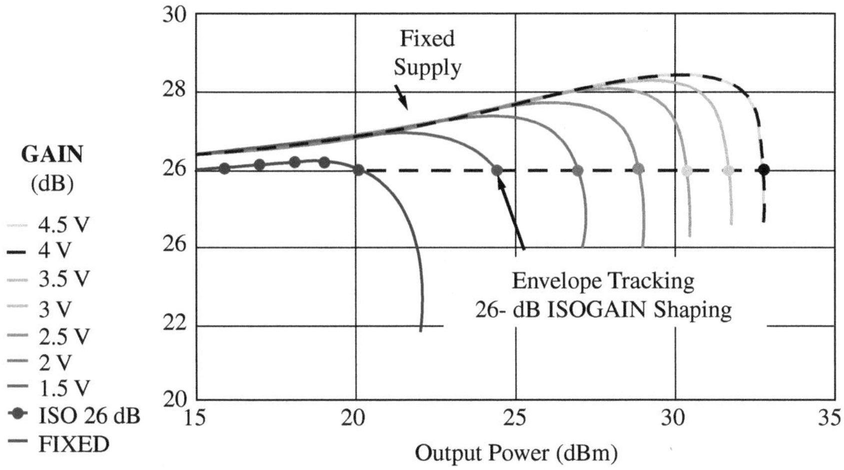

(*a*)

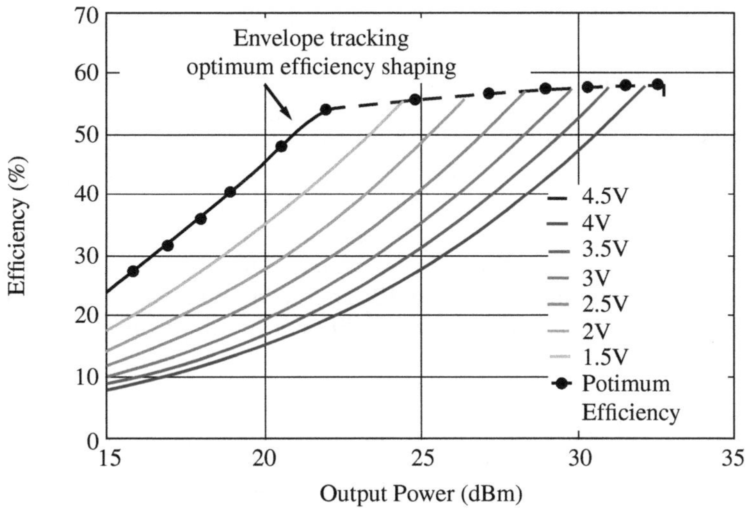

(*b*)

FIGURE 9.26 Envelope-tracking ISOgain and optimum efficiency shaping.

the envelope shaping table does not directly control *AM-PM* distortion, which ideally should be flat across the entire output power range, especially in a compressed region. To define the shaping table, it is necessary to measure the basic power amplifier characteristics such as output power, efficiency, gain, and phase over the full range of supply voltage, input power, and varying load impedance.

For an OFDM application at 2.4 GHz, the overall efficiency of the complete wideband ET system with SiGe HBT power amplifier is achieved as high as 28% with an error vector magnitude (EVM) of 5% at an output power of 20 dBm [78]. In this case, the gain of 11 dB was measured for this ET-based power amplifier, which is lower by 2.7 dB than that of the similar power amplifier with fixed dc-supply voltage. This is a common feature of the supply-modulated power amplifiers and is a result of their gain compression at low dc-supply voltages. Because the wideband envelope-tracking system has a significant nonlinearity associated with the collector modulation, as well as the intrinsic nonlinearity of the power amplifier, off-chip baseband DPD with an LUT scheme was implemented in this case to improve the system linearity, in particular EVM.

9.5 Switched-Path and Variable-Load Power Amplifiers

The power amplifiers in wireless communications systems operating in CDMA2000, WCDMA, or LTE standards are required to cover a dynamic range of the transmitting output powers up to 80 dB satisfying the corresponding linearity requirements. As a result, being designed for the highest power level with maximum achievable efficiency, the power amplifier tends to operate less efficiently at lower power levels, which leads to shortening the battery life time. Figure 9.27(*a*) shows the transmitter architecture that includes a variable gain amplifier, a power amplifier to provide a high output power level, a bypass line for bypassing the smaller output power level, and a two-pole switch between two signal paths [79]. Usually, the power amplifiers for wireless handset transmitters are designed to achieve the transducer power gain of about 25 to 30 dB. Therefore, it is very important to provide an efficient operation condition around maximum probable

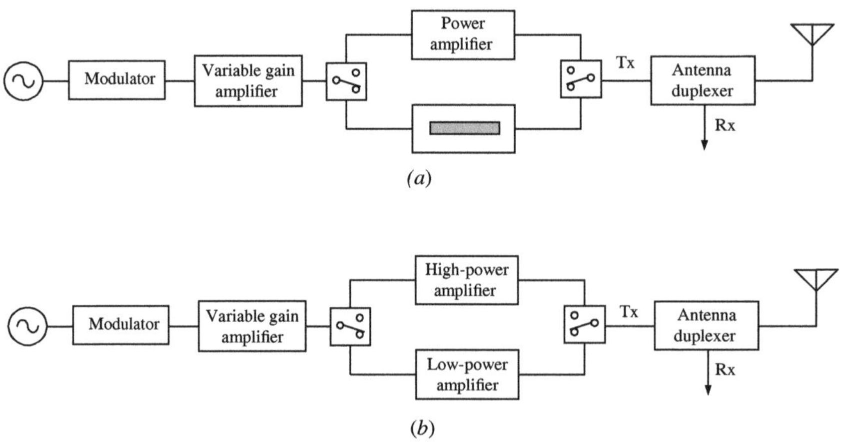

FIGURE 9.27 Transmitter architectures with dual-path power amplifier.

transmitting power required to deliver signal to the base station, which is of about 12 to 15 dB less than the maximum peak power. The output power of the widely used variable gain amplifiers is usually less than 10 dBm; otherwise, it is difficult to realize their linear operation. Besides, the variable gain amplifiers usually have a sufficiently high value for their noise figure. This contributes to the degradation in nonlinear distortion and excessive noise level in receiver bandwidth, which can only be improved by additional filtering that increases the cost and size of the transmitter.

A possible solution to improve the performance of the cellular handset transmitter is to use two power amplifier paths with different output power levels. This can result in a significant reduction of power consumption, because the low-power amplifier provides higher efficiency at output power level corresponding to maximum PDF. The block schematic of a dual-path transmitter architecture is shown in Fig. 9.27(*b*). When it is required to transmit the signal with an output power between the maximum level P_{max} and the statistically averaged power P_{avg}, the low-power amplifier is turned off. When it is enough to transmit the signal with the output power equal or less than P_{avg}, the high-power amplifier is turned off. Consequently, at any moment the only one power amplifier is turned on.

To further improve the efficiency of the multistage power amplifier system, generally it may need to provide more than two power amplifier paths with different output power levels when either each amplifying stage is connected to the three-pole switch or all amplifying stages are connected in parallel to the multipole switch. In this case, an efficiency improvement can be achieved by bypassing the power amplifier stages. For example, the power amplifier topology shown in Figure 9.28(*a*) provides the possibility of bypassing the second stage [80].

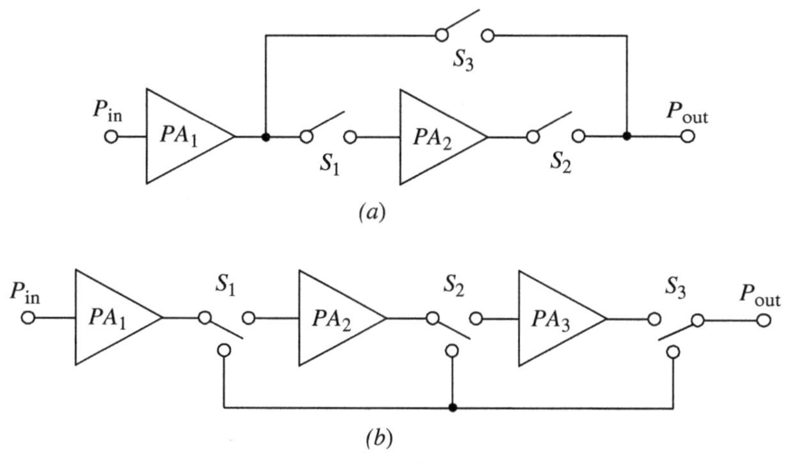

Figure 9.28 Switched-stage power amplifier configurations.

At lower output power levels, the signal amplification can be achieved using only the first amplifying stage PA_1 with the switch S_1 being turned on, whereas the maximum output power level is achieved using a two-stage power amplifier configuration with the switches S_2 and S_3 being turned on. To eliminate any additional impedance matching, both stages should be designed to operate in a 50-Ω environment at the input and output. An improvement in average efficiency of greater than four times at backoff output power levels (compared with the conventional two-stage structure) was achieved for a CDMA power amplifier operating over the frequency bandwidth of 825 to 849 MHz with a maximum output power of 29 dBm [81].

The power amplifier structure shown in Fig. 9.28(*b*) includes three amplifying stages with each stage having its own configuration so that the selected number of desired output power levels may be obtained directly from the selected amplifying stage [82]. The amplifying stages also can be configured in tandem to deliver maximum output power when an output of the first amplifying stage PA_1 provides an input to the second amplifying stage PA_2, while an output of the second amplifying stage PA_2 provides an input to the third amplifying stage PA_3. As a result, a three-stage power amplifier can provide three output power levels with maximum achievable efficiency.

Figure 9.29(*a*) shows the block schematic of a switched-path three-stage power amplifier configuration, which represents a separate connection of the amplifying stages with an output switchplexer [83].

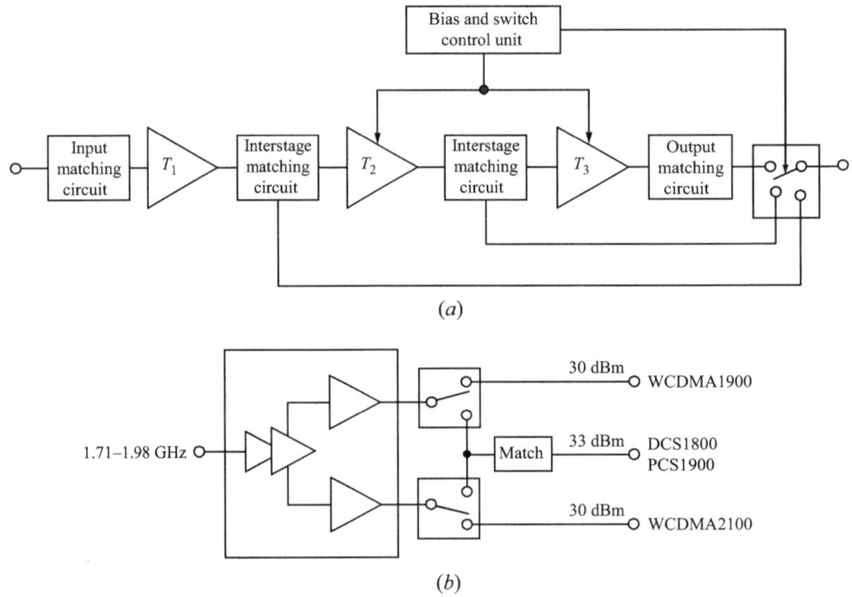

FIGURE 9.29 Switched-path power amplifier configurations.

Here, the interstage matching circuits represent the three-port networks with a single input and two outputs each, and the switchplexer is configured to provide communication between its corresponding input and output. This approach allows easy MCM (multichip module) implementation using separate technologies for the power amplifier and switchplexer. The other solution of a highly efficient reconfigurable power amplifier module for cellular-phone transmitter applications having reduced size and high efficiency at different output power levels, whose block schematic is shown in Fig. 9.29(*b*), can provide the signal transmission corresponding to constant-envelope DCS/PCS and varying-envelope WCDMA/LTE cellular modes in a high-frequency bandwidth of 1710 to 1980 MHz [84, 85]. The three-stage power amplifier module consists of common first and second amplifying stages and final stage with two parallel identical amplifying paths to provide separately equal saturated output powers of about 30 to 31 dBm to obtain a linear power of 27 dBm for WCDMA1900/2100 signals (or 26 dBm for LTE1900/2100), and a matching circuit for DCS/EDGE1800 and PCS/EDGE1900 modes to provide the required saturated output power of 33 dBm. When the top and bottom switches are connected together, the signals from two amplifying paths flow through the matching circuit required to match the resulting 25 Ω to the standard 50-Ω load and to provide an additional harmonic suppression. Otherwise, each separate amplifying path can be used to transmit either WCDMA/LTE1900 or WCDMA/LTE2100 signals, as well as WCDMA/LTE1800 mode.

As an alternative to switching of the power amplifier paths, it is possible to improve efficiency of the power amplifier operated at different output power levels with a fixed supply voltage by providing the impedance transformation between the load and the transistor output using switched-circuit arrangements of the load network [86]. The load network may include the series transmission line and shunt capacitors that can properly be connected by the turning on or off the corresponding *p-i-n* diode to provide the impedance matching at maximum and specified reduced power levels. For practical implementation, it is necessary to choose the *p-i-n* diodes with minimal series resistance and minimize an influence of the diode biasing circuitry on RF performance. The operational principle of the diode-switched or variable-load network configuration is illustrated in Fig. 9.30. In order to maximize efficiency of the power amplifier, the load resistance for different output power levels should be different, so as to provide a collector voltage amplitude close to the value of the dc-supply voltage V_{cc} in accordance with Eq. (9.27). This means that the load-line angle at lower output power levels becomes smaller, so that the smaller collector current amplitude corresponds to approximately the same collector voltage amplitude as for the higher output power level. Moreover, for lower power levels, the saturation voltage becomes smaller, as seen from the collector voltage amplitude corresponding to

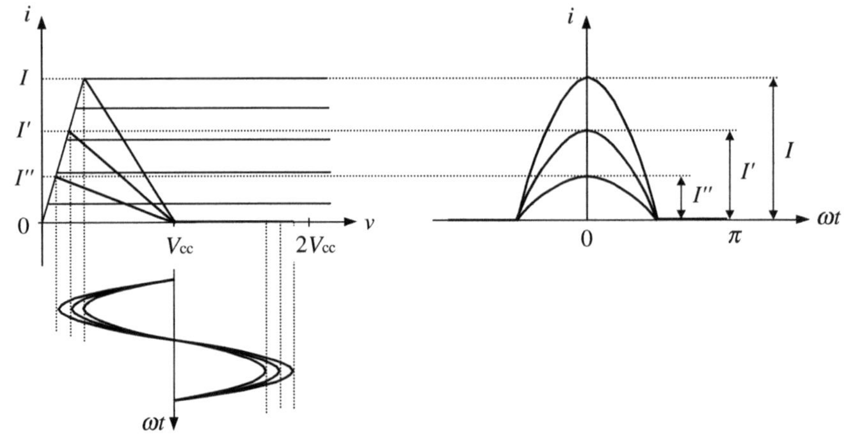

Figure 9.30 Collector voltage and current waveforms for different load lines.

the collector current amplitude I'' in Fig. 9.30, and peak collector voltage becomes even higher. This smaller load-line angle corresponds to the higher value of the load resistance seen by the device collector. For a dual-mode DCS/EDGE power amplifier implemented in a 0.18-μm CMOS technology where a tunable output matching network with a high-power switched capacitor and on-chip magnetic-coupled transformer is used, a saturated *PAE* of 45% at a 32 dBm was achieved for DCS1800 mode, whereas a modulated *PAE* of 28.1% at 27.5 dBm was optimized for EDGE1800 mode at 1.76 GHz [87].

To provide higher efficiency over a wide range of output power levels, the load-network configurations with variable elements can be used [88]. Figure 9.31(*a*) shows the high-pass π-type configuration of the output matching network with two shunt variable inductors. A π-type matching network has two variable elements, which in principle allows a load termination to be transformed to any impedance point on the Smith chart. The variable inductance can be practically implemented by means of the series connection of a quarterwave transmission line and a varactor diode, as shown in Fig. 9.31(*b*). In this case, the characteristic impedance of a quarterwave transmission line must be sufficiently low, so that the voltage swing across the varactor capacitance is substantially reduced. By using commercially available varactor diodes, it is possible to achieve an approximately constant value of the maximum *PAE* over the 5-dB dynamic range and more than double improvement in efficiency at 15-dB backoff output power level. The reverse-bias voltage applied to the varactor diodes should be high enough to minimize the insertion loss and parasitic phase distortion.

As an alternative, a ladder-type two-stage matching network can provide impedance transformation that covers the entire Smith chart.

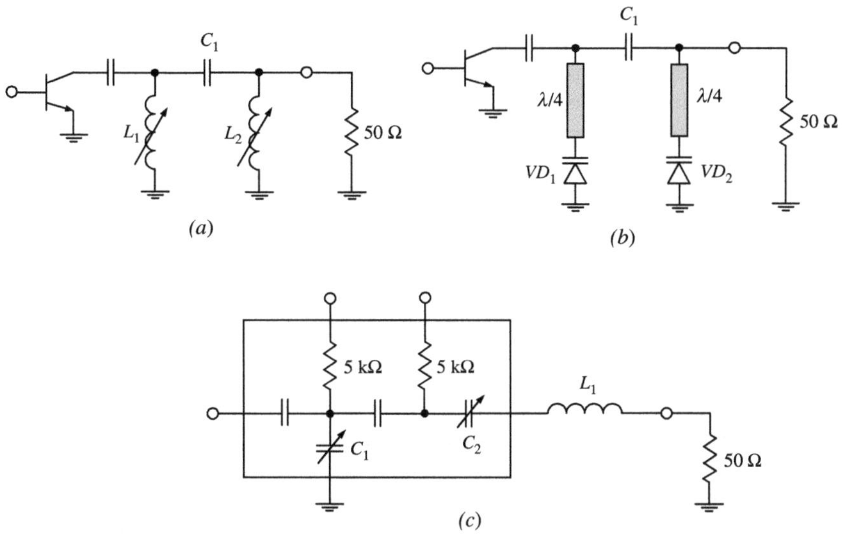

FIGURE 9.31 Load network configurations with variable load-network elements.

The advantage of the ladder matching network over the π-type network is that, for higher impedance transformation ratios, the loaded Q-factor of a two-stage ladder network is significantly lower, which improves the broadband capability of the power amplifier. In monolithic implementation, each tunable capacitor can be composed of an antiseries connection of two varactors independently controlled through its center-tap voltage and the inductors can be realized either by coplanar waveguide or bondwires [89]. In this case, the collector efficiencies of 30 to 55% are achieved over a 10-dB range of the output powers from 28 dBm for the 1800-, 1900-, and 2100-MHz bands. The varactor double-stub matching network can be used as an adaptive impedance antenna tuner for a WCDMA handset transmitter, with the goal of increasing overall efficiency by reducing the reflected power when the antenna is mismatched [90].

The variable capacitors using a thin-film $Ba_xSR_{1-x}TiO_3$ (BST) high-permittivity dielectric material with a voltage-dependent dielectric constant is a promising candidate for a voltage-controlled tunable device with a remarkably large variation in dielectric constant [91]. The parallel-plate BST capacitors can achieve as much as a 6:1 tuning range in dielectric constant. Unlike varactor diodes, the BST capacitors do not have a forward conduction region and hence can support a large RF voltage swing at low bias voltages, which makes them attractive for applications in cellular frontends, for example, in the antenna tuners and matching circuits of the power amplifiers. Figure 9.31(c) shows a simple matching circuit designed to transform a 50-Ω load to lower impedances in the frequency range of 850 to 950 MHz [92]. Here, an inductor L_1 is realized with a high-Q off-chip SMT air-core inductor, and the rest elements including BST capacitors are monolithically implemented. In this case, the insertion loss of an entire circuit is below 0.3 dB over all bias voltages and the output impedance varies continuously between 13 and 29 Ω at 900 MHz. Similar tunable matching network with low distortion and low operating bias voltages using as an output matching network of the power amplifier for PCS band demonstrated an *ACLR* better than –48 dBc at an output power of 28 dBm and a center bandwidth frequency of 1880 MHz [93]. The thick-film BST capacitors can be used as tunable interdigital varactors in a π-type low-pass output matching network of the high-power GaN HEMT amplifiers for base station applications at 2 GHz [94].

An electronically tunable load network with one or more variable elements can be used as an alternative load modulation technique, in which amplitude-modulate signals are produced with high efficiency by dynamic variation of the load impedance of the power amplifier, and the average efficiency can be twice that achieved in a linear operation of the same power amplifier for a multicarrier envelope with a 10-dB *PAR* and modulation bandwidths up to 50 kHz at high frequencies [95]. The load modulation technique can be extended to use in microwave high-power applications when, for a 1-GHz 7-W LDMOSFET power amplifier, the *PAE* can be increased by 10% at 10-dB power backoff compared to the linear power amplifier with a fixed 50-Ω load [96]. Under 3GPP WCDMA modulation, a dynamically modulated power amplifier with a varactor-tunable matching network achieves up to 5% improvement in drain efficiency at a maximum output power of approximately 15 dBm maintaining an *ACLR* below –33 dBc at 5-MHz offset [97].

Using the diode switches in the load network, or varactor-based tunable matching networks, results in increased size, cost, circuit complexity, and additional power losses. To improve the performance of the wireless handset transmitter, it might be possible to use its architecture with two amplifying paths with different output power levels and a single three-port nonswitchable output load network [98]. Such an approach provides high-efficiency operation at low and medium output power levels with a significant reduction of the overall transmitter power consumption. The basic dual-path two-stage power amplifier schematic with a three-port output matching circuit is shown in Fig. 9.32(*a*), which also includes a common first stage, a common three-port interstage matching circuit, and a dual-path second stage. When it is necessary to transmit a signal with output power between maximum output power P_{max} and some averaged backoff output power P_{avg}, the low-power amplifying stage with active device Q_3 is turned off. When it is enough to transmit the signal with output power equal or less than P_{avg}, the high-power amplifying stage with active device Q_2 is turned off. Both transistors are biased in

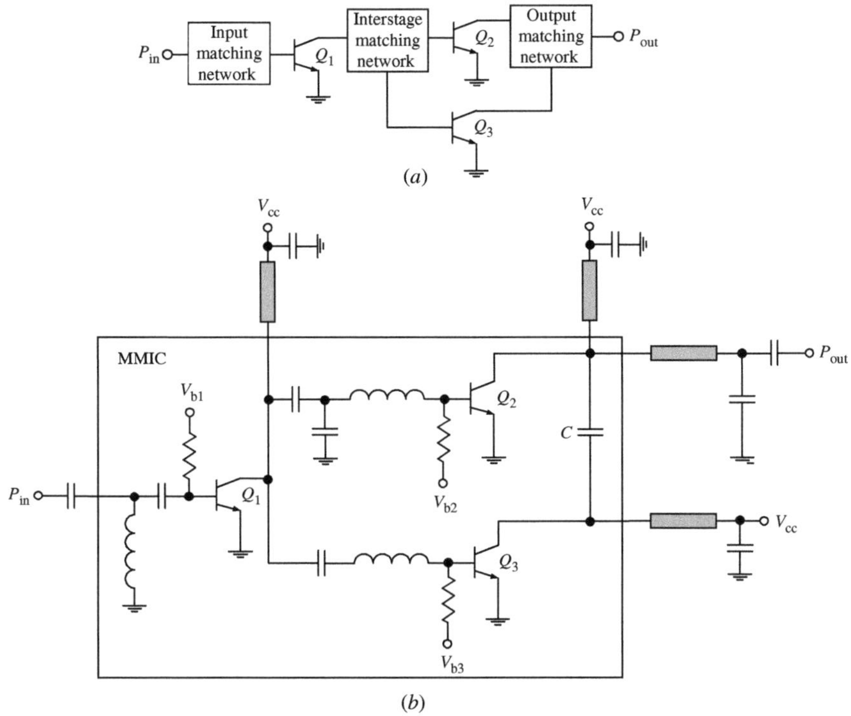

FIGURE 9.32 Three-port load network configurations of dual-path power amplifier.

Class AB with a small quiescent current to provide a linear operation. The area of the smaller device Q_3 corresponds to the output power P_{avr} required to deliver to the antenna. The three-port output matching circuit should be configured so that it provides a lower load impedance seen by the collector of the transistor Q_2 to deliver maximum output power P_{max} with maximum achievable collector efficiency and a higher load impedance seen by the collector of the transistor Q_3 to maximize collector efficiency at the most probable output power P_{avr}. Because such a dual-path configuration with a single three-port load network does not require additional components rather than transistors, it is very practical for single-chip integration without serious increasing in manufacturing cost and size.

Figure 9.32(*b*) shows the circuit schematic (without bias circuits) of a two-stage InGaP/GaAs HBT MMIC power amplifier intended for WCDMA handset applications [98]. The emitter areas of the transistors for driver stage Q_1, power stage Q_2, and dual stage Q_3 were chosen of 480 μm^2, 1920 μm^2, and 480 μm^2, respectively. The output impedance-transforming circuit is constructed with a series capacitor *C*. The current-mirror circuits located within the MMIC were used for transistor biasing. As a result, by using a Class-AB mode with quiescent

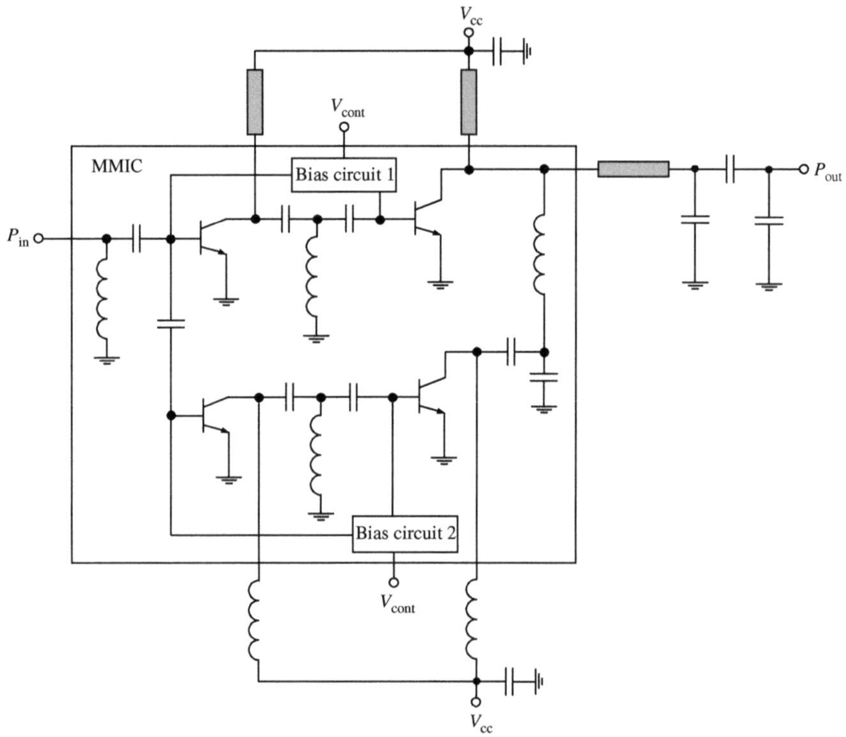

FIGURE 9.33 Schematic of dual-chain MMIC power amplifier.

currents of 10 mA for the devices Q_1 and Q_3 and 40 mA for the device Q_2 at a dc-supply voltage V_{cc} = 3.3 V, the *PAE* of 16.4% was obtained in a low-power mode with P_{1dB} = 16.7 dBm. In a high-power mode with P_{1dB} = 27.6 dBm, the *PAE* was 34.2%. Generally, an overall efficiency improvement of at least 1.81 times over a wide power range was provided, compared to a conventional two-stage Class-AB power amplifier.

Figure 9.33 shows an alternative approach to realize the dual-path power amplifier using a chain configuration of the output impedance-transforming circuit [99]. The dual-chain two-stage InGaP/GaAs HBT MMIC power amplifier with common input and output matching circuits is implemented by parallel connection of two amplifying chains having different output powers. Either the low-power amplifying chain with P_{1dB} = 16 dBm or the high-power amplifying chain with P_{1dB} = 28 dBm is activated through the bias selection. The matching circuit between the collectors of the output transistors, with a series inductor and a shunt capacitor composing a simple low-pass *L*-type matching circuit, allows the power-added efficiency at backoff output powers to be increased by more accurate impedance matching of the higher output impedance of the low-power device to the 50-Ω

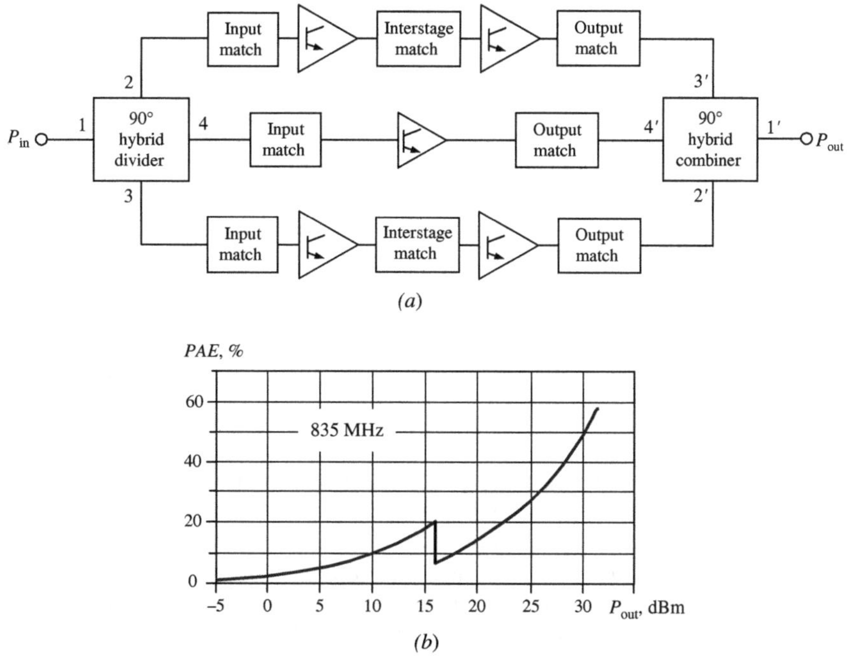

FIGURE 9.34 Block diagram and test result of balanced switched-path MMIC power amplifier.

load. Using such a dual-chain power amplifier configuration, it is possible to obtain the *PAE* of 21% at P_{1dB} = 16 dBm in a low-power mode and the *PAE* of 40% at P_{1dB} = 28 dBm in a high-power mode.

Figure 9.34(*a*) shows the block diagram of the power amplifier with a high efficiency at different output power levels, which incorporates an additional separate low-power amplifying path connected between the isolated ports of a quadrature input hybrid power divider and a quadrature output hybrid power combiner [100]. In a high-power mode, the bias control unit provides proper biasing of the transistors in both balanced paths and switches off the biasing in a low-power path 4-4′. In this case, the output powers from both high-power balanced paths 3-2′ and 2-3′ are combined in the output quadrature hybrid combiner and the low-power path is isolated from them, thus providing a high efficiency at high output power. When all ports of a hybrid branch-line combiner are matched, the incident power entering the input port 1 is divided between the output ports 2 and 3 with a phase shift of 90° between these outputs and no power is delivered to the isolated port 4. In a low-power mode, the bias control unit switches off the biasing of the transistors in both balanced paths and provides proper biasing for a low-power amplifying path.

A quadrature hybrid has an important advantage compared to the in-phase dividers so that, at equal values of reflection coefficients from the loads connected to the output ports 2 and 3, the reflected wave is absent at the input port 1, and consequently the input *VSWR* of a quadrature hybrid does not depend on the equal-load mismatch level. In this case, all reflected power in a quadrature divider is dissipated in a 50-Ω ballast resistor connected to the isolated port 4. Similarly, for a quadrature combiner, the reflected waves from two inputs will combine and flow to the load. As a result, the isolated port of the input hybrid becomes the input for a low-power amplifying path, whereas the isolated port of the output hybrid operates as its output 50-Ω port. Hence, the high efficiency at lower output power levels can be achieved, and no need to include any additional switches to isolate the high-power balanced paths and low-power path from each other.

Practical results for high-efficiency linear multiband and multimode two-stage balanced switched-path SiGe HBT MMIC power amplifiers intended to operate across DCS1800/PCS1900 and WCDMA850/900/1900/2100 frequency bands had demonstrated the $PAE \geq 50\%$ at maximum output power and $PAE \geq 20\%$ at backoff output power of 16 dBm. For example, the saturated output power of 31.5 dBm was achieved with a maximum *PAE* of 58% at operating frequency of 835 MHz, whereas, for an average WCDMA power of 28.5 dBm, a *PAE* of 41.5% was measured. In a low-power mode at P_{out} = 16 dBm with an *ACLR* of –34 dBc, the efficiency is increased by greater than a factor of 2 reaching 20% by turning off the bias current for the high-power balanced paths and turning on the low-power path, as shown in Fig. 9.34(*b*) [100].

To provide an efficient linear power amplifier operation, it is necessary to minimize the quiescent current at backoff output powers because maximum of the PDF for a WCDMA standard occurs at an output power of about 25 dB below its maximum level. As a current-controlled device, the bipolar transistor at RF operation requires the dc base driving current, whose value depends on the output power and device parameters. From technology viewpoint, because the bipolar device represents a parallel connection of the basic cells, important issue is to use the ballasting series resistors to avoid the current imbalance and possible device collapse at higher current density levels. The basic current-mirror and emitter-follower bias circuits can provide the temperature compensation over a wide range of ambient temperatures, with very small reference current for the latter case [101]. However, an adaptive bias circuit can additionally control the dc power consumption with varying output power, thus greatly improving a power-added efficiency when the output power is low and maintaining high linearity of the power amplifier when its output power is high [102].

9.6 Monolithic HBT and CMOS Power Amplifiers for Handset Applications

In monolithic microwave integrated circuit (MMIC) design, it is very important to minimize the die area in order to reduce cost. However, the design of the high-power monolithic power amplifiers can take advantage of the large device area, which is a large gate width for FET (field-effect transistor) devices or emitter area for bipolar transistors, available on an MMIC. As a result, the required high output power level is realized by combining the elementary device cells. For example, it can be done by using a simple parallel connection, as shown in Fig. 9.35(*a*), which requires too long device structure, or by using the transmission-line divider and combiner, as shown in Fig. 9.35(*b*), where it is necessary to keep the same length to provide an in-phase operation of all device cells requiring additional MMIC space. The in-phase operation is very important for a high-power device in terms of preventing excessive heating of any device cell, providing maximum output power delivery to the load with higher efficiency. The latter is very important when designing the wireless handset transmitter where the main requirements are to provide the long-term operation and low cost. Such an efficient operation condition can be achieved using GaAs HBT transistors having low on-saturation resistance r_{sat} and high transition frequency f_T. However, in this case, it is necessary to take care about equal current distribution through each HBT cell. Even small differences in cells or their placement can cause a heating imbalance, which can lead to the cell failure causing at the same time a chain reaction failure of the other cells.

For parallel connection of the device cells shown in Fig. 9.36(*a*), this problem can be eliminated by using the segmented capacitor with one layer connected to the RF input and the other segmented layer where individual segments are dc isolated and connected to each HBT cell base terminal, as shown in Fig. 9.36(*b*) [103]. To provide an identical dc current density for each cell, the series base resistors,

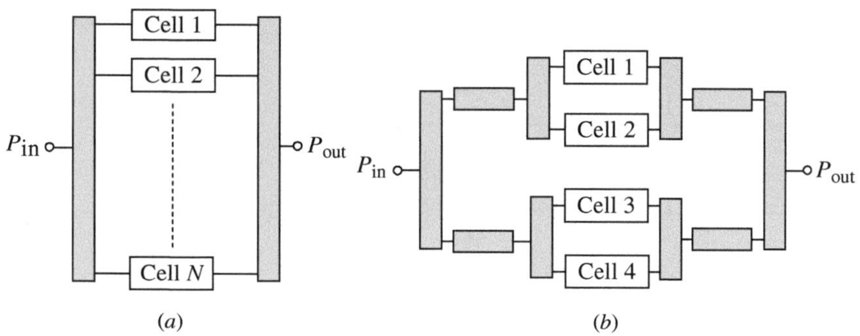

FIGURE 9.35 Different device cell connections for power combining.

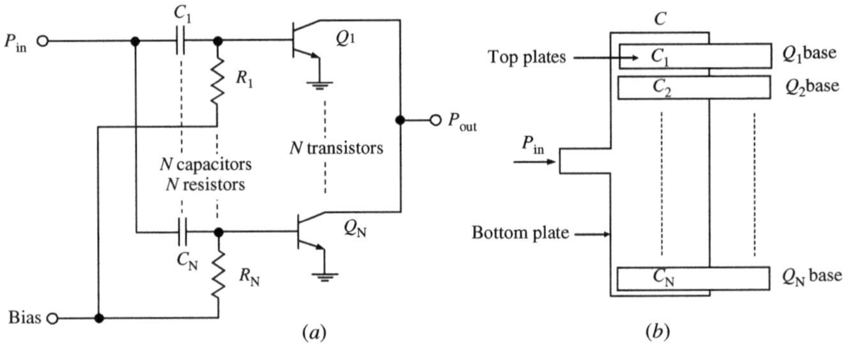

FIGURE 9.36 Parallel on-chip connection of device cells.

whose values are optimized over a wide temperature range, are connected to each cell.

However, when it is necessary to design a power amplifier with two or more amplifying stages, it is necessary to arrange additional input and interstage matching circuits, which can include several capacitors and inductors. Figure 9.37 shows the electrical schematic of a two-stage InGaP/GaAs HBT power amplifier MMIC designed to operate in a wide frequency bandwidth of 1.71 to 1.98 MHz [104]. Without any tuning, this power amplifier can provide $PAE \geq 51\%$ and $P_{out} \geq 30$ dBm over the entire frequency range at saturation, as well as $PAE \geq 38\%$ and $ACLR \leq -37$dBc at $P_{out} = 27$ dBm for a WCDMA signal in a frequency bandwidth of 1920 to 1980 MHz. The series resistors R_1 and R_2 are required to provide the unconditionally stable (without any parasitic oscillations) operation conditions. The input matching circuit representing a high-pass LC-transformer provides an input $VSWR$ better than 2:1 over the entire frequency range. Using two high-pass LC matching sections in the interstage matching circuit

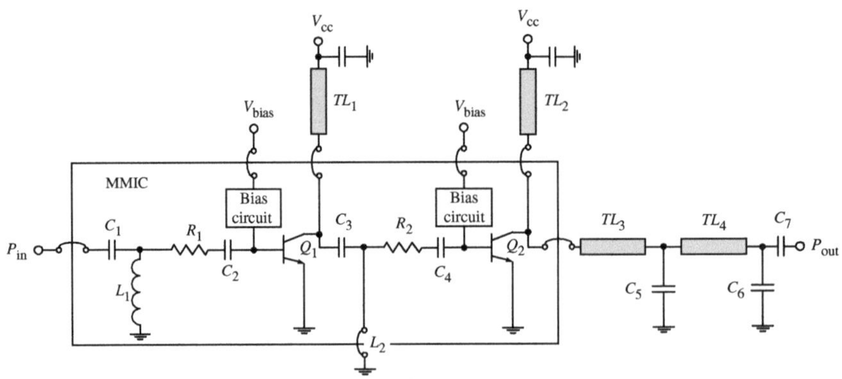

FIGURE 9.37 Two-stage InGaP/GaAs HBT power amplifier MMIC.

provides broadband operation by minimizing the impedance sensitivity to its parameter variations. In this case, the first shunt inductor represents a series connection of the bondwire and short-length transmission line TL_1, whereas the second shunt inductor can fully be represented by the inductive bondwire shorted outside of the chip. The designed load network approximates the parallel-circuit Class E structure (see Fig. 7.58 in Chap. 7), which includes the shunt inductor consisting of the bondwire and short-length transmission line TL_2 (shunt capacitance is fully represented by the device collector capacitance) and a two-section low-pass LC-transformer with a series 50-Ω microstrip line ($TL_3 + TL_4$) and two shunt chip capacitors C_5 and C_6. All microstrip lines are implemented by using FR4 substrate.

The emitter area of the HBT transistor is large enough for output powers of 1 W or greater. In order to minimize the MMIC size, it makes sense to split the overall device in several segments (rows) and then to combine them. Usually the input and output terminals of the segments are connected in parallel. However, in this case, the distances from the segment terminals to the input or output are different that causes some phase imbalance. For output terminals, this problem is easily solved by using several bondwires from different segments connected in parallel to off-chip element of the output matching circuit. However, for input segment connection, it is difficult to provide the same phase lengths, because it is necessary to insert additional equal-length transmission lines directly on MMIC. Besides, generally it is impossible to use the segmented realization of the series interstage capacitor (capacitor C_3 in Fig. 9.37) without MMIC size extension, because it should provide the equal segments connected to each device base. However, the capacitor profile may differ from square form and the output terminal of the device from previous stage can be located sufficiently far from the device of the output stage.

Therefore, an alternative approach shown in Fig. 9.38(*a*) is to connect all inputs of the output device segments directly to the second layer of the interstage capacitor C_{int} without segmentation (capacitor C_3 in Fig. 9.37). This approach allows the additional transmission lines to be eliminated (the phase difference between the most far-off cells is $< 5°$), thus providing approximately equal input powers flowing into the output device cells and significantly minimizing the overall MMIC size. The interstage inductor L_{int} represented by the bondwire (inductor L_2 in Fig. 9.37) can be connected to any point of the first layer of the capacitor C_{int}. The input shunt inductor L_{in} is a typical spiral inductor of square geometry having 3.5 turns. The overall die size of a two-stage power amplifier MMIC shown in Fig. 9.38(*b*) was of 0.9×1.0 μm^2 with emitter areas of the first and second stages equal to 540 μm^2 and 3600 μm^2, respectively. Such an approach can be applied to any type of the MMIC power amplifier.

Implementation into an MCM can reduce significantly the overall size of the power amplifier. For example, using a multilayer substrate

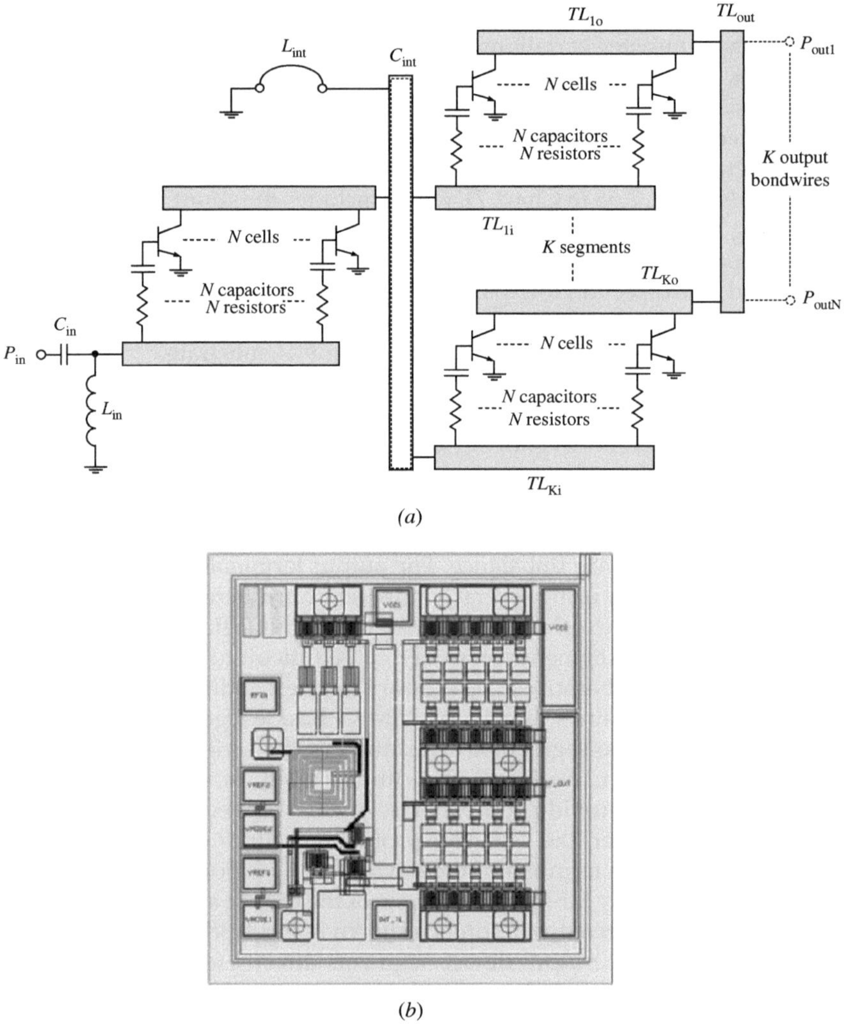

FIGURE 9.38 Monolithic implementation of two-stage HBT power amplifier.

consisting of three resin and four conductor layers having thermal via holes results in the MCM size of 7×7 mm^2 [105]. All chip elements and RF lines were located on the top resin layer with a dielectric constant of 10.5. The dc bias voltage was supplied through the third conductor layer. An HBT cell with an emitter area of 3×20 μm^2 showed a dc current gain of about 150 at a collector current density of 104 A/cm^2 and a high collector-emitter breakdown voltage of 20 V. A two-stage InGaP/GaAs HBT power amplifier with emitter areas of the first and second devices of 480 μm^2 and 2880 μm^2, respectively, exhibited WCDMA power performance with a *PAE* of 46% and an *ACLR* of

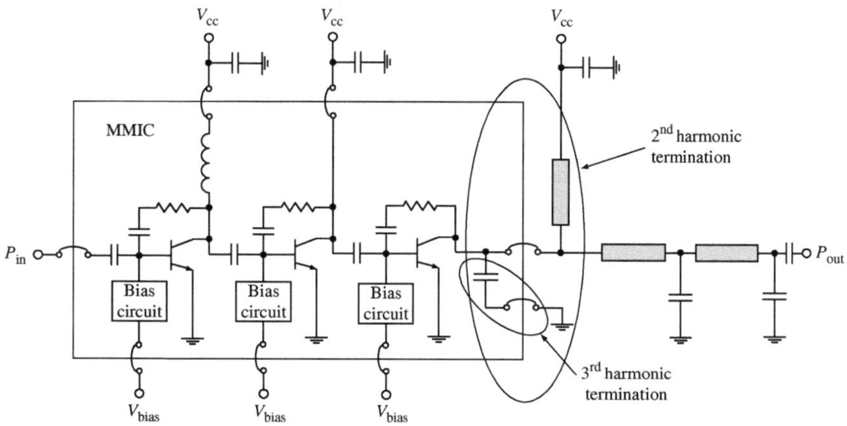

FIGURE 9.39 Three-stage GaAs HBT power amplifier MMIC.

–37 dBc at the measured output power of 26 dBm. Using SMT components (chip capacitors) of the smallest size and providing more compact dc and RF routing can reduce the overall MCM size to 3 × 3 mm^2 for a dual-mode WCDMA power amplifier with an output power of 27.5 dBm in a frequency bandwidth of 1920 to 1980 MHz [106].

Figure 9.39 shows the circuit schematic of a three-stage GaAs HBT power amplifier module operating in an inverse Class-F mode to achieve an output power of not less than 35 dBm with a *PAE* of about 60% in a frequency bandwidth of 824 to 915 MHz [107]. Here, the third-harmonic short-circuit termination is achieved by using a third-harmonic series resonant circuit composed of an on-die MIM capacitor and a grounded bondwire, and the reactance of these two elements, the collector bondwires, and the microstrip bias choke provide the second-harmonic open-circuit termination. The final two stages are subject to a high level of overdrive, which allows for an approximately constant output power under a wide range of input power levels to achieve flat saturated output power (within 0.2 dB) and high efficiency (within 2%) across the required frequency bandwidth. Because the frequency bandwidth is not very wide (about 10%), each interstage matching circuit represents a single high-pass *LC* matching section with a series MIM capacitor and a shunt inductor (bondwire in the second stage and on-die spiral inductor followed by a bondwire in the first stage), also including the device input and output parasitic reactances. Stability of operation is provided by using a parallel *RC* feedback circuit in each amplifying stage.

Recent progress in CMOS technology has shown their promising future for RF power applications. Much progress has been achieved at the research and technology levels, and the obvious possibility to minimize cost and size of the integrated circuits for cellular handset transmitters, especially power amplifier MMICs, makes CMOS technology very feasible bringing considerable economic benefits. However, realizing high-efficiency operation of CMOS power amplifiers is limited by some technology issues, such as higher value of the device saturation resistance, availability of TSVs (through silicon vias), low value of breakdown voltage, and lossy silicon substrate. Therefore, it is very vital to apply a high-efficiency technique into the design of new-generation CMOS power amplifiers. For example, a 900-MHz power amplifier based on a 0.25-μm CMOS technology with active die area of 2×2 mm², whose circuit schematic is shown in Fig. 9.40(*a*),

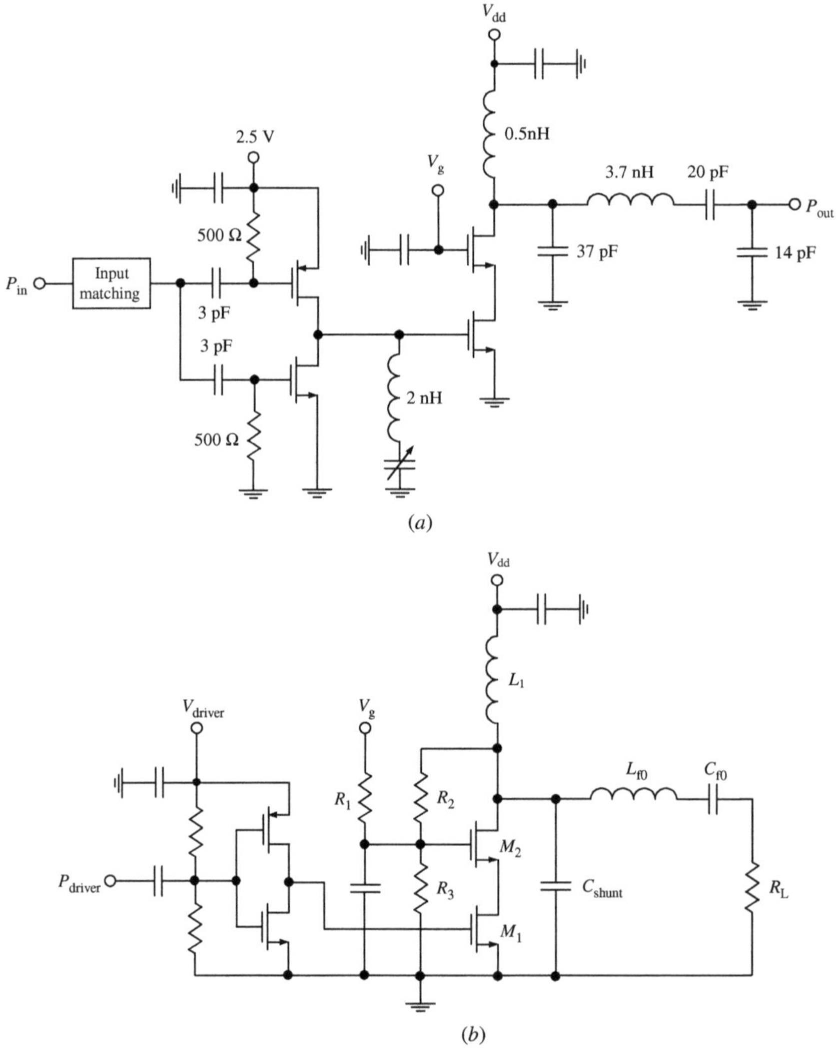

FIGURE 9.40 Cascode CMOS Class-E power amplifiers with finite dc-feed inductance.

can provide an output power of 0.9 W and a *PAE* of 41% using a Class-E load network with shunt capacitance and finite dc-feed inductance [108].

In this case, to minimize the voltage stress on the switching transistor and maximize the allowable supply voltage, the cascode connection of two nMOS devices is used, which allows the supply voltage to be as high as 1.8 V. Because a cascode switch has higher on-resistance per unit channel width than a single common-source switch during on-state mode, wider devices of 15-mm gate widths are used. The interstage bondwire inductor of 2 nH and external capacitor are used to resonate out the gate capacitance of the cascode device. Because the quality factor of the on-chip spiral inductors in a typical CMOS technology is low because of a large loss in the silicon substrate and thin metal layers, the bondwires can be successfully used instead, having less than 5% of inductance variation and less than 6% of *Q*-factor change as a result of machine-bonding process. As a result, the complete power amplifier load network consists of two aluminum bondwire inductors and one on-chip (37 pF) and two off-chip (20 pF and 14 pF) capacitors. The implemented power amplifier is differential, and baluns are used at both input and output to combine two single-ended paths. In order to further increase the efficiency of the CMOS power amplifier, a high-*Q* integrated passive device (IPD) based on a high-resistivity substrate to implement an output matching network can be used, which also makes the power amplifier module easier to manufacture. As a result, for an input power of 0 dBm, an output power of 34.5 dBm with a *PAE* of 55% was achieved in a low-frequency band of 824 to 915 MHz, whereas an output power of 32.5 dBm with a *PAE* of 52% was measured in a high-frequency band of 1.71 to 1.98 GHz at a 3.4-V power supply [109].

The circuit schematic of a 30-dBm Class-E power amplifier with a self-biasing technique implemented in a 65-nm CMOS technology is shown in Fig. 9.40(*b*) [110]. Here, the final cascode stage is formed by a standard thin-oxide (18 Å) device and a dedicated high-voltage extended-drain thick-oxide (50Å) device (ED-NMOS), which has an offset breakdown voltage of 15 V and operates safely up to at least 10 V in the on-state. The devices measured f_T and f_{max} exceed 30 GHz and 50 GHz, respectively, enabling its usage at microwave frequencies up to 5 GHz. As a result, such a cascode topology combines the fast switching capability of the thin-oxide device M_1 in a common-source configuration and the high voltage sustainability of the thick-gate M_2 device in a common-gate configuration. Additionally, the device M_1 provides a high-gain capability and presents lower values of parasitic capacitances at the input of the device M_2, and proper choice of its width results in a better matching of its output impedance to the input impedance of the common-gate device. The self-biased cascode configuration was introduced to optimally divide the voltage swing across the common-gate and common-source devices so as to optimize the efficiency and output power within the breakdown limits of the device. Because the linearity and efficiency degrade as the device M_2 is turned off at lower supply voltages V_{dd}, which is required for envelope tracking, the resulting poor performance can be restored by dc-biasing the gate of the cascode device M_2 independently with a fixed gate supply voltage V_g and optimum values of the bias resistors R_1, R_2, and R_3, so that the transistor is always turned on. To operate the cascode stage as a switch optimally, an inverter-based driver utilizing thin-oxide devices with a sufficient driving capability and an input impedance of 50 Ω to minimize the return loss was used. By using an active load-pull measurement system for power amplifier characterization, a *PAE* over 60% when delivering a 30-dBm output power was achieved in a frequency range of 0.8 to 2 GHz at $V_{dd} = 5$ V.

Figure 9.41 shows the circuit schematic of a Class-F power amplifier implemented in a 0.2-μm CMOS technology [111]. The advantage of using a Class-F operation mode is a substantially less drain voltage peak factor compared with a Class-E mode. This helps overcome the problem of low oxide breakdown voltage limiting the maximum output power and efficiency of the CMOS power amplifier because of the lower supply voltage required for the device protection. A Class-F operation is achieved by using a short-circuited quarterwave transmission line, having low impedances at its input for even harmonics, and a series resonant circuit in the load network, having high impedances at the second and higher-order harmonics. In a cascode configuration of the final stage, the thin-gate device M_1 is protected by a thick-oxide (80Å) device M_2 with no threat to oxide breakdown under supply voltage of 3 V. Using a pMOS device M_4 as an inverter in the driver stage eliminates the problem of the negative voltage swing

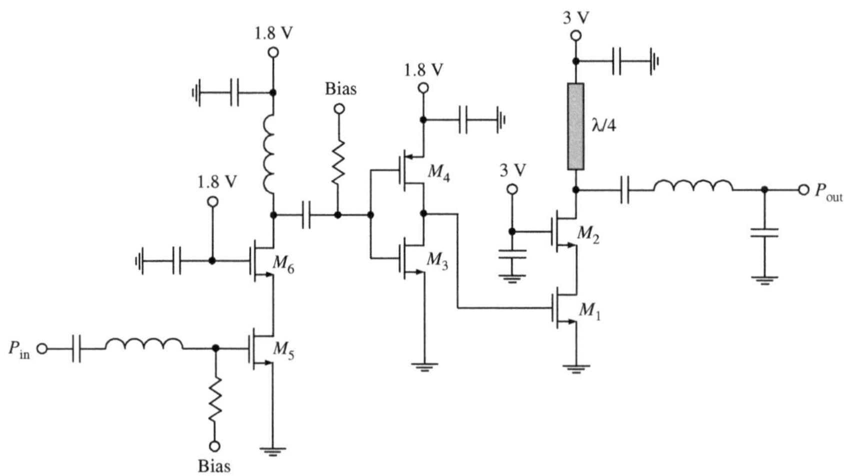

FIGURE 9.41 Schematic of Class-F power amplifier with quarterwave transmission line.

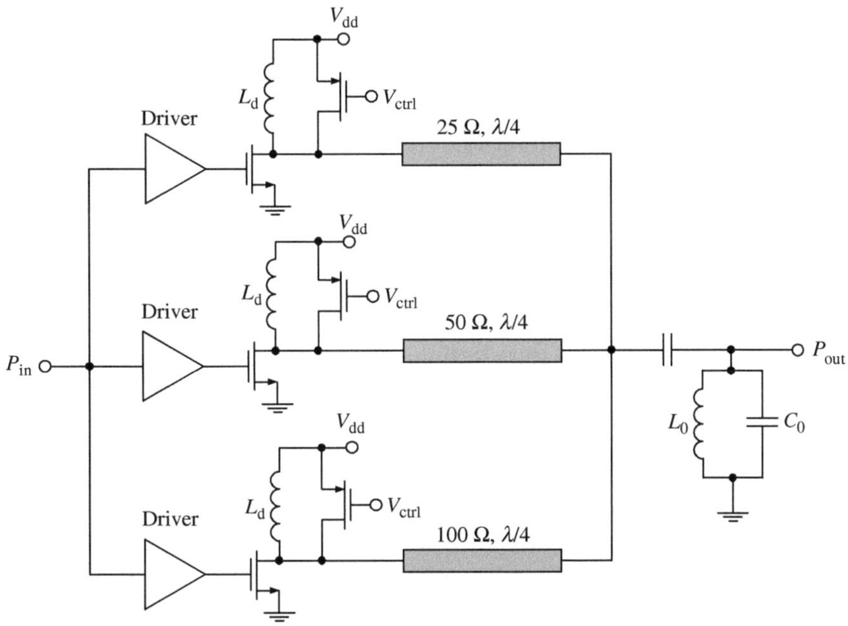

Figure 9.42 Class-F parallel power amplifier architecture.

across the gate of the cascode nMOS device M_1, thus preventing the forward biasing of its drain junction diode. Such a CMOS Class-F power amplifier was able to deliver a maximum output power of 1.5 W with a *PAE* of 43% at 900 MHz.

Figure 9.42 shows the simplified schematic of a parallel power amplifier architecture implemented in a 0.25-μm CMOS technology and intended to provide high operation efficiency at backoff output power levels [112]. This architecture uses three binary weighted Class-F power amplifiers, whose output powers are combined in a power-combining network using the quarterwave transmission lines loaded on the parallel resonant circuit tuned to the fundamental. The capability to turn off completely each individual power amplifier without interfering with the operation of other individual power amplifiers is provided by the addition of pMOS shorting switches, which result in high impedance at the end of the corresponding transmission line. The power amplifier architecture operating at 1.4 GHz from a 1.5-V supply occupies an active die size of 0.43 mm^2 and achieves a *PAE* of 49% at maximum output power of 300 mW, while maintaining a *PAE* greater than 43% over a lower output power range down to 100 mW. The transmission lines are implemented using PCB microstrip lines. On-chip transmission-line fabrication in CMOS technology using an equivalent *LC* ladder circuit makes it significantly shorter.

For example, the frequency response of 10 *LC* ladder sections, containing a series inductor and a shunt capacitor each, can approximate the frequency response of the transmission line within 5%, occupying an area about 14 times shorter. In this case, it is enough to use a spiral inductor to implement both series inductance and shunt capacitance, which can be obtained with the bottom-plate parasitic capacitance of the spiral. However, the maximum *PAE* of the parallel on-chip power amplifier architecture degrades by 10 to 15%.

To implement CMOS power amplifiers in modern communication systems using GSM/EDGE, CDMA2000, or WCDMA/LTE signals with nonconstant envelope when high linearity requirements need to be satisfied, some linearization techniques such as envelope tracking or DPD can be used to obtain simultaneously highly efficient and linear operation [113, 114]. Basically, the two main sources of CMOS device nonlinearity are the nonlinear behavior of its gate-source capacitance and drain current source. For example, for a 0.6-μm CMOS technology, when the transistor turns on and off, the gate-source capacitance C_{gs} changes from approximately of 1 pF/mm to almost of 0 pF/mm, whereas the variation of the gate-drain capacitance C_{gd} is negligible [115]. This nonlinearity can be reduced by introducing a parallel inverse nonlinearity at the final stage of the CMOS power amplifier through the use of a pMOS device connected in parallel to the nMOS device, as shown in Fig. 9.43.

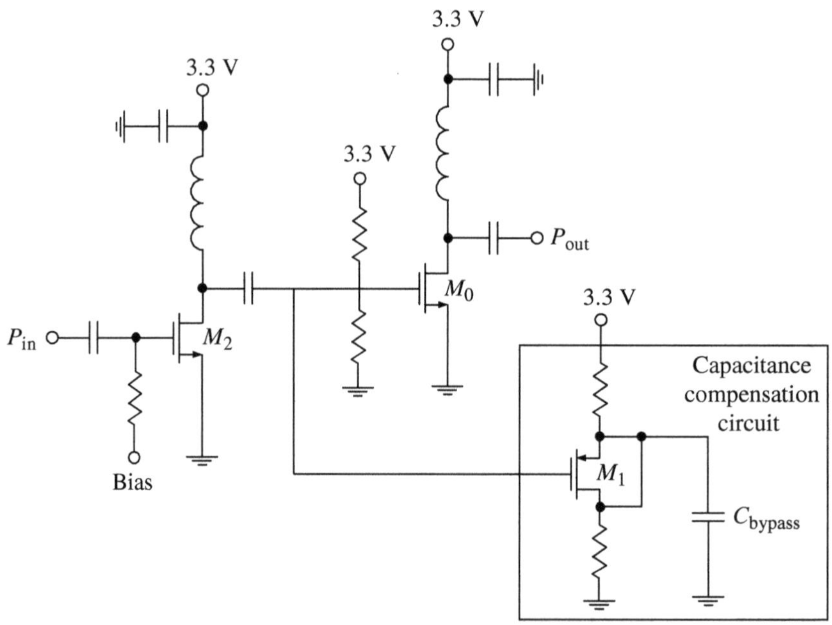

FIGURE 9.43 Schematic of two-stage Class-AB power amplifier.

In this case, no matter what state the transistor is in, the overall input capacitance is always approximately constant. By connecting the drain and source of the pMOS device together, only capacitive current flows into the pMOS device. To minimize intermodulation distortion due to the nonlinear current source, it is necessary to apply a gate bias voltage corresponding to the sweet spot where IM_3 is minimal. Note that, because of the specific nonlinear behavior of the device transfer function, there are two sweet spots where IM_3 is minimal [116]. The first sweet spot appears because of the turn-on knee region contribution, whereas the second sweet spot close to the output power compression point is due to the combined effect of the quadratic-to-linear and compression transitions of the device transfer function. To keep good linearity, the driver stage can be used in a Class-A operation mode. As a result, this technique was capable of improving IM_3 by 10 dB and *ACPR* by 6 dB for the Class-AB power amplifier implemented in a 0.6-μm CMOS technology, achieving a maximum output power of 20 dBm and a drain efficiency of 40% at the operating frequency of 1.9 GHz [115].

To obtain a high-efficiency and linear operation of the CMOS power amplifier at a high power, the gate of a common-source device should be biased in a deep Class AB. However, this bias mode results in a severe nonlinear behavior near turn-on point, thereby generating the large IM_3 products in a low-power region. In this case, a linearity improvement without a significant reduction in the efficiency can be achieved by using a signal envelope injection at the gate of a common-source device when the power approaches high levels, which occurs only during certain fractions of the signal periods [117]. To generate the injecting signal, a Class-D envelope amplifier is used in the bias circuit, whose circuit schematic is shown in Fig. 9.44(*a*). Such a bias circuit consists of two parts: the envelope detector of the input signal that uses an nMOS transistor M_1 with an *RC* network and an envelope amplifier. When an increasing voltage envelope signal enters the input of the bias circuit, the output voltage of the first stage decreases from V_{DD} to a lower value, and the pMOS transistor M_2 begins to charge the output node from the initial V_{BIAS} value to its higher value. However, when the input envelope signal decreases, the V_{BIAS} is decreased to a lower value. The injection angle and the shape of the injection signal are determined by the size ratio of the pMOS transistor M_2 and nMOS transistor M_3 in the second stage. Because the total width of the transistors in the bias circuit is much smaller than that of the power amplifier, there is no degradation in the overall efficiency. To minimize the sideband asymmetry and second-order nonlinear components, the values of the parallel resistor R_2 and capacitor C_1 in the bias circuit for the common-gate transistor shown in Fig. 9.44(*b*) were optimized to provide the optimum impedances at low frequencies and at the second harmonic.

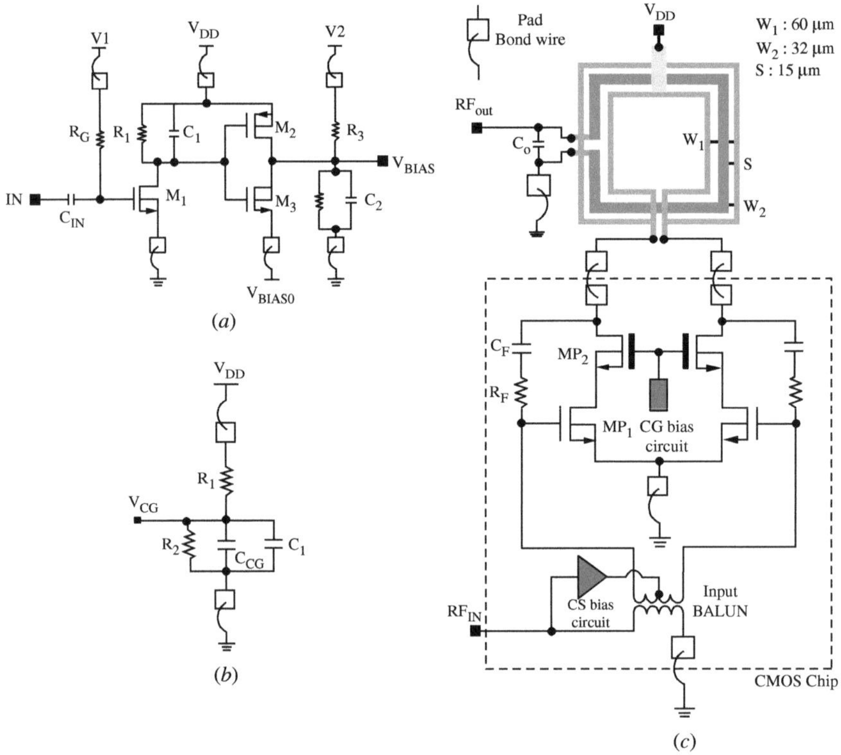

FIGURE 9.44 Schematic of differential CMOS power amplifier with adaptive bias circuits.

Figure 9.44(*c*) shows the complete schematic of a differential CMOS power amplifier, where the CMOS chip with an input balun is implemented in a 0.18-μm CMOS process and the output differential transmission-line transformer with a matching capacitor is completed using an IPD technology having a 10-μm thick copper metal and high-resistivity substrate. A feedback network with the series resistor (R_F = 140 Ω) and capacitor (C_F = 1 pF) is used between the drain of a common-gate device to the gate of a common-source device in each transistor pair to improve stability and linearity of the CMOS power amplifier. As a result, for an uplink WCDMA signal, an output power of 26.8 dBm with a *PAE* of 43.3% and an *ACLR* of –37 dBc at a 5-MHz offset at a dc-supply voltage of 3.5 V was achieved at 1.75 GHz [117].

In another adaptive bias circuit implementation, the Class-D envelope amplifier can be replaced by a resistive dividing circuit with an nMOS resistor, where the increased nMOS resistance for the increased envelope signal lowers the bias voltage of the common-gate devices, thus improving an *ACLR* by about 7 dB at a medium power region and by 2.5 dB at a high power of 26.4 dBm for a 10-MHz LTE signal with a *PAR* of 7.5 dB at 1.85 GHz for a fully integrated differential cascode power amplifier based on a 0.18-μm CMOS process [118]. Because these gate bias circuits generate memory effects, the second-harmonic control circuits at the source of the common-source stage and the gate of the common-gate stage can be used to minimize the sideband asymmetry for *ACLR* that may result in a 1.5-dB difference at an average output power of 27.5 dBm [119].

The gate oxide breakdown and hot-carrier effect are the two most critical issues of deep submicron CMOS device reliability. For example, for 0.18-μm and 0.25-μm CMOS technologies, the breakdown voltages are of about 5.7 V and 6 V, respectively [120, 121]. For a thermally grown silicon dioxide (SiO_2) layer, a field of 7×10^6 V/cm generally leads to irreversible breakdown [122]. Therefore, a safety margin should correspond to approximately 6 V across a 30-nm oxide. The electrical field near the corner of silicon-silicon dioxide ($Si\text{-}SiO_2$) interface, where the drain junction is directly under the gate, is the largest in the device and results in the following hot-carrier effects: electron injection into the SiO_2 layer producing a gate current; an avalanche hole-electron pair production process, increasing the substrate potential and resulting in the built-in bipolar transistor and current source between the drain and substrate; or punchthrough for shortened channel lengths of less than 1 μm contributing to a drain-source leakage current [122, 123]. Consequently, for high-efficiency operation modes with a large drain voltage peak factor, the device is liable to a hot-carrier effect when it is turned off. On the other hand, the device is under significant gate oxide stress when it is turned on having a zero drain voltage. The results of the accelerated stress condition with $V_g = V_{dd} = 5$ V applied to a 0.18-μm nMOS device demonstrate the significant degradation in the transition frequency f_T, transconductance g_m, and third-order intermodulation distortion IM_3, and drift of the gate-source capacitance C_{gs} [120]. If the gate breakdown is a catastrophic phenomenon, the hot-carrier effect contributes to a long-term power amplifier performance. Because of this hot-carrier effect, an output power of the CMOS power amplifier decreases exponentially and the slope of this decrease becomes sufficiently small after about 70 to 80 hours of its operation, suggesting that most of the created trap sites at the $Si\text{-}SiO_2$ interface have been filled by electrons. The recovery of the overall degradation in the output power of about 0.7 dB from its nominal value of 23.1 dBm can be achieved by increasing the gate bias voltage by 0.2 V [121]. This indicates that the performance degradation is mainly due to the increase in threshold voltage.

To overcome the low breakdown voltage limit of CMOS devices and minimize the output impedance transformation ratio for output power levels ranging from 26 to 35 dBm required at the power amplifier output in modern 2/3/4G cellular phone transmitters, a stacked-FET

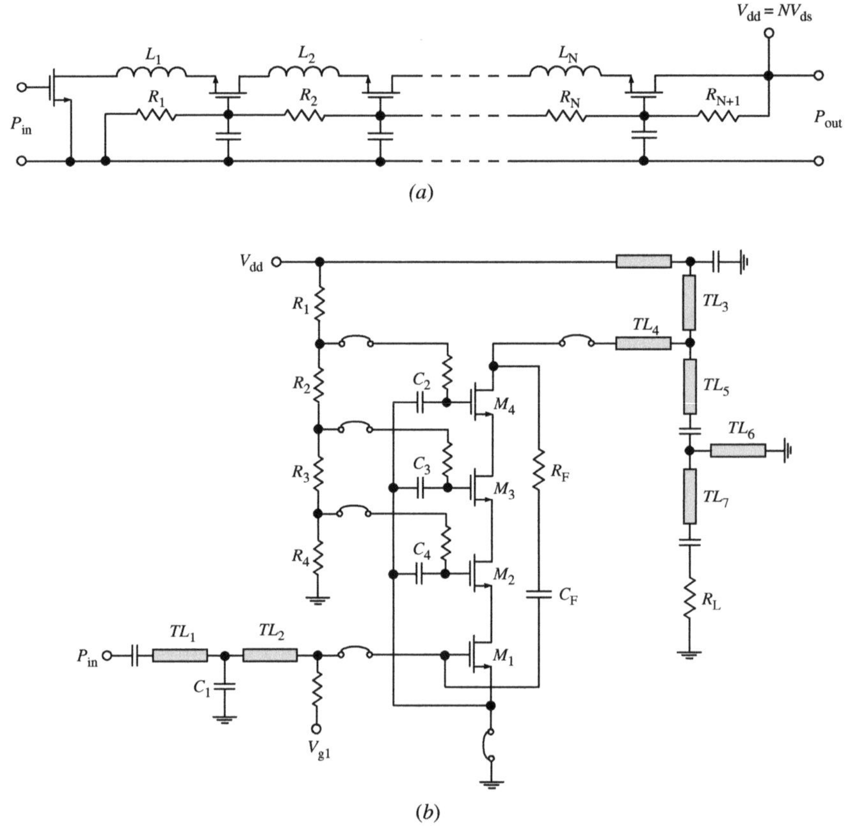

FIGURE 9.45 Circuit schematics of stacked CMOS power amplifiers.

structure can be used, whose simplified circuit diagram is shown in Fig. 9.45(*a*) [124]. The circuit is composed of a common-source input transistor and stacked transistors connected in series so that their output swings are added in phase. Here, all transistors are of equal size and the circuit elements are designed in such a way that each transistor operates under the same dc and RF conditions, where the resistors $R_1, \ldots, R_{N+1}$ form a dc bias network to provide proper bias to the series connected transistors. Unlike in a cascode configuration where the gate of the common-gate transistor is RF grounded, relatively small external capacitances $C_1, \ldots, C_N$ are introduced to allow the proper RF swing at the gate of each stacked transistor. In this case, the input and output impedances for the same total device size are N times and N^2 time higher, respectively, than that for a parallel configuration. In addition, the series configuration has higher gain because each transistor can be considered an amplifying stage followed by another transistor, where the inductances $L_1, \ldots, L_N$ are the matching elements.

Figure 9.45(*b*) shows the circuit schematic of a single-stage stacked CMOS power amplifier, which was designed to operate at 1.9 GHz using 0.28-μm 2.5-V devices available in a 0.13-μm silicon-on-insulator (SOI) CMOS process [125]. Here, each transistor has the total gate width of 5 mm so that the total device gate width in the amplifier is 20 mm. To avoid the device breakdown, the drain-source voltage of each transistor was limited to 4.5 V, allowing a maximum voltage swing of 18 V at the top drain node with a 9-V drain bias. The external capacitances C_2, C_3, and C_4 are 9, 2.6, and 2 pF, respectively, setting the optimum load impedances seen by each transistor. In this case, the gate bias voltages of each stacked transistor are implemented by an off-chip resistive voltage divider and applied through 1-kΩ on-chip resistors. The parallel feedback circuit consisting of the series connected resistor R_F and capacitor C_F is introduced to improve stability. The optimum load impedance is equal to 11.5 Ω, which can be conveniently matched with 50-Ω load over wide frequency bandwidth with high efficiency. In this case, the input and output transmission-line matching networks are implemented off-chip to allow opportunities for tuning capability. Special consideration is needed when setting the gate bias voltages of the stacked devices when the dc current is determined by the common-source device increasing with the RF power level and the source bias voltages of the stacked devices must decrease to accommodate higher dc current. As a result, a saturated output power of 32.4 dBm with a *PAE* of 47% was achieved at 1.9 GHz with a 6.5-V supply. For an uplink WCDMA modulated signal, an average output power of 28.5 dBm with a *PAE* of 38.7% and an *ACLR* of −38 dBc was measured. The saturated output power above 31 dBm and drain efficiency above 40% are maintained across the frequency bandwidth of 1.71 to 1.98 GHz. As an alternative, a stacked cascode configuration comprising four MOSFET devices can be used, which can provide an output power of 29.4 dBm at a maximum *PAE* of 51.2% operating from 3.4-V supply at 1.78 GHz with an off-chip output matching circuit using grounded coplanar waveguides and chip capacitors [126, 127]. In this case, a two-section interstage matching network was used to achieve a wide bandwidth of greater than 300 MHz, where the amplifier shows a high *PAE* of greater than 45%.

References

1. H. S. Black, "Translating System," U.S. Patent 1,686,792, Oct. 1928.
2. W. D. Lewis, "Self-Correcting Amplifier," U.S. Patent 2,592,716, Apr. 1952.
3. H. Seidel, "Feed-Forward Amplifier," U.S. Patent 3,471,798, Oct. 1969.
4. H. Seidel, "A Feedforward Experiment Applied to an L-4 Carrier System Amplifier," *IEEE Trans. Commun. Technol.*, vol. COM-19, pp. 320–325, Jun. 1971.
5. H. Seidel, "A Microwave Feedforward Experiment," *Bell Syst. Tech. J.*, vol. 50, pp. 2879–2916, Nov. 1971.
6. R. G. Meyer, R. Eschenbach, and W. M. Edgerley, "A Wide-Band Feedforward Amplifier," *IEEE J. Solid-State Circuits*, vol. SC-9, pp. 422–428, Dec. 1974.

7. C.-C. Hsieh and S.-P. Chan, "A Feedforward *S*-Band MIC Amplifier System," *IEEE J. Solid-State Circuits*, vol. SC-11, pp. 271–278, Apr. 1976.
8. R. J. Wilkinson and P. B. Kenington, "Specification of Error Amplifiers for Use in Feedforward Transmitters," *IEE Proc.-G*, vol. 139, pp. 447–480, Aug. 1992.
9. S.-G. Kang and I.-K. Lee, "Cancellation Performance of a Linearisation Loop of a Feedforward Amplifier," *Electronics Lett.*, vol. 33, pp. 444–446, Mar. 1997.
10. K. Konstantinou and D. K. Paul, "Analysis and Design of Broadband High Efficiency Feedforward Amplifiers," *1996 IEEE MTT-S Int. Microwave Symp. Dig.*, vol. 2, pp. 867–870.
11. P. B. Kenington, "Efficiency of Feedforward Amplifiers," *IEE Proc.-G*, vol. 139, pp. 591–593, Oct. 1992.
12. P. B. Kenington, *High Linearity RF Amplifier Design*, Artech House, 2000.
13. E. E. Eid, F. M. Ghannouchi, and F. Beauregard, "Optimal Feedforward Linearization System Design," *Microwave J.*, vol. 38, pp. 78–86, Nov. 1995.
14. H. Choi, Y. Jeong, J. S. Kenney, and C. D. Kim, "Cross Cancellation Technique Employing an Error Amplifier," *IEEE Microwave and Wireless Components Lett.*, vol. 18, pp. 488–490, Jul. 2008.
15. R. M. Bauman, "Adaptive Feed-Forward System," U.S. Patent 4,389,618, Jun. 1983.
16. B. Shi, W. Shan, and L. Sundstrom, "An Analog Adaptive Feedforward Amplifier Linearizer," *Proc. 34th Europ. Microwave Conf.*, pp. 1065–1068, 2004.
17. S. J. Grant, J. K. Cavers, and P. A. Goud, "A DSP Controlled Adaptive Feedforward Amplifier Linearizer," *5th IEEE Int. Conf. Record. Universal Personal Commun.*, pp. 788–792, 1996.
18. S. G. Randall, J. G. McRory, and R. H. Johnston, "Broadband DSP Based Feedforward Amplifier Linearizer," *Electronics Lett.*, vol. 38, pp. 1470–1471, Nov. 2002.
19. E. Peterson, "Control of Distortion," U.S. Patent 2,233,061, Feb. 1941.
20. T. Nojima and T. Kohno, "Cuber Predistortion Linearizer for Relay Equipment in 800 MHz Band Land Mobile Telephone System," *IEEE Trans. Vehicular Technol.*, vol. VT-34, pp. 169–177, Nov. 1985.
21. R. C. Tupynamba and E. Camargo, "MESFET Nonlinearities Applied to Predistortion Linearizer Design," *1992 IEEE MTT-S Int. Microwave Symp. Dig.*, pp. 955–958.
22. H. S. Read, "Electric Circuits," U.S. Patent 1,464,111, Aug. 1923.
23. Y. Ikeda, K. Mori, S. Shinjo, F. Kitabayashi, A. Ohta, T. Takagi, and O. Ishida, "An L-Band High Efficiency and Low Distortion Multi-Stage Amplifier Using Self Phase Distortion Compensation Technique," *IEICE Trans. Electron.*, vol. E85-C, pp. 1967–1972, Dec. 2002.
24. K. Yamauchi, K. Mori, M. Nakayama, Y. Itoh, Y. Mitsui, and O. Ishida, "A Novel Series Diode Linearizer for Mobile Radio Power Amplifiers," *1996 IEEE MTT-S Int. Microwave Symp. Dig.*, vol. 2, pp. 831–834.
25. K. Yamauchi, K. Mori, M. Nakayama, Y. Mitsui, and T. Takagi, "A Microwave Miniaturized Linearizer Using a Parallel Diode with a Bias Feed Resistance," *IEEE Trans. Microwave Theory Tech.*, vol. MTT-45, pp. 2431–2435, Dec. 1997.
26. M. Nakayama, K. Mori, K. Yamauchi, Y. Itoh, and Y. Mitsui, "An Amplitude and Phase Linearizing Technique for Linear Power Amplifiers," *Microwave J.*, vol. 39, pp. 96–104, Mar. 1996.
27. S. Ogura, K. Seino, T. Ono, A. Kamikokura, and H. Hirose, "Development of a Compact Broadband FET Linearizer for Satellite Use," *1997 IEEE MTT-S Int. Microwave Symp. Dig.*, vol. 3, pp. 1195–1198.
28. J. Boix, "Ku-Band Solid State Power Amplifiers for the New European Satellites," *Proc. 23rd Europ. Microwave Conf.*, pp. 11–14, 1993.
29. J. Yi, Y. Yang, M. Park, W. Kang, and B. Kim, "Analog Predistortion Linearizer for High-Power RF Amplifiers," *IEEE Trans. Microwave Theory Tech.*, vol. 48, pp. 2709–2713, Dec. 2000.
30. A. Grebennikov, *RF and Microwave Transmitter Design*, John Wiley & Sons, 2011.
31. K. Fayed, A. Ezzeddine, H. Huang, and A. Zaghloul, "Linear Power Amplifier Uses Mirror Predistortion," *High Frequency Electronics*, vol. 11, pp. 18–26, Jun. 2011.

32. I. Kim, J. Cha, S. Hong, Y. Y. Woo, J. Kim and B. Kim, "Predistortion Power Amplifier for Base-Station Using a Feedforward Loop Linearizer," *Proc. 36th Europ. Microwave Conf.*, pp. 141–144, 2006.
33. F. Zavosh, D. Runton, and C. Thron, "Digital Predistortion Linearizes CDMA LDMOS Amps," *Microwaves & RF*, vol. 39, pp. 55–61, Mar. 2000.
34. L. Sundstrom, M. Faulkner, and M. Johansson, "Quantization Analysis and Design of a Digital Predistortion Linearizer for RF Power Amplifiers," *IEEE Trans. Vehicular Technol.*, vol. VT-45, pp. 707–719, Nov. 1996.
35. A. R. Mansell and A. Bateman, "Adaptive Predistortion with Reduced Feedback Complexity," *Electronics Lett.*, vol. 32, pp. 1153–1154, Jun. 1996.
36. N. Nascas and Y. Papananos, "A New Non-Iterative, Adaptive Baseband Predistortion Method for High Power RF Amplifiers," *2003 IEEE Circuits and Systems Symp. Dig.*, vol. I, pp. 413–416.
37. H. Chireix, "High Power Outphasing Modulation," *Proc. IRE*, vol. 23, pp. 1370–1392, Nov. 1935.
38. D. C. Cox, "Linear Amplification with Nonlinear Components," *IEEE Trans. Commun.*, vol. COM-22, pp. 1942–1945, Dec. 1974.
39. F. H. Raab, "Average Efficiency of Outphasing Power-Amplifier Systems," *IEEE Trans. Commun.*, vol. COM-33, pp. 1094–1099, Oct. 1985.
40. S. A. Hetzel, A. Bateman, and J. P. McGeehan, "A LINC Transmitter," *Proc. 41st IEEE Vehicular Technol. Conf.*, pp. 133–137, 1991.
41. R. Langridge, T. Thornton, P. M. Asbeck, and L. E. Larson, "A Power Re-Use Technique for Improved Efficiency of Outphasing Microwave Power Amplifiers," *IEEE Trans. Microwave Theory Tech.*, vol. MTT-47, pp. 1467–1470, Aug. 1999.
42. C. P. Conradi, R. H. Johnston, and J. G. McRory, "Evaluation of a Lossless Combiner in a LINC Transmitter," *Proc. 1999 IEEE Canadian Conf. Electrical and Computer Eng.*, pp. 105–110.
43. F. Casadevall and J. J. Olmos, "On the Behavior of the LINC Transmitter," *Proc. 40th IEEE Vehicular Technol. Conf.*, pp. 29–34, 1990.
44. S. Tomisato, K. Chiba, and K. Murota, "Phase Error Free LINC Modulator," *Electronics Lett.*, vol. 25, pp. 576–577, Apr. 1989.
45. L. Sundstrom, "Automatic Adjustment of Gain and Phase Imbalances in LINC Transmitters," *Electronics Lett.*, vol. 31, pp. 155–156, Feb. 1995.
46. X. Zhang, L. E. Larson, P. M. Asbeck, and P. Nanawa, "Gain/Phase Imbalance-Minimization Techniques for LINC Transmitters," *IEEE Trans. Microwave Theory Tech.*, vol. MTT-49, pp. 2507–2515, Dec. 2001.
47. C. P. Conradi and J. G. McRory, "Predistorted LINC Transmitter," *Electronics Lett.*, vol. 38, pp. 301–302, Mar. 2002.
48. A. Birafane and A. B. Kouki, "On the Linearity and Efficiency of Outphasing Microwave Amplifiers," *IEEE Trans. Microwave Theory Tech.*, vol. MTT-52, pp. 1702–1708, Jul. 2004.
49. I. Hakala, D. K. Choi, L. Gharavi, N. Kajakine, J. Koskela, and R. Kaunisto, "A 2.14-GHz Chireix Outphasing Transmitter," *IEEE Trans. Microwave Theory Tech.*, vol. MTT-53, pp. 2129–2138, Jun. 2005.
50. T.-P. Hung, D. K. Choi, L. E. Larson, and P. M. Asbeck, "CMOS Outphasing Class-D Amplifier with Chireix Combiner," *IEEE Microwave and Wireless Lett.*, vol. 17, pp. 619–621, Aug. 2007.
51. S. Lee and S. Nam, "A CMOS Outphasing Power Amplifier with Integrated Single-Ended Chireix Combiner," *IEEE Trans. Circuits and Systems–II: Express Briefs*, vol. CAS-II-57, pp. 411–415, June 2010.
52. A. Huttunen and R. Kaunisto, "A 20-W Chireix Outphasing Transmitter for WCDMA Base Stations," *IEEE Trans. Microwave Theory Tech.*, vol. MTT-55, pp. 2709–2718, Dec. 2007.
53. M. Helaoui, S. Boumaiza, and F. M. Ghannouchi, "On the Outphasing Power Amplifier Nonlinearity Analysis and Correction Using Digital Predistortion Technique," *Proc. 2008 IEEE Radio and Wireless Symp.*, pp. 751–754.
54. W. Gerhard and R. Knoechel, "Improvement of Power Amplifier Efficiency by Reactive Chireix Combining, Power Back-Off and Differential Phase Adjustment," *2006 IEEE MTT-S Int. Microwave Symp. Dig.*, pp. 1887–1890.

55. R. Beltran, F. H. Raab, and A. Velazquez, "HF Outphasing Transmitter Using Class-E Power Amplifiers," *2009 IEEE MTT-S Int. Microwave Symp. Dig.*, pp. 757–760.
56. R. Beltran, F. H. Raab, and A. Velazquez, "An Outphasing Transmitter Using Class-E PAs and Asymmetric Combining," *High Frequency Electronics*, vol. 10, pp. 18–26, Apr., pp. 34–46, May 2011.
57. M. C. A. van Schie, M. P. van der Heijden, M. Acar, A. J. M. de Graauw, and L. C. N. de Vreede, "Analysis and Design of a Wideband High Efficiency CMOS Outphasing Amplifier," *2010 IEEE RFIC Symp. Dig.*, pp. 399–402.
58. M. P. van der Heijden, M. Acar, J. S. Vromans, and D. A. Calvillo-Cortes, "A 19 W High-Efficiency Wide-Band CMOS-GaN Class-E Chireix RF Outphasing Power Amplifier," *2011 IEEE MTT-S Int. Microwave Symp. Dig.*, pp. 1–4.
59. D. A. Calvillo-Cortes, M. P. van der Hejden, M. Acar, M. de Langen, R. Wesson, F. van Rijs, and L. C. N. de Vreede, "A Package-Integrated Chireix Outphasing RF Switch-Mode High-Power Amplifier," *IEEE Trans. Microwave Theory Tech.*, vol. MTT-61, pp. 3721–3732, Oct. 2013.
60. D. J. Perreault, "A New Power Combining and Outphasing Modulation System for High-Efficiency Power Amplification," *IEEE Trans. Circuits and Systems–I: Regular Papers*, vol. CAS-I-58, pp. 1713–1726, Oct. 2011.
61. T. W. Barton, J. L. Dawson, and D. J. Perreault, "Experimental Validation of a Four-Way Outphasing Combiner for Microwave Power Amplification," *IEEE Microwave and Wireless Components Lett.*, vol. 23, pp. 28–30, Jan. 2013.
62. G. Hannigton, P.-F. Chen, V. Radisic, T. Itoh, and P. M. Asbeck, "Microwave Power Amplifier Efficiency Improvement with a 10 MHz HBT DC-DC Converter," *1998 IEEE MTT-S Int. Microwave Symp. Dig.*, pp. 313–316.
63. G. Hannigton, P.-F. Chen, P. M. Asbeck, and L. E. Larson, "High-Efficiency Power Amplifier Using Dynamic Power-Supply Voltage for CDMA Applications," *IEEE Trans. Microwave Theory Tech.*, vol. MTT-47, pp. 1471–1476, Aug. 1999.
64. D. F. Kimball, J. Jeong, C. Hsia, P. Draxler, S. Lanfranco, W. Nagy, K. Linthicum, et al., "High-Efficiency Envelope-Tracking W-CDMA Base-Station Amplifier Using GaN HFETs," *IEEE Trans. Microwave Theory Tech.*, vol. MTT-54, pp. 3848–3856, Nov. 2006.
65. J. Jeong, D. F. Kimball, M. Kwak, P. Draxler, C. Hsia, C. Steinbeiser, T. Landon, et al.,"High-Efficiency WCDMA Envelope Tracking Base-Station Amplifier Implemented with GaAs HVHBTs," *IEEE J. Solid-State Circuits*, vol. SC-44, pp. 2629–2639, Oct. 2009.
66. M. Hassan, L. E. Larson, V. W. Leung, and P. M. Asbeck, "A Combined Series-Parallel Hybrid Envelope Amplifier for Envelope Tracking Mobile Terminal RF Power Amplifier Applications," *IEEE J. Solid-State Circuits*, vol. SC-47, pp. 1185–1198, May 2012.
67. M. Hassan and U. Saeed, "A Digitally-Controlled Hybrid Envelope Amplifier for LTE Envelope/Polar Transmitters," *2013 IEEE MTT-S Int. Microwave Symp. Dig.*, pp. 1–3.
68. M. Ranjan, K. H. Koo, G. Hannigton, C. Fallesen, and P. M. Asbeck, "Microwave Power Amplifiers with Digitally-Controlled Power Supply Voltage for High Efficiency and High Linearity," *2000 IEEE MTT-S Int. Microwave Symp. Dig.*, vol. 1, pp. 493–496.
69. D. F. Kimball, J. Jeong, C. Hsia, P. Draxler, S. Lanfranco, W. Nagy, K. Linthicum, et al., "High-Efficiency Envelope-Tracking W-CDMA Base-Station Amplifier Using GaN HFETs," *IEEE Trans. Microwave Theory Tech.*, vol. MTT-54, pp. 3848–3856, Nov. 2006.
70. J. Hoversten, S. Schafer, M. Roberg, M. Norris, D. Maksimovic, and Z. Popovic, "Codesign of PA, Supply, and Signal Processing for Linear Supply-Modulated RF Transmitters," *IEEE Trans. Microwave Theory Tech.*, vol. MTT-60, pp. 2010–2020, Jun. 2012.
71. J. Staudinger, B. Gilsdorf, D. Newman, G. Norris, G. Sadowniczak, R. Sherman, and T. Quach, "High Efficiency CDMA RF Power Amplifier Using Dynamic Envelope Tracking Technique," *2000 IEEE MTT-S Int. Microwave Symp. Dig.*, vol. 2, pp. 873–876.

72. S. Abedinpour, I. Deligoz, J. Desai, M. Figiel, and S. Kiaei, "Monolithic Supply Modulated RF Power Amplifier and DC-DC Power Converter IC," *2003 IEEE MTT-S Int. Microwave Symp. Dig.*, vol. 1, pp. 89–92.
73. B. Sahu and G. A. Rincon-Mora, "A High-Efficiency Linear RF Power Amplifier with a Power-Tracking Dynamically Adaptive Buck-Boost Supply," *IEEE Trans. Microwave Theory Tech.*, vol. MTT-52, pp. 112–120, Jan. 2004.
74. T. Gillenwater and M. Schindler, "Technology Trends in Mobile Handsets," *2013 IEEE Int. Wireless Symp. Dig.*, pp. 1–4.
75. B. Kim, Ju. Kim, D. Kim, J. Son, Y. Cho, Jo. Kim, and B. Park, "Push the Envelope," *IEEE Microwave Mag.*, vol. 14, pp. 68–81, May 2013.
76. N. Maslennikov, M. Gurvich, and A. Rabinovich, "Constant Gain Nonlinear Envelope Tracking High Efficiency Linear Amplifier," U.S. Patent 7,440,733, Oct. 2008.
77. G. Wimpenny, "Understand and Characterize Envelope-Tracking Power Amplifiers," *Nujira Ltd.*, 2012.
78. F. Wang, D. F. Kimball, D. Y. Lie, P. M. Asbeck, and L. E. Larsson, "A Monolithic High-Efficiency 2.4-GHz 20-dBm SiGe BiCMOS Envelope-Tracking OFDM Power Amplifier," *IEEE J. Solid-State Circuits*, vol. SC-42, pp. 1271–1281, Jun. 2007.
79. Y. Aihara, "Transmitter Power Varying Device Having a Bypass Line for a Power Amplifier," U.S. Patent 5,909,643, Jun. 1999.
80. S. Brozovich, "High Efficiency Multiple Power Level Amplifier Circuit," U.S. Patent 5,661,434, Aug. 1997.
81. J. Staudinger, "Applying Switched Gain Stage Concepts to Improve Efficiency and Linearity of Mobile CDMA Power Amplification," *Microwave J.*, vol. 43, pp. 152–162, Sep. 2000.
82. H. X. Wu, "High-Efficient Configurable Power Amplifier for Use in a Portable Unit," U.S. Patent 5,758,269, May 1998.
83. A. Grebennikov, "Efficient Power Amplification System," U.S. Patent 7,345,534, Mar. 2008.
84. G. Itkin and A. Grebennikov, "Transmitter Arrangement," U.S. Patent 7,853,290, Dec. 2010.
85. G. Klemens, N. M. Pletcher, B. Nejati, N. L. Frederick, and T. A. Myers, "Output Circuit with Integrated Impedance Matching, Power Combining and Filtering for Power Amplifiers and Other Circuits," U.S. Patent 8,432,237, Apr. 2013.
86. A. V. Grebennikov, H. Jaeger, and E. P. Heaney, "High-Efficiency Cellular Phone Transmitters with Reduced Power Consumption," *Proc. 2002 Europ. Conf. Wireless Technol.*, pp. 115–118.
87. H. Kim, Y. Yoon, O. Lee, K. H. An, D. H. Lee, W. Kim, C. H. Lee, and et al., "A Fully Integrated CMOS RF Power Amplifier with Tunable Matching Network for GSM/EDGE Dual-Mode Application," *2010 IEEE MTT-S Int. Microwave Symp. Dig.*, pp. 800–803.
88. G. Leuzzi and C. Micheli, "Variable-Load Constant-Efficiency Power Amplifier for Mobile Communications Applications," *Proc. 11*th *Europ. GAAS Symp.*, pp. 481–484, 2003.
89. W. C. E. Neo, Y. Lin, X. Liu, L. C. N. de Vreede, L. E. Larson, M. Spirito, M. J. Pelk, et al., "Adaptive Multi-Band Multi-Mode Power Amplifier Using Integrated Varactor-Based Tunable Matching Networks," *IEEE J. Solid-State Circuits*, vol. SC-41, pp. 2166–2176, Sep. 2006.
90. L. Sankey and Z. Popovic, "Adaptive Tuning for Handheld Transmitters," *2009 IEEE MTT-S Int. Microwave Symp. Dig.*, pp. 225–228.
91. R. A. York, A. S. Nagra, P. Periaswamy, O. Auciello, S. K. Streiffer, and J. Im, "Synthesis and Characterization of $(Ba_xSr_{1-x})Ti_{1+y}O_{3+z}$ Thin Films and Integration into Microwave Varactors and Phase Shifters," *Int. J. Integrated Ferroelectrics*, vol. 34, pp. 177–188, Aug. 2001.
92. L. V. Chen, R. Forse, D. Chase, and R. A. York, "Analog Tunable Matching Network Using Integrated Thin-Film BST Capacitors," *2004 IEEE MTT-S Int. Microwave Symp. Dig.*, pp. 261–264.

93. H. Katta, H. Kurioka, and Y. Yashima, "Tunable Power Amplifier Using Thin-Film BST Capacitors," *2006 IEEE MTT-S Int. Microwave Symp. Dig.*, pp. 564–567.
94. H. Maune, O. Bengtsson, F. Golden, M. Sazegar, R. Jacoby, and W. Heinrich, "Tunable RF GaN-Power Transistor Implementing Impedance Matching Networks Based on BST Thick Films," *Proc. 42nd Europ. Microwave Conf.*, pp. 1206–1209, 2012.
95. F. H. Raab, "High-Efficiency Linear Amplification by Dynamic Load Modulation," *2003 IEEE MTT-S Int. Microwave Symp. Dig.*, vol. 3, pp. 1717–1720.
96. H. M. Nemati, C. Fager, U. Gustavsson, R. Jos, and H. Zirath, "Design of Varactor-Based Tunable Matching Networks for Dynamic Load Modulation of High Power Amplifiers," *IEEE Trans. Microwave Theory Tech.*, vol. MTT-57, pp. 1110–118, May 2009.
97. J. S. Fu and A. Mortazawi, "Improving Power Amplifier Efficiency and Linearity Using a Dynamically Controlled Tunable Matching Network," *IEEE Trans. Microwave Theory Tech.*, vol. MTT-56, pp. 3239–3224, Dec. 2008.
98. H-M. Park, S.-H. Cheon, J.-W. Park, and S. Hong, "Demonstration of On-Chip Appended Power Amplifier for Improved Efficiency at Low Power Region," *2003 IEEE MTT-S Int. Microwave Symp. Dig.*, vol. 2, pp. 691–694.
99. J. H. Kim, J. H. Kim, Y. S. Noh, and C. S. Park, "An InGaP-GaAs HBT MMIC Smart Power Amplifier for W-CDMA Mobile Handsets," *IEEE J. Solid-State Circuits*, vol. SC-38, pp. 905–910, Jun. 2003.
100. A. Grebennikov, B. Zogl, H. Hermann, C. Roth, and W. Thomann, "High-Efficiency Balanced Switched-Path Monolithic SiGe HBT Power Amplifiers for Wireless Applications," *Proc. 37th Europ. Microwave Conf.*, pp. 1189–1192, 2007.
101. T. Sato and C. Grigorean, "Design Advantages of CDMA Power Amplifiers Built with MOSFET Technology," *Microwave J.*, vol. 45, pp. 64–78, Oct. 2002.
102. Y. S. Noh and C. S. Park, "An Intelligent Power Amplifier MMIC Using a New Adaptive Bias Control Circuit for W-CDMA Applications," *IEEE J. Solid-State Circuits*, vol. SC-39, pp. 967–970, Jun. 2004.
103. W. Pratt, "HBT Power Amplifier," U.S. Patent 5,629,648, May 1997.
104. H. Jaeger, A. V. Grebennikov, E. P. Heaney, and R. Weigel, "Broadband High-Efficiency Monolithic InGaP/GaAs HBT Power Amplifiers for Wireless Applications," *Int. J. RF and Microwave Computer-Aided Eng.*, vol. 13, pp. 496–510, Nov. 2003.
105. T. B. Nishimura, M. Tanomura, K. Azuma, K. Nakai, Y. Hasegawa, and H. Shimawaki, "A 50% Efficiency InGaP/GaAs HBT Power Amplifier Module for 1.95 GHz Wide-Band CDMA Handsets," *2001 IEEE RFIC Symp. Dig.*, pp. 31–34.
106. G. Hau, A. Hussain, J. Turpel, and J. Donnenwirth, "A $3 \times 3 mm^2$ LTE/WCDMA Dual-Mode Power Amplifier Module with Integrated High Directivity Coupler," *Proc. 2011 IEEE Bipolar/BiCMOS Circuits and Technology Meeting*, pp. 138–141.
107. M. J. Franco, "An Efficient, 35 dBm, Inverse Class-F, UHF RF Power Amplifier Module on a 12 mm^2 Footprint Designed in First Pass through Accurate Modeling and Simulation," *2010 IEEE MTT-S Int. Microwave Symp. Dig.*, pp. 928–931.
108. C. Yoo and Q. Huang, "A Common-Gate Switched 0.9-W Class-E Power Amplifier with 41% PAE in 0.25-μm CMOS," *IEEE J. Solid-State Circuits*, vol. SC-36, pp. 823–830, May 2001.
109. C. H. Lee, J. J. Chang, K. S. Yang, K. H. An, I. Lee, K. Kim, J. Nam, et al., "A Highly Efficient GSM/GPRA Quad-Band CMOS PA Module," *2009 IEEE RFIC Symp. Dig.*, pp. 229–232.
110. M. Apostolidou, M. P. van der Heijden, D. M. W. Leenaerts, J. Sonsky, A. Heringa, and I. Volokhine, "A 65 nm CMOS 30 dBm Class-E RF Power Amplifier with 60% PAE and 40% PAE at 16 dB Back-Off," *IEEE J. Solid-State Circuits*, vol. SC-44, pp. 1372–1379, May 2009.
111. T. C. Kuo and B. B. Lusignan, "A 1.5-W Class-F RF Power Amplifier in 0.2-μm CMOS Technology," *2001 IEEE Int. Solid-State Circuits Conf. Dig.*, pp. 154–155.

112. A. Shirvani, D. K. Su, and B. Wolley, "A CMOS RF Power Amplifier with Parallel Amplification for Efficient Power Control," *IEEE J. Solid-State Circuits*, vol. SC-37, pp. 684–693, Jun. 2002.
113. P. Asbeck, L. Larson, D. Kimball, and J. Buckwalter, "CMOS Handset Amplifiers: Directions for the Future," *Proc. 2012 IEEE Custom Integrated Circuits Conf.*, pp. 1–6.
114. D. Kang, B. Park, C. Zhao, D. Kim, J. Kim, Y. Cho, S. Jin, et al., "A 34% PAE, 26-dBm Output Envelope-Tracking CMOS Power Amplifier for 10-MHz BW LTE Applications," *2012 IEEE RFIC Symp. Dig.*, pp. 1–3.
115. C. Wang, L. E. Larson, and P. M. Asbeck, "A Nonlinear Capacitance Cancellation Technique and Its Application to a CMOS Class AB Power Amplifier," *2001 IEEE RFIC Symp. Dig.*, pp. 39–42.
116. C. Fager, J. C. Pedro, N. B. Carvalho, H. Zirath, F. Fortes, and M. J. Rosario, "A Comprehensive Analysis of IMD Behavior in RF CMOS Power Amplifiers," *IEEE J. Solid-State Circuits*, vol. SC-39, pp. 24–34, Jan. 2004.
117. B. Koo, Y. Na, and S. Hong, "Integrated Bias Circuits of RF CMOS Cascode Power Amplifier for Linearity Enhancement," *IEEE Trans. Microwave Theory Tech.*, vol. MTT-60, pp. 340–351, Feb. 2012.
118. S. Jin, B. Park, K. Moon, Y. Cho, D. Kim, H. Jin, et at. "Enhanced Linearity of CMOS Power Amplifier Using Adaptive Common Gate Bias Control," *2013 IEEE MTT-S Int. Microwave Symp. Dig.*, pp. 1–4.
119. S. Jin, B. Park, K. Moon, M. Kwon, and B. Kim, "Linearization of CMOS Cascode Power Amplifiers through Adaptive Bias Control," *IEEE Trans. Microwave Theory Tech.*, vol. MTT-61, pp. 4534–4543, Dec. 2013.
120. Q. Li, J. Zhang, W. Li, J. S. Yuan, Y. Chen, and A. S. Oates, "RF Circuit Performance Degradation due to Soft Breakdown and Hot-Carrier Effect in Deep-Submicrometer CMOS Technology," *IEEE Trans. Microwave Theory Tech.*, vol. MTT-49, pp. 1546–1551, Sep. 2001.
121. V. R. Vathulya, T. Sowlati, and D. Leenaerts, "Class 1 Bluetooth Power Amplifier with 24 dBm Output Power and 48% PAE at 2.4 GHz in 0.25 μm CMOS," *Proc. 27th Europ. Solid-State Circuits Conf.*, pp. 57–60, 2001.
122. R. S. Muller and T. I. Kamins, *Device Electronics for Integrated Circuits*, New York: John Wiley & Sons, 1986.
123. Y. Tsividis, *Operational and Modeling of the MOS Transistors*, McGraw-Hill, 1999.
124. M. Shifrin, Y. Ayasli, and P. Katzin, "A New Power Amplifier Topology with Series Biasing and Power Combining of Transistors," *1992 IEEE Microwave and Millimeter-Wave Monolithic Circuits Symp. Dig.*, pp. 39–41.
125. S. Pornpromlokot, J. Jeong, C. D. Presti, A. Scuderi, and P. M. Asbeck, "A Watt-Level Stacked-FET Linear Power Amplifier in Silicon-on-Insulator CMOS," *IEEE Trans. Microwave Theory Tech.*, vol. MTT-58, pp. 57–64, Jan. 2010.
126. S. Leuschner, S. Pinarello, U. Hodel, J. E. Mueller, and H. Klar, "A 31-dBm, High Ruggedness Power Amplifier in 65-nm Standard CMOS with High-Efficiency Stacked-Cascode Stages," *2010 IEEE RFIC Symp. Dig.*, pp. 385–398.
127. S. Leuschner, J. E. Mueller, and H. Klar, "A 1.8GHz Wide-Band Stacked-Cascode CMOS Power Amplifier for WCDMA Applications in 65 nm Standard CMOS," *2011 IEEE RFIC Symp. Dig.*, pp. 1–4.

CHAPTER 10

High-Efficiency Doherty Power Amplifiers

This chapter describes the historical aspect of the Doherty approach to the power amplifier design and modern trends in Doherty amplifier design techniques using multistage and asymmetric multiway architectures. To increase efficiency over the power-backoff range, the switchmode Class-E, conventional Class-F, or inverse Class-F operation mode by controlling the second and third harmonics can be used in the load network. The Doherty amplifier with a series connected load and inverted Doherty architectures are also described and discussed. Finally, examples of the lumped Doherty amplifier implemented in monolithic microwave integrated circuits, digitally driven Doherty technique, and broadband capability of the two-stage Doherty amplifier are given.

10.1 Historical Aspect and Conventional Doherty Architectures

A new power amplifier technique for amplitude-modulated (AM) radio-frequency signals was introduced by William H. Doherty in broadcasting in the mid-1930s as a more efficient alternative to both conventional amplitude-modulation techniques and Chireix outphasing [1, 2]. This new technique achieves plate circuit efficiencies of up to 60 to 65% independent of modulation by means of a combined action of the variation of load distribution of the vacuum tubes, and the variation of the circuit impedance over the modulation cycle. When Doherty joined the radio development department of the Bell Telephone Laboratories in June 1929, he was engaged in the development of high-power radio transmitters for transoceanic radiotelephony and broadcasting. As a result, in 1936 he invented a means to greatly improve the efficiency of radio-frequency power amplifiers, quickly termed the "Doherty amplifier."

It was first used in a 50-kW transmitter with audio-frequency feedback providing a resulting distortion level from less than 1% at lower frequencies to a few percent at high audio frequencies. In this case, the power amplifier was operated at an efficiency of 60% representing a reduction of nearly one- half in the all-day power consumption as compared with the power required in the conventional type of linear power amplifier operating at 33% efficiency [3]. The IRE Morris Liebmann Memorial Award was voted to Doherty in May 1937 for his improvement in the efficiency of radio-frequency power amplifiers [4]. By 1940, the Doherty amplifiers had been incorporated in 35 commercial radio stations worldwide, at powers up to 50 kW.

In subsequent years, the Doherty amplifiers continued to be used in a number of medium- and high-power low-frequency (LF) and medium-frequency (MF) vacuum-tube AM transmitters [5, 6]. In August 1953, a 1-MW vacuum-tube transmitter in which the outputs of the two 500-kW Doherty amplifiers were joined in a bridge-type combiner began regular operation in the long-wave band in Europe. The Doherty amplifiers had also been considered for use in solid-state MF and high-frequency (HF) systems, as well as in high-power ultra–high-frequency (UHF) transmitters [7–9]. The practical implementation of a classic triode-based Doherty scheme was restricted by its substantial nonlinearity for both linear amplification of AM signals and grid-type signal modulation that required complicated envelope correction and feedback linearization circuits. At the same time, the Doherty amplifiers using tetrode transmitting tubes could improve their overall performance when the modulation was applied to the screen grids of both the carrier and peaking tubes, while the control grids of both tubes are fed by an essentially constant level of RF excitation [10]. This resulted in the peaking tube being modulated upward during positive half of the modulating cycle and the carrier tube being modulated downward during negative half of the modulating cycle.

The Doherty amplifier still remains in use in very high-power AM transmitters, but for lower-power AM transmitters, vacuum-tube amplifiers in general were eclipsed in the 1980s by solid-state power amplifiers due to their advantages of smaller size and cost, lower operating voltages, higher reliability and greater physical ruggedness, insensitivity to mechanical shock and vibration, and possibility to use fully automated manufacturing processes and high level of integration. However, the transistor (bipolar or field-effect) as a three-electrode solid-state device is characterized by significant nonlinearity of its transfer characteristic and intrinsic capacitances that requires complicated linearization schemes, which was difficult to realize in analog domain at that time. Interest in the Doherty configuration has been later revived due to significant progress in radio communication systems based on complex digital modulation schemes such as WiMAX in OFDM (orthogonal frequency division multiplexing) systems, CDMA2000, WCDMA or LTE (long-term evolution) enhancement to the UMTS wireless standard, where the sum of several constant-envelope signals creates an aggregate AM signal with high peak-to-average ratio (*PAR*) and digital linearization techniques can be applied to reduce the Doherty amplifier distortion. Recently, Doherty amplifiers of different architectures have found widespread application in cellular base station transmitters operating at GHz frequencies.

10.1.1 Basic Structures

Generally, a Doherty amplifier system combines the outputs of two (or more) linear RF power amplifiers (PAs) through an impedance-inverting network composed of the lumped elements or represented by a quarterwave transmission line. The two fundamental forms of the Doherty amplifier are shown in Fig. 10.1, with a shunt connected load in Fig. 10.1(*a*) and with a series connected load in Fig. 10.1(*b*) [2]. In the former case, the load impedance used is $R_L = R/2$, which is the same as would be used if the tubes were to be connected in parallel in the conventional type of power amplifier, where R is the load impedance seen by each tube at maximum output power. In the latter case, as long as the right-hand or *peaking* tube does not conduct, the impedance-inverting network provides zero impedance being terminated as open circuit, and the left-hand or *carrier* tube operates into a load impedance $R_L = 2R$. However, when the peaking tube is permitted to conduct, each tube is operating into the impedance R at the peak of modulation and delivering twice the carrier power (or power of the transmitted unmodulated signal), so that the total instantaneous output is the required value of four times the carrier power. The shunt connection appears to be more advantageous for most practical applications because the load circuit is grounded, while the load is neither grounded nor balanced to ground in the series arrangement.

The ideal anode voltage and current behavior in the carrier and peaking tubes as the amplitude of the grid excitation is varied is shown in Fig. 10.1(*c*). Here, for the classic Doherty amplifier with equal-power tubes, the transition voltage is half the peak-envelope point (PEP), and the total output power of the amplifier comes from the carrier tube for input amplitudes less than or equal to the transition point. The region between the transition-point and PEP values represents the load modulation region and the voltage on the carrier tube remains constant at the PEP level. At the same time, the voltage across the peaking tube continues to rise linearly, with its current commencing and rising twice as fast as the current in the carrier tube in order to reach its PEP value at maximum output power. Thus, at low output power levels, the carrier amplifier operates linearly, reaching saturation that corresponds to maximum efficiency at some transition voltage below the system peak-output voltage. However, at higher output power levels, the carrier amplifier remains saturated while the peaking amplifier operates linearly.

The corresponding shapes of the envelopes of anode current and voltages during complete load variation are shown in Fig. 10.2 [2].

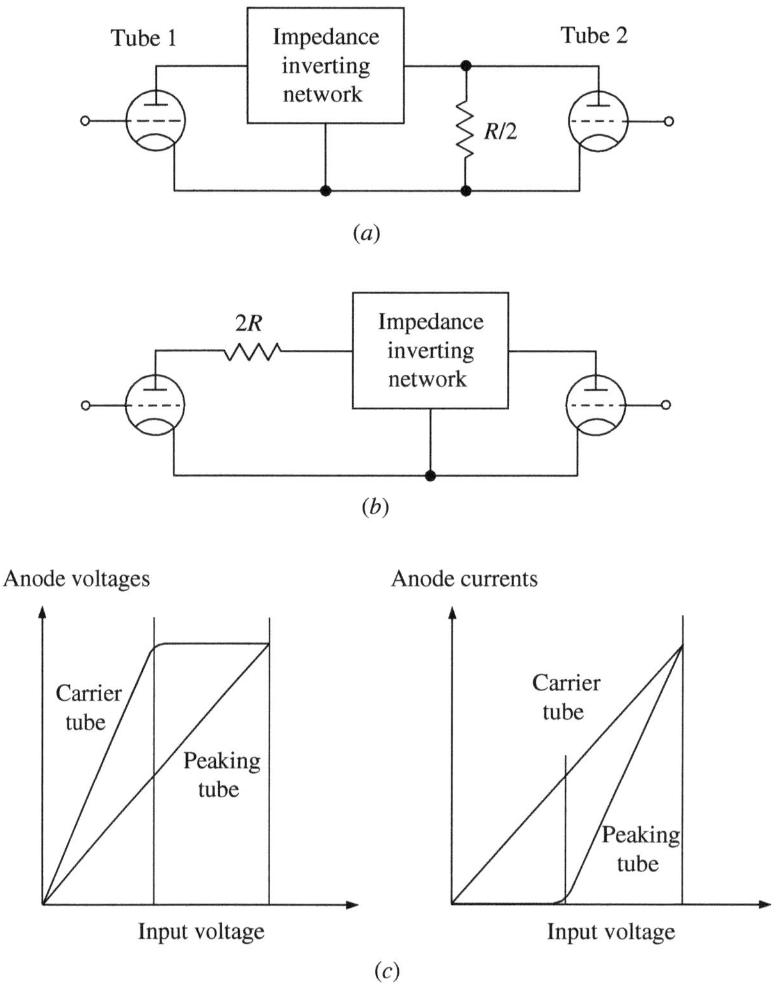

FIGURE 10.1 Doherty fundamental load-network structures and their ideal voltage and current behavior.

In a simple case of a sinusoidal modulating signal $v_m(t) = V_m \cos\Omega t$, where V_m is the modulating amplitude and Ω is the modulating frequency, the average output power P_{avr} for the modulated signal with a time-varying amplitude $V = V_0 + v_m(t) = V_0(1 + m\cos\Omega t)$, where V_0 is the carrier amplitude and $m = V_m / V_0$ is the modulation index, can be found as

$$P_{avr} = \frac{V_0^2}{4\pi R_L} \int_0^{2\pi} (1 + m\cos\Omega t)^2 \, d\Omega t = P_0 \left(1 + \frac{m^2}{2}\right) \tag{10.1}$$

where $P_0 = V_0^2/2R_L$ is the carrier power without modulation and m denotes the ratio of the variation of the modulated carrier amplitude

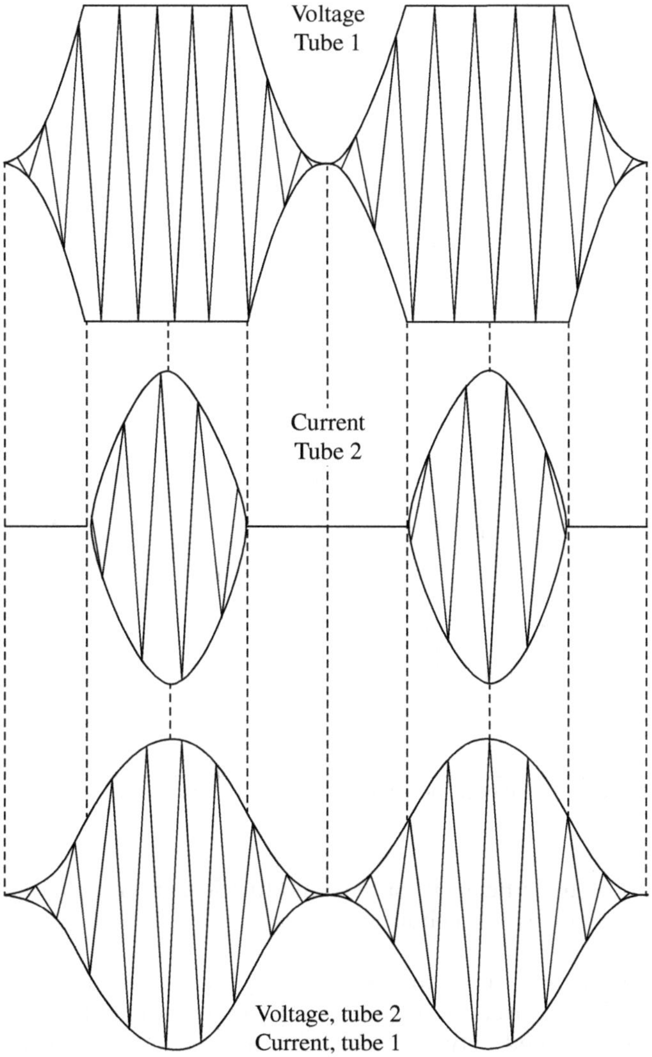

FIGURE 10.2 Envelopes of anode currents and voltages during load variation.

to the unmodulated carrier amplitude. Because the average anode current in the carrier amplifier remains constant for all cycle of load variation but the duration of anode current in the peaking amplifier increases gradually with the amplitude, the average efficiency through the modulation cycle is then found to be

$$\eta_{\text{avr}} = \eta_0 \frac{1 + \dfrac{m^2}{2}}{1 + \dfrac{2mq}{\pi}} \tag{10.2}$$

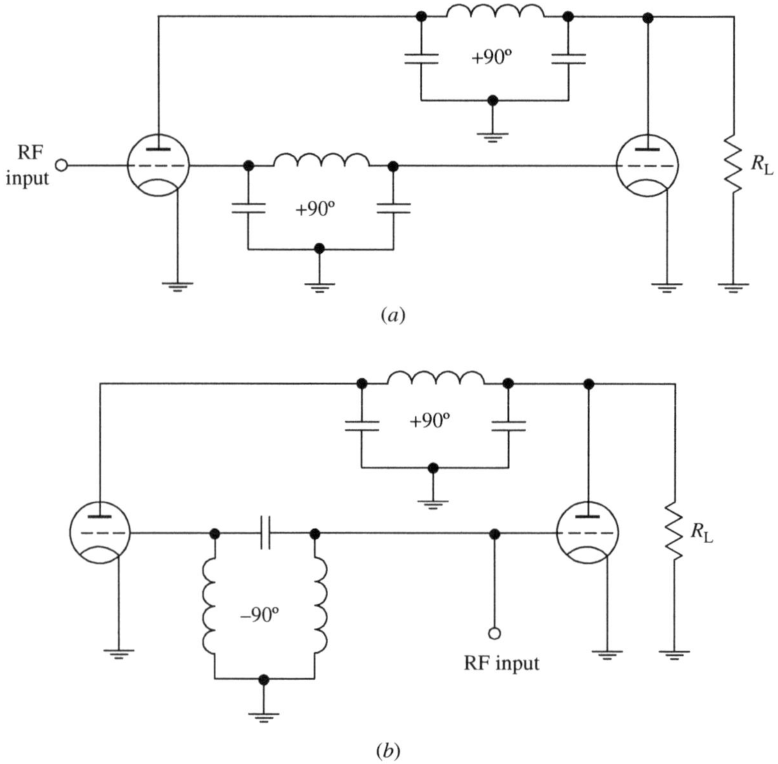

FIGURE 10.3 Doherty amplifier basic schematics with lumped elements.

where η_0 is the efficiency for zero modulation and q is the variable factor that ranges from about 0.7 for zero modulation to 0.93 for full modulation [1].

Figure 10.3(*a*) shows the Doherty amplifier schematic, where the anodes of the carrier and peaking tubes are connected together by a π-type 90° lumped network, which introduces a lagging phase shift of 90° from the anode of the carrier tube to the anode of the peaking tube [2, 10]. Such a 90° lumped network is used due to its impedance-inverting characteristics. This means that if the terminating impedance at the point in the network where the peaking tube is located is reduced, the impedance seen by the carrier tube will increase. In this case, because the load network of the high-power transmitter was approximately of 35 Ω, the 90° lumped network was set to provide an impedance at the carrier tube of about 140 Ω. As a result, the series inductance and two shunt capacitors are each selected to have a reactance equal to $\sqrt{35 \times 140} = 70\ \Omega$. To compensate for the output phase shift of 90°, the input to the grid of the peaking tube is delayed by 90° by similar means. If the input excitation is applied to the grid of the

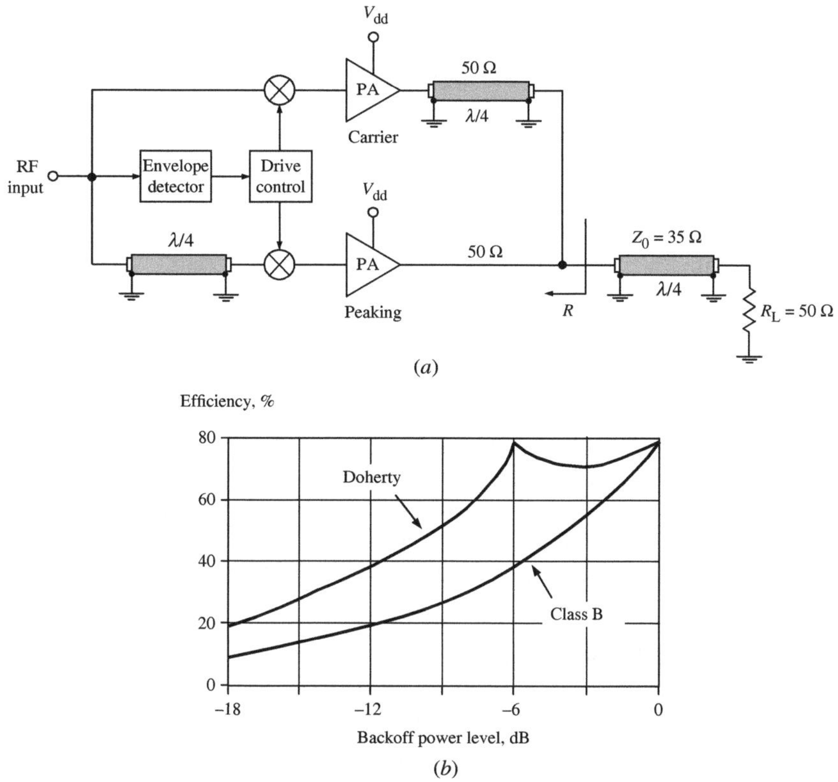

Figure 10.4 Doherty amplifier architecture with quarterwave lines and collector efficiencies.

peaking tube, a π-type lumped network consisting of the series capacitor and two shunt inductances is added between the grids of the carrier and peaking tubes to compensate for the output phase shift of 90°, as shown in Fig. 10.3(*b*).

The simplified two-stage transmission-line Doherty power-amplifier architecture shown in Fig. 10.4(*a*) incorporates the carrier and peaking power amplifiers, separated by a quarterwave transmission line in the carrier amplifier path [11, 12]. Such a section of line, known as a *quarterwave transformer*, has the ability to invert impedances according to

$$Z_0 = \sqrt{Z_{\text{in}} Z_{\text{out}}} \tag{10.3}$$

where Z_0 is the characteristic impedance of the transmission line. This property can be seen more clearly by rewriting Eq. (10.3) as

$$Z_{\text{out}} = \frac{Z_0^2}{Z_{\text{in}}} \tag{10.4}$$

from which it follows that the output impedance Z_{out} increases inversely with the input impedance Z_{in} for constant Z_0. The quarter-wave transmission line at the input of the peaking amplifier is required to compensate for the 90° phase shift caused by the quarter-wave transmission line at the output of the carrier amplifier. The output quarterwave line with $Z_0 = \sqrt{25 \times 50} = 35\ \Omega$ is required to match the standard load impedance of 50 Ω when both carrier and peaking amplifiers deliver maximum power, each of which designed in a 50-Ω environment.

An input drive controller is used to turn on the peaking amplifier (bias control) when the carrier amplifier starts to saturate because it is assumed that the carrier and peaking amplifiers are biased in Class B for idealized system analysis. However, in practice, the carrier amplifier is biased in a Class-B mode, whereas the peaking amplifier is biased in a Class-C mode. At a backoff power level of –6 dB, the saturated output power of the carrier amplifier is four times lower than the peak output power $P_{PEP.}$ This indicates that its collector (or drain) efficiency, when operated in an ideal Class-B mode, is twice than that of a conventional Class-B power amplifier, achieving a maximum efficiency of 78.5%, as shown in Fig. 10.4(*b*).

10.1.2 Operation Principle

The basic operation principle of a conventional Doherty power-amplifier architecture shown in Fig. 10.5(*a*) can be analyzed for low, medium, and peak output power regions separately [12]. Figure 10.5(*b*) shows the current and voltage behavior for ideal transistors and lossless matching circuits, where V_L is the load voltage and I_L is the load current. The condition of power conservation for a lossless output transmission line results in

$$I_3 = I_1 \sqrt{\frac{R_1}{R_3}} \tag{10.5}$$

while the current division ratio β is defined by

$$\beta = \frac{I_3}{I_2 + I_3} \tag{10.6}$$

As a result, the overall output power P_{out} is the sum of the carrier (main) amplifier output power $P_1 = \beta P_{out}$ and peaking (auxiliary) amplifier output power $P_2 = (1 - \beta)P_{out}$. The impedance seen at the output of the transmission line in the carrier-amplifier path is

$$R_3 = \frac{I_2 + I_3}{I_3} \frac{Z_1^2}{R_L} = \frac{Z_1^2}{\beta R_L} \tag{10.7}$$

and the impedance seen by the peaking power amplifier is

$$R_2 = \frac{I_2 + I_3}{I_2} \frac{Z_1^2}{R_L} = \frac{Z_1^2}{(1-\beta) R_L} \tag{10.8}$$

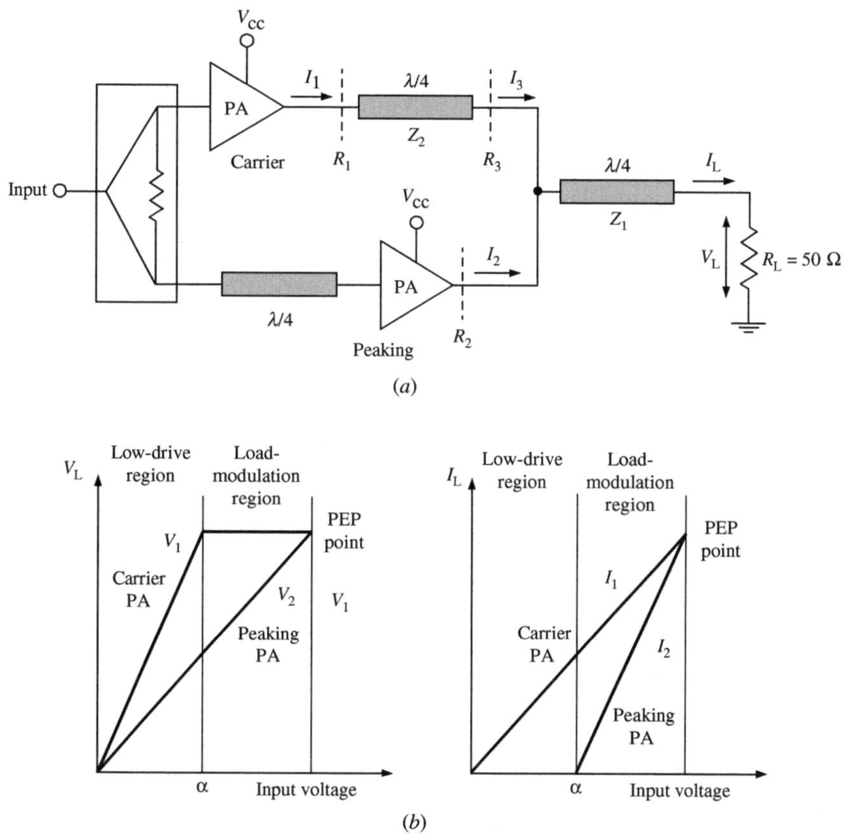

FIGURE 10.5 Basic two-stage Doherty amplifier architecture and its operation principle.

At peak output power P_{PEP} when both carrier and peaking amplifiers are saturated, the resultant collector efficiency is equal to the maximum achievable efficiency $\eta = \pi/4 \approx 78.5\%$ for an ideal Class-B operation. For the conventional Doherty power-amplifier architecture shown in Fig. 10.5(*a*) with the current and power division ratios $\beta = \alpha = 0.5$, respectively, when both carrier and peaking amplifiers produce equal output powers, their load impedances are equal to $R_1 = R_3 = R_2 = Z_2 = 2Z_1^2 / R_L$. If the characteristic impedance of the output transmission line is chosen to be $Z_1 = 35\ \Omega$, then $R_1 = R_3 = R_2 = Z_2 = R_L = 50\ \Omega$.

At lower power levels in a low-drive region, the peaking amplifier is turned off because the instantaneous amplitude of the input signal is insufficient to overcome the negative Class-C bias and appears as an open circuit, whereas the carrier amplifier operates in the active region. In this case, the load impedance seen by the carrier amplifier is

$$R_1 = \left(\frac{Z_2}{Z_1}\right)^2 R_L \tag{10.9}$$

resulting in $R_1 = 2R_L = 100\ \Omega$ when $Z_1 = 35\ \Omega$ and $Z_2 = R_L = 50\ \Omega$. Because the output power of the carrier amplifier in saturation is four times less than the peak output power P_{PEP}, the collector efficiency of the carrier amplifier in an ideal Class-B mode will be twice than that of a conventional Class-B power amplifier, achieving maximum of 78.5% at backoff power level of –6 dB, as shown in Fig. 10.4(*b*).

At medium power levels in a load-modulation region, the carrier amplifier is saturated, whereas the peaking amplifier is turned on and operates in the active region. Because the output voltage of the carrier amplifier $V_1 = I_1R_1$ is constant under saturation conditions, from Eq. (10.5) it follows that the current I_3 is constant in the medium power region as well. The collector efficiency of the carrier amplifier remains at its maximum value, whereas the collector efficiency of the peaking amplifier increases up to its maximum value for Class-B operation at peak output power P_{PEP}. As a result, the Doherty amplifier architecture achieves maximum efficiency at both the transition –6 dB backoff point and the peak output power, and remains relatively high in between, as seen from Fig. 10.4(*b*).

For high-quality transmission, it is important to obtain in vacuum-tube power amplifiers a linear relation between grid exciting voltage and output anode voltage. The high impedance used for the carrier tube over the lower half of the modulation envelope causes the dynamic characteristic to be quite straight in this region. To obtain linearity from the point where curvature begins on the carrier tube, up to point representing the peaks of modulation, is a matter involving both the point, at which the peaking tubes comes into operation, and the rate, at which its contribution increases with drive. For securing a satisfactory adjustment, there are two variables: the bias on the carrier tube and the amplitude of the excitation on this tube, whose careful selection may allow the power amplifier to operate with low distortion [2]. Besides, in a low-power region, linearity of the Doherty amplifier is entirely determined by the carrier amplifier that should be highly linear even though the load impedance is high. In a high-power region, linearity can be improved by the harmonic cancellation from the carrier and peaking amplifiers using appropriate gate bias voltages. For example, in terms of gain characteristics of each amplifier using LDMOSFET devices, a late gain expansion of the Class-C biased peaking amplifier compensates the gain compression of the Class-AB biased carrier amplifier, thus improving the third-order intermodulation (IM_3) level [13].

In the mid-1990s, it was found that by using existing microwave design techniques and implementing a few modifications, a microwave version of the Doherty amplifier could be realized [14]. In this case, an efficiency of 61% was achieved at 1-dB gain compression point and this level of efficiency was maintained through a 5.5-dB reduction in output power at an operating frequency of 1.37 GHz. A few years later, a possibility to achieve a high-efficiency performance of the microwave monolithic Doherty amplifier was demonstrated in the *Ku*-band and *K*-band frequencies using pHEMT and InP DHBT technologies [15, 16]. The first fully integrated Doherty amplifier MMIC with a chip size of 2 mm^2 operating in a frequency range of 38 to 46 GHz was developed using a 0.15-μm GaAs HEMT process [17]. By using modern high-voltage HBT and GaN HEMT technologies, a high average efficiency greater than 50% can be achieved for multicarrier WCDMA signals using a high-power two-way symmetrical Doherty amplifier [18, 19].

10.1.3 Offset Lines

At high frequencies, it is necessary to take into account that the transistor input impedance is varied with bias voltage and that the transistor output reactance should be compensated to provide the required open-circuit condition by using so-called *offset lines* with optimized electrical lengths [13, 20]. Figure 10.6 shows the basic schematic diagram of a fully matched microwave Doherty amplifier with offset lines in the output circuit and phase-compensating circuits in the input circuit required to reduce amplitude modulation/phase modulation (AM/PM) variations because the output power level can change significantly with phase variations between carrier and peaking paths [20, 21]. At high frequencies, it is enough to use an input phase-compensating transmission line (offset line) at the input of the peaking amplifier. Because of different biasing of the carrier (Class AB) and peaking (Class C) amplifiers, an offset line at the input of the peaking amplifier can be used to optimize linearity and efficiency for multicarrier applications [22]. To closer approximate ideal Doherty amplifier performance, an adaptive power-dependent input power distribution between the carrier and peaking amplifier can be provided so as to deliver more power to the carrier amplifier in the low-power region and to the peaking amplifier in the high-power region, which can result in a linearity improvement of 5 to 7 dB over a wide range of output powers and an increased efficiency up to 5% for WCDMA signals at 2.4 GHz [23].

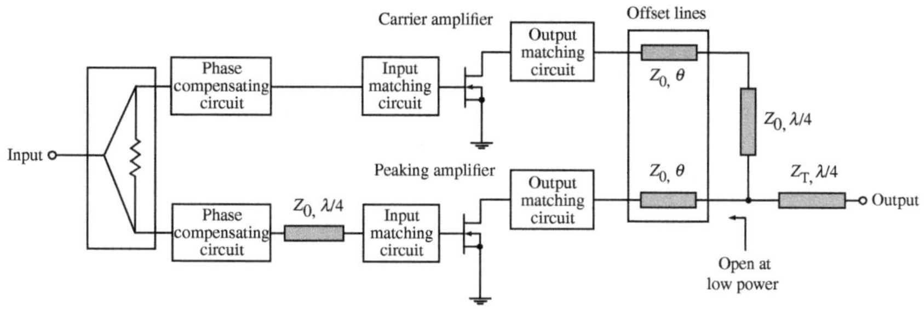

FIGURE 10.6 Block diagram of microstrip LDMOSFET Doherty amplifier.

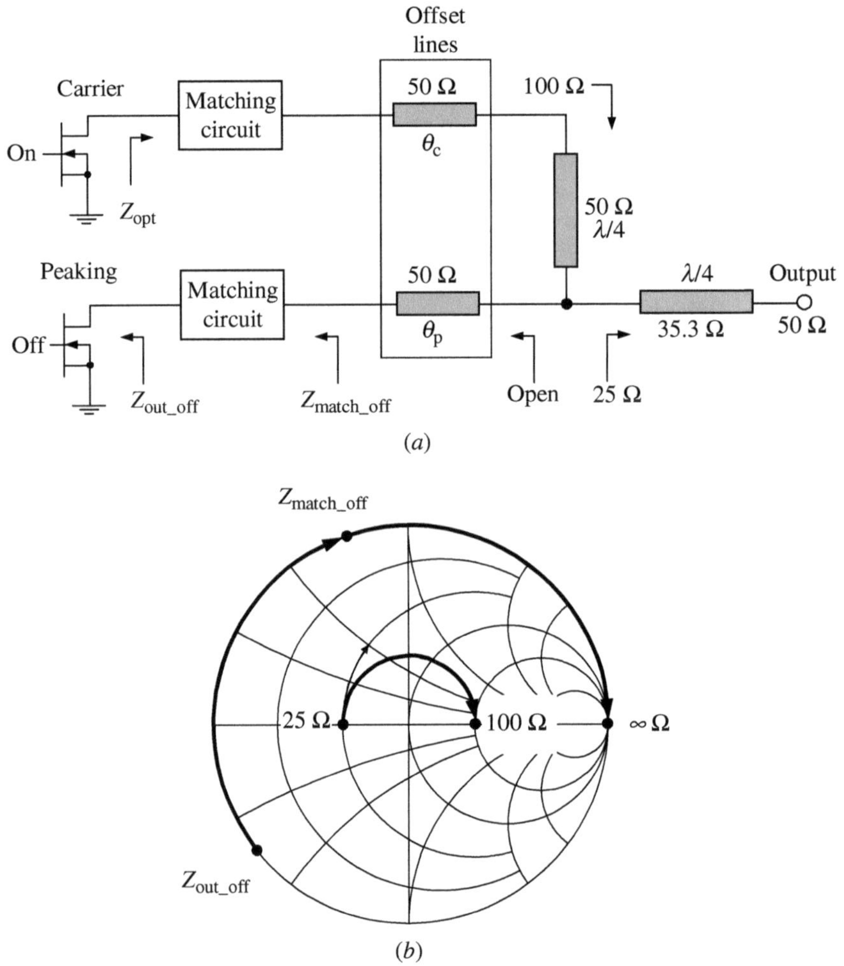

FIGURE 10.7 Load-network schematic and impedances.

The additional offset lines with the characteristic impedance of 50 Ω and fixed electrical lengths θ are connected after the matching circuits of the carrier and peaking amplifiers, as shown in Fig. 10.7(*a*). In a low-power region, the phase adjustments of the offset lines cause the peaking amplifier to be open-circuited and the load impedance seen by the carrier amplifier is doubled to 100 Ω due to an impedance-transforming property of a 50-Ω quarterwave transmission line, assuming that the matching circuit in conjunction with offset line provides the required impedance transformation to the optimum high impedance Z_{opt} seen by the device output at the 6-dB power backoff, as shown in Fig. 10.7(*b*). At the same time, the offset line of the peaking amplifier is adjusted to provide a high impedance (ideally infinite) from the matching-circuit impedance Z_{match_off} so that it prevents power leakage to the peaking path when the peaking transistor is turned off. In real device, the effect of the knee voltage should be considered due to nonzero value of the device on-resistance, which may be smaller or larger depending on the transistor implementation technology. To maximize efficiency at the 6-dB backoff power, the carrier amplifier with nonzero knee voltage should have a load impedance larger than 100 Ω, or larger than $2Z_0$ if $Z_0 \neq 50$ Ω, otherwise the carrier amplifier does not reach the saturation region where maximum efficiency can be achieved [24]. In this case, the offset-line length θ_c of the carrier amplifier is optimized to increase the load impedance, while the offset-line length θ_p of the peaking amplifier is adjusted to block the output-power leakage, and the phase-compensating line with electrical length of 90° + ($\theta_c - \theta_p$) is used at the input of the peaking amplifier.

10.1.4 Linearity

Generally, the nonideal power gain and phase performance in a high-power nonlinear region can cause a significant linearity problem for transmitting signals with nonconstant envelope in wireless communication transmitters. To solve this problem, an improved Doherty amplifier architecture can be used by using an envelope tracking technique to control the gate bias voltage of the peaking amplifier in accordance with the input signal envelope. Such an approach can also provide a higher efficiency with a lower bias voltage for higher output powers. This ensures both high efficiency and good linearity requirements over a wide range of output powers. Figure 10.8(*a*) shows the block diagram of a 2.14-GHz LDMOSFET microstrip Doherty power amplifier for WCDMA applications with adaptive gate bias control [25]. For the same average output powers of 32.7 dBm, such a two-stage Doherty power amplifier demonstrates an improvement in a *PAE* of 15.2% at an *ACLR* of –30 dBc compared to its Class-AB counterpart. This is because the quiescent current in a Doherty architecture is maintained constant only for the carrier amplifier, whereas the bias point of the peaking amplifier is varied according to the input signal envelope. However, it should be noted that the wider the modulation bandwidth of the transmitting signal, the more problematic to implement such a technique is in practice to achieve significant linearity improvement.

The linearity of the power amplifier can be improved by using a digital signal processing to provide more accurate gate-bias control alongside with digital predistortion (DPD) needed for the simultaneous correction of the gain and phase characteristics in a high-power region. Figure 8(*b*) shows the block schematic of an 840-MHz MESFET Doherty power amplifier for CDMA applications with the digital signal processor (DSP) implemented externally on a board controlled

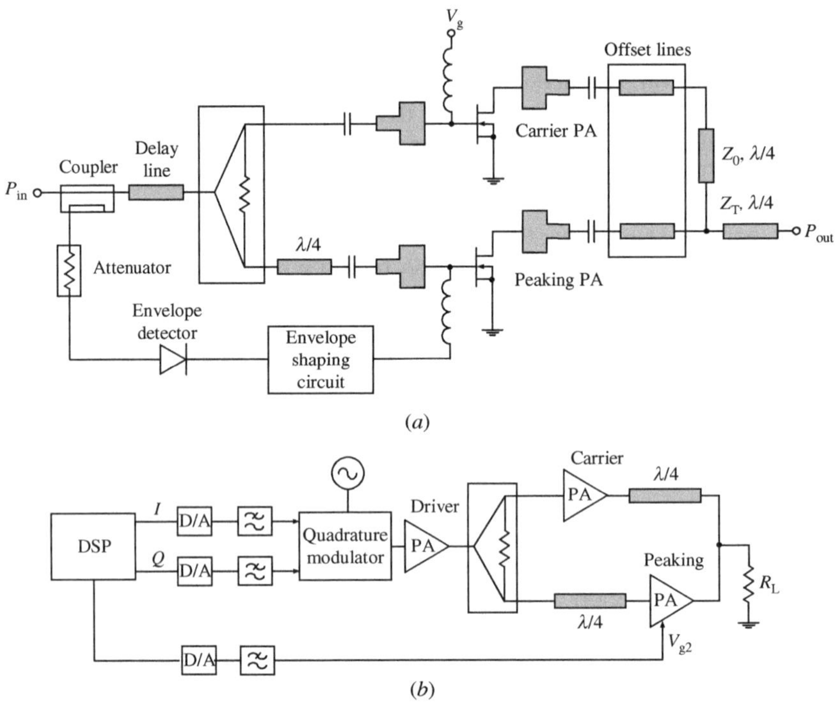

FIGURE 10.8 Block diagrams of Doherty amplifier architectures with adaptive control.

by a personal computer [26]. In this case, the DSP generates both the baseband in-phase (I) and quadrature (Q) signals, which are upconverted to form an RF signal using the quadrature modulator. The DSP unit also generates the voltage signal V_{g2}, which is applied to the peaking amplifier as the gate bias. This results in an efficiency improvement by the dynamic gate biasing of the peaking amplifier according to the instantaneous envelope of the input signal. At the same time, the phase performance is corrected by the phase predistortion at baseband level based on the dynamic gate bias-voltage values from the gain correction, thus resulting in a linearity improvement. An overall improvement of *PAE* from 3 to 5% and an *ACPR* of about 10 dB at an average output power of 23 dBm can be achieved by using such a DSP technique. Besides, a simple bias-switching technique can be used in Doherty-type amplifiers, so that they can satisfy the linearity requirements of the power amplifiers for CDMA handset applications over the entire dynamic power range [27]. In addition to bias-switching technique, a dual-mode matching approach can be used to optimally design a dual-mode Doherty amplifier operated simultaneously in HPSK (hybrid phase shift keying) mode and OFDM 64-QAM (quadrature amplitude modulation) mode for mobile terminals [28].

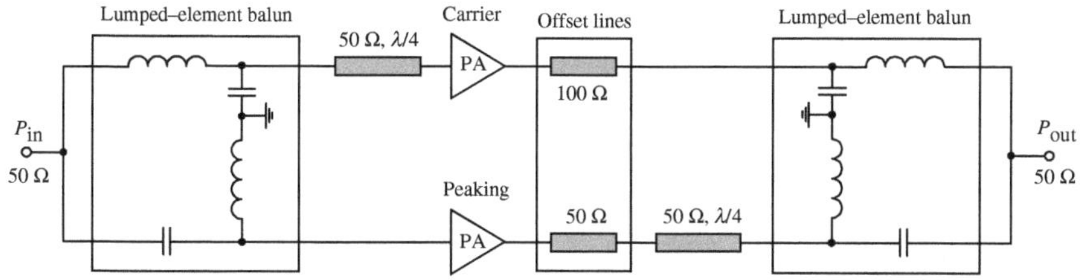

FIGURE 10.9 Block diagram of Doherty amplifier with series-connected load.

10.1.5 Series-Connected Load

In comparison with the shunt-connected load type, the series-connected load combines the output powers of the carrier and peaking amplifiers in a manner similar to that of the push-pull amplifiers, having the same capability to suppress even harmonic components in the output signal spectrum. Figure 10.9 shows the block diagram of a Doherty amplifier with a series-connected load, where the input and output baluns are implemented with the lumped inductances and capacitors [29]. The lumped-element balun can be designed for an arbitrary unbalanced-load value by proper selecting its element values. In this case, an unbalanced load of 50 Ω and a balanced port load of 100 Ω were chosen. At low-power levels when the peaking amplifier is turned off, one balanced port of the balun connected to the open-circuited quarterwave transmission line is shorted that results in a load of 100 Ω for carrier amplifier. At high power level when the peaking amplifier becomes active, both carrier and peaking amplifiers are operated in a near push-pull mode. For such a GaN HEMT Doherty amplifier operating at 1.8 GHz, high efficiencies of 31% and 56% were achieved at 24-dBm and 31-dBm saturated output powers, respectively.

10.2 Efficiency Improvement

Further efficiency improvement of the vacuum-tube Doherty amplifiers was achieved by creating biharmonic modes at the input and load networks of the carrier and peaking amplifiers. Specifically, additional parallel resonant circuits tuned to the third harmonic were used, which have resulted in an efficiency of over 80% [30]. Ideally, an infinite number of odd-harmonic resonators results in an idealized Class-F mode with a square voltage waveform and a half-sinusoidal current waveform, whereas an infinite number of even-harmonic resonators results in an idealized inverse Class-F mode with a half-sinusoidal voltage waveform and a square current waveform at the device output terminal. However, in practice at microwave frequencies, it is enough to control the second- and third-harmonic components to provide a high operation efficiency of the power amplifier, and it is

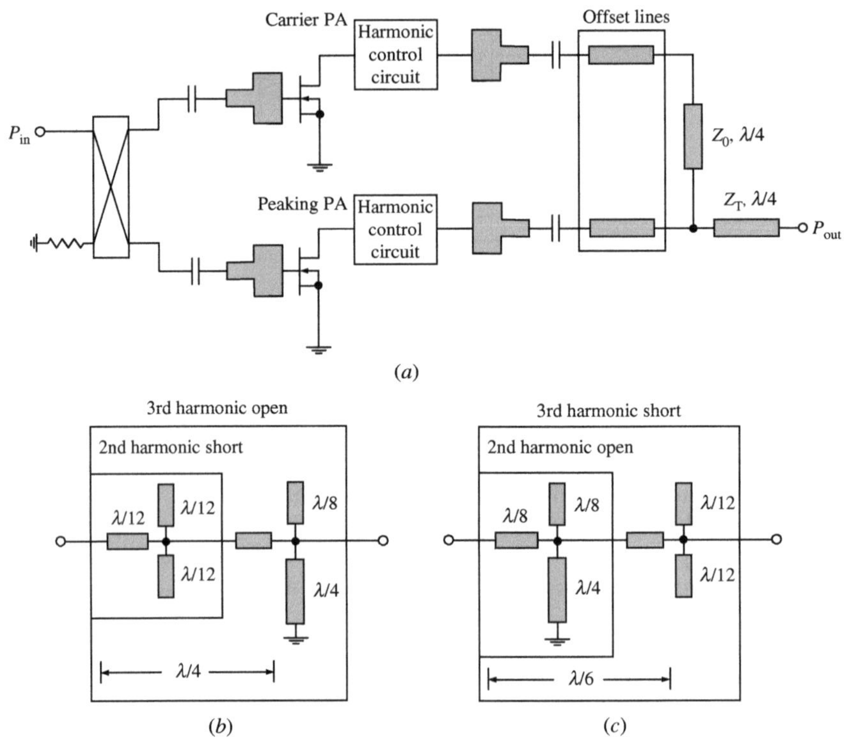

Figure 10.10 Block diagram of Doherty amplifier with harmonic-control circuits.

preferable to use short- and open-circuit stubs instead of lumped capacitors in the load network.

Figure 10.10(*a*) shows the block schematic of a fully matched microwave Doherty amplifier, including the harmonic-control circuits used in front of the output matching circuits and offset lines. To approximate a Class-F operation mode, the harmonic-control circuit includes both arm shunt stubs for a better harmonic trap and a series tuning line to compensate for the output device parasitic elements, as shown in Fig. 10.10(*b*). Despite the fact that the Doherty amplifier with harmonic control provides worse linearity than the conventional Doherty amplifier in terms of *AM*/*AM* and *AM*/*PM* performance, applying a digital feedback predistortion linearizer allows the *ACLR* to be significantly improved. For example, the *ACLR* of a 2.14-GHz two-stage Class-F GaN HEMT Doherty amplifier with an average output power of 36 dBm and a drain efficiency of 52.4% was improved by greater than 20 dB [31]. A two-stage inverse Class-F LDMOSFET Doherty amplifier using the harmonic-control circuits shown in Fig. 10.10(*c*) is capable of providing a drain efficiency of 54.7% at an average output power of 32 dBm for a 1-GHz forward-link WCDMA signal [32].

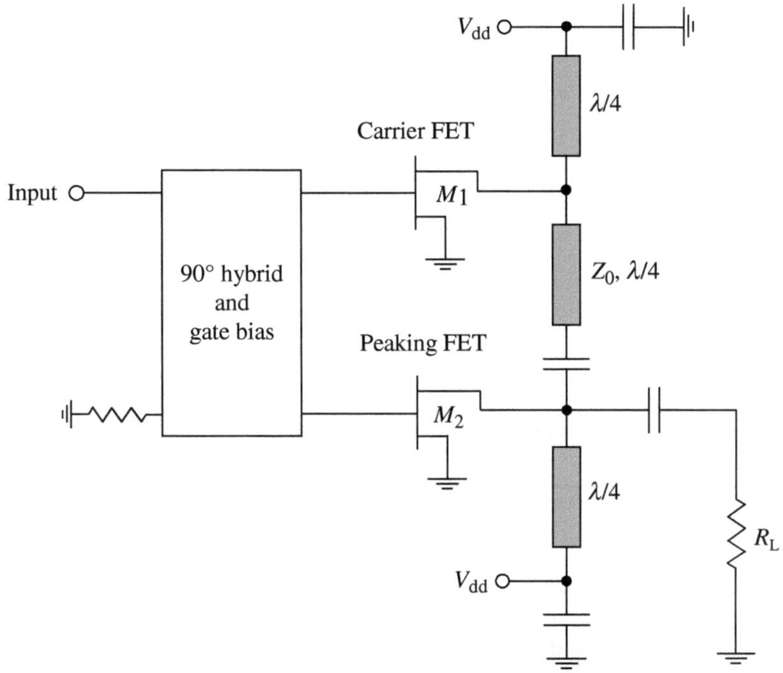

Figure 10.11 Schematic of Class-F/Class-C Doherty amplifier.

In some cases, a Class-F design strategy can be applied to the carrier amplifier only, whereas the peaking amplifier operates in a conventional Class-C mode [33]. Figure 10.11 shows the simplified circuit schematic of a Class-F/Class-C Doherty amplifier, in which the drain bias supplies are connected to the corresponding device drain terminals through the short-circuited quarter-wavelength transmission lines providing inherent even-harmonic suppression [34, 35]. In this case, the drain current of the Class-B biased carrier FET (field-effect transistor) contains the dc, fundamental-frequency, and even-harmonic components. On the other hand, the drain current of the Class-C biased peaking FET is purely sinusoidal ideally because its odd-harmonic components are shunted by the short-circuited bias-feed quarterwave line connected to the carrier FET, thus resulting in no harmonics appearing at the peaking FET output. However, the effect of output parasitics of the carrier FET reduces the quality of the odd-harmonic short-circuit termination at the peaking FET, resulting in power-amplifier performance degradation when measured efficiency was slightly above 60% in saturation at an operating frequency of 770 MHz for a supply voltage of 3.5 V.

The test board of a highly efficient 2.14-GHz 2.5-W transmission-line two-stage GaN HEMT Doherty amplifier with an input branch-line coupler operating in an inverse Class-F mode and implemented in a

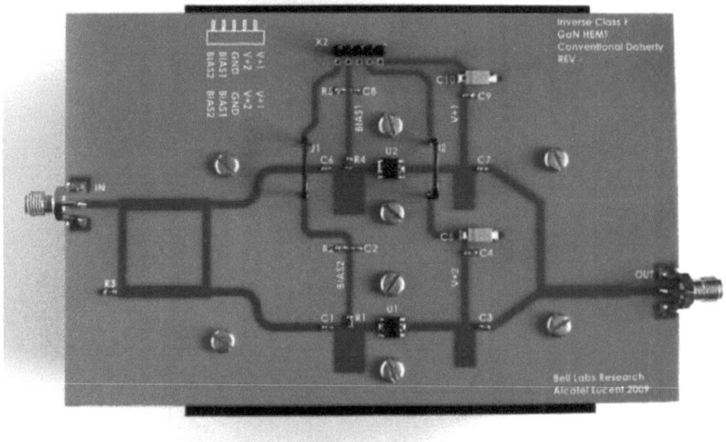

FIGURE 10.12 Test board of inverse Class-F GaN HEMT Doherty amplifier.

30-mil RO4350 substrate is shown in Fig. 10.12. Both output matching circuits of the carrier and peaking amplifiers having the same structure provide the load matching at the fundamental frequency and the corresponding second- and third-harmonic control. Here, high impedance at the second harmonic and low impedance at the third harmonic are created at the output of each device using a short-length series microstrip line short-circuited at the second and third harmonics by a quarterwave short-circuited transmission line (RF grounded by a bypass capacitor) and an open-circuit stub with an electrical length of 30°, respectively [36]. The exact electrical lengths of the microstrip lines depend on the values of the device output shunt capacitance and series inductance. As a result, the drain efficiency at maximum output power and 6-dB backoff point exceeded 65% using two 5-W NPTB00004 GaN HEMT devices. Efficiency enhancement of a Doherty amplifier can also be provided with a combination of Class-F and inverse Class-F schemes for the carrier and peaking amplifiers, respectively, and an efficiency of 45% can be achieved at 10-dB backoff [37].

A high efficiency of the Doherty amplifier can also be achieved using a Class-E mode when the device output shunt capacitance can be considered as an internal element of the Class-E load network [38]. If the device shunt capacitance is larger than the nominal Class-E capacitance, an additional compensation circuit with a series capacitor and a shunt inductance needs to be added to the device output. For example, this configuration was used in the design of a 2.14-GHz Class-E Doherty power amplifier using 25-W GaN HEMT devices in the carrier and peaking amplifiers, resulting in a *PAE* of 44.8% at an average output power of 37 dBm for a single-carrier WCDMA signal [39].

The symmetrical Doherty configuration using packaged devices in the carrier and peaking amplifiers can also be designed in a Class-E mode when the device output capacitance is used to compose a π-section Class-E load network [40]. In this case, the ideal high-Q series resonant circuit required for the Class-E mode is replaced by a low-pass matching circuit with the series inductance, implemented as a microstrip line on a ceramic substrate, and a shunt ceramic chip capacitor.

10.3 Asymmetric Doherty Amplifiers

There is a possibility to extend the region of high efficiency over a wider range of output powers if the carrier and peaking amplifiers are designed to operate with different output powers: smaller for the carrier amplifier and larger for the peaking amplifier. For instance, for a power division ratio $\alpha = P_{\text{carrier}}/(P_{\text{peaking}} + P_{\text{carrier}}) = 0.25$, the transition point with maximum drain efficiency corresponds to the backoff power level of −12 dB from peak output power [12]. At peak output power when the carrier and peaking amplifiers shown in Fig. 10.5(*a*) are saturated, it follows from consideration of their output powers that $R_1 = Z_2 = R_3 = 3R_2$. As a result, $I_2 = 3I_3$ and $\beta = 0.25$. The output impedances R_2 and R_3 as functions of the load resistance R_L and characteristic impedance Z_1 can be obtained from Eqs. (10.7) and (10.8). For example, if one can choose the characteristic impedance of the output transmission line and load resistance equal to $Z_1 = 15\ \Omega$ and $R_L = 50\ \Omega$, respectively, then the characteristic impedance of the quarter-wave transformer and load impedance for the carrier amplifier are $Z_2 = R_1 = 18\ \Omega$, whereas the output impedance of the peaking amplifier is equal to $R_2 = R_1/3 = 6\ \Omega$.

Because from Eqs. (10.7) and (10.8) it follows that

$$R_1 = \frac{Z_2^2}{\beta R_3} \tag{10.10}$$

hence, at lower power levels when the peaking amplifier is turned off, the output impedance R_1 is four times higher than that at peak output power where $R_1 = Z_2 = R_3$. In this case, the power ratio between the peaking and carrier amplifiers should be 3:1. However, because of availability issues of the power GaAs HBT devices, a scaling ratio of 4:1 was chosen, with total emitter areas of 3360 μm² and 840 μm² for the peaking and carrier amplifiers, respectively, to implement the extended Doherty technique into the monolithic power amplifier developed for CDMA handset applications. As a result, the *PAE* of 45% and 23% were measured at the highest output power of 25 dBm and at 10-dB backoff level, respectively [41]. The conventional Class-AB power amplifiers designed for the same application normally have the power-added efficiency of about four times lower at this backoff power.

To optimize linearity and efficiency of the asymmetric Doherty amplifier, it is very important to optimize the biasing conditions for the carrier and peaking amplifiers [42].

For the packaged devices when it is difficult to choose the proper power ratio between the devices, it is convenient to use the identical power amplifiers that can compose ideally the N-way Doherty architecture where one carrier amplifier is in parallel with $(N-1)$ numbers of the peaking amplifiers. This is the simplest hybrid approach to acquire an $(N-1)$ times larger-sized peaking amplifier compared with the carrier amplifier for an asymmetric two-way Doherty amplifier configuration [43]. Figure 10.13(*a*) shows the schematic diagram of an N-way Doherty amplifier with a parallel connection of one carrier amplifier and $(N-1)$ identical peaking amplifiers. The ideal drain efficiencies of the N-way Doherty amplifier (DA) architectures with peak values at –6 dB, –9.5 dB, –12 dB, and –14 dB power backoff points according to $P_{\text{backoff}} = 20 \log_{10} N$ for the two-, three-, four-, and five-way structures, respectively, and the conventional Class-B power amplifier are shown in Fig. 10.13(*b*).

Generally, the N-way Doherty amplifier is composed of the N-way power splitter, identical fully matched carrier and $(N-1)$ peaking amplifiers, N offset lines, and an output combiner representing a quarterwave impedance transformer. The characteristic impedance of a quarterwave transmission line for converting the load of the carrier amplifier is $Z_0 = R_0/\sqrt{\alpha}$ and the common load resistance is $R_L = R_0/(\alpha + 1)$, where R_0 is the matched load for both carrier and peaking amplifiers, usually equal to 50 Ω [44]. For asymmetric two- and three-way Doherty amplifiers with optimized individual bias conditions and load matching for the carrier and peaking amplifiers, applying an uneven drive results in more linear operation and produces more power than an even drive [45, 46]. For example, a three-way Doherty amplifier based on Class-F load networks and fabricated using 10-W GaN HEMT devices achieved a *PAE* of 45.9% and an *ACLR* of –49.2 dBc for a single-carrier WCDMA signal with a *PAR* of 10 dB at 2.14 GHz using a digital feedback predistortion technique [47].

In spite of the efficiency improvements offered by the asymmetric N-way Doherty amplifier over its symmetric two-way Doherty amplifier counterpart, its total power gain, which significantly depends on the power gain of the carrier amplifier, will be reduced due to the corresponding insertion loss in the required input N-way power splitter. This issue is circumvented by using distributed amplification. In particular, distributed amplification is a technique whereby power combining is performed directly at the transistor level without the need for an N-way power combiner. Figure 10.14 shows the simplified schematic of a distributed Doherty amplifier, where the powers of both N-carrier and N-peaking amplifiers are combined using half-wave and quarterwave microstrip lines [48, 49]. The desired

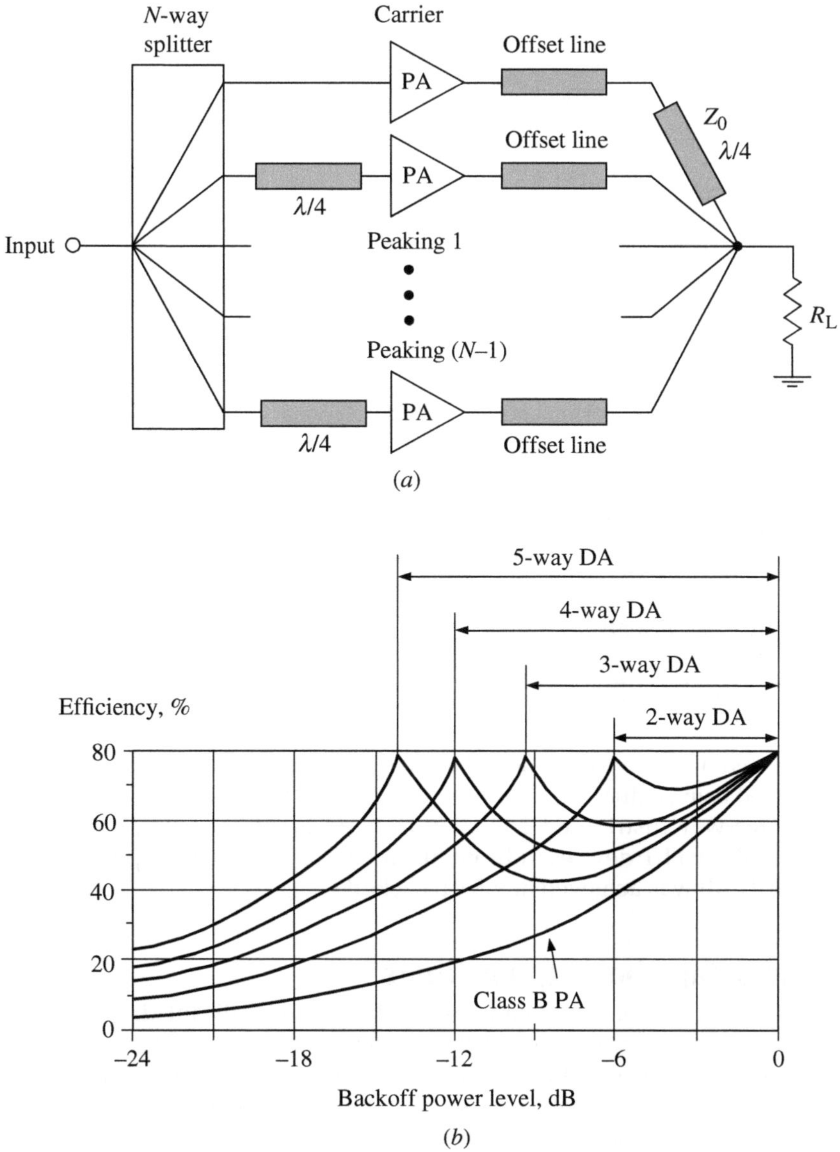

FIGURE 10.13 Asymmetric *N*-way Doherty amplifier and efficiencies.

location of peak efficiency points of such a distributed *N*-way Doherty amplifier can be given in decibels by

$$P_{\text{backoff}} = 20 \log_{10} \left(\frac{K}{M} + 1 \right) \tag{10.11}$$

where *K* and *M* are the numbers of the peaking and carrier amplifiers, respectively. Practically, in order to design a high-efficiency three-way

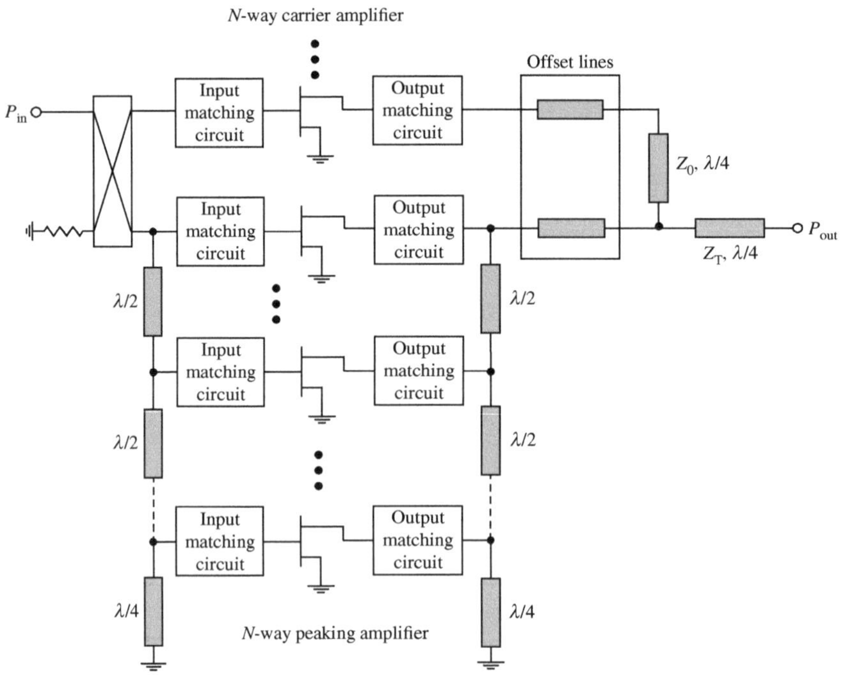

Figure 10.14 Distributed *N*-way Doherty amplifier.

distributed Doherty amplifier, the two peaking amplifiers can be combined using a dual-fed distributed structure. The measured results of a three-way distributed 2.14-GHz Doherty amplifier using three 45-W LDMOSFET devices indicate that a *PAE* of 39.5% with a power gain of 11 dB was achieved at 9.5-dB backoff.

10.4 Multistage Doherty Amplifiers

An asymmetric Doherty architecture exhibits a significant drop in efficiency in the region between the efficiency peaking points, especially for large power ratios between the carrier and peaking amplifiers. However, it is possible to use more than two power amplifiers to prevent significant deterioration of efficiency at backoff output power levels. This can be provided by the so-called multistage Doherty amplifiers, whose operation is somewhat similar to that of a two-stage Doherty amplifier as having a low-power region when only the carrier amplifier is turned on and a high-power region when all peaking amplifiers are turned on [12]. In this case, unlike the asymmetric Doherty amplifier when all peaking amplifiers are turned off simultaneously, the peaking amplifiers in a multistage Doherty amplifier are turned off consecutively at the corresponding peak efficiency points, beginning from (N–1) peaking amplifier.

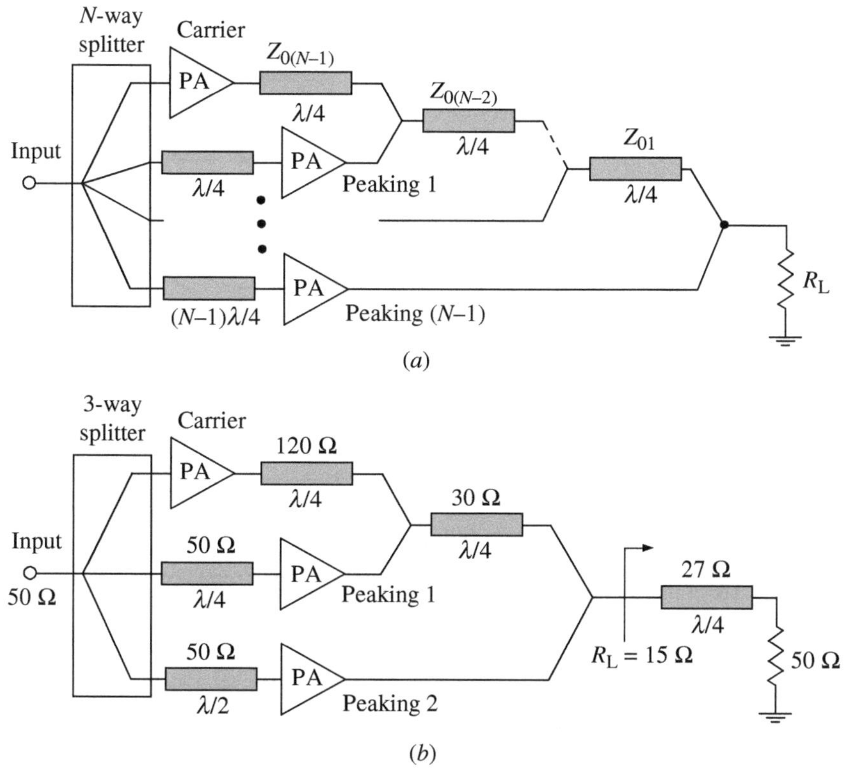

FIGURE 10.15 Multistage Doherty amplifier architectures.

The basic multistage Doherty power amplifier architecture shown in Fig. 10.15(*a*) comprises more than one peaking amplifier, with quarterwave transmission lines to combine their output powers [50]. The characteristic impedances of each output quarterwave transmission line depend on the levels of backoff power and can be calculated from

$$Z_{0i} = R_L \prod_{j=1}^{i} \gamma_j \tag{10.12}$$

$$\prod_{j=k}^{(i+k)/2} \gamma_{(2j-k)} = 10^{(B_i/20)} \tag{10.13}$$

where $i = 1, 2, \ldots, N-1$, $k = 1$ (for odd i) or 2 (for even i), N is the total number of amplifier stages, and B_i is the backoff level (positive value in decibels) from the maximum output power of the system, at which the efficiency peaks. The maximum level of backoff B_{N-1} is set by the carrier amplifier, while the number of efficiency peaking points is directly proportional to the number of amplifier stages used in the design.

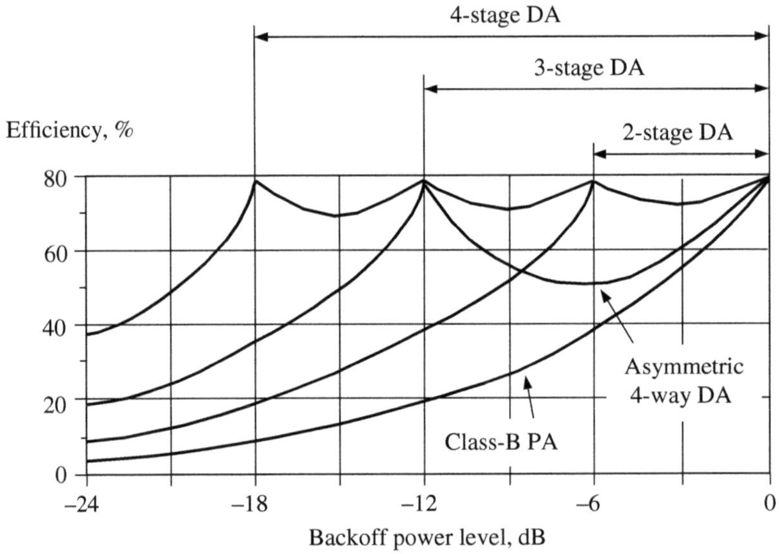

FIGURE 10.16 Efficiencies of different Doherty amplifier architectures.

Figure 10.16 shows the instantaneous drain efficiencies of the multistage Doherty amplifier (DA) architectures for the two, three, and four stages, having maximum efficiencies at the transition points of –6 dB, –12 dB, and –18 dB backoff output power levels, respectively. From Fig. 10.16, it follows that the multistage architecture provides higher efficiencies at backoff levels between the efficiency peaking points compared with an asymmetric Doherty architecture and significantly higher efficiency at all backoff output power levels compared with the conventional Class-B power amplifiers. For the most practical case of a three-stage Doherty amplifier, whose block schematic is shown in Fig. 10.15(*b*), the characteristic impedances of each output quarterwave transmission line can be obtained from Eqs. (10.12) and (10.13) to be

$$Z_{01} = \gamma_1 R_L \tag{10.14}$$

$$Z_{02} = \gamma_1 \gamma_2 R_L \tag{10.15}$$

where

$$\gamma_1 = 10^{(B_1/20)} \tag{10.16}$$

$$\gamma_2 = 10^{(B_2/20)} \tag{10.17}$$

where $B_1 = 6$ and $B_2 = 12$ for peak efficiencies at –6 dB and –12 dB backoff points, respectively, resulting in $Z_{01} = 30\ \Omega$ and $Z_{02} = 120\ \Omega$ for $R_L = 15\ \Omega$.

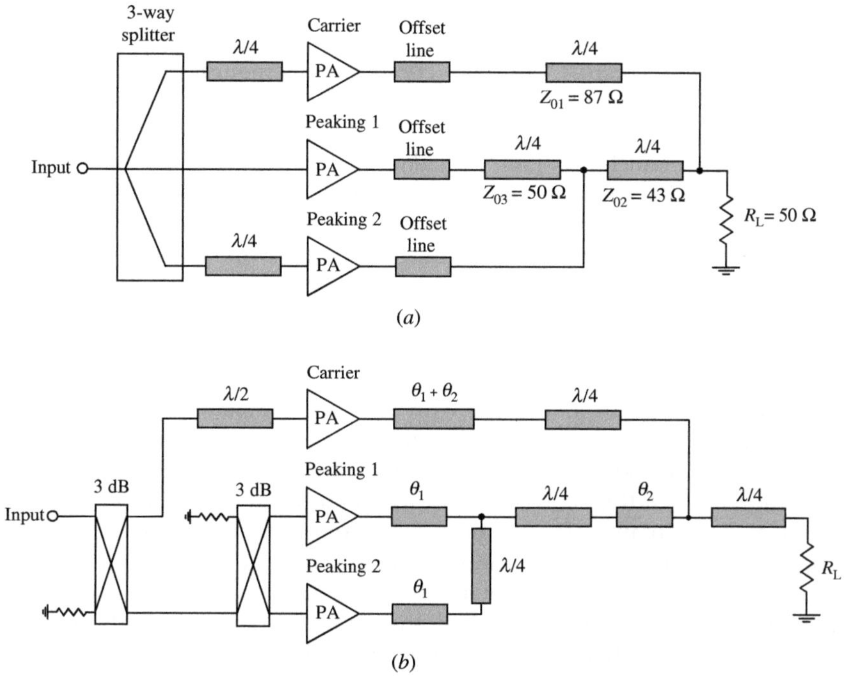

Figure 10.17 Modified three-stage Doherty amplifier architectures.

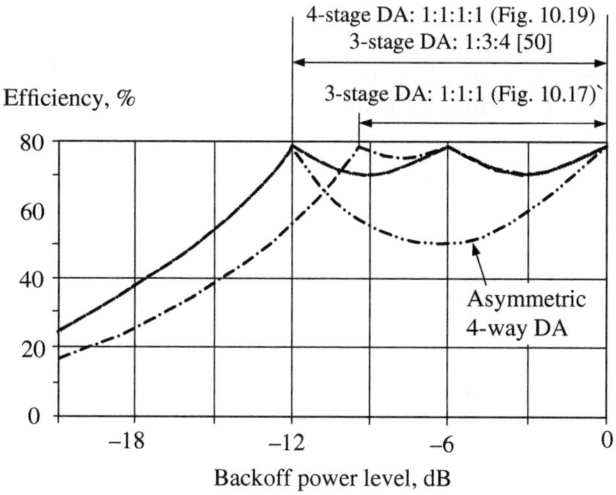

Figure 10.18 Efficiencies of different Doherty amplifier architectures.

For a 1.95-GHz WCDMA application, a three-stage Doherty amplifier structure using GaAs MESFET devices with the device periphery ratio of 1:2:4 and microstrip power combining elements provides a *PAE* of 48.5% and a power gain of 12 dB at P_{1dB} = 33 dBm. The peak power-added efficiencies of 42% and 27% were measured at –6 dB and –12 dB backoff levels [50]. Efficiencies at backoff points can be increased by optimizing the input drive conditions for the peaking amplifiers [51]. Moreover, further efficiency improvement of a three-stage Doherty amplifier at maximum output power and backoff points can be achieved by using the highly effective GaN HEMT devices and applying a DPD technique for linearization. In this case, the drain efficiency at –12 dB output power backoff point can be increased to be higher than 60% [52].

A typical problem associated with the conventional three-stage Doherty amplifier is that the load-line modulation of the carrier stage stops at a certain power level, leaving the carrier amplifier in deep saturation and leading, consequently, to a significant degradation of its linear performance. In addition, when the carrier and peaking amplifiers have equal configurations with the same device periphery sizes, similar performance is obtained with regards to the symmetrical two-stage Doherty amplifier, with the efficiency peaking points at –3.5 dB and –6 dB backoff output powers. These problems can be partially solved by using a modified three-stage Doherty amplifier architecture with a parallel combination of one carrier and one Doherty amplifier used as a peaking amplifier, as shown in Fig. 10.17(*a*) [53]. In this case, a novel way of combining enables high instantaneous efficiencies at –6 dB and –9.5 dB backoff output powers with a single device size. The characteristic impedances of the transforming quarterwave transmission lines are calculated as $Z_{01} = \sqrt{3}R_L$, $Z_{02} = (\sqrt{3}/2)R_L$, and $Z_{03} = R_L$, where R_L is the load resistance [54, 55].

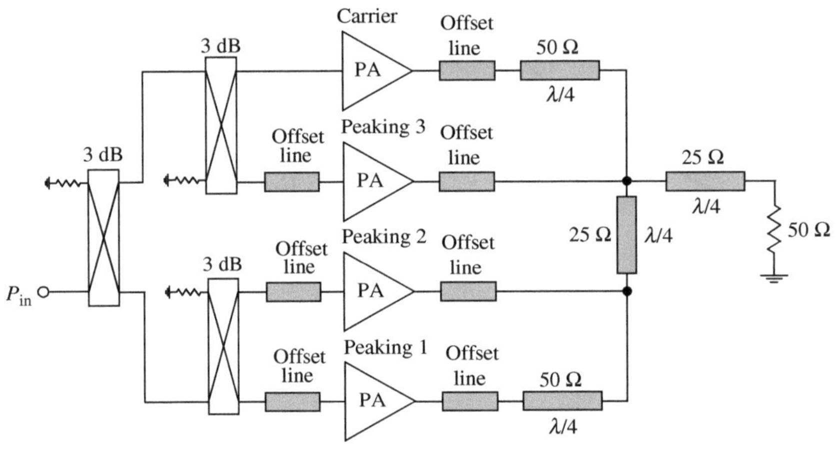

FIGURE 10.19 Modified four-stage Doherty amplifier architecture.

Figure 10.17(*b*) shows the other modified three-stage Doherty architecture operating in a 2.14-GHz WCDMA system with a 36-dBm average output power provided by the three 10-W GaN HEMT transistors where both peaking amplifiers represent in turn the conventional two-way Doherty configuration [56]. In this case, only the carrier amplifier is turned on at low-power region, the carrier amplifier is saturated and the first peaking amplifier is turned on at medium-power region, and the carrier and first peaking amplifier are both in saturation and the second peaking amplifier is turned on at high-power region. The optimum electrical lengths of the required 50-Ω offset lines are $\theta_1 = 0.28\lambda$, $\theta_2 = 0.36\lambda$, and $\theta_1 + \theta_2 = 0.64\lambda$, respectively. As a result, the carrier amplifiers are turned on at near –9 dB and –6 dB backoff output power levels. The efficiency and linearity can be optimized by using the two driving amplifiers connected to the output ports of the input 3-dB coupler and biased in Class B and Class C, respectively.

Figure 10.18 shows the theoretical instantaneous drain efficiencies of the multistage (three and four stages) and four-way asymmetric DA architectures for different power (or device size) ratios, with peak efficiencies ranging from −12 dB power backoff levels. From Fig. 10.18, it follows that the four-way or any asymmetric multiway Doherty architecture provides significantly lower efficiency between the corresponding peak efficiency points. However, for a multistage Doherty configuration, the peak efficiency points at lower power backoff levels can be achieved using an optimum device size ratio. For example, a peak efficiency at the lowest backoff point of –12 dB is achieved for a device periphery ratio of 1:3:4 in a three-stage Doherty amplifier, whereas the lowest backoff of about –9.5 dB corresponds to the peak efficiency for an equal device periphery size of 1:1:1 in the modified three-stage Doherty amplifiers shown in Figs. 10.17(*a*) and 10.17(*b*) [50, 55, 56].

In a classic four-stage Doherty power amplifier with the corresponding peak efficiencies at –6 dB, –12 dB, and –18 dB backoff output power points, the maximum ratio between the characteristic impedances of the quarterwave transmission lines is equal to 16 [50]. For example, for $R_L = 6$ Ω, the characteristic impedances of the consecutive quarterwave transmission lines are $Z_{01} = 12$ Ω, $Z_{02} = 48$ Ω, and $Z_{03} = 192$ Ω, respectively. These values are difficult to correctly implement using microstrip lines on a single substrate with a fixed thickness and dielectric permittivity. In a modified four-stage Doherty configuration with the device size ratio of 1:1:1:1 shown in Fig. 10.19, where the two conventional two-stage Doherty amplifiers are combined in a final four-stage Doherty configuration (Doherty-in-Doherty), the maximum ratio between the transmission-line characteristic impedances is equal to 50 Ω / 25 Ω = 2 only [57].

Figure 10.18 shows the three efficiency peaking points provided by the modified four-stage Doherty amplifier with equal gate bias

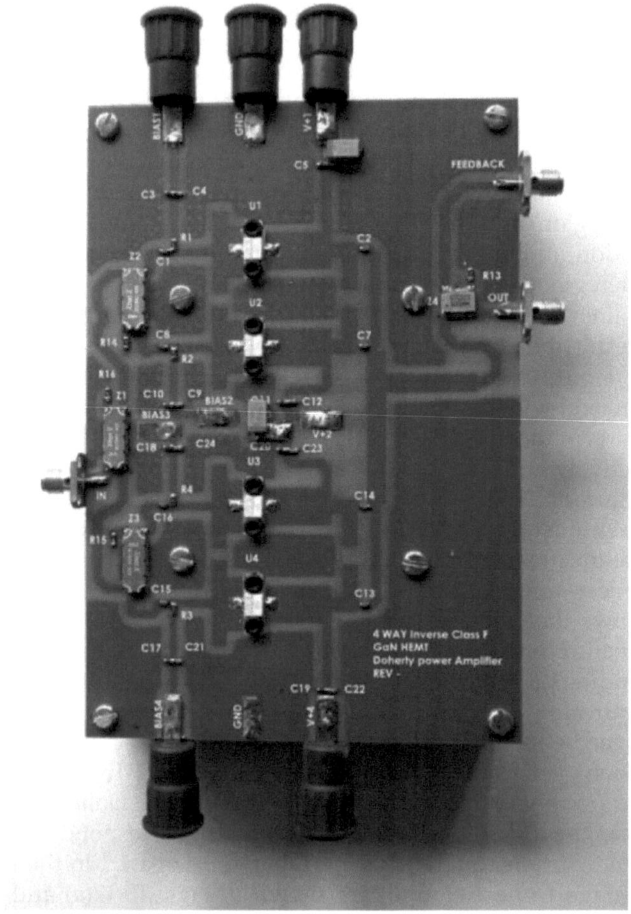

Figure 10.20 Test board of 2.14 GHz 100-W four-stage Doherty GaN HEMT amplifier.

voltages for the second and third peaking amplifiers. Furthermore, the optimization of their gate bias voltages can change the efficiency profile between the peak power and –6 dB backoff points and contribute to linearity improvement. Because of the device input and output parasitics such as the gate-source and drain-source capacitances, additional input offset lines are implemented at the input of the peaking amplifiers and identical output offset lines that introduce the compensating inductive reactances are connected in series to each output circuit. This is a very practical version of a four-stage Doherty amplifier, capable of achieving high output powers with high drain efficiency and having three 90° hybrid couplers at the input and four quarterwave microstrip lines at the output.

Figure 10.20 shows the test board of the modified four-stage GaN HEMT Doherty power-amplifier architecture based on four 25-W Cree CGH40025F devices and fabricated using a 30-mil RO4350 substrate [57]. The carrier and peaking amplifiers are designed to operate in an inverse Class-F mode with the second- and third-harmonic control by using a transmission-line load-network technique. The input dividing network includes three commercial 90° hybrid couplers, while a 30-dB directional coupler required to sampling output power for linearization loop needs to be connected to the output port. In a CW operation mode when all transistors are biased with the same gate voltage of –3.4 V, an output power of 50 dBm (100 W) and a drain efficiency of 77% were achieved at a supply voltage of 34 V. In a single-carrier 2.14-GHz WCDMA operation mode with a *PAR* of 6.5 dB, a drain efficiency of 61% was achieved at an average output power of 43 dBm (20 W) with an *ACLR* of –31 dBc.

10.5 Inverted Doherty Amplifiers

Figure 10.21 shows the schematic diagram of an inverted Doherty amplifier configuration with an impedance transformer based on a quarterwave line connected to the output of the peaking amplifier. Such architecture can be very helpful if it is easier in a low-power region to provide a short circuit rather than an open circuit at the output of the peaking amplifier, which depends on the characteristic of the transistor. The quarterwave line can be implemented in a compact form suitable for use in mobile applications [58]. In this case, at low-power levels, a quarterwave line is used to transform very low output impedance after the offset line to high impedance seen from the load junction. In particular, by taking into account the device package parasitic elements of the peaking amplifier, an optimized output matching circuit and a proper offset line are designed to provide the maximum output power from the carrier device [59]. At a high power level, for the matched phase difference between identical carrier and peaking amplifiers, the load impedance seen from each amplifier after the offset lines is equal to the standard 50-Ω load impedance.

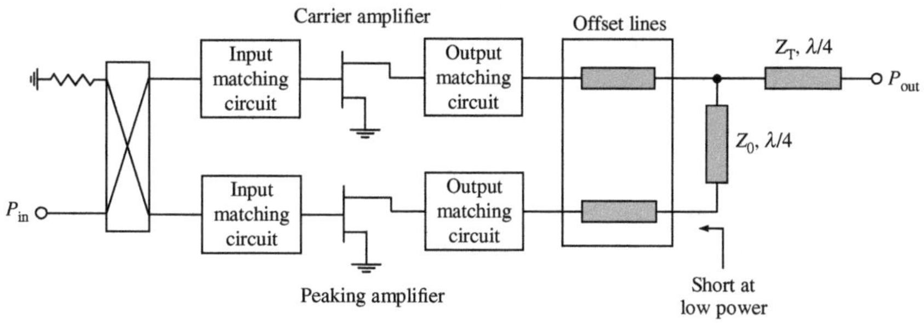

Figure 10.21 Schematic diagram of inverted Doherty amplifier.

When using a four-carrier WCDMA signal, a *PAE* of 32% with an *ACLR* of –30 dBc at an average output power level as high as 46.3 dBm was achieved for an inverted 2.14-GHz LDMOSFET Doherty amplifier. This provides a 9.5% improvement in efficiency and 1-dB improvement in the output power under the same *ACLR* conditions as for the balanced Class-AB operation using the same devices [60]. For a 64-QAM modulated signal with 24-MHz channel bandwidth, a *PAE* higher than 31% with a 0.5-dB output-power flatness at 27-dBm was achieved across a frequency bandwidth from 2.4 to 2.5 GHz with an *ACPR* better than –40 dBc [61].

To better understand the operation principle of an inverted Doherty amplifier, consider separately the load network shown in Fig. 10.22(*a*), where the peaking amplifier is turned off. In a low-power

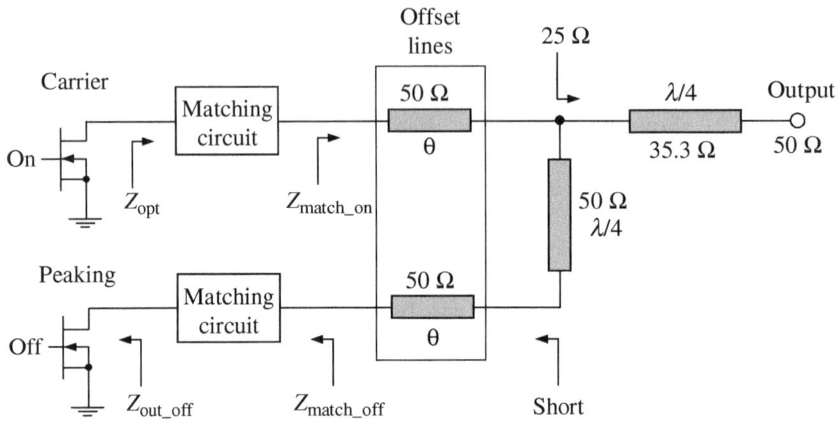

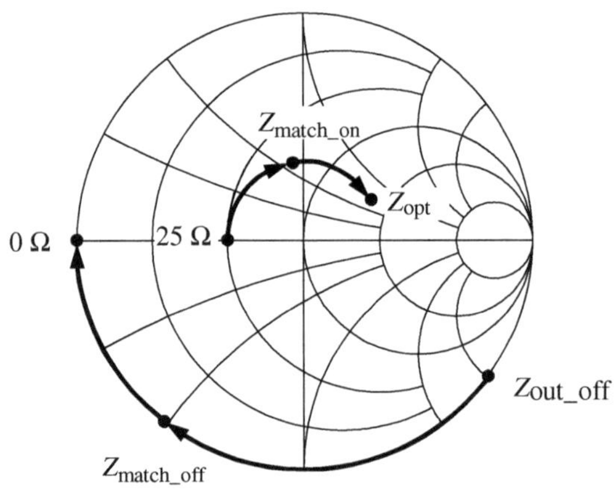

Figure 10.22 Load-network schematic and impedances.

region, the phase adjustment of the offset line with electrical length θ causes the peaking amplifier to be short-circuited (ideally equal to 0 Ω), and the matching circuit in conjunction with offset line provides the required impedance transformation from 25 Ω to the optimum high impedance Z_{opt} seen by the carrier device output at the 6-dB power backoff, as shown in Fig. 10.22(*b*). In this case, the short circuit at the end of the quarterwave line transforms to the open circuit at its input so that it prevents power leakage to the peaking path when the peaking transistor is turned off. In a high-power region, both carrier and peaking amplifiers are operated in a 50-Ω environment in parallel, and the output quarterwave line with the characteristic impedance of 35.3 Ω transforms the obtained 25 Ω to the required 50-Ω load.

In a Doherty configuration, both the Class-AB carrier amplifier and Class-C peaking amplifier are not fully isolated from each other. This can result in a serious problem to robustly design the optimum load impedance shift presented to both transistors for high efficiency and low distortion [62]. From the load-pull measurements for a unit-cell 28-V GaAs HJFET device, it was observed that in order to obtain high efficiency and low distortion, the carrier amplifier load impedance should change from the maximum efficiency point to the maximum output power point at Class AB, while the peaking amplifier load impedance should vary from the small-signal gain point to the maximum output power point in Class C [63]. In this case, the load impedance corresponding to the maximum efficiency point is lower than the load impedance corresponding to the maximum output power point. An inverted Doherty architecture can be suitable to realize the carrier amplifier load impedance variation from lower impedance to higher impedance in accordance with the increase in the input power level. The external input and output matching circuits are necessary to optimize the load impedance shift presented to both carrier and peaking amplifiers as a function of the input power level. As a result, a drain efficiency of 42% at an output power of 49 dBm around the 6-dB backoff level was achieved for a two-carrier WCDMA signal of 2.135 GHz and 2.145 GHz with an IM_3 of −37 dBc.

Figure 10.23 shows the three-stage inverted Doherty amplifier configuration, where the quarterwave transmission lines are added in the outputs of the carrier and peaking amplifiers to provide a proper load modulation ratio [64]. The half-wave transmission line in the input path of the carrier amplifier is used to compensate for the delay provided by the output load network. The characteristic impedances of the quarterwave transmission lines are optimized to provide a high efficiency over wide output power backoff range. If the device size ratio of the carrier first peaking and second peaking amplifiers is $1{:}m_1{:}m_2$, respectively, the characteristic impedances of the quarterwave transmission lines at the full power loading condition can be obtained by

$$Z_T = Z_1\sqrt{\frac{1}{1+m_1+m_2}} \tag{10.18}$$

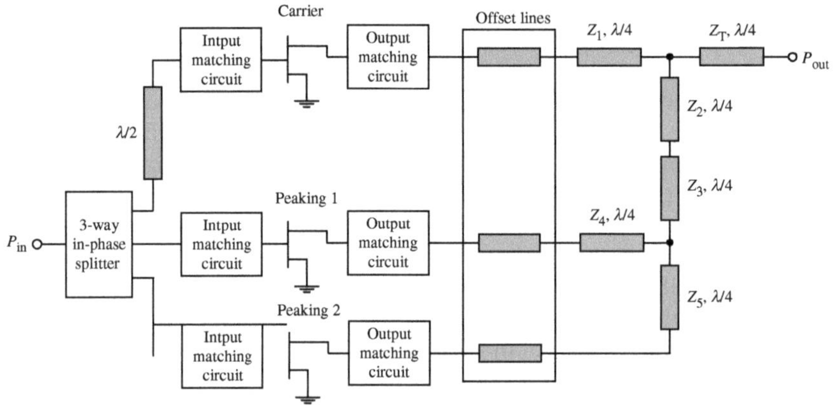

FIGURE 10.23 Schematic diagram of three-stage inverted Doherty amplifier.

$$Z_4 = Z_1 \frac{Z_3}{Z_2} \sqrt{\frac{1}{m_1}} \tag{10.19}$$

$$Z_5 = Z_1 \frac{Z_3}{Z_2} \sqrt{\frac{1}{m_2}} \tag{10.20}$$

assuming the same 50-Ω loading conditions for the standard load and the carrier and peaking amplifiers at full loading conditions. As a result, for the same device sizes for the carrier and peaking amplifiers when $m_1 = m_2 = 1$, $Z_2 = Z_3 = 50\ \Omega$, and $Z_1 = 70\ \Omega$, from Eqs. (10.18) to (10.20), it follows that $Z_T = 40.4\ \Omega$ and $Z_4 = Z_5 = 50\ \Omega$, respectively. In this case, the drain efficiency for a single-carrier 2.14-GHz WCDMA signal with a *PAR* of 10.5 dB can be improved by 5% over wide range of output powers.

10.6 Integration

The transmission-line two-stage Doherty amplifier can easily be implemented into the monolithic microwave integrated circuit (MMIC) by using a pHEMT or CMOS process. For example, a fully integrated *Ku*-band MMIC Doherty amplifier using a 0.25-μm pHEMT technology achieved a two-tone *PAE* of 40% with a corresponding IM_3 of –24 dBc at 17 GHz, whereas a single-tone *PAE* of 38.5% at 1-dB compression point was measured for a 20-GHz MMIC Doherty amplifier implemented in a 0.15-μm pHEMT process for use in digital satellite communication (DSC) systems [65, 66]. Furthermore, by using a 0.13-μm RF CMOS technology, a transmission-line MMIC Doherty amplifier based on cascode configuration of the carrier and peaking amplifiers achieved a saturation output power of 7.8 dBm from a supply voltage of 1.6 V at an operating frequency of

60 GHz for use in wireless personal area network (WPAN) transceivers [67]. On the other hand, the efforts to directly apply the Doherty technique to the design of power-amplifier integrated circuits with a high level of integration at lower frequencies face difficulties, because the physical size of the quarterwave transmission lines is too large in this case. For example, for an FR4 substrate with effective dielectric permittivity of $\varepsilon_r = 3.48$, the geometrical lengths of the quarterwave transmission lines are 48 mm, 19 mm, and 8.7 mm at the operating frequencies of 900 MHz, 2.4 GHz, and 5.2 GHz, respectively. Therefore, one of the acceptable solutions for the fabrication of small-size Doherty amplifier MMICs intended to operate in WLAN or WiMAX transmitter systems is to replace each quarterwave line in the input combining circuit and output impedance transformer by its low-pass π-type lumped-distributed equivalent with a short-length series transmission line and two shunt capacitors connected to its both ends [68, 69]. Additionally, simple and small-size second-harmonic termination circuits can be realized with integrated MIM capacitors and bondwires at the end of the carrier and peaking amplifier collectors [68].

In order to minimize the inherently high substrate loss and increase the level of integration to implement the Doherty amplifier in a CMOS process, the branch-line coupler and quarterwave transformer in the amplifier input and output circuits are fully substituted by their lumped equivalents [70]. By considering the transmission *ABCD*-matrices for a quarterwave transmission line shown in Fig. 10.24(*a*) and a π-type low-pass lumped circuit consisting of a series inductance and two shunt capacitors shown in Fig. 10.24(*b*) and equating the corresponding elements of both matrices, the ratio between the circuit elements can be written as

$$Z_0 \omega C = \frac{Z_0}{\omega L} = 1 \tag{10.21}$$

where Z_0 is the characteristic impedance of the quarterwave transmission line. A high-power Doherty amplifier MMIC can be integrated with lumped elements in a standard discrete package, where the compensation series circuits (each consisting of an inductance and a capacitor) are connected to the drain terminals of the carrier and

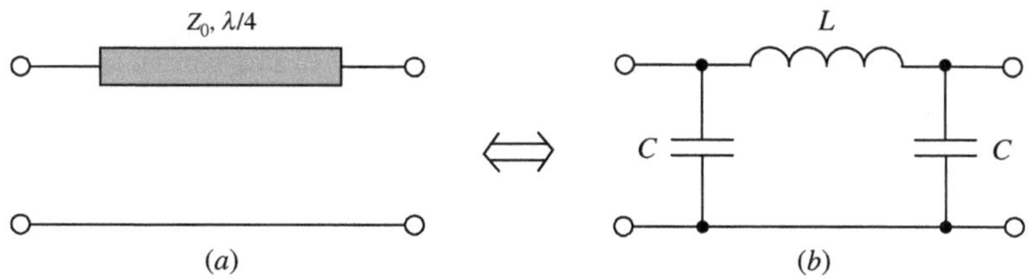

Figure 10.24 Quarterwave transmission line and its single-frequency lumped equivalent.

peaking transistors to compensate for their output capacitances [71]. For example, an integrated solution based on four 10-W MMIC Doherty amplifier cells combined in parallel achieves a drain efficiency of 39.8% at an average output power of 7.5 W with an *ACLR* of –50 dBc using a DPD technique for a two-carrier 2.14-GHz WCDMA signal with a *PAR* of 7.6 dB [72].

Similarly, the input in-phase transmission-line two-way Wilkinson divider can be replaced by its lumped equivalent, where a π-type low-pass *LC* circuit is used in its each branch. Figure 10.25(*a*) shows an example of the simplified schematic of a two-stage lumped Doherty amplifier, where the output quarterwave transmission line connected to the carrier amplifier output and the input phase-shifting quarterwave transmission line connected to the peaking amplifier input are replaced by equivalent π-type low-pass lumped circuits. In addition, the output quarterwave transformer is replaced by an *L*-type high-pass matching circuit, whereas two *L*-type low-pass matching circuits are used to provide the input matching of the carrier and peaking amplifiers. At the peaking amplifier input path, the right-hand shunt capacitor as a part of the equivalent quarterwave phase

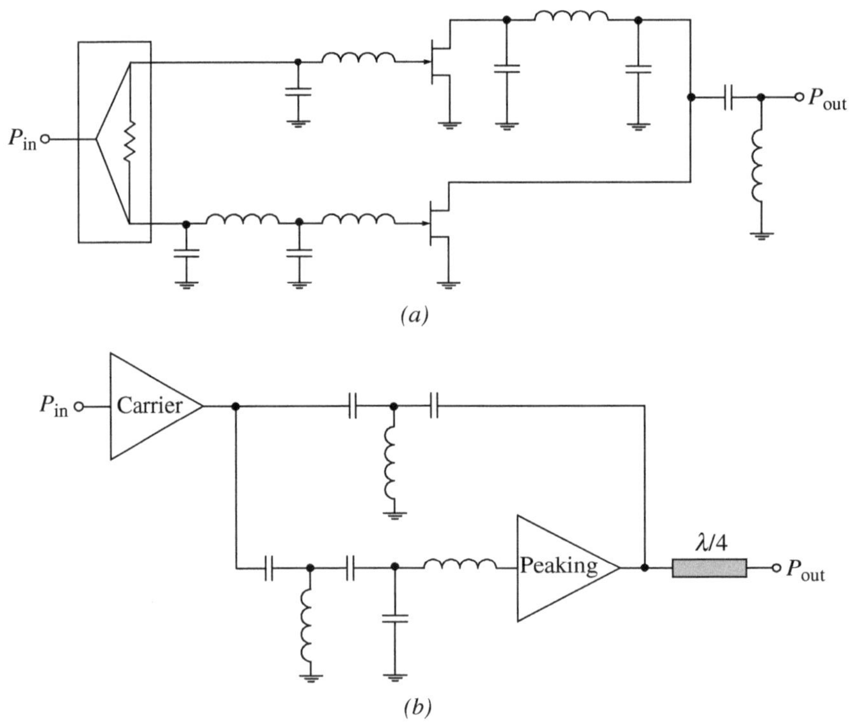

FIGURE 10.25 Circuit schematics of lumped Doherty amplifiers for handset applications.

shifter and the shunt capacitor as a part of the input *L*-type low-pass matching circuit can be combined into a single shunt capacitor.

To remove a 3-dB hybrid input divider, the conventional Doherty amplifier can be rearranged to be more suitable for handset applications. Figure 10.25(*b*) shows the circuit schematic of a "series-type" Doherty architecture, where the subcircuit of the peaking amplifier and impedance transformer is connected to the output of the carrier amplifier in series rather than in parallel configuration [73, 74]. The impedance transformers in both paths are composed of a high-pass lumped-element *T*-network each. This is because the high-value inductances or relatively long series microstrip lines are required for the low-pass *T*-networks, making chip-level integration impractical. The shunt inductance in the high-pass *T*-network can be implemented using a high-impedance microstrip line instead of a lumped-element inductor. Besides, a high-pass *T*-network helps prevent occasional LF oscillations. The output quarterwave transformer fabricated externally can be replaced by the equivalent low–pass π-type matching circuit with a series short-length microstrip line and two shunt chip capacitors. In a low-power region, the input impedance from the peaking amplifier path that is turned off due to deep Class-C bias mode is sufficiently high. However, in a high-power region, it reduces significantly when the peaking amplifier is turned on, so the load seen by the carrier amplifier reduces significantly, by about three times for a practical case of a series-type MMIC Doherty amplifier with optimum device sizes based on a 2-μm InGaP HBT technology [74]. As a result, for a 1.9-GHz IS-95A CDMA signal, a *PAE* of 18% and 42.8% were achieved at 16 dBm and 28 dBm, respectively.

As the Doherty amplifier for handset applications should be compact, a direct input-dividing circuit considering the impedance variations of the carrier and peaking amplifiers can be used instead of the Wilkinson power combiner. In this case, because the input impedance of the carrier amplifier remains almost constant, whereas that of the peaking amplifier changes significantly because of the Class-C bias, this effect can be utilized for the uneven input dividing [45, 75]. As a result, large power is delivered to the carrier amplifier at the low-power region, and the power gain at the low-power region becomes much higher than that at the high-power region, deteriorating the gain flatness and linearity of the Doherty amplifier. Figure 10.26 shows the full circuit schematic of a lumped Doherty amplifier with two-stage carrier and peaking amplifiers [75]. Here, the output matching circuit takes a role of a quarterwave transformer including parasitics, with the phase compensation network used at the input of the carrier path, and the offset line used at the output of the peaking path. The second and third harmonics are properly controlled to enhance the efficiency of both carrier and peaking amplifiers. Moreover, the second- and third-harmonic control circuits are also utilized for the quarterwave transformer by connecting the capacitor C_q,

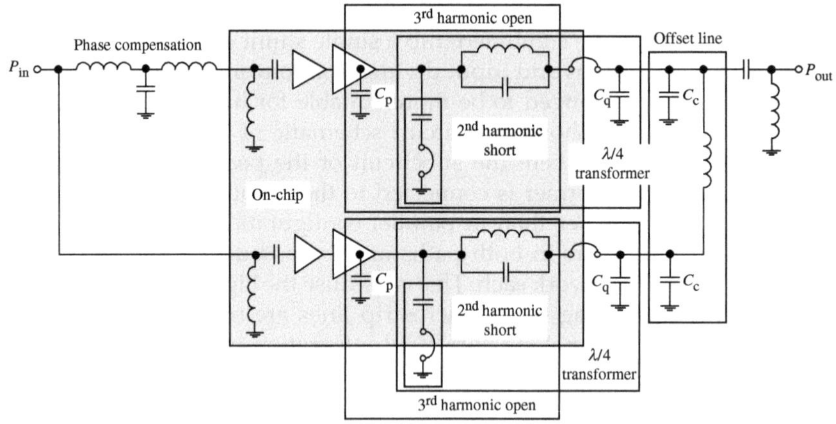

FIGURE 10.26 Schematic of Doherty amplifier with harmonic control for handset applications.

forming a π-network where the device output capacitance C_p and second-harmonic control circuit are considered as one capacitor and the parallel resonant *LC*-circuit is inductive at the fundamental. The capacitors C_c of the offset line can be combined with capacitors C_q to reduce the number of components. As a result, the Doherty amplifier MMIC implemented in a 2-μm InGaP/GaAs HBT process presents a *PAE* of 40.2% at an output power of 26 dBm with an error vector magnitude (EVM) of 3% for a 16-QAM m-WiMAX signal having a 9.54-dBc crest factor and 8.75-MHz bandwidth.

Because a frequency-dependent quarterwave transformer and output matching circuits generally provide a narrowband operation of a conventional Doherty amplifier, a quarterwave transmission line as an additional matching element can be added at the output of the peaking amplifier in series with the offset line to minimize the loaded quality factor for broader operation by increasing the impedance at the output junction of the carrier and peaking amplifiers [76]. In a handset monolithic application, the transmission-line quarterwave impedance transformer and offset line are implemented with lumped elements, representing the equivalent π-type low-pass and π-type high-pass *LC* networks, respectively, as shown in Fig. 10.27(*a*). In this case, the network parameters are optimized to provide an open-circuit condition at the output of the peaking branch over broadband frequency range when the peaking amplifier is turned off. To simplify the load-network structure, the values of the inductances L_1 and L_2 are chosen so as to merge them with the corresponding capacitances C_1 and C_2. Figure 10.27(*b*) shows the circuit schematic of a 2-μm GaAs HBT Doherty amplifier, where all of the components are fully integrated on a chip. In this case, the inductors are implemented using bondwires and slab inductors, the input dividing circuits are broadband

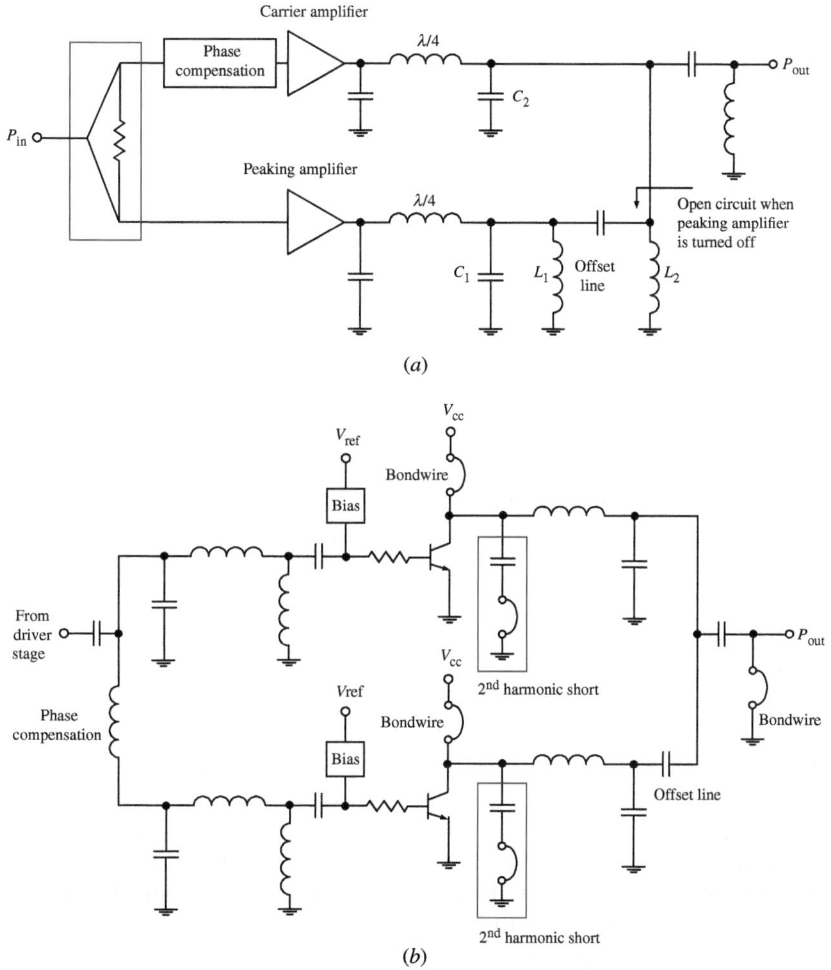

FIGURE 10.27 Schematics of Doherty amplifiers for handset applications with bandwidth enhancement.

based on low-Q matching networks, and the second- and third-harmonic impedances are controlled for high efficiency across the bandwidth. The open-circuit conditions are achieved by optimizing all load-network elements, including drain bondwires and device output capacitances. For a mobile 8.75-MHz 16-QAM m-WiMAX application with a 9.6-dB crest factor, such a lumped Doherty amplifier exhibits a *PAE* of over 27% and an output power of over 23.6 dBm across 2.2-2.8 GHz using a DPD technique. Similar Doherty amplifier with broadband lumped networks can provide a *PAE* over 30% and an output power of over 28 dBm across 1.6 to 2.1 GHz for a 10-MHz LTE signal with a *PAR* of 7.5 dB [77].

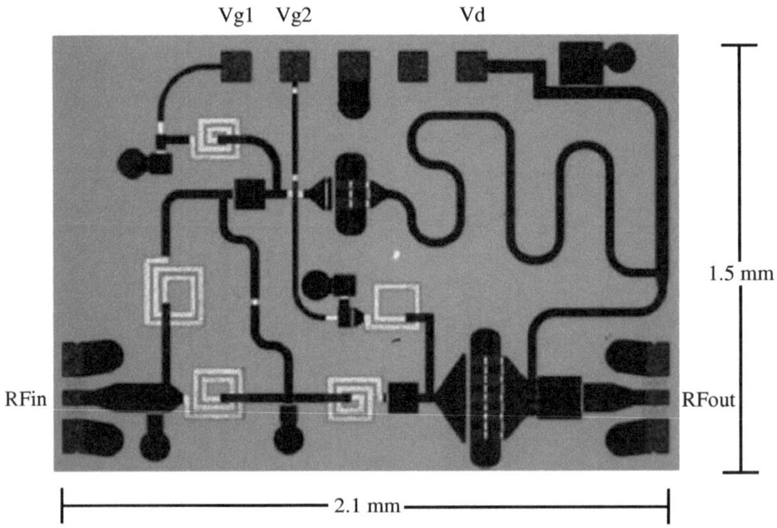

FIGURE 10.28 Test chip of wideband monolithic GaN HEMT Doherty power amplifier. (*Courtesy of Chalmers Institute of Technology.*)

Figure 10.28 shows the test chip of a wideband monolithic GaN HEMT asymmetric Doherty power amplifier, where the input network consists of a lumped-element Wilkinson power divider and phase shifter, whereas the load network represents a *T*-line impedance inverter with optimized characteristic impedances and electrical lengths of microstrip lines [78]. The input matching and stability of operation is provided with a series resistor at the input of the carrier device and with two shunt *LR* circuits connected at the inputs of the carries and peaking devices, respectively. The Doherty power amplifier was implemented in a 0.25-μm GaN HEMT process, having a 100-μm thick SiC substrate, a relative permittivity of 9.7, a maximum drain current density of 900 mA/mm, and a maximum power density of 5 to 7 W/mm, with a total chip size of 2.1 mm × 1.5 mm. To obtain the highest possible output power with model verified device sizes, the total gate widths of the carrier and peaking devices were chosen as 4 × 100 μm and 10 × 100 μm, respectively. As a result, a power-added efficiency of greater than 30% at 9-dB power backoff within the frequency range of 6.7 to 7.8 GHz and a maximum output power of 35 ± 0.5 dBm from 6.6 to 8.5 GHz were achieved. For a 10-GHz MMIC Doherty power amplifier using Class-E approximation in both carrier and peaking amplifiers that are based on 140-nm GaN HEMTs of a 1-mm gate width each, the simulated two-tone results demonstrated a *PAE* of 40.4% at peak output power of 25.6 dBm and a *PAE* of 24% at 6-dB backoff [79].

10.7 Digitally Driven Doherty Amplifier

In a digitally driven dual-input Doherty amplifier architecture, the input signal of each branch is digitally preprocessed and supplied separately to each branch of the Doherty amplifier to optimize its overall performance. In this case, digital signal processing is applied to reduce the performance degradation due to phase impairment in the Doherty amplifier branches achieved by adaptively aligning the phases of the carrier and peaking paths for all power levels after the peaking amplifier is turned on. Generally, a DSP includes a DPD system to improve linearity, which can be configured to provide a carrier signal component along a carrier amplifier path and a peaking signal component along a peaking amplifier path from a digital input signal [80]. In this case, the carrier and peaking amplifiers can amplify the signal components according to the programmable proportions of the split input signal, and not based on a saturation condition of the carrier amplifier, thus resulting in a higher efficiency. Because the signals are isolated prior to being input to the Doherty amplifier, the Doherty amplifier need not include an asymmetric splitting with input phase-matching delay, and input impedance-matching circuitry is simplified. The DPD system also performs phase and gain adjustments to each of the signal components. To further improve efficiency performance of a two-stage Doherty amplifier, the separated amplitude and phase modulated signals, produced by the DSP, drive through the corresponding quadrature upconverters both the carrier and peaking amplifiers, each operated in a Class-E mode [81].

Figure 10.29(*a*) shows the block diagram of a dual-input digitally driven Doherty amplifier with digital signal processing and dual-channel upconverter [82]. In this case, direct access and software control of the individual inputs can bring an improvement in efficiency of a Doherty amplifier between the two efficiency peaking points at maximum and 6-dB backoff powers, as shown in Fig. 10.29(*b*). The Doherty amplifier design is performed by deriving the offset line with electrical length θ_p to be inserted at the output of the peaking branch to ensure a quasi-open circuit condition and prevent leakage from the carrier amplifier to the output of the peaking amplifier at the low-power region. The offset line with electrical length θ_c was optimized to maximize efficiency around the turn-on point of the Doherty amplifier. Because of the different bias conditions for the carrier amplifier (Class AB) and the peaking amplifier (Class C), the degradation in output power due to phase imbalance condition can be as high as 40% after the peaking amplifier is fully turned on, which directly translates into significant deterioration in a drain efficiency of the Doherty amplifier. However, the dual-input digitally driven Doherty architecture, allowing for the adoption and implementation of a power adaptive phase-alignment mechanism, can minimize the adverse effects of phase imbalance between the carrier and peaking branches.

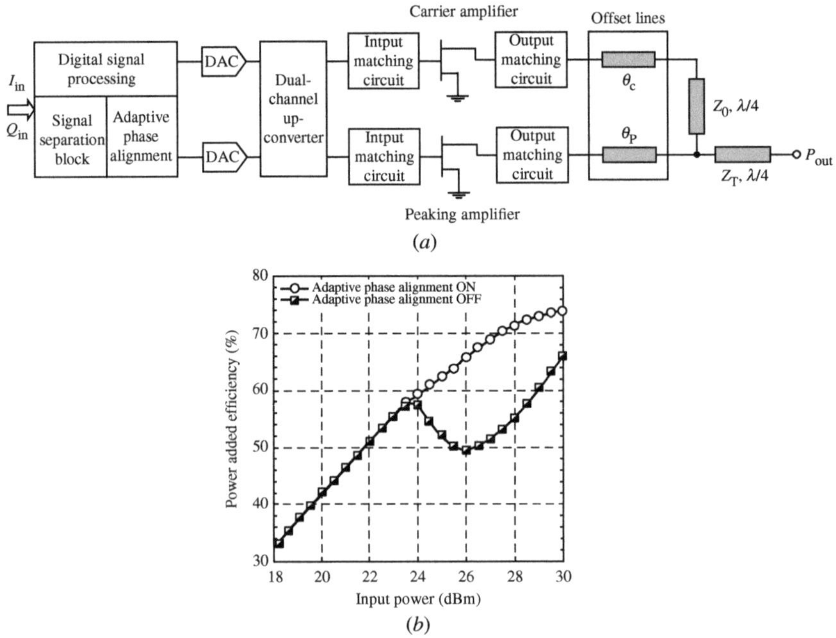

FIGURE 10.29 Block diagram and simulated performance of dual-input digital Doherty amplifier.

The power-dependent phase offset is adjusted using a power-indexed lookup table (LUT) to correct for the phase disparity at all power levels, where both the carrier and peaking amplifiers contribute to the total output power of the Doherty amplifier. As a result, the phase difference between the carrier and peaking branches is reduced to 0° over the input power range spanning from the turn-on of the peaking amplifier until the saturation of the Doherty amplifier. The phase-aligned Doherty amplifier based on two 10-W GaN HEMT transistors demonstrates a *PAE* higher than 50% over an 8-dB output-power backoff range and a *PAE* of 57% at an average output power of 37 dBm for a single-carrier WiMAX signal with a *PAR* of 7 dB, thus resulting in an improvement of 7% in *PAE* and 1 dB in average output power with similar linearity performance corresponding to an *ACPR* of −22 dBc compared to the fully analog Doherty amplifier [82].

Efficiency enhancement in a digital Doherty amplifier over wide power range can also be achieved by using a digitally controlled dynamic input power distribution scheme to minimize the drive power waste into the peaking branch at backoff power levels [83]. In this case, the carrier amplifier should get significantly more input power in comparison to the peaking amplifier at low-power drive, whereas the carrier amplifier should get slightly less input power in comparison to the peaking amplifier after turn-on point. As a result, the efficiency can be improved by 7% compared to the conventional fully analog symmetrical Doherty amplifier based on two 10-W GaN HEMT devices and operating at 2.14 GHz for a single-carrier WiMAX signal with a 9-dB *PAR* and 10-MHz bandwidth.

10.8 Multiband and Broadband Capability

A multiband capability of the conventional two-stage Doherty amplifier can be achieved when all of its components are designed to provide their corresponding characteristics over the required bands of operation, as shown in Fig. 10.30(*a*) [84]. In this case, the carrier and peaking amplifiers should provide broadband performance when, for example, their input and interstage matching circuits are designed as broadband and the load network generally can represent a low-pass structure with two or three sections tuned to the required frequencies. Some bandwidth extension can be achieved by simply optimizing the characteristic impedances of the quarterwave impedance transformer and quarterwave output combiner in a combining load network [85, 86]. For a multiband operation with the center frequency ratio at each of the frequency bands of 2 or greater, the input divider can be

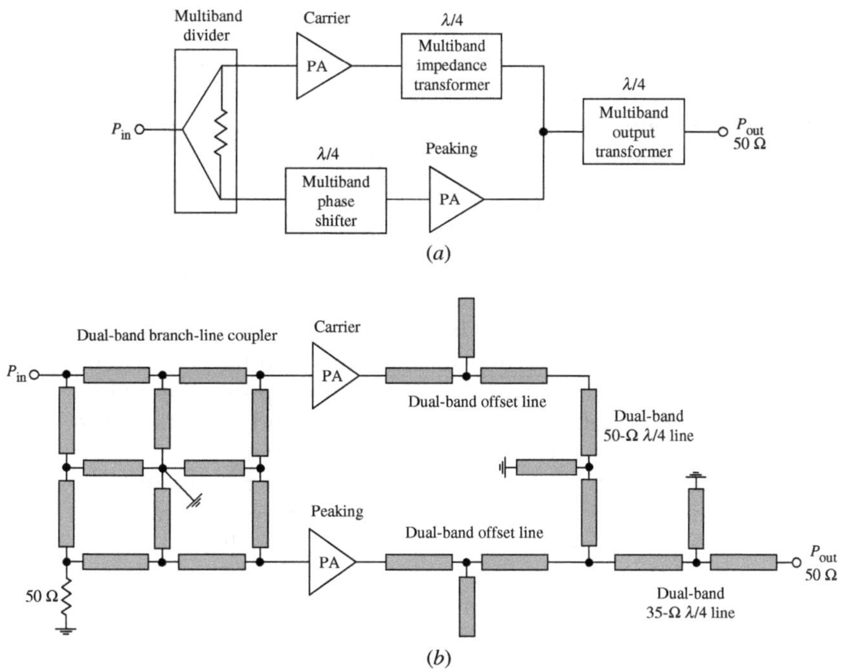

Figure 10.30 Block diagrams of multiband Doherty amplifiers.

configured by a multisection Wilkinson power divider or coupled-line directional coupler. In a dual-band operation mode, a dual-frequency Wilkinson power divider can represent a structure, where each quarterwave branch of a conventional Wilkinson power divider is substituted by the two transmission-line sections with different characteristic impedances and electrical lengths [87, 88]. In practical applications, especially if the operating frequencies of one of the frequency band are sufficiently low, the miniaturized version of a dual-band Wilkinson power divider can be designed based on the concept of slow wave periodic structure [89].

A dual-band input power splitter can also represent a π-shape or T-shape stub-tapped branch-line coupler, as well as an impedance transformer network, which introduces a 90° phase shift. Similarly, the offset lines and an output quarterwave transformer can be based on a π-type or T-type transmission-line impedance-inverting section with proper selected transmission-line characteristic impedances and electrical lengths, where the shunt element is realized by an open- or short-circuit stub, as shown in Fig. 10.30(*b*) [90, 91]. However, it should be noted that it is not easy to design a multiband impedance transformer that should adequately provide two separate matching options simultaneously: first, to operate in a 50-Ω environment without affecting the power amplifier performance in a high-power region, and second, to provide an impedance matching from 25 to 100 Ω in a low-power region. In this case, as an alternative, it is also possible to switch between two quarterwave transmission lines in a dual-band operation when each of the quarterwave transmission line is tuned to the corresponding center bandwidth frequency. However, it may not be so simple in practical implementation because of the load-network complexity and additional power losses.

10.8.1 Dual-Band Parallel Doherty Architecture

The classic two-stage Doherty amplifier has limited bandwidth capability in a low-power region as it is necessary to provide an impedance transformation from 25 to 100 Ω when the peaking amplifier is turned off, as shown in Figs. 10.31(*a*) and 10.32(*a*), thus resulting in a loaded quality factor $Q_L = \sqrt{100 / 25 - 1} = 1.73$ at 3-dB output-power reduction level, which is sufficiently high for broadband operation. The parallel architecture of a two-stage Doherty amplifier with modified modulated load network, whose block schematic is shown in Fig. 10.31(*b*), can improve bandwidth properties in a low-power region by reducing the impedance transformation ratio by a factor of two [92].

In this case, the load network for the carrier amplifier consists of a single quarterwave transmission line required for impedance transformation, the load network for the peaking amplifier consists of a 50-Ω quarterwave transmission line followed by another quarterwave transmission line required for impedance transformation, and

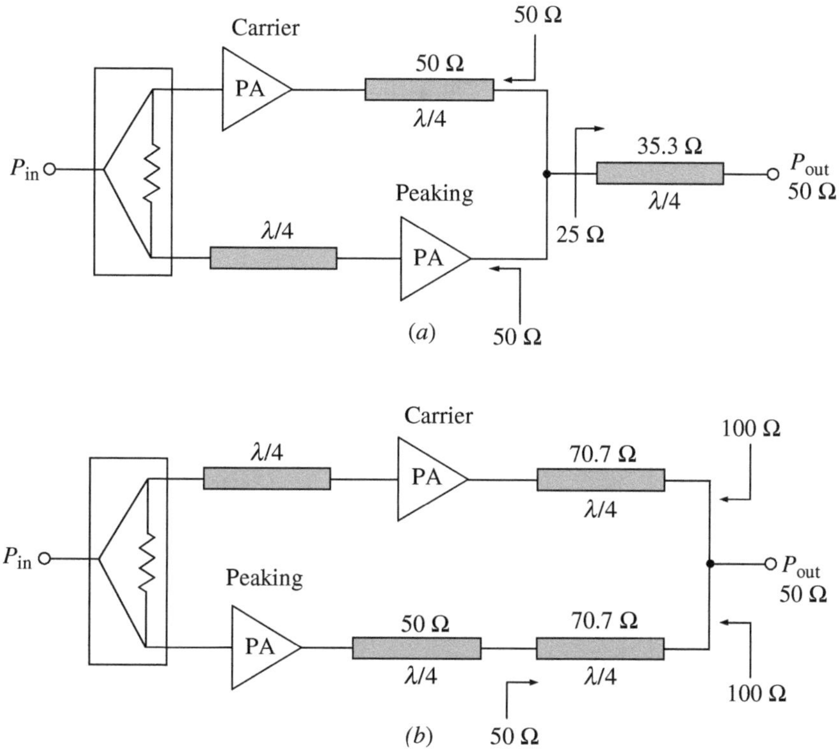

FIGURE 10.31 Block diagram of conventional and modified two-stage Doherty amplifiers.

the quarterwave transmission line at the input of the carrier amplifier is necessary for phase compensation. Both impedance-transforming quarterwave transmission lines, having a characteristic impedance of 70.7 Ω each, provide a parallel connection of the carrier and peaking amplifiers in a high-power region by parallel combining of the two 100-Ω impedances at their output into a 50-Ω load, with 50-Ω impedances at their inputs seen by each amplifier output. In a low-power region below output-power backoff point of –6 dB when the peaking amplifier is turned off, the required impedance of 100 Ω seen by the carrier-amplifier output is achieved by using a single quarterwave transmission line with the characteristic impedance of 70.7 Ω to match with a 50-Ω load, as shown in Fig. 10.32(*b*).

This provides a loaded quality factor $Q_L = \sqrt{100 / 50 - 1} = 1$, resulting in a 1.73 times wider frequency bandwidth, as shown in Fig. 10.32(*c*) by curve 1 compared with a conventional case (curve 2). Because the load network of the peaking amplifier contains two quarterwave transmission lines connected in series, an overall

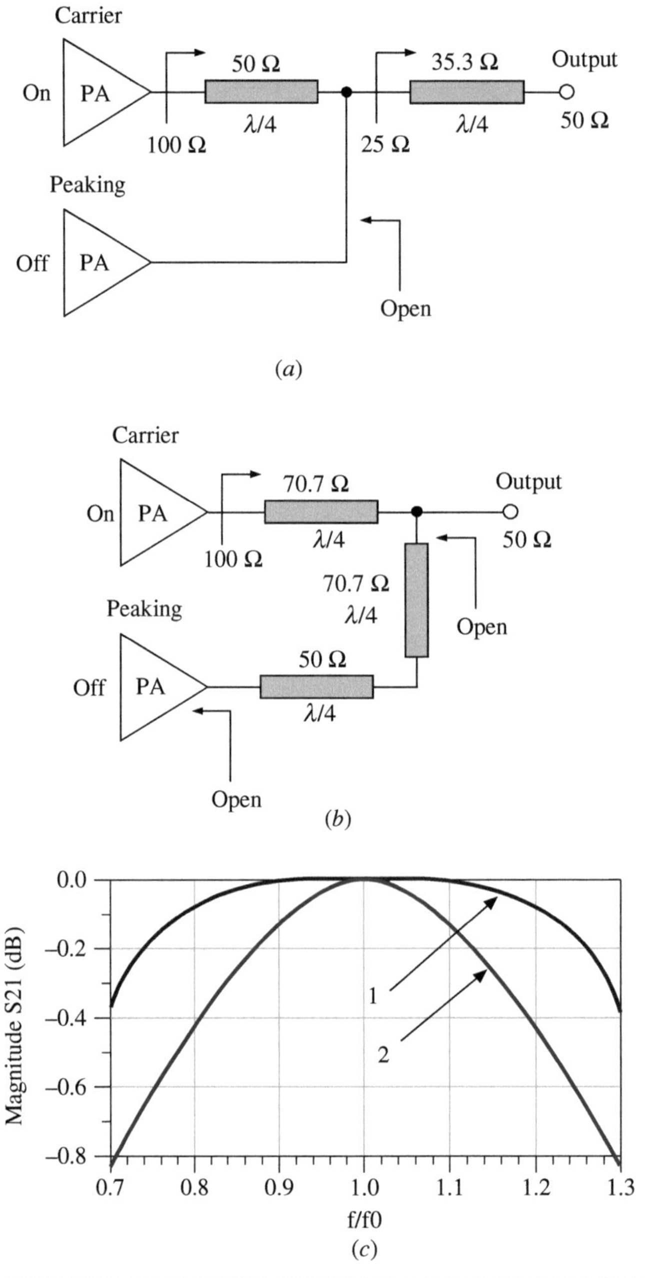

FIGURE 10.32 Load-network schematics and broadband properties.

half-wavelength transmission line is obtained, and an open circuit at the peaking-amplifier output directly translates to the load providing a significant isolation of the peaking-amplifier path from the carrier-amplifier path in a wide frequency range. The input in-phase divider and phase-compensating transmission line can be replaced by a broadband coupled-line 90° hybrid coupler.

From Fig. 10.32(*c*), it follows that use of a parallel Doherty architecture can provide a broadband operation within 25 to 30% around center bandwidth frequency with minimum variation of the load-network transfer characteristic. As a result, a dual-band operation can be easily provided by this architecture, for example, in 1.8-GHz (1805–1880 MHz) and 2.1-GHz (2.11–2.17 GHz) or in 2.1-GHz and 2.6-GHz (2.62–2.69 GHz) WCDMA/LTE frequency bands, respectively.

Figure 10.33 shows the simulated circuit schematic of a dual-band parallel GaN HEMT Doherty architecture, where the carrier and peaking amplifiers are based on broadband transmission-line Class-E power amplifiers, whose circuit structure is shown in Fig. 8.45 (see Chap. 8). Here, the input matching circuits and output load network are based on microstrip lines with their parameters corresponding to a 20-mil RO4360 substrate. The ideal broadband 90° hybrid coupler is used at the input to split signals between the carrier and peaking amplifying paths. The electrical lengths of both offset and combining

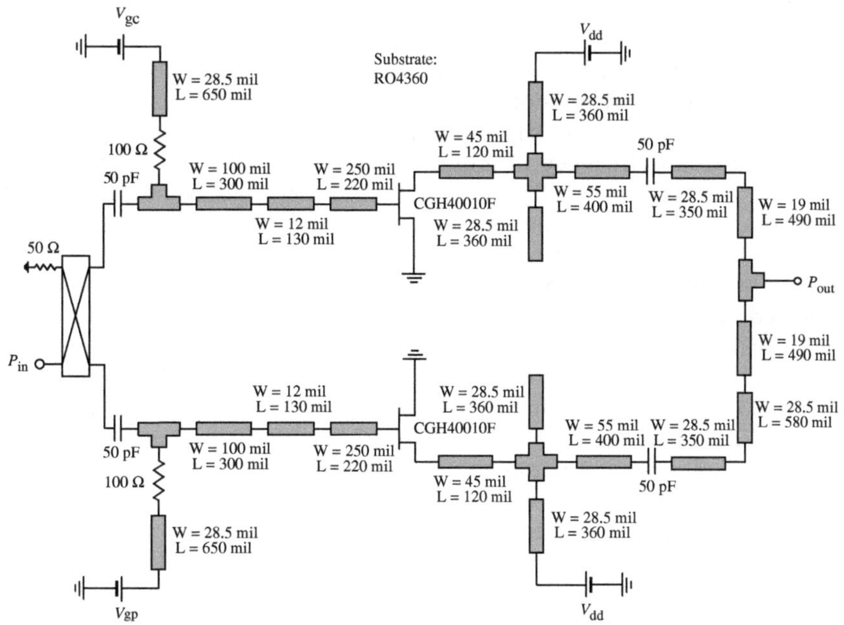

FIGURE 10.33 Circuit schematic of dual-band parallel GaN HEMT Doherty amplifier.

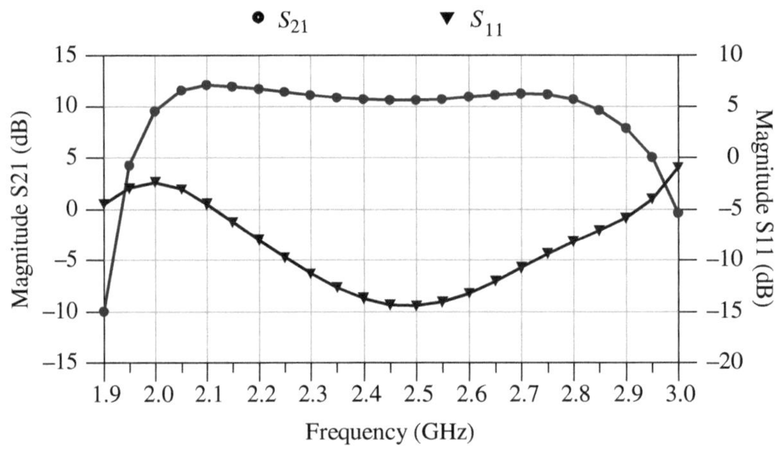

FIGURE 10.34 Simulated small-signal S_{11} and S_{21} versus frequency.

microstrip lines were optimized to maximize efficiency at saturated and backoff output power levels.

Figure 10.34 shows the simulation results for the small-signal S_{21}-parameters versus frequency demonstrating the bandwidth capability of a parallel transmission-line GaN HEMT Doherty amplifier covering a frequency range from 2.0 to 2.8 GHz with a power gain over 10 dB. In this case, an input return loss defined from the magnitude of S_{11} is less than 5 dB over the frequency bandwidth of 2.1 to 2.9 GHz.

Figure 10.35 demonstrates the broadband capability of a parallel Doherty structure, where the carrier and peaking amplifiers are based on a broadband transmission-line reactance compensation Class-E technique. In an amplifier saturation mode with an input power of 36 dBm, a drain efficiency of around 70% with an average output power of greater than 43 dBm and a gain variation of about 1 dB was simulated across the frequency range of 2.0 to 2.8 GHz, as shown in Fig. 10.35(*a*). At the same time, high drain efficiencies over 50% at backoff output powers of 5 to 6 dB from saturation can potentially be achieved across the frequency range of 2.1 to 2.7 GHz, as shown in Fig. 10.35(*b*). This means that the practical implementation of a parallel Doherty power amplifier, the simulation setup of which is shown in Fig. 10.33, can provide a highly efficient operation in two cellular bands of 2.11 to 2.17 GHz and 2.62 to 2.69 GHz without any tuning of the amplifier load-network parameters, either with separate or simultaneous dual-band transmission of WCDMA or LTE signals.

The large-signal simulations versus input power have been done at two center bandwidth frequencies of 2.14 GHz and 2.655 GHz with optimized circuit parameters to achieve maximum performance. Figure 10.36 shows the simulated large-signal power gain and drain efficiencies of a dual-band parallel transmission-line GaN HEMT Doherty

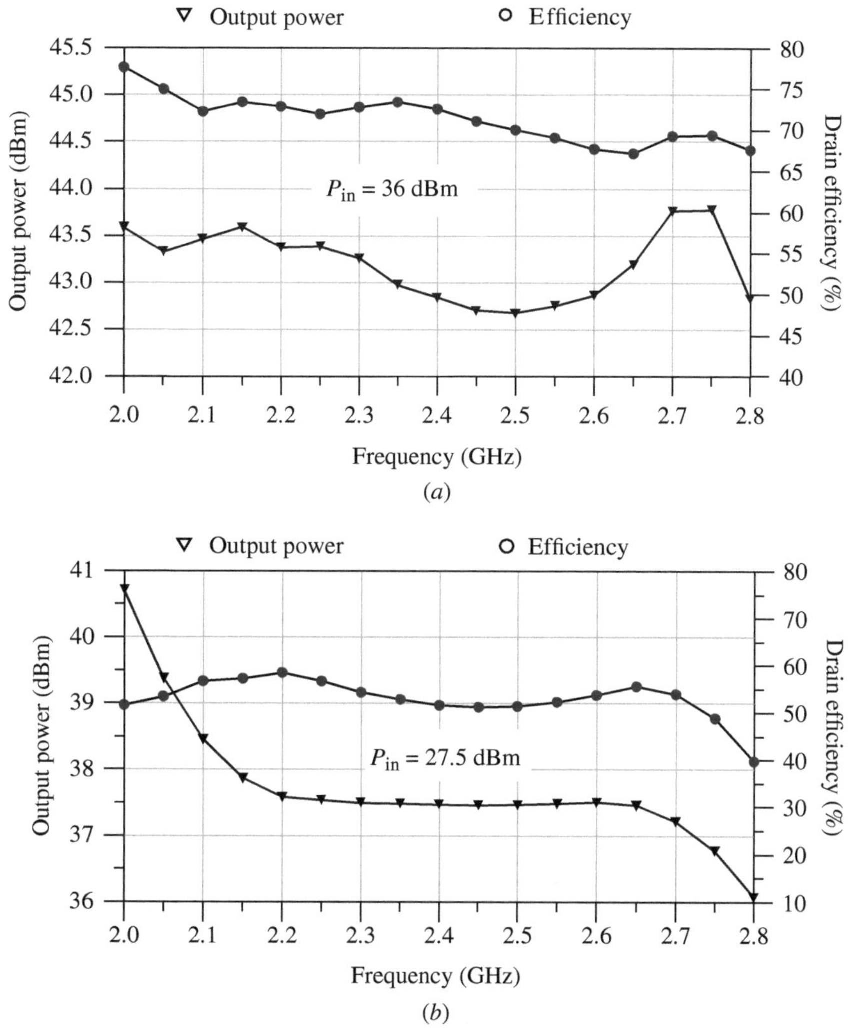

FIGURE 10.35 Broadband capability of parallel Doherty amplifier.

amplifier, with the carrier gate bias $V_{gc} = -2.45$ V, peaking gate bias $V_{gp} = -7$ V, and dc supply voltage $V_{dd} = 28$ V. In this case, a linear power gain of about 11 dB was achieved at an operating frequency of 2.655 GHz, whereas a slightly higher linear power gain of about 12 dB was achieved at lower operating frequency of 2.14 GHz. At the same time, the drain efficiencies of 64% and 53% were simulated at backoff output powers of 39 dBm (4-dB backoff from saturated power of 43 dBm) and 37 dBm (6-dB backoff) at both center bandwidth frequencies, respectively.

The dual-band transmission-line GaN HEMT Doherty amplifier was fabricated on a 20-mil RO4360 substrate. An input power splitter

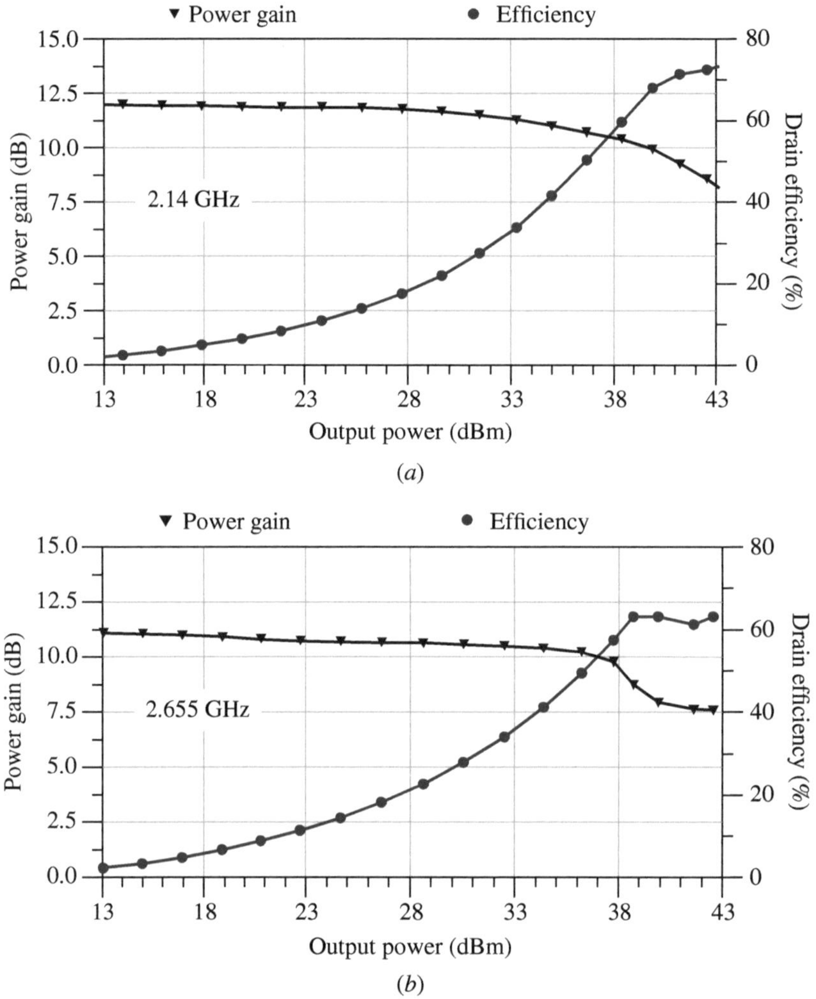

FIGURE 10.36 Simulated results of dual-band parallel Doherty amplifier.

represents a broadband coupled-line coupler from Anaren, model 11306-3, which provides maximum phase balance of ± 5° and amplitude balance of ± 0.55 dB across the frequency range of 2 to 4 GHz. Figure 10.37 shows the test board of a dual-band parallel Doherty amplifier based on two 10-W Cree GaN HEMT power transistors CGH40010P in metal-ceramic pill packages. The input matching circuit, output load network, and gate and drain bias circuits (having bypass capacitors on their ends) are fully based on microstrip lines of different electrical lengths and characteristic impedances according to the simulation setup shown in Fig. 10.33. Special care should be taken for device

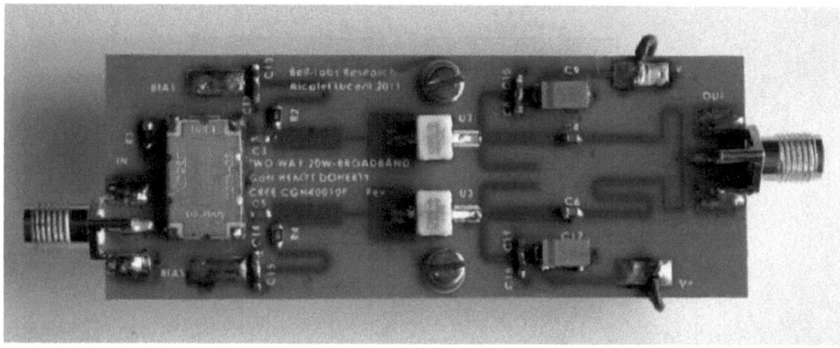

Figure 10.37 Test board of dual-band GaN HEMT parallel Doherty amplifier.

implementation process in order to minimize the input and output lead inductances of the packaged GaN HEMT device, which can significantly affect the amplifier performance such as power gain, output power, and drain efficiency.

For a single-carrier 5-MHz WCDMA signal with a *PAR* of 6.5 dB, a drain efficiency of 45% with a power gain of about 10 dB and *ACLR* (at 5-MHz offset) lower than –30 dBc at 2.14 GHz and a drain efficiency of 40% with a power gain of about 11 dB and *ACLR* around –30 dBc at 2.655 GHz were achieved at an average output power of 39 dBm. In both cases, optimization of the gate bias voltages for the carrier (Class-AB mode) and peaking (Class-C mode) amplifiers were provided.

10.8.2 Tri-Band Inverted Doherty Configuration

Generally, the multiband impedance transformer can represent a configuration with N ($N \geq 2$) cascade-connected transmission lines with different characteristic impedances. In this case, a simple two-stepped transmission-line impedance transformer can provide a two-pole response with different characteristic impedance ratio and different electrical lengths of the transmission-line sections [93]. It can be used as a dual-band output transformer as it is necessary to provide an impedance transformation from the output impedance of 25 Ω to the standard 50-Ω load in both low- and high-power regions [91].

As an example, the dual-band output transformer can be realized using a two-section transmission line, where the characteristic impedance of the first quarterwave transmission-line section is equal to 30 Ω and the characteristic impedance of the second quarterwave transmission-line section is set to 42 Ω, as shown in Fig. 10.38(*a*). In this case, the amplitude variations of ±0.5 Ω shown in Fig. 10.38(*b*) by curve 1 and phase variations of ±1° shown in Fig. 10.38(*c*) by curve 1 can be achieved across the frequency range from 2.0 to 2.8 GHz covering simultaneously 2.1-GHz (2.11–2.17 GHz) and 2.6-GHz (2.62–2.69 GHz) WCDMA/LTE bands. It was shown that by using a

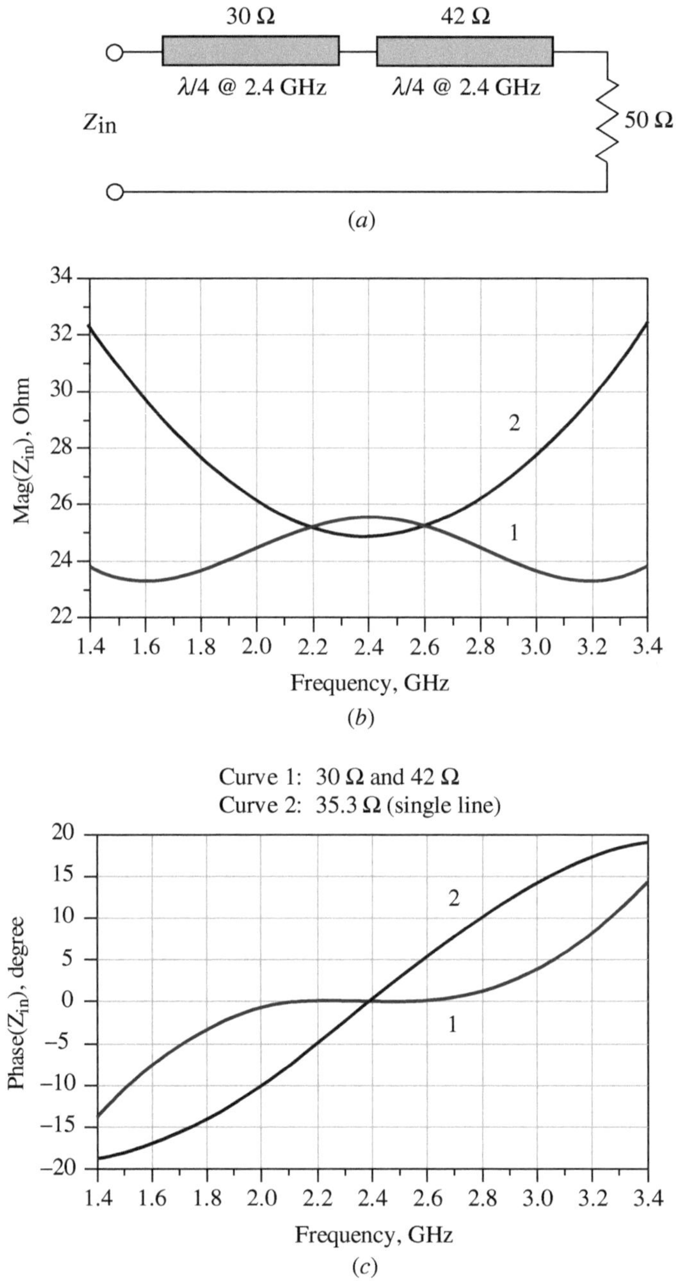

FIGURE 10.38 Stepped transmission-line transformer and its input impedances.

multisection output transmission line with different characteristic impedances, the frequency range from 2.2 to 2.96 GHz can be covered [94]. In this case, the broadband matching is realized by applying the simplified real frequency technique with the desired frequency-dependent optimum impedances. For comparison, the narrowband amplitude and phase responses of a quarterwave single-line impedance transformer are shown in Figs. 10.38(*b*) and 10.38(*c*) by curves 2, respectively. At the same time, from Figs. 10.38(*b*) and 10.38(*c*), it follows that the amplitude variations of ±1.0 Ω and phase variations of ±2° can be achieved with a 1-GHz bandwidth from 1.9 to 2.9 GHz, which means that reducing the midband frequency to 2.3 GHz can result in a simultaneous tri-band operation with inclusion of an additional 1.8-GHz (1805-1880 MHz) WCDMA/LTE bandwidth.

Figure 10.39(*a*) shows the modified broadband load network of an inverted Doherty amplifier, which consists of a two-section

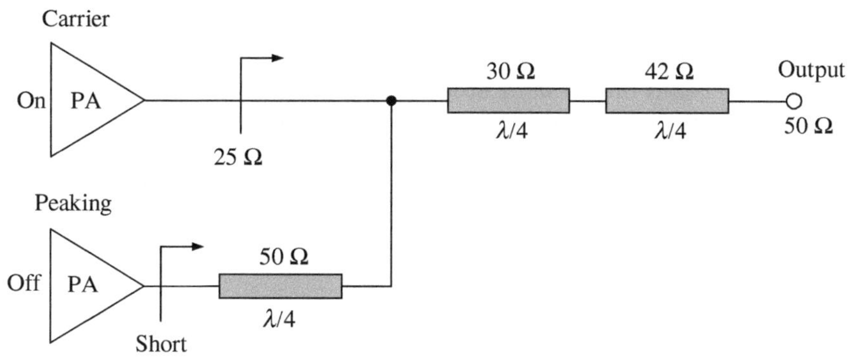

(*a*)

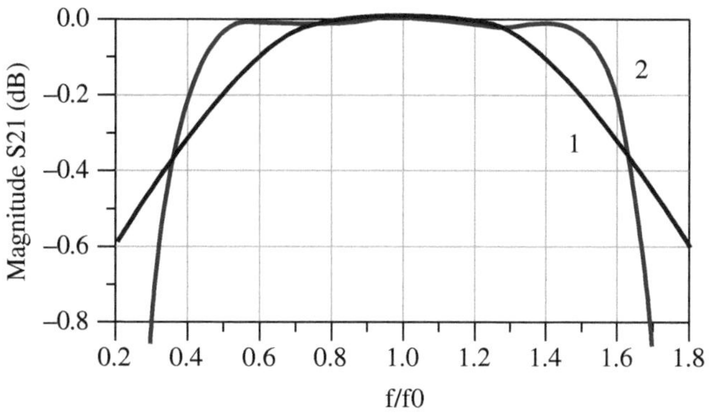

(*b*)

FIGURE 10.39 Load-network schematic and broadband properties.

transmission-line output impedance transformer, where each quarterwave transmission line has a different characteristic impedance to match first the initial 25 Ω to intermediate 35.3 Ω and then to 50-Ω load. Such 25- to 50-Ω transformer provides a wide frequency range, as shown in Fig. 10.39(*b*) by curve 1. However, broader frequency range with flatter frequency response can be achieved with a quarterwave open-circuit stub connected at the input of the two-line transformer when the peaking amplifier is turned off, as shown in Fig. 10.39(*b*) by curve 2, resulting in more than octave bandwidth in a low-power region at output power levels less than –6 dB backoff point. In this case, it is assumed that the output matching circuit of the carrier amplifier provides ideally a broadband impedance transformation from 25 to 100 Ω or close seen by the device multiharmonic current source. At the same time, broadband performance is also provided in a high-power region when both carrier and peaking amplifiers are turned on.

Figure 10.40 shows the simulated circuit schematic of a tri-band inverted GaN HEMT Doherty amplifier configuration, where the carrier and peaking amplifiers using Cree CGH40010 GaN HEMT devices are based on the same broadband transmission-line Class-E power amplifiers, whose idealized circuit structure is shown in Fig. 8.45 (see Chap. 8), and the broadband load network corresponds to the impedance-transforming structure shown in Fig. 10.39(*a*). The input matching circuits and output load network are based on microstrip lines with their parameters corresponding to a 20-mil

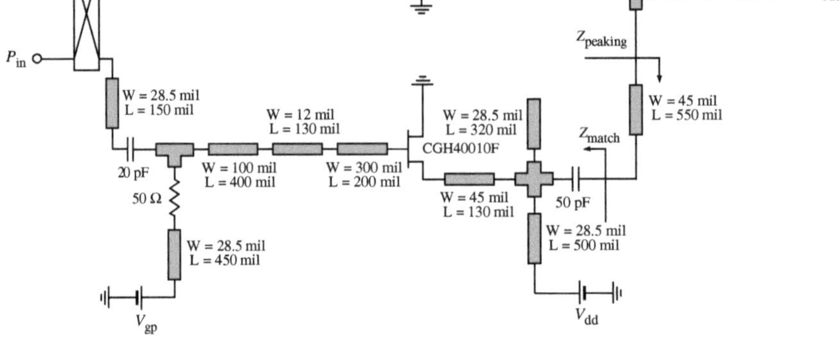

FIGURE 10.40 Circuit schematic of tri-band inverted GaN HEMT Doherty amplifier.

RO4360 substrate. In this case, it was found that, when the broadband Class-E power amplifier as a peaking amplifier is turned off, a short-circuit condition is achieved at the input of a series 35-Ω transmission line shown in Fig. 8.45 (see Chap. 8), which has a quarter wavelength at high bandwidth frequency to match with a 50-Ω load. Therefore, such a quarterwave transmission line was removed from the load networks of both the carrier and peaking amplifiers.

As result, the overall combining load-network is significantly simplified, and only a small optimization of the electrical lengths of the short- and open-circuit stubs in the load networks of each carrier and peaking amplifier is required to achieve broader frequency response. The broadband 90° hybrid coupler is used at the input to split signals between the carrier and peaking amplifying paths and to provide a 90° phase shift at the input of the carrier amplifier across the entire frequency bandwidth.

The impedance conditions at different points of the load network of the peaking amplifier when it is turned off are shown in Fig. 10.41, where Z_{match} shown in Fig. 10.41(*a*) indicates a low reactance at the output of a Class-E load network with short- and open-circuit stubs across the required frequency range from 1.8 to 2.7 GHz, having nearly zero reactance at high bandwidth frequency of 2.7 GHz and increasing capacitive reactance when the operating frequency reduces to 1.8 GHz. At the same time, by using the series transmission line of a quarter-wavelength long at high bandwidth frequency, an open-circuit condition is provided at higher bandwidth frequencies with sufficiently high inductive reactances at lower bandwidth frequencies, indicating by Z_{peaking} shown in Fig. 10.41(*b*). Hence, the broadband performance of such an inverted Doherty structure can potentially be achieved in a practical realization.

Figure 10.42 shows the frequency behavior of the impedance Z_{carrier} seen by the carrier device, as shown in Fig. 10.40, which has an inductive reactive component required for a high-efficiency Class-E operation and its real component slightly varies between 17 and 22 Ω. This means that by taking into account the device output shunt capacitance of 1.3 pF and series bondwire inductor of about 1 nH, the impedances seen by the device multiharmonic current source at the fundamental across the entire frequency bandwidth of 1.8 to 2.7 GHz can be increased up to around 50 Ω, which is high enough to achieve high efficiency at backoff output power levels. In this case, the device output capacitance and bondwire inductor constitute a low-pass *L*-type matching section to increase the load impedance seen internally by the device multiharmonic current source at the fundamental.

Figure 10.43 shows the simulation results for the small-signal S_{21}-parameters versus frequency, demonstrating the bandwidth capability of a modified inverted transmission-line GaN HEMT Doherty amplifier, which potentially can cover a wide frequency range of 1.6 to 3.0 GHz with a power gain over 11 dB.

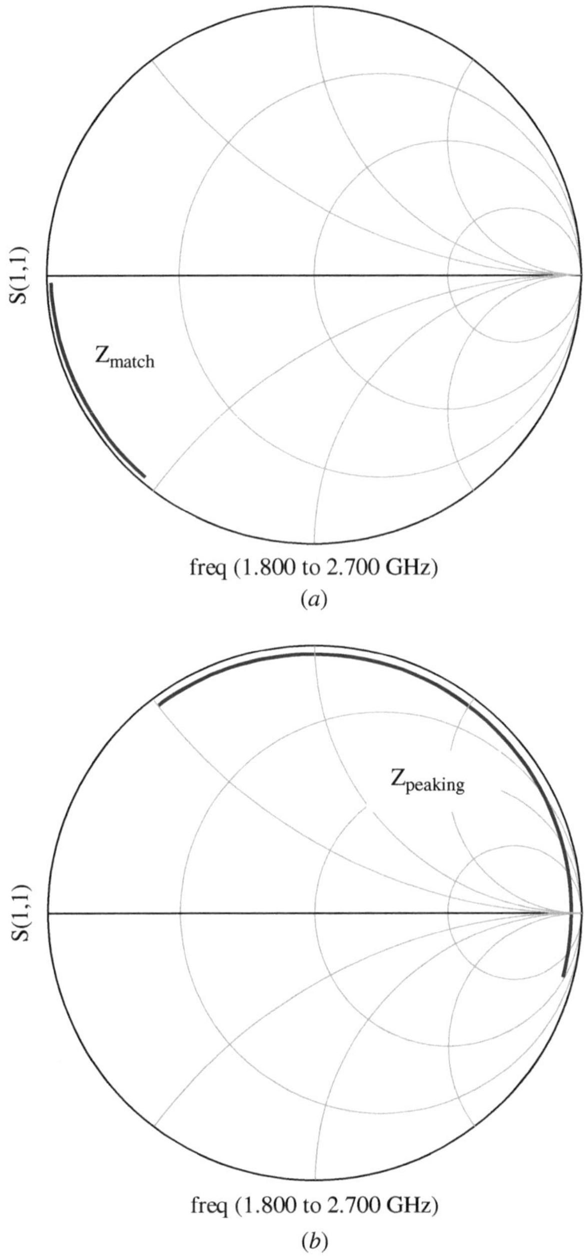

FIGURE 10.41 Impedances for peaking amplifier.

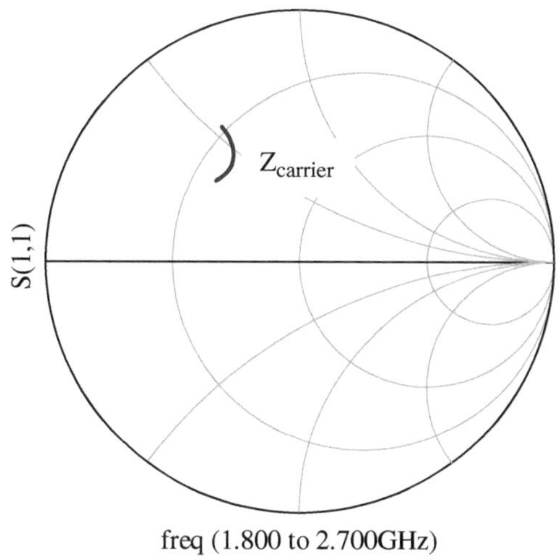

Figure 10.42 Impedance for carrier amplifier.

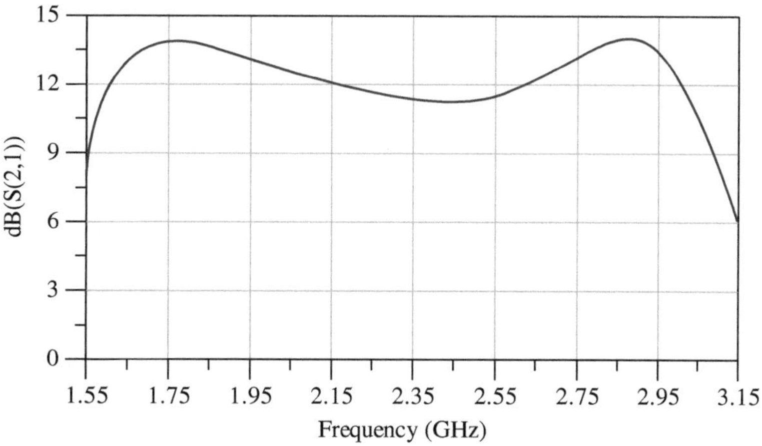

Figure 10.43 Simulated small-signal S_{21} versus frequency.

Figure 10.44 shows the simulated large-signal power gain and drain efficiencies of a transmission-line tri-band inverted GaN HEMT Doherty amplifier, with the carrier gate bias $V_{gc} = -2.45$ V, peaking gate bias $V_{gp} = -7.45$ V, and dc supply voltage $V_{dd} = 28$ V. In this case, a linear power gain of about 11.5 dB was achieved at higher bandwidth frequencies of 2.655 GHz and 2.14 GHz, whereas a slightly higher linear power gain of about 13 dB was achieved at lower bandwidth frequency of 1842.5 MHz. At the same time, the drain efficiencies of 71.5%, 69.0%, and 64.0% at backoff output powers of 40 dBm (4-dB backoff from saturated power of 44 dBm) and 59.0%, 57.0%,

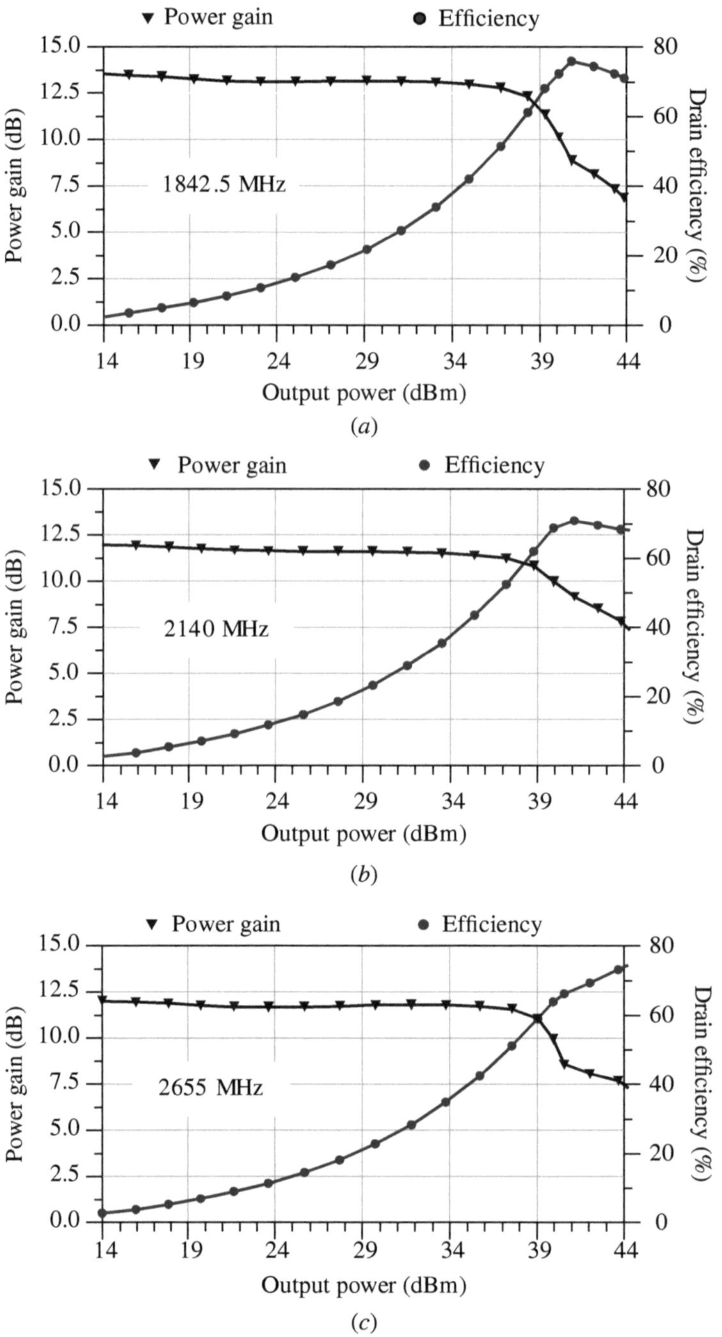

FIGURE 10.44 Simulated power gain and drain efficiencies of tri-band inverted Doherty amplifier.

and 53.5% at backoff output powers of 38 dBm (6-dB backoff) were simulated at center bandwidth frequencies of 1842.5 MHz, 2140 MHz, and 2655 MHz, respectively. Here, the peak drain efficiency peaks near 4-dB backoff output power at low bandwidth frequency of 1842.5 MHz and at medium bandwidth frequency of 2.14 GHz are clearly seen, while high efficiency maintains almost constant at high output powers at high bandwidth frequency of 2.655 GHz.

The tri-band transmission-line GaN HEMT Doherty amplifier was fabricated on a 20-mil RO4360 substrate. An input splitter represents a broadband coupled-line coupler from Anaren, model X3C17A1-03WS, which provides maximum phase balance of ±5° and amplitude balance of ±0.5 dB across the frequency range of 690 to 2700 MHz.

Figure 10.45 shows the test board of a tri-band inverted Doherty amplifier based on two 10-W Cree GaN HEMT power transistors CGH40010P in metal-ceramic pill packages. The input matching circuit, output load network, and gate and drain bias circuits (having bypass capacitors on their ends) are fully based on microstrip lines of different electrical lengths and characteristic impedances according to the simulation setup shown in Fig. 10.40. Special care should be taken for device implementation process in order to minimize the input and output lead inductances of the packaged GaN HEMT device. Additional tuning has been done in the input matching circuits to maximize power gain over the entire frequency range.

FIGURE 10.45 Test board of tri-band inverted GaN HEMT Doherty amplifier.

For a single-carrier 5-MHz WCDMA signal with a *PAR* of 6.5 dB, the drain efficiencies of 58%, 50%, and 42% at an average output power of 38 dBm with a power gain of more than 11 dB were achieved at the operating frequencies of 1.85 GHz, 2.15 GHz, and 2.65 GHz, respectively, with the *ACLR* (at 5-MHz offset) measured from –32 dBc at 1.85 GHz to –37 dBc at 2.65 GHz. The gate bias voltages for carrier (Class-AB mode with a quiescent current of 100 mA) and peaking (Class-C mode) amplifiers were the same for all three frequencies.

References

1. W. H. Doherty, "Amplifier," U.S. Patent 2,210,028, Aug. 1940 (filed Apr. 1936).
2. W. H. Doherty, "A New High Efficiency Power Amplifier for Modulated Waves," *Proc. IRE*, vol. 24, pp. 1163–1182, Sep. 1936.
3. W. H. Doherty and O. W. Towner, "A 50-Kilowatt Broadcast Station Utilizing the Doherty Amplifier and Designed for Expansion to 500 Kilowatts," *Proc. IRE*, vol. 27, pp. 531–534, Sep. 1939.
4. "William H. Doherty," *Proc. IRE*, vol. 25, p. 922, Aug. 1937.
5. C. E. Smith, J. R. Hall, and J. O. Weldon, "Very High Power Long-Wave Broadcast Station," *Proc. IRE*, vol. 42, pp. 1222–1235, Aug. 1954.
6. J. B. Sainton, "A 500 Kilowatt Medium Frequency Standard Broadcast Transmitter," *Cathode Press* (Machlett Company), vol. 22, no. 4, pp. 22–29, 1965.
7. V. M. Rozov and V. F. Kuzmin, "Use of the Doherty Circuit in SSB Transmitters," *Telecommunications and Radio Eng.*, 1970–1971.
8. P. Y. Vinogradov, N. I. Vorobyev, E. P. Sokolov, and N. S. Fuzik, "Amplification of a Modulated Signal by the Doherty Method in a Transistorized Power Amplifier," *Telecommunications and Radio Eng.*, part. 1, vol. 31, pp. 38–41, Oct. 1977.
9. "Development of Circuitry for Multikilowatt Transmitter for Space Communications Satellites," Report No. CRI19803, (NASA N71–29212), General Electric Company, Space Systems Division, Feb. 1971.
10. A. Mina and F. Parry, "Broadcasting with Megawatts of Power: The Modern Era of Efficient Powerful Transmitters," *IEEE Trans. Broadcast.*, vol. BC-35, pp. 121–130, Jun. 1989.
11. G. Clark, "A Comparison of Current Broadcast Amplitude-Modulation Techniques," *IEEE Trans. Broadcast.*, vol. BC-21, pp. 25–31, Jun. 1975.
12. F. H. Raab, "Efficiency of Doherty RF Power-Amplifier Systems," *IEEE Trans. Broadcast.*, vol. BC-33, pp. 77–83, Sep. 1987.
13. B Kim, J. Kim, I. Kim, and J. Cha, "The Doherty Power Amplifier," *IEEE Microwave Mag.*, vol. 7, pp. 42–50, Oct. 2006.
14. R. J. McMorrow, D. M. Upton, and P. R. Maloney, "The Microwave Doherty Amplifier," *1994 IEEE MTT-S Int. Microwave Symp. Dig.*, pp. 1653–1656.
15. C. F. Campbell, "A Full Integrated Ku-Band Doherty Amplifier MMIC," *IEEE Microwave and Guided Wave Lett.*, vol. 9, pp. 114–116, Mar. 1999.
16. K. W. Kobayashi, A. K. Oki, A. Gutierrez-Aitken, P. Chin, L. Yang, E. Kaneshiro, P. C. Grossman, et al., "An 18–21 GHz InP DHBT Linear Microwave Doherty Amplifier," *2000 IEEE RFIC Symp. Dig.*, pp. 179–182.
17. J.-H. Tsai and T.-W. Huang, "A 38–46 GHz MMIC Doherty Amplifier Using Post-Distortion Linearization," *IEEE Microwave and Wireless Compon. Lett.*, vol. 17, pp. 388–390, May 2007.
18. C. Steinberser, T. Landon, C. Suckling, J. Nelson, J. Delaney, J. Hitt, L. Witkowski, et al.,"250 W HVHBT Doherty with 57% WCDMA Efficiency Linearized to -55 dBc for 2c11 6.5 dB PAR," *IEEE J. Solid-State Circuits*, vol. SC-43, pp. 2218–2228, Oct. 2008.
19. H. Deguchi, N. Ui, K. Ebihara, K. Inoue, N. Yoshimura, and H. Takahashi, "A 33 W GaN HEMT Doherty Amplifier with 55% Drain Efficiency for 2.6 GHz Base Stations," *2009 IEEE MTT-S Int. Microwave Symp. Dig.*, pp. 1273–1276.

20. Y. Yang, J. Yi, Y. Y. Woo, and B. Kim, "Optimum Design for Linearity and Efficiency of a Microwave Doherty Amplifier Using a New Load Matching Technique," *Microwave J.*, vol. 44, pp. 20–36, Dec. 2001.
21. G. K. Wong, T. R. Shah, and K. Titizer, "Doherty Power Amplifier with Phase Compensation," U.S. Patent 7,295,074, Nov. 2007 (filed Mar. 2005).
22. S.-C. Jung, O. Hammi, and F. Ghannouchi, "Design Optimization and DPD Linearization of GaN-based Unsymmetrical Doherty Power Amplifiers for 3G Multicarrier Applications," *IEEE Trans. Microwave Theory Tech.*, vol. MTT-57, pp. 2105–2113, Sep. 2009.
23. M. Nick and A. Mortazawi, "Adaptive Input-Power Distribution in Doherty Power Amplifiers for Linearity and Efficiency Enhancement," *IEEE Trans. Microwave Theory Tech.*, vol. MTT-58, pp. 2764–2771, Nov. 2010.
24. J. Moon, Ja. Kim, Ju. Kim, I. Kim, and B. Kim, "Efficiency Enhancement of Doherty Amplifier through Mitigation of the Knee Voltage Effect," *IEEE Trans. Microwave Theory Tech.*, vol. MTT-59, pp. 143–152, Jan. 2011.
25. Y. Yang, J. Cha, B. Shin, and B. Kim, "A Microwave Doherty Amplifier Employing Envelope Tracking Technique for High Efficiency and Linearity," *IEEE Microwave and Wireless Compon. Lett.*, vol. 13, pp. 370–372, Sep. 2003.
26. Y. Zhao, M. Iwamoto, L. E. Larson, and P. M. Asbeck, "Doherty Amplifier with DSP Control to Improve Performance in CDMA Operation," *2003 IEEE MTT-S Int. Microwave Symp. Dig.*, pp. 687–690.
27. S. Bae, J. Kim, I. Nam, and Y. Kwon, "Bias-Switching Quasi-Doherty-Type Amplifier for CDMA Handset Applications," *2003 IEEE RFIC Symp. Dig.*, pp. 137–140.
28. T. Kato, K. Yamaguchi, Y. Kuriyama, and H. Yoshida, "An HPSK/OFDM 64-QAM Dual-Mode Doherty Power Amplifier Module for Mobile Terminals," *IEICE Trans. Electron.*, vol. E90-C, pp. 1678–1684, Sep. 2007.
29. S. Kawai, Y. Takayama, R. Ishikawa, and K. Honjo, "A GaN HEMT Doherty Amplifier with a Series Connected Load," *Proc. 2009 Asia-Pacific Microwave Conf.*, pp. 325–328.
30. D. F. Bowers, "HEAD — A High Efficiency Amplitude-Modulation System for Broadcasting Transmitters," *Communication and Broadcasting*, vol. 7, pp. 15–23, Feb. 1982.
31. J. Kim, J. Moon, Y. Y. Woo, S. Hong, I. Kim, J. Kim, and B. Kim, "Analysis of a Fully Matched Saturated Doherty Amplifier with Excellent Efficiency," *IEEE Trans. Microwave Theory Tech.*, vol. MTT-56, pp. 328–338, Feb. 2008.
32. J. Kim, B. Kim, and Y. Y. Woo, "Advanced Design of Linear Doherty Amplifier for High Efficiency Using Saturation Amplifier," *2007 IEEE MTT-S Int. Microwave Symp. Dig.*, pp. 1573–1576.
33. P. Colantonio, F. Giannini, R. Giofre, and L. Piazzon, "Theory and Experimental Results of a Class F AB-C Doherty Power Amplifier," *IEEE Trans. Microwave Theory Tech.*, vol. MTT-57, pp. 1936–1947, Aug. 2009.
34. K. W. Eccleston, K. J. I. Smith, P. T. Gough, and S. I. Mann, "Harmonic Load Modulation in Doherty Amplifiers," *Electronics Lett.*, vol. 44, pp. 128–129, Jan. 2008.
35. K. W. Eccleston, K. J. I. Smith, and P. T. Gough, "A Compact Class-F/Class-C Doherty Amplifier," *Microwave and Optical Technology Lett.*, vol. 53, pp. 1606–1610, Jul. 2011.
36. A. Grebennikov, "High-Efficiency Transmission-Line GaN HEMT Inverse Class F Power Amplifier for Active Antenna Arrays," *Proc. 2009 Asia-Pacific Microwave Conf.*, pp. 317–320.
37. S. Goto, T. Kunii, A. Inoue, K. Izawa, T. Ishikawa, and Y. Matsuda, "Efficiency Enhancement of Doherty Amplifier with Combination of Class-F and Inverse Class-F Schemes for S-Band Base Station Application," *2004 IEEE MTT-S Int. Microwave Symp. Dig.*, pp. 839–842.
38. G. W. Choi, H. J. Kim, W. J. Hwang, S. W. Shin, J. J. Choi, and S. J. Ha, "High Efficiency Class-E Tuned Doherty Amplifier Using GaN HEMT," *2009 IEEE MTT-S Int. Microwave Symp. Dig.*, pp. 925–928.
39. Y.-S. Lee, M.-W. Lee, and Y.-H. Jeong, "Highly Efficient Doherty Amplifier Based on Class-E Topology for WCDMA Applications," *IEEE Microwave and Wireless Compon. Lett.*, vol. 18, pp. 608–610, Sep. 2008.

40. E. Takahashi, T. Ishikawa, K. Kashimura, and N. Adachi, "High-Efficiency Four-Stage Class-E Doherty Amplifier for W-CDMA Base Stations," *Proc. 38th Europ. Microwave Conf.*, pp. 234–237, 2008.
41. M. Iwamoto, A. Williams, P-F. Chen, A. G. Metzger, L. E. Larsson, and P. M. Asbeck, "An Extended Doherty Amplifier with High Efficiency over a Wide Power Range," *IEEE Trans. Microwave Theory Tech.*, vol. MTT-49, pp. 2472–2479, Dec. 2001.
42. J. Kim, B. Fehri, S. Boumaiza, and J. Wood, "Power Efficiency and Linearity Enhancement Using Optimized Asymmetrical Doherty Power Amplifiers," *IEEE Trans. Microwave Theory Tech.*, vol. MTT-59, pp. 425–434, Feb. 2011.
43. Y. Yang, J. Cha, B. Shin, and B. Kim, "A Fully Matched N-Way Doherty Amplifier with Optimized Linearity," *IEEE Trans. Microwave Theory Tech.*, vol. MTT-51, pp. 986–993, Mar. 2003.
44. Y. Takayama, T. Harada, T. Fujita, and K. Maenaka, "Design Method of Microwave Doherty Power Amplifiers and Its Application to Si Power MOSFET Amplifiers," *Electronics and Communications in Japan*, part 2, vol. 88, pp. 9–17, Apr. 2005.
45. J. Kim, J. Cha, I. Kim, and B. Kim, "Optimum Operation of Asymmetrical-Cells-Based Linear Doherty Power Amplifiers—Uneven Power Drive and Power Matching," *IEEE Trans. Microwave Theory Tech.*, vol. MTT-53, pp. 1802–1809, May 2005.
46. I. Kim, J. Cha, S. Hong, J. Kim, Y. Y. Woo, C. S. Park, and B. Kim, "Highly Linear Three-Way Doherty Amplifier with Uneven Power Drive for Repeater System," *IEEE Microwave and Wireless Compon. Lett.*, vol. 16, pp. 176–178, Apr. 2006.
47. J. Moon, Ja. Kim, I. Kim, Ju. Kim, and B. Kim, "Highly Efficient Three-Way Saturated Doherty Amplifier with Digital Feedback Predistortion," *IEEE Microwave and Wireless Compon. Lett.*, vol. 18, pp. 539–541, Aug. 2008.
48. K. J. Cho, W. J. Kim, S. P. Stapleton, J. H. Kim, B. Lee, J. J. Choi, J. Y., et al., "Design of *N*-Way Distributed Doherty Amplifier for WCDMA and OFDM Applications," *Electronics Lett.*, vol. 43, pp. 577–578, May 2007.
49. W. J. Kim, K. J. Cho, S. P. Stapleton, and J. H. Kim, "N-Way Doherty Distributed Power Amplifier," U.S. Patent 7,688,135, Mar. 2010 (filed Apr. 2008).
50. N. Srirattana, A. Raghavan, D. Heo, P. E. Allen, and J. Laskar, "Analysis and Design of a High-Efficiency Multistage Doherty Power Amplifier for Wireless Communications," *IEEE Trans. Microwave Theory Tech.*, vol. MTT-53, pp. 852–860, Mar. 2005.
51. W. C. E. Neo, J. Qureshi, M. J. Pelk, J. R. Gajadharsing, and L. C. N. de Vreede, "A Mixed-Signal Approach towards Linear and Efficient N-Way Doherty Amplifiers," *IEEE Trans. Microwave Theory Tech.*, vol. MTT-55, pp. 866–879, May 2007.
52. M. J. Pelk, W. C. E. Neo, J. R. Gajadharsing, R. S. Pengelly, and L. C. N. de Vreede, "A High-Efficiency 100-W GaN Three-Way Doherty Amplifier for Base-Station Applications," *IEEE Trans. Microwave Theory Tech.*, vol. MTT-56, pp. 1582–1591, Jul. 2008.
53. J. Gajadharsing, W. C. E. Neo, M. Pelk, L. C. N. de Vreede, and J. Zhao, "3-Way Doherty Amplifier with Minimum Output Network," U.S. Patent 8,022,760, Sep. 2011 (filed Dec. 2008).
54. B Kim, I. Kim, and J. Moon, "Advanced Doherty Architecture," *IEEE Microwave Mag.*, vol. 11, pp. 72–86, Aug. 2010.
55. I. Kim, J. Moon, S. Jee, and B. Kim, "Optimized Design of a Highly Efficient Three-Stage Doherty PA Using Gate Adaptation," *IEEE Trans. Microwave Theory Tech.*, vol. MTT-58, pp. 2562–2574, Oct. 2010.
56. Y.-S. Lee, M.-W. Lee, S.-H. Kam, and Y.-H. Jeong, "Advanced Design of a Double Doherty Power Amplifier with a Flat Efficiency Range," *2010 IEEE MTT-S Int. Microwave Symp. Dig.*, pp. 1500–1503.
57. A. Grebennikov, "A High-Efficiency 100-W Four-Stage Doherty GaN HEMT Power Amplifier Module for WCDMA Systems," *2011 IEEE MTT-S Int. Microwave Symp. Dig.*, pp. 1–4.

58. R. F. Stengel, W.-C. A. Gu, G. D. Leizerovich, and L. F. Cygan, "High Efficiency Power Amplifier Having Reduced Output Matching Networks for Use in Portable Devices," U.S. Patent 6,262,629, Jul. 2001 (filed Jul. 1999).
59. G. Ahn, M. Kim, H. Park, S. Jung, J. Van, H. Cho, S. Kwon, et al., "Design of a High-Efficiency and High-Power Inverted Doherty Amplifier," *IEEE Trans. Microwave Theory Tech.*, vol. MTT-55, pp. 1105–1111, Jun. 2007.
60. S. Kwon, M. Kim, S. Jung, J. Jeong, K. Lim, J. Van, H. Cho, et al.,"Inverted-Load Network for High-Power Doherty Amplifier," *IEEE Microwave Mag.*, vol. 10, pp. 93–98, Feb. 2009.
61. S. Jin, J. Zhou, and L. Zhang, "A Broadband Inverted Doherty Power Amplifier for IEEE 802.11b/g WLAN Applications," *Microwave and Optical Technology Lett.*, vol. 53, pp. 636–639, Mar. 2011.
62. J. Sirois, S. Boumaiza, M. Helaoui, G. Brassard, and F. M. Ghannouchi "A Robust Modeling and Design Approach for Dynamically Loaded and Digitally Linearized Doherty Amplifiers," *IEEE Trans. Microwave Theory Tech.*, vol. MTT-53, pp. 2875–2883, Sep. 2005.
63. I. Takenaka, K. Ishikura, H. Takahashi, K. Hasegawa, T. Ueda, T. Kurihara, K. Asano, et al., "A Distortion-Cancelled Doherty High-Power Amplifier Using 28-V GaAs Heterojunction FETs for W-CDMA Base Stations," *IEEE Trans. Microwave Theory Tech.*, vol. MTT-54, pp. 4513–4521, Dec. 2006.
64. M.-W. Lee, S.-H. Kam, Y.-S. Lee, and Y.-H. Jeong, "Design of Highly Efficient Three-Stage Inverted Doherty Power Amplifier," *IEEE Microwave and Wireless Compon. Lett.*, vol. 21, pp. 383–385, Jul. 2011.
65. C. F. Campbell, "A Fully Integrated *Ku*-Band Doherty Amplifier MMIC," *IEEE Microwave and Guided Wave Lett.*, vol. 9, pp. 114–116, Mar. 1999.
66. C. P. McCarroll, G. D. Alley, S. Yates, and R. Matreci, "A 20GHz Doherty Power Amplifier MMIC with High Efficiency and Low Distortion Designed for Broad Band Digital Communication Systems," *2000 IEEE MTT-S Int. Microwave Symp. Dig.*, pp. 537–540.
67. D. Yu, Y. Kim, K. Han, J. Shin, and B. Kim, "A 60-GHz Fully Integrated Doherty Power Amplifier Based on 0.13-mm CMOS Process," *2008 IEEE RFIC Symp. Dig.*, pp. 69–72.
68. D. Yu, Y. Kim, K. Han, J. Shin, and B. Kim, "Fully Integrated Doherty Power Amplifiers for 5 GHz Wireless-LANs," *2006 IEEE RFIC Symp. Dig.*, pp. 177–180.
69. M. Elmala, J. Paramesh, and K. Soumyanath, "A 90-nm CMOS Doherty Power Amplifier with Minimum AM-PM Distortion," *IEEE J. Solid-State Circuits*, vol. SC-41, pp. 1323–1332, Jun. 2006.
70. C. Tongchoi, M. Chongcheawchamnan, and A. Worapishet, "Lumped Element Based Doherty Power Amplifier Topology in CMOS Process," *2003 IEEE Int. Circuits and Systems Symp. Dig.*, vol. 1, pp. 445–448.
71. I. I. Blednov, "High Power Doherty Amplifier," U.S. Patent 7,078,976, Jul. 2006 (filed Oct. 2005).
72. I. I. Blednov and J. van der Zanden, "High Power LDMOS Integrated Doherty Amplifier for W-CDMA," *2006 IEEE RFIC Symp. Dig.*, pp. 1–4.
73. J. Jung, U. Kim, J. Jeon, J. Kim, K. Kang, and Y. Kwon, "A New "Series-Type" Doherty Amplifier for Miniaturization," *2005 IEEE RFIC Symp. Dig.*, pp. 259–262.
74. C. Koo, U. Kim, J. Jeon, J. Kim, and Y. Kwon, "A Linearity-Enhanced Compact Series-Type Doherty Amplifier Suitable for CDMA Handset Applications," *2007 IEEE Radio and Wireless Symp. Dig.*, pp. 317–320.
75. D. Kang, J. Choi, D. Kim, and B. Kim, "Design of Doherty Power Amplifiers for Handset Applications," *IEEE Trans. Microwave Theory Tech.*, vol. MTT-58, pp. 2134–2142, Aug. 2010.
76. D. Kang, D. Kim, and B. Kim, "Broadband HBT Doherty Power Amplifiers for Handset Applications," *IEEE Trans. Microwave Theory Tech.*, vol. MTT-58, pp. 4031–4039, Dec. 2010.
77. D. Kang, D. Kim, Y. Cho, B. Park, J. Kim, and B. Kim, "Design of Bandwidth-Enhanced Doherty Power Amplifiers for Handset Applications," *IEEE Trans. Microwave Theory Tech.*, vol. MTT-59, pp. 3473–3483, Dec. 2011.

78. D. Gustafsson, J. C. Cahuana, D. Kuylenstierna, I. Angelov, N. Rorsman, and C. Fager, "A Wideband and Compact GaN MMIC Doherty Amplifier for Microwave Link Applications," *IEEE Trans. Microwave Theory Tech.*, vol. MTT-61, pp. 922–930, Feb. 2013.
79. J. S. Moon, H. Moyer, P. Macdonald, D. Wong, M. Antcliffe, M. Hu, P. Willadsen, et al.,"High Efficiency X-band Class-E GaN MMIC High-Power Amplifiers," *Proc. 2012 IEEE Topical Conf. Power Amplifiers for Wireless and Radio Appl.*, pp. 9–11.
80. R. Sperlich, G. C. Copeland, and R. Hoppenstein, "Hybrid Doherty Amplifier System and Method," U.S. Patent Appl. 2008/0111622, May 2008 (filed Nov. 2007).
81. D. R. Pehlke, "Class E Doherty Amplifier Topology for High Efficiency Signal Transmitters," U.S. Patent 6,396,341, May 2002 (filed Dec. 2000).
82. R. Darraji, F. M. Ghannouchi, and H. Hammi, "A Dual-Input Digitally Driven Doherty Amplifier for Performance Enhancement of Doherty Transmitters," *IEEE Trans. Microwave Theory Tech.*, vol. MTT-59, pp. 1284–1293, May 2011.
83. R. Darraji and F. M. Ghannouchi, "Digital Doherty Amplifier with Enhanced Efficiency and Extended Range," *IEEE Trans. Microwave Theory Tech.*, vol. MTT-59, pp. 2898–2909, Nov. 2011.
84. Y. Suzuki and S. Narahashi, "Multiband Doherty Amplifier," U.S. Patent 7,602,241, Oct. 2009 (filed Jun. 2007).
85. K. Bathich, A. Z. Markos, and G. Boeck, "Frequency Response Analysis and Bandwidth Extension of the Doherty Amplifier," *IEEE Trans. Microwave Theory Tech.*, vol. MTT-59, pp. 934–944, Apr. 2011.
86. K. Bathich, D. Gruner, and G. Boeck, "Analysis and Design of Dual-Band GaN HEMT Based Doherty Amplifier," *Proc. 6th Europ. Microwave Integrated Circuits Conf.*, pp. 248–251, 2011.
87. L. Wu, Z. Sun, H. Yilmaz, and M. Berroth, "A Dual-Frequency Wilkinson Power Divider," *IEEE Trans. Microwave Theory Tech.*, vol. MTT-54, pp. 278–284, Jan. 2006.
88. K. M. Cheng and F. Wong, "A New Wilkinson Power Divider Design for Dual Band Applications," *IEEE Microwave and Wireless Compon. Lett.*, vol. 17, pp. 664–666, Sep. 2007.
89. K. Rawat and F. Ghannouchi, "A Design Methodology for Miniaturized Power Dividers Using Periodically Loaded Slow Wave Structure with Dual-Band Applications," *IEEE Trans. Microwave Theory Tech.*, vol. MTT-57, pp. 3380–3388, Dec. 2009.
90. W. Chen, S. A. Bassam, X. Li, Y. Liu, K. Rawat, M. Helaoui, F. M. Ghannouchi, et al.,"Design and Linearization of Concurrent Dual-Band Doherty Power Amplifier with Frequency-Dependent Power Ranges," *IEEE Trans. Microwave Theory Tech.*, vol. MTT-59, pp. 2537–2546, Oct. 2011.
91. P. Colantonio, F. Feudo, F. Giannini, R. Giofre, and L. Piazzon, "Design of a Dual-Band GaN Doherty Amplifier," *Proc. 18th Int. Microwave Radar and Wireless Commun. Conf.*, pp. 1–4, 2010.
92. A. Grebennikov and J. Wong, "A Dual-Band Parallel Doherty Power Amplifier for Wireless Applications," *IEEE Trans. Microwave Theory Tech.*, vol. MTT-60, pp. 3214–3222, Oct. 2012.
93. C. Monzon, "A Small Dual-Frequency Transformer in Two Sections," *IEEE Trans. Microwave Theory Tech.*, vol. MTT-51, pp. 1157–1161, Apr. 2003.
94. G. Sun and R. H. Jansen, "Broadband Doherty Power Amplifier via Real Frequency Technique," *IEEE Trans. Microwave Theory Tech.*, vol. MTT-60, pp. 99–111, Jan. 2012.

Index

Note: Page numbers followed by *f* denote figures; page numbers followed by *t* denote tables.

B

BJTs, HBTs
- bipolar device, small-signal *T*-model, 116*f*
- equivalent circuit elements determination, 112–116
- intrinsic π-, *T*-type bipolar device typologies, 117*f*
- intrinsic π-, *T*-type typologies equivalence, 116–117
- large-signal Ebers-Moll model, 118*f*
- large-signal Gummel-Poon model, 120*f*
- nonlinear bipolar device modeling, 117–122
- nonlinear BJT and HBT model, with HBT structure, 111*f*, 112*f*
- parasitic pad capacitances, lead inductances models, 114*f*
- small-signal equivalent circuit, 110–112
- small-signal model at low frequencies and cutoff operation mode, 115*f*

broadband power amplifiers. *See also* high-efficiency power amplifiers; power amplifiers
- Bode-Fano criterion, 414–416, 415*f*, 416*f*

broadband power amplifiers (*Cont.*):
- broadband LDMOSFET power amplifier circuit schematic, performance, 430*f*
- broadband lumped-element microwave FET amplifiers, schematics, 426*f*
- broadband microwave GaN HEMT power amplifier, 438*f*
- capacitive impedance-transforming circuits, 417*f*
- five-step transformer, theoretical frequency bandwidth, 441*f*
- four-section transformers, optimal parameters, 443*t*
- impedance-transformer design procedure using low-pass filter-prototype, 424*f*
- impedance-transforming bandpass filter, frequency response, 426*f*
- Kuroda identities, 434*f*
- loaded lossless matching circuits, 414*f*
- low-pass Chebyshev filters-prototypes parameters, 421*f*
- low-pass to microstrip transforming filter, transformation design, 436*f*

broadband power amplifiers (*Cont.*):
lumped *L*-, p-, and *T*-type impedance-transforming circuits, 421*f*
matched networks with transmission lines, 431–443, 433*f*
matching circuit with ideal transformer, 416*f*
matching networks with lumped elements, 416–427
matching networks with mixed lumped, distributed elements, 427–431, 429*f*
maximum ripple level *vs.* frequency bandwidth, 422*f*
stepped transmission line transformers, equal length sections, 442*f*
stepped transmission line transformers, schematic structures, 440*f*
two-section broadband low-pass matching circuit, frequency response, 420*f*
two-section broadband matching circuit, 435*f*
two-section impedance-transforming circuit, 418*f*
two-section low-pass Chebyshev filter parameters, 419*t*
broadband power amplifiers, Class-E:
broadband cascode Class-E power amplifier with compensating inductor, 488*f*
broadband Class-E CMOS power amplifiers, client schematics, 490*f*
broadband Class-E pHEMT power amplifier, circuit schematic, 487*f*
broadband CMOS, 487–489
broadband high-efficiency microstrip LDMOSFET power amplifier, 483*f*
broadband parallel-circuit Class-E, 477–484
broadband power amplifiers (*Cont.*):
broadband two-stage *X*-band Inp DHBT Class-E power amplifier, 486*f*
broadband *X*-band Inp DHBT Class-E power amplifier, impedance conditions, 485*f*
circuit schematics of broadband GaN HEMT Class-E power amplifier, 475*f*
conductance, susceptance, 471*f*
double-reactance compensation circuit, impedances, 464*f*
double-reactance compensation circuits, 462*f*
8.49 broadband performance of Class-E LDMOSFET power amplifier, 480*f*
load network with bandpass filter and impedance, 472*f*
monolithic microwave, 484–487
output power, power gain, efficiency *vs.* frequency, 476*f*
reactance compensation circuit with lumped elements and transmission line, 473*f*
reactance compensation technique, 456–469
simulated broadband Class-E LDMOSFET power amplifier, 479*f*
single-, double-susceptance compensation circuits, 478*f*
single-reactance compensation circuits, 457*f*
single-susceptance compensation circuit and admittances, 460*f*
single-susceptance compensation circuit and impedances, 461*f*
stunt capacitance, 469–477, 469*f*
susceptance compensation circuit with quarter-wavelength transmission line, 467*f*
transmission-line Class-E load network with susceptance compensation, 482*f*

broadband power amplifiers (*Cont.*):
transmission-line reactance compensation circuits, design equations, 466*t*
transmission-line single-frequency equivalence technique, 468*f*
transmission-line susceptance compensation circuit, 465*f*, 481*f*
broadband power amplifiers, lossy compensation networks, 443–444, 444*f*
bipolar broadband high-power amplifier for VHF FM transmitters, 452*f*
broadband LDMOSFET power amplifier, schematic and performance, 451*f*
device input impedance, equivalent circuits, 448*f*
input, output, and interstage broadband matching circuits, 453*f*
lossy gain-compensation circuits, 446*f*
lossy match design techniques, 444–450, 448*f*
microstrip lossy match MESFET amplifiers, 447*f*
microstrip two-stage lossy match MESFET power amplifier, 454*f*
microwave broadband GaN HEMT power amplifiers, schematics, 455*f*
practical examples, 450–456

D

Doherty power amplifiers
anode currents and voltages during load variation envelopes, 577*f*
asymmetric Doherty amplifiers, 591–594
asymmetric *N*-way Doherty amplifier and efficiencies, 593*f*

Doherty power amplifiers (*Cont.*):
basic schematics with lumped elements, 578*f*
basic structures, 575–580
basic two-stage architecture, operation principles, 581*f*
Class-F/Class-C Doherty amplifier, schematic, 589*f*
conventional, modified two-stage Doherty amplifiers block diagram, 615*f*
digitally driven Dohery amplifiers, 611–613
distributed *N*-way Doherty amplifier, 594*f*
Doherty fundamental load-network structures, ideal voltage and current behavior, 576*f*
dual-band GaN HEMT parallel Doherty amplifier, test board, 621*f*
dual-band parallel Doherty architecture, 614–621, 615*f*, 616*f*, 620*f*
dual-band parallel GaN HEMT Doherty amplifier, circuit schematic, 617*f*
dual-input digital Doherty amplifier, block diagram and simulated performance, 612*f*
efficiencies, different Doherty amplifier architectures, 596*f*, 597*f*
for handset applications with bandwidth enhancement, schematic, 609*f*
with harmonic control circuits for handset applications, schematic, 608*f*
with harmonic-control circuits, block diagram, 588*f*
historical aspect, conventional Doherty architectures, 573–575
integration, 604–610

Doherty power amplifiers (*Cont.*):
inverse Class-F GaN HEMT Doherty amplifier, test board, 590*f*
inverted, 601–604, 601*f*, 602*f*, 623*f*
load-network schematic, impedances, 584*f*
lumped Doherty amplifiers for handset applications, circuit schematics, 606*f*
microstrip LDMOSFET Doherty amplifier, block diagram, 583*f*
modified four-stage Doherty amplifier architectures, 598*f*
modified three-stage Doherty amplifier architectures, 597*f*
multiband, broadband capability, 613–614, 613*f*
multistage Doherty amplifier architectures, 595*f*
multistage Dohery amplifiers, 594–601
operation principle, 580–583, 581*f*
parallel Doherty amplifier, broadband capability, 619*f*
quarterwave transmission line, single-frequency lumped equivalent, 605*f*
series connected load, 587
with series-connected load, block diagram, 587*f*
simulated small-signal *S*11 and *S*21 *vs*. frequency, 618*f*
stepped transmission-line transformer, input impedances, 622*f*
three-stage inverted, schematic diagram, 604*f*
tri-band inverted Doherty configuration, 621–630, 622*f*, 623*f*, 626*f*, 627*f*, 628*f*
tri-band inverted GaN HEMT Doherty amplifier, circuit schematic, 624*f*

Doherty power amplifiers (*Cont.*):
tri-band inverted GaN HEMT Doherty amplifier, test board, 629*f*
2.14 GHz 100-W four-stage Doherty GaN HEMT amplifier, 600*f*
wideband monolithic GaN HEMT Doherty power amplifier, test chip, 610*f*

E

efficiency enhancement techniques. *See* linearization, efficiency enhancement techniques

F

frequency-domain analysis
Bessle functions, 45–47, 46*f*
bipolar transistor simplified equivalent circuit, 54*f*
coefficients $\gamma_n(\theta)$ for dc, fundamental, higher-order current components, 44*f*
conduction angle schematic, 42*f*
device junction capacitance, large-signal behavior, 56*f*
DUT amplitude, phase characteristics, 67*f*
functions, harmonic superposition, 64*f*
harmonic balance method, 58–62, 58*f*
harmonic balance method, application to three-terminal MESFET device, 59*f*
MESFET power amplified equivalent circuit, 61*f*
MESFET simplified equivalent circuit, 56*f*
Newton-Raphson algorithm, 50–54
nonlinear parallel circuit schematic, 49*f*
piecewise-linear approximation, 40–45, 41*f*

frequency-domain analysis (*Cont.*):
quasilinear method, 54–58
resistor, diode, voltage source circuit, 53*f*
three-step iteration procedure, 54*t*
time-domain analysis, 47–50
trigonometric identities, 38–40
X-parameters, 62–68

H

HBTs. *See* BJTs, HBTs
HEMTs. *See* MSEFTs, HEMTs
high-efficiency power amplifiers. *See also* Doherty power amplifiers
idealized inverse Class-F mode, 356–359, 357*f*, 358*f*
optimum-efficiency Class-B collector current, voltage waveforms, 327*f*
overdriven Class-B collector current, voltage waveforms, 324*f*
power-added efficiencies for different values of power gain, 326*f*
high-efficiency power amplifiers, Class-E shunt capacitance, 373–374
Class-C power amplifier, with detuned resonant circuit, 375*f*
Class-E power amplifier with shunt capacitance, basic circuits, 375*f*
Class-E power amplifiers with transmission lines, equivalent circuits, 389*f*
equivalent Class-E load network at fundamental frequency, 379*f*
high-power high-frequency Class-E MOSFET amplifiers, 390*f*
high-power VHF Class-E amplifiers, 392*f*
load network with transmission lines, 387–390

high-efficiency power amplifiers (*Cont.*):
normalized (*a*) load current, and collector (*b*) voltage (*c*), current waveforms for idealized optimum Class E with shunt capacitance, 378*f*
optimal, nominal, off-nominal Class-E operations, 385–387
optimum load impedance and two-harmonic Class-E voltage waveform, 388*f*
optimum load-network parameters, 374–381
practical Class-E power amplifiers, 390–394
saturation resistance, finite switching, time, nonlinear shunt capacitance, 381–385, 381*f*
suboptimum operation above maximum frequency, 385–387, 387*t*
transmission-line Class-E power amplifiers, circuit topologies, 393*f*
high-efficiency power amplifiers, Class-E with finite DC-feed inductance, 394
Class-E power amplifiers with generalized load network, equivalent circuits, 396*f*
Class-E power amplifiers with transmission-line matching, schematics, 408*f*
idealized optimum parallel-circuit Class E, normalized (*a*) load current and collector (*b*) voltage and (*c*) current waveforms, 404*f*
load networks with transmission lines, 405–408
normalized optimum Class-E load network parameters, 400*f*
optimal circuit parameters, general analysis, 395–401

high-efficiency power amplifiers (*Cont.*):
optimal load-network parameters, 401*f*
optimum Class-E parameters p and φ *vs.* q, 399*f*
parallel-circuit Class-E power amplifier, 402–405, 402*f*
parallel-circuit Class-E power amplifier for handset application, transmission-line load network, 407*f*
parallel-circuit Class-E power amplifier, with lumped matching circuit, 405*f*
transmission-line parallel-circuit Class-E power amplifier, equivalent circuits, 406*f*
high-efficiency power amplifiers, Class-F circuit design, 327–329
biharmonic and polyharmonic power amplifiers, Fourier voltage, current waveforms, 328*f*
Class-F power amplifier, with series quarterwave transmission line, 338*f*
Class-F power amplifier, with stunt quarterwave transmission line, 339*f*
Class-F with maximally flat waveforms, 333–338
Class-F with quarterwave transmission line, 338–342, 342*f*
current waveforms for *n*th-harmonic peaking, 336*f*
design examples, 348–354
drain efficiency, power gain, and voltage waveform, 351*f*
drain voltage waveform, efficiency, power gain, 353*f*
idealized Class-F mode, 329–333, 329*f*, 332*f*
idealized collector current, voltage waveforms with nonzero on-resistance, 343*f*
high-efficiency power amplifiers (*Cont.*):
load network with parallel resonant circuit, frequency response, 346*f*
load networks, with lumped and distributed parameters, 344–348
load networks with parallel, series resonant circuits, 345*f*
microstrip impedance-peaking load network, frequency response, 348*f*
parasitic resistance, shunt capacitance, 342*f*
resultant efficiencies, voltage and current harmonic components, 337*t*
saturation resistance effect, 342–344
simulated lumped LDMOSFET Class-F power amplifier, 350*f*
simulated microstrip LDMOSFET Class-F power amplifier, 352*f*
small-signal LDMOSFET equivalent circuit, output *I-V* curves, 349*f*
transmission-line Class-F GaN HEMT power amplifier, 353*f*
transmission-line impedance-peaking circuit, 347*f*
voltage waveforms for *n*th-harmonic peaking, 334*f*
high-efficiency power amplifiers, inverse Class-F mode, 354–356, 355*f*, 356*f*
design examples, 363–372
drain efficiency, power gain, voltage, current waveforms, 367*f*
equivalent representations of load network at fundamental frequency, 371*f*
load networks seen by the device output at corresponding harmonics, 369*f*

high-efficiency power amplifiers (*Cont.*):
  load networks with lumped, distributed parameters, 361–363, 361*f*, 363*f*
  microstrip impedance-peaking circuit, 364*f*
  with quarterwave transmission line, 359–361, 360*f*
  simulated 500-MHz microstrip power amplifier, 366*f*
  simulated 500-MHz microstrip power amplifier, with *T*-transformer, 364*f*
  tranmission-line impedance-peaking circuit, 363*f*
  tranmission-line inverse Class-F amplifier, equivalent circuit, 368*f*
  transmission-line 10-W inverse Class-F GaN HEMT power amplifier, 372*f*

I

impedance matching
  equivalent circuits with (*a*) voltage and (*b*) current sources, 128*f*
  main principles, 127–131
  Smith chart, 131–133, 134*f*, 135–136, 135*f*
impedance matching, broadband UHF high-power amplifier, 170–174
  complete broadband input two-port network, 172*f*
impedance matching, equivalence between circuits with lumped and distributed parameters, 165–168
  lumped element, transmission line equivalence, 165*f*
  matching-circuit equivalence, lumped and transmission line, 167*f*
impedance matching, lumped elements
  analytic design technique, 136–148
impedance matching, lumped elements (*Cont.*):
  impedance parallel, series equivalent circuits, 137*f*
  loaded two-port network, input impedance, 138*f*
  L-transformer, additional *LC* filter, 141*f*
  L-type matching circuits, relevant equations, 139*f*
  matching circuits, *L*-, π-, and *T*-transformers, 136*f*
  matching circuits, two *L*-transformers connected, 141*f*
  parallel resonant circuit input impedance, frequency plot, 140*f*
  parallel resonant circuit, loaded *L*-transformer, 139*f*
  π-transformer, additional *L*-filter, 143*f*
  π-transformer, with series capacitor, 144*f*
  transformers, relevant equations, 146*f*
  transformers, with series and shunt capacitors, 147*f*
impedance matching, MOSFET VHF high-power amplifier, 153–156
  complete broadband matching circuit, 154*f*
  complete broadband matching circuit, Smith chart, 155*f*
  complete broadband output circuit, 155*f*
  complete broadband output circuit, Smith chart, 155*f*
impedance matching, narrow-band microwave amplifier, 169–170
  complete output two-port network circuit, 169*f*
  complete output two-port network circuit, Smith chart, 171*f*

impedance matching, transmission lines types
coaxial line, 174–176, 175*f*
conductor materials, electrical resistivity, 182*t*
coplanar waveguide (CPW), 183–184, 184*f*
coplanar waveguide (CPW) characteristic impedance *vs.* s/(s + 2W), 185*f*
microstrip line, 178–181, 179*f*
microstrip line characteristic impedance *vs.* *W/h*, 180*f*
slotline, 181–183, 182*f*
slotline characteristic impedance *vs.* *W/h*, 183*f*
stripline, 176–178, 176*f*
stripline characteristic impedance *vs.* *W/b*, 178*f*
substrate materials, electrical and thermal properties, 179*t*
impedance matching, UHF bipolar power amplifier, 149–152
complete input circuit, 149*f*
complete input circuit, Smith chart, 150*f*
complete output network circuit, 151*f*
complete output network circuit, Smith chart, 151*f*
impedance matching, with transmission lines
analytic design technique, 156–165
*L-t*ransformer, with series transmission line, 160*f*
with π-, *T*-transformers, 163*f*
transmission-line impedance transformer, 157*f*
transmission-line *L*-transformer, calculation nomographs, 162*f*
transmission-line π-transformer, relevant equations, 164*f*
transmission-line transformer, any source and load impedances, 160*f*
transmission-line *T*-transformer, relevant equations, 164*f*

L

linearization, efficiency enhancement techniques
adaptive analog, digital feedforward amplifier linearizer, 509*f*
asymmetric Chireix outphasing systems, Class-E power amplifiers, 524*f*
balanced feedforward amplifier topologies, 508*f*
balanced switched-path MMIC power amplifier, block diagram and test result, 549*f*
cancellation as function of amplitude, phase imbalance, 505*f*
cascode CMOS Class-E power amplifiers with finite dc-feed inductance, 556*f*
Chireix outphasing power amplifier system and instantaneous, 521*f*
Chireix outphasing power amplifier system and instantaneous efficiencies, 531*f*
Class-F parallel power amplifier architecture, 559*f*
Class-F power amplifier with quarterwave transmission line, 558*f*
CMOS Chireix outphasing systems, Class-E power amplifiers, 526*f*
collector voltage, current waveforms for different load lines, 544*f*
device cell connections for power combining, 551*f*
device cells parallel on-chip connection, 552*f*
differential CMOS power amplifier with adaptive bias circuits, 562*f*
digital predistortion (DPD) system, 516*f*
dual-chain MMIC power amplifier, 548*f*

linearization, efficiency (*Cont.*):
dual-path power amplifier, three-port network configurations, 547*f*
envelope tracking (ET), 532–540
envelope-tracking (ET) ISOgain and optimum efficiency shaping, 539*f*
envelope-tracking (ET) power amplifier architecture with analog control, 533*f*
envelope-tracking (ET) power amplifier architecture with digital control, 534*f*
envelope-tracking (ET) power amplifier dc supply (*a*) voltage and (*b*) current, 536*f*
envelope-tracking (ET) power amplifier, varying supply voltage, 538*f*
envelope-tracking (ET) power amplifier, with Class-S, 535*f*
feedforward amplifier architecture, 501–509, 503*f*
four-stage outphasing architecture and instantaneous efficiencies, 529*f*
LINC transmitter with phase error compensating loop, 520*f*
load network configurations, variable load-network elements, 545*f*
modified three-path predistortion linearizer, 514*f*
monolithic HBT, CMOS power amplifiers for handset applications, 551–565
outphasing power amplifier system with hybrid combiner, 519*f*
outphasing power amplifiers, 517–531
power amplifier linearizers with input power splitting, block diagrams, 513*f*
power amplifier module with linearizer, 515*f*

linearization, efficiency (*Cont.*):
power amplifier with predistortion linearizer, block diagram, 510*f*
predistortion linearization, 509–517
simple diode-based predistortion linearizers, 511*f*
simple outphasing power amplifier system, 518*f*
stacked CMOS power amplifiers, 564*f*
switched-path power amplifier configurations, 542*f*
switched-path, variable-load power amplifiers, 540–550
switched-stage power amplifier configurations, 541*f*
three-stage GaAs HBT power amplifier MMIC, 555*f*
transistor-based linearizers, 512*f*
transmitter architectures with dual-path power amplifier, 540*f*
two-stage Class-AB power amplifier, schematic, 560*f*
two-stage HBT power amplifier, monolithic implementation, 554*f*
two-stage InGaP/GaAs HBT power amplifier MMIC, 552*f*

lumped elements
capacitor topologies, different series, 28*f*
capacitors, 27–29
inductors, 23–27
parallel capacitor topology, equivalent circuit, 28*f*
shaped spiral inductor, equivalent circuit, 25*f*
spiral inductor layouts, 24*f*

lumped elements, impedance matching
analytic design technique, 136–148
impedance parallel, series equivalent circuits, 137*f*

lumped elements (*Cont.*):
loaded two-port network, input impedance, 138*f*
L-transformer, additional *LC* filter, 141*f*
L-type matching circuits, relevant equations, 139*f*
matching circuits, *L*-, *π*-, and *T*-transformers, 136*f*
matching circuits, two *L*-transformers connected, 141*f*
parallel resonant circuit impute impedance, frequency plot, 140*f*
parallel resonant circuit, loaded *L*-transformer, 139*f*
π-transformer, additional *L*-filter, 143*f*
π-transformer, with series capacitor, 144*f*
transformers, relevant equations, 146*f*
transformers, with series and shunt capacitors, 147*f*

M

MSEFTs, HEMTs
capacitance equivalent circuits consistent with charge conservation, 98*f*
Chalmers (Angelov) nonlinear model, 105–108, 105*f*
Curtice quadratic nonlinear model, 102–104, 103*f*
device intrinsic *Z*-parameters, extraction method, 101*f*
distributed *RC* channel network schematic under device gate, 102*f*
equivalent circuit elements determination, 99–102
IAF (Berroth) nonlinear model, 108–109, 108*f*
Materka-Kacprzak nonlinear model, 104–105, 104*f*
measured and modeled $I_{ds}(V_{ds})$ curves of low-voltage LDMOSFET, 95*f*

MSEFTs, HEMTs (*Cont.*):
model selection, 109–110
nonlinear GaN HEMT model with electrothermal elements, 97*f*
nonlinear MESFET and HEMT model with HEMT physical structures, 96*f*
small-signal equivalent circuit, 94–98
small-signal MESFET circuit, zero drain bias voltage, 102*f*

N

network parameters, traditional, 1–6
loaded two-port transmission system, 5*f*
parameters relationships, 6*t*
two-port nonautonomous transmission system, 2*f*

P

power amplifiers, design fundamentals
A, AB, B, and C operation classes, 267–274
adaptive bias circuit, performance, 312*f*
basic power-amplifier structure, classes of amplification, 279*f*
bias circuits, 306–313
bipolar power amplifier stage, with current mirror bias circuit and its performance, 310*f*
bipolar power amplifier stage, with emitter follower bias circuit, 311*f*
bipolar power amplifier, with lineaerizing bias resistor, 296*f*
bipolar tuned power amplifier, circuit schematics, 284*f*
Class-A operation, voltage and current wave forms, 269*f*
Class-AB, Class-C collector current waveforms, 275*f*

power amplifiers (*Cont.*):
Class-B operation, voltage and current wave forms, 271*f*
classes of operation, finite number of harmonics, 278–281
collector current waveforms for the device operating in saturation, active, and pinch-off regions, 277*f*
harmonic, intermodulation components straight-line dependencies, 292*f*
high-efficiency bipolar mixed-mode Class-C power amplifier, 282*f*, 283*f*
high-power UHF LDMOSFET push-pull amplifier, circuit schematic, 316*f*
linearity, 290–297
load line, output impedance, 274–278
load lines for (*a*) inductive, and (*b*) capacitive load impedances, 278*f*
matched bipolar amplifier, simple equivalent circuit, 246*f*
matched FET power amplifier, simple equivalent circuit, 246*f*
MESFET power amplifier intermodulation distortions, different quiescent currents, 294*f*
microwave power amplifier, typical topology, 313*f*
mixed-mode Class B, nonlinear effect of collector capacitance, 281–286
monolithic microwave GaAs MESFET power amplifier, circuit schematic, 316*f*
overview, main characteristics, 241–248
passive tuner, reflective coefficient position, 289*f*
practical aspect, RF and microwave power amplifiers, 313–319

power amplifiers (*Cont.*):
single-stage power amplifier, block schematic, 242*f*
3.0 to 3.5 GHz GaAs MESFET power amplifier, circuit schematic, 314*f*
300-W VHF MOSFET power amplifier, circuit schematic, 317*f*
TV applications, bipolar VHF power amplifier, 315*f*
2.5 to 2.7 GHz GaAs MESFET power amplifier, circuit schematic, 314*f*
two-port loaded amplifier networks, 243*f*
two-stage *X*-band MMIC driver amplifier, schematic and photo, 318*f*
two-tone driving signal, 297*f*
two-tone excitation, typical output power spectrum, 294*f*
typical bipolar Class-AB bias circuit for high-power amplifier, 308*f*
typical bipolar Class-AB circuits, 308*f*

power amplifiers, push-pull and balanced power amplifiers
balance power amplifiers, 303–306
balanced high-power GaAs MESFET amplifier with branch-line impedance-transforming hybrids, 306*f*
basic configurations, 298–303
branch-line hybrids, power amplifier, 305*f*
power amplifiers with quadrature hybrid couplers, schematics, 303*f*
push-pull operation, basic operation, 299*f*
push-pull with compact balanced-to-unbalanced transformers, 302*f*

power amplifiers (*Cont.*):
push-pull with compact unbalanced-to-balanced transformers, 302*f*
single-ended and balance transistors, matching techniques, 301*f*
power amplifiers, stabilization circuit technique
BJT potential instability, frequency domains, 252–258
MOSFET potential instability, frequency domains, 258–261
parasitic bipolar oscillators, equivalent circuits, 257*f*
parasitic MOSFET oscillators, equivalent circuits, 261*f*
simplified bipolar π-hybrid equivalent circuit, with emitter lead inductance, 255*f*
stabilization circuits, examples, 262–267
power MOSFETS
charge conservation, 90–91
device intrinsic *Z*-parameters, extraction method, 76*f*
drain current *vs.* gate-source voltage, 81
equivalent circuit elements determination, 75–79
forward gate biasing condition, device equivalent circuit, 77*f*
gate-source capacitance *vs.* gate-source voltage, 87*f*
gate-source resistance, 91–92
high-voltage LDMOSFET, $I_{ds}(V_{ds})$ curves, 83*f*
high-voltage LDMOSFET, measured and modeled $C_{ds}(V_{gs})$ dependencies, 89*f*
high-voltage LDMOSFET, measured and modeled $I_{ds}(V_{gs})$ and $g_m(V_{gs})$ curves, 85*f*
power MOSFETS (*Cont.*):
high-voltage LDMOSFET, simulated *I-V* model parameters, 84*t*
intrinsic MOSFET equivalent circuit, first-, second-order channel approximation, 74*f*
low-voltage MOSFET, measured and modeled $g_m(V_{gs})$ curves, 86*f*
low-voltage MOSFET, measured and modeled $I_{ds}(V_{gs})$ and $g_m(V_{gs})$ curves, 86*f*
low-voltage MOSFET, simulated *I-V* model parameters, 86*t*
modeled temperature dependencies of V_{th} and g_m, 93*f*
MOSFET distributed channel structure, schematic, 73*f*
nonlinear *I-V* models, 79–84
small-signal equivalent circuit, 72–75
small-signal equivalent circuit, with extrinsic linear elements, 75*f*
temperature dependence, 92–94
zero-drain bias condition, device equivalent circuit, 78*f*
power transformers, combiners, couplers
1:1 coaxial cable transformer, low-frequency model, 195*f*
4:1 coaxial cable transformer, schematic configurations, 197*f*
9:1 coaxial cable transformer, schematic configurations, 199*f*
balanced power amplifier topology, with 45° delay lines, 220*f*
baluns, 204–210
baluns, 1:1 balun circuit configurations, 205*f*

power transformers (*Cont.*):
baluns, circuit arrangement with two cable transformers for push-pull operation., 207*f*
baluns, coupled-line Marchand balun, 209*f*
baluns, Marchand balun schematic configurations, 208*f*
basic properties, 187–188
branch-line hybrid couplers, 221–229
broadband microwave branch-line quadrature hybrid, 224*f*
coaxial cable combiners, fully matched and isolated, 204*f*
coaxial cable combiners, with increased isolation, 203*f*
coaxial cable transformer, schematic configurations, 194*f*
compact microstrip three-way Wilkinson power divider, 220*f*
coupled-line directional couplers, 229–236, 230*f*, 233*f*, 234*f*, 236*f*
equal-delay 2.25:1 unun, schematic configuration, 199*f*
four-port networks, directional coupler schematic diagram, 189–191, 189*f*
fractional 1:2.25 impedance transformer, schematic configurations, 200*f*
Guanella 1:1 and 4:1 transformers, schematic configuration, 192*f*
Gysel high-power in-phase planar combiner/divider, 214*f*
hybrid combiners, with one coaxial cable transformer, 201*f*
lumped hybrid with capacitive and inductive coupling, equivalent circuits, 228*f*

power transformers (*Cont.*):
lumped *LC*-type hybrid coupler, equivalent circuits, 227*f*
microstrip branch-line quadrature hybrid coupler, 222*f*
microstrip three-way recombinant divider, with improved isolation, 216*f*
N-way in-phase combiners/dividers, circuit topologies, 210*f*
N-way Wilkinson divider, frequency performance, 212*f*
practical four-way microstrip Wilkinson power combine/divider, 215*f*
reduced-size branch-line quadrature hybrid, 225*f*
Ruthroff 1:4 impedance transformer, schematic configurations, 197*f*
single-section branch-line hybrid coupler, bandwidth performance, 223*f*
split-tee power divider, 217*f*
three-port networks, 188–189
three-port networks, power divider/combiner schematic diagrams, 188*f*
three-way power divider, new type, 219*f*
transmission-line transformers and combiners, 191–204
two-cable hybrid combiner, 202*f*
two-way Wilkinson divider, microstrip realization, 213*f*
VSWR = 2, Smith-chart impedances, 221*f*
Wilkinson power dividers/combiners, 210–221

practical broadband RF, microwave power amplifiers, 490–497
bipolar high-power UHF amplifier for TV transmitter, 493*f*

practical broadband RF (*Cont.*):
broadband GaN HEMT MMIC power amplifier, circuit schematic, 495*f*
broadband high-power VHF bipolar amplifier, circuit schematic, 491*f*
microstrip broadband 15-W GaAs MESFET power amplifier, 492*f*
reactively matched two-stage broadband GaN HEMT MMIC power amplifier, circuit schematic, 496*f*

S

scattering parameters, 7–11
parameter two-port network, 9*f*

T

three-port networks, common terminal, 20–23, 21*f*
bipolar transistors, different common terminals, 22*f*
Y- and *Z*-parameters, active device with different common terminal, 23*f*
transmission lines, 29–34. *See also* impedance matching, with transmission lines
coaxial line, 174–176, 175*f*
conductor materials, electrical resistivity, 182*t*
loaded transmission line, 31*f*
microstrip line, 178–181, 179*f*
microstrip line characteristic impedance *vs.* *W/h*, 180*f*
schematics, 30*f*
slotline, 181–183, 182*f*
slotline characteristic impedance *vs.* *W/h*, 183*f*
stripline, 176–178, 176*f*
stripline characteristic impedance *vs.* *W/b*, 178*f*
substrate materials, electrical and thermal properties, 179*t*
transmission lines, impedance matching
analytic design technique, 156–165
L-transformer, with series transmission line, 160*f*
with π-, *T*-transformers, 163*f*
transmission-line impedance transformer, 157*f*
transmission-line *L*-transformer, calculation nomographs, 162*f*
transmission-line π-transformer, relevant equations, 164*f*
transmission-line transformer, any source and load impedances, 160*f*
transmission-line *T*-transformer, relevant equations, 164*f*
two-port networks, interconnections, 11, 13*f*, 14–16
parameter conversions, 12*t*
two-port networks, practical
π- and *T*-circuit equivalence, 19*f*
π- and *T*-circuit parameters, 20*f*
π- and *T*-type networks, 17–20, 17*f*
single-element networks, 16–17, 16*f*

Printed by Libri Plureos GmbH in Hamburg,
Germany

9 780071 828628